13TH INTERNATIONAL SYMPOSIUM ON INDUSTRIAL ROBOTS AND ROBOTS 7

Conference Proceedings

Future Directions
Volume 2

April 17-21, 1983
Chicago, Illinois

Sponsored by:
Robotics International of SME
One SME Drive
P.O. Box 930
Dearborn, Michigan 48121

in cooperation with:
Robot Institute of America

PROCEEDINGS
13TH ISIR/ROBOTS 7

Volume 2
Future Directions

First Edition

First Printing

Library of Congress Catalog Number 83-060355

International Standard Book Number: 0-87263-114-1

Manufactured in the United States of America

PREFACE

The 13th International Symposium on Industrial Robots/Robots 7 will be the largest gathering ever assembled for the purpose of exchanging robotic technology.

"The Emerging Challenge", the theme for this year's conference, certainly describes the present industrial movement towards the application of robotics and programmable manipulators. The challenge to produce products faster, with higher quality, requiring less investment, with larger and larger variations has resulted in the realization of technologies far beyond the imagination of our forefathers.

Presently:

- Production Engineers have the challenge to keep abreast with the exploding robotic equipment market.
- Robot manufacturers on the other hand because of the fierce competition within their industry have a challenge to build smarter, faster and lower cost equipment.
- Management has to provide for competent staffing capable of integrating technology into production systems to meet the challenge for market position.
- Educators most importantly are challenged to train men and women in these new technologies during a period of decreasing financial support and increasing educational costs.
- Government's continuing challenge is to allocate resources for technological advances all so vital to each country's survival.

These are only a few of the real challenges facing the world's technical community. How these challenges are met will continue to affect every man, woman and child throughout the world.

The 13th ISIR/Robot 7 Conference is meeting these challenges through four special forums (Educators, Aerospace, Design, Department of Defense), 72 technical and three basic sessions with over 180 manufacturers exhibiting high technology equipment. This conference will demonstrate and discuss the latest in robotic technology.

Robotics International (RI) of the Society of Manufacturing Engineers (SME) sponsors the conference. RI is a user organization which has grown in three and one-half years to a membership of over 7,000. Its members are engineers, scientists, technicians and educators who are devoted to the application of high technology. Each member is invited and encouraged to become an active member of RI by joining technical action groups such as: human factors, assembly, aerospace, foundry, welding, R&D, education, machine loading, finishing and others. These groups provide the basis for technology transfer.

Finally the 13th ISIR/Robots 7 Conference will meet the challenge because of many dedicated hard-working men and women who have so unselfishly worked for months on various committees. With the professional guidance of SME the overall conference is well planned, hundreds of abstracts and papers have been reviewed, safety regulations have been established, special events will happen, communications among foreign dignitaries will be possible and the facilities will be comfortable.

Conference Committee
13th ISIR/Robots 7

RI/SME

A member organization founded in 1980, Robotics International of SME is both applications and research-oriented—covering all phases of research, design, installation, operation and maintenance of industrial robots. In addition, RI/SME serves as a liaison between industry, government and education to determine areas where technological development is needed. RI/SME is the vital link in the chain that connects the individual to the many aspects of robotic-oriented manufacturing.

Members of RI/SME are involved in any of these three catagories:

- Planning, selecting and applying robots.
- Research and development leading to the creation or improvement of robots.
- Other creative robotic activities in administration, education or government.

The members of RI/SME update their knowledge and skills through educational programs, chapter meetings, conferences, expositions and publications including *Robotics Today.*

The largest American scientific and educational association exploring the expanding use of robots, RI/SME is an official association of the Society of Manufacturing Engineers and is headquartered in Dearborn, Michigan. RI/SME addresses the informational needs of the engineer invloved in the application of robots and provides a forum for discussion among concerned individuals. It is, in a very real sense, industry's partner and a major force in an engineer's professional advancement.

TABLE OF CONTENTS
VOLUME 2
FUTURE DIRECTIONS

CHAPTERS

ROBOT PROGRAMMING LANGUAGES

INTELLIGENT ROBOTS/ADAPTIVE CONTROL SYSTEMS

ADVANCED VISION AND TACTILE SENSORS

ROBOTICS PROGRAMS IN EDUCATION COMPLEXES

KINEMATICS

ADVANCEMENTS IN NON-BINARY VISION

END EFFECTORS

ROBOTIC EDUCATION AND TRAINING FORUM

ROBOT DESIGN FORUM

TABLE OF CONTENTS
VOLUME 1
APPLICATIONS WORLDWIDE

CHAPTERS

ROBOTIC OVERVIEW

ROBOT MATERIAL HANDLING SYSTEMS

PROCESS SYSTEMS

IMPLEMENTATION

CAD/CAM

7

FINISHING

HUMAN FACTORS AND SAFETY

APPLICATION DEVELOPMENT

APPLICATIONS GENERAL

SUNDAY, APRIL 17, 1983

Numbers listed in parentheses indicate the location of the paper in these proceedings.

Conference Registration Begins at the Conrad Hilton Hotel

Registration for the 13th ISIR/Robots 7 Conference will begin at 1:00 p.m. Sunday, April 17, at the Conrad Hilton Hotel. This is an ideal time to register at your leisure for the week of activities.
After registering for the conference, begin your week by attending these Sunday p.m. programs.

4:00-6:00 P.M.—ROBOTIC BASICS WORKSHOP
Chair: *Robert Day*, CMfgE, Sr. Industrial Engineer, Eastman Kodak, Rochester, New York, U.S.A.
This two-hour Basics Workshop will be jointly conducted by two well-known consulting firms. *Ronald Potter*, Robot Systems, Inc., will discuss the fundamentals of robotics, and *William Higgins*, Williams & Higgins Industrial Sales, will cover basic implementation and applications.

6:00-10:30 P.M.—ROBOTIC EDUCATION AND TRAINING FORUM
Co-Chairs: *Jack Lane*, CMfgE, P.E., Director, GMI Robotics Center, General Motors Engineering and Management Institute, Flint, Michigan, U.S.A., and *C. W. Burckhardt*, Professor, Ecole Polytechnique, Lausanne, Switzerland
"Career Opportunities in Robotics and Educational Requirements". . .*Bill Weisel*, Director of Educational Relations, Prab Robots, Inc., U.S.A. **(19-1)**
"State of the Art in Robotic Education". . .*Frederick Sitkins*, Assistant Professor of Mechanical Engineering, Western Michigan University, U.S.A.
"Automated Manufacturing Technology—A Systems Training Concept". . .*Jim Rehg*, Director, Robotics Center, and *Richard Upchurch*, Dept. Head Mathematics, Piedmont Technical College, U.S.A. **(19-10)**
"Education and Training: Comments from Japan". . .*Souji Inagaki*, Nagoya Municipal Industrial Research Institute, Japan
"U.S. Robotics Educational Efforts". . .*Dr. Lois Galbiati*, Dean, College of Technology, State University of New York-Utica, U.S.A.
"Bridging the Gap in Robotics Education". . .*Dr. Winifred Warnat*, Director, National Center on Teaching and Learning, Eastern Michigan University, U.S.A.

6:00-10:00 P.M.—DEPARTMENT OF DEFENSE FORUM
The U.S. Government's Role in Robotics Research, Development and Applications: An Overview of Current Programs and Plans for the Future
Co-Chairs: *William F. Holden*, Naval Material Command, and *George M. Peace*, Environmental Research Institute of Michigan, U.S.A.

• Congressional Perspective:
• Manufacturing Technology:
Frederick J. Michel, Director, Manufacturing Technology, U.S. Army Material Development & Readiness Command, Alexandria, Virginia, U.S.A.
Richard Gamble, Head, Manufacturing Technology Office, Naval Ocean Systems Center, San Diego, California, U.S.A.
Lt. Gordon Mayer, Project Manager, Robotics, Computer Integrated Manufacturing Branch, U.S. Air Force Materials Laboratory, Dayton, Ohio, U.S.A.
Dr. James S. Albus, Chief, Industrial Systems Division Center for Manufacturing Engineering, National Bureau of Standards, Washington, D.C., U.S.A.
Dr. William M. Spurgeon, Program Director, Production Research, Division of Mechanical Engineering & Applied Mechanics, National Science Foundation, Washington, D.C., U.S.A.
• Robotics In Space:
Capt. Robert F. Freitag, Deputy Director, Space Station Task Force, National Aeronautics and Space Administration, Washington, D.C., U.S.A.
• Military R & D Program Plans:
Charles Shoemaker, Special Projects Office, U.S. Army Human Engineering Laboratory, Aberdeen Proving Grounds, Maryland, U.S.A.
Marvin Denicoff, Division Leader, Information Sciences Division, Office of Naval Research, Arlington, Virginia, U.S.A.
Dr. Thomas E. Walsh, Director, Electronic and Material Sciences, Air Force Office of Scientific Research, Bolling Air Force Base, Washington, D.C., U.S.A.

10:30-11:30 P.M.—You are invited to join other 13th ISIR/Robots 7 conferees at a cash-bar reception.

MONDAY, APRIL 18, 1983

Numbers listed in parentheses indicate the location of the paper in these proceedings.

A Good Day To Visit The Show

To maximize your visit to the 13th ISIR/Robots 7 meeting, conference registrants are urged to visit the Exposition at McCormick Place East on Monday, April 18. The full schedule of conference sessions on Tuesday and Wednesday may not permit you to thoroughly view the robot exhibits on these two days.

Exposition hours on Monday, April 18, are 10:00 a.m. to 8:00 p.m.; on Tuesday, April 19, 10:00 a.m. to 6:00 p.m., and on Wednesday, April 20, 10:00 a.m. to 5:00 p.m. We suggest the following schedule for your convenience:
1—Register for the conference at the Conrad Hilton Hotel on Sunday, April 17, from 1:00 p.m. to 8:00 p.m., or on Monday, April 18, from 8:00 a.m. to 8:00 p.m. Pick up your registration credentials including your show admission and shuttle bus ticket to McCormick Place East.
2—Visit the show at McCormick Place East on Monday, April 18. Shuttle busses will connect the Conrad Hilton and McCormick Place during the exposition hours Monday through Wednesday.
3—Following your Monday tour of the show, return to the Conrad Hilton for the Robotic Basics Session at 4:00 p.m. or 7:00 p.m., and the Robot Design Forum at 7:00 p.m.

4:00-6:00 P.M.—ROBOTIC BASICS WORKSHOP #2 (repeat)
Chair: *Thomas Blunt*, CMfgE, Manager, Advanced Production Technology Operation, Technology Division, General Electric, Louisville, Kentucky, U.S.A.
This two hour Basics Workshop will be jointly conducted by two well-known consulting firms. *Ronald Potter*, Robot Systems, Inc., will discuss the fundamentals of robotics, and *William Higgins*, Williams & Higgins Industrial Sales, will cover basic implementation and applications.

7:00-9:00 P.M.—ROBOTIC BASICS WORKSHOP #3 (repeat)
Chair: *Dale Hendrixson*, Director of Marketing, Cincinnati Milacron, Lebanon, Ohio, U.S.A.
Please see above description for details.

7:00-10:00 P.M.—ROBOT DESIGN FORUM
Chair: *Lou R. Paul*, Professor, School of Electrical Engineering, Purdue University, West Lafayette, Indiana, U.S.A.
"Types of Kinematic Chains Used in Robots". . .*James C. Colson*, Graduate Student, and *N. Duke Perriera*, Assistant Professor, Mechanical Engineering, University of Texas, Austin, Texas, U.S.A. **(20-1)**
"Development of a New Type of Mobile Robot TO-ROVER". . . *Masahaw Takano*, Professor, and *Gotoro Odawara*, Professor, University of Tokyo, Japan **(20-81)**
"Designing Custom Robots for In-Plant Use". . .*Robert A. Brown*, Sr. Mechanical Engineer Staff, Titlist Golf Division, Acushnet Company, Mattapoisett, Massachusetts, U.S.A. **(20-45)**
"Straightline Motion of a Manipulator through Hybrid Computation". . . *J. F. Steinman*, Research Assistant, and *R. D. Klafter*, Associate Professor, Dept. of Electrical and Computer Engineering, Drexel University, Philadelphia, Pennsylvania, U.S.A. **(20-69)**
"Travelling Assembly Robot". . .*Hisaaki Hirabayashi*, Research Engineer, Hitachi, Ltd., Yokohama, Japan **(20-19)**
"TBD"

TUESDAY, APRIL 19, 1983

Numbers listed in parentheses indicate the location of the paper in these proceedings.

DAY CHAIR: *RONALD POTTER*, VICE PRESIDENT, DIRECTOR OF OPERATIONS, ROBOT SYSTEMS, INC., NORCROSS, GEORGIA, U.S.A.
7:00-9:00 A.M.—CONFERENCE REGISTRATION
9:00 A.M.-12:00 NOON—CONCURRENT SESSIONS

APPLICATIONS: ROBOT MATERIAL HANDLING SYSTEMS

Co-Chairs: *K. Yonemoto*, Executive Director, Japan Industrial Robot Association, Tokyo, Japan, and *Robert Green*, National Sales Manager, Prab Robots, Inc., Kalamazoo, Michigan, U.S.A.

"Robotic Applications at Texas Instruments, Sherman Facility". . .*Kerry Arnold*, Manager, FAB Engineering, and *Rick Edwards*, Robot Engineer, Texas Instruments, Sherman, Texas, U.S.A. **(2-1)**

"Simulation of Multi-Machine Service by Industrial Robots". . .*H. J. Warnecke*, Professor, Dr.-Ing., *U. Schmidt-Streier*, Dr.-Ing., *M. Schweizer, Dr.-Ing., and A. Altenhein*, Dipl.-Ing., Fraunhofer Institute, Stuttgart, Federal Republic of Germany **(2-10)**

"Improving Your Machine Loading Robot Application Savvy". . .*Jeff Balcom*, Application Engineer, Prab Robots, Inc., Milford, Ohio, U.S.A.

"A Simple Robot System for Loading/Unloading Internal Grinders". . .*Terence G. Gandy*, Undergraduate Student, Omaha, Nebraska, U.S.A. **(2-25)**

RESEARCH: ROBOT PROGRAMMING LANGUAGES

Co-Chairs: *Carl Page*, Professor of Computer Science, Michigan State University, East Lansing, Michigan, U.S.A., and *A. Arnstrom*, The Swedish Institute for Production Engineering Research, Stockholm, Sweden

"What Can't Robot Languages Do?". . .*Barry I. Soroka*, Electrical Engineering, University of Southern California, Los Angeles, California, U.S.A. **(12-1)**

"Hierarchical Control of Integrated Manufacturing Systems". . .*R. Cassinis*, Sr. Researcher, Milan Polytechnic Artificial Intelligence Project, Milano, Italy **(12-9)**

"The SRI Robot Programming Systems (RPS)". . .*William T. Park*, Sr. Research Engineer, SRI International, Menlo Park, California, U.S.A. **(12-21)**

"Using ADA as a Robot System Programming Language". . .*R. A. Volz*, Professor; *T. N. Mudge*, Professor; *D. A. Gal*, University of Michigan, Ann Arbor, Michigan, U.S.A. **(12-42)**

Alternate Speaker:
"Evaluation of Commercially Available Robot Programming Languages". . .*William A. Gruver*, Vice President, LTI Robotic Systems; *Barry I. Soroka*, Professor of Electrical Engineering, University of Southern California; *John J. Craig*, Consultant, Stanford University; *Timothy L. Turner*, Systems Engineer, LTI Robotic Systems, Torrance, California, U.S.A. **(12-58)**

ROBOTIC OVERVIEW

Co-Chairs: *A. Morecki*, International Federation for the Theory of Machines and Mechanism, Warszawa, Poland, and *John DiPonio*, CMfgE, Manufacturing Program Specialist, Ford Motor Company, Dearborn, Michigan, U.S.A.

"Trends in the Robotic Industry (revisited): Where Are We Now?". . .*Laura Conigliaro*, Vice President, Prudential-Bache Securities, Inc., New York, New York, U.S.A. **(1-1)**

"The Challenge of Robotic Education". . .*Jack Lane*, Professor, GM Engineering and Management Institute, Flint, Michigan, U.S.A.

"Delphi Forecast of the Future of Robotics in the United States". . .*Don Smith*, Professor, University of Michigan, Ann Arbor, Michigan, U.S.A.

"The Unattended Factory—Fanuc's New Flexibility Automated Manufacturing Plant Using Industrial Robots". . .*Doug Urbaniak*, GM Fanuc Robotics Corporation, Troy, Michigan, U.S.A. **(1-18)**

12:00 NOON-1:30 P.M.—GROUP LUNCHEON • LUNCHEON SPEAKER: "TBD"
1:30-4:30 P.M.—CONCURRENT SESSIONS

APPLICATIONS: PROCESS SYSTEMS

Co-Chairs: *Bill Willis*, Sr. Administrator, Process Development, Central Foundry, General Motors Corporation, Saginaw, Michigan, U.S.A., and *Daniele L. Fabrizi*, President, SIRI, Bergamo, Italy

"Recent Proliferation in Industrial Artificial Vision Applications". . .*Philippe Villers*, President, Automatix, Inc., Billerica, Massachusetts, U.S.A. **(3-1)**

"Applications of Machine Vision to Parts Inspection and Machine Control in the Piece Part Manufacturing Industries". . .*Timothy Pryor*, President, Diffracto Ltd., Windsor, Ontario, Canada **(3-21)**

"Routing and Drilling with an Industrial Robot". . .*Thony Molander*, M.Sc., Linkoping Institute of Technology, Linkoping, Sweden **(3-37)**

"Visual Sensing System for 3-D Object Using High Light Pattern". . .*K. Takashima*, *T. Inari*, *M. Tanaka*, Product Development Laboratory, Mitsubishi Electric Corporation, Hyogo, Japan, and *Y. Tomioka* and *Y. Matsushita*, Sakai Works, Nippon Steel Corporation, Sakai, Japan **(3-48)**

Alternate Speaker:
"A New Robot with Voice, Hearing, Vision, Touch, Grasping, Controlled by One Microprocessor, with Mechanical and Electronic Integrated Design". . .*A. Rovetta*, Professor, Politechnico Milano, Italy **(3-57)**

RESEARCH: INTELLIGENT ROBOTS/ADAPTIVE CONTROL SYSTEMS

Co-Chairs: *Ichiro Kato*, Professor, Waseda University, Tokyo, Japan, and *Dennis Derfiny*, Communications Group, Motorola, Inc., Schaumburg, Illinois, U.S.A.

"Adaptive Control of the T3 Robot". . .*Lloyd Clark*, Undergraduate Student, Massachusetts Institute of Technology, Cambridge, Massachusetts, U.S.A. **(13-1)**

"Adaptive Control of Manipulators in the Task Oriented Coordinate Space". . .*Albo Balestrino*, Professor; *Giuseppe DeMaria*, Associate Professor; and *Lorenzo Sciavicco*, Professor, Instituto Elettrotecnico, University of Naples, Naples, Italy **(13-13)**

"Hierarchical Control for Robots in an Automated Factory". . .*James Albus*, *Charles McLean*, *Anthony Barbera*, and *M. L. Fitzgerald*, National Bureau of Standards, Washington, D.C., U.S.A. **(13-29)**

"Application of Sensor Controlled Robots for Fettling of Castings". . .*R. D. Schraft*, Dr.-Ing.; *M. Schweizer*, Dr.-Ing.; *E. Abele*, Dipl.-Ing., and *W. Sturz*, Dipl.-Ing., Fraunhofer Institute for Manufacturing Engineering and Automation, Stuttgart, Federal Republic of Germany **(13-44)**

Alternate Speakers:
"A Guidance System for a Mobile Robot". . .*M. Julliere*, Professor; *L. Marce*, Professor; *H. Place*, Professor, Institut National Des Sciences Appliquees, Rennes, France **(13-58)**

"Some Problems of Mechanics and Control for Pneumatic Industrial Robots". . .*A. Yu Ishlinsky*, Academician/Director; *F. L. Chernousko*, Professor/Department Chief, and *V. G. Gradetsky*, Dr., Robotics Lab Chief, Institute for Problems in Mechanics, USSR Academy of Sciences, Moscow, U.S.S.R. **(13-69)**

IMPLEMENTATION

Co-Chairs: *Robert Fish*, CMfgE, Supervisor, Industrial Engineering-Special Projects, Hydra-Matic Division, General Motors Corporation, Ypsilanti, Michigan, U.S.A., and *Goran Lundstrom*, Technical Editor, NY TEKNIK, Stockholm, Sweden

"Effective Interfacing through End Effectors". . .*Rudolph F. Mutter*, Sr. Robotics Engineer, ASEA, Inc., White Plains, New York, U.S.A. **(4-1)**

"Two Hands Are Better Than One". . .*Chris Reed*, Sr. Applications Engineer, Prab Robots, Inc., Kalamazoo, Michigan, U.S.A. **(4-12)**

"Strategic Economic Models: The Future of Robotic Justification". . .*John Van Blois*, Manager of Robotics Education, IBM Corporation, Boca Raton, Florida, U.S.A. **(4-24)**

"Robotic Implementation—Do It Right". . .*Pat John Rosato*, Director, Technical Training/Publications, Unimation, Inc., Danbury, Connecticut, U.S.A. **(4-32)**

Alternate Speaker:
"Electricity Use Impact of Industrial Robots". . .*Alan Letzt*, Program Vice President, and *Scott E. Fenton*, Sr. Manager, *Robert L. Hershey*, P.E., Vice President, and *Alan M. Letzt*, P.E., Program Vice President, Science Management Corporation, Washington, D.C., U.S.A. **(4-50)**

4:30-6:00 P.M.—WINE AND CHEESE RECEPTION
The RI/SME Technical Council cordially invites the 13th ISIR/Robots 7 Conference attendees to join them for a social gathering
6:30 P.M.-12:00 MIDNIGHT—INTERNATIONAL BANQUET.

WEDNESDAY, APRIL 20, 1983

Numbers listed in parentheses indicate the location of the paper in these proceedings.

DAY CHAIR: *BRIAN FORD*, CONSULTANT, FORD INDUSTRIES, MAHOPAC, NEW YORK, U.S.A.
8:00-9:00 A.M.—CONFERENCE REGISTRATION
9:00 A.M.-12:00 NOON—CONCURRENT SESSIONS

APPLICATIONS: ROBOTIC ASSEMBLY

Co-Chairs: *Q. Huang*, Deputy Director, Machine Design Division, Peking Institute of Aeronautics and Astronautics, Peking, China, and *Gordon VanderBrug*, Product Line Manager, Automatix, Inc., Billerica, Massachusetts, U.S.A.

"Robotic Assembly of Floppy Disk Drives" . . .*Ronald T. Albo*, Manager, Advanced Production Technology, Shugart Associates, Sunnyvale, California, U.S.A. **(5-1)**

"Robotic Assembly: Design, Analysis and Economic Evaluation". . .*Peter B. Scott*, Roboticist, Centre for Robotics, and *Tom M. Husband*, Professor of Engineering Manufacture, Imperial College, London, England **(5-12)**

"Application of Industrial Robots for Assembly Operations in the Automobile Industry" . . .*H. J. Warnecke*, Professor, Dr.-Ing.; *D. Schraft*, Dr.-Ing.; *M. Schweizer*; *J. Walther*, Dipl.-Ing.; Fraunhofer Institute for Manufacturing Engineering and Automation, Stuttgart, Federal Republic of Germany **(5-30)**

"Robotic Assembly of Keycaps to Keyboard Arrays". . .*Donald Wilt*, Sr. Mechanical Engineer, and *Gordon VanderBrug*, Product Line Manager, Automatix, Inc., Billerica, Massachusetts, and *Jim Davis*, Principal Manufacturing Engineer, Digital Equipment Corporation, Phoenix, Arizona, U.S.A. **(5-40)**

Alternate Speaker:
"A Modular Programmable Assembly Station" . . .*Randall Smith*, Research Engineer, and *David Nitzan*, Director, Robotics Department, SRI International, Menlo Park, California, U.S.A. **(5-53)**

RESEARCH: ADVANCED VISION AND TACTILE SENSORS

Co-Chairs: *Roger Nagel*, Professor and Director, Institute for Robotics, Lehigh University, Bethlehem, Pennsylvania, U.S.A., and *Yow-Lam Oh*, Executive Director, Centre for Robotics, Chisholm Institute for Technology, Melbourne, Victoria, Australia

"Problems of Vision-Directed Robot Manipulators in an Unstructured Parts Handling Environment". . .*Chuck Raymond*, Graduate Student; *Max Donath*, Assistant Professor, University of Minnesota, and *Richard Olson*, Manager, Honeywell, Inc., Minneapolis, Minnesota, U.S.A. **(14-1)**

"Development of a Machine-Vision Based Robotic Arc-Welding System". . .*Jan Kremers*, *Carl Blahnik*, *Alfred Brain*, *Ronald Cain*, *Jeff DeCurtins*, *Jose Mesequer*, and *Norman Peppers*, SRI International, Menlo Park, California, U.S.A. **(14-19)**

"Piezoelectric Polymers: New Sensor Materials for Robotic Applications". . .*P. Dario*, Dr., *C. Domenici*, *R. Bardelli*, *D. DeRossi*, and *P. C. Pinotti*, Centro "E. Piaggio", Faculty of Engineering, Pisa, Italy **(14-34)**

"Robot Vision for a Hierarchical Control System". . .*M. Shneier*, *E. Kent*, *J. Albus*, *P. Mansbach*, *M. Nashman*, *L. Palombo*, *W. Rutkowski*, and *T. Wheatley*, National Bureau of Standards, Washington, D.C., U.S.A. **(14-50)**

Alternate Speaker:
"Programming a Vision System". . .*G. Cividini*, Assistant Researcher; *G. Villa*, Assistant Researcher, and *M. Gini*, Research Associate, Department of Electronics, Politecnico, Milano, Italy **(14-67)**

RESEARCH: KINEMATICS

Co-Chairs: *Bernard Sallot*, Executive Director, Robot Institute of America, Dearborn, Michigan, U.S.A., and *P. Lal Tummala*, Professor, Electrical Engineering & Systems Science Department, Michigan State University, East Lansing, Michigan, U.S.A.

"A Geometric Approach in Solving the Inverse Kinematics of PUMA Robots". . .*C.S.G. Lee*, Assistant Professor of Electrical and Computer Engineering, and *M. Ziegler*, University of Michigan, Ann Arbor, Michigan U.S.A. **(16-1)**

"Interactive Procedure for Computer-Aided Design of Industrial Robot Mechanisms". . . *Potkonjak Veljko*, Ph.D., Assistant Professor, Electrotech Faculty; *Miomir Vukobratovic*, D.Sc., Head of Robotics Dept., and *Dragen Hristic*, Ph.D., Leading Researcher, Mihailo Pupin Institute, Beograd, Yugoslavia **(16-85)**

"A New Approach to Coordinate Transformation: Parallel Computation". . .*L. Mancini*, *N. Nicolosi*, *S. Scarselli*, and *G. Zoly*, Associate Researchers, Politecnico di Milano, Milano, Italy **(16-19)**

"Kinematics of Major Robot Linkage". . .*V. Milenkovic*, Principal Research Engineer, and *B. Huang*, Sr. Research Engineer, Manufacturing Process Lab., Ford Motor Company, Redford, Michigan, U.S.A. **(16-31)**

Alternate Speaker:
"A Design of a Controller as a Component of a Robotic Manufacturing System". . .*Paul Fussell*, Research Assistant; *Paul K. Wright*, Associate Professor, Mechanical Engineering, and *D. A. Bourne*, Project Leader, Robotics Institute, Carnegie-Mellon University, Pittsburgh, Pennsylvania, U.S.A. **(16-48)**

12:00 NOON-1:30 P.M.—GROUP LUNCHEON
Luncheon Program: The 13th ISIR/Robots 7 marks the official "Changing of the Guard" for RI/SME Officers and Board Members. The 1982 President, Walter Weisel, will welcome the incoming President, Thomas Blunt, who will preview the highlights of RI/SME activity for 1983-84.
1:30-4:30 P.M.—CONCURRENT SESSIONS

APPLICATIONS: MIG AND TIG WELDING WITH ROBOTS

Co-Chairs: *Chuck Clark*, Regional Sales Manager, Industrial Robot Division, Cincinnati Milacron, Inc., Lebanon, Ohio, U.S.A., and *J. B. Eijlers*, Drs., Director, Meininger Automation b.v., Rijswijk, Holland

"Feasibility Studies for Small Batch Robotic Arc Welding". . .*P. K. Wadsworth*, *R. G. Thorne*, *J. E. Middle*, Loughborough University, Loughborough, England **(6-1)**

"Integration of a Robotic Welding System with Existing Manufacturing Processes". . . *Richard Wolke*, Manufacturing Engineer, General Electric, Lynchburg, Virginia, U.S.A.

"Robotic Welding at Honeywell". . .*John A. Kallevig*, P.E., Principal Application Engineer—Robotics, Honeywell Corporation, Golden Valley, Minnesota, U.S.A. **(6-15)**

"Flexible Automation in the Shipbuilding Industry". . .*James B. Acton*, Manager, Research and Development, Todd Pacific Shipyards Corporation, San Pedro, California, U.S.A.

Alternate Speaker: (6-26)
"Position Sensing with an Electric Arc". . . *George E. Cook*, Manager, Research and Development, CRC Welding Systems, Inc., Nashville, Tennessee, U.S.A. **(6-46)**

RESEARCH: CAD/CAM

Co-Chairs: *A. J. Niemi*, Department of Electrical Engineering, Helsinki University of Technology, Finland, and *Ronald J. Meyer*, CAD/CAM Project Manager, Rockwell International, Downey, California, U.S.A.

"A Graphical Emulation System for Robot Design and Program Testing". . .*Rudiger Dillmann*, Dr.-Ing., Universitat Karlsruhe, Karlsruhe, Federal Republic of Germany **(7-1)**

"Parametric Design of Robotic Grippers on an Interactive Graphics System". . .*R. Scott Leckie*, Managing Director, SDI: Computer Aided Technologies, Inc., Cleveland, Ohio, U.S.A. **(7-16)**

"Geometric Modelling: A Contribution Towards Intelligent Robots". . .*A. De Pennington*, Sr. Lecturer, *Dr. Susan Bloor*, Lecturer, and *Mazine Balila*, Research Student, The University of Leeds, Leeds, United Kingdom **(7-35)**

"Evaluating Robotic Systems through Simulation". . .*T. L. Welch*, Sr. Systems Analyst, Pritsker & Associates, Inc., West Lafayette, Indiana, U.S.A. **(7-55)**

Alternate Speaker:
"Analysis of Robot Performance Operation" . . .*Nicholas Dagalakis*, Research Engineer, National Bureau of Standards, Washington, D.C., U.S.A. **(7-73)**

ROBOTIC PROGRAMS IN EDUCATIONAL COMPLEXES

Co-Chairs: *Edward Konopka*, CMfgE, Professor, Robotics Technology, Oakland Community College, Auburn Heights, Michigan, U.S.A., and *P. Rabischong*, Unite de Recherches Biomechaniques, Montpellier, France

"Training—The Key to Success in the Utilization of Robots". . .*David Osborne*, Instructor, ASEA, Inc., Troy, Michigan, U.S.A. **(15-1)**

"How Should the University Teach Robotics?" . . .*Barry I. Soroka*, Electrical Engineering, University of Southern California, Los Angeles, California, U.S.A. **(15-12)**

"The Use of Intelligent Chips in Educational Robot Systems". . .*B. Schroer*, *G. Workman*, and *L. Fullerton*, The University of Alabama in Huntsville, Huntsville, Alabama, U.S.A. **(15-22)**

"Education in Robotics in Poland". . .*A. Morecki*, Professor, *Kaczmarczyk*, and *Strzelecki*, Technical University of Warsaw, Warsaw, Poland **(15-33)**

Alternate Speaker:
"Robot Education for Engineers in Manufacturing". . .*Gary Hammond*, GM Engineering and Management Institute, Flint, Michigan, U.S.A. **(15-51)**

7:00-10:15 p.m.—THE AEROSPACE CHALLENGE FORUM.

7:00-10:15 P.M.—**ROBOTIC TECHNOLOGY TRANSFER: THE AEROSPACE CHALLENGE FORUM**

Chair: *Philip Barone*, CMfgE, Sr. Development Engineer, Grumman Aerospace Corporation, Bethpage, New York, U.S.A.

7:00-8:00 P.M.—Part I, Advanced Robotic Applications For Aerospace

Chair: *Clint Burdick*, Staff Scientist, Littom Industries, Beverly Hills, California

"Populating a Printed Circuit Board with Non-Standard Electronic Components". . .*Keith Kerstetter*, Applications Engineer, IBM, Systems Products Division, Boca Raton, Florida, U.S.A.

"Light Machining and Assembly with Robots in Aerospace". . .*Robert L. Scherrer*, Applications Engineer, Cincinnati Milacron, Industrial Robot Division, Lebanon, Ohio, U.S.A.

"Robotic Drilling and Routing Applications in Aerospace". . .*Thomas Bopp*, Applications Development Engineer, ASEA, Inc., Industrial Robot Division, White Plains, New York, U.S.A.

"Future Robotic Applications in Aerospace". . .*Klaus W. Nielson*, Manager of Product Engineering, Unimation, Danbury, Connecticut.

"To be Announced". . .*Harry D. Mackie*, Director-Marketing, GCA Corporation, Industrial Systems Group, Naperville, Illinois, U.S.A.

"Application of Hydraulic Robots in Aerospace". . .*John Fleckenstein*, Regional Sales Manager, Prab Robots, Inc., Kalamazoo, Michigan, U.S.A.

"Wire Harnessing with Robots in the Aerospace Industry". . .*William F. Rooney*, Vice President—Marketing, American Robot Corporation, Clinton, Pennsylvania, U.S.A.

8:00-8:30 p.m.—Part II, Interactive Group Discussion Led by a Panel of Robot Manufacturers.

9:00-9:30 p.m.—Part III, Software for Robotic/Aerospace Applications and Future Needs

Chair and Speaker: *Robert B. Thornhill*, Principal Computer Systems Designer—Robotics, Martin-Marietta Data Systems, Orlando, Florida, U.S.A.

Topics to be Covered: On-Line Programming of Robots. . .Off-Line Programming of Robots. . .Animation of Robots and Robot—Tended Manufacturing Work Cells. Presentation is based on an extensive industry survey.

9:30-9:45 p.m.—Part IV, Current and Future Aerospace Robotic Applications

Chair and Speaker: *Wallace J. Plumley*, Group Engineer of Automation Sciences and Manufacturing Research, Lockheed-Georgia, Marietta, Georgia, U.S.A.

Presentation is based on a recent extensive survey of robot users and manufacturers in aerospace and related industries.

9:45-10:15 p.m.—Part V, Interactive Group Discussion Led by a Panel of Users.

THURSDAY, APRIL 21, 1983

Numbers listed in parentheses indicate the location of the paper in these proceedings.

DAY CHAIR: *ANTHONY CLARKE*, MARKETING AND SALES MANAGER, NORDSON CORP., AMHERST, OHIO, U.S.A.
8:00-9:00 A.M.—CONFERENCE REGISTRATION
9:00 A.M.-12:00 NOON—CONCURRENT SESSIONS

APPLICATIONS: FINISHING

Co-Chairs: *T. E. Brock*, Publisher & Executive Secretary, *Industrial Robot* Magazine, IFS Ltd., & British Robot Association, Bedford, United Kingdom, and *Stephen T. Kostelec*, Manager, Manufacturing Services, GR Manufacturing Company, Grand Rapids, Michigan, U.S.A.
"Automatic Grinding". . .*Asbjorn Mortensen*, Researcher, Rogaland Research Institute, Stavanger, Norway **(8-1)**
"Robots Towards 100% Uptime". . .*John Goodwin*, Tool and Process Engineer, General Motors of Canada, Ontario, Canada **(8-12)**
"Cleaning of Castings—A Typical Job for a Robot". . .*Lars Gustafsson*, Systems Design Engineer, ASEA, Inc., White Plains, New York, U.S.A. **(8-24)**
"Thermal Spray Robots". . .*Gary Irons*, Manager, Systems Engineering, and *Lou George*, Manager, Systems Marketing, METCO, Inc., Westbury, New York, U.S.A. **(8-34)**
Alternate Speaker:
"Development of a Robot-Integrated Workingplace for the Brushing of Rubber-Metal Pats". . .*G. Spur*, *G. Seliger*, and *W. Felsing*, Fraunhofer Institute for Manufacturing Engineering and Automation, Berlin, West Germany **(8-52)**

RESEARCH: ADVANCEMENT IN NON-BINARY VISION

Co-Chairs: *James West*, CMfgE, Vice President, Perceptron, Inc., Farmington Hills, Michigan, U.S.A., and *H. J. Warnecke*, Professor, Dr.-Ing., Fraunhofer Institute for Productivity and Automation, Stuttgart University, Federal Republic of Germany
"A Hierarchical Vision System for Object Identification and Localization". . .*A. Riad*, Dr., and *M. Briot*, LAAS du CNRS, Toulouse, France **(17-1)**
"An Interactive Software System for Developing Robot Vision Algorithms". . .*Peter G. Selfridge*, Member of Technical Staff, Bell Laboratories, Holmdel, New Jersey, U.S.A. **(17-11)**
"Industrial Computer Vision by Mathematical Morphology". . .*Stanley Sternberg*, Ph.D., President, Cytosystems Corporation, Ann Arbor, Michigan, U.S.A. **(17-24)**
"Grey Scale Robot Vision for Real Time Inspection and Assembly". . .*Arkady Makhlin*, Advisory Engineer, and *Glenn Tisdale*, Manager, Information Technology, Westinghouse Electric Corporation, Pittsburgh, Pennsylvania and Baltimore, Maryland, U.S.A. **(17-36)**
Alternate Speaker:
"3D Camera Based on Synchronized Scanning". . .*Marc Rioux*, Research Officer, National Research Council, Ontario, Canada **(17-48)**

HUMAN FACTORS AND SAFETY

Co-Chairs: *Y. Hasegawa*, Professor, Waseda University, Systems Science Institute, Tokyo, Japan, and *Sandra Pfister*, President, R.T.S.I., New Baltimore, Michigan, U.S.A.
"The R Factor of Industrial Robot Safety". . . *Joseph P. Ziskovsky*, Robotics Manager, GCA PaR Systems, St. Paul, Minnesota, U.S.A. **(9-1)**
"Fault-Tree Analysis of Hazards Created by Robots". . .*Kunitomo Kawagucki*, Director, and *Noboru Sugimoto*, Researcher, The Research Institute of Industrial Safety, Tokyo, Japan **(9-13)**
"Human Factors Effort in Robotic System Design". . .*Michael Olex*, Sr. Project Engineer, Nordson Corporation, Amherst, Ohio, and *Harvey Shulman*, Associate Professor of Psychology, Ohio State University, Columbus, Ohio, U.S.A. **(9-29)**
"Human Factors Engineering in Robotics" . . .*H. McIlvaine Parsons*, Ph.D., and *Greg Kearsley*, Ph.D., Human Resource Research Organization, Alexandria, Virginia, U.S.A. **(9-40)**
Alternate Speaker:
"Retrofitting Safety Devices on Industrial Robots". . .*Ronald Potter*, Vice President, Application Development and Engineering, Robot Systems, Inc., Norcross, Georgia, U.S.A.

12:00 NOON-1:30 P.M.—GROUP LUNCHEON—"The Advent of Personal Robots"
LUNCHEON SPEAKER: *Douglas Bonham* and *James Lytle*, Heath Company, St. Joseph, Michigan, USA

1:30-4:30 P.M.—CONCURRENT SESSIONS

APPLICATION DEVELOPMENT

Co-Chairs: *George Munson*, CMfgE, Vice President, Systems Division, Unimation, Inc., Danbury, Connecticut, U.S.A., and *F. Denis*, Consultant, TEAM CONSULT, Brussels, Belgium
"Robotic Fastener Installation in Aerospace Subassembly". . .*D. Mark Lambeth*, Manufacturing Technology Engineer, General Dynamics, Fort Worth, Texas, U.S.A. **(10-1)**
"Visually Guided Arc-Welding Robot with Self-Training Features". . .*W. Baudot*, *G. Clermont*, *P. Gaspart*, and *P. Lecocq*, Engineers, C.R.I.F., Brussels, Belgium **(10-13)**
"Analytical Balance Operation with an Assembly Robot". . .*H. N. Nunnally*, Research Engineer, and *E. G. Craze, Jr.*, Research Engineer, Philip Morris Research Center, Richmond, Virginia, U.S.A. **(10-31)**
"A Robotic Cell for Application Evaluation and Training". . .*Alexander Holzer*, Research Scientist, Commonwealth Scientific and Industrial Research Organization, Fitzroy, Victoria, Australia **(10-42)**
Alternate Speaker:
"The Australian Robotic Sheep Shearing Research and Development Programme". . . *P. C. Wong*, and *P. R. W. Hudson*, Australian Wool Corporation, Sydney, Australia **(10-56)**

RESEARCH: END EFFECTORS

Co-Chairs: *M. S. Konstantinov*, Professor, Head of Central Laboratory for Manipulators and Robots, Higher Institute for Mechanical & Electrical Engineering, Sofia, Bulgaria, and *J. L. Nevins*, Division Leader, Robotic and Assembly Systems Division, Charles Stark Draper Laboratory, Inc., Cambridge, Massachusetts, U.S.A.
"End Effector Technology and Programmed Automatic Exchange". . .*Allen J. Wright*, Mechanical Engineer—R & D, Intelledex, Inc., Corvallis, Oregon, U.S.A. **(18-1)**
"Robotic Drilling End Effector Test Equipment for Research and Development and Production Support". . .*Jon M. Heywood*, Manufacturing Technology Engineer, General Dynamics, Fort Worth, Texas, U.S.A. **(18-14)**
"A Sensory Gripper for Handling Textiles". . . *D. R. Kemp*, *G. E. Taylor*, *P. M. Taylor*, and *A. Pugh*, Department of Electronic Engineering, University of Hull, Humberside, England **(18-23)**
"Advanced Composite Material Handling Applications". . .*Kenneth B. Bubeck*, Project Manager, Rensselaer Polytechnic Institute, Troy, New York, U.S.A. **(18-34)**

APPLICATIONS: GENERAL

Co-Chairs: *Paul A. Misegades*, CMfgE, Manager, Equipment & Tool Services, Advance Production Technology Operation, General Electric Company, Louisville, Kentucky, U.S.A., and *H. Knudsen*, Manager, Production Engineering & Quality Assurance Office, ASEA A/S, Odense C, Denmark
"Automated Body Systems from the Ground Up". . ., *Frank A. DiPietro*, Director, Production Engineering Activity, Fisher Body Division, General Motors Corporation, Warren, Michigan, U.S.A. **(11-1)**
"Robots in Aerospace: A Multi-Tasking Robot for the Manufacture of Precision Guided Projectile Structures". . .*Larry L. Michaud*, Sr. Staff Engineer, Martin Marietta Aerospace, Orlando, Florida, U.S.A. **(11-14)**
"NASA's Use of Robotics in Building the Space Shuttle". . .*Ken Fernandez*, *C.S. Jones*, and *M. L. Roberts*, Marshall Space Flight Center, Huntsville, Alabama, U.S.A. **(11-35)**
"Product Design for Robotic Assembly". . .*J. Ronald Bailey*, Sr. Systems Engineer, Systems Products Division, IBM Corporation, Boca Raton, Florida, U.S.A. **(11-44)**
Alternate Speaker:
"Simple Robotics Systems for Industrial Applications". . .*James B. Canner*, Vice President, Sterling Detroit Company, Detroit, Michigan, U.S.A. **(11-58)**

CHAPTER 12

ROBOT PROGRAMMING LANGUAGES

What Can't Robot Languages Do?

Barry Irvin Soroka
University of Southern California
Los Angeles, California

Abstract

Today's robot programming languages are deficient in a variety of ways. They separate the control level from the user interface so that new algorithms cannot be easily tested. They communicate only stumblingly with external devices. They achieve completeness by requiring complexity: simple things are often hard to do. They cannot adapt to changes in architecture such as adding an arm or adding a processor. They emphasize the procedural component of robot programming making it difficult to program multiple arms working together. Some solutions to these difficulties are proposed.

The Current Scene.

Numerous robot programming languages (RPLs) have appeared since Unimation first introduced VAL in 1979. Here is a catalog of the current commercial offerings:

language	designer
AL	Stanford, Robot Technology Inc
AML	IBM
HELP	GE
MCL	McDonnell-Douglas, USAF
RAIL	Automatix
VAL	Unimation

Despite their profusion, RPLs have had little impact on the factory. Most robots are still taught by guiding: you lead the arm thru the motions and then ask the robot to repeat that sequence for a very long time.

If almost nobody uses them, then why do RPLs keep appearing? The answer is that a robot programming language makes a robot more marketable, since the vendor can claim up-to-the-minute software as well as up-to-the-minute hardware and mechanisms.

The sales brochures are full of pronouncements about what robot programming languages *can* do. This paper examines some of the things these languages *can't* do today, and aren't likely to do without extensive rethinking.

The One-Robot, One-Terminal Syndrome.

The typical RPL-equipped robot comes today as a robot plus a computer box. The box hides two things: a controller and a language processor. The controller generates trajectories and servos the joints. The language processor accepts the robot program from the user and translates the language into commands for the controller. This is not an economical use of resources since (hopefully) the robot spends more time doing work than the user spends programming. The language processor is idle while the robot is working.

Some companies are taking steps to make better use of computer resources by sharing one language processor among several robot controllers. This is a step in the right direction, but it is still too low a level of integration, as we shall see later on.

Robots and CAD/CAM.

Today, each robot stands in splendid isolation --- equipped with its keyboard, its teach-box, its CRT, its arm, and its processors --- majestically ignorant of other computing nearby. This paradigm has limited utility for the factory of today or the factory of the future. Robots must be integrated into existing and projected CAD/CAM facilities. Why can't we use our existing CAD databases to speed up the robot programming process? Why can't we collect production statistics from a robot for inclusion in CAM reports and plans? Serious thought and many experiments are needed before promulgating standards for information interchange among factory components, robots included.

Understanding the Factory.

Today's RPLs do not understand the other machines in the factory. Robot languages are full of small atomic statements which let you do practically anything, but they do not include concepts describing typical factory components. Here are a number of examples:

In starting a machine like a lathe, the typical robot program executes something like the following:

(1) send the signal to close the chuck,
(2) check that the chuck has closed.

Why can't we have robot language primitives which understand this primitive dance between machines as a single construct? GE's HELP language has taken a step in the right direction with its STROKE statement:

```
stroke(feeder,sensor,kicker,tries).
```

This commands the program to pulse the line called FEEDER, then check the SENSOR line to see if a part has been delivered. If not, then pulse the line called KICKER to try dislodging a jammed part; TRIES is the number of times to try this before causing an error and summoning human assistance.

Conveyors are another source of problems in robot programming. Many robots are equipped with software for following a moving line, but the line must be straight and it must be parallel to the X or Y axis of the robot. This is a simplification of what we actually find in factories, where conveyors don't have to be straight lines. What we need is the software which lets you teach a task while the parts are not moving, accepts a description of the arbitrary path of the conveyor, and relocates the task automatically while the line is in motion.

Pallets are one of the prime reasons for using a robot programming language, since it is cumbersome to teach manually every position on a pallet. Typically, an RPL lets you compute the position of the (i,j)-th element on a pallet using a construct like:

base + (i-1)*dRow + (j-1)*dCol,

where BASE is the position of element (1,1) and dRow and dCol are the vector directions and spacings along the row and column directions. But programming applications with pallets can be made much simpler. When the user declares a pallet to have M rows and N columns, the robot should ask for positions it needs to know:

Please teach the position of element (1,1).
Please teach the position of element (M,1).
Please teach the position of element (1,N).
Please teach the position of element (M,N).

The position (M,N) would be used as a check. This sort of primitive understanding of the pallet concept would be easy to build into a robot language, and some work on this has been done at Stanford. Further work is needed to incorporate an understanding of 3-D pallets and dunnage into RPLs.

Flexible Programs.

Current RPLs are installation-specific. Programs describe motions of specific individual arms rather than tasks to be performed. Adding an extra robot to the workstation means you have to rewrite the robot program, because the assignment of tasks to arms is static and fixed in advance. In a workspace with two robots, the loss of one robot usually means the program cannot continue, even if the first robot is able to reach the locations serviced by its partner. If a robot becomes idle, there is no way it can move to assist other arms in their work.

RPLs cannot handle the foreground-background problem, where a robot might service a lathe (foreground) but perform some other task (background) at times when the lathe is operating. Current programs enforce a lock-step discipline to interaction among robots and machines, and programming such activities explicitly is very hard to do, even when the language contains constructs for concurrent actions. RPLs today force the user to think procedurally, explicitly anticipating all situations which might occur. What we need is a declarative language with notions of priority and interrupts. There should be mechanisms for describing parallel activities: At what points can a motion sequence be interrupted safely? What is the relative priority of two tasks? The exact decision of which arm does what task can then be left to an intelligent scheduling program.

Obstacles.

Despite considerable research, we seem to be years away from programs which can efficiently plan paths around obstacles in the workspace. Nonetheless, today's robot systems could easily detect imminent collisions by checking the proposed next micro-motion against a list of known obstacles.

The difficulty is: how do we tell the robot system about the obstacles? This is something which current RPLs could handle if the obstacles were modelled by some small set of primitives, say half-planes and spheres. Teaching or textually specifying such obstacles is very simple, and testing for imminent collision is computationally fast. More complex obstacles might be composed of boolean combinations of these primitives: AND, OR, and NOT.

Why a Robot System is like Babel.

The software of a modern robot contains a variety of subsystems, including: a parser which checks the syntax of user statements; an interpreter which executes the user's commands; a trajectory generator which computes the path the tool should take; an arm solver which converts from motion of the tool tip to motion of the joints; a servo system for efficiently driving each joint to where it's told to be. These processes are typically distributed over a number of processors, the current trend being one central processor plus a smaller processor at each joint. Each computer in the system offers a different complement of languages in which to program. The RPL of the system is yet another language, but it's not powerful enough to describe itself.

Currently, each component of a robot system is programmed in a fairly low-level language, insuring speed, but also insuring non-portability. The cost of moving a robot system from one set of hardware to another is enormous. Vendors must maintain an in-house stable of programmers familiar with a whole host of computer programming languages. It is in the vendors' interest to find a single high-level language capable of hosting all of the software of a robot system. Current robot software is heavily tied to the hardware architecture for which it was developed. You can upgrade an LSI-11/03 to an 11/23, and you can upgrade a 6502 to a 6503, but heaven help you if you try to move your robot software from one vendor's computer to another's.

Limits to Growth.

Great effort is currently required to expand a robot system by adding another processor to its current architecture, yet this is exactly what will soon be happening. Microcomputers are getting cheaper and more powerful, and it makes sense to "farm out" some of the computation now done centrally. Similarly, it will soon be economically feasible to design and fabricate special-purpose chips for robot algorithms.

Vast amounts of recoding will be required for current robot systems to grow by adding these new processors. Part of the bottleneck is architectures, and part is languages. We need robot system architectures which are flexible and evolutionary. The RS-232 interface is about the only standard we have for one machine talking to another, and this is exactly how most robots talk to add-on vision systems. But different, more efficient interfaces will be needed for a host of other devices: smart sensors, chips, and new processors.

Current languages --- neither RPLs nor the languages in which they are implemented --- simply fail to supply the concepts required in a future robot system. The ideal language for robot programming and implementation must contain facilities for handling:

- description of the processors available to the system;
- description of the interconnections available;
- estimation of how long it takes a given processor to run a given program;
- distributing processes among available processors;
- graceful degradation (surviving the loss of a processor);
- process migration (moving a running process from one processor to another);
- fault tolerance (detecting and correcting errors);
- optimization (locating performance bottlenecks and taking action to widen them).

Languages with these facilities are still in the research stages, but robot implementors should aim towards software of this flexibility.

Robot Independence.

Robot languages and systems are heavily dependent on the robots with which they are marketed. Languages for cartesian robots (those with XYZ-sliding joints) require that users describe the motions in terms of the available joints; programs in such languages are useless for robots of other types. Worse, these languages make the user think about the joints and not about the task. It makes no sense to learn a new robot language each time a new robot comes into the shop. The software investment in robot programs is too great to have to reprogram continuously from language to language. Companies with many robots from many vendors are already feeling pressure on this point.

What we need are vendors of robot-independent software: languages and systems which can interface to many robots, of many types, from many vendors. The reason this has been so difficult in the past is that robot vendors regard the interface to their robots as proprietary information; this often means that the interface is undocumented, prone to bugs, and ugly. But I think such independent language vendors will appear eventually, when the time is ripe for robot interface standards. After all, the robot builder is an expert on mechanisms and manufacturing, not necessarily an expert at software, which is a completely different exercise. And only a small portion of profits from robot sales comes from the software.

Robot-independent languages will need a standard language for describing robots, so that the system knows enough about the robot to adapt the programs written for one robot to run without textual change on another. This robot description language must be rich enough to answer a number of questions about the robot:

How many joints does it have?
What types of joints?
What are the joint limits?
What is its reach?

What are its nominal and actual parameters?
What is the shape of the links?

Already, the vendors of robot simulators feel the need for such a robot description language. With new robots appearing twice a month, how else can they keep their library of robot models up to date?

A robot system needs calibration data on each individual arm it controls. Today you cannot even move a robot program from one robot to another which is theoretically identical: manufacturing errors make each robot different. The specs may say 90 degrees on paper, but in practice it's plus or minus 0.01 degrees, and that makes for considerable error if the robot has a reach of six feet. How can we ship a robot program across the country to a different make of robot if we can't even move it across the hall to its "twin"?

Completeness and Complexity.

RPLs have been getting more ornate as vendors try to make their languages complete: they should be able to do everything. (One vendor boasted that his RPL was so complete that his best programmer was able --- in an afternoon --- to write an interpreter for one of the other RPLs.) This completeness has cost a considerable price in complexity: sure, this language can do everything, but it takes terrific work to get even the simplest task programmed. An old rule of software engineering says that the effort should match the gain: simple things should be easy to do, and complex things should take a little more work. Robot programmers are not all alike, and a robot programming language should offer a variety of levels of expression, ranging from simple teaching-by-guiding to complex textual programming.

Basic Issues.

One of the reasons why current RPLs are used so little is that there are certain practical questions to which we have no answers, and on which little research has been done.

Who will be programming the robots and using these languages? Will it be

- assembly-line workers?
- factory foremen?
- CAD operators?
- NC part programmers?
- manufacturing engineers?
- Computer Science PhDs?

Nobody knows. What type of language is optimal for each of these types of programmers? And how can we design languages of the appropriate level if we don't know who our ultimate user is going to be?

Another problem involves the many modalities available for human-robot interaction:

- text (RPLs);
- graphics (off-line simulators);
- motion (teaching-by-guiding);
- audio (voice input/output).

What mix of these modalities is optimal for programming a robot task? Many more human factors experiments are needed in this area.

Conclusions.

We have described some of the defects of current robot programming languages, things they do badly or not at all. Some tentative areas for improvement have been identified based on knowledge we have today. Finally we set forth some basic problems on which experiments are needed before robot programming languages will be widely and effectively used.

Hierarchical Control of Integrated Manufacturing Systems

Riccardo Cassinis
Milan Polytechnic Artificial Intelligence Project
Milano, Italy

Abstract

Although many robot programming languages are available, very few efforts have been made for connecting different machines (robots and tool machines) that cooperate in the same technological process. No standard connections systems are available, and synchronization means provided by manufacturers are usually inadequate. This paper proposes a simple connection solution based on a centralized controller that coordinates the activities of the machines, and provides a common programming language for operations that involve more than one robot. The solution preserves the possibility of programming each robot in its own language.

1. INTRODUCTION.

It is quite obvious that a single robot cannot perform useful tasks in an industrial environment, unless it is provided with some additional equipment that allows the machine to correctly grasp, handle and dispose of the mechanical parts onto which technological operations are to be performed. This is true even when the process does not require physical contact between the robot's end effector and the part being processed (a typical example is spray painting). Moreover, in order to avoid the need of manually loading and unloading the robot, it is usually desirable to integrate the robot unit in the production line, that also includes special purpose and tool machines, conveyor belts, vision systems, etc.

The aim of the paper is to illustrate a method that allows easy programming of a system that includes different robots and a variety of auxiliary devices.

Section 2 gives an overall view of the problem to be solved, and the main standpoints of the research that was carried on.

Section 3 emphasizes the need of such integrated programming system, discussing a practical example of cooperation between two robots, ans section 4 shows the practical implementation of the proposed system.

2. THE PROBLEM.

The problem of creating a universal robot programming language or automatic translators from a language to another one has been studied by many authors [1,2,3,4]. All the proposed systems concern however an "off-line" translation, while, in the case being discussed here, a real time translation is needed.

Synchronizing a deterministic process is not a conceptually difficult task, if each machine **can** be synchronized, i.e. it has suitable input and output signals, and suitable programming primitives. Things are not so easy in practice, specially when the evolution of the process depends on unpredictable parameters (presence or absence of pieces, down-times of some components of the system, varying length operations, etc.).

As an example, let us consider an experiment that was carried on at Milan Polytechnic Robotics Laboratory. Two robots (Supersigma and PUMA 500) were faced to each other, and a program was written that made PUMA hand out a piece to Supersigma, who had to grasp it, change its orientation and give it back to PUMA.

In order to synchronize the robots, a link composed of two logic signals (one input and one output line for each robot) was established, and these signals were used as flags by each program.

Although the task was extremely simple, it took about eight hours to program it, and, during debugging, there was constant danger of collisions between the two machines.

Should the same program have been written on a single robot with three arms, it would require no more than half an hour, if a suitable language were available.

Another difficulty comes from the fact that each robot has its own programming system, and it is often undesirable to alter or to substitute it with something else. Besides cost considerations, this is because each robot programming language is tailored to the machine it has to control, and it would be simply impossible, for instance, to obtain a good performance from an articulated robot using a language designed for a cartesian one.

The decision was then taken to develop a programming system that would provide the following characteristics:

- All activities of the global system are supervised by a single controller, so that the process schedule may be easily laied down, inspected and corrected;
- Each robot may be programmed in its own language, in order to maintain top performance of each machine;
- When strict cooperation between two machines is needed (this only happens a limited number of times in the whole process) the same language may be used to program operations of all the involved robots.

3. SINGLE CONTROLLER PROGRAMMING.

When two or more concurrent processes must be loosely synchronized, primitives such as flags or messages may be used. This method does not imply a centralized controller, since synchronizing signals may be physically interchanged between separate control units. On the other side, some operations, such as passing pieces from one robot to another one, are best described by a set of sequential instructions, all written in the same language. The method being introduced here allows the robot programmer to choose either philosophy, according to the specific needs of the task being programmed.

Let us consider the example given in section 2 again. It should be noted that, in order to allow PUMA to enter the working space of Supersigma, the right arm of the latter must be moved all the way to the left. In the following, the required clearance area will be referred to as "exchange area". The sequence the two robots must follow to exchange a piece from each other is then:

1. Remove Supersigma arms from the exchange area;
2. Move PUMA arm along the entrance path (three straight movements are required);
3. Move Supersigma right arm to a position where the piece may be exchanged and exchange it;
4. Remove Supersigma arms from the exchange area;

5. Remove PUMA arm from the exchange area.

Block diagrams of the programs that were developed are shown in figure 1. Horizontal arrows represent the points where the two programs must be synchronized.

If a common programming language were available, all synchronizations could be eliminated, and the program flow chart would be as it appears in fig. 2. The first two operations may be done simultaneously in order to increase speed, and all the other, that are strictly sequential, follow in a very traditional way. In each block a letter indicates whether the operation is to be performed by Puma (P) or Supersigma (S).

4. IMPLEMENTATION.

The experimental setup that was used includes the Supersigma robot with its own control system and a PUMA 500 (fig. 3).

The two robots are mechanically linked by a conveyor belt, that is driven by Supersigma control system. They are no more directly facing to each other, because of the already mentioned problems related to the mechanical structure of the machines, and because such placement implies severe limitations of PUMA moving capabilities. Moreover, keeping robots apart during experiments helps preventing collisions on such delicate devices as the robot hands are.

A final reason for choosing this layout was that a vision system can be placed over the belt to inspect and to recognize parts being carried from a robot to the other one.

Since the computing power of the CCU of Supersigma is much greater than is needed for controlling the robot, it was decided to use its computer (an LSI11/2) as the global process controller.

Two languages are available for programming Supersigma: MAL [5] and VML [6]. There are no conceptual differences between the two languages as long as the problem being discussed is concerned: actually, the system was first implemented in VML and is now being applied to MAL.

Instructions concerning Supersigma and the conveyor belt are already included in VML, and do not require any additional processing.

As it was previously said, the main goal of the research was to allow PUMA programming by Supersigma CCU without limiting the performance of the former robot, and without altering the structure of its control system.

This means that no modifications to PUMA were allowed, neither to hardware nor to software. The only possibility was then to feed data to PUMA through its terminal input, that is the only connection of PUMA to the external world, besides the digital I/O lines, that are not suitable for this purpose.

PUMA program will then be a sequence of commands that must be executed **immediately** as they are received. This obviously limits the execution speed, since immediate commands execution is slower than normal execution of a VAL program. In order to minimize this problem, it

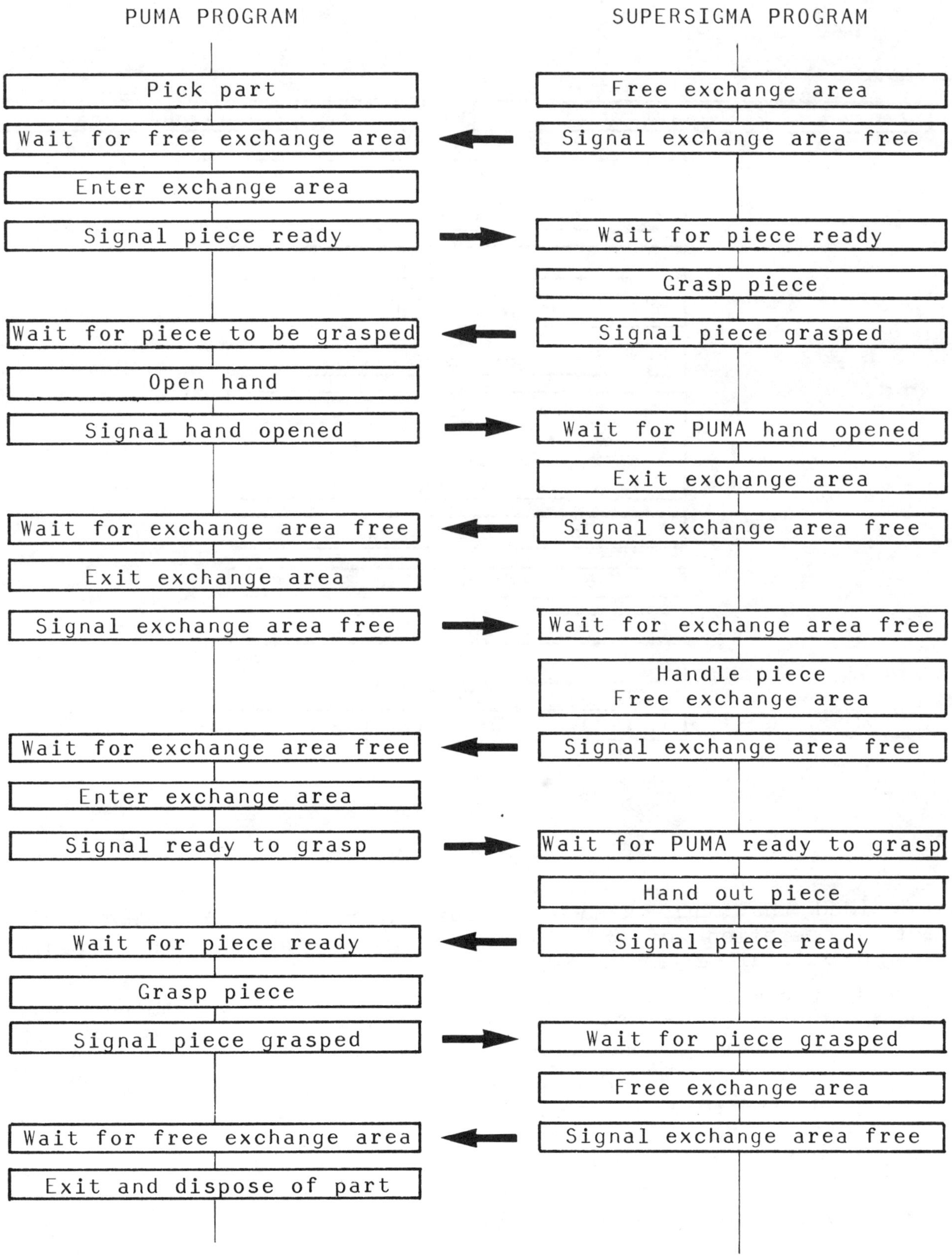

Fig. 1. Separate programs.

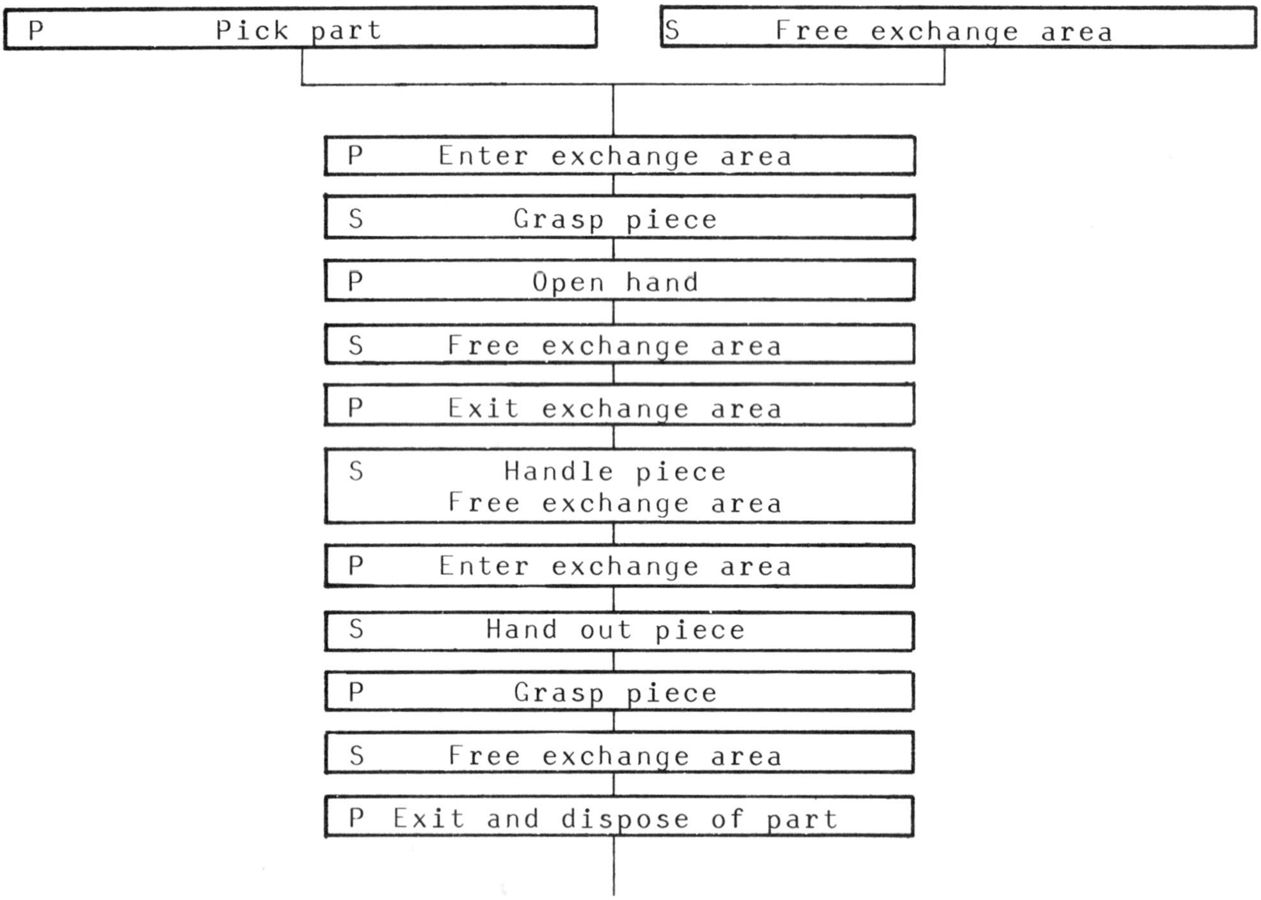

Fig. 2. Single program.

was decided that this kind of programming should be used only when strict cooperation of the machines is needed, using traditional programming in all other cases.

Two features must then be added to Supersigma programming languages: movement instructions for PUMA, and the possibility of sending alphanumeric strings of any kind to PUMA.

Since Supersigma is controlled by an array of microcomputers, each one being devoted to a specific task, the same philosophy should be used for connecting PUMA to the system. A single microcomputer is devoted to the communication between the CCU and PUMA, that takes place on a serial line operated at 9600 baud.

The block diagram of the control system is shown in fig. 4.

The functions of PUMA control microcomputer are:

1. If strings are to be sent to PUMA, transfer them from MICROBUS (that is a parallel bus) to PUMA, accomplishing the protocol of the latter machine.

Fig. 3. Experimental setup.

2. If movement instructions are to be sent to PUMA, translate them from the standard format used for Supersigma in suitable alphanumeric strings.
3. At any time, keep trace of PUMA status and issue appropriate status signals towards the CCU.

Points 2 and 3 are the most interesting from the programming standpoint.

4.1. Control system structure.

A full description of Microbus is given in [7]: a short summary will however be sufficient for understanding of the problem being discussed.

The communication between the CCU and microcomputers takes place in the form of commands from the CCU towards microcomputers and of data and status signals in the opposite direction.

Only "free" microcomputers may receive commands. This means that microcomputers that are executing a previously received command will be virtually disconnected from the array until the execution of the command has come to an end.

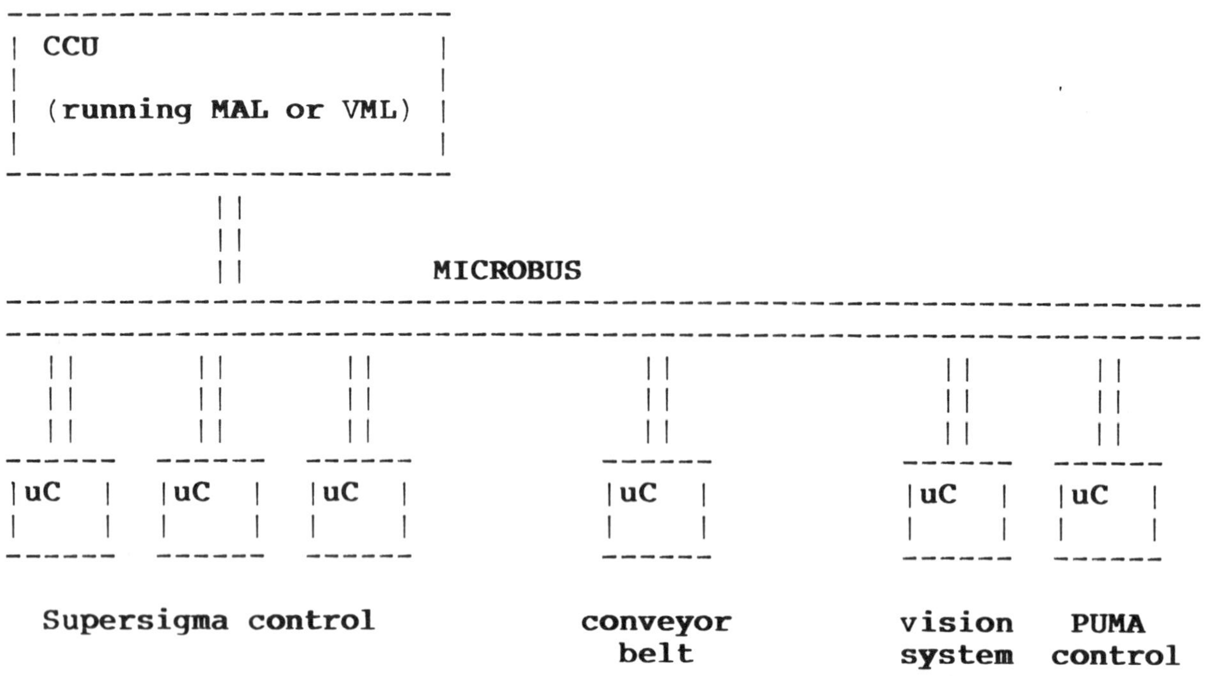

Fig. **4**. Block diagram of the control system.

The quantity of information carried by each command depends on the nature of the command itself. Five 8-bit words are sufficient to contain the longest possible command. These words are transmitted on an 8-bit bus with an asynchronous protocol. This requires the use of the additional line to provide initial synchronization.

In order to let the CCU know if microcomputers are free or busy, some **"status lines"** are included. These lines (typically two per microcomputer) should directly reach an input port of the CCU. In order to minimize wiring and I/O ports on the CCU, status lines are multiplexed in a "status bus".

Two status lines per microcomputer were chosen instead of one because, for motor microcomputers, it is important to state not only if the device is free or busy, but also if the previous command was correctly executed.

The meaning of status lines is shown in fig. 5.

Since data are transmitted over the same bus that carries commands, a protocol has been stated in order to avoid conflicts (for the sake of simplicity, no hardware bus arbiter was implemented).

This protocol is based on a "speak when you are requested to" principle. In other words, microcomputers must receive a special command from the CCU before they can send data over the bus. It is then the CCU responsibility to interrogate one microcomputer at a time, and to send no commands until the answer is complete.

Under normal operating conditions, answers flow is only a small fraction of the total data flow on Microbus: the communication protocol may then be slower and simpler than the one used for commands. Moreover, the number of bytes for each answer is not yet known, since new sensors are being studied. Motor microcomputers may answer with only two bytes, since sixteen bits are sufficient to represent the position of a motor,

LINE A	LINE B	MEANS
0	0	Free.
0	1	Busy (previous command still under execution).
1	0	Free, but previous command was aborted because there has been an attempt to overcome rail limits.
1	1	Free, but previous command was aborted because force acting on the hand was excessive (possibly because the hand has hit something unexpected).

Fig. 5. Status lines attribution.

but complex sensors may require a greater amount of information to be sent for each answer.

4.2. Sending strings to PUMA.

This feature allows the programmer complete control of PUMA: since Supersigma is connected to PUMA's terminal input, any kind of command can be sent over the line, including those commands that cause the execution of complete programs previously stored in PUMA's mass memory.

4.3. Sending movement instructions to PUMA.

The system must provide a means of translating movement instructions written in MAL (or VML) to VAL language.

Programming philosophies of Supersigma and PUMA are quite different from this point of view, and some details must be added.

In MAL, only joint-oriented movement instructions are available. This means that, in order to reach a point P(X1, Y1, Z1), the user must use the instruction

MOVE XL=X1, YL=Y1, ZL=Z1

for the left arm, or

MOVE XR=X1, YR=Y1, ZR=Z1

for the right one. It should be noted that, due to the cartesian structure of Supersigma, joint movements always take place along cartesian axes.

In VML, both joint-oriented and point-oriented instructions are available. This means that the point may be defined as a set of three separate coordinates and reached by three movement instructions, or a triple of coordinates may be used, thus allowing the use of a single movement instruction.

At Microbus level, since each microcomputer controls a single axis, a movement instruction of the kind previously stated is always translated into the set

```
MOVE XL=incr1
MOVE YL=incr2
MOVE ZL=incr3
WAIT XL
WAIT YL
WAIT ZL
```

where incr1, incr2 and incr3 are the number of elementary steps each axis has to perform. Since transmission time of the commands is quite small, the three movements are virtually executed at the same time. If no restrictions are placed on the execution time of each movement, the motors microcomputers will optimize each axis movement separately, and the path followed by the end effector will be quite unpredictable. It has been noted however that in cartesian machines this is not a drawback, if linear interpolation, that is only needed in a very limited number of cases, is available.

On the other side,, PUMA allows three kinds of movement instructions: joint-oriented, along cartesian axes and point-oriented. Moreover, cartesian axes may belong to either a fixed reference system or to a system that is dynamically defined by the position of the end effector.

4.4. Translating movement instructions

The most immediate way of translating CCU movement commands into PUMA instructions is then to use instructions that allow movements along cartesian axes.

A problem arises because the movement command to PUMA may only be issued **after** all the elementary movements have been defined. This is because PUMA will not accept any movement instruction while another movement instruction is being executed.

This requires a special treatment of movement instructions that refer to PUMA. The previous sequence

```
MOVE XL=incr1
MOVE YL=incr2
MOVE ZL=incr3
WAIT XL
WAIT YL
WAIT ZL
```

should be translated into

```
MOVE XL=incr1
MOVE YL=incr2
MOVE ZL=incr3
START
WAIT_END_MOVE
```

in order to allow the microcomputer that connects the two robots to know when the movement information is complete, and the movement may be started.

This of course implies some additional processing at VML (or MAL) level but, if a complete uniformity is desired, the same protocol may be used for Supersigma movements also.

In the present layout, since the two robots are far from each other, a common coordinates system cannot be implemented. The presence of the conveyor belt suggests the creation of a "coordinate shifting mechanism" that would allow an automatic conveying of objects that are beyond the reach of the robot that has to grasp or to work on them.

This solution, that is presently being investigated, requires a careful modeling of the objects on the belt, since moving one of them means moving the whole belt, and, if no care is taken, some of the objects might fall from the belt or cause other kinds of undesirable side effects.

For the sake of simplicity, in all the previous examples only three degrees of freedom were considered. If the robot (as it happens for PUMA) also allows rotations of its end effector, the system must be expanded accordingly, but no additional problems arise.

4.5. Keeping trace of PUMA's status.

Since the global control system is connected to PUMA's terminal input, the only available information about the state of the latter machine is the answer that PUMA issues after each command is executed (or rejected, if it was wrong or unexecutable).

All possible answers from PUMA were examined, and divided into six classes, that are:

1. No answer yet.
2. Command execution terminated, VAL program running;
3. Command execution terminated, VAL program stopped;
4. Command execution rejected, VAL program running;
5. Command execution rejected, VAL program stopped;
6. PUMA is waiting for an answer (Yes/No);
7. PUMA fatal error.

All messages from PUMA are stored in a ring buffer in PUMA control microcomputer, and, as soon as they come to an end (carriage return is detected) they are examined, and status lines towards the CCU are set accordingly.

Since Microbus status lines only allow four configurations, a grouping was necessary:

PUMA status	**Status lines: A**	**B**
1 (No command issued from the CCU), or 6	**0**	**0**
1 (Command issued from the CCU)	**0**	**1**
2, 3, 4, 5 (non-fatal errors)	**1**	**0**
7 (fatal error)	**1**	**1**

This information is sufficient in most cases: if required, the complete answer from PUMA may be transferred to the CCU via Microbus data lines.

5. EXPERIMENTAL RESULTS.

The fundamental functions of the system have already been tested, and results are encouraging. In order to make debugging simpler, in the first implementation no PUMA control microcomputer was used, and all the necessary procedures were directly included in VML. The link between the CCU and PUMA was established using one of the serial ports of the CCU. This method can be used if complete modularity is not required.

Programming tasks that involve both robots has become much easier and straightforward.

Although the execution speed of immediate commands in PUMA is not very high, it is not noticed if the number of commands issued from the CCU is small (i.e. the method of common programming is only used when required). During normal operation, PUMA executes VAL routines stored in its own memory, and the execution speed is the normal one.

6. CONCLUSIONS.

A method was shown that allows programming different robots using a common supervisor. The method may be easily expanded to drive other machines, if a command language similar to VAL is available, or, changing the structure of the program running on the microcomputer that provides communication between the supervisor and the slave machine, to different programming systems.

REFERENCES

[1] W.T. Park, D.J. Burnett - An interactive incremental compiler for more productive programming of computer-controlled industrial robots and flexible automation systems - Proc. IX ISIR - Washington, 1979.

[2] J. Courtney, G. Herlin - A portable robot software system - Proc. XI ISIR - Tokyo, 1981.

[3] M. Weck, W. Eversheim - ROBEX - An off-line programming system for industrial robots - Proc. XI ISIR - Tokyo, 1981.

[4] J.C. Latombe, E. Mazer - LM: A high-level programming language for controlling assembly robots - Proc. XI ISIR - Tokyo, 1981.

[5] G. Gini, M. Gini, R. Gini, D. Giuse - Introducing software systems in industrial robots - Proc. IX ISIR - Washington, 1979.

[6] P. Bison, G. Lorenzin, E. Pagello - The formal definition of VML and a proposed portable implementation - - Proc. XI ISIR - Tokyo, 1981.

[7] R. Cassinis - An example of a distributed intelligence discrete process controller - Digital Systems for Industrial Automation - Vol. 1, N. 1.

The SRI Robot Programming System (RPS)

William Thomas Park
SRI International
Menlo Park, California

Abstract

A robot programming system developed at SRI International consists of a partial compiler and interpreter for a simple robot programming language. Debugging facilities speed-up development of control programs for small to moderately sized robotic manufacturing systems not requiring parallel processing. Powerful, general-purpose, and easy to use, it runs on an LSI-11 under the RT-11 disk operating system. A library of fully-compiled subroutines currently supports FORTRAN-like control structures, operator communication, integer and floating-point computations, and file I/O. It operates equipment such as the SRI Vision Module, the Unimation PUMA 600 robot manipulator, and other devices.

I INTRODUCTION

The purpose of this paper is to describe SRI's Robot Programming System (RPS). We describe the organization of the system software, give an example of the Robot Programming Language (RPL) that it supports, and illustrate some of the debugging features in RPS that we think should be included in contemporary robot control systems.

RPS controls robots, machine vision systems, and many other kinds of industrial automation equipment. RPS was designed primarily to help people develop small to moderately sized automatic manufacturing systems by speeding up the development, testing, and debugging of control algorithms for those systems. It is general-purpose and easy to use; users can extend its capabilities; and it runs in an inexpensive minicomputer.

The RPS software consists of two programs for a Digital Equipment Corporation (DEC) PDP-11 or LSI-11 computer with 56K bytes (28K words) of memory. The programs are a compiler and an interpreter for robot control programs in a simple computer language called RPL. The programs run under the single-job monitor of DEC RT-11 operating system.

Using RPS is a four-step process:

(1) Write an RPL program that describes what the automatic manufacturing equipment should do.

(2) Compile the RPL program.

(3) Start the interpreter and specify the compiled RPL program to be run.

(4) Run the RPL program, using the interpreter to control its execution in order to test and debug it.

RPS has been used at SRI in three demonstrations to date:

* The use of tactile feedback in automatic pop riveting of sheet metal boxes [6].

* The use of machine vision in orienting large parts by controlled tumbling, so that, e.g., a limited-sequence robot can handle them [8].

* The use of machine vision to replace hard tooling in a bowl feeder [9].

RPS was developed by the author as part of SRI's project, "Machine Intelligence Research Applied to Industrial Automation."* Dr. David Nitzan is principal investigator. One of the main goals of this project is the timely transfer of technology such as the RPS system to over 30 companies that are affiliated with the project. A series of SRI reports describes the research performed [1-11].

----

* Supported by National Science Foundation Grants GI38100X1, APR75-13074, DAR78-27128, and DAR80-23130, with supplementary support from approximately 30 industrial affiliates.

Recent work on RPS has been generously supported by the Manufacturing Research Laboratory of Lockheed Aircraft Corporation, under the supervision of Mr. Richard Movich. Lockheed uses RPS in experimental robotic drilling and riveting of sheet metal assemblies for aircraft [12,13,14].

The remainder of this paper discusses the following topics:

* Design goals for the RPL language and RPS system
* Final design of the RPL compiler and interpreter programs
* Performance of the compiler and interpreter
* An example of a simple RPL program
* Use of the interpreter to run the example program
* Command menus in the interpreter
* Debugging aids in the interpreter
* Evaluation and directions for future work.

II DESIGN GOALS

The original design goals for the RPS software were the following:

(1) An inexpensive operating environment that our industrial affiliates could duplicate easily for low-risk experimentation in advanced sensor-directed, computer-controlled robotic manufacturing.

(2) Ease of use by people who are not skilled programmers.

(3) Debugging facilities to speed testing of robot-control programs.

(4) A short turnaround time for program revisions.

(5) Ease of adaptation to new automatic manufacturing equipment, robotic or conventional.

Goal (1), an inexpensive computer environment, led us to select a popular 16-bit computer environment (Digital Equipment Corporation PDP-11 and LSI-11 series with the minimum of special options). Low-cost versions of the machine were about to become available and a powerful system development language (BLISS-11 [15]), whose compiler had excellent optimization capabilities, was available to us and our affiliates from Carnegie-Mellon University.

To minimize the required run-time environment, we decided to compile RPL programs completely before executing them. Thus, a compiler program first produces a file containing a compact description of the RPL program, known as interpretable object code. Whenever the RPL program is to be run, an interpreter program reads its object file into memory, decodes the instructions therein, and carries them out. Thus, the compiler and interpreter do not have to fit into memory at the same time, as with Unimation's VAL system [16], for example.

Goal (2), ease of use, led us to decide that the RPL language should be very simple, rather than a more powerful but complex one such as Stanford's AL [17]. We felt that the majority of production personnel who might use the system would be more skilled in manufacturing methods rather than in computer programming. Also, we felt that a manufacturing specialist who had received a little training in programming would be able to make the equipment operate more productively than a computer programmer who had received a little training in manufacturing methods. Finally, supporting a complex language would have required a more costly operating environment. We settled on a language whose power and complexity lie somewhere between those of BASIC and FORTRAN. In fact, RPL programs can easily be converted to FORTRAN programs. This might be worthwhile in order to obtain the maximum execution speed--e.g., for a rapid inspection task.

Goal (3), debugging facilities, led us to decide upon an interpretive mode of execution rather than direct execution of fully compiled code. The interpreter now supports the following debugging facilities: single-stepping, tracing, breakpointing, and manual interruption.

Goal (4), short turnaround time, was achieved by designing the RPL language for one-pass compilation. For example, labels and subroutine names that may be forward referenced have a distinguishing prefix of L$ and S$, respectively (e.g., L$LOOP, S$PICKUP). Furthermore, no lookahead is required during parsing of tokens to detect syntax errors. Compilation is sufficiently rapid that text editing, even for minor revisions, accounts for the major portion of turnaround time with typical RPL programs.

Goal (5), ease of adaptation to new equipment, was met by restricting all equipment-specific code to a set of library routines written in either BLISS-11 or assembly language. The user can therefore extend the library using only the RT-11 text editor, assembler, and linker on the PDP-11 or LSI-11. Industrial users who have manipulators, cameras, tools, or sensors similar to those used at SRI can operate them with available library routines.

III FINAL DESIGN

The compiler parses by the method of recursive descent [18]. It reads the RPL source text from a file, writes statically allocated user data values to a temporary file, and keeps interpretable instructions and the symbol table in core. It optionally writes a listing file that is merely a copy of the RPL source text file with line numbers prefixed to each line. A printout of this file is useful during debugging because run-time error messages, breakpoint locations, and trace reports all refer to these line numbers. Any error during compilation prevents generation of an interpretable object file, for safety. Thus, only RPL programs that are free of syntax errors can be run.

The interpreter is a skeleton program that calls routines in a library according to instructions in the RPL program. The interpreter and compiler program are completely general-purpose and can be used with any kind of equipment. Two of our project's reports describe a basic set of library routines in detail [9,11].

For speed, library routines are directly executed rather than interpreted. The basic library can be augmented with new device-control routines to suit different automatic equipment from that in SRI's laboratory testbed.

One can also write subroutines in the RPL language if maximum speed and efficiency is not required. These subroutines will be interpreted. RPL programs for most applications consist of a main program and several subroutines. Only occasionally does an application require new library routines, too.

The library itself consists of general-purpose data processing routines together with device-specific routines. The general-purpose routines perform such functions as arithmetic computation, conditional branching, and conventional I/O operations to keyboard/display terminals, the file system, and a cassette drive. Device-specific routines have been written to operate Unimate 2000 and PUMA 560 robot manipulators, the SRI vision module, stepper motors, relays, and analog-to-digital and digital-to-analog converters made by DEC and Datel Corporation. Lately we have been developing library routines to operate Machine Intelligence Corporation's VS-100 vision module.

In the remainder of this document we will use the terms "run," "execute," and "interpret" interchangeably to refer to carrying out the instructions in the user-written RPL program. All mean "interpret the instructions in the core-resident copy of the compiled RPL program."

IV PERFORMANCE

Compilation is reasonably fast--about 20 simple statements per second in a PDP-11/40. Interpretation proceeds at about 100 statements per second. This is adequate for all but rapid visual inspection tasks or high-bandwidth servoing. In typical application programs, the CPU spends most of its time waiting for a manipulator to complete a motion. One factor that contributes to execution speed is that the compiler fully resolves all symbolic references in the RPL program so that the interpreter never has to search a symbol table to find data or statements.

Execution speed could have been even faster had we not decided to make the interpreter perform a number of run-time consistency checks on each statement. We felt the checks were worthwhile because the minimal-configuration CPUs that we use have no memory protection mechanisms, so programming errors could corrupt the object program. This could lead to a system crash or incorrect operation of the manufacturing equipment. Some of the run-time checks are designed to detect such problems and interrupt execution before harm is done. These checks are not foolproof, however.

A minimal-configuration PDP-11 or LSI-11 can address 64K bytes of memory. Of these, the top 4K bytes are used as memory-mapped device control registers and are unavailable for storage, leaving 60K of available memory. Of this, the compiler and RT-11 occupy about 83% (50K bytes). The interpreter, together with RT-11 and a typical library that includes routines for operating a PUMA 760 manipulator and an SRI or MIC vision module take up about 90% of memory (54.5K bytes). Thus the compiler can translate a program whose object code and symbol table total 10K bytes or less, and the interpreter can load up to 5.5K bytes of object code. Because a compiled RPL statement occupies only 8 bytes on average, an RPL program could be as large as 700 lines. However, the object code also includes any data items used, and the interpreter needs some space for a run-time stack. So, 400 to 500 lines is probably a practical upper limit on RPL program size. This does not, however, include comments, which are not stored, so they may be used liberally for documentation.

V RPL EXAMPLE

Figure 1 shows the listing file for an RPL program with line numbers on the left, as they would be produced by the RPL compiler. This program repeatedly asks for a number, then prints out its square. A conditional branch in line 19 stops the program when the squared value equals nine.

```
0001 ! Example of an RPL program.
0002 ! Asks for a number, prints its square.
0003 ! Repeats until squared value equals nine.
0004
0005 ! ... this whole line is a trailing comment ...
0006
0007         COMMON                     ! Common data area
0008 REAL X, % ... this is an embedded comment ... % Y, Z
0009         INTEGER N                  ! Integer variable
0010         CHAR NNEQL = "N*N = "      ! Character string
0011         END
0012
0013         PROGRAM P$TEST           ! Main program
0014         INTEGER TEMP               ! Local dynamic variable
0015         ENDLOCALS                  ! End of local declarations
0016 L$LOOP: S$ASK ("N = ", N)          ! User subroutine
0017         ITIMES ( TEMP, N , N )! Lib. subr. to multiply
0018         S$PRINT ( NNEQL, TEMP ) ! User subroutine
0019         IFNEQ ( TEMP , 9 , L$LOOP ) ! If (TEMP .NE. 9)
0020         END                                !  go to L$LOOP
0021
0022
0023 ! Asks question, reads number as decimal integer value.
0024
0025         SUBROUTINE S$ASK ( QUESTN, NUMBR )
0026         TYPEL QUESTN  ! Type string on new line
0027         DECINV NUMBR  ! Input decimal integer to variable
0028         END
0029
0030
0031 ! Prints message, then value as decimal integer.
0032
0033       SUBROUTINE S$PRINT ( MESSG, VALUE )
0034       TYPEL MESSG  ! Type string on new line
0035       DECOUT VALUE ! Output decimal number
0036       END
```

Figure 1 RPL Program to Square a Number

Because the purpose of this example is to illustrate RPL syntax and interpreter operation, it does not illustrate operation of robot manipulators, vision systems, or the like. Meaningful examples of these activities would require a much longer sample program and would only confuse the issues. The program would appear much the same, except for the use of other library routines for operating those devices.

The following sections use the example program to illustrate the overall structure of RPL programs and the data types that can be used in them.

A. Overall Program Structure

A complete RPL program consists of:

* An optional COMMON block (as in lines 7-11 of Figure 1) in which the user defines and names any data items that are to be available to all parts of the program.

* A mandatory named PROGRAM block (lines 13-20), which is the starting point of the entire program.

* Optional SUBROUTINE blocks (lines 25-28 and 33-36).

B. Comments

RPL allows two different forms of comment--trailing and embedded. Trailing comments (lines 1-3, 5, 7, 9, etc.) are introduced by an exclamation point ("!"), and extend to the end of the line. Embedded comments (line 8) are bounded by percent signs ("%"). They may appear entirely within a statement wherever a space might normally appear, or they can span several lines.

C. Data Types and Forms

Data items may be literal constants (e.g., "N*N = " in line 10, or "9" in line 19) or symbolic variables (e.g., "N" in line 9) of type REAL, INTEGER, or CHARACTER. A literal constant must be a scalar, but a REAL or INTEGER variable may be either a scalar or an array with up to three dimensions. A CHARACTER variable can only be one character long, but a CHARACTER constant can be of any length. A CHARACTER data item is stored as a byte string (i.e., a sequence of ASCII character codes, including carriage return, tab, bell, etc.) terminating in a zero byte.

A variable declared in the COMMON block (such as "X" in line 8) has global scope (i.e., it is accessible to the PROGRAM block and any SUBROUTINE block). The scope of a variable declared in a PROGRAM or SUBROUTINE block (such as "TEMP" in line 14) is just that block.

D. PROGRAM and SUBROUTINE Blocks

1. Block Format

PROGRAM and SUBROUTINE blocks are composed of an optional sequence of local declarations (terminated by "ENDLOCALS" if present), followed by an optional sequence of executable statements, followed by "END." Each SUBROUTINE block describes a recursively callable subroutine subprogram that may take up to seven parameters.

2. Parameters

Parameters to SUBROUTINE blocks are passed by name (i.e., pointers to the locations of the data items are passed, rather than their values) so that subroutines can return values to callers in a structured way. The names of the SUBROUTINES themselves have global scope so that they can be used as actual parameters in subroutine calls.

3. Local Variables

The PROGRAM and SUBROUTINE blocks may also contain declarations of local variables that are known only within the block in which they are declared. For example, "TEMP" in line 14 can only be used by name in the main program. PROGRAM and SUBROUTINES blocks have the same format as declarations in the COMMON block, except that initial values are not allowed. Any local declarations must be followed by the word "ENDLOCALS," and both must precede any executable statements. Local variables with the same name in different blocks are treated as different quantities, and they are dynamically allocated on each entry to the subroutine in which they are declared, and deallocated on exit. This permits recursive calls to a subroutine without overwriting local variable values in other incarnations of that subroutine.

4. Statement Types

The most unusual feature of the RPL language is that it has only one type of executable statement--the subroutine call. In this respect, it superficially resembles LISP, which has only function calls. This restriction greatly simplified the design and operation of the parser and code generator in the compiler as well as the entire interpreter.

We have in a sense moved the greater part of the complexity of a general-purpose programming language from its syntax to its semantics. The library mechanism, which effectively defines the semantics, also makes it possible to extend the language easily--by revising or adding to the library. Neither requires any change to the compiler or interpreter, other than to update the list of library routine names in each.

The RPS system also bears some resemblance to P-code interpreter systems, such as UCSD Pascal [19], in the sense that it it uses a library of routines to perform functions such as computation, assignment, transfer of control, and so on. Pascal [19]. For people familiar with programming concepts, such systems are much easier to use because the languages they support (such as PASCAL) have many different specialized types of statements. These allow complex procedures to be described in a few lines of code. As noted above, we decided to address the needs of people with less experience of programming (and typing!) by minimizing the number of different kinds of statment that have to be learned and by avoiding concepts such as nested block structures. In fact, RPL programs could easily be constructed by a menu-pick system so that the user would have to type little or nothing to program a complex manufacturing procedure.

An RPL statement may call either an RPL-coded SUBROUTINE block (as in line 16, which calls subroutine S$ASK) or a built-in library routine (as in line 17, which calls library routine ITIMES). There are library routines to perform most of the common functions such as assignment, branching, computation, and I/O, that would be performed by specialized statement types in a conventional programming language such as FORTRAN. Others operate the robot arms, cameras, etc.

Any RPL statement may have one or more labels (e.g., label "L$LOOP" on the statment in line 16). The scope of a label is the block in which it occurs. A statment may pass up to seven parameters to the subroutine or library routine that it calls. Any of these may be a variable name or literal constant, and the callee receives their addresses rather than their values. This allows the callee to return values to the caller and compensates of the lack of functions in the RPL language. If a statement calls a library routine, however, it may also pass as a parameter the name of a library routine, the name of a user-defined RPL subroutine, or a label name within the scope of the statement.

5. Coding Expressions

Because there are no operators in RPL, expressions must be evaluated by a sequence of calls to library routines that perform arithmetic and logical operations. Because there are no functions in RPL, these calls cannot be nested. These two characteristics make RPL programs take up many more lines on a listing than a FORTRAN or LISP program to do the same thing. Skilled programmers will find this annoying, but it makes it easier for a novice to follow the flow of control in a complex program during testing and debugging. This is one example of our overall concern in designing the RPS to cater to the manufacturing specialist rather than the skilled programmer.

VI INTERPRETER USAGE

Figure 2 is a transcript of the dialogue between the user and the interpreter program during a run of the RPL program in Figure 1. The program has previously been compiled without errors.

Numbers have been appended to the left of certain lines in the transcript so that we may refer to them in this discussion. The transcript illustrates the following procedures:

* Starting the interpreter (lines 1-4)

* Loading the program (lines 5-7)

* Starting the user program (lines 8-9)

* Program execution (lines 10-13)

* Program termination (line 14)

* Exit from the interpreter to the RT-11 monitor (lines 15-17).

The following sections explain these procedures in more detail.

A. Starting the Interpreter

The RT-11 monitor command "RUN" in line 1 loads the interpreter from file RPSINT.SAVE on device DK1: (an RK05 disc), and starts it. The interpreter displays (line 2) its version number, its creation date, and (in decimal in line 3) the number

NOTE: User type-ins are underlined.

```
   +------------------------------------------------+
01 | .RUN DK1:RPSINT.SAV                            |
   |                                                |
02 | RPL Interpreter, Version RT5.0  June 15, 1982  |
   |                                                |
03 | Free words: 2937                               |
04 | LOAD RPL program from file ...                 |
05 | RPL*TEST                                       |
   |                                                |
06 | LOADED                                         |
07 | Free words: 2733                               |
   |                                                |
08 | TOP>INITIALIZE                                 |
09 | TOP>START                                      |
10 | N = 2                                          |
11 | N*N = 4                                        |
12 | N = 3                                          |
13 | N*N = 9                                        |
   |                                                |
14 | TOP>EXIT                                       |
   |                                                |
15 | Exit to monitor [Y/N] ... Y                    |
16 | FINISHED                                       |
   |                                                |
17 | .                                              |
   +------------------------------------------------+
```

Figure 2 Interpreting the Example Program TEST.SRC

of bytes of memory available for storage of the user's compiled program. It then prompts the operator (line 4-5) for the file name of a compiled RPL program.

B. Loading the RPL Program

The user types the file name, "TEST," of the compiled RPL program that is to be run. If the compiled file is too large to fit into available free memory, the interpreter will say "Program too big," and ask for another file name. In the example, the report "LOADED" (line 6) indicates that there was enough room for the program.

A second display of available free memory (line 7) shows how much space now remains. Some space must be left over in order to store the run-time stack. The interpreter uses a stack to store local variables, pointers to parameter values, and other calling linkage information. Most user programs need about 50 to 100 words of stack space, but the actual requirement may be much larger if a program calls subroutines to many levels. Heavy use of recursion (i.e; RPL subroutines that call themselves) can also consume large amounts of stack space.

C. Starting the Program

The prompt "TOP>" (line 8) requests the user to type in one of several commands from the menu shown in Table 1. This is called the "top-level menu." Selecting certain of the options, such as "BPT," will activate other, lower-level command menus. Section VII describes menu usage in more detail.

Table 1

Top-Level Interpreter Commands

COMMAND	FUNCTION
INITIALIZE	Initialize equipment
LOAD	Load compiled RPL program
START	Start program
CONTINUE	Continue interrupted program
RETRY	Retry failed statement
STOP	Stop program
EXIT	Exit to RT-11 monitor
ODT	Call Octal Debugging Tool program
DUMP	Dump RPL program to new file
CLEAR	Erase RPL program from core.
ENABLE	Enable selected debugging options
DISABLE	Disable selected debugging options
BPT	Set/clear breakpoints in RPL program
VAL	Talk to VAL system console TTY
VALERR	System debugging for RPS-VAL I/O
DOWNLOAD	Download VAL controller with a VAL program
SAVE	Save a VAL program on a file

In line 8 the user types the INITIALIZE command. This resets the automatic equipment by means of a call to a certain library routine. The system programmers can easily modify that routine to suit any new automatic equipment that may be installed.

The START command (line 9) then starts the robot control program running. In this example, the RPL program in Figure 1 immediately prompts the user to supply a number as input (line 10). The first value that the user types is "2," followed by a space or carriage return to terminate the number. The RPL program computes its square, four, and prints it out in line 11. Next the user types "3," the program prints "9," and consequently the program terminates (14). The user then types "EXIT" to leave the interpreter and go back to the monitor. Before quitting, however, the interpreter asks the user to confirm that he is really finished (15). When the user replies "Y" for "yes," the interpreter transfers control back to RT-11 (lines 16-17).

VII MENU USAGE

The interpreter's prompting message "TOP>" indicates that the user may type in one of the top-level commands listed in Table 1. Such a list of commands is called a menu. There are several different menus, and some commands in a menu automatically invoke another menu. Each menu has its own unique 3-character prompt followed by a ">", such as "TOP>", to indicate to the user the particular set of commands from which he can select at the moment.

A. How to Use a Menu

In general, when presented with a menu prompt the user has the following options:

* In response to the prompt for any given menu, he may type "?" to get a printout of all the commands in that menu.

* Typing a "#" will get back to the next-higher menu in the hierarchy, until the TOP> menu is reached again.

* The user need only type enough characters of a command word to uniquely distinguish it from the others in the menu.

* Type a space or carriage return to end a command word, or type an ESC character to make the system finish typing the whole command word. (On some keyboards, the ESC key is labeled ALT MODE.)

* A menu may have a default command for the user's convenience, which may be selected by typing just an ESC, blank, or carriage return character. The default command, if any, is indicated at the bottom of the display of menu options that results from typing a "?."

B. Menu-Based Command Language

Menu sequences are be set up so that the experienced user need only type the command words that he wants from each menu, sequentially, in the same line, and separated by blanks. This feature of the menu-driven command processor effectively implements a simple command language, using what amounts to an augmented transition network technique. However, if at any point in typing in such a command phrase the user should make a mistake, the processor will automatically display an appropriate error message and prompt him for the correct command word. The user may then type either the correct command word, "?" to remind himself of his options, or "#" to escape to the next higher-level menu. "#" will usually back up one word at a time through the command phrase that was being entered. Some menu chains, however, are set up so that a "#" aborts the entire command phrase.

C. VAL-Specific Commands in the Top-Level Menu

The commands VAL, VALERR, DOWNLOAD, and SAVE in Table 1 are specific to the VAL controller for the Unimation PUMA manipulator. VAL is used to communicate directly with the VAL controller from the user's keyboard. VALERR is used to diagnose problems in communicating with the VAL controller and operating the PUMA manipulator through it. DOWNLOAD loads the VAL controller with manipulation programs written in the VAL language and stored on an RT-11 file. SAVE extracts VAL programs from the VAL controller and stores them on the RT-11 file system. The usage of these four commands, too complex to explain here, is described in our eleventh project report [11].

VIII TESTING AND DEBUGGING AIDS

System programmers can issue the ODT command in the top-level menu to use DEC's Octal Debugging Tool (ODT) program when developing new library routines. To use ODT requires familiarity with the machine language, architecture, and operation of the PDP-11 or LSI-11 computer family. We will not explain how to use ODT here, but merely refer the reader to the manufacturer's documentation [20]. Because most users will not be expert enough to use ODT, RPS provides four other facilities for testing and debugging RPL programs, as follows.

* Run-time error messages
* Manual interrupts
* Execution controls
* Error-recovery procedures
* Saving and reloading procedures.

A. <u>Run-Time Error Messages</u>

Errors detected during execution result in an error report that contains the following information:

* The line number of the statement in which the error occurred.
* The type of error.
* The line numbers of the preceding ten statements that were executed.

A typical run-time error message might look like this:

```
+------------------------------------------------------------+
| ERROR IN LINE 123: TOO MANY/FEW ARGUMENTS                  |
| LINE HISTORY: ... 88, 89, 90, 105, 106, 107, 105, 106, 107, 123 |
+------------------------------------------------------------+
```

B. <u>Manual Interrupts</u>

Whenever an RPL program is running, typing a ^X character will suspend operation at the completion of the statement being executed at the time. The interpreter prints the line number of the next statement that will be executed, when and if execution resumes, and reports that a manual interrupt occured. It then offers the user the top-level command menu. For example,

```
+--------------------------+
| TOP>START                |
| .                        |
| . ... program runs       |
| .                        |
| ^X                       |
| @00057 MANUAL INTERRUPT  |
| TOP>                     |
+--------------------------+
```

Here the user typed the ^X character just before the program reached line 57.

C. Execution Controls

Testing a robot control program is much riskier than testing a conventional computer program that does not control automatic machinery. An error in the program can damage equipment or injure people. The RPS software provides means for closely controlling RPL programs so the user can detect problems and prevent accidents before they occur. The primary means are:

* Trace printouts

* Single-step execution mode

* Breakpoints.

1. Tracing Execution

Figure 3 illustrates the use of the top-level menu command ENABLE TRACE to trace program execution. The top-level command ENABLE TRACE makes the interpreter display the line number of each statement it executes. DISABLE TRACE turns off this display.

```
+-----------------------------+
| TOP>ENABLE TRACE            |
| TOP>START                   |
| @00053                      |
| @00054                      |
| @00055                      |
| @00056 ^X                   |
|                             |
| @00057 MANUAL INTERRUPT     |
|                             |
| TOP>DISABLE TRACE           |
| TOP>CONTINUE                |
|    .                        |
|    . --execution continues  |
|    .   without trace prints |
|                             |
+-----------------------------+
```

Figure 3 Tracing Program Execution

The trace printout indicates the actual flow of control, which may not be obvious from inspection of the RPL program text. Because execution can proceed rapidly, limited only by the speed of the printout or display device, the trace is more useful for diagnosing the cause of a failure than preventing it. Single-stepping is safer when an error can have serious consequences.

2. Single-Stepping

Figure 4 illustrates the single-step mode of operation. ENABLE SINGLE-STEP allows the user to run the RPL program one step at a time so that he can detect problems before they become serious. At each statement, the interpreter prints its line number and pauses. Hitting the space bar executes the statement. If the user decides something is wrong, he can type "T" instead to invoke the top-level command menu. He can then use any of the other testing and error-recovery features described in this section to correct matters. Alternatively, he can give the command DISABLE SINGLE-STEP to continue execution at full speed if he thinks it is safe.

```
+-----------------------------+
| TOP>ENABLE SINGLE-STEP      |
| TOP>START                   |
| @00055: <blank>             |
| @00056: <blank>             |
| @00057: ?                   |
|  SPACE  Proceed             |
|  T      TOP> command menu   |
|                             |
| @00057: T                   |
| TOP>DISABLE SINGLE-STEP     |
| TOP>CONTINUE                |
|     .                       |
|     . --execution continues |
|     .   without single-step |
|         printouts           |
|                             |
+-----------------------------+
```

Figure 4 Single-Stepping the Program

3. Breakpointing

"Breakpointing" means specifying one or more points in the RPL program at which execution should be interrupted automatically. When such a breakpoint is reached, the interpreter displays a message describing the line number and the kind of breakpoint and then offers the user the top-level command menu.

a. Types of Breakpoint

There are two kinds of breakpoints--programmed and interactive. A programmed breakpoint is a statement "BREAK" in the RPL program (a call to library routine BREAK). Any number of these may be present.

Up to ten interactive breakpoints (named "A" through "J") can be specified (or "set") by the user using the BPT command in the top-level menu. The same command will display a list of those currently set and also remove (or "clear") them.

b. Setting and Clearing Interactive Breakpoints

Figure 5 shows how the BPT command is used. "BPT" invokes the breakpoint submenu, which prompts for a command with "BPT>." The user types "?" to review the options in that menu (namely, SET, CLEAR, and DISPLAY), and the last option, DISPLAY, is seen to be the default. On the next "BPT>" prompt, the DISPLAY command shows all current breakpoint settings. Breakpoint "A" is set at line 12, "E" at 6, and "J" at 17. The others are unassigned. Next, the CLEAR command removes breakpoint "J" from the program, and "?" shows the result. The user then removes all by typing "X" instead of a breakpoint name. The system checks to make sure that he really wants to remove them all, asking "ALL BPTS?", which the user confirms by typing "Y." The next display, again requested by "?", shows that now no interactive breakpoints are set. A bare carriage return then indicates that the user is through clearing breakpoints, and the "BPT>" prompt reappears.

The "SET" command allows the user to place some interactive breakpoints in the program. He puts breakpoint "A" on line 123 and "D" on 456, then types "?" to check the result. Satisfied, he types a bare carriage return to indicate he is done setting breakpoints. Finally, he types the general menu-escape character "#" to leave the BPT menu entirely and return to the top-level menu.

c. Breakpoints in Action

Figure 6 illustrates the effect of breakpoints on an RPL program. In addition to the two interactive breakpoints set up in the previous example (A on 123 and B on 456), we assume that line 43 in the RPL program is a BREAK statement.

Whenever the interpreter reaches a breakpoint in the program, it reports the location and the type of breakpoint. If a programmed breakpoint, it offers the top-level command menu. If interactive, it acts as if single-stepping were in effect--that is, it waits until the user types a blank and then continues. If he types a "T," he gets the top-level menu.

In Figure 6, when execution reaches line 43 the interpreter stops and prompts the user to type a top-level command. He simply continues execution. Interactive breakpoint "D" is reached next at line 456. This time the interpreter prompts for the same type-ins as it would in trace mode--space, "T", or "?." The user types "?" to review these choices and then elects to continue execution by typing a blank instead of going to the top-level menu. However, at interactive breakpoint A, line 123, he does request the top-level menu. After using some of the top-level commands, he gives the CONTINUE command to proceed again.

D. Error-Recovery Procedures

In general, whenever an RPL program cannot proceed, the interpreter offers the user the top-level command menu. This can happen for any of the following reasons:

* An error occured.

* The user typed ^X to cause a manual interrupt.

```
TOP>BPT
BPT>?
 SET
 CLEAR
 DISPLAY (DEFAULT)

BREAKPOINT-CONTROL COMMAND
BPT>DISPLAY

                    BREAKPOINTS

    A   12 | B       | C       | D       | E    6 |
    F      | G       | H       | I       | J   17 |

BPT>CLEAR

BPT I.D.? [A-J/X/?/CR]: J at line 17 O.K.? [Y/N]... Y
BPT I.D.? [A-J/X/?/CR]: ?

                    BREAKPOINTS

    A   12 | B       | C       | D       | E    6 |
    F      | G       | H       | I       | J      |

BPT I.D.? [A-J/X/?/CR]: X ALL BPTS? [Y/N] ... Y
BPT I.D.? [A-J/X/?/CR]: ?

                    BREAKPOINTS

    A      | B       | C       | D       | E      |
    F      | G       | H       | I       | J      |

BPT I.D.? [A-J/X/?/CR]:
BPT>SET
BPT I.D.? [A-J/X/?/CR]: A at line ? [1-9999]: 123
BPT I.D.? [A-J/X/?/CR]: D at line ? [1-9999]: 456
BPT I.D.? [A-J/X/?/CR]: ?

                    BREAKPOINTS

    A  123 | B       | C       | D  456 | E    6 |
    F      | G       | H       | I      | J   17 |

BPT I.D.? [A-J/X/?/CR]:
BPT>#
TOP>
```

Figure 5 Setting, Clearing, and Displaying Breakpoints

```
+----------------------------------+
| TOP>START                        |
| @00043 PROGRAMMED BREAK          |
| TOP>CONTINUE                     |
| @00456 BREAKPOINT D:?            |
|  SPACE  Proceed                  |
|  T      TOP> command menu        |
|  ? [BLANK/T/?]: <blank>          |
| @00123 BREAKPOINT A:T            |
| TOP> ...                         |
| .                                |
| . --user uses top-level commands |
| .                                |
| TOP>CONTINUE                     |
+----------------------------------+
```

Figure 6 Example of Breakpoint Usage

* The user typed "T" at an interactive breakpoint or while in single-step mode.

* The interpreter came to a BREAK or QUIT statement, or the END statement of the main program.

The following top-level commands are provided to recover from whatever condition interfered with execution of the RPL program:

* RETRY--Retries the current statement, if an error occured during it. Otherwise, same as CONTINUE.

* CONTINUE--executes the next statement. Used to proceed with execution and to skip a statement in which an error occured.

* START--Will restart the program (must first stop the program with the STOP command if it is currently running, for safety).

* LOAD--Clears old program (which must first be stopped if it is running), then loads a new one.

* EXIT--Returns control to RT-11. Stops any program currently running. Cannot re-enter the interpreter from RT-11.

E. Saving and Reloading Procedures

The DUMP command in the top-level menu allows one to save an RPL program on a specified file, along with the current values of all variables in the COMMON area. The LOAD command will reload them later.

LOAD and DUMP can be used to speed up testing and debugging. For example, suppose that the first activity in an RPL program is a lengthy training or calibration procedure. Once that part of the program has been debugged and works properly, it is a waste of time to repeat it. To save time, the program can be saved after the training or calibration has been performed, and each subsequent test run can skip it.

Alternatively, one "generic" program can be particularized to many slightly different applications by adjusting some of the data values in the program and then dumping it. For example, one might write a generic pallet-packing program in RPL, train it several times to pack different shapes of pallet, and save each version as a different program. This approach is especially useful in applications in which a part-coding scheme is used to classify workpieces or group technology methods are used to organize manufacturing operations.

IX EVALUATION AND DIRECTIONS FOR FUTURE WORK

The RPS system has met all of the design goals that were set for it about five years ago. However, the present version has the following weaknesses:

* The BLISS language used for system implementation has not been as widely adopted by our industrial affiliates as we had hoped. Consequently, only a few have been able to use RPS [14]. In fact, in the future SRI's Robotics Department intends to write most of its software in the C language, which is now much more widely accepted.

* Differences in the operation of various models of computer in the PDP-11 and LSI-11 line have impeded transfer to some of our affiliates. Another impediment is that some affiliates use a different operating system, such as RSX-11.

* Several desirable features, such as arithmetic expressions and source-level debugging, have never been implemented.

* In the last two years, the SRI laboratory testbed equipment has been modularized and interconnected through a local network. The RPS system is currently unable to communicate over this network to operate the equipment.

Because of the availability of inexpensive computers, programming skill levels in industry have risen more rapidly than we had anticipated. Commercial manipulation and vision systems with much more complex programming languages than RPL are already being used by industry. Industry is even beginning to demand more advanced language features than those systems offer, such as parallelism. The need for simplicity in the RPL language is not so strong today.

Another important development is that some programming systems for conventional general-purpose languages, such as PASCAL, now provide interactive debugging facilities. At least one robot-control system (JARS) is based on a general-purpose PASCAL top-level program that calls upon a specialized library of low-level routines to operate the automatic equipment [21,22]. This is a very attractive approach to robot software development today.

Current work consists of development of library routines to operate an MIC vision module and comprehensive documentation on the whole RPS system. The first priority for further work would probably be to interface the system to the existing local network in our robotics testbed laboratory. A method for accomplishing this task has been developed. Reimplementation in the C language or in a PASCAL development environment is also possible. A capability for parallelism would be desirable.

However, that can cause problems having to do with deadlock of processes, process starvation, and simultaneous updating of a data base. We would want to look closely at methods for guaranteeing that the top-level control software for a task is always correct and free of such problems.

REFERENCES

1. C. Rosen et al., "Exploratory Research in Advanced Automation," First Report, NSF Grant GI38100X, SRI Project 2591, Stanford Research Institute, Menlo Park, California (December 1973).

2. C. Rosen and D. Nitzan, et al., "Exploratory Research in Advanced Automation," Second Report, NSF Grant GI38100X1, SRI Project 2591, Stanford Research Institute, Menlo Park, California (August 1974).

3. C. Rosen et al., "Exploratory Research in Advanced Automation," Third Report, NSF Grant GI38100X1, SRI Project 2591, Stanford Research Institute, Menlo Park, California (December 1974).

4. C. Rosen and D. Nitzan, et al., "Exploratory Research in Advanced Automation," Fourth Report, NSF Grant GI38100X1, SRI Project 2591, Stanford Research Institute, Menlo Park, California (June 1975).

5. C. Rosen and D. Nitzan, et al., "Exploratory Research in Advanced Automation," Fifth Report, NSF Grant GI38100X1, SRI Project 4391, Stanford Research Institute, Menlo Park, California (January 1976).

6. C. Rosen and D. Nitzan, et al., "Machine Intelligence Research Applied to Industrial Automation," Sixth Report, NSF Grant APR75-13074, SRI Project 4391, Stanford Research Institute, Menlo Park, California (November 1976).

7. C. Rosen and D. Nitzan, et al., "Machine Intelligence Research Applied to Industrial Automaton," Seventh Report, NSF Grant APR75-13074, SRI Project 4391, SRI International, Menlo Park, California (August 1977).

8. C. Rosen and D. Nitzan, et al., "Machine Intelligence Research Applied to Industrial Automation," Eighth Report, NSF Grant APR75-13074, SRI Project 4391, SRI International, Menlo Park, California (September 1978).

9. D. Nitzan et al., "Machine Intelligence Research Applied to Industrial Automation," Ninth Report, Grants APR75-13074 and DAR78-27128, SRI Projects 4391 and 8487, SRI International, Menlo Park, California (August 1979).

10. D. Nitzan et al., "Machine Intelligence Research Applied to Industrial Automation," Tenth Report, NSF Grant DAR78-27128, Project 8487, SRI International, Menlo Park, California (November 1980).

11. D. Nitzan et al., "Machine Intelligence Research Applied to Industrial A NSF Grant DAR80-23130, Project 2996, SRI International, Menlo Park, California (January 1982).

12. R. C. Movich, "Sensory Feedback Computer System for Automation of Aircraft Manufacturing," Proc., AUTOFACT West, Vol. 1, CAD/CAM 8, Anaheim, CA, U.S.A., Nov. 17-20, 1980, pp. 629-648.

13. R. C. Movich, "Robotic Drilling and Riveting using Computer Vision," SME Tech. Paper Series, MS 80-712, 21 p. (1980).

14. R. C. Movich, "A Vision-Controlled Robotic Cell," Proc., S.P.I.E., Vol. 336 (1981).

15. W. A. Wulf, D. . Russell, and A. N. Habermann, "BLISS: A Language for Systems Programming," Communications of the ACM, Vol. 14, No. 12, pp. 780-790 (December 1971).

16. B. Shimano, "VAL: A Versatile Robot Programming and Control System," Proc. IEEE Third Int. COMPSAC79, pp. 878-883, Chicago, Illinois (November 1979).

17. R. Goldman and M. S. Mujtaba, "AL User's Manual," Artificial Intelligence Laboratory Memo, Third Edition, Stanford University, Stanford, California (December 1981).

18. D. Gries, Compiler Construction for Digital Computers, pp. 97-100, (John Wiley & Sons, Inc., New York, New York, 1971).

19. K. Bowles, "UCSD Pascal: A (Nearly) Machine Independent System (for Microcomputers and Minicomputers)," BYTE, Vol. 3, No. 5, p. 46 (May 1978).

20. Introduction to RT-11, Order No. DEC-11-ORITA-A-D, DN1, pp. 14-4 through 14-14, Digital Equipment Corporation, Maynard, Massachusetts (March 1978).

21. JARS User's Guide and Manual, Jet Propulsion Laboratory, Pasadena, California (1980).

22. W. A. Gruver, B. I. Soroka, J. J. Craig, and T. L. Turner, "Evaluation of Commercially Available Robot Programming Languages," to appear in Proc., 13th International Symp. on Industrial Robots, April 18-21, 1983, Chicago, Illinois

Using ADA as a Robot System Programming Language

R.A. Volz
Center for Robotics and
Integrated Manufacturing
Ann Arbor, Michigan

T.N. Mudge
Center for Robotics and
Integrated Manufacturing

D.A. Gal
Center for Robotics and
Integrated Manufacturing

Abstract

The ADA programming language was developed for the Department of Defense in 1978. ADA is intended as the DoD's sole system implementation language, in particular, one of the primary aims of ADA is for the programming of real-time embedded systems. The logical structure of an ADA-based multiprocessing system used for the real-time control of a robot system is discussed. The robot system is comprised of a PUMA 600 arm and a vision subsystem that uses a GE TN2500 camera as an input device. Multiprocessing tasks in the robot system can be easily achieved using ADA running on a dual processor machine. Using multiprocessing to meet real-time constraints is discussed. Research is associated with the vision subsystem.

I. INTRODUCTION[1]

With the advent of flexible automated manufacturing environments, the need for a standard implementation language to program manufacturing cells has become important. The present practice of designing new robot languages for nearly every new robot is certainly counterproductive in the long run [1]. Moreover, current high level languages seem insufficient for the real-time requirements of production systems, while standards for the sake of readability, flexibility and portability are absent from most manufacturing languages. Design requirements for the Department of Defense's (DoD's) future system implementation language, Ada [2], assumes all of these properties. Though fully validated compilers are not currently commercially available, DoD's strong support of the language guarantees a large scale presence in the future. Therefore, a serious inquiry into the feasibility of using Ada as a standard implementation language for manufacturing is warranted.

This paper describes the logical structure of an Ada-based manufacturing cell, exploiting the data-abstraction and multitasking features of the language. Specifically, the discussion will focus on the vision subsystem developed in Ada and the limitations and improvements encountered in the process.

II. SOFTWARE ISSUES IN MANUFACTURING CELLS

To illustrate some of the problems which arise in flexible manufacturing cells consider the simple cell illustrated in Figure 1 (figures are after the bibliography). This cell is a simple machine loading/unloading system consisting of an NC milling machine, an input conveyor, a robot for loading, unloading and tool changes, and an output conveyor. For purposes of generality we assume that the incoming stock arrives in a disoriented fashion and must be located and identified by a machine vision system before being grasped. The goal is for the cell to be fully automatic and flexible in the sense that it can be used to manufacture any part within some reasonable class. There is to be no loss of production time due to the training of either the robots or the machine vision system as occurs presently [3], [4]. It is presumed that all the parts to be produced have been designed via a CAD system.

A hierarchical computer system controls the cell as shown in Figure 2. Dedicated microcomputers interface with the various physical process machines and sensors in the cell. These communicate with a central cell control computer which manages the overall behavior of the cell and handle communications with the CAD database. The principle computer issues are:

- extraction of information from the CAD system to assist cell operations
- development and management of the complex real-time software system to operate the cell
- the architecture of the computer hardware/software system to support the cell
- computational speed

This paper addresses principally the second of these problems. The last two are beyond the scope of this paper and are treated elsewhere [5]-[7]. Exploration of the use of CAD information for cell operations beyond automatic production of NC programs is only beginning, and is also treated elsewhere [8]-[10], though portions of our work on automatic training of vision systems impinge upon the current discussion.

As the cell complexity grows, so too does the software system which must manage it. In similar cases it has been shown that over 70% of the total cost of developing embedded software goes

[1] This work was supported in part by the Zimmer Foundation and in part by the Air Force Office and Scientific Research under contract F49620-82-C-0089.

into software maintenance [11]. It thus becomes critical to treat the software development functions very carefully. Ada was designed to address such problems through its support of abstract data types, separate compilation, multi-tasking, and extensibility.

The use of Ada for cell control is explored through a case study of a portion of the cell: the operation of the CAD driven vision subsystem. The underlying philosophy of Ada is centered upon the use of objects for program design. **Objects** are essentially data structures[2] [12]. The key element is that they maybe embedded in **packages** and made visible to program parts that would use them only through procedure and function references. That is, object and packages together hide data structure implementation and are an implementation of abstract data types. However, since objects are only visible through the operations performed on them, we will use the term object oriented loosely to imply the object/package combination. **Generic** procedures, an attempt at polymorphic function implementation [14], [15] allow operations to be defined over a set of data types, thus providing a broader use of objects. **Tasks** provide a means of dividing a program into logically concurrent operations with possible synchronization between them [16]. In addition to forming the basis for real-time operations, in a parallel processor environment, they may also provide a means of increasing processing efficiency. While there are many other features in Ada, these are the principal ones used in the vision system.

III. DESIGN OF THE VISION SYSTEM

The goal of the vision module project was to write a vision subsystem in Ada to provide a vision interface for our experimental manufacturing cell and allow recognition of nonoverlapping parts. VISUAL (Vision Interface System Using the Ada Language) was not intended to produce new algorithms nor was it intended to be a simple translation of existing vision systems into comparable Ada code. Rather, the goals of VISUAL were to implement the SRI (Stanford Research Institute) vision algorithms [4], taking advantage of the facilities provided in the Ada programming language, both to explore the use of Ada and the use of CAD information to replace the vision training phase.

The main implementation issue behind the development of VISUAL was to create a vision subsystem which could be used extensively for the variety of future projects concerning research with manufacturing cells. It was not only important to make VISUAL transportable to a variety of systems and configurations, but also extensible for a variety of vision tasks to be performed. Therefore, clear modularization, documentation and data abstraction were built into the system to allow future users to easily tailor VISUAL to their particular problems.

The overall vision module can be divided into three main submodules and a vision driver which communicates with an external program or an operator at a terminal (see Figure 3). The following are brief descriptions of the modules to familiarize the reader with the techniques used in solving the vision problem, and to set the stage for discussions on the use of Ada.

Feature Extraction

The feature extraction module is based on binary input from a video camera [4]. Raster scan lines are converted from an array of white and black pixels to a smaller sequence of bands which represent a run-length encoded sets of pixels which are of the same color and in contiguous locations on the line. If silhouettes are used, dark bands are interpreted as projections of a part while light bands are considered to be the absence of a part (i.e., a hole or empty space).

---

[2] The definition of "objects" is problematic, and our usage is fairly narrow. See [13] for a discussion of various viewpoints.

VISUAL analyzes consecutive scanlines, linking overlapping dark bands to form blobs and overlapping gap bands to form holes. Concurrently, values corresponding to the number of pixels and horizontal and vertical coordinate summations are calculated. Then when the links are made, these values are added into appropriate locations in an intermediate feature vector. In addition, concurrent computation of perimeter pixels are calculated. Thus when all the bands for a blob have been linked together, the intermediate feature vector and perimeter structure already contain the information which is needed to calculate the final feature values. These features, which usually contain values for area, perimeter length, number of holes, etc., are then used to identify the part in the scene.

Decision Tree Builder

Prior to the actual operation of the vision system, it is necessary to train the system to recognize the parts expected. Because of the quantization of the image, distortion through perspective and deviations within the limits of manufacturing tolerances, a given part will usually have mean and variance values for each feature measured [17], [18]. These values can be used to construct a decision strategy to follow at run-time.

To generate this mean and variance information, the vision system need only to be given a series of views of a part from a variety of locations in the viewing area. Traditionally, manual training of the vision system is required before each run-time situation [3], [19]. The work being done at the University of Michigan (as is also being done elsewhere [10]) generates the images from CAD information, calculating the necessary statistics off-line.

For each class of parts to be expected, a set of subclasses can be formed on the basis of the single feature which possesses the widest variance between mean values of the feature over the class of parts (measured in units of standard deviations). Thresholds can be calculated which divide the given set into subsets which should be disjoint and cannot be equivalent. Repeating this process on each subclass of parts produced until no classes can be further divided (i.e., each consists of only one part) yields a decision strategy for the run-time operation.

This decision strategy is then implemented through a decision tree data-structure. Each subtree's root node consists of the identity of the feature to be tested and then the thresholds which direct the routine to the corresponding offspring nodes. Each leaf node contains the part identified by the respective path through the structure. This essentially formulates a decision at run-time with roughly log(n) scalar comparisons.

Decision Mechanism

This module processes a given feature vector which has been calculated from an input image through the decision tree created prior to run-time as described above. It then sets up a resulting data structure which can return such information as the identity of the part, coordinates of its centroid and angle of orientation as well as possible pick points for robot acquisition.

IV. Object-Oriented Construction of VISUAL

One of the most crucial problems facing the designers of embedded computer systems is the management of complex software systems. As manufacturing becomes more and more integrated with sophisticated technologies, the level of software complexity dramatically increases. The need for such extensive software will require techniques to improve software organization, reliability and maintainability as programs are modified and expanded to fit changing needs. It has long been established that the solution for these problems is in efficient program modularization.

Conventional modularization

Good programmers initially factor problems into general functions to be performed by the solution program. These operations usually constitute logical processing units within the program and may later become subroutines or functions which are called in specified sequences by a driver routine. Typically these processes act on subsets of the global data structures, generating desired output from a stream of input. The global data-structures correspond to COMMON blocks in a FORTRAN program.

The feature extraction module of the vision system, may be broken down in the following way:

SLE ::= Scan-Line Encoding
CNA ::= Connectivity Analysis
FFC ::= Final Feature Calculation

Thus a single run through the program would be the processing of one image by each module in succession.

One way of defining the set of data structures operated on by the program is the following:

Frame ::=
Camera's 256X256 pixel array.
Band_Line ::=
Expanded run-length encoding represented as a linked-list of blob and gap bands.
Intermediate Feature Vectors ::=
Arrays containing summation values of area, coordinate summations, perimeter values, etc., for each hole and blob in the scene.
Perimeter ::=
Doubly linked-list which contains all the exterior pixels for a given part.
Final_Feature_Vector ::=
Vector which contains the scaler values for each feature requested.

Each process of the program takes subsets of these data-structures as parameters for their operation. A data-structure used as input for a process will be prefaced by the word **in** and a data-structure output from a process will be prefaced by the word **out.** Thus the above program becomes:

program(**in** Frame, **out** Final_Feature_Vector) ::=

SLE(**in** Frame, **out** Band_Line vector)
CNA(**in** Band_Line, **out** Intermediate_Vector list, **inout** Perimeter)
FFC(**in** Intermediate_Feature_Vector, **in** Perimeter, **out** Final_Feature_Vector)

As we can see in Figure 4, both the CNA and FFC processes must have direct access to both Intermediate_Feature_Vector and Perimeter structures. Therefore, design decisions as to how Intermediate_Feature_Vector and Perimeter were implemented must be known by both. In the case where this design needs to be changed, which often occurs in the lifetime of a well used piece of software, we are faced with the task of patching each process in the program which uses the structures to correct the situation. Obviously, in very large systems with extensive interdependence of data-structures, this could be a tremendous task.

Object modularization

The problems encountered with conventional methods of modularization had little to do with the algorithms for calculating the information. They lay with the storage and retrieval of the information once it was generated. In addition, debugging a conventional program is made more difficult when a single data structure is accessed and modified by the entire program; memory can be *clobbered* from anywhere and tracing offending statements is a formidable proposition. To avoid these problems, we must hide these design decisions from the processes which only need to know values and not necessarily how these values are stored and retrieved.

Program modularization based upon Ada objects allows us to create a module which has sole access to a given data structure. All external processes must access data through rigidly defined storage and retrieval subroutine calls. Then when modifications are required in the data structure, the only code which must be modified are the local routines which access the data. For example it is unimportant for the program to know the actual implementation of the Frame object, as long as there is a means of finding the coordinates of where the image changes from white to black and black to white. Similarly, the program does not need to know how the band lines or final objects are implemented, only how to store and retrieve information.

In addition we would like to have processes interact with objects at a uniform level of abstraction. Instead of processes knowing that an intermediate feature vector and its perimeter, as in Figure 4, are distinct, we treat the perimeter as another attribute of the intermediate feature object by embedding a perimeter module with in an Intermediate Feature Module, as shown in Figure 5. Thus the feature extraction data-structures can be factored into the following object modules.

Frame Module:
dump_new_frame()--command camera to take a picture and return results
next_black_pixel() return coordinates
next_white_pixel() return coordinates

Band_Line Module:
store_blob(<band-info>)--storage details not needed by user
store_gap_band(<band-info>)
next_band(band) return band--starting coordinate to length
retrieve_<attr>(band) return <attr>--<attr>=hole or blob

Intermediate Feature Module:
store_features(<feature-info>)
sum_<attr>(feature_vector,value)
retrieve_<attr>(feature_vector)return<attr>
start_perimeter(<pixel_coordinates>) returns perim_pixel
link_perimeter(feature_vector,perim_pixel) returns length
retrieve_<attr>(perim_pixel) return <attr>

Final Feature Module:
store_<feature>(<feature-info>)--e.g. area, perimeter, centroid
retrieve_<feature>(final_vector) return <feature>

The interface is now clearly defined and hence easier to use by the program, and, most importantly more easily maintained and modified (compare Figures 4 and 5).

Objects in Ada

Object oriented programs are a rather difficult to implement in traditional high-level languages. Data structures which must be accessed in separate blocks must be global to the block containing both in most highly structured languages. In less restrictive languages, the memory locations can be made available to the subroutines which need access to them, but then we are faced with problems of information security. In Ada, modules for both objects and processes are clearly defined with the package constructs. A package, as the name implies, contains a collection of routines and related data structures, composing a syntactic and semantic unit [2]. A package definition consists of a specification section and a package body. The specification formally defines visibility and scope rules for variables and data objects as well as the subprograms which may be called. It creates a global environment for both the package body and any routines which need to access the module. The body and any modules which reference the package can then be compiled separately from the specification as long as they are able to recreate the environment of the specification section during their own compilation. The package body elaborates the code for each subroutine and function in the package specification. In this way, strong static-semantic checking can be achieved concurrently with separate compilation. Packages are then used to create data and process modularization.

As illustrated above, we abstracted the data-structures into data-objects to hide their actual implementation. Each object contains a set of routines which perform storage operations and access information contained in the objects. By making the data structures exclusive to the object module, using the Ada construct **private**, user processes are forced to use only these routines when accessing the object. The inefficiencies produced by excessive subroutine calls can be reduced with the INLINE pragma (a compiler directive which expands the subroutine source code inline wherever called).

The most immediate feature of such large scale data modularization aside from facilitating program development is tremendous program flexibility. By compiling a package body separately from its specification, we can fix the interface of processes with a module without elaborating any code for the actual implementation. Thus, when entire programs are linked, different package bodies can be substituted to provide a variety of interfaces. To illustrate the power of this flexibility we will elaborate the work accomplished with the frame module.

The frame buffer contains a single load routine which conceptually causes a new image to be dumped into the frame object and two access operations which return the row and column of the next occurrence of a black or white pixel, essentially returning a binary run-length encoded output of an image. In the run time operation of VISUAL, the first operation causes a GE-TN2500 camera to dump an image into a frame grabber, and then an attached processor translates the resulting matrix into a run length encoding. The encoded image is then dumped into the execution machine and the access instructions return the encoded stream as called.

However, because the actual implementation is hidden from the rest of the system, we can replace the package body for the frame buffer and have entirely new processing modes. In debugging the system, it is necessary to enter images interactively. Through a simple package substitution, the frame buffer access routines request column and row information from an operator's terminal. This allows us to bypass the need to perform elaborate calculations on entire frames of information initially and test specific calculation problems directly. In addition, experiments with replacing manual training using real images of an object by images generated from CAD information can be easily performed. By substituting another body for the frame buffer, we could read images created from CAD information on a host mainframe.

Tailoring VISUAL with Generics

An undersirable attribute of strongly typed languages is the inability of type independent operations to be used on a variety of conflicting data-types. An example which is often mentioned is a storage management module. Objects of the same data-type are to be dynamically pushed and popped in the manner of a queue or stack. The manager does not need to know any information about the object being stored, it only places a new object in the first free location and retrieves the front-most or top-most object. The generic facilities of Ada allow one to define such a module and then expand the Ada source code necessary to implement it for each data type desired. One merely writes generic subroutines or packages, leaving the data-types manipulated by the package as formal parameters. One can then create an instance of the package which is specific to a given data-type in the compilation of a package. This allows repetitive software to be written once and then copies which are specific to a given data-type can be created for each instance.

Generics are not limited, however, to operations which are totally independent of the data-types involved. Though not as flexible as polymorphic functions [14], Ada generics can adjust for broad classes of objects due to the **with** clause [2]. Because certain operations which are specific to the data-type given as a generic formal parameter can be inherited, we can extend the use of generics to more elaborate software systems such as robot vision.

In the process of developing the vision module, it was unclear as to what features should be calculated to best suit our problems. Contemporary vision systems [3], [20] utilize a list of nearly forty features which could be useful in distinguishing between objects, inspecting parts or guiding specific assembly operations. Most vision systems calculate subsets of four to twelve features from this list. Because of the limited amount of time, no vision system calculates the entire list at run-time. Selection of the proper subset is, therefore, crucial for an effective solution. Furthermore, different operations require different subsets. General part recognition usually focuses on total area, perimeter and first or second moments of inertia while vision systems tailored to operations such as valve inspection [21] are obviously more concerned with hole information.

We wanted a vision module which could be easily adapted to the variety of expected vision tasks in our manufacturing cell. It was important to build-in a mechanism which would facilitate modifying the feature set to be calculated while at the same time allowing for fast run-time execution. We, therefore, explored the possibility of using Ada generics to instantiate (create an instance of) vision systems tailored to a specific subset of features, thereby placing most of the code generation burden on the compiler.

A library of routines to calculate all features from given object vectors was established. To create an instance of the vision system using a subset of the features, one creates a feature vector module which contains the definition of only the feature vector to be used and a feature calculation driver which makes calls to the necessary feature calculation routines. This is done through generic formed parameters. In this way, relatively specific and efficient vision systems can be produced at a minimal programming cost.

When this technique is generalized to the entire manufacturing cell, a great deal of extensibility is added to cell programs. Not only are the programs transportable, pending advent of Ada compilers for many systems, but also reconfigurable to the various machines if their interfaces are implemented in an object-oriented manner. Thus a manufacturing cell program which performs a given cell task with a generic robot arm, a comparable program can be developed for a system with an Asea or Puma robot as long as the interface has been designed according to a generic specification, simply by generating an instance of the program specific to the robot type. This, of course, requires a greater amount of effort at development time but may payoff if the program will be widely used.

Speed Up Through Multitasking

With ever diminishing costs for hardware, multiprocessor systems will become more and more cost-efficient. Not only does parallel processing expand the complexity of problems which can be solved in real-time, but also allows for greater fault tolerance in operation. The manufacturing cell's central computer already must synchronize activities in attached processors controlling motion in robot arms, performing low level vision operations from camera input and executing tooling programs in an NC device. In addition, the cell's computer could also have a multiprocessor architecture, (e.g. the Intel 432)·providing more powerful computer control and computation. Finally, processes which absorb exorbitant processing time may need to be down loaded to special purpose hardware made possible through VLSI technology. An implementation must make such a decision transparent to the user program.

Unfortunately, conventional programming languages lack formal constructs to facilitate programming in a multiprocessor environment. Defining separate processes and then establishing communication and synchronization is a nontrivial programming operation. Plus, most solutions require so much knowledge of the hardware targeted for execution that transportability is virtually impossible.

Conversely, the Ada **task** construct provides a relatively clear means of dividing a program into processes which can be executed in parallel on one processor or concurrently on several different processors [16]. Tasks are declared in a similar manner as subroutines. Communication is formally defined with message passing facilities. Whether or not this leads to any real ease in parallel programming depends upon both the computer implementation and the underlying computer architecture. In our experimental cell an Intel 432 system is used which automatically handles parallel processing of tasks.

Again using the feature extraction module as an example, suppose the actual operation was to continually read images from the camera and execute the program on each. If we define each process as a task instead of as a subroutine, we can take advantage of the extra processing power available in our multi-processing system. The scan-line encoding process can read images from the camera and then send Band_Lines to the connectivity process. Now, instead of waiting for the rest of the scan lines to be entirely encoded, the connectivity analysis begins while the scan-line encoding process fetches the second raster scan line. Thus each process performs its own loop, fetching information from the previous process and sending information to following processes (see Figure 6). This technique, identified as "serial concurrency," shows a reduction in processing time which has a lower time-bound of the time to process a single record once plus the amount of time for the slowest process to work on k-1 records where k is the total number of records processed.

Let T(k) be the time to run a program on k data records, then:

$$Tsc(k)=time(\ P1+P2+...+Pn\ ) + (k-1)*time(max(P1,P2,...,Pn)) <= T(k)$$

where n is the number of processes in the program, assuming sufficient processors are available. This, of course, is an ideal situation which assumes uniformity in the processing of records and little or no overhead to manage the extra-processors, but programs which cycle through large data sets on systems with enough processors, should gravitate to this lower bound.

This particular technique, aside from maximizing the use of processors also provides a degree of flexibility in the run-time implementation. The scan-line encoding task must be performed on every raster line in the image. The transmission and calculations on this array prove to be to be quite exorbitant. Therefore, instead of having the image dumped directly into the central cell computer, the major portion of the work is done in an attached processor which works directly with the low level frame grabber. Because the program had been divided into tasks originally, the final implementation required no special interface with the rest of the processes.

The calculations in the third process, the final feature calculations, easily separate into independent subtasks such as bounding_box_calculations, hole_calculations, perimeter_calculations, etc., all requiring the same input, the Object_Buffer, and all outputting to unique locations in the final feature vector. These independent calculations can be grouped into parallel subtasks within the process. The parent process merely passes chunks of work to each subtask and waits for them to finish (see Figure 6). This method, identified as "parallel concurrency," shows a reduction in processing time with a lower bound of the time necessary to process each record through the longest subtask.

$$Tpc(k) = time(max(Sp1, Sp2, Sp3,...,Spn) <= T(k)$$

where Spi is the ith parallel sub-process out of n processes. Again, this is in an ideal situation, but extensive processing using this technique should reflect significant speed-up. Not only do we reduce the processing but with enough processors we could add additional features to our chosen subset without increasing processing time.

Further, by defining the library of subtasks to be used with the feature calculation module, we have established an efficient mechanism for implementing the generic tailoring of the vision system. For a given subset of features, we write a feature calculation process which spawns the sub-processes necessary to calculate the given feature set. Then when the vision system is instantiated via the appropriate feature vector as a formal parameter, this process, specific to the feature vector, is inherited via the Ada **with** clause.

V. SUMMARY AND CONCLUSIONS

The future development of automated manufacturing cells will be increasing linked to the integration of cell components amongst themselves and with higher level computer aided engineering functions. This integration will depend upon increasing complex and sophisticated computer systems. Ada was developed specifically for large complex real-time embedded software systems. This paper has explored its use as the basis for developing manufacturing cell software and illustrated this with the implementation of a computer vision module via Ada.

The use of Ada packages to implement abstract data types (objects) was shown to provide a clean, easily read implementation. The well defined interfaces to the data and hiding of implementation details make it easy to shift from one implementation to another, either for debugging purposes, or to accommodate new developments in associated hardware. The use of generics helps one to write general systems and then take tailor them to specific configurations for given problems. When Ada is used in conjunction with an object based multiprocessing hardware system, the use of tasking in unconventional ways can yield a parallel programming implementation of some algorithms, thus increasing the overall through put. Though not explicitly discussed here, the ability to define one's own data types and overload (that is redefine) operators on datatypes can be a significant aid in developing readable programs.

One of the principal goals of the use of Ada as the base of programming for the manufacturing cell is to achieve highly readable and easily maintained code. While it is too soon to assess the maintainability of the code, it can be noted that the use of Ada substantially enchanced the code production process. The programmer was totally unfamiliar with Ada, SRI vision work, and the hardware system upon which the code was to run at the beginning of that period of time. Moreover, a significant portion of time was lost due to bugs in both the hardware and software of the system. In spite of these difficulties the program which constituted 5,000-6,000 lines of code was produced in less than four months.

Another issue, one that requires considerable additional research, is that of debugging. Ada is intended for compile based translation and almost all robot programming languages are interpretive. From the point of view of the cell programmer, however, the robot program may be a separately prepared and debugged entity. Moreover, what is really necessary is a fast interactive translate/debug system, and this does not preclude compile translation, particularly if used in conjunction with a simulator [22].

Recent programming language research has yielded a number of new concepts which will aid the program development process. A number of these are incorporated into Ada. Future languages will undoubtably encompass more of these concepts. However, at present, the considerable resources being put into the Ada effort by DOD coupled with its orientation toward real-time embedded systems make it a candidate for manufacturing cells worth investigating. Based on the work outlined here it is our conclusion that from a programming point of view the use of Ada will be a highly attractive alternative.

VI. BIBLIOGRAPHY

[1] K. G. Shin, "Comparative Study of Robot Programming Languages," Center for Research in Integrated Manufacturing Report, (to appear).

[2] *Reference Manual for the Ada Programming Language (Revision)* Washington, D. C.: United States Department of Defense, July 1982.

[3] J. W. Hill, "Survey of commercial vision systems," *Industrial Automation Group*, May 1980.

[4] G. J. Gleason, "Vision Module Development," Ninth Report, NSF Grants APR75-13074 and DAR78-27128, SRI Projects 4391 and 8487, Stanford Research Institute, Menlo Park, CA, pp. 9-16, Aug. 1979.

[5] J. L. Turney and T. N. Mudge, "VLSI implementation of a numerical processor for robotics," in *Proc. of the 27th Int. Instrumentation Symp.*, Indianapolis, IN, pp. 169-175, April 1981, also presented at the Instrument Soc. of America Anaheim Conf., Oct. 1981.

[6] T.N. Mudge, "Special Purpose VLSI Processors for Industrial Robots," in *Proc. of the IEEE Computer Soc. 5-th Int. Computer Software & Application Conf.*, pp. 270-271, Nov. 1981.

[7] T. N. Mudge, R. A. Volz, and D. E. Atkins, "Hardware/Software transparency in robotics through object level design," in *Proc. of the Soc. of Photo-optical Instrumentation Engineers Technical Symp. West*, Aug. 1982.

[8] J. Wolter, T. C. Woo, and R. A. Volz, "Gripping position for 3D objects," *Industrial Applications Soc. Meeting*, Oct. 1982.

[9] E. W. Baumann, "Model Based Vision and the MCL Language," *IEEE SMC Conference*, pp. 433-438, Oct. 1981.

[10] E. W. Baumann, "CAD Model Input for Robotic Sensory Systems," *Proc. Autofact IV*, Nov. 1982.

[11] J. D. Cooper, "Why a DoD Standard Programming Language," in *Ada Conference*, Boston, Mass. and Washington, D.C., pp. 1-6, June 17-18 and June 28-29.

[12] E. I. Organick, *A Programmer's View of the Intel 432 System* Santa Clara, CA 95051: Intel Corp., 1982.

[13] T. Rentsch, "Object Oriented Programming," *Sigplan Notices*, vol. 17, no. 9, pp. 51-57, Sept. 1982.

[14] R. Milner, "Theory of Type Polymorphism in Programming," *Journal of Computers and System Sciences*, vol. Vol. 17, pp. 348-375, 1978.

[15] D. I. Good and W. D. Young, "Generics and Verification in Ada," *Sigplan Notices*, vol. l5, pp. 123-127, Nov. 1980.

[16] E. S. Roberts, A. Evans, Jr., C. R. Morgan, and E. M. Clarke, "Task Management in Ada - A Critical Evaluation for Real-Time Multiprocessors," *Software - Practice and Experience*, vol. 11, pp. 1019-1051, 1981.

[17] R. Duda, J. Kremers, and D. Nitzan, "Automatic Part Classification," Fifth Report, NSF Grant G138100X1, SRI Project 4391, Stanford Research Institute, Menlo Park, CA, pp. 7-22, Jan. 1976.

[18] R. Duda and D. Nitzan, "Error Analysis for Automatic Part Recognition," Fifth Report, NSF Grant G138100X1, SRI Project 4391, Stanford Research Institute, Menlo Park, CA, pp. 65-81, Jan. 1976.

[19] W. A. Perkins, "A Computer Vision System that Learns to Inspect Parts," General Motors Research Laboratories, Research Publication GMR-3650, June 1981.

[20] H. G. Barrow and J. M. Tenenbaum, "Computational Vision," *Artificial Intelligence Center, SRI Int'l.*, vol. 69, no. 5, pp. 572-595, May 1981.

[21] W. A. Perkins, D. B. Juliette, R. Dewar, and J. H. West, "KEYSIGHT: A Computer Vision System for Inspecting Value Spring Assemblies," General Motors Research Laboratories, Research Publication GMR-3601, April 1981.

[22] Stuart J. Kretch, "Robotic Animation," *Proc. Robots* 6, Mar. 1982.

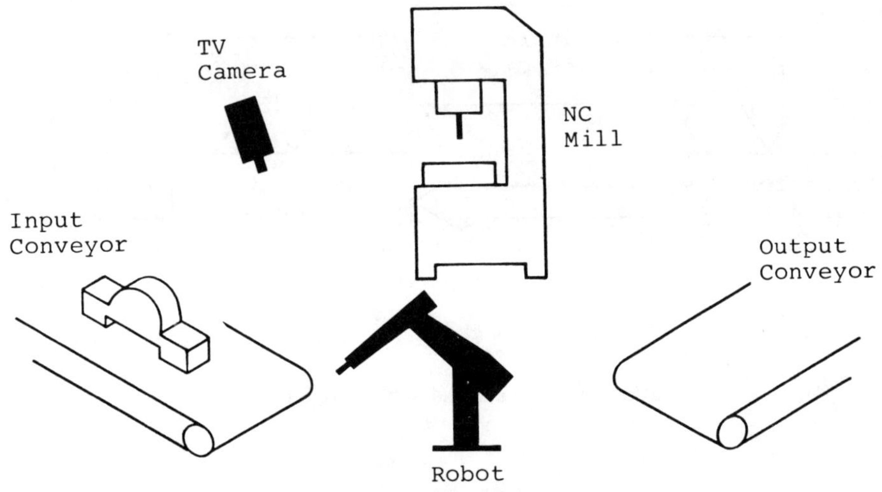

Figure 1. Hypothetical integrated manufacturing cell.

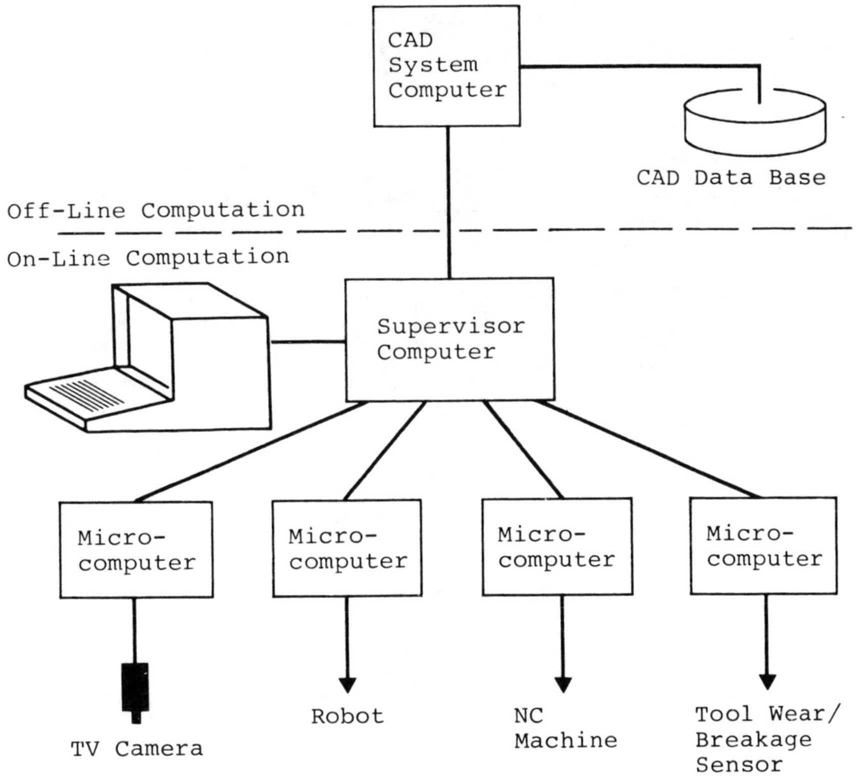

Figure 2. Hierarchical organization of hypothetical manufaturing cell.

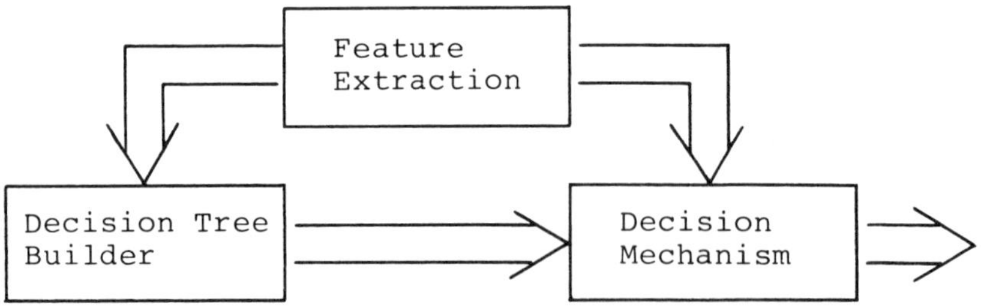

Figure 3. Interrlation of three main vision system modules.

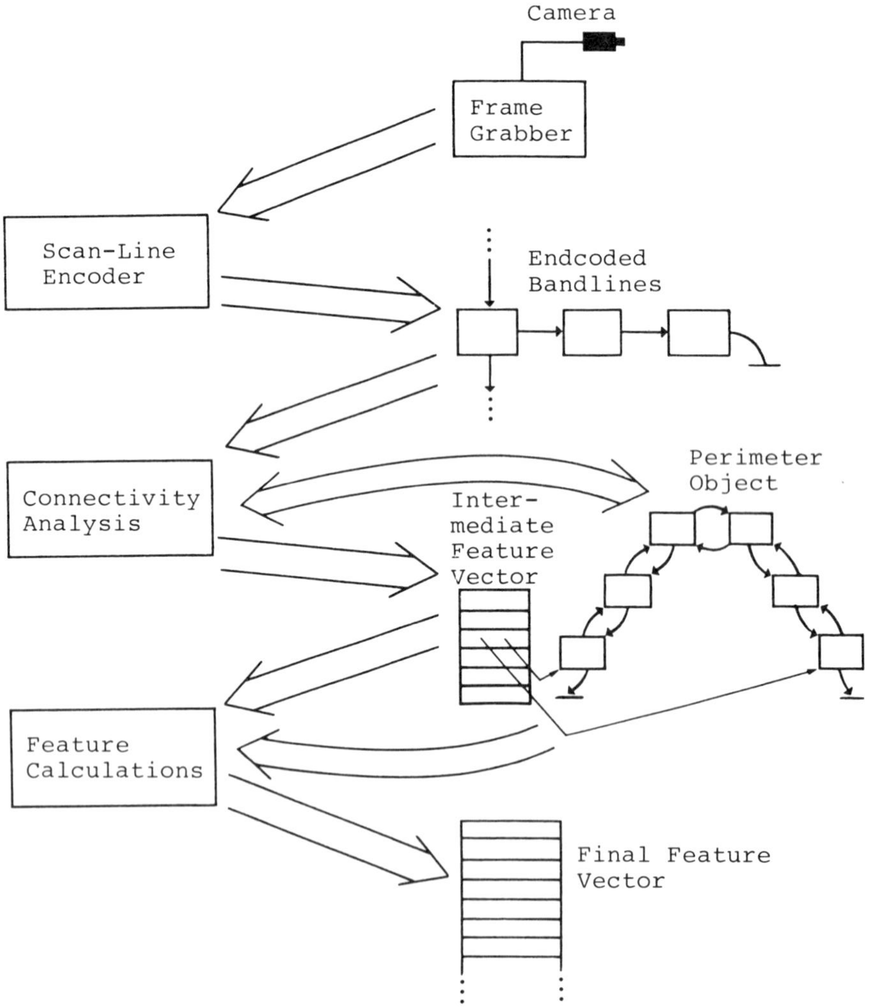

Figure 4. Conventional modularization of vision system.

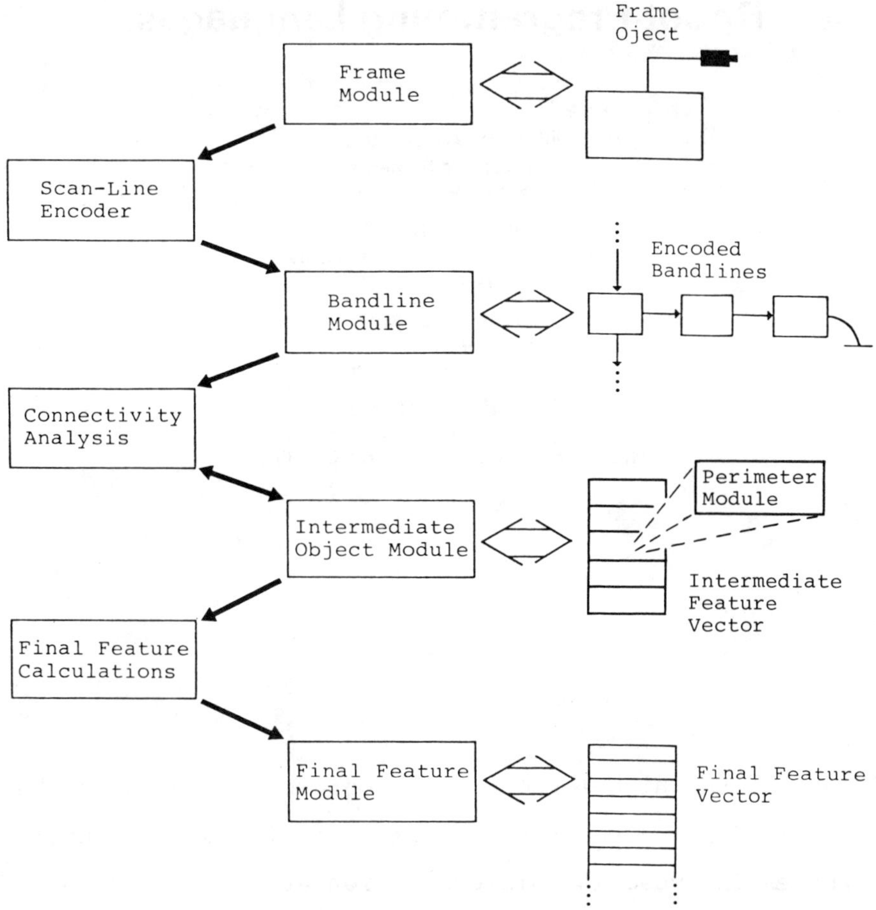

Figure 5. Object-based modularization of vision system.

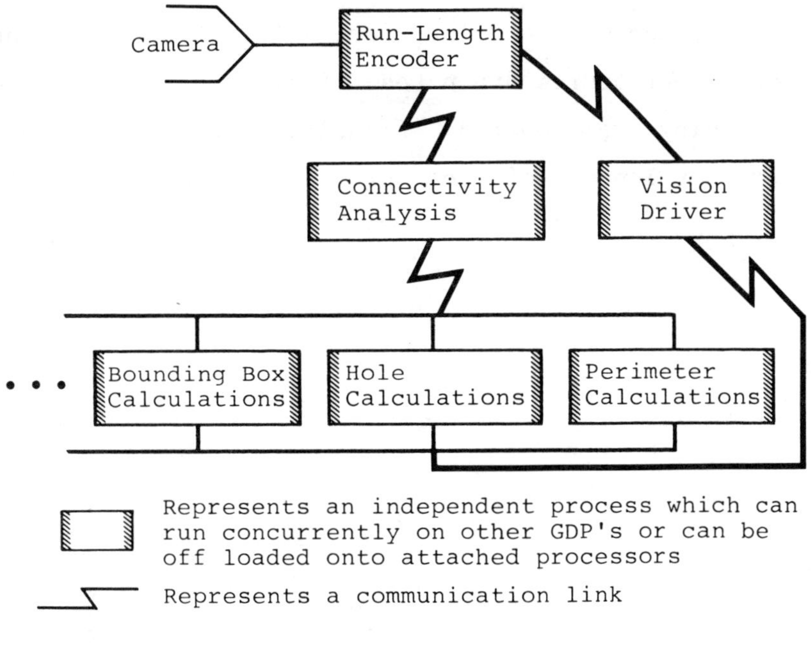

Figure 6. Multiprocess organization of vision system.

Evaluation of Commercially Available Robot Programming Languages

William A. Gruver
LTI Robotic Systems
Palos Verdes, California

Barry I. Soroka
University of Southern California
Los Angeles, California

John J. Craig
Stanford University
Palo Alto, California

Timothy L. Turner
Research Triangle Institute
Research Triangle Park, North Carolina

Abstract

An evaluation of most of the high-level robot programming languages that have been developed by robot builders and research laboratories for manufacturing applications is reviewed. Languages included in this evaluation are: AL (Stanford University), AML (IBM Corporation), HELP (General Electric Corporation), JARS (Jet Propulsion Laboratory), MCL (McDonnell Douglas Corporation), RAIL (Automatix Inc), RPL (SRI International), and VAL (Unimation Inc). Historical background of the languages is reviewed. Languages are evaluated on the basis of data types, control structures, motion specification, use of sensors, and interface with external machines and devices. Common and uncommon features are noted. Facilities for program editing, entry, debugging, and teaching are compared. Finally, conclusions regarding the present status and future development of robot programming languages are given.

INTRODUCTION

Simple push-button teach-and-repeat schemes are sufficient for robot programming in a wide variety of tasks, especially where a robot will perform a single task for a long run of workpieces. Reasons for going beyond use of the push-button approach to a textual robot programming language are as follows:

- Teaching points by use of a teach-box can be cumbersome for operations such as palletizing, where each new position could be computed rather than taught.
- A textual language permits robots to be programmed off-line, allowing the actual robot to continue to perform useful work during the programming process.
- Textual robot programs can be moved from one robot to another provided that sufficient accuracy is available.
- CAD/CAM can be integrated with robot programming.

For these reasons, high-level robot programming languages will become increasingly important in the future.

This paper presents results from an extensive evaluation of robot programming languages performed by LTI Robotic Systems. Eight robot programming languages were selected: AL, AML, HELP, JARS, MCL, RAIL, RPL, and VAL. Historical background, description, and evaluation matrices are summarized for each new language. Conclusions are given regarding the present and future development of robot programming languages.

HISTORICAL BACKGROUND

The earliest robots during the 1960s were used only in the teach-and-repeat mode without a textual programming language. This mode of operation was sufficient for applications such as spray painting and spot welding, where a single task was endlessly repeated.

WAVE was developed in 1973 as an experimental language for programming robots for research purposes at the Stanford Artificial Intelligence Laboratory. Stanford offered excellent computer facilities for software development, and their research was directed at finding the limitations of robotics theory rather than using robots to perform manufacturing tasks. The WAVE system was interfaced to a vision system under which successful hand-eye coordination was demonstrated.

In 1974, the Stanford group began the development of a second robot programming language, AL, based on ALGOL and incorporating advanced constructs for controlling multiple arms in parallel and cooperative tasks. AL was designed to facilitate programming assembly operations using the extensive force sensing capabilities at Stanford. An interactive teaching module was added to an existing compiler, leading to the mostly interpretive system in use today.

Another major research effort occurred at the IBM T. J. Watson Research Center at Yorktown Heights, New York. This group has developed a number of languages since 1976 for controlling a Cartesian manipulator. The first languages were EMILY and ML, which were successfully used in several assemblies. Later, AUTOPASS was developed as a high-level language for semi-automatic programming of assembly tasks using a proprietary geometric modeling system. Recently, the AML language emerged as a commercial offering from IBM based on the work of the Yorktown Heights group.

Two laboratories used existing computer programming languages as hosts for subroutines for controlling robots. SRI International developed RPL, a robot programming language for in-house research and distribution to members of SRI's Industrial Affiliates Program. RPL emphasizes the use of special purpose subroutines that are called by a general purpose program. RPL is designed to facilitate control of the varied machines which comprise a robot work-cell: robots, machine tools, and vision systems. The Jet Propulsion Laboratory developed JARS, a robot system based on PASCAL, in connection with contracts with NASA for controlling robots to be used in the assembly of solar cell arrays.

The first commercially available robot programming language was VAL, released in 1979 by Unimation. VAL was developed by a Stanford alumnus quite familiar with AL, but constrained by the need to use an available minicomputer for implementation. VAL is an extension of the BASIC programming language and runs on a DEC LSI-11/03. At first, VAL was supplied with the PUMA manipulators, but VAL controllers are now available for many of Unimation's robots. VAL has been upgraded to run on an LSI-11/23, and a new version, VAL-II, that corrects some of its predecessor's deficiencies, is now in beta test.

Although it does not offer a true textual language, the Cincinnati Milacron T3 robot provides the user with features that resemble a language. This robot system has become very popular since its inception during the late 1970s, particularly in the aerospace industries. Recently, an option has also been developed for the T3 that permits the user to obtain a hard copy of the sequences that have been programmed from the teach-box and control panel. However, the T3 programming system has not been included in this paper because it is not a textual language.

1981 witnessed the introduction of two more commercial robot programming languages. Automatix developed RAIL to control visual inspection, and robots for assembly and arc-welding tasks. Under U.S. Air Force ICAM funding, McDonnell Douglas developed MCL as an extension of the NC machine tool programming language APT for the off-line programming of robots from a CAD data base; it is now being offerred commercially to U.S. firms.

Two robot programming languages were commercially announced in 1982, AML by IBM Corporation and HELP by General Electric Company (under license to the Italian firm DEA). New robot programming languages will undoubtedly be developed with increasing frequency as the market matures and robot builders offer advanced programming features as part of their total marketing strategy.

LANGUAGE DESCRIPTIONS

The languages selected for evaluation satisfied the following criteria: the language had been in actual use, either in industry or in the laboratory; the language had a textual form; and manuals or other detailed documentation were available to LTI Robotic Systems.

AL

AL is the second generation robot programming language produced at the Stanford Artificial Intelligence Laboratory, an early leader in robot research. Based on concurrent PASCAL, it provides constructs for control of multiple arms in cooperative motion. Commercial arms have been integrated into the AL system. AL has been copied by several research groups around the world. The current AL implementation requires a large main frame computer, but a stand-alone portable version has been announced. In the present AL system, programs are developed and compiled on a PDP-10. The resulting p-code is downloaded into a PDP-11/45 where it is executed at run-time. High-level code is written in SAIL (Stanford Artificial Intelligence Language). The run-time system is written in PALX. The 11/45 has floating point, no cache, a single terminal, and 128 Kbyte memory. Two PUMA 600s and two stanford Scheinman arms are controlled. AL can control all four arms at the same time. The stand-alone AL runs completely on the 11/45. It is written almost entirely in OMSI PASCAL. This new version is being marketed for industrial applications.

AML

A Manufacturing Language (AML) was designed to be a well-structured, semantically powerful interactive language that would be well adapted to robot programming. The central idea was to provide a powerful base language with simple subsets for use by programmers with a wide range of experience. An interpreter implements the base language and defines the primitive operations, such as the rules for manipulating vectors and other "aggregate" objects, that are naturally required to describe robot behavior. A major design point of the language was that these rules should be as consistent as possible, with no special-case exceptions. Such a structure provides a ready growth path as programmers and applications grow more sophisticated. The computer controller on the RS/1 assembly robot consists of an IBM Series/1 mini-computer with a minimum of 192 Kbyte memory. Peripherals include disk and diskette drives, matrix printer, and keyboard/display terminals. AML is being used to control the RS/1 assembly robot, a Cartesian arm having linear hydraulic motors and active force-feedback from the end-effector. A subset of AML is employed on the Model 7535 robot that is controlled by the IBM personal computer; however, the features of this version are not included in this evaluation.

HELP

HELP is a commercial offering of General Electric Company's Model A12 Allegro Robot System designed for control of multiple Cartesian arms in assembly tasks. It is an interpreter, based roughly on PASCAL. There are no geometric data types or constructs whatsoever; motions are in terms of actual robot joints, which also define the xyz coordinate system. It includes provisions for defining and activating multiple tasks simultaneously; the tasks communicate by means of global flags. A DEC LSI-11/2 serves as the system controller. An Intel 8080A controls the joints. Peripherals include a keyboard, display screen, floppy disk, and a teach box. The Allegro system can be configured with up to four arms having at most 12 degrees of freedom for all arms.

JARS

JARS was developed at the Robotics and Teleoperators Group of NASA's Jet Propulsion Laboratory in 1979. Its purpose is to provide a high-level programming environment to support basic research in manipulation, control, and robot programming systems. The intended users are members of the Robotics and Teleoperators Group at JPL, although the language may be desirable for other sites as well. The JARS language is PASCAL with many robot specific types, variables, and subroutines added. All of JARS runs on a single DEC PDP-11/34 having floating point, no cache, and 28K bytes of useable memory. OMSI PASCAL and MACRO-11 are used in the implementation. The system runs under the RT-11SJ monitor. A Tektronix graphics terminal provides the capability for real-time graphing operations. Currently, JARS controls a modified Stanford Scheinman arm and it has also been interfaced to a PUMA 600.

MCL

MCL is an extension of APT. It provides statements for robot motion specification, vision system operation, image modeling, real-time conditional logic, and compile-time language extensions. MCL was written for the programming of work-cells in which a number of devices, including a robot, are under the control of one or more computers. When writing programs, the MCL programmer describes the cell. The cell contains standard names which are given exact definition by the work-cell control system. The ICAM developed version of APT/MCL utilizes an IBM 370, with 1 Mbyte memory, batch and TSO. A McAuto proprietary version (MCL/11) for real-time operation utilizes a DEC PDP-11 with 128 Kbyte memory, and an interactive compiler. Input is via a terminal. APT/MCL uses BAL and FORTRAN. MCL/11 uses FORTRAN. Currently, only the Cincinnati Milacron T3 robot is controlled using MCL.

RAIL

RAIL was developed by Automatix Inc as a high-level language for the control of both vision and manipulation. It is an interpreter, loosely based on PASCAL. Many constructs have been incorporated into RAIL to support inspection and arc-welding systems which are a major product of Automatix. The central processor of the RAIL system is a Motorola 68000. Peripherals include a terminal and a teach-box. RAIL is being supplied with three different systems: vision only, no arm; a custom-designed Cartesian arm for assembly tasks; and a Hitachi process robot for arc welding.

RPL

RPL was developed at SRI International to facilitate development, testing, and debugging of control algorithms for modest automatic manufacturing systems that consist of a few manipulators, sensors, and pieces of auxiliary equipment. It was designed for use by people who are not skilled programmers, such as factory production engineers or line foremen. RPL may be viewed as LISP cast in a FORTRAN-like syntax. The SRI Robot Programming System (RPS) consists of a compiler that translates RPL programs into interpretable code, an an interpreter for that code. RPS is written mostly in Carnegie-Mellon's BLISS-11 and cross-compiles from a DEC PDP-10 to a PDP-11 or LSI-11. The RPS programs run under RT-11 with floppy or hard disks. The RPL language is implemented as subroutine calls. The user sets up the subroutine library and documents it for people who must write RPL programs. Present efforts at SRI are concentrated on the control of a Unimation PUMA 500 arm and the MIC vision module. Previously, SRI operated the Unimate 2000A and 2000B hydraulic arms and the SRI vision module.

VAL

VAL is a robot programming language and control system designed for use with the Unimation robots. Its stated purpose is to provide the ability to easily define robot tasks. The intended user of VAL will typically be the manufacturing engineer responsible for implementing the robot in a desired application. Prior programming knowledge is helpful but not essential. VAL has the structure of BASIC, with many new command words added for robot programming. It also has its own operating system, called the VAL Monitor, which contains the user interface, editor and file manager. The central monitor contains a DEC LSI-11/03, or more recently, the LSI-11/23. Each of the joints is controlled by a separate 6503 microprocessor. The monitor communicates with the user terminal, the floppy disk, the teach-box, a discrete i/o module, and an optional vision system from Machine Intelligence Corporation. C language and 6503 assembly language are involved in the implementation. VAL has been released for use with all PUMA robots and with the Unimate 2000 and 4000 series.

LANGUAGE COMPARISONS

Language Modalities

	AL	AML	HELP	JARS	MCL	RAIL	RPL	VAL
textual	X	X	X	X	X	X	X	X
menu		X						

Language Type

	AL	AML	HELP	JARS	MCL	RAIL	RPL	VAL
subroutines				X			X	
extension					X			
new language	X	X	X			X		X

Geometric Data Types

	AL	AML	HELP	JARS	MCL	RAIL	RPL	VAL
frame (pose)	X			X	X	X		X
joint angles		X		X				X
vector	X	X		X	X			
transformation	X			X	X			X
rotation	X			X	X			
path						X		

Display and Specification of Rotations

	AL	AML	HELP	JARS	MCL	RAIL	RPL	VAL
rotation matrix	1			4				
angle about a vector	x			4				
quaternions								
Euler angles	2	x		x		x		x
roll-pitch-yaw	3		x		x			

1 AL displays rotations as a rotation matrix.
2 AL accepts directly the specification of an orientation by three Euler angles (or by an angle about a vector).
3 AL orientations could also be specified by roll-pitch-yaw angles.
4 Normally JARS does not display these forms, however, the user may write a routine to print them because JARS has the forms available internally.

Ability to Control Multiple Arms

	AL	AML	HELP	JARS	MCL	RAIL	RPL	VAL
multiple arms	x		x		x			

Control Structures

	AL	AML	HELP	JARS	MCL	RAIL	RPL	VAL
statement labels		x	x	x	x		x	x
if-then	x	x	x	x	x	x	x	x
if-then-else	x	x	x	x	x	x	x	
while-do	x	x	x	x	x	x	x	
do-until	x	x		x		x	x	
case	x			x		x	x	
for	x	x		x		x	x	
begin-end	x	x		2	1			
cobegin-coend	x		3		1			
procedure/function/subroutine	x	x	x	x	x	x	x	x

1 MCL can invoike tasks in parallel using INPAR.
2 Since it is a language based on subroutines added to PASCAL, JARS has all the structures of PASCAL.
3 HELP permits the simultaneous activation of several tasks.

Control Modes

	AL	AML	HELP	JARS	MCL	RAIL	RPL	VAL
position	x	x	x		x	x	x	x
guarded moves	1	1	1					
bias force	x							
stiffness/compliance	x			2				
visual servoing	3			3		3	3	3
conveyor tracking				x	x			

1 Using force control or limit switch action.
2 Currently being implemented at Jet Propulsion Laboratory.
3 Uses visual inputs to determine set points but does not specifically perform visual servoing.

Motion Types

	AL	AML	HELP	JARS	MCL	RAIL	RPL	VAL
coordinated-joint between two points	x	x	x	x		x	2	x
straight-line between two points	1			x	x	x	2	x
splined through several points	x		x	x		x	2	x
continuous path ("tape recorder" mode)								
implicit geometry circles					x			
implicit geometry patterns					x			

1 Currently being implemented at Stanford University.
2 Relies on the VAL controller.

Signal Lines

	AL	AML	HELP	JARS	MCL	RAIL	RPL	VAL
binary input	0	64	*	0	242	6	32	32
binary output	0	64	*	2	242	10	32	32
analog input	64	0	*	0	242	0	32	0
analog output	4	0	0	0	242	0	64	0

* Custom for each system.

Successful Sensor Interfaces

	AL	AML	HELP	JARS	MCL	RAIL	RPL	VAL
vision	x	1	x	x	x	x	x	x
force	x	x	x					
proximity								
limit switch	x	x		x	x	x	x	x

1 Reported by the IBM T. J. Watson Research Center, Yorktown Heights, NY; not commercially available.

Support Modules

	AL	AML	HELP	JARS	MCL	RAIL	RPL	VAL
text editor	1	x	3	3		x	x	x
file system	1	x	3	3		x	x	x
hot editor		x						
interpreter	x	x	x			x	1	x
compiler	x			x	x		x	
simulator	x	2			x			
MACROs	x		x		x			
INCLUDE statement	x	x						
command files				x			x	
logging of sessions	x							
error logging	x							
HELP functions	x	x						
tutorial dialog		x						

1 AL uses the support features of the PDP-10 operating system.
2 A simulator has been developed at the IBM T. J. Watson Research Center, Yorktown Heights, NY.
3 JARS and HELP use the systems support features of the RT-11 operating system.

Debugging Features

	AL	AML	HELP	JARS	MCL	RAIL	RPL	VAL
single stepping		x	x			x	x	
break points	x	x				x	x	
trace		x	x		x		x	
dump	x		x		x		x	

CONCLUSIONS

Although most of the commercially available robot programming languages have been included in this evaluation, notable omissions are those products developed by Bendix, Olivetti, and Phillips. These and other languages are being studied by LTI Robotic Systems for comparison in a subsequent paper.

This study has resulted in a number of conclusions regarding the present status and future development of robot programming languages. Even though most industrial robots are still programmed using guide-through with a teach-box and subsequent playback of learned points, complex assembly and machine loading tasks will demand the best features of textual languages.

It can be therefore expected that more and better robot programming languages will be introduced to meet this demand. New languages should include attention to the following items:

- Deficiencies in existing languages should be corrected. For example, some languages provide inadequate names for variables and deficient mechanisms for passing arguments to subroutines.

- A language should include the common features of existing languages. For example, most languages provide binary i/o lines, and some provide an interface to commercial visions systems.

- A language should provide room for expansion. Force sensing may not be needed in initial commercial systems, but will probably become important in a few years; the language should not require extensive alteration to support it. Similarly, multiple arms should be controlled by one language system, even if most initial customers will want only a single arm.

- A language should provide a natural interface for a wide variety of users, both novices and experts. It should be possible to program simple tasks simply, but more advanced constructs also should be available for difficult tasks.

ACKNOWLEDGEMENTS

LTI Robotic Systems wishes to thank the following individuals for providing valuable assistance relating to the evaluation of the robot programming languages described in this paper: Phillip Summers and Russel Taylor from IBM T. J. Watson Research Center, Yorktown Heights, NY; Ewald Heer and Raymond Eskenazi from Jet Propulsion Laboratory, Pasadena, CA; Stuart Kretch from McDonnell Douglas Corporation, St. Louis, MO; William Park from SRI International, Menlo Park, CA; and Frank Weigand from LTI Robotic Systems, Palos Verdes, CA.

REFERENCES

AL User's Manual, Third Edition, by Shahid Mujtaba and Ron Goldman, Stanford University, Palo Alto, CA, Report No. STAN-CS-81-889, December 1981.

IBM Robot System/1: AML Reference Manual, Second Edition, SC34-0410, IBM Corporation, Boca Raton, FL, September 1981; IBM Robot System/1: AML Concepts and User's Guide, First Edition, SC34-0411-0, IBM Corporation, Boca Raton, FL, September 1981; IBM Robot System/1: General Information Manual and User's Guide, Third Edition, GA34-0180-2, IBM Corporation, Boca Raton, FL, March 1982.

Allegro Operator's Manual (A12 Assembly Robot), General Electric Company, Bridgeport, CN, 1982.

JARS: JPL Automomous Robot System, by John J. Craig, Jet Propulsion Laboratory, Pasadena, CA, 1980 (unpublished report).

Robotic System for Aerospace Batch Manufacturing, Task B - High Level Language User Manual, Technical Report AFML-JR-79-4202, Wright Patterson Air Force Base, OH, October 1981.

RAIL Software Reference Manual (ROBOVISION and CYBERVISION), Rev. 3.0, MN-RB-07, Automatix Inc., Burlington, MA, January 1982.

The SRI Robot Programming System (RPS) - An Executive Summary, by William T. Park, SRI International, Menlo Park, CA, March 19, 1981; How to Generate a New Version of the SRI Robot Programming System (RPS), by William T. Park, SRI International, Menlo Park, CA, August 6, 1979.

User's Guide to VAL, Version 12, #398-H2A, Unimation Inc., Danbury, CN, June 1980; VAL Univision Supplement, Version 13 (VSN), Second Edition, July 1981.

LMAC: A Language Generator System for the Command of Industrial Robots

A. Haurat
Centre Microsystemes et Robotique
France

M.C. Thomas
Centre Microsystemes et Robotique

Abstract

LMAC is a software produced for the programming of flexible robot systems. LMAC proposes a group of fundamental software tools: data processing modularity, two methods of parameterization of modules and synchronization mechanisms, both explicit and implicit. LMAC was produced with an aim to minimize the problems associated with transportability to make it adaptable to a very wide range of material, both of a data processing and of a mechanical nature.

INTRODUCTION

The development of a new language for the programming of robots does not seem to us to be desirable at the present time. We are not looking for a specialized language, which inevitably implies choices to be made ; there are already in existence a certain number of these languages which permit the programming of robots, for example, in terms of elementary movements of the machine or in terms of actions to carried out the object, etc ...

Nor are we looking for a so-called universal language, since the high cost of such languages, particularly in terms of memory centention , does not allow for adaptation to small data-processing materials.

We only need make the parallel with the programming languages : the multiplicity of specialized software in data-processing has led to the disappearance of several of their number and the classic universal languages no longer have a future.

The solution was to define a group of software tools easily employable from an elementary language based upon the structured programming.

This sytem known as LMAC ([1]), is simple to transport from one environment (both of a data-processing and of a mechanical nature) to another, and allows the programming of any type of application of the industrial robot, from the work of a simple robot (mechanical arm) to the direction of a fully automised production line.

LMAC has been devised upon the basis of technics which are currently prevalent in data-processing for the writing of software : thus transportability, maintenance, readability, etc ... as much of LMAC , as of the software which it allows to be programmed are assured.

In the same way, certain basic software tools of our system have been developed from advanced concepts of programming : for instance, modularity as in ADA or Dijkstra guarded phrases. These software tools have been adapted to meet our needs and others have been created in order to achieve our objectives.

The complete set of these tools, expounded in this article, comprises of :

- the concept of a module, a module being an <u>abstract type</u> as in ADA
- the parameterization of the modular type by means of the <u>clause "employ"</u> which permits the creation of a new type based upon others,
- the parameterization of the objects of a given modular type by means of the <u>clause "import"</u> which allows us in particular to give the physical characteristics of these objects,
- a mechanism of explicit synchronization of processes, programmed with the help of a syntax inspired by Dijkstra guarded phrases and which correspond to the cooperation of robots,
- a mechanism of implicit synchronization, i. e. a certain number of real times tools (control of a clock, semaphores, ...) grouped together in a executive real time exclusive to our system.

MODULARITY WITHIN LMAC

The notion of module in general

The study of a particular problem leads to the elaboration of a sequence of actions, known as algorithm, and the data used by this algorithm. If the algorithm is represented with a programming language such as Fortran or Basic, then the data must be of different types : real, array, etc ...

The type is a data set, which may or may not be structured, upon which is defined a set of pre-defined operations, that is to say pre-existant with the use of the language. The use of a type data is like the use of an operation of this type without having to be acquainted with the manner in which the operation is realized within the computer. On the other hand, in actual languages, if more complex data is required such as a queue, a stack, etc ..., the structure of this data and the operations specific to this data will have to be created.

Everything is therefore visible to the user which can lead to non desired operations. For example, in a stack represented by a array, a user acquainted with this representation cannot be prevented from having access to any element whereas only top of the list should be accessible. In order to partly prevent this visibility and therefore give protection against improper use of the data as well as the algorithms, the module notion has been introduced into certain languages ([2], [3], [4], [5]). It is not within our scope to present the different types of modules which exist within the languages, but, after having given a functional definition of the module, we will explain the approach which we have undertaken.

A module is an algorithm or data description unit grouping together the manipulation functions, that is to say the operations. Only a part of the module is visible to the user. This part contains objects what is necessary for using the module, for example the list of operations, the data syntax, etc ...

Hereunder is a diagrammatic description of a data module. It would be the same description for a treatment module.

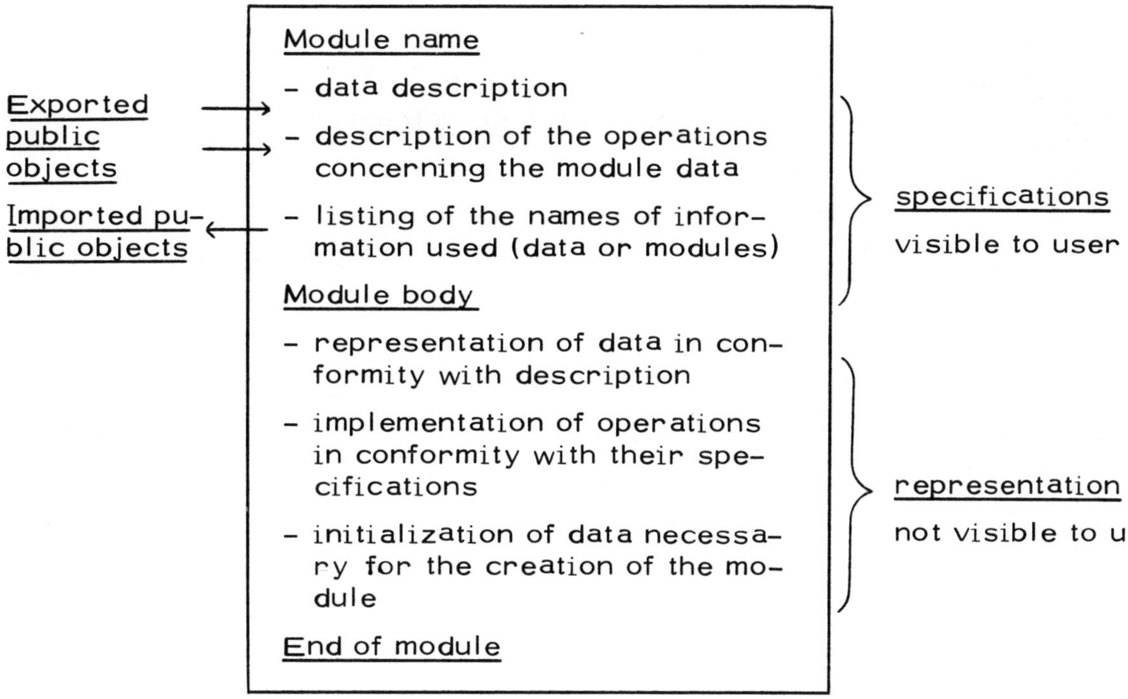

The execution of a non modular program is a temporal sequence of instruction executions using the data. The execution of a modular program is a temporal sequence of module utilization. A module is used when an operation of this module is employed and when data originating from other data module utilizations are employed.

As a conclusion we may state a module as being s structuration tool which :

. protects the user from an improper utilization of data or treatments,

. clarifies the interdependances between object modules,

- strengthens the structuration of the programs because the writing, the debugging and the maintenance of these programs are done by bit,
- increases the language possibilities by the creation of new modules. Generally speaking, a subset of a programming language constitues the writing language of the modules.

Modularity in LMAC

Amongst the different types of existing modules, the one which seems to us to be the most suitable for robot command is the one which covers the abstract type notion such as exists in CONCURENT PASCAL ([5]). The modules are abstract sets of data structures or parts of programs. To command these robots, the user must write programming unit which, on the one hand declare the elements of these modules, and, on the other hand, use the operations.

Example :

This example is in two parts :

- the definition of a module which describes the transformations which may be applied to the references, that is to say the translation and rotation compositions. These translations and rotations are represented by means of quaternions.
- the utilization of a module in a programmation unit. During execution this unit realizes a transformation which is the composition of a translation according to a direction vx and a composition of the translation according to the direction vy.

- Definition of a module

```
        Module transf is transl ∘ rot
     ⎧ Op (transl-along) v : direction (of) x : real return transf ;
 (1) ⎨ Op (compound) t1 : transf (then) t2 : transf return transf ;
     ⎩ Op (rot-axis) v : direction (angle) t : real return transf ;

     ⎧ body transf is
     ⎪    record (transl : quaternion, rot : quaternion) ;
     ⎪ transl-along : build-trans (v, x) ;
     ⎪ compound : begin
 (2) ⎨                transl:=add-quat (t1. transl, t2. transl) ;
     ⎪                rot:=mul-quat (t1. rot, t2. rot)
     ⎪            end ;
     ⎪    rot-axis : build-rot (v, t) ;
     ⎩ end transf ;
```

(1) : specification part

(2) : representation part.

The specification part of this module, only part visible to the user, contains :

- the data description, that is to say transl ∘ rot,
- the list of the operations which may be applied to this data and the type of operands,
- the utilization syntax (encircled).

The operations which appear in this example use the "direction" module and the real pre-defined type.

The representation part, corresponding to the body part, clarifies the implementation of the module "transf", that is to say the manner in which its data and operations are represented in the computer. Within this structure are used exported operations via the "quaternion" module :

. the operation build-trans (v, x) which is a function which gives the constructed quaternion with the components of v and the real x (it is the same system for fab-rot).

. the operation add-quat which is the addition of two quaternions.

. the operation mul-quat which is the multiplication of two quaternions.

- Utilization of module within programming unit

```
program p ;
def
   transf : t1, t2, t3 ;
   direction : vx, vy ;
body
   t1 := transl-along vx of 100 ;
   t2 := transl-along vy of 100 ;
   t3 := compound t1 then t2 ;
end p
```

This programming unit contains :

. a heading section "p" specifying the name of the unit,

. a declaration section (beginning with def and ending with the beginning of body) which specifies the names t1, t2, t3, vx and vy of the module utilization occurences "transf" and "direction",

. a body section containing the actual programming that is to say the utilization of compound and transl-along operations.

***

The command of a robot implies, on the one hand, calculations independent from the command, for example the calculation of a trajectory or of a position, and on the other hand, the physical actions of the robot which depend upon the environment. It is useful, within LMAC, to distinguish between two categories of data module utilization :

. the utilization referred to as static : the operations issue the calculations or the actions realized in open loop by the robot,

. the utilization referred to as dynamic : the actions are realized in a closed loop, that is with a return of information.

This distinction appears in the body of the module, it is therefore transparent to the user. This distinction may be justified because it enables to separate the manipulation part of the robot realizing a repetitive task from the "intelligent" part which may alter the realization of the task with relation to an acquaintance with the universe. In this manner the operation of a robot only requires a minimum of calculations expressed in real time.

A module equally enables to describe a set of data and its operations independent from the environments within this module could be used. In this way we are going to be able to express the flexibility of a robotised system, all the more so if it is a mechanically modular system.

LMAC enables the writing of new modules. The module, corresponding to the part of the system which we wish to describe, is written with the same syntax as that used for the programming units. This creation of the module is the same thing as the creation of a set of operations which may be realized on the new architecture. In this way can be commanded the robotised systems, the different components of which are easily interchangeable.

Finally a module being a unit of compilation, most of the type verifications are realized independent from the command and improve therefore the operating security of the robot.

THE PARAMETERIZATION OF MODULES

Parameterization of types : employ clause

Within certain programming languages such as ADA, a module may be parametered, that is to say the data belonging to a module has its type specified at the time of the module utilization. For example, in a "stack" data module, the type of the components which compose the stack is specified only at the time of utilization of this stack. This parameterization is known as genericity. Within the LMAC it is necessary to have this possibility of genericity if one considers the robot, or more generally speaking a robotised system as a modular mechanical structure in which it is necessary to be able to clarify the liaisons between the different mechanical parts. With this condition, we have a command software which is always adapted to the mechanical structure.

Example :

If a robot can choose and assemble its tool at the beginning of the task, then the software must enable to realize a displacement of the carrier (the tool not yet being assembled) as well as of the tool once assembled.

Within LMAC the genericity clause is clarified by "employ". In the programming unit body, the sequence which enables to realize displacements of the carrier without tool and displacements of the carrier with its tool (in this case the tool being a gripper), may be written :

```
body
move-along path1 the carrier1 ;
robot employ gripper1 on carrier1 ;
move-along path2 the robot ;
end prog ;
```

"move-along" is a displacement module operation.

"path1" and "path2" are trajectories.

***

The genericity enables to reinforce the operating security of a robot, because a failure of any one part of the mechanical structure may be taken into consideration. For this it will be necessary to have anticipated, via the genericity clause, the parameterization of the module corresponding to the defective part of the robot. The system then enables to regain the task after having changed the faulty part. The task is regained at the data which has been protected at the time of the incident.

Parameterization of objects : import clause

Within a program, the usage of the "import" clause allows the parameterization of objects as defined by the module. This parameterization is achieved by means of values or of algorithms which are found in a so-called physical file. This clause gives us the possibility of specifying the values which are directly linked to the mechanical functions of the flexible robotised system.

In the same example, it is necessary to initialize the occurence of the "gripper" module gripper1 utilization and the occurence of the "carrier" module. For this, in the declaration part of the programming unit it will be necessary to introduce initial physical data.

decl

gripper : gripper1 import terminal tool ;

carrier : carrier1 import arm ;

"terminal tool" and "arm" are two files which contain the set of physical variables which give the characteristics of the tool and of the arm.

SYNCHRONIZATION AND COMMUNICATION

At a given work position, it is necessary to control either the functions of one machine or those of several which must work in cooperation. This cooperation is achieved by the establishment of a means of communication between the various elements of the system and indeed between the system and the exterior. Thus, for example, one must coordinate the actions of various robots arms in order to carry out assembly operations or to control the movements of a robot within an environment subject to change. In order to solve such problems, a programming system must possess both explicit and implicit tools of synchronization.

Explicit synchronization

These tools use the Dijkstras notion of guarded command ([6]) to achieve the synchronization and the communication between processes. The carrying of a program corresponds to the execution of a series of actions. Within the program unit, these guarded commands are expressed in the following manner :

P1 then [guard → P2]

P1 and P2 signify processes and "guard" is a condition relative to process P1. If the guard becomes true during the execution of process P1, then process P2 goes ahead.

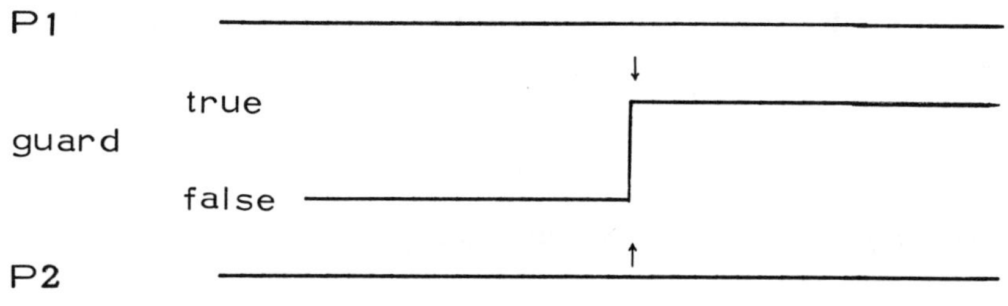

The operands of the guard may be not only boolean conditions but also the state of the process (ready or not to transmit) at the time of attempts of communication.

As far as the exclange mechanism between processes is concerned, this is initiated by the simultaneous execution of input/output commands within each of the processes.

In a flexible system, the make-up of the "real time" command must be organized within a local network of computers. We believe that, for such a make-up, the implementation of the preceding tools must be carried out by a protocol of interprocess exchanges, broken down into several hierarchical layers :

. the first of these layers is the application protocol, made up of procedures called upon by the processes wishing to communicate with one another,

. upon the lower layer, the session protocol breaks down the input/output commands sent out in elementary communication signals by the application protocol and memorized the state of advancement in the exchange,

. the transfer protocol is composed of procedures called upon by the session protocol of the same computer. Its function is to localise the receiving process and to transmit the signals sent out by the session protocol to the network protocol if the corresponding processes are located upon distinct processors,

. the final layer is constituted of the network protocol which is charged with the sending and the reception of messages along the lines of communication which link the calculators to each other.

There are two things to be noticed about such an exchange mechanism ([7]) :

. the layers evolve in parallel by sending signals to one another. It is therefore necessary to have upon each processor an executive real time which possesses elementary mechanisms of synchronization (semaphore, for example),

. the transmission of data between processes is carried out without intermediary memorisation, i.e. directly from the memory space of the source process into the memory space of the receiving process.

In conclusion, communication between two processes happens in the following manner :

. a "rendez-vous" phase, where each process is ready to receive or to send out information,

. an exchange phase, where the data are transmitted from the memory space of the source process to the memory space of the receiving process,

. a phase of reactivation of the communicant processes.

Implicit synchronization

Since real time applications are characterized by the interaction between the program and external events, it is important to take into account the notion of time and the place of the events in time.

Since the specific character of real time problems concern synchronism, the necessary software tools appear to be :

- the wait dog,
- the control of the clock,
- the control of hierarchically organized interruption,
- semaphores,

These mechanisms need not be directly accessible to the user (contrary to guarded phrases).

These real time tools are regrouped into an executive real time, invisible to the user, but used at the time of execution of LMAC program ; it increases further the independance of our system in relation to a given material, i.e. its adaptability.

Synchronization example :

The part of the programming unit hereunder enables to command a robot which takes a test-tube with its gripper, then transports this test-tube causing it to vibrate in order to prevent a precipitate.

```
program p ;
def
   vibrating-gripper : gripper1 import terminal tool ;
   displacement : dep ;
   object : test-tube ;
body
   closed-with  0. 1N the gripper1 ;                                   (line 1)
   if space-apart gripper1 < 1 then write ('test-tube absent')         (line 2)
      else
        move-along dep the test-tube and-then                          (line 3)
          [start → shake gripper1 // stop → end-shake gripper1]
   endif
end p
```

The object module describing the gripper is named "vibrating-gripper". A declaration, gripper1 utilization accurence is made. The operators which authorize the utilization are closed-with, space-apart, shake, end-shake.

At line 1 the gripper is closed until the tightening force represents 0. 1 newton,

At line 2 the tightening is finished and the space between the jaws is measured to determine whether the test piece has been correctly held,

From line 3 onwards the following two processes are synchronized :

- P1 : the displacement which is realized according to the trajectory "dep",
- P2 : the action of shaking the test-tube.

The two processes must start and end together. For this, the following phrase is written :

P1 and-then [guard→P2] (this writing is inspired from [6]) in which "guard" is a condition relative to process P1. If "guard" is true then the process P2 evolues. In the example, is used a predicate function "start" which becomes true once the process P1 begins. It is the same for the "stop" function when P1 stops. Finally the symbol // indicates the asynchronism between the stop and the start actions.

CONCLUSION

The software tools introduced in this article have been conceived to simplify and unify the guidance of both flexible robotised system and simple manipulator-robots. Obviously, LMAC must also take into account the direction of the transfer of components and the cooperation of the automatisms making up a given work position ([8]) ; the integration of such interfaces poses no problem in LMAC.

Besides, one of our objectives being the simplicity of programming, we offer to the user, among other things, direct help in the writing of programs, upon a visual console. LMAC is at present implemented on LSI 11/23 at the "Centre microsystèmes et robotique" at Besançon.

BIBLIOGRAPHY

[1] Haurat A., Thomas M.C. LMAC : a modular command language for industrial robots. 1er colloque génie logiciel, Paris, juin 1982.

[2] Liskov B., Zilles S. "Programming with abstract data types". Proceedings of A.C.M. Sigplan Conference on very high level languages, Sigplan Notices, Vol.9, No 4, (1974).

[3] Ichbiah, J.C. Reference manual for the green programming language, publication CII-HB, 1979.

[4] Wirth, N. Modula 2, ETH Zurich, rapport No 27, 1977.

[5] Brinch, P. Hansen. The programming language Concurent Pascal, Pascal I.E.E.E. software engineering, 1975.

[6] Hoare, C.A.R. Communicating sequential processes, Com. A.C.M., August 1978, Vol.21, No 9, p. 666-677.

[7] Elloy, J.P., Molinaro, P. Implantation de CSP dans un réseau local par protocole. IASTED International Symposia and Course, Davos, March 1982.

[8] Haurat A. Des outils logiciels nécessaires au pilotage d'ateliers flexibles. Congrès IFAC, décembre 1982.

Voice Communications with Robots

A. Levas
The University of Connecticut
Storrs, Connecticut

M. Selfridge
The University of Connecticut

Abstract

Planning ability provides robots with the needed flexibility to cope with complex and uncertain environments but requires that robots be able to learn new plans. Robots should acquire new plans by accepting spoken natural language input which specifies the steps in the plan as instances of previously known general planning knowledge. A robot manipulator system learns the plan for assembling an electric switch by understanding voice input specifying the assembly steps. The success of this system suggests that this method of teaching robots new plans may prove a useful technique for robots of the future.

I. Introduction

Robots are increasingly being used in situations requiring them to exhibit flexible behavior. In such situations point-to-point motion sequences are inadequate [1]. Instead, robots will probably use planning mechanisms, coupled with sensors of various kinds, to carry out their tasks [1]. Such mechanisms allow robots to perform in situations where object position cannot be exactly specified in advance and in which unanticipated events can occur. However, training robots which use plans requires teaching them new plans. This paper describes one method, telling the robot the steps in the new plan using voice input to natural language understanding programs.

To see why teaching plans to a robot is important, consider a possible robot application, that of a shop specializing in short production runs producing things from fairly standard parts. Such a shop cannot use special purpose equipment, since runs are short, therefore using robots may be the only way to automate. They would use planning mechanisms, for the reasons outlined above. However, short production runs require frequent retraining, and the robot must acquire new plans frequently. Teaching new plans by using voice input to natural language programs to tell the robot the steps in the plan appears to be an effective solution to this problem.

It must be stated that the research reported here does not attempt to break new ground in the difficult and complex area of speech understanding. Rather, the approach uses existing technology to explore other issues. Clearly the ability to teach robots new plans using voice input can only improve as the state-of-the-art of speech recognition improves.

The rest of this paper will describe a system which learns a new plan by being told the plan steps via voice input. This description will encompass the system's hardware, its world knowledge and planning knowledge, and its learning mechanisms.

A sample of the system's performance will then be presented, and finally conclusions and suggestions for future research.

II. The System

This system has several hardware and software components. The robot manipulator is a five degree of freedom Microbot Minimover-5, having a parallelogram-type hand with a touch sensor. The arm is controlled by an Apple II microcomputer which is in turn connected to a DEC VAX-11/780 via a serial line. In addition, the Apple II drives a Heuristics 2000 voice input device and a Votrax TNT speech generator for user communication with the robot. All Apple software is written in BASIC, while all VAX software is written in LISP. In operation, a planning program running on the VAX sends manipulator movement commands to be executed by the Apple, thus moving the manipulator.

The system uses a control program running on the VAX to coordinate its operation. This program must communicate with the Apple for two different purposes: generating speech and recognizing speech, and also directing the Apple to move the manipulator. In addition, it must also run natural language, planning, and learning programs on the VAX. System operation proceeds when the control program tells the Apple to prompt the user using speech generation, and to accept a spoken command (see [4] for a review of the recognition process). This command is received from the Apple and passed to a natural language understanding program [2] which produces a representation of the command's meaning. An inference program completes any unspecified parts of the meaning, verifying the completed meaning with the user by generating a natural language paraphrase of the meaning [3], speaking the paraphrase to the user with speech synthesis via the Apple, and receiving voice confirmation from the Apple. Given a representation of the meaning of the command, the control program then calls a planning program [5] to plan the series of manipulator motions needed to carry out the command. The sequence of motions is passed to the Apple for execution and when execution is complete the cycle begins again.

The need for learning occurs when the planning program cannot retrieve any plans to carry out the command. In this circumstance, the control program operates differently. It informs the user that no plans were available and tells him to specify the steps in the plan. The user then has the opportunity to use voice input to issue a series of commands which when obeyed will carry out the original goal. The system accepts and obeys these commands just as it accepts top-level commands. The control program saves the meaning of each of these commands, and when the sequence is over, combines them with the meaning of the original command to produce a new planning rule. This new planning rule specifies that if the system has the goal obeying a command of the original form, then it should use the plan specified by the series of sub-commands.

III. An Example

This section presents an example of the system being taught to assemble a simple momentary contact electric switch. This switch has three parts: a base, a post, and a contact. The base is one inch square by three long, and on its top has a hole for the post and a slot for the contact. Assembling this switch requires that the post be inserted into the hole and the contact be inserted into the slot. When assembled, the switch can be closed by depressing the contact in the manner of a telegraph key.

For this example, the system has been supplied with the kind of knowledge that would be available from a high-performance vision system. This includes knowledge of object shape, location, orientation and the handling of all know parts. The objects that are in the workspace for this example are a switch base, a contact, and two posts of different lengths used to demonstrate the system's handling of ambiguity. In addition, the speech recognition system has been previously trained to recognize all the words appearing in the example.

In this example the system is taught the plan for making the electrical switch by being given voice commands for the steps in that plan. The example includes the voice verification by the system of the intended command where the command given was ambiguous. The last part of the example shows the system using the learned plan to make a second switch. This example is taken from an actual run of the system, but has been edited for clarity and compactness.

```
% lisp loadvoiceI/O

Franz lisp, Opus 36

-->(runvoiceio)
SPEECH INPUT: "make a switch"
RECOGNIZED: MAKE A SWITCH
CONFIRM: "yes"
SEARCHING PLAN LIST...NO PLANS FOUND FOR SWITCH

ENTER TRAINING MODE (YES/NO)?
SPEECH INPUT: "yes"
TYPE GOAL NAME: switch

ENABLING VOICE TRAINING MODE

SPEECH INPUT: "put the post in the block hole"
RECOGNIZED: PUT THE POST IN THE BLOCK HOLE
CONFIRM: "yes"
SPECIFY: DO YOU MEAN: PUT THE THREE INCH POST IN THE BLOCK HOLE
CONFIRM: "yes"
SPEECH INPUT: "put the contact in the block slot"
RECOGNIZED: PUT THE CONTACT IN THE BLOCK SLOT
CONFIRM: "yes"
SPEECH INPUT: "end training"
RECOGNIZED: END TRAINING
CONFIRM: "yes"

TRAINING CONCLUDED

BUILDING NEW PLANNING RULE:
GOAL: SWITCH
PLAN: 1) INSERT POST HOLE
      2) INSERT CONTACT SLOT

-->(runvoiceio)

SPEECH INPUT: "make a switch"
RECOGNIZED: MAKE A SWITCH
CONFIRM: "yes"
```

```
SEARCHING PLAN LIST...PLAN RETRIEVED

[...planning program proceeds to execute the new learned
 plan which accesses the previously known planning rules
 and assembles the switch...]

-->(exit)
%
```

IV. Conclusions

Future development of this system will proceed in several directions. First, a top-down vision system is being added to the system to allow it to actually see the objects in the workspace. Second, interaction between the natural language analyzer and the speech recognition device is being expanded to improve speech recognition accuracy. Third, the capabilities of the system are being expanded to include implementation on a full-size industrial robot system, and behaviors such as product testing as well as manufacture.

The successful performance of the system described in this paper suggests that using natural language and voice input and output can be an important part of teaching robots new actions in real-world situations, and that teaching robots new plans by voice input is a practical technique.

BIBLIOGRAPHY

[1] Birk, J. and Kelly, R. (1980) Proceedings of the Workshop on the Research Needed to Advance the state of Knowledge in Robotics, printed by the NSF and the Univ. or R.I.

[2] Birnbaum, L. and Selfridge, M. (1981) Conceptual Analysis of Natural Language. In Inside Computer Understanding: 5 Programs Plus Miniatures, Lawrence Erlbaum Assoc., Hillsdale, New Jersey.

[3] Cullingford, R.E., Krueger, M.W., Selfridge, M. and Bienkowski, M. (1981). Automated Explanations as a Component of a CAD System. To appear in IEEE Trans. SM&C. Special Issue on Human Factors and User Assistance in CAD, November, 1981.

[4] Reddy, D.R. (1975) Speech Recognition. Academic Press, New York.

[5] Winston, P. (1977). Artificial Intelligence. Addison-Wesley Publishing Company, Reading, Mass.

MCL, The Manufacturing Control Language

Brian O. Wood
The Pennsylvania State University
University Park, Pennsylvania

Mark A. Fugelso
The Pennsylvania State University

Abstract

Under contract within the United States Air Force ICAM program, the McDonnell Douglas Corporation has developed the Manufacturing Control Language (MCL) for use with robotic work cells. An extension of the numerical control language APT, MCL contains control words for real-time decision making and vision processing. These facilities, along with several other features, make this language a versatile off-line programming tool. This paper gives a basic overview of MCL's capabilities.

I. INTRODUCTION

The programming of robots for manufacturing tasks is a major industrial concern. In the past, robotic motions have been programmed through a procedure known as "teaching". By physically leading the manipulator through the motions manually or with a control box, the programmer "teaches" the robotic controller a series of desired motions. These motions are then stored in the computer's memory for later use. Although very little operator skill is needed to teach a robot, valuable production time is wasted whenever a new job must be taught. In a batch production situation, this wasted time could mean the difference between profit and loss. A better solution to the robot programming problem is to create control programs independently of the manufacturing hardware. This procedure, known as off-line programming, uses a high level computer language to construct a set of control commands. By creating control programs off-line, unproductive "teaching" time can be eliminated.

To meet the needs of off-line programming, the Manufacturing Control Language (MCL) has been developed by the McDonnell Douglas Corporation under contract with the United States Air Force. Based on the numerical control language for machine tools, APT (Automatic Programmed Tool) , MCL can control robots, sensory devices, machine tools, parts feeders, and all other devices used in manufacturing work cells. APT was chosen as the basis for this new language because of its geometry specification abilities and its widespread use and popularity in the manufacturing community.

In recent years, new developments in robotic and manufacturing technology, such as improvements in sensory and vision systems, have created a need for a more versatile control language. Although many robotic control languages have been developed in the past, few have the ability to control all modern day cell functions. MCL is an attempt to combine all needed control features into one flexible programming language. In this paper, after briefly discussing the evolution of off-line programming languages, MCL's control facilities are more carefully examined. To illustrate the features described, a sample program is also included.

II. EVOLUTION OF ROBOTIC PROGRAMMING LANGUAGES

The need for an off-line programming language for industrial robots was first realized in the late 1960's. At that time, research work, such as MIT's hand-eye project, Paul (1) was begun to find ways of controlling robotic manipulators without physically teaching. Since that time, several robotic control languages have been developed, each with the unique advantages and disadvantages. The block diagram in Figure 1 illustrates this evolutionary development. Some of the languages shown in this diagram are further described in the following discussion.

In 1970, the first substantial off-line programming language was developed. Known as WAVE, this language provided for force and motion control using homogeneous transformations, Paul (1). WAVE's stumbling block to further development, however, was its inability to provide simultaneous interactive control of two or more devices.

Five years after the introduction of WAVE, a language called AL was created to control motions of objects and the assemblies of objects, Paul (1). This language enabled the control system to realize when objects affixed to each other and to the robotic end effector. All positions of affixed parts were automatically updated as the assembly was moved within the work volume. One major problem with AL was its limited ability to use sensory data.

After AL, researchers at Purdue University began working on PAL. All tasks in PAL were represented in terms of structured Cartesian Coordinates. Motion statements

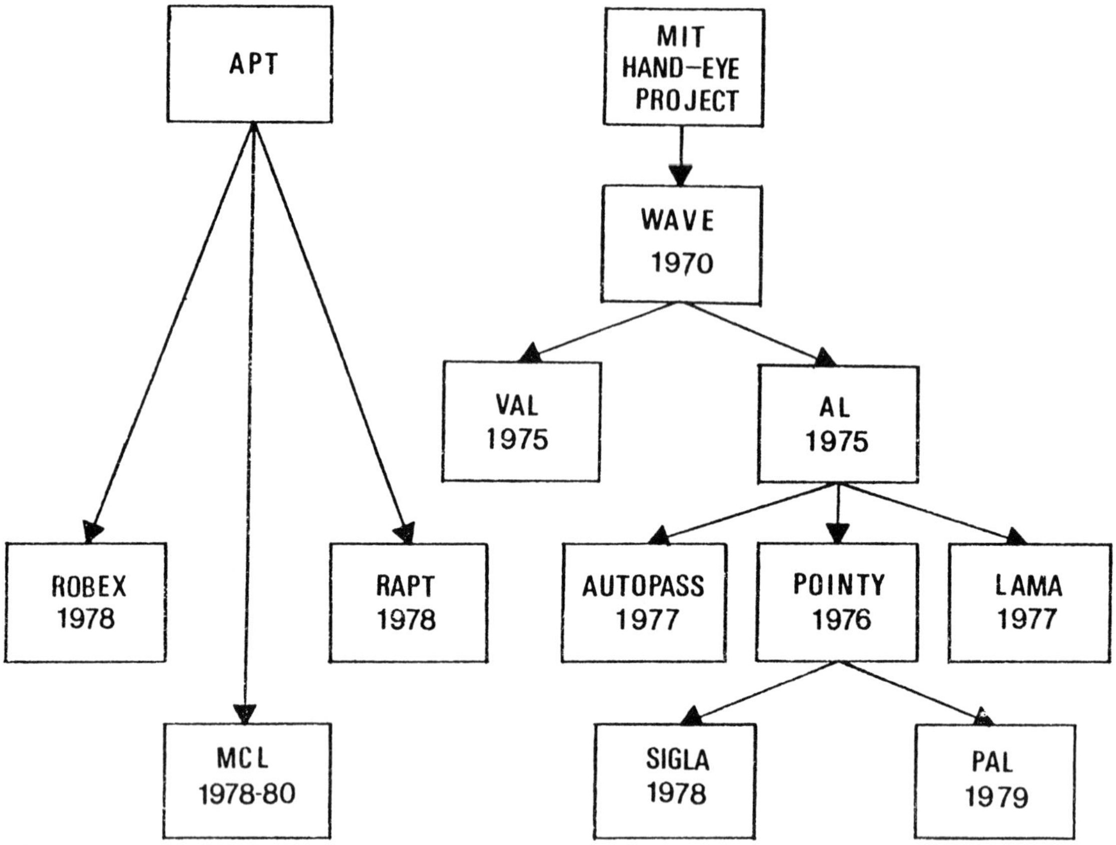

Figure 1 Evolution of Robotic Control Languages

were requests to position the manipulator such that the position equation, representing a closed kinematic chain of homogeneous transformations, was satisfied, Berg et al. (2). Using sensory feedback, PAL had the ability to allow for the continuity of motion and compliance to eliminate all unnecessary stopping of the manipulator. This feature was advantageous for spraying and welding applications.

The next language developed, VAL, was created by Unimate, Inc. to control their PUMA robot. Working interpretively, VAL gave the robotic system the ability to work in unpredictable or moving frames of reference, Shimano (3).

IBM contributed to off-line programming software with the development of AUTOPASS. Used primarily in the laboratory for assembly research, AUTOPASS used motion commands that were generated from a geometric data base called the world model, Lieberman et al. (4). Recently, IBM has developed another control language called AML for use with their new robotic systems.

In 1978, McDonnell Douglas began their work on MCL. After examining the existing control languages for positive and negative characteristics, they chose the well known numerical control language, APT, to be the basis for their new language. MCL consists of the original APT source program plus a set of newly defined control words. With the addition of statements for cell control and decision making to APT's existing geometrical and motion specification capabilities, this new language is a powerful programming tool with many desirable features.

III. FEATURES OF MCL

In order to be effective, an off-line programming language must be able to control functions needed by modern day robotic work cells. This fact was taken into account during the creation of MCL. As a result, this language offers a wider range of features than most other control languages, MCL capabilities include:

1) Ability to control any robot. As previously mentioned, MCL sends instructions to lower level computers which in turn send commands to their respective devices. To control a different kind of robot, a new postprocessing routine must be used by the controller to change MCL Code into a form understandable to the robot.

2) Ability to control entire manufacturing cells. Unlike most other off-line programming languages, a single MCL program can control the activities of every device in a work cell.

3) Real time decision making. MCL programmed controllers can process real time sensory data and make logical decisions based on this information.

4) Vision processing. Two dimensional binary views of parts can be modeled in an MCL controlled system. With this model, vision systems for locating or inspecting parts can be used.

5) Program verification. The Verifier Postprocessor feature will indicate whether a given program can be used on any of the cells described in the database.

6) Ability to specify convenient frames of reference. To aide in programming, different frames of reference can be used to define points. The lower level control system interprets these points in relation to the cell hardware.

A more detailed description of some of these features is contained in the following sections.

IV. HIERARCHIAL CONTROL STRUCTURE

The MCL control system uses a hierarchial structure to operate and control cell hardware. All levels of this hierarchy are integrated to work together as a single system. Figure 2 is a block diagram of this structure.

At the top of the control hierarchy is a main frame computer system, such as the IBM 3033. Using terminals linked to this computer, programmers create control programs independently of cell equipment. To aide in program writing and debugging, various databases are available here. For example, the Vision Data Generator database computes all vision system information needed for the modeled workpiece. Such as

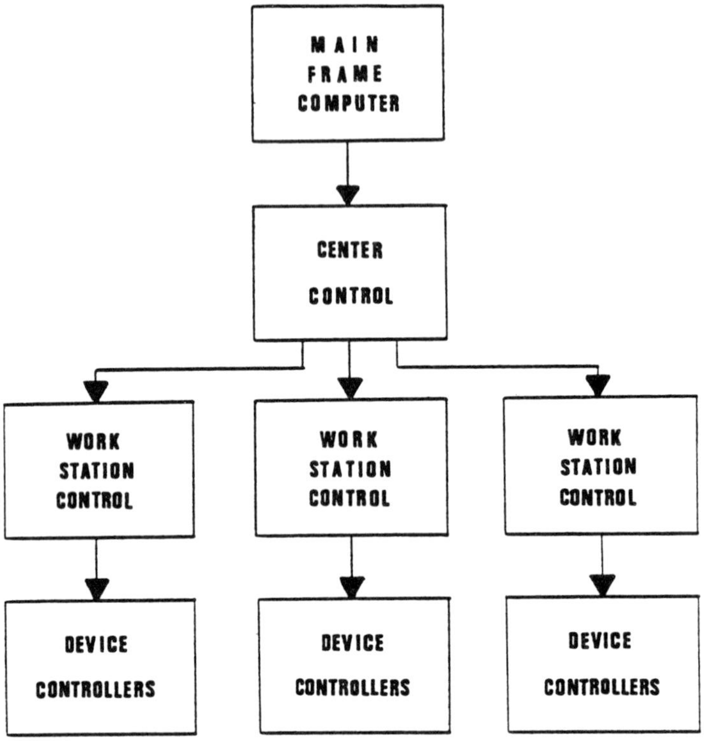

Figure 2 Hierarchial Control Structure

x- and y-coordinates, centroids, and area. Also, verification that a given program can be used with a previously defined work cell is performed at this level by the database containing the Verifier Postprocessor. Completed programs are stored by this main frame computer system.

After being successfully processed, control programs are then moved down to the next hierarchial level, Center Control. Here, the computer system acts as a link between the main frame computer and the factory's manufacutring hardware. Postprocessing of programs for specific cells, containing a unique collection of robots, machine tools, sensory devices, and other manufacturing equipment, is performed at this level.

From Center Control, programs are distributed to the individual cells. At this level, known as Workstation Control, language semantics are changed into signals understandable to the lowest level controllers for each device. For example, the command SEND/HAND, CLOSE is converted by workstation control into electrical impulses which make the mechanical system close the robotic gripper. These signals are the final link between the high level language and physical movement.

V. REAL TIME CONTROL AND DECISION MAKING

The words real time refer to actions and events that occur as programs run at the shop floor level. Using information from sensory and vision devices, the control system can make logical decisions regarding the operation of the cell hardware. MCL defined statements for decision making, looping, tasking, and assignment, use data in the form of real-time variables and their operators to alter cell operation from the shop floor. Real time variables fall into three different categories: STRING, REAL, or LOGIC. To be used in a program, a real time variable must be declared as one of these types.

Assigning values or expressions to declared variables is accomplished using the LET/[assignment] command. In this statement, [assignment] represents a) real variable = real expression, or b) logic variable = logic expression, or c) string variable = string expression. Some examples of real time assignment statements are shown in Figure 3.

```
STRING:  LET/S1 = 'ENTER # OF PARTS'
         LET/S2 = 'RESET CELL'

REAL:    LET/R1 = NUM
         LET/R2 = R1 + 1.0

LOGIC:   LET/L1 = S1.EQ.S2
         LET/L2 = R1.NE.R2
```

Figure 3 Real Time Assignment Examples

Real time decision making in MCL is done through a WHEN-ELSE structure. The syntax for this structure is as follows:

```
WHEN (logical expression)
  { First block of statements
ELSE
  { Second block of statements
ENDOF/WHEN
```

If the logical expression in the WHEN statements is true, the first statement block is executed. The second block of statements following the work ELSE are executed if the logical expression is false. It should be noted that the ELSE statement and its following statement block are optional.

One control work often used within a WHEN-ELSE structure is the ABORT statement. Whenever ABORT is encountered in the execution of an MCL program, all cell activity ceases. By stoping all devices in abnormal situations, damage to expensive cell hardware can be avoided.

MCL's facility for real time looping is the WHILE statement. A WHILE loop structure is similar to that of a WHEN structure:

```
WHILE (logical expression)
    { Block of statements
ENDOF/WHILE
```

The block of statements continue to execute as long as the logical expression is true. If the expression is false, the program execution continues after the ENDOF/While statement.

Often, the simultaneous execution of two or more cell devices is necessary. The TASK statement allows various control subprograms to be defined which are later run in parallel. To execute a set of TASKs, a statement of the form INPAR/[task name 1], [task name 2],... is employed. The listed TASKs will simultaneously be performed whenever this command is encountered.

The ability of a control system to make real time decisions adds a great deal of versatility and flexibility to robotic manufacturing cells. Because of the increasing number of technological advances in the area of sensory devices, "intelligent" control systems for manufacturing are a must. For more information on MCL's real time programming abilities, please see reference 5.

VI. VISION PROCESSING

Robotic visuals is beginning to play an important role in the operation of manufacturing cells. An off-line programming language, therefore, must be able to effectively process real time visual information. Applying a model based approach, Baumann (6), MCL uses vision data to make decisions and adjustments regarding device movements within each cell.

Vision processing in MCL is a two step process. First, a 2-dimensional view of the workpiece is modeled in the program. As the program executes, the model description enters an appropriate database. During the second step, MCL compares the actual camera view with the model. This comparison information is then used for real time decision making.

To create a 2-dimensional model, the outlines of various geometric part features are traced using standard APT motion statements. Assuming a zero cutter radius, these outlines enclose areas that are combined to form a binary representation of the entire workpiece.

Specific part features are defined with the REGION statement and structure:

```
[region name] = REGION/[non-geometric property]
    { Region definition statements
ENDOF/REGION
```

The region definition statements are a set of APT motion commands (i.e. GOFWD, GORGT, GOLFT,...) which trace a closed path. To describe the area inside this perimeter, a non-geometric property modifier is given. The three most common modifiers are SOLID, HOLE, and WINDOW. If the area inside a REGION is filled with material, the modifier

SOLID is used. Similarly, the work HOLE indicates that the defined area is empty. WINDOW describes special square REGIONs used for fine resolution vision systems. Within a WINDOW square, a completely solid view, a completely empty view, a view with a straight edge, or a corner view must appear.

Using a PROJEC/[region 1], [region 2],... statement, the previously defined REGIONs are grouped to form the 2-dimensional model. Figure 3 shows a 2-dimensional view of a workpiece. This workpiece is modeled in the sample program in Figure 4.

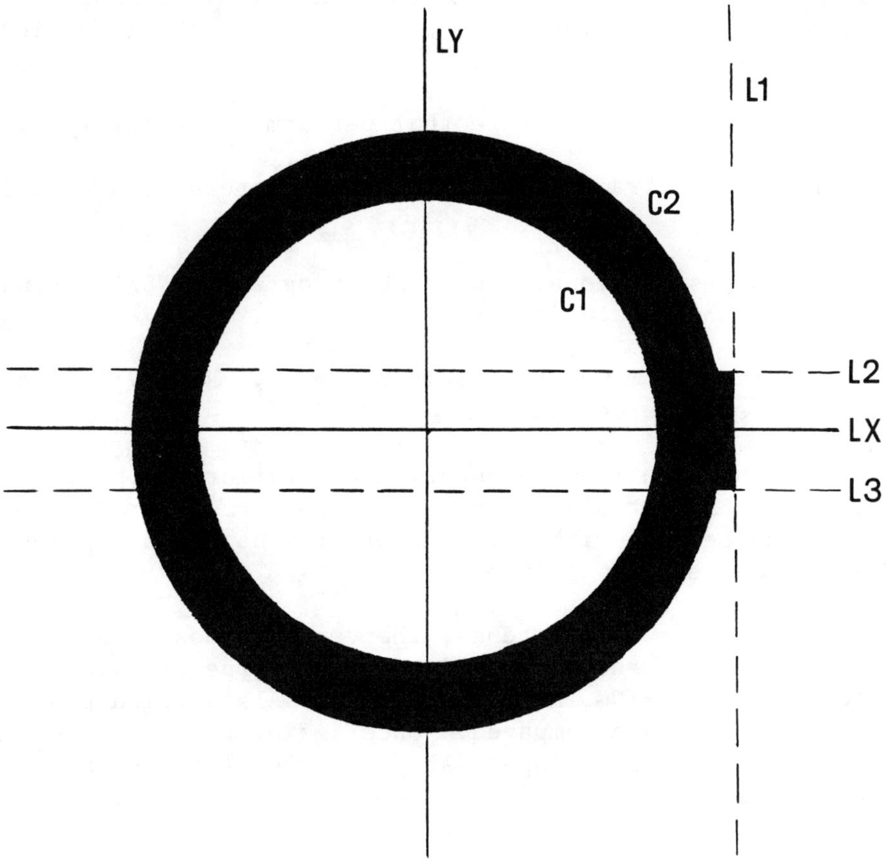

Figure 4 2-Dimensional Binary Model of Sample Workpiece

Once a PROJEC has been defined, the vision system can compare the camera's binary view with the part model. MCL has two facilities for making these comparisons. First, the LOCATE command can be used to verify the identity of a part. Additionally, LOCATE can determine the position and orientation of a workpiece. This information is then relayed to the control system where real time decisions regarding adjustments in the movement of cell devices can be made.

Part inspection is the second function MCL can perform. Using the INSPEC command, specific geometrical properties of a workpiece can be compared to expected property values. Once again, real time decisions can be made on whether a part should be used or scrapped. Further information on MCL vision processing can be obtained from references 5 to 6.

VII. VERIFIER POSTPROCESSOR

MCL's verifier postprocessor is a facility for determining if a given program can run on any of the database described cells. Via a tree structured input code, new cell descriptions can be entered into this database. A FORTRAN translator program then converts this input data into information readable by the verifier postprocessor. The first check made by the postprocessor is a comparison of standard names used in the program to standard names on the database. A standard name is a label given to particular cell components. These names are recognizable to the lower level controllers. The four types of standard names are:

1) DEVICE - refer to pieces of hardware that perform functions within a cell.

2) FRAME - label specific frames of reference interpretable by the lower level control system.

3) DEST - name data destinations. In an MCL program, SEND/[destination], [data] indicates where data is sent.

4) SOURCE - name sources from which data is recieved. RECIEV/[source], [data] references a source name.

When a program is run on the mainframe computer, the verifier postprocessor is involved to compare program and database standard names. The postprocessor then outputs all cells that the program can be used with. If the program cannot be run on any cells, reasons for failure in each cell are printed.

In addition to checking for name matches, the verifier postprocessor also checks the compatibility of data to be sent and received to the type of data that can be handled by the hardware (i.e. STRING, LOGIC, or REAL). Also vision processing algorithms needed and available are compared. Once again, the postprocessor will output any discrepancies. An example of input data for several cells is shown in Figure 5. Reference 5 and 7 contain further information on this facility.

VIII. SAMPLE PROGRAM

Figure 6 is a sample MCL program for a robotic pallet loading application. This program runs a manufacturing cell equipped with material handling conveyors and a vision system for part locating. A schematic diagram of the cell layout is illustrated in Figure 7.

The parts to be loaded are transported on conveyor belt 1 to an area that is in the field of vision of the camera system. When a part enters this area, the belt stops moving. Using the part's location data obtained from the vision system, the control computers make the necessary manipulations to enable the robot arm to move to the required position for part pickup. The robot then moves the workpieces to a desired position on the pallet. This procedure repeats until 100 parts have been stacked on the pallet. After the pallet is completely loaded, the second belt moves one increment to position an empty pallet in place of the full pallet.

```
0001CENTER     00
0002FACTRY1   -01
0003CELL       00
0004TEST5     -01
0005MEMORY     00
000632768     -01
0005FRAME      00
0006DRIL      -01
0006ROUT      -01
0006ROBOT     -01
0006STAGE     -01
0005DEVICE     00
0006ROBOT     -01
0007TYPE       00
0008MOTION    -01
0009AXIS       00
00106         -01
0006DRIL      -01
0007TYPE       00
0008POSTP     -01
0009PPWORD     00
0010CYCLE     -01
0006ROUT      -01
0007TYPE       00
0008POSTP     -01
0009PPWORD     00
0010CYCLE     -01
0005DEST       00
0006CONSOL    -01
0007DATATYP    00
0008STRING    -01
0006POSIT     -01
0007DATATYP    00
0008STRING    -01
0006ROBOT     -01
0007DATATYP    00
0308STRING    -01
0005SOURCE     00
0006CONSOL    -01
0007DATATYP    00
0008REAL      -01
0004TEST6     -01
0005MEMORY     00
000632768     -01
0005FRAME      00
0006PALLET    -01
0006ROBOT     -01
0006HAND      -01
0005DEVICE     00
0006ROBOT     -01
0007TYPE       00
0008MOTION    -01
0009AXIS       00
00106         -01
0006VSS       -01
0007TYPE       00
0008VISION    -01
0009ALGORM     00
0010COARS1    -01
0006DUMMY     -01
0007TYPE       00
0308MOTION    -01
0009AXIS       00
00101         -01
0005DEST       00
0006CONSOL    -01
0007DATATYP    00
0008STRING    -01
0008REAL      -01
0006BELT1     -01
0007DATATYP    00
0008MINOR     -01
0006HAND      -01
0007DATATYP    00
0008MINOR     -01
0005SOURCE     00
0006CONSOL    -01
0007DATATYP    00
0008REAL      -01
```

Figure 5 Input Data for the Verifier Postprocessor

```
PARTNO  MCL PALLET LOADING PROGRAM
CLPRNT
MACHIN/CELL
SYN/CIRCLE,C,POINT,P,VECTOR,V,LINE,L
REAL/NUMPAL,FINDP
FRAME/PRT
USEFRM/PRT
PRTP=P/0,0,0
C1=C/0,0,4
C2=C/0,0,5
LX=L/XAXIS
LY=L/YAXIS
L1=L/PARLEL,LY,XLARGE,5.25
L2=L/PARLEL,LX,YLARGE,1
L3=L/PARLEL,LX,YSMALL,1
PRTVEC=V/0,1,0
PRTROL=V/1,0,0
OUTLIN=REGION/SOLID
   DNTCUT
   FROM/PRTP
   GO/ON,LY
   AUTOPS
   CUT
   INDIRV/0,1,0
   GOFWD/LY,ON,C2
   GORGT/C2,ON,L2
   GOLFT/L2,ON,L1
   GORGT/L1,ON,L3
   GORGT/L3,ON,C2
   GOLFT/C2.ON,LY
   GOFWD/C2,ON,LY
ENDOF/REGION
INNER=REGION/HOLE
   DNTCUT
   FROM/PRTP
   GO/ON,LX
   AUTOPS
   CUT
   INDIRV/1,0,0
   GOFWD/LX,ON,C1
   GORGT/C1,ON,LX
   GOFWD/C1,ON,LX
ENDOF/REGION
PART1=PROJEC/OUTLIN,INNER
USEFRM/PALLET
RESERV/Z,100,Y,100,X,100,PP,100
M=1
J=1
K=1
N=1
LOOPST
1A)Z(M)=5.0+5.0*(M-1)
   2A)Y(J)=6.0+10.0*(J-1)
      3A)X(K)=6.0+10.5*(K-1)
         PP(N)=P/X(K),Y(J),Z(M)
         K=K+1
         N=N+1
         IF(K-5)3A,3A,3B
      3B)J=J+1
         K=K-5
         IF(J-5)2A,2A,2B
   2B)M=M+1
      J=J-5
      IF(M-4)1A,1A,1B
1B)LOOPND
PALVEC=V/0,1,0
PALROL=V/1,0,0
USEFRM/HAND
HP=P/0,0,0
HVEC=V/0,-1,0
HROL=V/1,0,0
MULTAX/ROLL
SEND/CONSOL,'ENTER NUMBER OF PALLET LOADS'
RECEIV/CONSOL,NUMPAL
WHILE(NUMPAL.GT.0)
   LL=1
   LOOPST
   6A)SEND/BELT1,ON
      DEVICE/VSS
      LET/FINDP=-1
      WHILE(FINDP.EQ.-1)
        LOCATE/PROJEC,PART1,STATUS,FINDP
      ENDOF/WHILE
      SEND/BELT1,OFF
      WHEN(FINDP.EQ.-2)
         SEND/CONSOL,'TOO MANY PARTS UNDER CAMERA'
         ABORT
      ENDOF/WHEN
      DEVICE/ROBOT
      SEND/HAND,CLOSE
      WORKPT/HP,HVEC,ROLL,HROL
      USEFRM/PRT
      GOTO/0,0,5
      GOTO/PRTP,PRTVEC,ROLL,PRTROL
      SEND/HAND,OPEN
      CONECT/ROBOT,PRT
      GODLTA/0,0,5
      USEFRM/PALLET
      GOTO/0,0,10
      GOTO/PP(LL),PALVEC,ROLL,PALROL
      SEND/HAND,CLOSE
      DISCON/ROBOT,PRT
      GODLTA/0,0,5
      LL=LL+1
      IF(LL-100) 6A,6A,6B
   6B)LOOPND
   USEFRM/PRT
   GOTO/24,0,0
   SEND/CONSOL,'PALLET LOADED'
   LET/NUMPAL=NUMPAL-1
ENDOF/WHILE
FINI
```

Figure 6 Sample Program

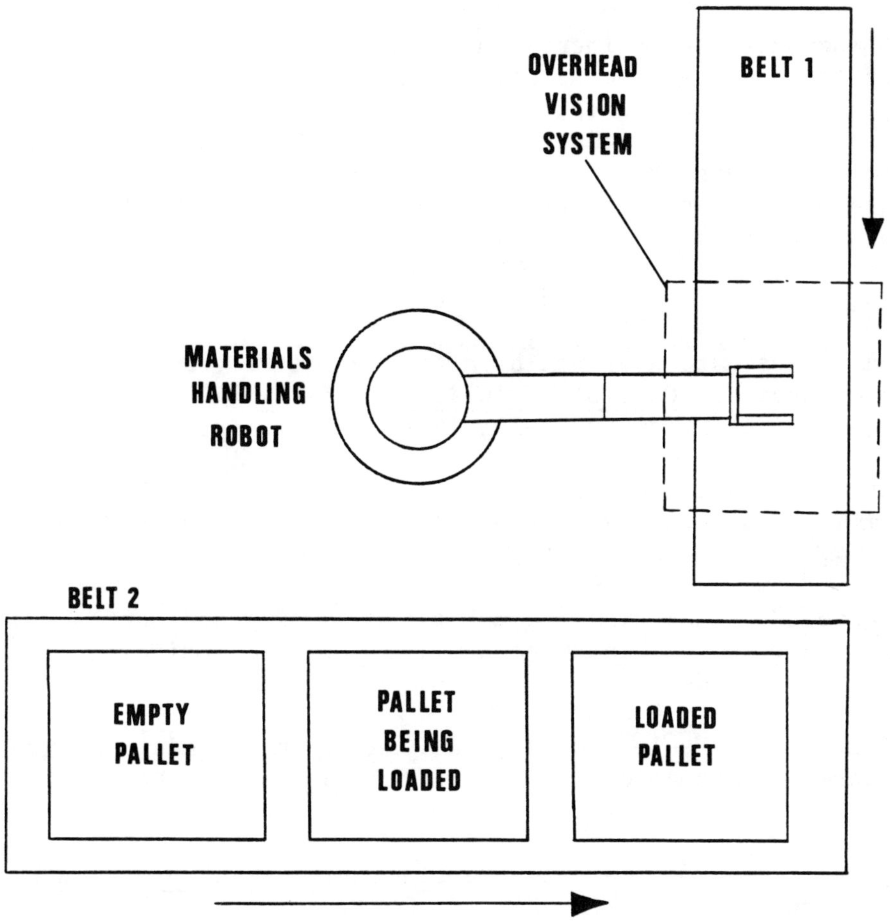

Figure 7 Layout of Sample Work Cell

The workpiece loaded by this manufacturing cell is a cylindrical object with a small rectangular protrusion on the external diameter as illustrated in Figure 4. The sample MCL program given models this workpiece using a set of REGION definitions. These REGION's are then combined using a PROJEC statement to form the 2-dimensional model. The vision system uses this model to locate a part as it enters the camera's field of view.

Each pallet is loaded with parts in a 5x5x4 geometric arrangement (5 across, 5 back, 4 high). This means that the robot end effector must move to 100 different positions. These positions are defined by the LOOPST-LOOPND structure. A GOTO statement references these defined points one at a time within another loop.

When this cell is ready to begin functioning the operator is prompted on the control consol with the command 'ENTER NUMBER OF PALLET LOADS'. An appropriate number is then entered and relayed to the control computers. The control program then directs the work cell to perform the desired movements this given number of times using a WHILE loop. When all pallets are loaded, all cell activity ceases.

IX. CONCLUSION

Because of Industry's ever increasing need for greater productivity, robotic off-line programming languages are being more widely used. MCL's ability to make real time decisions, process vision data, and specify convenient frames of reference make it advantageous over most other control languages. In addition, features such as the verifier postprocessor make this language easy to use from a programmer's point of view. As the use of robotic work cells continues to grow, MCL has the potential to be a valuable programming tool.

BIBLIOGRAPHY

1. Paul, Richard, "Evaluation of Manipulator Control Programming Languages," Proc. IEEE Conference on Decision Control Including Symposium on Adaptive Processes 18th, Fort Lauderdale, Florida, December 12-14, 1979, vol. 1, pp. 252-256.

2. Berg, Eilert J., Paul, Richard P., and Takose, Kumikatsu, "A Structured Approach to Robot Programming and Teaching," Trans. IEEE Systems, Man, and Cyber., April 1981, vol. SMC-11, no. 4, pp. 274-289.

3. Shimano, Bruce, "VAL: A Versatile Robot Programming and Control System," Proc. IEEE Comput. Soc. Int. Comput. Software Appl. Conf., 3rd, Chicago, Illinois, November 6-8, 1979, pp. 878-883.

4. Lieberman, L. I. and Wesley, M. A., "AUTOPASS: An Automatic Programming System for Computer Controlled Mechanical Assembly," IBM Journal of Research and Development, July 1977, pp. 321-333.

5. Robotic System for Aerospace Batch Manufacturing, Task B - High Level Language User Manual, ICAM Project 812-8, October 17, 1981.

6. Baumann, Erwin W., "Model Based Vision and the MCL Language," presented at The 1981 International Conference on Cybernetics and Society, Atlanta, Georgia, October 26-28, 1982.

7. Robotic System for Aerospace Batch Manufacturing, Task B - System User Manual, ICAM Project 812-8, October 1, 1981.

CHAPTER 13

INTELLIGENT ROBOTS/ ADAPTIVE CONTROL SYSTEMS

Adaptive Control with the T^3 Robot

Lloyd Clark
International Harvester
Cambridge, Massachusetts

Neil Webber
International Harvester

Terry Sutton
International Harvester

Abstract

A Cincinnati Milacron T^3 robot has been equipped with binary sensors. The sensors and the robot itself have been interfaced to a PDP-11/34 computer. An experiment was conducted in which the robot had to locate gear forgings within a pallet which partially constrains their locations.

INTRODUCTION

The Cincinnati Milacron T^3 is a large, hydraulically activated industrial robot capable of manipulating a payload of up to 120 pounds. International Harvester installed ten T^3's in a typical machine loading operation at its Farmall plant in East Moline, Illinois. Each robot has a turning center associated with it. The task of the robot is to unload gear blank forgings from forklift-delivered pallets and load them into the turning center. When the turning center completes its operation, the robot unloads the forging and sends it on for further processing. The entire operation is programmed on a strict point-to-point basis.

This paper is concerned only with the very first step of the process--unloading the forklift-delivered pallets. There are two problems that Farmall experiences with its current set-up regarding the pallets. The first problem is that there are many different sizes of gear forgings to be processed. Because the T^3's are programmed to pick up the forgings at fixed locations in the work space, each size of forging requires a specially built pallet which is unusable for any other size of forging. The main disadvantage of this is that it is expensive and wasteful to manufacture and store the various sizes of pallets required.

The second problem that Farmall experiences is that the relative position between the robot and the pallet fixture (into which the pallet is delivered and secured) tends to drift. This shows up as differences between each of the ten robot/turning-center stations, and as occasional failures in a particular station, requiring adjustment of the point-to-point program. Because of this drift, it is necessary to maintain ten different versions of the point-to-point programs; programs cannot be exchanged between stations.

The Automation Technology Department of International Harvester undertook a research project to address the major problems at Farmall, namely the drift of the fixtures and the requirement of special pallets. The goal of the project was to demonstrate cost-effective techniques that would be feasible to implement in a manufacturing environment. This goal placed constraints on the cost of the solution, as well as on its complexity and reliability.

In the following section an overall view of the proposed solution is presented; the section following that details the actual laboratory implementation of the solution, and the final section concludes with a discussion of the results of the implementation.

SOLUTION OVERVIEW

Forgings

The forgings processed at the Farmall plant are doughnut shaped, with outer diameters ranging from about 4"(10 cm) to 6"(15 cm), and inner diameters ranging from about 2.5"(6 cm) to 3"(8 cm). The dimensions of a typical forging are shown in Figure 1.

Pallet

To address the first problem at the Farmall plant, namely the requirement of a special pallet for each type of forging, it was decided to use a pallet with square cells capable of holding the largest forging. Figure 2 depicts such a pallet. Using this type of pallet would allow any type of forging to be processed, without having to have special pallets for each forging type. However, because the forgings are not held at fixed positions in space by such a pallet, a new problem is introduced: it becomes necessary to search for the forging using some form of sensor.

Sensors

The sensor chosen would have to be capable of determining the forging's location in a plane. There is no question of object orientation since the forgings are rotationally symmetric. In addition, it is desirable to minimize the length of time required to search for the forging.

Although these requirements are within the capabilities of available vision systems(1), vision was ruled out as being impractical in this application for a variety of reasons. Vision systems are too expensive at the present time to be cost effective. More importantly, binary vision systems require controlled lighting and high contrast between object and background in order to perform, and these conditions cannot be met in the factory environment at the Farmall plant.

The type of sensor most likely to be effective in the given environment is a simple binary sensor such as an LED emitter/detector pair. This type of sensor can be used either in a beam-break mode of operation, or in a reflected-beam mode of operation. Because of the limited clearance around the parts in the pallet, the reflected-beam mode was chosen. The sensor would be mounted directly onto the robot gripper. In order to use the sensor to locate the forging center, it is necessary to pass over the forging with the sensor in some predefined search pattern.

Search Strategy

To find the (x,y) coordinate of the center of the forging requires two sweeps with the sensor: one parallel to the x axis and one parallel to the y axis. Each sweep gives at least two data points, one at the start of the forging and one at the end. Averaging the two will give the distance to the center of the forging in the given direction (Figure 3A). It is possible to get more data points if the sweep crosses the center of the forging; however, no new information is obtained from those readings (Figure 3B). Conversely, it is possible to get only one data point if the sweep only just crosses the outer edge of the forging (Figure 3C), in which case the center coordinate is found directly. In the worst case, it is possible to miss the forging entirely, yielding no data (Figure 3D).

It was considered necessary to minimize the amount of time taken to locate the forging; thus the situation depicted in Figure 3D should be avoided because it will be necessary to repeat the scan at a new starting position in order to find the forging. Fortunately,

---

(1) Roger N. Nagel, et al., "Connecting the PUMA Robot with the MIC Vision System and Other Sensors".

if the size of the pallet cell is constrained to be less than twice the diameter of the smallest forging that can be processed, it is impossible to miss the forging if the scan is down the center of the cell. This has the added advantage of eliminating the special case situation of Figure 3C (assuming the cell size is <u>less</u> than twice the smallest forging diameter).

Computer Interface

It is necessary for the sensor to be connected to a device which can accept raw binary sensor data, uses this raw data to compute the center of the forging, and move the T^3 robot to the computed forging center. A PDP-11/34 digital computer was used for this purpose. The PDP-11/34 and the T^3 robot were connected via an RS232 communications link. This connection allows the PDP-11/34 to send the T^3 robot new coordinates when the T^3 is in "ADJUST" mode.

Shortly after receiving a set of new coordinates from the computer, the T^3 robot begins to move toward the new position at a velocity that is constant. Thus, in order to do a line search, the T^3 is first sent coordinates which correspond to one endpoint of the line segment. Next, the T^3 is sent new coordinates which correspond to the other endpoint of the line segment. This causes the T^3 robot to scan the line segment at constant velocity. At the same time the computer samples the binary output of the sensor.

In order to compute the positions of the forging edges from the raw sensor data, the computer must know the robot's position as a function of time. The robot's starting position and the robot's velocity are known, but the time at which the robot begins moving in response to the new coordinates is not known. Therefore, some kind of start signal is needed to signal the computer when the robot begins movement. Without a start signal, the PDP-11/34 does not have enough information to allow the coordinates of the forging center to be computed. Fortunately, a useable start signal was obtained from the servo circuitry of the T^3 robot.

ACTUAL SOLUTION

Gripper

A gripper capable of picking up forgings by their inner diameter was designed and attached to the T^3 robot. The gripper consists of two parallel, air-actuated jaws that are inserted into the center hole of a forging and opened. The force of the jaws against the inner diameter of the forging allows the forging to be picked up and transported by the gripper. Closing the gripper jaws releases the forging. Actuation of the gripper is controlled by the T^3 robot.

The gripper contains mountings for three sensors so that the performance of various sensors could be easily evaluated. A small optical sensor attaches to the tip of one of the gripper jaws. Another optical sensor and an eddy-current sensor are mounted elsewhere on the gripper since they are not small enough to be integrated into the tips of the gripper jaws.

The gripper also contains a mechanism which is supposed to allow the gripper to go down and check to determine if the gripper is positioned over the center hole of the forging. The mechanism is spring-loaded and is designed to close a switch if it contacts the forging as it moves down. This would make accidental ramming of the forging with the gripper less likely. It would also allow the gripper to check the centering of the forging before it attempted to search for the forging. If the gripper happened to be close to the center of the forging, the verify feature would enable the gripper to pick up the forging quickly without wasting time searching. Unfortunately, the verify feature was unusable due to mechanical problems.

Adjust

Adaptive control of the T^3 was achieved with the optional ADJUST feature. Normally, the T^3 is programmed on a point-to-point basis. At each point one function, such as "open gripper", "turn on device #1", etc., may be programmed. ADJUST is such a function; with it it is possible for an external computer to gain control of the T^3 in order to command new positions.

When the T^3 comes to a point programmed as an ADJUST function point, the coordinates describing the position of the T^3 are transmitted over the serial interface outlined below. The external computer, which in this case was a PDP-11/34, must then do one of two things. It may simply transmit an acknowledge message back to the T^3, in which case the robot continues with its point-to-point program, or, new coordinates may be sent to the T^3. When the T^3 receives these new coordinates it will move to the position they describe, and repeat the ADJUST function (i.e., transmit the new coordinates back to the external computer).

It is important to note that only the position of the T^3 is controllable via the ADJUST function. It is not possible to alter the speed of the robot, the status of the gripper (open, closed) or the state of any device connected to the T^3 controller.

Interface

The T^3 controller contains an RS232 serial interface port when the optional ADJUST feature is present. It is through this port that an external computer is interfaced.

The protocol used is a modified version of the Digital Equipment Corporation DDCMP protocol. Messages are formatted into header and data packets, with cyclic redundancy checks for each. In addition, there is an interlock scheme involving keeping track of message numbers which prevents the two computers from getting out of sync.

A complete discussion of the interface is beyond the scope of this paper. Readers are referred to the Cincinnati Milacron External Function Manual for the T^3 robot.

One detail that is of importance is that transmission errors (presumably detected by the CRC) will cause the T^3 to ask for a retransmission. This means that the time required for a given transmission to complete is not predictable. There is an upper bound, because after seven retransmission requests the serial link is considered to be broken by the T^3 software.

The method used for interpreting sensor data requires knowledge of the point in time at which the T^3 began to move; which in turn requires knowledge of the actual transmission time. The uncertainty of the transmission time is one factor that led to the use of a servo error signal for synchronization, as described in a later section on sensors.

Software Support

Higher level control of the T^3 was implemented on the PDP-11/34 through three FORTRAN routines and a transformation package. This combination allowed the programmers to specify transformations to be used when receiving or transmitting T^3 coordinates. In this way programs could be written independently of which sensor head was actually being used at the time, as long as the appropriate transformations were supplied.

The three basic routines provided were WAIT, ADJUST, and CONTINUE. The WAIT routine caused the PDP-11/34 to wait for the T^3 to come to an Adjust point. WAIT returned the coordinates of the T^3 relative to the currently selected transformation.

The ADJUST procedure took coordinates as input arguments and transformed them according to the currently selected transformation. The transformed coordinates were then sent to the T^3.

It was presumed by the software that WAIT and ADJUST were interleaved appropriately by the programmer. Using ADJUST before a WAIT, or WAITing twice without an intervening ADJUST would cause a DDCMP protocol error and a break of the link between the PDP-11/34 and the T^3. This extra responsibility was placed on the programmers to allow for greater parallelism in the processing algorithms; i.e., only WAITing when nothing else remains to be computed.

The final routine is CONTINUE, which sends an acknowledge message to the T^3, causing it to resume execution of its point-to-point program.

Transformations

It was necessary to implement a transformation package for two reasons. The first reason is that the T^3 can only transmit and receive coordinates in the world system when using ADJUST. Although hand-based coordinates are available through the T^3's teach pendant, they are not available through the ADJUST function. In most cases it is advantageous to move in a coordinate frame relative to the hand (gripper) of the T^3. This is especially true in executing the search patterns.

The second reason for implementing a transformation package is that the gripper used in the experiment has a number of different sensors mounted in different locations. By using the appropriate transformation matrices (2) it was possible to program the higher levels independently of the actual sensor in use.

---

(2) Richard P. Paul, Robot Manipulators: Mathematics, Programming, and Control (Cambridge: MIT Press, 1981).

One disadvantage of this approach is that certain optimizations are lost. For example, to switch from using the eddy-current sensor to using the gripper requires raising the T^3 in order to clear the pallet. However, to switch from using the finger mounted incandescent sensor to using the gripper requires only a small shift along the axis in the direction of the gripper opening/closing motion. As implemented, the software always drew the T^3 up to a safe position before switching from one tool to another. We did not get a chance to explore ways to avoid unnecessary moves without sacrificing program generality.

Sensors

The objective of sensor selection for the T^3 robot was to obtain simple binary sensors that could sense the edges of a gear blank forging accurately enough to enable the location of the center of the forging to be reliable computed. The sensors had to be accurate enough to insure that the calculated center of the forging was never far enough from the actual center to cause the gripper to ram the forging as it came down. It was initially decided that the sensors had to be capable of sensing a forging within a variation of $\pm 3/16$ inches of its nominal sensing height, since variations in the parts delivery system and robot could cause sensing height deviations of that magnitude. The sensors had to be capable of detecting the forgings reliably even though the surfaces of the forgings varied in appearance from shiny silver to rusty brown. It was also necessary for the sensors to be unresponsive to the wooden pallet surface, which was 1.8 inches below the surface of the forging. It was desirable for the sensors to have good repeatability, low hysteresis, small beamwidth, and small size.

In accordance with the above requirements, three sensors were selected: an incandescent optical proximity sensor, a pulsed LED optical proximity sensor, and an eddy-current proximity sensor.

The incandescent sensor contains an incandescent lamp, a focusing lens, and a phototransistor in a package that measures 0.88x0.50x0.16 inches. It has a maximum range of 3.0 inches with white bond paper, an optimum range of 0.16 inches, a field of view 0.11 inches wide, and 0.015 inches of object definition. The chief advantage of the incandescent sensor over the other sensors selected is its small size, which enables it to fit in tight spots such as the tip of the gripper.

The pulsed LED sensor contains an infrared LED, a phototransistor, and two focusing lenses in a package that measures 3.25x3.10x1.00 inches. It has a detection range of 2.5 ± 0.25 inches and a spot size of 1/8 inch. The chief advantage of this sensor is its large sensing distance, which makes it less likely that it will accidentally be rammed into a forging.

The eddy-current sensor is contained in a cylindrical package that has a diameter of 0.65 inches and a length of 3.2 inches. It has a maximum sensing distance of 0.4 inches and 4% typical hysteresis. Its main advantage is its lack of dependence on the reflective properties of the forging.

Each of the three digital sensor outputs was connected to a multichannel analog-to-digital converter, which was interfaced to the

PDP-11/34 computer. This connection enabled PDP-11/34 programs to select a sensor and read data from that sensor.

Before the two photoelectric sensors could be used for detecting forgings, their sensitivity controls had to be adjusted to have as much sensitivity to the forging as possible without being sensitive to the pallet surface. The LED sensor was adjusted for maximum sensitivity and was still insensitive to the pallet surface. The incandescent sensor could not be set at maximum sensitivity since it would detect the pallet. Its sensitivity was set just below the point where it would detect the pallet at a height of 1.8 inches (the height of a forging). Thus, when scanned across the forging, the sensor will be farther than 1.8 inches from the pallet and will be unresponsive to the pallet surface.

The software associated with the sensors samples the output from a sensor at a 1kHz rate. The output from a sensor is either on or off, so it is a simple task for a program to determine the times at which the sensor output changes.

A start signal is obtained from the servo error circuitry of the T^3 robot. This is an analog signal which jumps to a new value when the T^3 robot begins to move in response to a command. The start signal is also sampled at a 1kHz rate. The computer detects the beginning of robot movement by determining the time at which the start signal exceeds a threshold level. Since the time at which the robot began moving, the velocity of the robot, the starting position of the robot, and the times of sensor transitions are known to the computer, it can calculate the coordinates of the positions where the forging was detected by the sensor.

High Level Control

In normal operation, the robot gripper is first positioned over the center of a pallet cell. At this point, had the verification capability worked, the robot would have gone straight down to determine if the forging was close enough to the center to allow it to be picked up immediately. Since the mechanical verification section of the gripper did not function properly, the verifcation operation was omitted. Next, the gripper is repositioned so that the sensor is correctly oriented for sensing. The X-Y search pattern is carried out, and the computer uses the resulting sensor data to compute the center of the forging. The computer instructs the robot to move to the computed location of the forging center. The robot then goes down and picks up the forging with the gripper.

RESULTS

Sensors

The sensors were very accurate in terms of their repeatability and hysteresis. Typical errors introduced by the sensors were on the order of ten thousandths of an inch. The chief problem of the sensors was sensitivity to variations in the vertical distance from the sensor to the forging. Robot wobble in the vetical direction (as a search pattern was being executed) tended to reduce the accuracy of the sensors and could even result in loss of data if the wobble was severe.

Adjust

It was demonstrated that the adjust function of the T^3 robot could be used to provide real-time adaptive control. The actual search and pick-up process was also sucessfully demonstrated.

CONCLUSION

In conclusion, it has been successfully demonstrated that a computer, sensors, and the T^3 robot can be interconnected so that the robot is able to search, find, and pick up forgings from a pallet cell which does not tightly fix the location of a forging. Such an approach eliminates the need for precision pallets which match one and only one forging size. Instead, one standard pallet can be used for many different sizes of forgings.

BIBLIOGRAPHY

1. Roger N. Nagel, et al., "Connecting the PUMA Robot with the MIC Vision System and Other Sensors," S.M.E. Tech Paper MS82-219, 1982.

2. Richard P. Paul, Robot Manipulators: Mathematics, Programming, and Control, MIT Press, 1981.

3. Roger N. Nagel, et al., "Experiments in Part Acquisition Using Robot Vision," S.M.E. Tech Paper Ms79-784, 1979.

4. J.L. Nevins and D.E. Whitney, "Computer-Controlled Assembly," Scientific American, Vol. 238, No. 2, February 1978, pp. 62-74.

5. D.E. Whitney and J.L. Nevins, "The Remote Center Compliance: What Can It Do?", Robotics Today, Summer 1979, pp. 19-23.

6. R. Bolles and R. Paul, "The Use of Sensory Feedback in a Programmable Assembly System," SRI Memo AIM-220, October 1973.

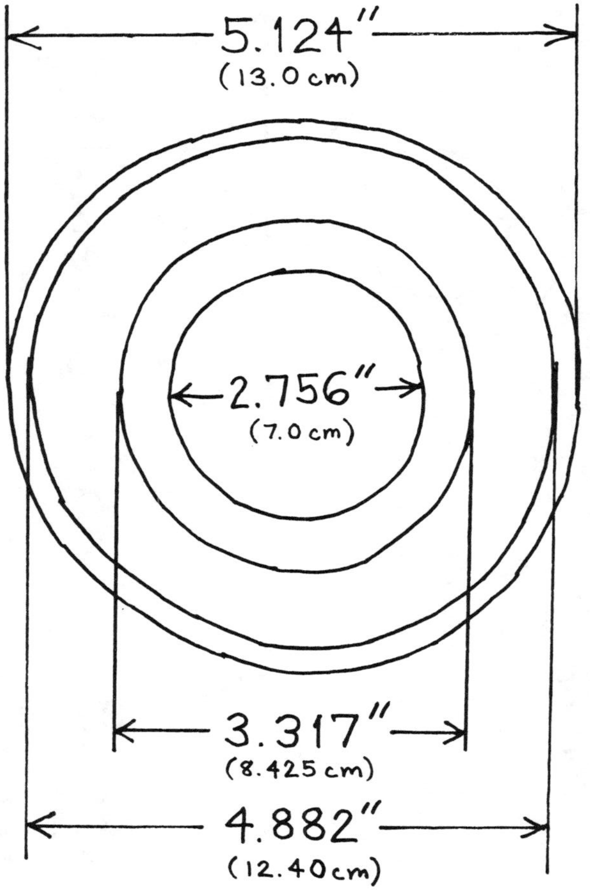

Figure 1. Typical Forging

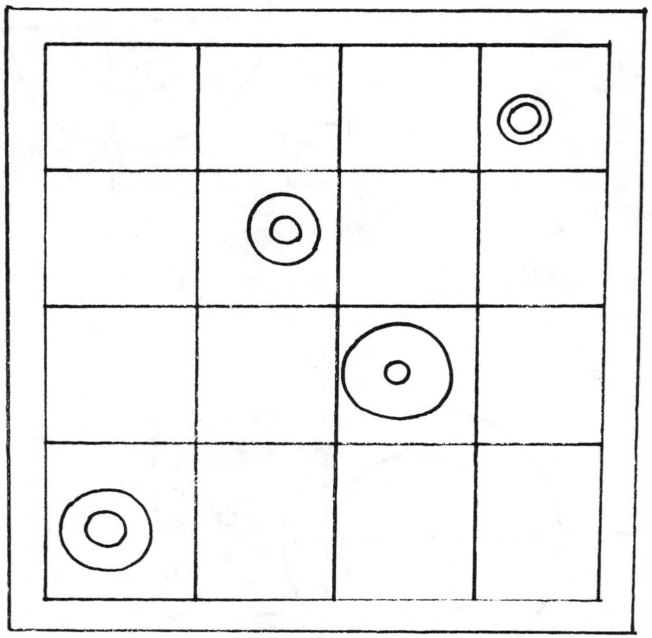

Figure 2. Pallet

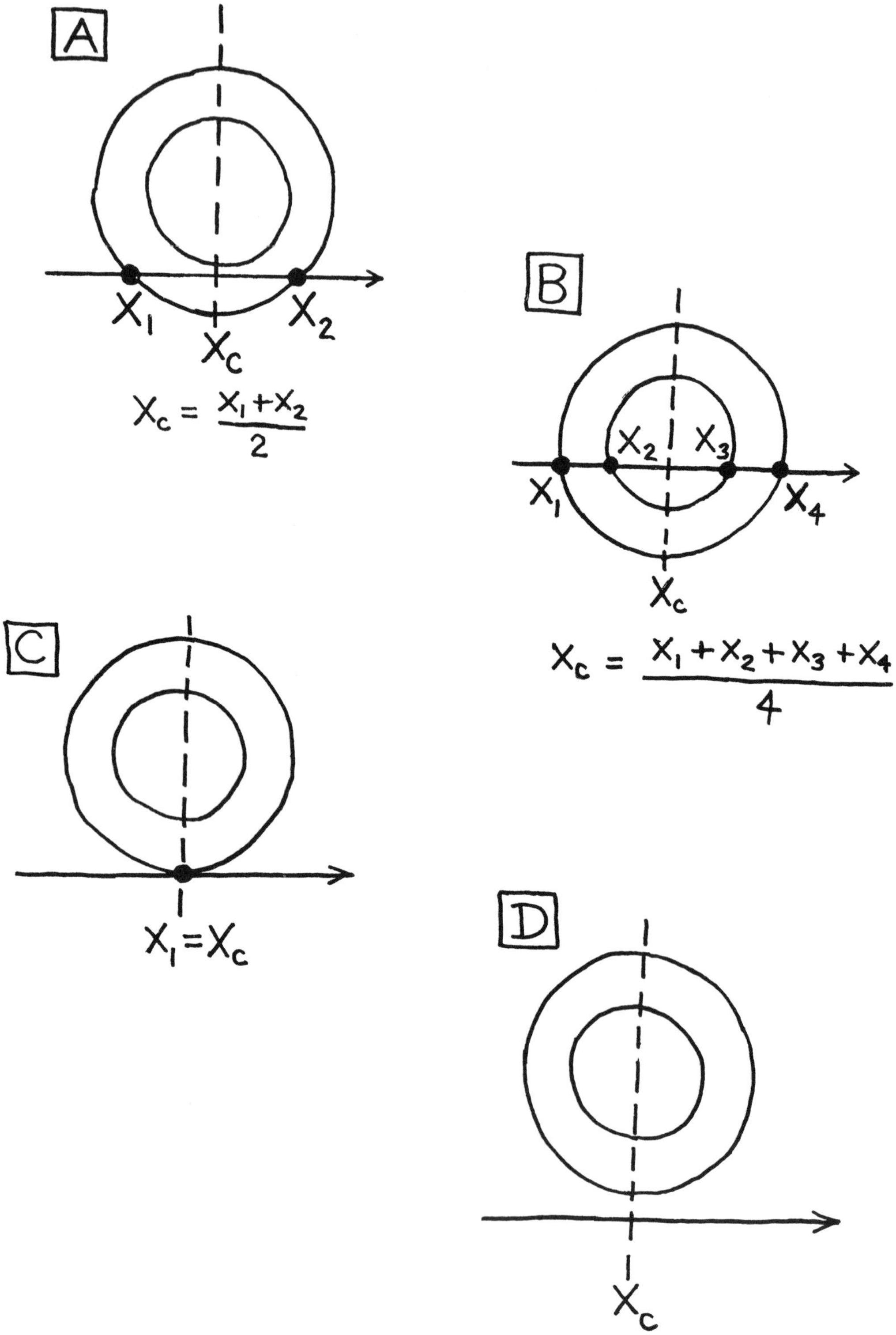

Figure 3. Search Results

Adaptive Control of Manipulators in the Task Oriented Space

A. Balestrino
University of Naples—Istituto Elettrotecnico
Naples, Italy

G. De Maria
University of Naples—Istituto Elettrotecnico

L. Sciavicco
University of Naples—Istituto Elettrotecnico

Abstract

The increased demand on robotic manipulator performances leads to the use of advanced control structures. An adaptive model, following control for robotic manipulators, has been previously developed in order to directly control the angular positions of the linkages. The target trajectory is assigned in the task-oriented coordinates in order to avoid involved computations to transform the task-oriented space into drive-oriented space. In this paper the above mentioned control technique is used to implement an adaptive controller which assures insensitivity and task-oriented coordinates decoupling.

INTRODUCTION

The task of a manipulator control system is to enable the manipulator terminal link to follow the prescribed motion trajectories, planned in a coordinate frame which suitably describes the work space. Simple joint servos are currently used when terminal body motion has low speed; the control of such servos has to be sufficiently robust, in order to compensate changing joint inertia and gravity loadings depending on the configurations assumed by the mechanical structure. More sophisticated controllers are currently studied in order to execute terminal body motion with high speed and precision. The design of these controllers requires the knowledge of a complete and accurate dynamical model of the mechanical chain, but this results in a great amount of computation during the motion.

The manipulator mechanical structure is a chain of rigid bodies (links) interconnected by joints; the generalized forces acting on the joints can be considered as inputs, whereas reciprocal positions of adjacent joints are representative of the outputs. It is generally agreed that dynamics of the spatial manipulator elements can be described by a set of coupled nonlinear second order differential equations [1,2] . Physically, the coupling terms represent gravitational torques, reaction torques, Coriolis and centrifugal torques.

The reciprocal positions of the joints θ_i define a space that we call drive - oriented space. With reference to a manipulator model described by a generalized coordinate vector $\theta = [\theta_1 \ldots . \theta_k]^T$ and by the input vector $u = [u_1 \ldots . u_k]^T$ of generalized forces acting on the joints, a number of methods for dynamical control of manipulators have been proposed in order to providing position control of the manipulator. They include resolved motion position control [3], resolved motion rate control [4] and resolved motion acceleration control [5]; other existing control algorithms emphasize nonlinear compensations of the interactions [6,7,8] . All these schemes require a detailed model of the manipulator and a good load forecasting. In order to relax these requirements alternative solutions to controller design have been given in [9,10,11] , which show good insensitivity properties with respect to changes of the manipulator characteristics and realize decoupled dynamics of the links by compensating the interactions among the various joints of the manipulator. In particular the approach followed in [11], based on a procedure proposed in [12], is simple and systematic, requires only approximate modelling and a limited computational effort. The simulation results have shown that the decoupling among the various joints is effective and the dynamic behaviour for each generalized coordinate can be specified in terms of a second order linear dynamical model.

In all the above mentioned control schemes the direct control of joint positions and torques is realized with the aim of minimizing the deviation of the executed trajectory from a target one. The target trajectory is usually planned in a coordinate frame characterizing the task-oriented space, different from the drive-oriented one. Hence a coordinate transformation is needed between task-oriented space and drive - oriented space, in order to specify the target trajectory in terms of joint generalized coordinates. In order to avoid these transformations,which remarkably increase the computational difficulties, it appears more convenient to implement a control algori thm which directly controls forces and positions of the terminal link in the task-oriented space, rather than torques and joint positions. In [13] a cartesian position control called resolved motion force control is presented; its basic concept is to determine the necessary joint torques to be applied when performing the cartesian motion control of the manipulator; a multielement strain gauge force wrist sensor is requested and the joint torques are determined by means of the Robbins-Monro stochastic approximation method. In [14] an adaptive control method is proposed; with the hypothesis that, during the adaptation, the system is time invariant, an adaptive control algorithm is deduced in form of differential equations able to track a desired trajectory planned in the task-oriented space.

The approach followed in the present paper is based on adaptive model following control (AMFC) system technique [15] and on hyperstability theory but it does away with the assumption that the plant is characterized by a time invariant linear model during the adaptation process. The unit vector adaptation law [12] is used to implement a cartesian position control of a k-degrees of freedom manipulator; the control is simple and does not require accurate modelling. The suggested adaptive controller generates discontinuous adaptation signals; their effects on the plant hardware should be taken into account. A mixed control strategy, by suitably combining nonlinear compensation and unit vector adaptation law, can be used in order to reduce the control effort generated by the adaptation mechanism [16] . If one assumes that the inertia matrix and the gravity torques can be approximately computed on line, the simulation results , for a 3-degrees of freedom manipulator , show the effectiveness of the proposed control technique with reference both to the trajectory tracking errors and to the control signal quality.

MANIPULATOR DYNAMICAL MODEL

We consider an open kinematical chain with k rotational joints; as generalized coordinate vector $\theta = [\theta_1, \theta_2, \ldots \theta_k]^T$ we denote the relative angular positions θ_i between two links at i-th joint. The analytical dynamical model can be obtained by applying Lagrange equation :

$$\frac{d}{dt}\frac{\partial T}{\partial \dot{\theta}_i} - \frac{\partial T}{\partial \theta_i} = Q_i^L + Q_i \qquad i = 1 \ldots k \tag{1}$$

where T is the kinematic energy , Q_i^L represents gravitational torque depending on position of the links and Q_i are generalized forces assumed as the input to the system. The kinematic energy of the manipulator system is :

$$T = \frac{1}{2}\dot{\theta}^T B(\theta)\dot{\theta} \tag{2}$$

where the inertia matrix $B(\theta)$ is symmetric and positive definite for all θ.

From Eqn(1) the manipulator dynamical model can be derived in the form :

$$B(\theta)\ddot{\theta} = - C(\theta,\dot{\theta}) + E(\theta) + M(t) \tag{3}$$

where

$$C(\theta,\dot{\theta}) = \frac{dB(\theta)}{dt}\dot{\theta} - \frac{\partial T}{\partial \theta}$$ takes into account Coriolis and centrifugal torques;

$$E(\theta) \triangleq [Q_1^L \; . \; . \; . \; Q_k^L]^T$$ is the $k \times 1$ vector of gravitational torques;

$$M(t) \triangleq [Q_1 \; . \; . \; . \; Q_k]^T$$ is the $k \times 1$ vector of torques acting on the joints .

The k variables θ_i define the drive-oriented space Θ . In this space an adaptive control technique has been proposed in [11] based on the positiveness property of the matrix $B(\theta)$.

Usually the performances and dynamical specifications about the target trajectories are given in the m-dimensional task oriented space X differing from the drive-oriented space Θ.

The tranformations between the coordinate frames in Θ and in X are given by :

$$x = H(\theta) , \tag{4}$$

$$\dot{x} = J(\theta)\dot{\theta} , \quad \text{where } J(\theta)= \frac{\partial H(\theta)}{\partial \theta} \text{ is the Jacobian matrix ,} \tag{5}$$

$$\ddot{x} = J(\theta)\ddot{\theta} + \frac{dJ(\theta)}{dt}\dot{\theta} . \tag{6}$$

The relationship between terminal link static forces F and static torques M is:

$$M = J^T(\theta)F . \tag{7}$$

$H(\theta)$ is a continuous function , rather involved , with known structure and parameters. The relationship (4) , (7) are one to one and allow the transformation from the Θ - space to the X - space. Note that the inverse transformation :

$$\theta = H^{-1}(x) \tag{8}$$

being solution of the nonlinear , coupled trascendental equation (4) , is not one to one , not always exists and hardly ever is an explicit function of x; therefore the computational effort required to specify the target trajectory in the Θ space is remarkable. In order to overcome these difficulties , the control law should be designed and implemented directly in terms of the task-oriented coordinates. To this end the manipulator dynamics must be rewritten in the task-oriented space. Assuming $m = k$, tha Jacobian matrix $J(\theta)$ is non singular almost everywhere in the Θ space.

Since θ can be regarded as a function of x , by considering matrices B and J to be function of x , the kinematic energy (2) can be written , taken into account Eqn. (5) ,

$$T = \frac{1}{2}\dot{x}^T J^{T^{-1}}(x)B(x)J^{-1}(x)\dot{x} . \tag{9}$$

In the task-oriented space the Lagrange equation becomes :

$$\frac{d}{dt}\frac{\partial T(x)}{\partial \dot{x}_i} - \frac{\partial T(x)}{\partial x_i} = F_i^L + F_i \tag{10}$$

where F_i^L and F_i represent the components in the task oriented space of the gravitational and generalized forces , respectively , applied to the terminal link. From Eqns. (9) and (10) the manipulator model in the task oriented space is described by :

$$J^{T^{-1}}(x)B(x)J^{-1}(x)\,\ddot{x} = \frac{\partial T}{\partial x} - \frac{d}{dt}(J^{T^{-1}}(x)B(x)J^{-1}(x))\dot{x} + F^L + F\,. \tag{11}$$

In this equation the Jacobian matrix J , the inertia matrix B and the gravity term F^L

$$F^L = J^{T^{-1}}(\theta)E(\theta) \tag{12}$$

are explicit functions of θ and can be computed in real time for a given θ , whereas they are too complicated to be expressed as explicit functions of x , because implicitly include the transformation (8) from θ to x. The other terms in the right hand side of (11) are rather involved. It must be noted that the matrix $J^{T^{-1}}BJ^{-1}$ exists almost everywhere in the task oriented space and is positive definite.

In order to evidentiate the explicit dependence on the coordinate frame , the model (11) can be rewritten starting from (3),(5),(6) and (7) as follows :

$$\ddot{x} = \frac{dJ(\theta)}{dt}\dot{\theta} - J(\theta)B^{-1}(\theta)C(\theta,\dot{\theta}) + J(\theta)B^{-1}(\theta)E(\theta) + J(\theta)B^{-1}(\theta)J^T(\theta)F\,. \tag{13}$$

A more compact form of Eqn.(13) can be obtained by putting :

$$z^T \triangleq [x^T, \dot{x}^T]\,, \tag{14}$$

so that we get :

$$\dot{z} = a_p(z) + B_p(z)F\,, \tag{15}$$

where :

$$a_p^T(z) \triangleq [\dot{x}^T,\ \bar{a}_p^T(z)]\,,\quad B_p^T(z) \triangleq [0 \mid J(x)B^{-1}(x)J^T(x)] \tag{16}$$

with

$$\bar{a}_p(z) \triangleq \frac{dJ(x)}{dt}J^{-1}(x)\dot{x} - J(x)B^{-1}(x)(C(x\ \dot{x}) - E(x)). \tag{17}$$

Eqn.(13) jointly with (3) suggests a representation of the manipulator system as in the block diagram of Fig. 1.

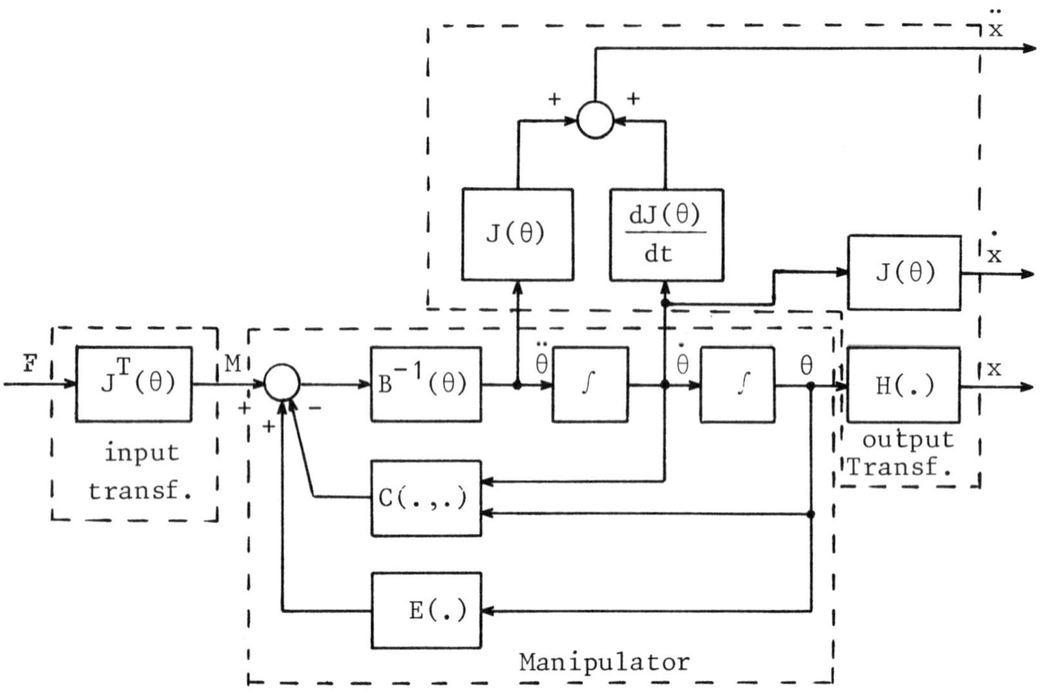

Fig.1. Manipulator system description.

The above considerations imply the impossibility of computing exactly the right hand side of the system equation in real time. Moreover, it should be remembered that there are always unknown deviations of the parameters from the true ones, due for instance to the presence of the payload. Hence it is necessary to implement robust control laws, which assure good performances as specified by means of a linear reference model, in spite of the plant nonlinearities and the parameter uncertainty. To cope with these difficulties, in the next section we propose an adaptive control scheme which does not require an exact knowledge of the plant model, but only the bounds of the plant parameters.

TASK ORIENTED ADAPTIVE CONTROL FOR MECHANICAL MANIPULATORS

In this section an adaptive control scheme for mechanical manipulators is considered in order to realize a system insensitive to changes of the manipulator parameters, and characterized by the decoupling among the motion components in the task-oriented space.

The adaptive control scheme (Fig.2) differs from the one suggested in [14] as it relies on the unit vector adaptation law specially developed for a class of nonlinear plants [12]. Referring to the block diagram of Fig.2 the linear and time invariant reference model

$$\dot{\mathbf{z}}_M = A_M z_M + B_M u_M \tag{18}$$

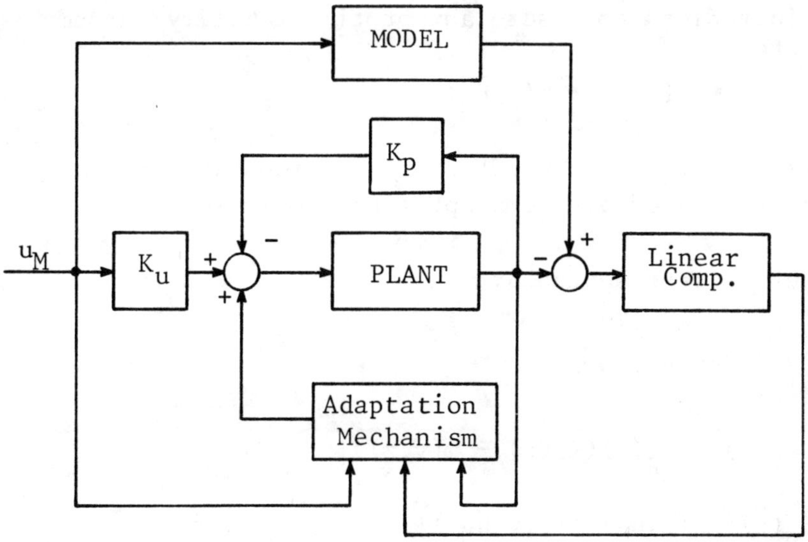

Fig.2. Adaptive Model Following Control System

provides a behaviour objective for the manipulator,as z_M represents the target trajectory; note that for a k-degrees of freedom manipulator we have $z_M^T = [x_M^T, \dot{x}_M^T] \in R^{2k}$ and $u_M \in R^k$.

The first step in the design of the adaptation mechanism is the choice of the reference model. In order to obtain from the nonlinear , coupled manipulator system well behaved linear , uncoupled characteristics , we choose a reference model which , for each coordinate , is represented by the linear , second order , time invariant differential equation :

$$\ddot{x}_{Mi} + a_{1i}\dot{x}_{Mi} + a_{0i}x_{Mi} = u_{Mi} , \qquad i = 1, \ldots, k . \tag{19}$$

Hence Eqn.(18) is characterized by

$$A_M = \begin{pmatrix} 0 & I_k \\ \Lambda_0 & \Lambda_1 \end{pmatrix} , \quad B_M = \begin{pmatrix} 0 \\ I_k \end{pmatrix} \tag{20}$$

where $\Lambda_0 = -\mathrm{diag}(a_{0i})$, $\Lambda_1 = -\mathrm{diag}(a_{1i})$, $a_{0i} > 0$, $a_{1i} > 0$, I_k is the $(k \times k)$ identity matrix.

Let

$$e = z_M - z \tag{21}$$

be the generalized state error , and

$$v = De , \quad v \in R^k \tag{22}$$

a linear compensator introduced to assure asymptotic stability; assume

$$F = K_u(z,u_M) - K_p(z) + \phi(v,z) + \psi(v,u_M) \tag{23}$$

where K_u and K_p are $(k \times 1)$ suitable feedforward and feedback actions, whereas ϕ and ψ are $(k \times 1)$ controls generated by the adaptation mechanism.

The error dynamics is represented by :

$$\dot{e} = A_M e + \begin{pmatrix} 0 \\ \hdashline I \end{pmatrix} w$$

$$w = (0 \vdots I)\left[A_M z - a_p(z) + B_p(z)(K_p(z) - \phi(v,z))\right]$$

$$+ \left[B_M u_M - B_p(z)(K_u(z,u_M) + \psi(v,u_M))\right] \tag{24}$$

The adaptation mechanism is asymptotically stable in the large [15] if the following conditions are satisfied

i) structure conditions :

$$B_p(z)B_p^+(z)(A_M z - a_p(z)) = A_M z - a_p(z)$$

$$B_p(z)B_p^+(z)B_M = B_M \tag{25}$$

where $B_p^+ \triangleq (B_p^T B_p)^{-1} B_p^T$. They essentially concern the structure of matrices B_p , B_M and vector $A_M z$, $a_p(z)$ rather than the actual values of their parameters. With the choice (20) they are satisfied.

ii) The transfer function

$$G(s) = D(sI - A_M)^{-1} \begin{pmatrix} 0 \\ \hdashline I \end{pmatrix} \tag{26}$$

is strictly positive real. By choosing

$$D = (0 \vdots I)P , \tag{27}$$

where the positive definite $(2k \times 2k)$ matrix P is a solution of the Lyapunov equation

$$A_M^T P + PA_M = - H , \quad H \text{ p.d.} \tag{28}$$

the positivity condition is satisfied.

iii) The integral inequality

$$\int_0^T v^T w \, dt \leq \gamma^2 \tag{29}$$

holds for all $T \geq 0$. This inequality is satisfied if ϕ and ψ are such that :

$$v^T(J(x)B^{-1}(x)J^T(x))\phi(v,z) \geq v^T m(z) \quad , \quad \forall z , \forall v \tag{30}$$

$$v^T(J(x)B^{-1}(x)J^T(x))\psi(v,u_M) \geq v^T n(z,u_M) , \quad \forall z , \forall u_M , \forall v \tag{31}$$

where

$$m(z) \triangleq \Lambda_0 x + \Lambda_1 \dot{x} - \bar{a}_p(z) + J(x)B^{-1}(x)J^T(x)K_p(z) \quad , \tag{32}$$

$$n(z,u_M) \triangleq u_M - J(x)B^{-1}(x)J^T(x)K_u(z,u_M) , \tag{33}$$

according to (16) , (17) , (20) and (24).

Assuming [12]

$$\phi(v,z) = q\frac{v}{\|v\|}\sum_{i=1}^{2k}|z_i| \quad , \qquad q > 0 , \tag{34}$$

$$\psi(v,u_M) = p\frac{v}{\|v\|}\sum_{i=1}^{k}|u_{Mi}| , \qquad p > 0 \quad , \tag{35}$$

by virtue of positive definiteness of the matrix $J(x)B^{-1}(x)J^T(x)$, the left hand side members of inequality (30) and (31) are positive.

Let assume

$$\|m(z)\| \leq \|M.z\| \ \forall z , \ \|n(z,u_M)\| \leq \|N\cdot u_M\| \ \forall z , \forall u_M \tag{36}$$

with M and N constant matrices of suitable dimensions. By denoting with $\Lambda(M)$ and $\Lambda(N)$ the square root of the maximum eigenvalues of M^TM and N^TN , and with $\lambda(J(x)B^{-1}(x)J^T(x))$ the minimum eigenvalues of $J(x)B^{-1}(x)J^T(x)$, inequalities (30) , (31) and then (29) are satisfied if :

$$q \geq \frac{\Lambda(M)}{\lambda(J(x)B^{-1}(x)J^T(x))} \quad , \tag{37}$$

$$p \geq \frac{\Lambda(N)}{\lambda(J(x)B^{-1}(x)J^T(x))} \quad , \tag{38}$$

we refer to (34) and (35) , jointly with (37) and (38) , as the unit vector adaptive control law [12] . The matrix H , the vectors K_p and K_u and the scalars p and q must be selected by the designer. As far as matrix H is concerned , a deeper insight in the error dynamics (24) is requested. The present approach leads to an asymptotically stable system with two distinct modes in error dynamics depending on the output v of the static compensator D [12]. Constraints imposed by the hyperstability condition (34) , (35) assure that all the trajectories reach subspace $v = 0$ in a finite time interval and then $e \to 0$ asymptotically. The subspace $v = 0$ is attractive and invariant; on this set the control is ideally switched at an infinite frequency; in reality the swi tching occurs at a finite frequency and the trajectories chatter with respect to the subspace $v = 0$. On this subspace the decaying speed of the error components only depends upon the matrices H and A_M. With the choice (20) we have :

$$e^T = [e_1, e_2, \ldots, e_k, \dot{e}_1, \dot{e}_2, \ldots, \dot{e}_k] = [e_p^T, \dot{e}_p^T] . \tag{39}$$

By partitioning the matrix P according to the structure of A_M , it results :

$$D = (0 \mid I) \begin{pmatrix} P_1 & P_2 \\ P_2^T & P_3 \end{pmatrix} = (P_2^T \mid P_3) , \tag{40}$$

if the trajectories belong to $v = 0$, we have :

$$\dot{e}_p = - P_3^{-1} P_2^T e_p . \tag{41}$$

As the reference model is decoupled , it is convenient to select H as a diagonal matrix of assigned trace

$$H = \begin{pmatrix} \mathrm{diag}(h_i) & 0 \\ 0 & \mathrm{diag}(h_{i+k}) \end{pmatrix} , \quad h_i > 0 , \quad h_{i+k} > 0 \quad i = 1, , \ldots , k . \tag{42}$$

Solving the Lyapunov equation (28) , computed (27) from (41) the error dynamics is :

$$\dot{e}_i = - \frac{a_{1i} h_i}{a_{0i} h_{i+k} + h_i} e_i , \qquad i = 1 , \ldots , k . \tag{43}$$

Hence the choice of the elements of H affects the decaying speed of the generalized state error when the system operates on $v = 0$. By noting that the i-th component of the position error goes to zero with the time constant $\frac{a_{0i} h_{i+k} + h_i}{a_{1i} h_i}$, the elements h_i , h_{i+k} are selected prescribing the error transient behaviour in the sliding mode. As regards the parameters q and p they are chosen in accordance with the inequality (37) and (38) only in order to assure the convergence properties of the adaptation mechanism and then the convergence of the error e(t) from the initial condition e(0) to the sliding set $v = 0$; whenever $v \neq 0$ an increase in the value of p and q forces the generalized error to reach the set $v = 0$ faster [12]. Since we are interested in trajectories tracking , we can assume the same initial state on the reference model and on the plant. Taking into account that the energy associated with the forcing signal generated by the adaptation mechanism depends on p and q values , it may be proved expedient to reduce the lower bounds in the inequalities (37) and (38) by a suitable design of feedforward and feedback actions K_u and K_p.

An interesting connection with the techniques based on the compensation of the nonlinearities may be here considered. In order to reduce p and q values it is necessary to lower the maximum eigenvalue of the matrices $M^T M$ and $N^T N$. These eigenvalues might be reduced to zero by choosing K_p and K_u such that m(z)=0 , n(z)=0 , but this should be equivalent to realize nonlinear feedback and feedforward blocks, which exactly compensate the manipulator nonlinearities and simultaneously achieve the decoupling. Also if the manipulator is accurately modeled, the exact nonlinear compensation is involved and of costly implementation; it seems expedient to assume only a partial knowledge of the plant dynamics with an amount of uncertainty to be balanced by UV adaptation law, considerably reducing the values p and q.

As previously noted the Jacobian matrix J, the inertia matrix B and the gravitational torques E are explicit functions of θ with a quite simple structure; assuming that θ is a measurable vector, they can be reasonably evaluated on line. Assuming:

$$K_p = J^{T^{-1}}(\theta)\,[E(\theta) - B(\theta)J^{-1}(\theta)(\Lambda_0 x + \Lambda_1 \dot{x})] \,, \qquad (44)$$

$$K_u = J^{T^{-1}}(\theta)B(\theta)J^{-1}(\theta)u_M \,, \qquad (45)$$

the balancing of the effects of gravity and inertia and the decoupling of the motion components with respect to the task-oriented space coordinates are left to the feedback and feedforward blocks. The task of the adaptive controllers is limited to compensate the terms $\frac{dJ(x)}{dt} J^{-1}(x)\dot{x}$ and $J(x)B^{-1}(x)C(x,\dot{x})$, which are rather involved, and to balance the parametric variations. In (44) and (45) different arguments θ, x are used in order to underline the explicit dependence. By using an inverse model technique the exact input u_M which eliminates tracking errors in nominal conditions can be easily computed. If the desired trajectory is assigned by specifying $\bar{x}(t)$, $\dot{\bar{x}}(t)$ and $\ddot{\bar{x}}(t)$ in the task-oriented space, by means of (19) we obtain:

$$u_M = \ddot{\bar{x}}(t) - \Lambda_1\dot{\bar{x}}(t) - \Lambda_0\bar{x}(t) \,. \qquad (46)$$

In Fig.3 a block diagram of the adaptive controller, resulting from the above considerations, is shown. From this control scheme the following appealing features are apparent.

i) The computer burden is greatly reduced by a suitable use of the state variables both in the task oriented space and in the drive oriented space. The nonlinear transformations from measured values of θ and $\dot{\theta}$ to x and $\dot{x}$ are one to one; the matrices B, J, J^{-1} and the vector E are explicit functions of θ.

ii) The adaptive controller balances neglected terms and parametric variations; as a consequence matrix B and vector E can be computed on line with poor accuracy.

iii) The nonlinear compensation E(θ) and the coupling term BJ^{-1} allow to reduce the effort required of the adaptation mechanism.

iv) The proposed adaptation law generates the control signal:
$u_a = \frac{v}{\|v\|}(q\,\Sigma|z_i| + p\,\Sigma|u_{Mi}|)$; only simple nonlinear operations are required, avoiding the on line solution of differential equations.

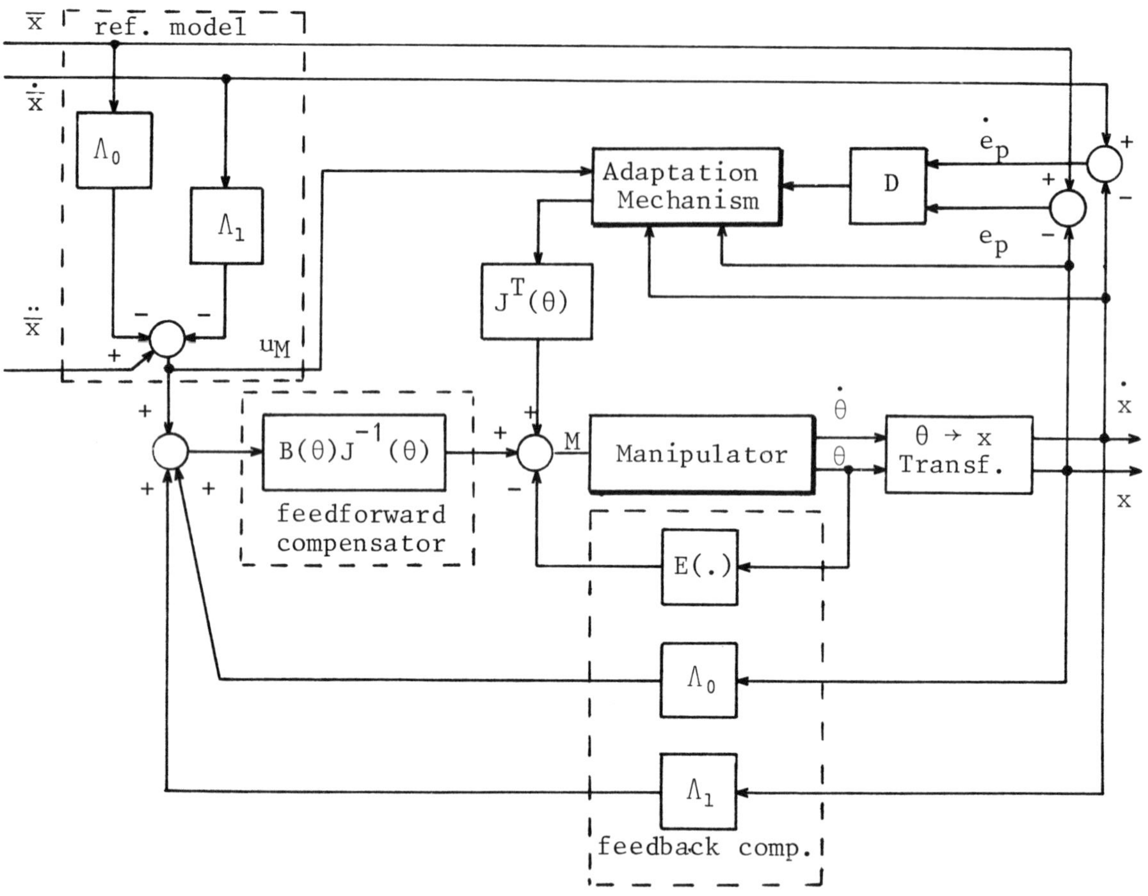

Fig.3. Adaptive manipulator control system with nonlinear feedback and feedforward compensations.

v) The adaptive model following control, with a linear decoupled model in the task-oriented space, allows the user to plan the desired trajectories in the most natural setting.

The major disadvantages are due to the discontinuous nature of the adaptation signal both with respect to its amplitude and with respect to its frequency. As regards the amplitude, saturations may occur, f.i. the torque saturation due to the actuators driving the joints; as regards frequency, the chattering phaenomenon occurs at some finite frequency so that the effects on the plant hardware must be taken into account. In any case, by considering the actuator dynamics, by means of simulation studies it has been shown [11] that the control signal ripple is cut down while the tracking characteristics remain satisfactory.

CASE STUDY AND CONCLUSIONS

A numerical simulation study has been performed to investigate the effectiveness of robotic manipulator adaptive control designed via the above described technique. A manipulator with three degrees of freedom is considered with the following set of parameters : m_1 = 30Kg , m_2 = 18.6Kg , l_1 = l_2 = 0.5m ; where m_i , l_i , i=1 , 2 denote the mass and the length of the i-th link.

The simulations have been carried out to study how the end point of the manipulator follows a desired trajectory given by a 1.5m straight line in the cartesian frame X (see Fig.4) (the task-oriented space) with a trapezoidal velocity law (Fig.5).

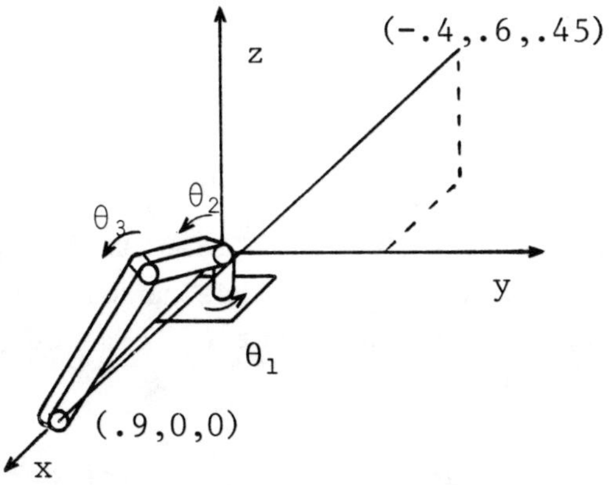

Fig.4. Robot and trajectory for case study.

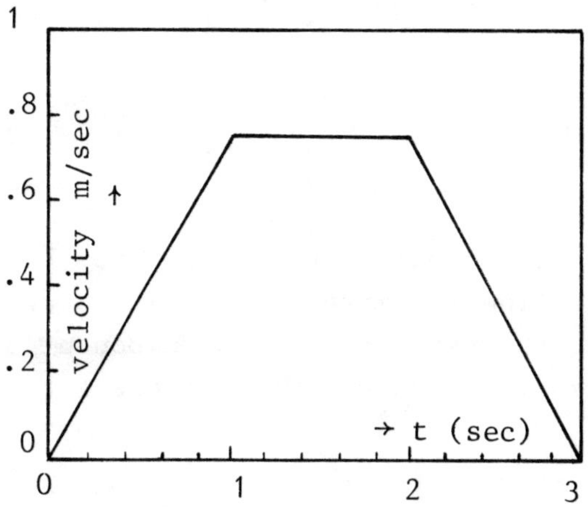

Fig.5. Velocity law.

The entries of the inertia matrix and the gravitational torques vector, obtained by using a code for the automatic generation of dynamical model of manipulators, are the following:

$$b_{11}(\theta) = 23.38 + 10.45\sin^2\theta_2 + 3.7\sin^2(\theta_2+\theta_3) + 7.9\sin\theta_2\sin(\theta_2+\theta_3)$$

$$b_{12}(\theta) = b_{13}(\theta) = 0.$$

$$b_{22}(\theta) = 84.9 + 7.9\cos\theta_3$$

$$b_{23}(\theta) = 3.87 + 3.95\cos\theta_3$$

$$b_{33}(\theta) = 27.03$$

$$Q_1^L(\theta) = 0.$$

$$Q_2^L(\theta) = 213.67\sin\theta_2 + 77.43\sin(\theta_2+\theta_3)$$

$$Q_3^L(\theta) = 77.43\sin(\theta_2+\theta_3) .$$

Note that the entries of B take also into account the contribution of the actuators rotor inertia.

Since we use the inverse model technique to synthesize the control signal u_M, the reference model has been chosen in order to guarantee a good trajectory tracking when the system operates without adaptation signal. We have chosen the same reference model for each linkage with: $\Lambda_0 = -I$ and $\Lambda_1 = -50I$.

In order to test the robustness of the proposed adaptive controller we have utilized, into the compensation and decoupling block of Fig.3, the inertia matrix $B(\theta)$ neglecting the coupling terms $b_{23}(\theta)$ and $b_{32}(\theta)$ and setting $b_{11}(\theta) = 23.38$, $b_{22}(\theta)=84.9$

and $b_{33}(\theta)=27.03$.

The matrix H of the Lyapunov equation (28) has been chosen as :

$$H = \begin{pmatrix} 30 & & & & & \\ & 90 & & & & \\ & & 30 & & & \\ & & & 5 & & \\ & & & & 15 & \\ & & & & & 5 \end{pmatrix},$$

where the values of the entries have been selected in order to give adequate gains to the linear compensator D. The choice of the matrix H and the reference model imposes for the first and third component of the state error a time constant of 23msec. , whereas the second component has a time constant of 14msec.

The effectiveness of the adaptive control may be evaluated by considering the trajectory tracking error computed as the distance σ of the end point from the straight line on which the target trajectory lies and the spatial delay Δ along the direction of the trajectory. In Fig.6 the trajectory tracking error is reported when the system operates without adaptation signal , whereas Fig.7 shows the tracking error when an adaptation signal is synthesized according to (34) and (35) with q=5 and p=8.

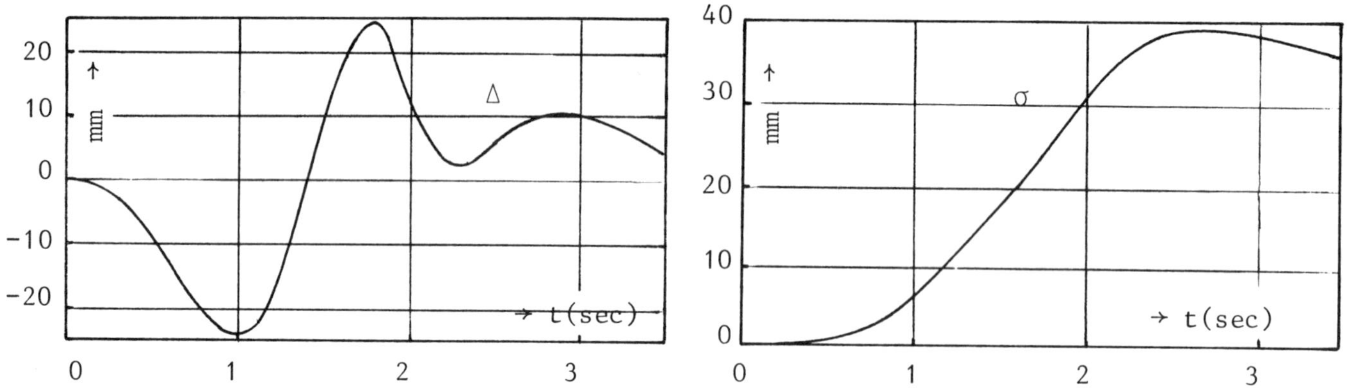

Fig.6. Tracking errors Δ,σ without adaptation algorithm.

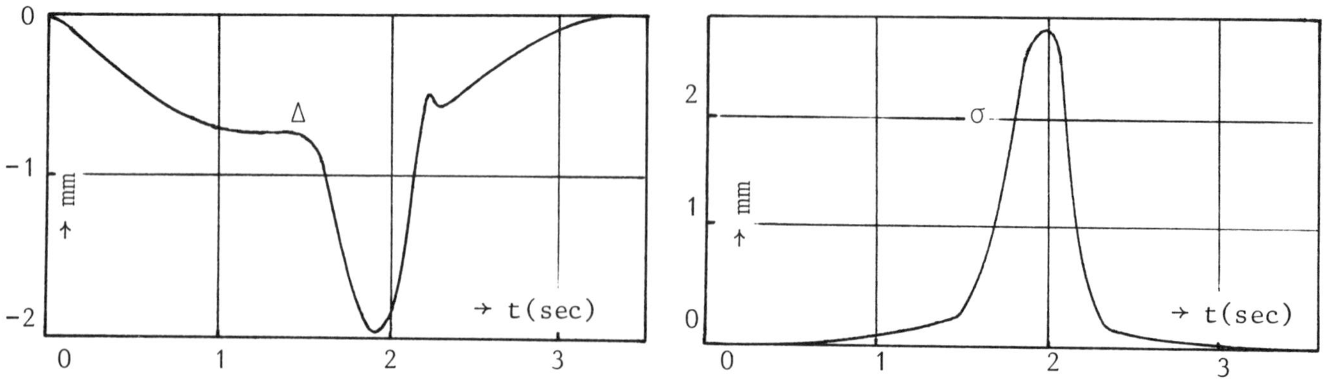

Fig.7. Tracking errors , with adaptation algorithm (p=8 , q=5)

The comparison between these results shows the effectiveness of the adaptive controller of the unit-vector type. As regards the errors in presence of the adaptation signal, they ought to be theoretically zero, since the reference model and the manipulator have the same initial conditions; but it must be taken into account that, whereas the adaptive control procedure has been developed for the continuous time system, actually the simulated system is a sampled data system as the simulations work with a sampling period equal to 1msec., so that the errors are to be ascribed to the sampling effects. It must be noted that such errors are in any case very small.

Fig.8 shows the control u_{p2} applied at the second joint, which consists of a low frequency (average) component, corresponding to the nominal torque required to follow the desired motion, and a high frequency (chatter) component, the envelope of which is drawn in black.

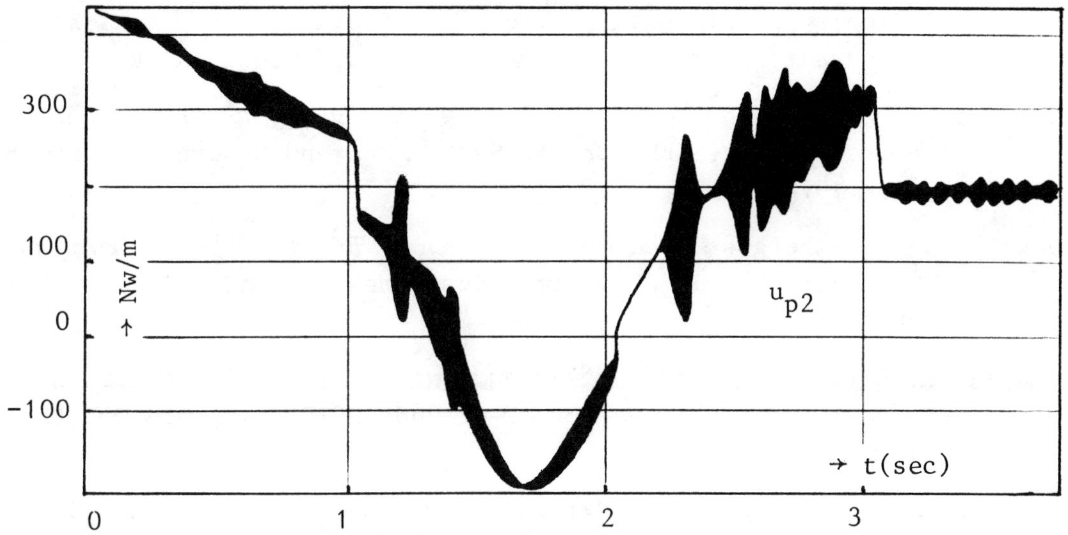

Fig.8. Driving torque at manipulator 2nd joint.

Finally the responses of the manipulator adaptive control system have been evaluated when a 5Kg payload is considered. Reference model response and manipulator response are indistinguishable, which implies that the system is insensitive to parametric variations.

From the above results the unit vector adaptive control is accurate in cartesian motion control for any kind of load, achieves the decoupling and automatically compensates for changing configurations and gravity loadings. As the scheme only uses simple matrices (the Jacobian matrix and an approximate inertia matrix and gravity loading vector) it is applicable to control manipulators of more than three degrees of freedom.

BIBLIOGRAPHY

[1] A.K. Bejczy, *Robot arm dynamics and control*, Jet Propulsion Lab. Tech. Memo. 33-669, Feb.1974.

[2] M. Vukobratovich, *Dynamics of active articulated mechanism and synthesis of artificial motion*, Mechanism and Machine Theory, vol.13, 1-18, 1978.

[3] R. Paul , *The mathematics of computer controlled manipulators*, Proc. JACC , vol.1, 124-131 , 1977.

[4] D.E. Whitney , *The mathematics of coordinate control of prosthesis arms and manipulators*, Trans. of ASME J. of Dynamic Syst. Measur. and Control , Dec. 1972.

[5] J.Y.S. Luh , M.W. Walker , R. Paul , *Resolved acceleration control of mechanical manipulators*, IEEE Trans. Autom. Contr. , vol.AC-25 , n.3 , June 1980.

[6] K.K.D. Young , *Control and optimization of robot arm trajectories*, IEEE Symp. on Autom. Computation and Control , Milwakee 1976.

[7] E. Freund , *A nonlinear control concept for computer controlled manipulators*, IFAC Symp. on Multivariable Control Systems , Toronto 1977.

[8] D. Lentini , F. Nicolò , S. Nicosia , L. Vecchio , *Decoupling and nonlinearities compensation in robot control systems*, Industrial Robot 1981.

[9] K.K.D. Young , *Controlled design for a manipulator using theory of variable structure systems*, IEEE Trans. Syst. , Man and Cybern., vol.SMC-8 , 108-119 , 1978.

[10] S. Dubowsky , D.T. DesForges , *The application of MRAS to robotic manipulator*, Trans of ASME J. of Dynamic Syst. Measur. and Contr. , vol.101 , 193-200 , 1979.

[11] G. De Maria , L. Sciavicco , *Performance evaluation of robotic manipulators with adaptive controllers*, Proc. of 2nd IASTED Symp. on Robotics , Davos 1982.

[12] A. Balestrino , G. De Maria , L. Sciavicco , *Hyperstable adaptive model following control of nonlinear plants*, Systems & Control Letters , vol.1 , n.4, 232-236 , 1982.

[13] C.H. Wu , R. Paul , *Resolved motion force control of robot manipulator*, IEEE Trans on Syst. , Man and Cybern. , vol.SMC-12 , n.3 , 1982.

[14] M. Takegaki , S. Arimoto , *An adaptive trajectory control of manipulators*, Int. J. on Control , vol.34 , n.2 , 1981.

[15] I.D. Landau , B. Courtiol , *Design of multivariable adaptive model following control system*, Automatica , vol.10 , 483-494 , 1974.

[16] A. Balestrino , G. De Maria , L. Sciaviccco , *Feedback and feedforward in AMFC of a class of nonlinear time varying plants*, Proc. of Workshop on Adaptive Control , Florence 1982.

Hierarchical Control for Robots in an Automated Factory

James S. Albus
National Bureau of Standards
Washington, District of Columbia

Charles R. McLean
National Bureau of Standards

Anthony J. Barbera
National Bureau of Standards

M.L. Fitzgerald
National Bureau of Standards

Abstract

A hierarchical architecture for real-time sensory-interactive control of robots, machine tools, inspection machines, and materials transport and inventory systems is described in this paper. Computer-aided design, computer-aided process planning, and management information systems make up the top level in the hierarchy, where the highest level goals are selected and the longest-range planning horizons exist. Commands and goals generated at this highest level are decomposed through a series of levels (SHOP, CELL, WORKSTATION, and EQUIPMENT) until at the lowest level there are generated a series of drive signals to individual actuators on robots, machine tools, and other equipment. Feedback from sensors and from the control hierarchy itself are used at all levels to produce real-time goal seeking behavior.

INTRODUCTION

The basic structure of a hierarchical control system is a tree, wherein each computational module has a single superior, and one or more subordinate modules. The top module is where the highest level decisions are made and the longest planning horizon exists. Goals and plans generated at this highest level are transmitted to the next lower level where they are decomposed into sequences of subgoals. In general, the decomposition at each level takes into account information derived from: (a) processed input data from sensors that measure the state of the environment, (b) reports from lower control levels as to the state of the control hierarchy itself, and (c) predictions (or expectations) generated by models, knowledge bases, or inference engines.

At each level, input commands from the next higher level are decomposed into sequences of output sub-commands to the next lower level in the context of the state of the environment, of the state of the control system, and the internal store of knowledge. At each level predictions and expectations are generated by the internal world model in the context of the state of the task, the goal of the system, and the best current hypothesis about the state of the environment. Also at each level, processed signals from the environment are compared against expectations from the world model. Correlations are computed and differences measured between observation and expectation. A high degree of correlation indicates that the task is proceeding according to plans, and expectations are being met. Differences represent error signals which can be used by the task decomposition module either to modify behavior, or change expectations so that the task goal is successfully accomplished. At the highest level, the input command represents the ultimate goal of the entire organism. At the lowest level, output drive signals are computed and sent to the physical actuators.

A hierarchical architecture for a factory control system is shown in Figure 1. A single chain of command from the bottom to the top of such a hierarchy is outlined by the dotted line in Figure 1. This chain of command can be further segmented, as shown in Figure 2, into three separate hierarchies: (1) a goal, or task decomposition, hierarchy (H); (2) a feedback processing hierarchy (G); and (3) a world model hierarchy (M). This has been discussed in a number of previous papers [2,3,4,5].

In general, each of the hierarchical levels shown in Figure 1 can be further partitioned into sublevels. For example the control hierarchy for a machining work station robot shown in Figure 2, decomposes the equipment level into three sublevels. An assembly robot might decompose the equipment level into four or more sublevels. The number of sublevels required depends on the complexity of the tasks that must be decomposed at that level.

At all levels, the H, G, and M modules are concurrent processes produced by real-time programs executing simultaneously in each module. Perhaps the simplest way to treat this conceptionally is to model each of the modules in the hierarchy as a finite-state automaton. The state-graph describing the activity of the entire hierarchy can then be described by a Petri diagram [1].

For each module in this architecture there are three concepts of time: the planning horizon, the response time, and the cycle time. The planning horizon is the interval over which a control module plans into the future. The response time is the delay between a change in a module's input and the generation of a new output. The cycle time is the period between sampling the input variables. In general, the response time will be slightly longer than the cycle time, and the planning horizon will be many times longer than the response time.

The response time of the finite-state automata at each level depends on the requirements for stability and dynamic response at the respective levels. The response time requirement is shorter at the lower levels, but the complexity of the control computations is less. The response time is longer at the higher levels, and the complexity of the computations is greater. Thus, the total computational power required at any level of the hierarchy is more or less constant.

Communication between the various modules in such a system can be accomplished by writing messages in a data base which is common to all modules which either compute or make use of those messages. Each message area (or mailbox) within the data base can be restricted so that only one system may write into it, although many can read its contents. If the cycles of the state-clock at all levels are synchronized, information transfer into and out of the common data base will occur at predictable time increments and each message can carry a time tag.

TASK DECOMPOSITION

In the robot control system architecture shown in Figure 2 the bottom (or first) level of the task decomposition hierarchy is where coordinate transforms and servo computations are made, and all joint motions are scaled to hardware limits on velocity and force.

At the second level, elemental movements such as <REACH TO (A)>, <GRASP>, <LIFT>, <ORIENT ON (B)>, <MOVE TO (X)>, <RELEASE>, etc. are decomposed into force and velocity trajectories in a convenient coordinate system. That coordinate system may be defined in the robot's work space, in the part, or in a coordinate frame in the robot's gripper.

At the third level, simple tasks such as <FETCH (A)>, <MATE (B) TO (A)>, <LOAD TOOL (C) WITH PART (D)>, etc. are decomposed into elemental movements which can be interpreted by the second level.

In the Automated Manufacturing Research Facility (AMRF) currently under construction at the National Bureau of Standards [8], a machining workstation robot will receive input commands to the third level of its hierarchical control system from a WORKSTATION CONTROLLER, which is the fourth level in the robot's hierarchy, as indicated in Figure 2.

A typical machining workstation in the AMRF will consist of a robot, a machine tool, a work tray buffer, and several tools and sensors that the robot can manipulate. Trays of parts and tools will be delivered to the workstation by a robot cart. The workstation controller will be given commands consisting of lists of operations to be performed on the parts in the trays. It is the task of the workstation controller to generate a sequence of simple task commands to the robot, the machine tool, and any other system under its control so that the set of operations specified by its input command list are carried out in an efficient sequence. For example, the workstation controller may generate a sequence of simple task commands to the robot to set up the clamping fixtures for the first part; to the machine tool to perform the specified machining operations; to the robot to modify the clamping fixtures for the next job, etc. The planning horizon for the workstation may vary from an hour or two up to about a day, depending on the complexity and number of parts that are being processed.

The fifth level of the robot control hierarchy in Figure 2 is the CELL CONTROLLER which is responsible for managing the production of a batch of parts within a particular group technology part family. The task of the cell is to group parts in trays and route the trays from one workstation to another. The cell generates dispatching commands to the material transport workstation to deliver the required tools, fixtures, and materials to the proper machining workstations at the appropriate times. The cell must have planning and scheduling capabilities to analyze the process plans for each part, the tooling and fixturing requirements, and the machineability time estimates for each operation. It will use these capabilities to optimize the make-up of trays and their routing from workstation to workstation. The planning horizon for the cell will depend on the size and complexity of the batch of parts in process, but may be on the order of a week.

The sixth level in the robot control hierarchy is the SHOP CONTROLLER which performs long term production planning and scheduling. It also manages inventory, and reports shortages in materials and tools to the next higher level (Facility Control) where orders are issued to outside vendors. The planning and scheduling functions are used to determine the workstation, robot, and material resources requirements for each cell. The shop then dynamically allocates workstations to, or reclaims them from the cells as necessary to meet the production schedule [7]. The control structure always satisfies the rules of a hierarchical tree at any instant in time, but the subtrees of the shop may shift from one moment to the next. For example, an individual robot may belong to Workstation #1 of Cell #26 at one moment, and to Workstation #16 of Cell #2 the next. The same logic can be applied to the passing of workstations between cells. This degree of flexibility becomes important in factories or construction sites where robots are mobile and rapidly move from one physical work site to another.

The seventh level is FACILITY CONTROL. It is at this level that engineering design is performed and the process plans for manufacturing each part, and assembling each system, are generated. Here also, management information is analyzed, materials requirements planning is done, and orders are processed for maintaining inventory. Because of the very long planning horizons at this level in the control hierarchy, the activities of the facility control module are not usually considered to be a part of a real-time control system. However, in the context of hierarchical control with exponentially increasing time horizons at each higher level, these facility control activities can be integrated into the real-time control hierarchy of the manufacturing system.

FEEDBACK PROCESSING

Each level of the task decomposition hierarchy is serviced by a feedback processing module which extracts the information needed for control decisions at that level from the sensory data stream and from the lower level control modules. The feedback processing modules at each level detect features, recognize patterns, correlate observations against expectations, and format the results to be used in the decisions and computational procedures of the task decomposition modules at that level.

At the lowest level of the robot hierarchy, the feedback processing modules extract and scale joint positions and force and torque data to be used by the servo and coordinate transformation computations.

At the second level, touch and proximity data, and simple visual measurements of distance and positions of grip points are extracted from the sensory input to be used in computing trajectory end points.

At the third level the three dimensional positions of visual features such as edges, corners, and holes are computed and combined to determine the position and orientation of surfaces and volumes of objects. Identities of objects may also need to be computed (or recognized) in order to generate the reaching and grasping commands at this level.

At the fourth (WORKSTATION) level, relationships between various objects need to be determined, in order to sequence simple task commands.

At the fifth (CELL) level, the location and composition of trays of parts and tools and the length of queues of parts needs to be determined. This may be derived from sensors which read coded tags on trays, or may be inferred from sensory input from lower level sensors on the robot or in the workstation.

At the sixth (SHOP) level, the condition of machines, tools, and the amount of inventory on hand must be determined in order to generate schedules, allocate resources, and evaluate and set priorities for production.

At the seventh (FACILITY) level, the requirements for changes in part design, or in process plans need to be recognized in order to make engineering changes, or redesign parts or processes.

THE WORLD MODEL

The world model hierarchy, made up of M modules in Figure 2, consists of a knowledge base containing all the information currently known about the task, the parts, or the workplace. The M modules also contain procedures that allow them, based on the state of the task and other contextual information from various places in the hierarchy, to compute expectations and predictions about what the sensory data to the corresponding G module should be. This allows the G modules at each level to compare expectations with observations, and to measure both the degree of correlation and the degree of difference. A strong degree of correlation means that the proper model is being matched with the incoming sensory data. It means that the observed object or situation has been correctly recognized, and that information contained in the model can be safely used for decision making even though it may not be directly observable by the sensory system.

A large degree of difference between expectations generated by the model and observations derived from sensors means that either an incorrect choice of models has been made, or the model has not been correctly transformed spatially or temporally so as to generate the proper set of expected feature relationships, or that the incoming sensory data is too noisy, or is being improperly processed and filtered. In this case, the computational problem for the task decomposition module is to decide which type of error is being encountered and what is required to remedy the discrepancy. In general, this type of problem can be solved either by a set of situation/action rules of an expert system, or a set of heuristic search procedures.

At low levels, the world model contains dimensional information describing the shapes and sizes of parts. The M modules use this to generate expected positions of image features such as edges, corners, holes, and surfaces.

At the facility control level the model contains information about machining processes, material properties, shop processing capabilities, and expected lead times for procurements which can be used to compute estimated completion times for various production plans.

At the shop level, the world model contains information about machine capabilities, machineability of materials, tool life, and inventory levels and is able to simulate the performance of various cell configurations.

At the cell level, the model contains information about workstation task times, and is able to simulate the performance of various hypothetical task sequences.

At the workstation level, the world model contains knowledge of tray layouts including the names of parts and their approximate positions, orientations, and relationships such as on-top-of, underneath, stacked N-deep, leaning-against, etc.

At the simple task level, the model contains knowledge of the geometrical size and shapes of three dimensional objects such as parts and tools and the relationships between coordinate systems based in the work space and the robot. These can be used to generate expected positions and orientations of three dimensional objects in a robot or machine tool coordinate system.

At the elemental move level, the model is able to generate expected positions and orientations of specific features of parts and tools, such as edges, corners, surfaces, holes, and slots [6].

At the coordinate transformation and servo level, the model generates windows or filter functions that are used to screen and track the incoming raw data stream.

PLANNING IN A HIERARCHICAL CONTROL SYSTEM

There can be three different modes of operation of a hierarchical control structure such as shown in Figure 2.

The first is the control execution mode. In this mode the task decomposition hierarchy is used to decompose tasks into subtasks in the generation of behavioral action. The world model hierarchy is used to generate expectations based on the state of the task. The sensory processing hierarchy compares these expectations against observed sensory data in evaluating the results of actions. Processed feedback information is used to servo the control system to successfully accomplish its goal.

A second mode of operation is the sensory recognition mode. In this mode the control hierarchy is used to generate hypotheses which drive the world model hierarchy in the creation of expectations. The sensory processing hierarchy compares these expectations against observed data in order to recognize and evaluate objects, situations, and events.

A third mode of operation is the planning mode. In this mode the control hierarchy is used to generate hypothesized tasks and task decompositions. These drive the world model hierarchy to create expected hypothetical results. The sensory processing hierarchy evaluates these hypothetical results and assigns a desirability (or cost) value to the hypothesized task decompositions. By this mechanism a search can be conducted over the space of potential task decompositions to find an optimal coarse of action prior to the initiation of behavioral output.

At each level of the control hierarchy, a computing module such as shown in Figure 3 can be used to execute the production rules that encode the control program at that level. A state-graph (or Petri diagram) corresponding to the rules in the state-transition table is the analog of the flow chart of a procedural program for the task decomposition module [1]. The left-hand side of the table consists of all the command, internal state, and feedback inputs that can be encountered at any tick of the state-clock. The right-hand side contains an output command (and/or a pointer to a procedure which computes an argument which becomes part of the output command) to the next lower level. It also contains a next internal state, and a report to the next higher level, or to other modules at the same level.

At the lowest hierarchical level, the left-hand side of the state-transition table consists of variables which select the type of coordinate transformation required and the type of servo computations needed.

At the second level, the left-hand side consists of variables which define the type of trajectories to be generated. The right-hand side contains pointers to procedures that compute forces, positions, accelerations, and velocities in the appropriate coordinate systems.

At the third level, the left-hand side consists of variables which specify the state of the environment as reported by sensors, and the right-hand side the names of appropriate elemental movements to be made for each state. Pointers to procedures are used to compute arguments and modifiers.

At the higher levels, the state-tables may be compared to production rules in expert systems. Procedures that are invoked by these state-tables may consist of heuristic search algorithms or linear programming techniques for generating plans, schedules, etc.

The response time and cycle time requirements grow longer for the finite-state automata at the upper levels of the hierarchy. Thus, the amount of computing power needed in the execution mode to execute state-transition tables decreases at higher levels in the hierarchy. On the other hand, there is much more need for planning at the upper levels. For example, the types of control decisions required at the upper levels of the factory control system shown in Figure 2 typicaly involve planning algorithms. The hierarchical control system proposed here thus provides the required planning capability. The upper levels of the control hierarchy can use the excess execution mode computing power to operate in the planning mode.

CONCLUSIONS

There appear to be a number of advantages of the table-driven hierarchical control system described above. The first is that it partitions the problem into simple, well-defined modules with clearly specified inputs, outputs, internal states, and rules for state-transitions. The control problem is partitioned vertically with respect to task complexity and abstraction, horizontally with respect to function (such as task decomposition, sensory processing, and world modeling), and along the time axis by the use of a state-clock. This simplifies the design and eases the synchronization of simultaneous processes in the many different computing modules.

The second advantage is that it facilitates the handling of error conditions. If additional states need to be defined to deal with unanticipated error conditions, they can simply be inserted into the state-graph and added to the state-transition table. There is very little interaction with other parts of the program code. A simple interpreter can insert the additional steps into the state-tables as fast as they are entered from the keyboard. Program executions can then continue from the point where the error was encountered.

Third, this approach formalizes the control problem into a very orderly structure. Each line in the state-transition table for any module is an IF/THEN production rule. <IF (the command is such, and the state is so, and the feedback conditions are thus) / THEN (the output is whatever is stored on the right hand side of the table, and the system steps to the indicated next state)>. The addition of each node or edge to the state-graph, and the corresponding lines added to the state-transition table is the equivalent of the addition of a new chunk of knowledge about how to deal with a specific control situation at a particular point in a problem domain at a unique phase in the task execution. This system architecture thus bridges the gap between servomechanisms and finite-state automata at the lower levels and expert system technologies at the upper levels. The fact that even very powerful expert systems typically do not contain more than a thousand production rules suggests that the size of the state transition tables for sensory interactive robot control systems will not become excessive, even at the higher levels.

Fourth, it allows the easy insertion of conditional tests of new sensory or feedback data. For example, if a new touch sensor is added, it is possible to merely insert a new column in the feedback portion of the state-transition table, and a new line, or lines, in the state-transition table for each of the new edges to be added to the state-graph.

Fifth, it facilitates debugging. Each state of the system is formally identified and the set of conditions that lead to and from that state are clearly specified. Diagnostic routines can be used to read these states from common memory and stop the state-clock when necessary. This makes it easy to perform traces, to set break points, and to reason backwards from error states. The system is completely deterministic and errors in logic are simple to reconstruct. Program bugs are therefore relatively easy to locate and correct.

Sixth, it makes it possible to build teaching and learning capabilities into sensory-interactive robot control systems. Simple programs with few or no error conditions can be defined first by state-graph flowcharts. The system can then be run until it encounters unanticipated problems or undefined conditions which cause an <ERROR STOP>. Then each problem condition can be dealt with specifically by adding lines to the state-transition table to address that particular problem state. Eventually the entire space of problem states will contain programmed solutions, and the frequency of <ERROR STOP>'s will decline. The rule-based approach facilitates the integration of an expert system which would have the capability to use its higher "learning" rules to modify the world model, sensory processing, or decision structures, as experience on different problem states is acquired.

There are, of course, many unanswered questions regarding this approach. The hierarchical control system architecture described in this paper is still largely a theoretical construct. There are many problems that remain to be addressed and many of the details of this method will undoubtedly change as more experience is acquired during the construction and testing of the experimental system.

BIBLIOGRAPHY

1. Albus, J.S., Barbera, A.J., Fitzgerald, M.L. (1982). Programming a Hierarchical Robot Control System. _Proc. 12th International Symposium on Industrial Robots,_ Paris, France, 9-11 June 1982.

2. Albus, J.S., Barbera, A.J., Fitzgerald, M.L. (1981a). Hierarchical Control for Sensory Interactive Robots. _Proc. 11th International Symposium on Industrial Robots,_ Tokyo, Japan, 7-9 October 1981.

3. Albus, J.S., Barbera, A.J., Fitzgerald, M.L., Nashman, M. (1981b). Sensory Interactive Robots. _Proc. of 31st General Assembly, International Institution for Production Engineering Research (CIRP)_, Toronto, Canada, September 1981.

4. Albus, J.S., Barbera, A.J., Nagel, R.N., (1981c). Theory and Practice of Hierarchical Control. _Proc. 23rd IEEE Computer Society International Conference_, September 1981, 18-39.

5. Barbera, A.J., Fitzgerald, M.L., Albus, J.S. (1982). Concepts for a Real-Time Sensory-Interactive Control System Architecture. _Proc. 14th Southeastern Symposium on System Theory_, Blacksburg, Virginia, April 1982, 121-126.

6. Kent, E.W. "A Hierarchical, Model-Driven, Vision System for Sensory-Interactive Robotics." _Compsac 82_, Chicago, IL, November 11, 1982.

7. McLean, C.R. (1982). The Virtual Manufacturing Cell. _Proc. 4th IFAC/IFIP Symposium on Information Control Problems in Manufacturing Technology_, Gaithersburg, Maryland, 26-28 October 1982.

8. Simpson, J.A., Hocken, R.J., Albus, J.S. (1982). The Automated Manufacturing Research Facility of the National Bureau of Standards. _Journal of Manufacturing Systems_, Society of Manufacturing Engineers 1:(1)17-31, 1982.

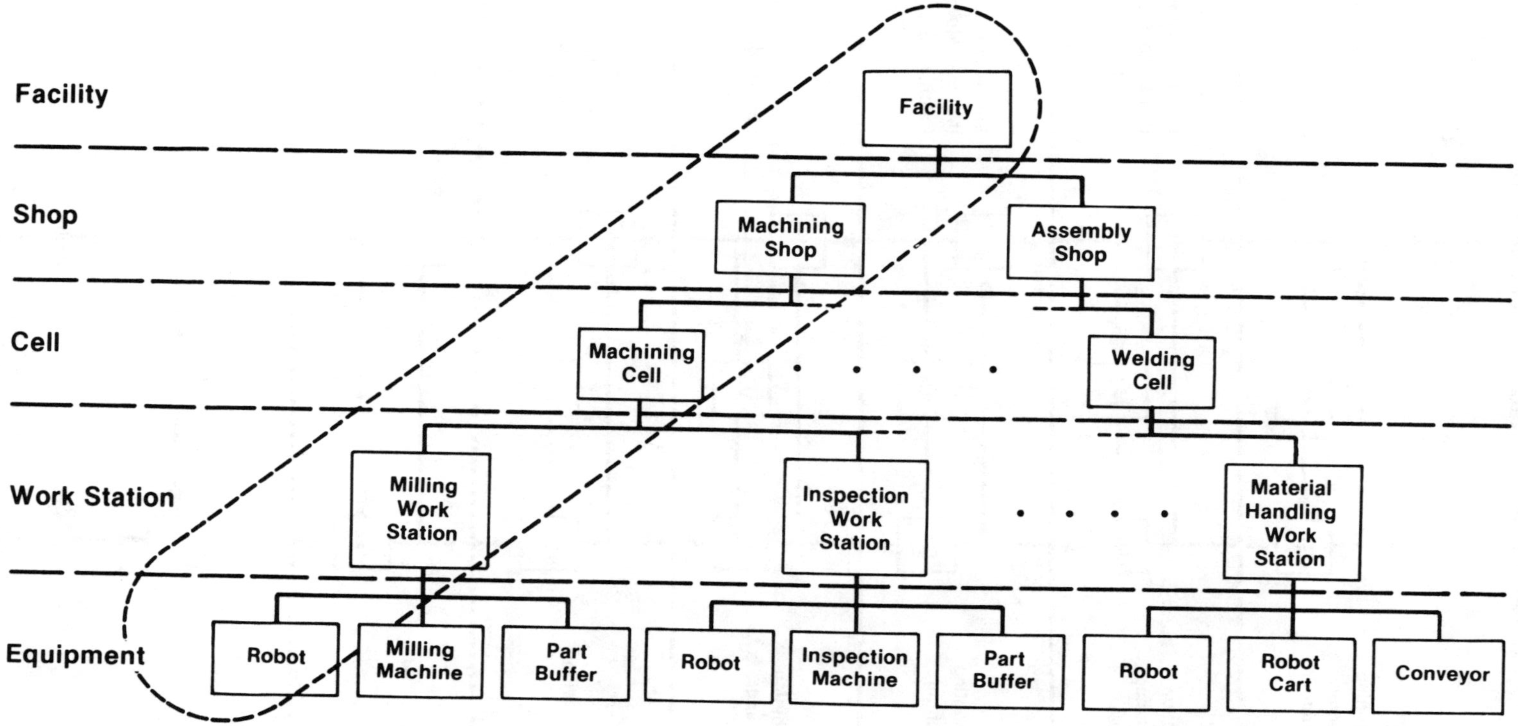

Figure 1. A hierarchical control system architecture for integration of robots, machine tools, and material transport facilities into an automatic factory. The chain of command enclosed in the dotted lines is detailed in Figure 2.

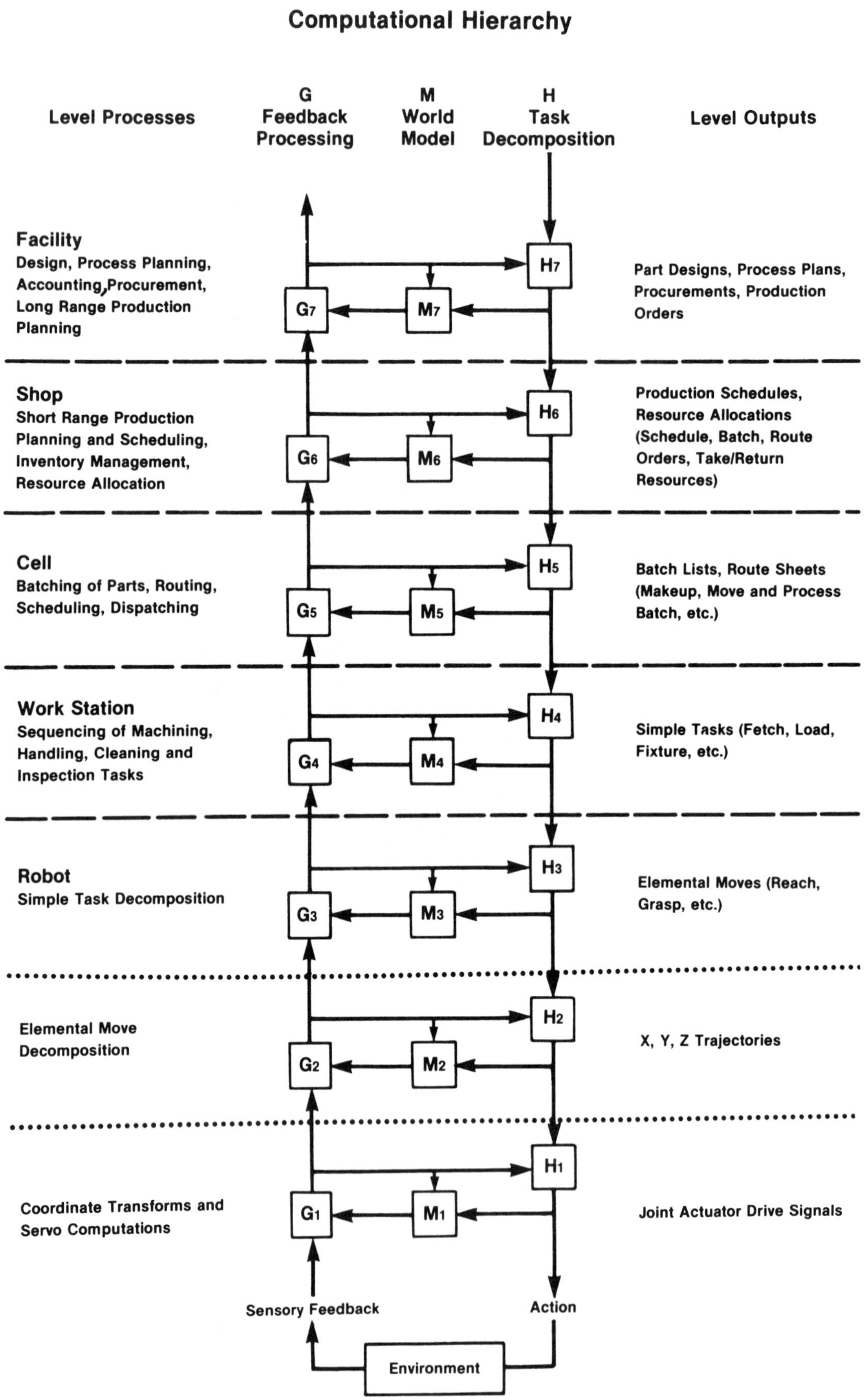

Figure 2. The computational hierarchy for a robot in a machining workstation. This hierarchy corresponds to the chain of command enclosed in dotted lines in Figure 1.

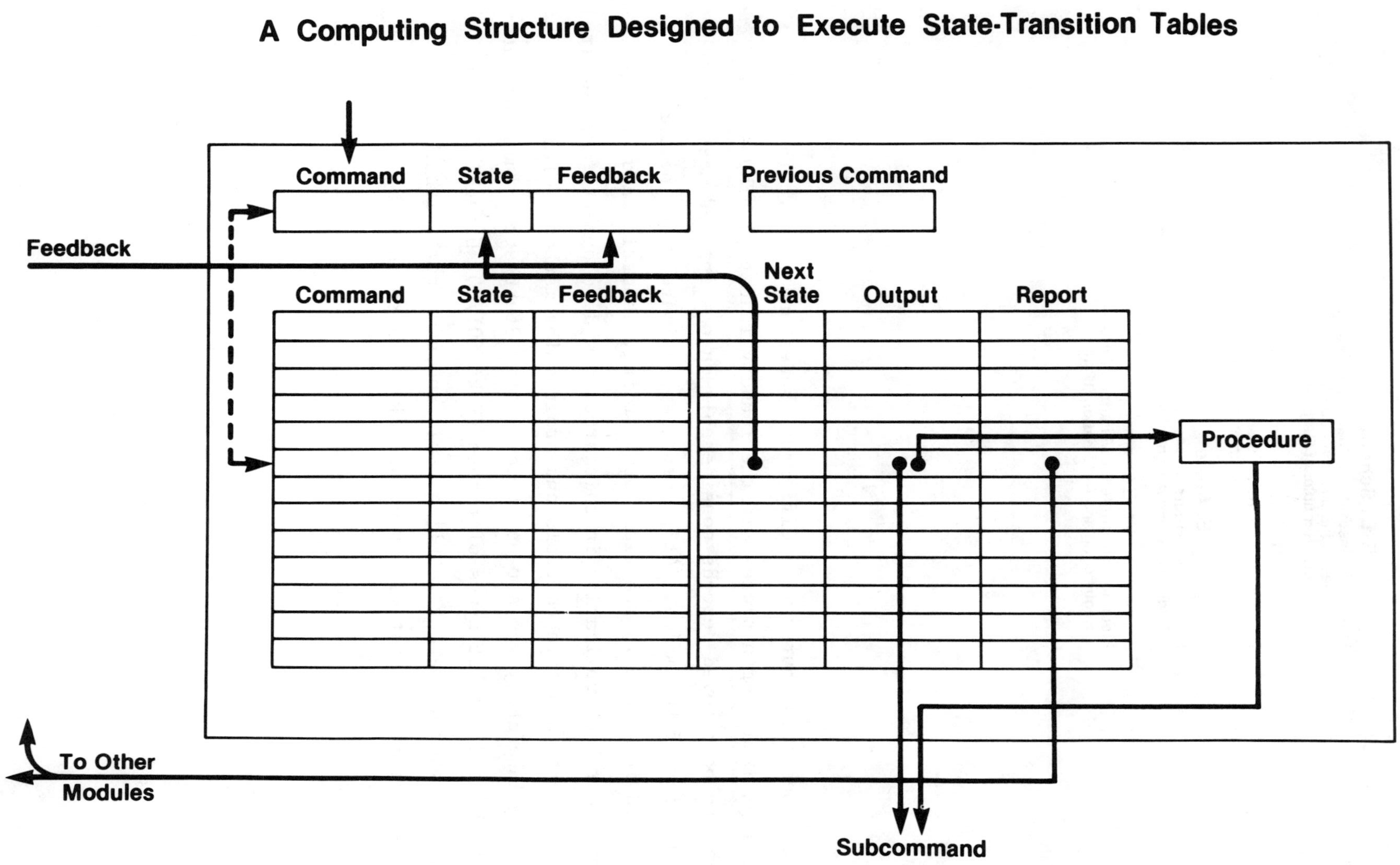

Figure 3. A computing structure to execute state-transition tables.

Application of Sensor Controlled Robots for Fettling of Castings

R.D. Schraft
Fraunhofer Institute for Manufacturing Engineering and Automation (IPA)
Federal Republic of Germany

M. Schweizer
Fraunhofer Institute for Manufacturing Engineering and Automation (IPA)

E. Abele
Fraunhofer Institute for Manufacturing Engineering and Automation (IPA)

W. Sturz
Fraunhofer Institute for Manufacturing Engineering and Automation (IPA)

Abstract

Many casting processes cause burrs with different shapes and sizes. Therefore, the cleaning of castings is a common process in the foundry industry. The size and form of the burrs to be removed depends considerably on the applied casting technology and the quality of the molds. At present, approximately 80 industrial robots are installed in Europe in the field of fettling, deburring and brushing. In order to increase robot application in this area, new developments are required in the field of sensors, adaptive controls and programming methods. For these reasons, a fettling laboratory was set up at the IPA in Stuttgart. This laboratory is equipped with three different types of robots (Kuka 601/60, Hitachi Process-Robot, Puma 600) and many different fettling tools and sensors. With this equipment, it was possible to test different adaptive control strategies and programming methods.

ABSTRACT

Almost all casting processes cause burrs with different shapes and sizes. Therefore, the cleaning of castings is a common process in the foundry industry. The size and form of the burrs to be removed depend considerably on the applied casting technology and the quality of the molds.
At present, approximately 8o industrial robots are installed in Europe in the field of fettling, deburring and brushing.

In order to increase the number of robot applications in this area, new developements are required in the field of sensors, adaptive controls and programming methods.

For these reasons, a fettling laboratory was set up at the IPA in Stuttgart. Right now, this laboratory is equipped with 3 different types of robots (Kuka 6o1/6o, Hitachi Process-Robot, Puma 6oo) and a considerable number of different fettling tools and sensors.
With this equipment, it was possible to test different adaptive control strategies and programming methods.

This contribution provides a survey about possible sensing methods and about the use of this sensor information for the different methods of controlling the cutting process (control of cutting speed, cutting depth, force etc.). Elastic suspensions for cutting tools or workpieces with adjustable and controlable elasticity will bei discussed.

In order to measure the height of the burr during grinding, a special tool-sensor unit was developed which allows

- easy programming of a contour by tracing a manually grinded workpiece (master part) and
- compensation of tool wear.

The function of this sensor, the signal processing as well as the range of applications will be described.

1. DISTURBANCES IN THE PROCESS OF FETTLING

Today, the use of industrial robots (IR) for the fettling of castings is still quite rare, compared to IR-applications in other fields /2/.

In Europe, it is estimated that 8o IR's are installed in the field of fettling and deburring (Jan. 1983). Main difficulties for the robot application in this field are:

- the variety of tasks (cutting burrs at the dividing line, cutting runners, grinding of irregular surfaces etc., see Fig. 1) /1/
- the variety of castings (small and medium-sized batches)
- castings differ in geometry and in the size of the burr. Since castings do not have an exact reference plane for clamping, the clamping tolerance is a great problem for the automation of the fettling process. The different types of tolerances are shown in Fig. 2.

An analysis of over 1oo representative workpieces in 11 foundries shows that with larger workpieces deviations of ± 1o mm at the edges and surfaces to be machined may occur, which are mainly caused by swelling of the mould, shrinkage of the casting and the type of heat treatment. In the case of GGG castings the appropriate standard /8/ in the nominal size range 125o - 16oo mm even allows tolerances of up to ± 19 mm.

The amount of shape clamping tolerances and the size of the cross-section of the burr depends on the applied moulding technique and usually increases with the rising weight of workpieces. Fig. 3 shows the results of measurements carried out on typical workpieces from three different weight groups. 21% of the workpieces had burr sizes of over 8 mm^2 and 3o% had burr sizes of 2-8 mm^2. Typical values are:

T_{wr} = o - 5o mm, and

T_v = o.o2 - o.o8 mm/N for elastic displacement-

These tolerances and disturbances, shown in Fig. 2 may be considered either by elastic suspension or by adaptive control (sensors).

2. ELASTIC SUSPENSIONS

In cases where the disturbances are relatively small deburring can be carried out without sensor or without adaptive control. For these cases the tool is fixed to the industrial robot by means of rubber elements or by pneumatic springs (Fig. 4). Using different programmable variables (e.g. stiffness) the tool-suspension can be optimized for different contour-sections. The flexibility of the suspension of the tool has to be adapted to the tool and the workpiece. It is important that a vibration of the tool can be avoided by special elements or damping elements. The advantage of this simple and cheap system is connected with the disadvantage that no reliable control of the process is possible.

The programming is complicated and laborious because all possible burrs with their largest possible size have to be taken into consideration. Since the cutting speed depends on the largest possible burr, the machining time for smaller burrs is not optimized.
Which such elastic suspensions workpieces can be deburred which are produced by die-casting or by pressure die-casting (Fig. 5) and which show small geometrical inaccuracies. Also, machining burrs can be removed. Workpieces cast in sand often show varying burrs and significant deformation so that simple systems like the one described above can hardly be applied.

3. ADAPTIVE SYSTEMS

3.1 Structure of Adaptive Systems

Starting from a certain size of the tolerances shown in Fig. 2 the cutting process has to be adapted to the respective conditions. Such an automatic control or optimization of the cutting process is called adaptive control. The basic set-up of an ACC-system (ACC = Adaptive Control Constraint) is shown in Fig. 6. Sensors control and describe the actual state of the process /2/. According to their targets, adaptive systems are divided into geometrical and technological AC-systems. Geometrical AC-systems effect the working process in order to achieve a certain geometry for the workpiece (e.g. a minimum height of the burr). Technological AC-Systems effect the process in order to reduce production time. During cleaning of castings and deburring with industrial robots geometrical sensors are required to provide for a minimum height of the burr as well as a smooth surface. Technological sensors are required to minimize the production time.

Similar to the different adaptive control developments in machine tools, different system quantities can be used in case of sensor controlled handling systems. For different applications, sensors were developed for measuring

- cutting forces (normal force and tangential force),
- geometric quantities (position of the workpiece, height and thickness of the burr),
- grinding capacity (power input of electrical and pneumatic drive).

The process can be controlled by changing the speed, the position and the number of repetitions. A systematical survey of the possible measuring quantities and variable quantities is shown in Fig. 7.

3.2 Sensors

3.2.1 Sensors for Geometry

If there are high demands regarding the smoothness of the surface, geometrical measuring quantities will provide for better results. Fig. 8 shows the possible geometrical measuring quantities and different concepts for sensor systems. Tactile sensors for scanning a line or a matrix were developed and tested in different research projects /4/. This type of sensor can monitor the complex contours during the grindung process. In order to measure the height of the burr during grinding, the measurement of the swivel angle of the tool (Fig. 8) seems to be a suitable concept.

Thus, a tool-sensor unit with the following features was built:

- easy programming of a contour by tracing a manually grinded master part,
- compensation of tool wear (decreasing diameter of grinding wheel),
- adaption of contact pressure depending on the size of the burr.

The tool-sensor-unit is shown in Fig. 9. The grinder can swivel around its centre of gravity and is pressed against the workpiece-surface by a pneumatic cylinder.
The swivel angle can be measured with a potentiometer. The signal of this potentiometer is digitized (12 bit).

The principle of operation of this sensor is explained in Fig. 1o. It shows a workpiece contour with a burr. Programming and grinding of the part is done in three steps. First, the different contours to be deburred are programmed in the teach-in mode by storing the starting points and the end points P1, P2 etc. Afterwards, the robot executes this initial programme without cutting. At this time the sensor processor, by means of the linear potentiometer, scans the contour of the already grinded master part. It is not to be expected that the actual path of the robot and the actual contour of the master part will be parallel. Therefore, the potentiometer output (U_{A1}) will vary. The value of this signal is stored at predetermined fixed distances of 1 mm.

After these two steps the programming is finished and the third step, the actual grinding of parts with burrs, can begin. During automatic deburring, the actual potentiometer signals are being compared with the stored values (U_{A1}). Depending on the deviation, the contact pressure of the grinding wheel can be increased or decreased by means of a governor valve which is controlled by an analogue output of the sensor-processor.

For the compensation of tool wear and clamping tolerances, a reference point next to the burr has to be programmed and the potentiometer signal U_{A1} (of the master part) has to be stored. Before grinding of a new part, this reference point is approached and the actual potentiometer signal can be compared with the stored value. The difference gives an indication of tool wear (T_{wr}) and the clamping tolerances (T_A). This value $T_{wr} + T_A$ can now be used to compensate tool wear and clamping tolerances by controlling the contact pressure.
The main advantage of this system is the easy programming of complicated contours by tracing a master part.
The adaptive control consists of the sensor, the processor and the governor-value and therefore the robot itself does not need any special sensor interface. Fig. 11 shows the interlinking of the Siemens-Robot-Control with the sensor processor (PDP-11/o3).

3.2.2 Sensors for Technological Measuring Quantities

The power input of the tool and the cutting forces /5/ acting on the tool are suitable measuring quantities for many machining processes. These technological measuring quantities allow an adaptive control of the deburring process if minimum production time (maximum speed) is required. In case where a minimum height of the burr must be achieved (high surface-quality) these type of sensor can not work at a optimum /9/.

3.2.3 Sensor Interface

The interlinking of sensor with IR-controls in order to build up an adaptive control is at a rather low level.
In particular, a better dynamic performance of the adapive system is desired, so that reliable machining is enabled even with high feed rates.
The manipulated variable "feed rate" may be influenced by an available interface which be applied for various industrial robots. Considerably more effort would be required to chance the path throughout the process. Fig. 12 shows various possibilities for path alterations. If the correction of the path is carried out through th IR-drive, there are two different possibilities:

- o correction vectors are programmed by the user and, with a threepoint control (switching sensor), change the programmed path /6/.
- o the corrected directions depend on the path vector or the orientation of the gripper. Here the input of correction vectors, which requires a considerable effort, is unnecessary /7/.

The latter method appears to be more suitable for machine tasks since, for technological reasons, the gripper or tool must be lead to the position of the workpiece surface with constant tool orientation. With this principle, the internally controlled coordination transformation, which is also necessary for the path interpolation, may be used for the calculation of the inividual corrections of the IR-axes. An example is shown in Fig. 13.

With these correction possibilities, the path velocity, the cutting depth (correction direction 3) as well as the position of the tool perpendicular to the longitudinal direction of the burr (correction direction 2) may be controlled.

The four corrective directions and changes of the velocity along the path shown in Fig. 13 may be activated by allocating binary inputs. Using analogue sensors, this may be achieved by use of a comparator.

The magnitude of the correction carried out per scanning cycle of the operation system of the control computer may be altered according to the given task before the start of the programme.

This principle of path- and speed- corrections is suitable for simple switching sensors as well as for multi-component transducers (MCT). If the MCT is installed parallel to the tool-related coordinate system, that is, in front of the sixth axis, simple signal processing without sensor-specific transformations may be realized. In this case, every axis of the sensor corresponds with a certain correction axis of the robot control. Since, especially during machining with high velocities along paths, increased demands are made on the dynamics of the controlling unit, a puls-width-modulator for analogical sensors is advantageous.

4. EXAMPLE

The preceding chapters describe geometry-processing and technological sensors and control systems. Now, the application of a technological sensor is to be illustrated by the example of a laboratory set-up at the IPA.

The task set for this laboratory set-up concerned wooden parts for the furniture industry, especially intricately formed legs of tables and chairs. The geometrical chape of these parts is produced by a milling machine (copy milling or form milling). After milling, to achieve the desired faultless surface, the wooden parts must be grinded. This was hitherto performed, completely manually, in two work cycles (rough grinding and fine grinding). So far, any attempts to automate this procedure have failed, amongst other things, because of the lack of flexibility and the small number of parts. Tests have been carried out at the IPA, in which an industrial robot takes the wooden parts from a magazine with a special gripper, to pass them on to the grinding tool for the grinding procedure. After the grinding procedure the wooden parts are deposited in another magazine (Fig. 14).

Since as much of the surface of the wooden parts as possible must be grinded, the industrial robot gripper must clamp the wooden part in a very small area. Because the wooden parts were very large, rather large clamping tolerances occurred when the small gripper was used. To achieve a good surface quality, these had to be picked up and compensated for with the aid of the sensor.

Since the geometrical shape of the wooden parts was predetermined, it seemed more appropriate to apply a technological- rather than a geometrical sensor. The power consumption of the grinding tool was selected as the measuring value.

The laboratory set-up was largely made up of the following components:

- special tool for wood grinding
- industrial robot KUKA IR 601/60 CP
- measuring device for continuous control of grinding force
- processing of measured values to adapt the measuring signal to the industrial robot control
- sensor interface for the industrial robot.

Aided by the sensor interface, correction data for the alteration of the fixed programmed path may be given to the industrial robot.

During the test runs it was found that such an adaptive control may be realized. As soon as the wooden part was pressed too strongly against the grinding tool, because of clamping tolerances, the grinding capacity exceeded its liminal value. Thereby, the programmed path of the industrial robot was automatically corrected (away from the grinding tool). If, however, the capacity remained under a particular minimum value, a correction of the programmed path into the direction of the grinding tool, automatically followed. In the investigation two contours of the wooden parts lying opposite each other were grinded. The consequence of clamping errors was that too much was grinded on one side and too little on the other side. By the application of the sensor system described, an approximately even grinding force on both sides, in spite of clamping tolerances could be achieved.

The investigations proved that such tasks may be solved today. Problems occur in the computing time necessary for the processing of measured values and in transferring the measured results to the industrial robot control. The correction of stored programmes has hitherto generally required a large amount of effort in order to modify that hardware and software of the industrial robot. Since more and more robots are now equipped with standard sensor interfaces for the correction of programmes, the use and application of the described sensors will surely be simplified quite considerably during the next few years.

5. REFERENCES

/1/ Autorenkollektiv: Rationalisierung in der Gußputzerei, Giessereiverlag GmbH, Düsseldorf, 1973.

/2/ Warnecke, H.-J.: Schraft, R.-D. Industrieroboter, Krausskopf-Verlag 1979.

/3/ N.N.: VDI-Richtlinien Nr. 3426. Numerisch gesteuerte Arbeitsmaschinen. Adaptive Control an spanenden Werkzeugmaschinen, Berlin, Köln: Beuth 1974.

/4/ Stute, G.: Erne, H.: The control design of an industrial robot with advanced tactile sensitivity. Proceedings of the 9th ISIR. Washington 1979.

/5/ Abele, E.: Entgraten von Werkstücken mit Industrieroboter. TÜ 2o (1979) 11, S. 413-418.

/6/ N.N.: ASEA Industrierobotersystem-Adaptive Steuerung. Info.Nr. YB 11o-318, Västeras 1979.

/7/ Wörn, H.: Anwendung der Mikroelektronik in der Handhabungstechnik. Düsseldorf: VDI-Verlag 198o S. 55-62. (VDI Bericht; Nr. 379).

/8/ DIN 1685: Gußrohteile aus Gußeisen mit Kugelgraphit, Allgemeintoleranzen. Berlin, Köln: Beuth, 1978.

/9/ Abele, E.: Adaptive controls for fettling of castings with industrial robots. Proceedings of the 1st International Conference on Robot-Vision and Sensory Controls, Stratford-upon-Avon-Kempston, Bedford: International Fluidics Services, 1981.

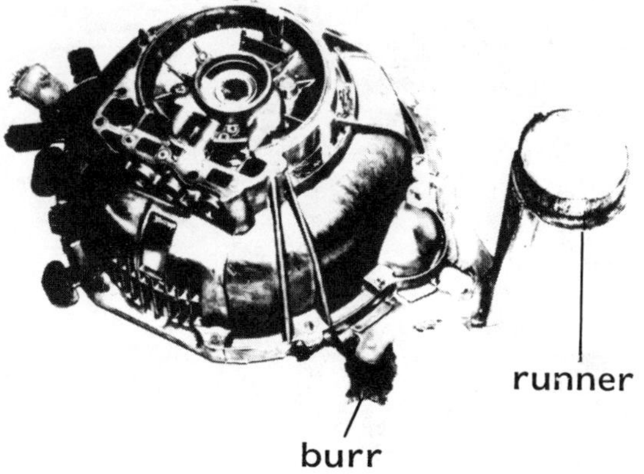

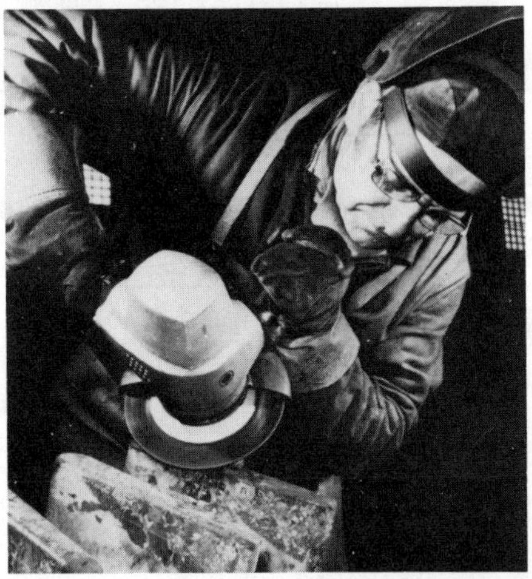

Figure 1:

Fettling of castings

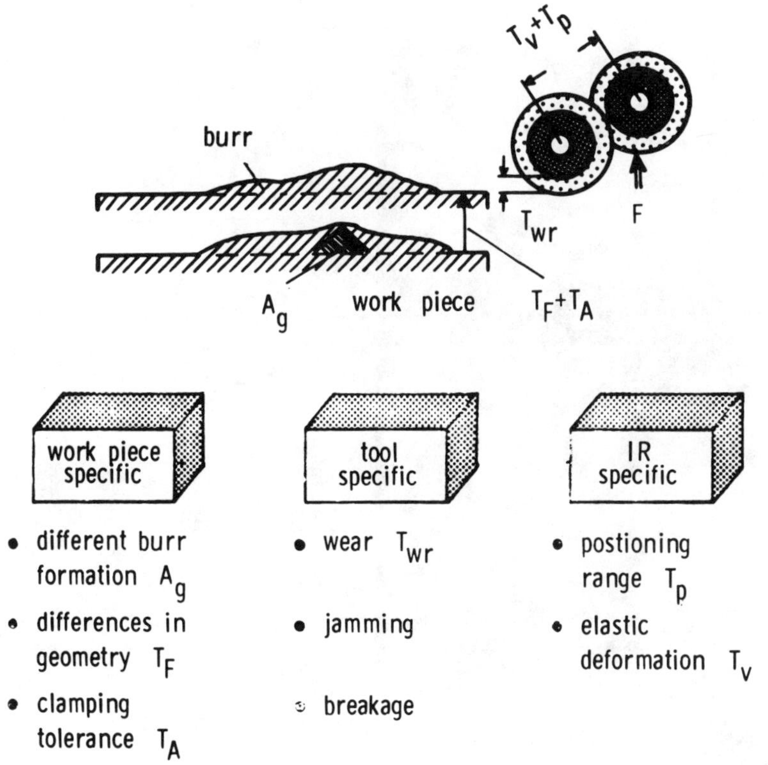

Figure 2:

Disturbances in automation of the fettling process

weight class \ max. size of T_A, T_F u. A_g	die casting		chill casting		sand casting	
	$T_A + T_F$*	A_g**	$T_A + T_F$*	A_g**	$T_A + T_F$*	A_g**
0,63 - 1 kg	<0,4	2	<0,6	2,8	<2,0	6
10 - 16 kg	<1,5	8	<2,1	11	<3,2	24
100 - 160kg	—	—	—	—	<10,0	55

*in mm **in mm^2 / without riser system

Figure 3:

Sectional area of a burr by different forming processes

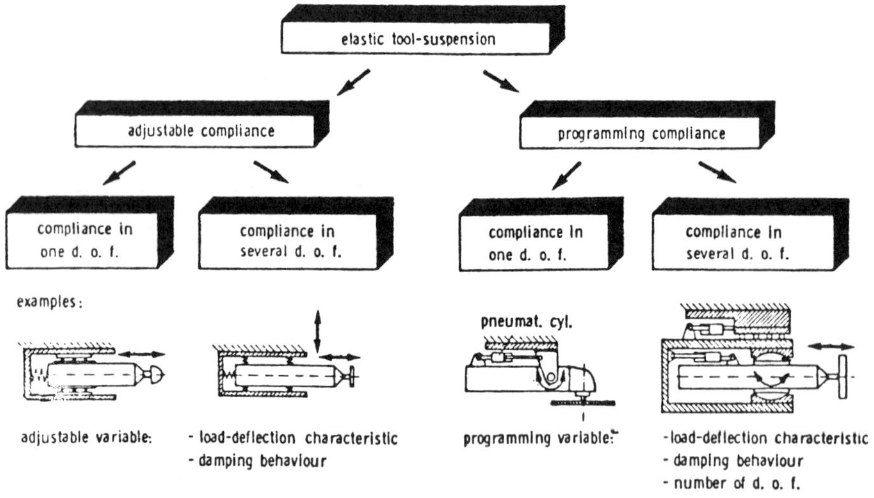

figure 4: Elastic tool suspensions for fettling

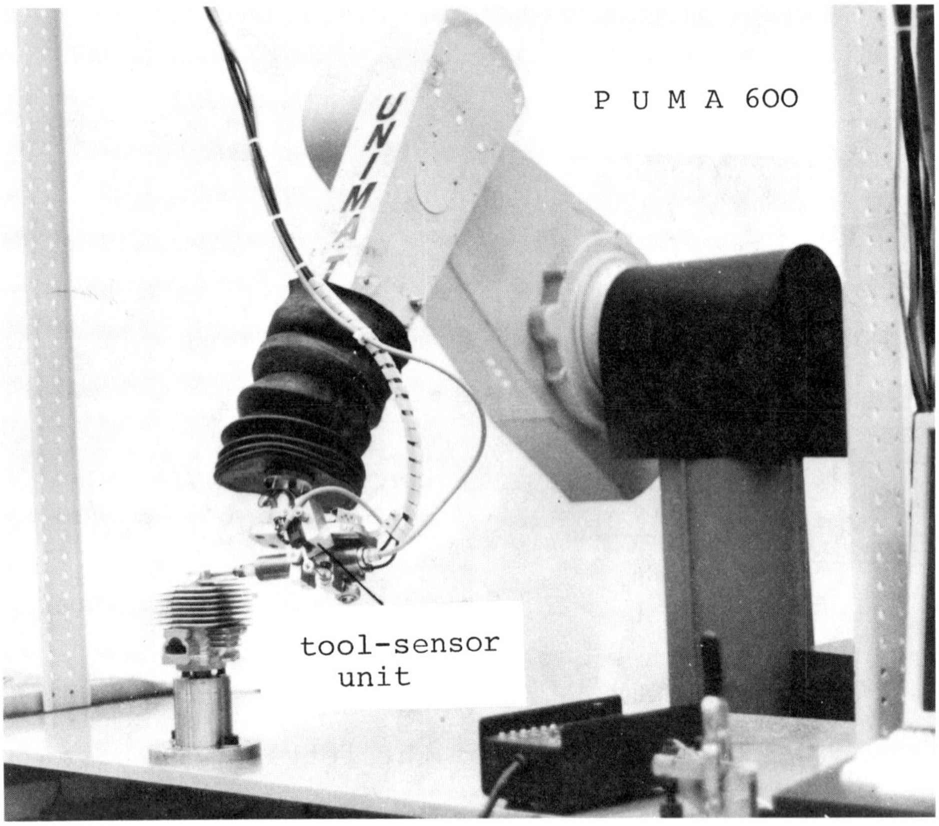

figure 5:

Application of a elastic tool suspension with programmable compliance

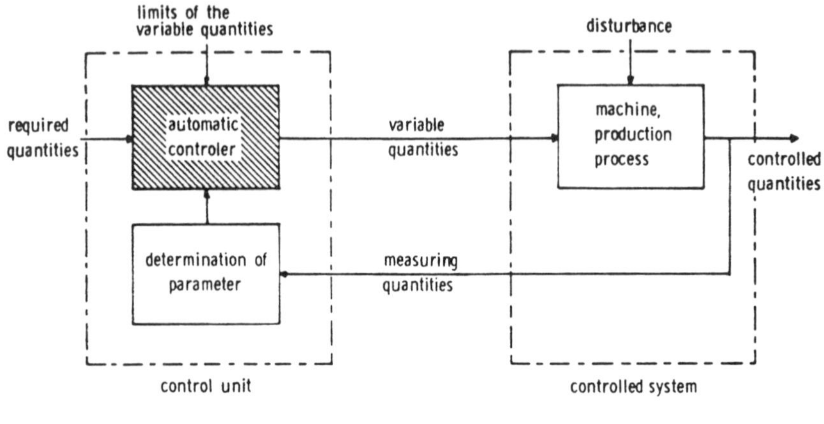

figure 6: Signal flow plan of an ACC-system

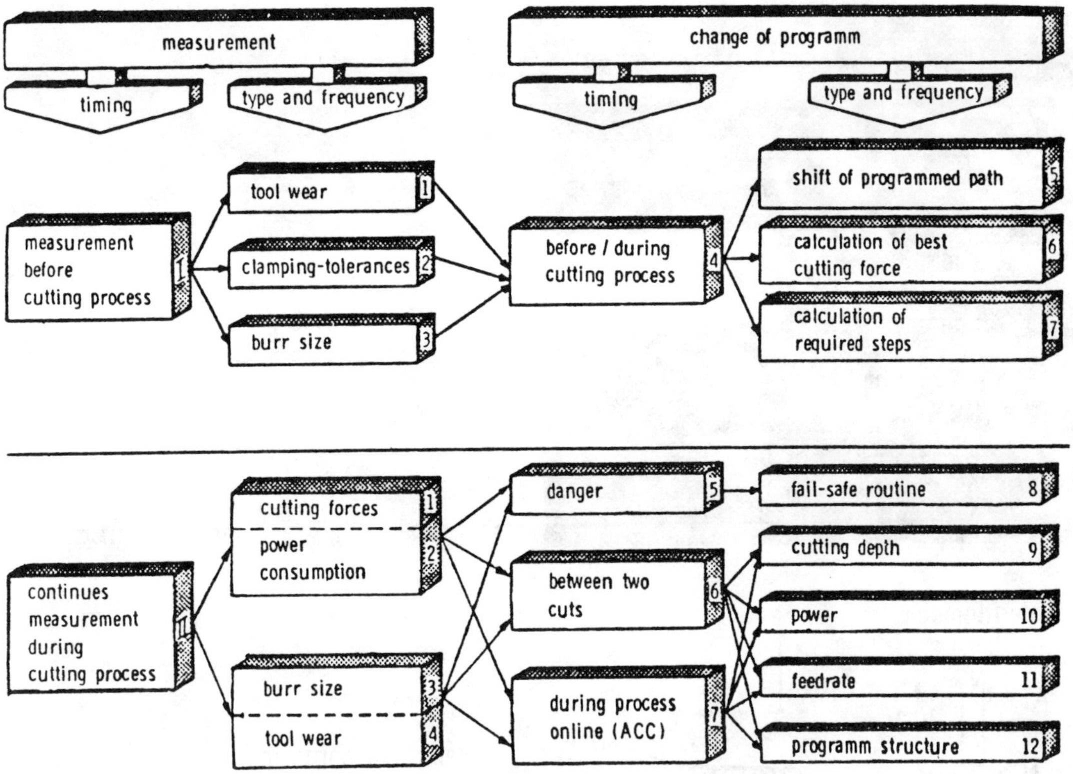

figure 7:

Strategies of sensor controled fettling

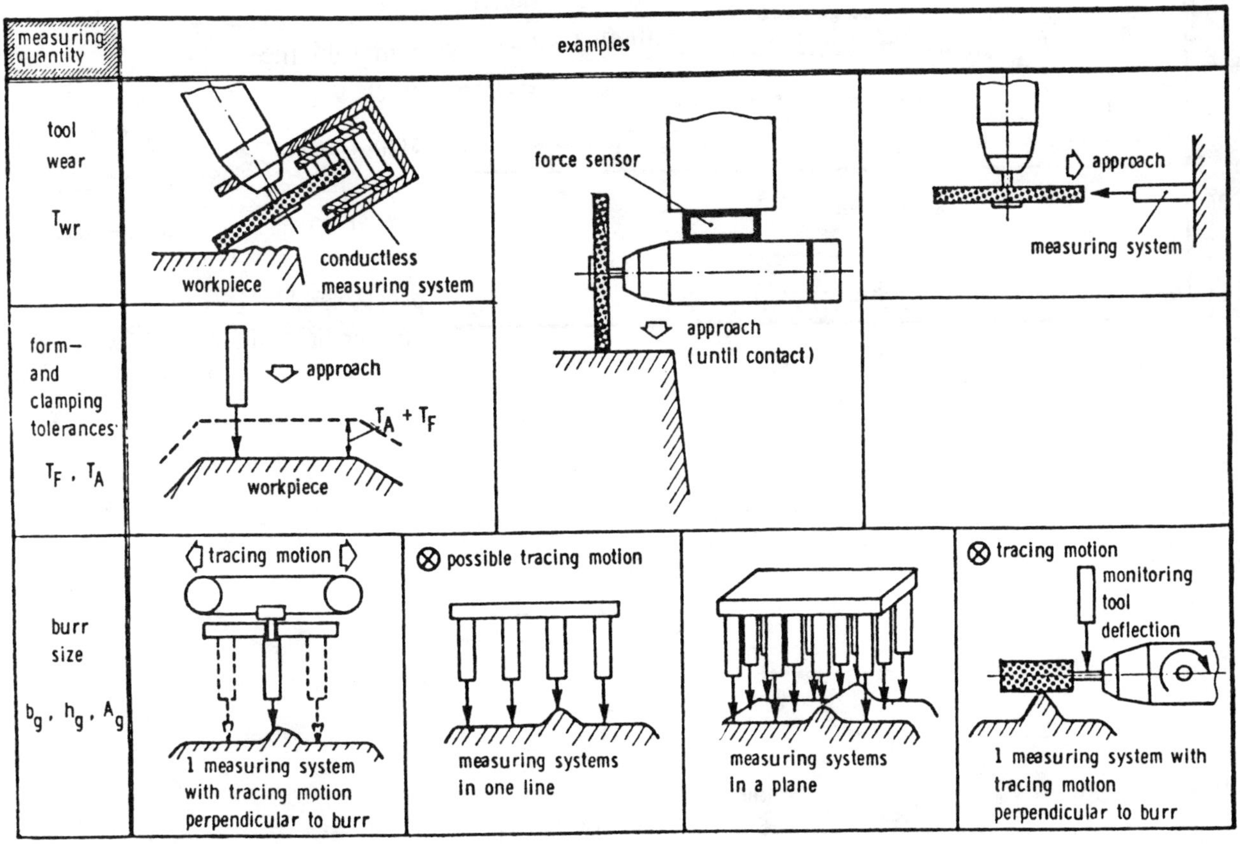

figure 8:

Measuring quantity and measuring principles of sensors for geometry

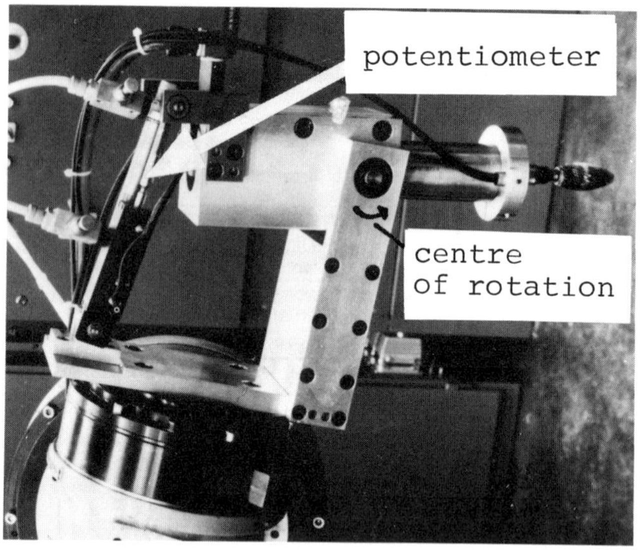

figure 9:
Example of a tool sensor unit

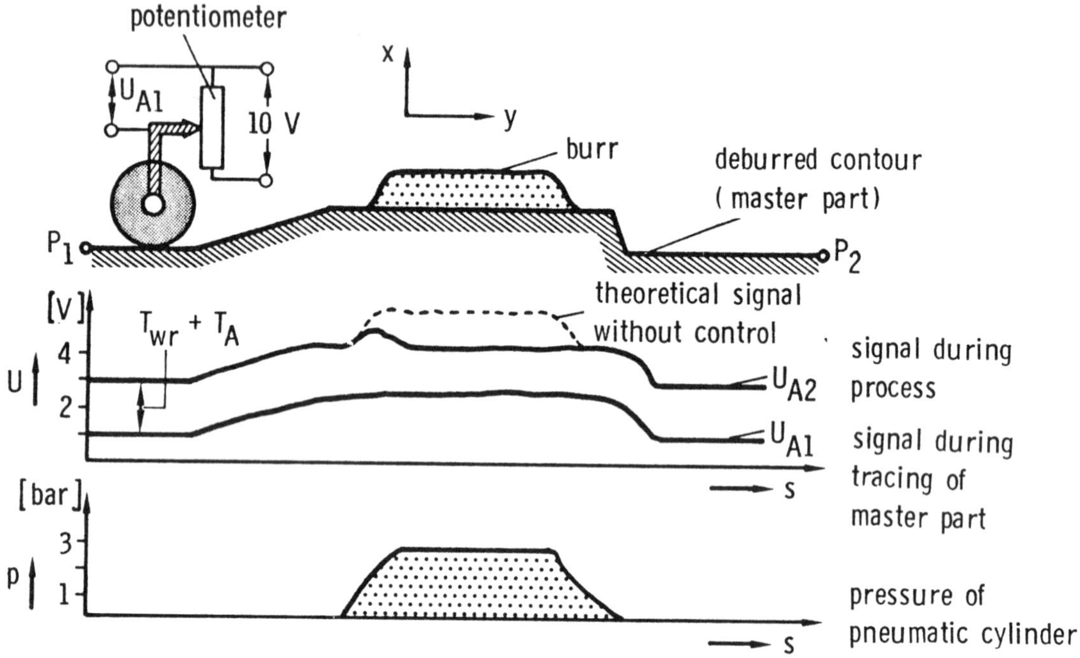

figure 1o:
Function of
a sensor for
measuring
the height
of the burr

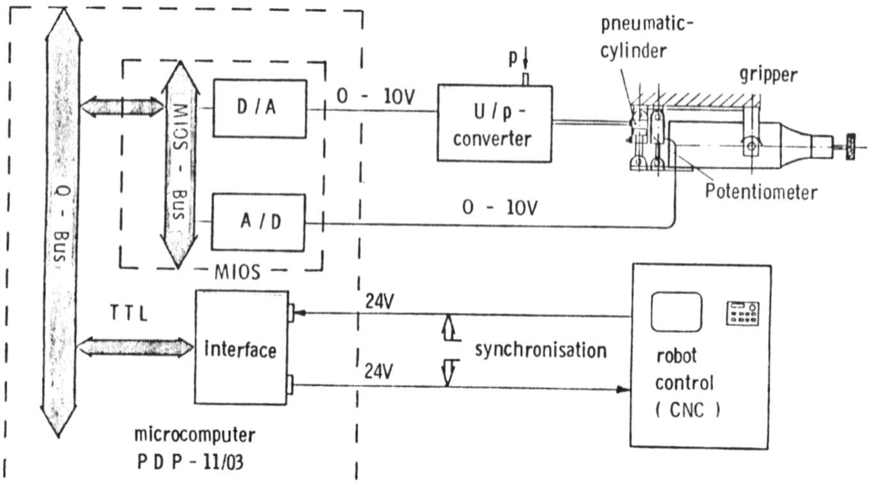

figure 11:
Sensor interfacing

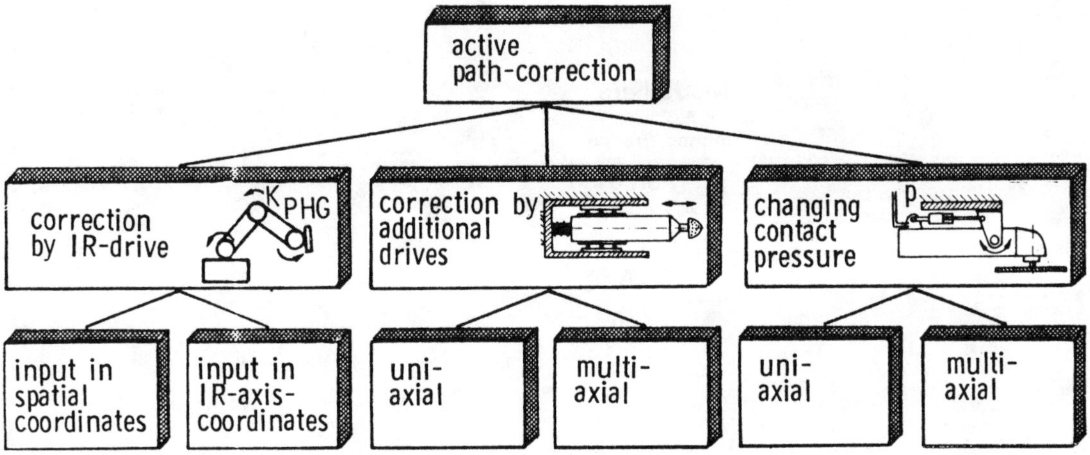

figure 12: Systematic of mani - pulated variables for path correction

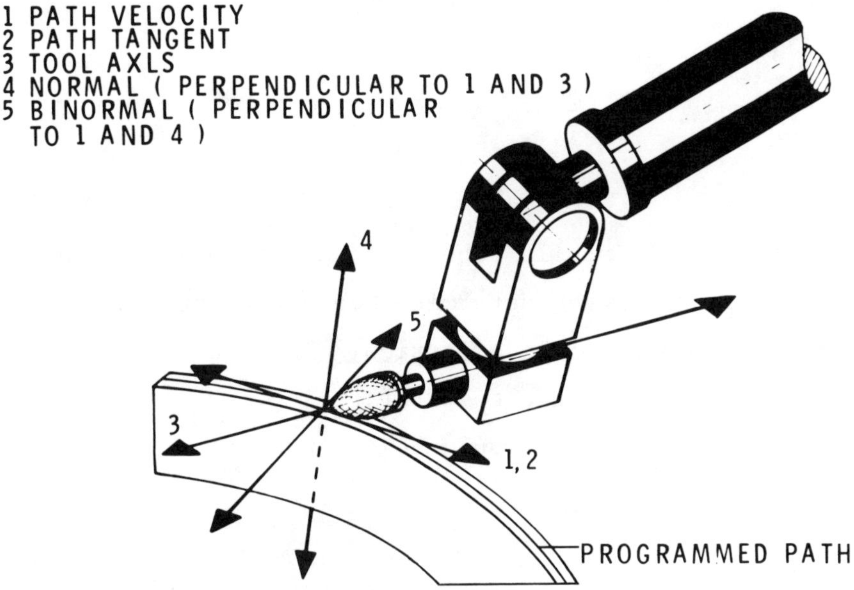

figure 13: Sensor control of programmed path with the KUKA IR 6o1/CP

figure 14: Grinding of wooden parts with an industrial robot

A Guidance System for a Mobile Robot

M. Julliere
L.A.T.E.A.
Rennes, France

L. Marce
L.A.T.E.A.

H. Place
L.A.T.E.A.

Abstract

The importance of knowing accurate position in mobile robot guidance is significant. The knowledge of accurate positioning describes the determination of an optimal route; the decomposition in segments; the means of localization. Robot control must perform an accurate and smoothed tracking of the computer trajectory. A guidance system is a basic element for a robot which would be autonomous.

I - INTRODUCTION

This paper describes the set of means developped to insure the displacements of the mobile robot VESA in an obstacles-dotted environment. Obstacles are assumed to be polygonals and known a-priori.

Determination of an optimal path to go from one point to another in such an environment, avoiding at best obstacles, will be described in first section. This determination is dependent on an upper level device, named central computer, which defines a reference trajectory the vehicle has to follow at best.

Good execution of this task implies that the vehicle has at its disposal at each instant an estimation of its coordinates (X,Y) and also of its attitude (i.e. orientation of the followed trajectory versus a reference direction). We will describe different systems which can be foreseen to resolve this problem and particularly those we use for VESA.

Finally, knowledge of vehicle coordinates must allow to minimize its distance from the reference trajectory. It is a control problem applied to vehicle driving and steering devices.

II - DETERMINATION OF AN OPTIMAL PATH

Robot is assumed to be moved in a plan environment dotted with obstacles, well characterised by polygons defined by their apexes. To take into account the robot size, obstacles are artificially enlarged of a quantity equal to a robot significant dimension, for example the radius in case of the circular robot VESA.

Determination of a trajectory between two points in such an environment reduces to searching a path in a graph formed of nodes (apexes of enlarged obstacles) and arcs (segments joining these nodes).

To minimize simultaneously searching cost and path cost (for distance or time), a heuristic method is used which permits to obtain in a very short time an optimal path between initial and final points. This method is based on an A^*-type algorithm (1). For that, nodes are classified in a file depending on the future of the node, expressed by an evaluation fonction f(n). - If $\hat{f}(n)$ is the value of this function at node n, defined as evaluation of cost $f^*(n)$ of the optimal path going through n, thus :

$$\hat{f}(n) = \hat{g}(n) + \hat{h}(n)$$

$$\text{and } f^*(n) = g^*(n) + h^*(n),$$

where $\hat{g}(n)$ is the estimation of cost for the minimal path from start-node d to node n, that is $g^*(n)$, and $\hat{h}(n)$ is the estimation of cost for the minimal path from node n to goal a, that is $h^*(n)$.

Figure 1 gives an example where this algorithm is applied to determine an optimal path in a workshop analog environment. Note that the determined trajectory is made of straight lines.

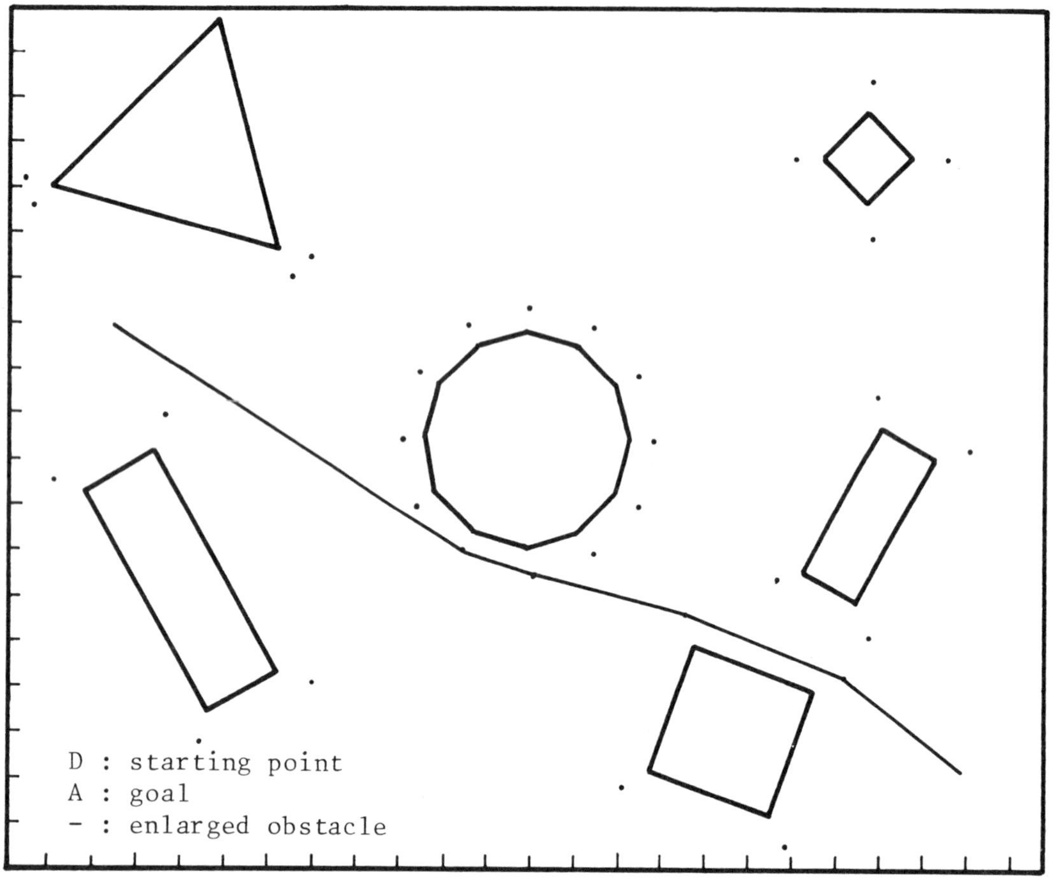

FIGURE 1 : EXAMPLE OF OPTIMAL TRAJECTORY

III - LOCALISATION MEANS OF THE VEHICLE

Here the problem is to estimate coordinates (X,Y) and attitude of vehicle. Two types of means can be foreseen, absolute and relative.

Absolute measurement systems give coordinates, and eventually attitude, measured versus fixed points or surfaces which occupy fixed and known positions in the robot world. These reference positions may be those of obstacles themselves, or those of artificial devices set there by the operator with the only goal to solve the localization problem (environment marking).

- The first solution (obstacles as reference) supposes that the robot has at its disposal in memory a world map to which it can compare the world map seen by its own sensors. The robot must be equipped by "vision" devices, such as one or more T.V. cameras (2), an infrared telemeter (3) or an ultrasonic telemeter (4).

- Environment marking can be realized with light-emitters, receptors or reflectors (generally I.R.) (3) or using other physical supports such as H.F. waves (5), painted marks on floor or obstacles (6).

Relative measurement systems estimate robot position by computing coordinate variations from an earlier instant. These variations can be obtained from acceleration measurement (inertial systems) or from measurement of the wheels displacements, more precisely displacements of contacts points between wheels and floor (odometry). In all cases, relative systems proceed by integration, therefore errors are cumulative : this fact limits the use of such systems to estimate short-term trajectory.

The two types of systems are not concurrent but complementary, each provides certain advantages :

- Relative systems (inertial or odometric) give a very good precision for short term measurements and can deliver data at each instant. Odometric systems are low cost, but it is not the case of inertial systems. On the other hand, long-term precision depends on parameters which are not easy to control.

- The precision given by absolute systems is better for long term and even excellent for short range systems (painted marks,...). But, in this case, measurements are only available in particular points. Finally, beacons setting up amounts to particularize the world and cannot be considered as universal solution (beacons screened by obstacles, for example).

From the foregoing discussion, generally, it is interesting to associate many systems. Therefore we have retained three of those :

- A triangulation system using I.R. beacons
- An odometric system
- An estimation of the trajectory actually followed in vicinity of the obstacles by means of telemetric measurements.

To execute a given trajectory at local level (vehicle), measurement is essentially odometric. It can be corrected ponctually, when passing near an obstacle, with the aid of telemetric measurements in direction of an identified part of the obstacle. This part is known by the localisation, considered precise, which is provided by the odometric measurement. The previous restrictions, particular obstacle for which a part is only used, permit a simple correlation i.e. being applied to a restricted zone of the world.

The triangulation system with receiving beacons takes place at an upper level : its job is to provide to the central computer an absolute measurement of the coordinates, which has not to be very precise. This computer linked by radio uses the information of position, in order to get the robot out of a bad situation, in the case of failure of the other systems (vehicle side-splipping, unpredicted obstacles,...). This latter system is seen by the central computer as one of its peripherals affected to a surveying function.

IV - TRIANGULATION MEASUREMENT USING I.R.

The vehicle is equipped, at its center, with an emitter built of 8 diodes (LED) each supplied by sinusoidal currents $\pi/4$ phase-shifted between them. Beyond certain distance, this source is seen as providing an I.R. rotating beam. The robot world is provided with three beacons (I.R. receptors) which are receiving a beam, the intensity of which is going periodically through a maximum.Measurement of time delays between maximums detected by two beacons gives the angle between the straight line joining the I.R. source to these beacons (Fig. 2). This no-mechanical solution has been adopted, because it permits to obtain a low period ($\simeq$ 1ms) for the rotating beam and to calculate a mean value of measurements on a time interval corresponding to low vehicle displacements. All the beacons are identical and measured angles are proportional to phase shift between received signals.

To calculate position, one proceeds as follows :

- The two lowest angles are first chosen among the three measured ones, allowing thus two pairs of beacons.

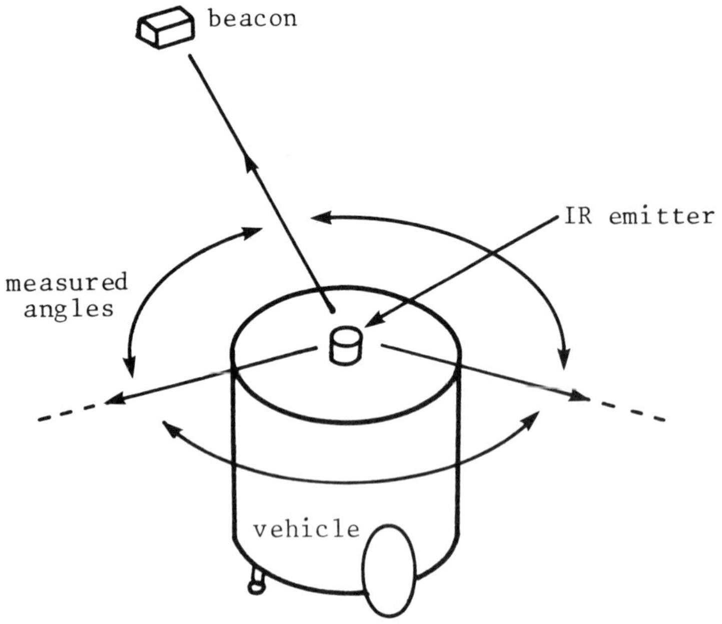

FIGURE 2 : PRINCIPLE OF I.R. TRIANGULATION

- For each of these pairs, the value of measured angle defines a circle passing through the beacons and vehicle center. Coordinates of the circle centers are thus calculated.

- Finally the intersection of the so defined circles gives the vehicles center coordinates.

V - ODOMETRIC MEASUREMENTS

This method uses the informations provided by optical encoder attached to the two vehicle wheels. These wheels have a common axis and are, for our vehicle, driving wheels. In other cases they are only steering ones. Calculation method, as described by TSUMURA and al. (5) gives :

$$\Delta\theta_n = \frac{1}{E} (\Delta U_L - \Delta U_R)$$

$$\Delta U_n = \frac{1}{2} (\Delta U_L + \Delta U_R)$$

$$\theta_n = \theta_{n-1} + \Delta\theta_n$$

$$X_n \simeq X_{n-1} + \Delta U_n \cos \frac{\theta_n + \theta_{n-1}}{2}$$

$$Y_n \simeq Y_{n-1} + \Delta U_n \sin \frac{\theta_n + \theta_{n-1}}{2}$$

where :

- θ_n, X_n, Y_n (figure 3) are, at time t_n, the attitude of vehicle and the coordinates of middle point of the wheels axis.

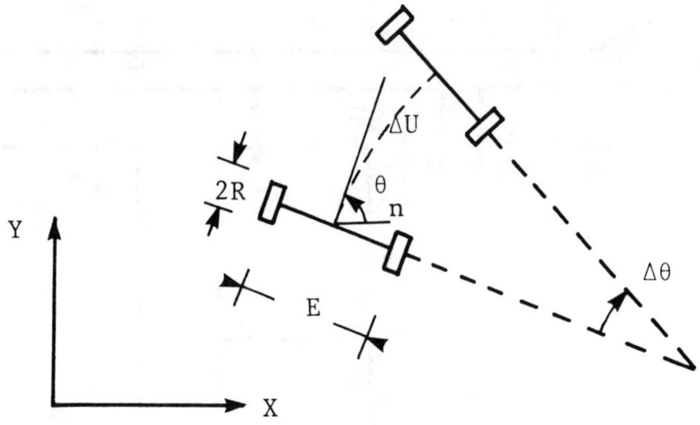

FIGURE 3 : PRINCIPLE OF ODOMETRY

- $\Delta\theta_n$ and ΔU_n are the variations, take place between instants t_{n-1} and t_n, of the attitude and the distance covered by the previous middle point.

- ΔU_L and ΔU_R are the covered distances, respectively by the left and right wheels and E the distance between these wheels.

The two wheels are identical, of radius R, and each drives an optical encoder giving 700 pulses during one wheel revolution (N=700). If ΔN_L and ΔN_R are the numbers of pulses provided by the encoders between t_{n-1} and t_n, then :

$$\Delta U_L = \frac{2\pi R}{N} \Delta N_L \qquad \Delta U_R = \frac{2\pi R}{N} \Delta N_R$$

which gives :

$$\Delta\theta_n = \alpha(\Delta N_L - \Delta N_R)$$

$$\Delta U_N = \beta(\Delta N_L + \Delta N_R)$$

with :

$$\alpha = \frac{2\pi R}{NE} \qquad \beta = \frac{\pi R}{N}$$

In order to simplify coordinates calculation, we have chosen to measure ΔU_L and ΔU_R , not with $\Delta t = t_n - t_{n-1}$ constant, but with ΔU or $\Delta\theta$ constant, depending on the signs of ΔU_L and ΔU_R. This condition is implemented with wired logic circuitry as :

$$|\Delta N_L| + |\Delta N_R| = 16 \qquad \text{(cf. fig. 4)}$$

If ΔN_L and ΔN_R are of the same sign ($\Delta U = \pm 16\beta$), the TSUMURA's relations become :

$$X_n = X_{n-1} \pm 16\ \beta \cos \frac{\theta_{n-1}+\theta_n}{2}$$

$$Y_n = Y_{n-1} \pm 16\ \beta \sin \frac{\theta_{n-1}+\theta_n}{2}$$

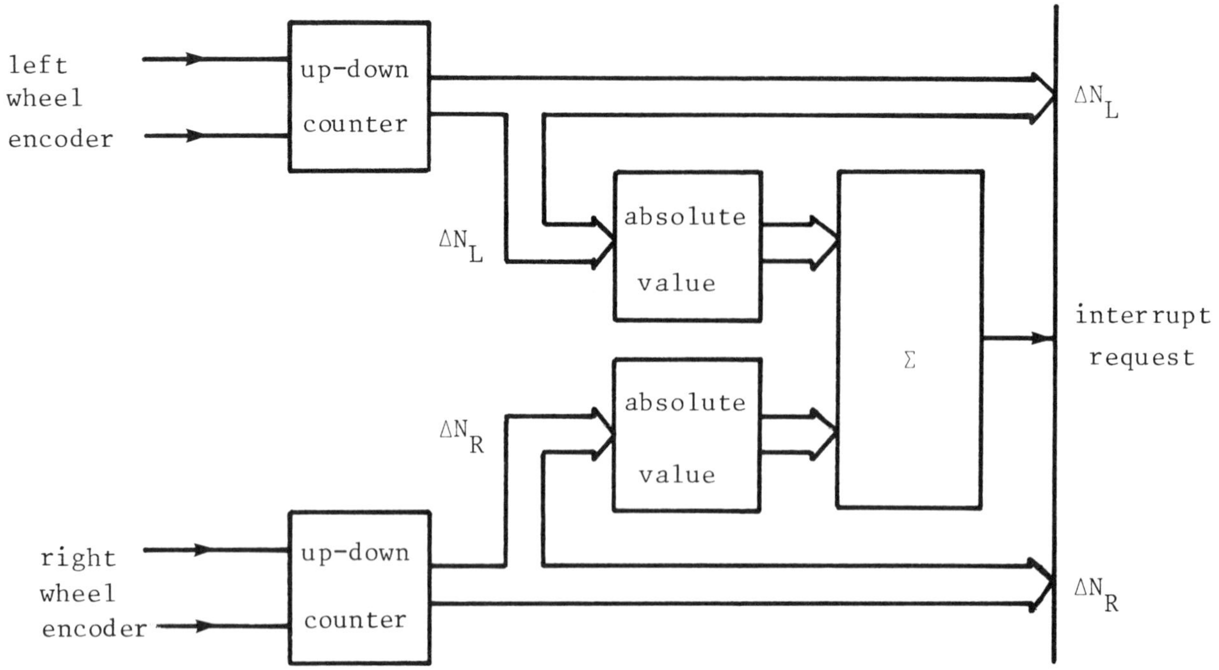

FIGURE 4 : INTERFACE BETWEEN ENCODERS AND MICROPROCESSORS

If ΔN_L and ΔN_R are of opposite sign ($\Delta\theta = \pm 16\alpha$) :

$$X_n = X_{n-1} + \Delta U_n \cos(\theta_{n-1} \pm 8\alpha)$$

$$Y_n = Y_{n-1} + \Delta U_n \sin(\theta_{n-1} \pm 8\alpha)$$

Calculations of θ_n, X_n, Y_n are now implemented on an Intel 8085 microprocessor and results are provided to the control device (microprocessor of the same type) or to the central computer on its request. These values can be replaced at any time by new values got by absolute systems and considered as more precise.

VI - FOLLOWING OF THE REFERENCE TRAJECTORY

The control of vehicle movement in a two dimensional space can be reduced to the control of two variables : the instantaneous speed of the axis middle point : $V = \frac{dU}{dt}$ and the attitude θ. In the case of own vehicle, for which wheels are also driving ones, this amounts to control the sum of their speeds $\omega_R + \omega_L = \frac{V}{\pi R}$ and their difference $\omega_L - \omega_R = \frac{E}{2R}\dot{\theta}$. Speed V will be maintained constant by an independent control loop, the input of which is received from the central computer. Performances of this loop are not critical. Control acts on the sum of voltages applied to motors.

The attitude control acts on the difference of these two voltages. Control law is inspired from the methods used for driving wire-guided vehicles. Many studies have been realized in this field concerning particularly guiding stability (7). This control law includes at first a term, the role of which is to minimize the distance error ε_d between the vehicle center and the reference trajectory : $C = K_1 \varepsilon_d$ (Figure 5a).

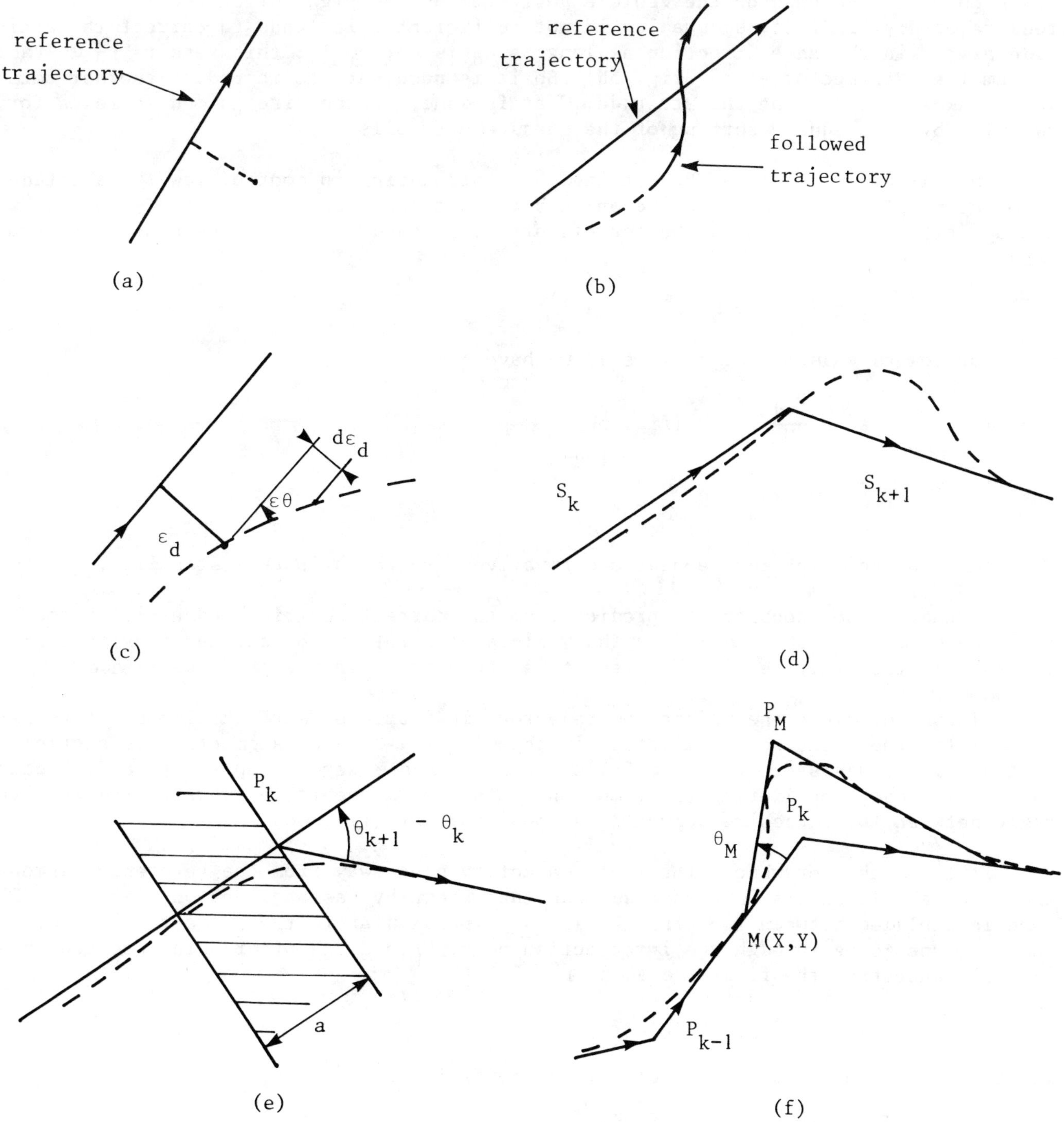

FIGURE 5

a) Distance error ε_d

b) Effect of a control law $K\varepsilon_d$

c) Relation between ε_d and ε_θ

d) Overstepping at the intersection os two reference segments

e) Succession of reference segments

f) Local modification of trajectory

The sign of ε_d depends on the vehicle position, on the right or on the left of oriented trajectory. In fact, such a law is not sufficient : it leads to correct the attitude always in the same direction as long as ε_d is not null ; this results in unstable system (see trajectories of fig. 5b). So it is necessary to introduce another term which takes into account the attitude. That is done, on the wire guided vehicles for example, by an adequate setting of the measurement coils (7).

An equivalent effet can be obtained by introducing in control law the attitude error $\varepsilon_\theta = \theta_R - \theta$, where θ_R is the angle between the segment of trajectory to follow and a reference axis and θ is the vehicle attitude measured versus the same reference axis :

$$C = K_1 \varepsilon_d + K_2 \varepsilon_0$$

For low ε_θ values ($\varepsilon_\theta \simeq \sin \varepsilon_\theta$), we have :

$$\varepsilon_\theta \simeq \frac{d\varepsilon_d}{dU} \qquad \text{(fig. 5c)}$$

and

$$C = K_1 \left(\varepsilon_d + K \frac{d}{dU} \varepsilon_d\right)$$

This control type lets appear a derivative term which insures stability.

Another method consists to predict from the current position and attitude the position which will be occupied by the vehicle at a latter instant and thus to correct accordingly the attitude (8). The result is the same as in the previous method.

Elsewhere, since the reference trajectory is formed of straight lines, it is important to smooth the real trajectory in the vicinity of each segments intersection. To do that, it is advisable to avoid following a reference segment up to its end, because this would lead, considering the time constants, to unacceptable overstepping when the angle between two succesive segments is too important (fig. 5d).

To avoid this drawback, the real trajectory goes away from the reference segment when the vehicle enters in a zone such as that shown by dashed lines on figure 5c. This zone is included between two straight lines perpendicular to the current reference segment S_k, one going through the intersection point P_k and the other being located ahead at a distance from the first one equal to :

$$a = \gamma V \left|\theta_{k+1} - \theta_k\right|$$

where : γ is an experimentally determined coefficient,

$V = \frac{dU}{dt}$ is the vehicle speed,

θ_{k+1} and θ_k are the reference attitude for the segments S_k and S_{k+1}.

Finally, the central computer can modify localy the trajectory which is to follow during execution. It has only to indicate to the control microprocessor a new direction to follow (fig. 5f) : the vehicle, at point M(X,Y), receives the order to follow a new segment passing through this point and a new point P_M. This approach offers to the central computer the way to modify a trajectory from informations given by sensors (impredicted obstacles, for example).

Figure 6 shows some example of guidance for a simulated vehicle (analog simulation).

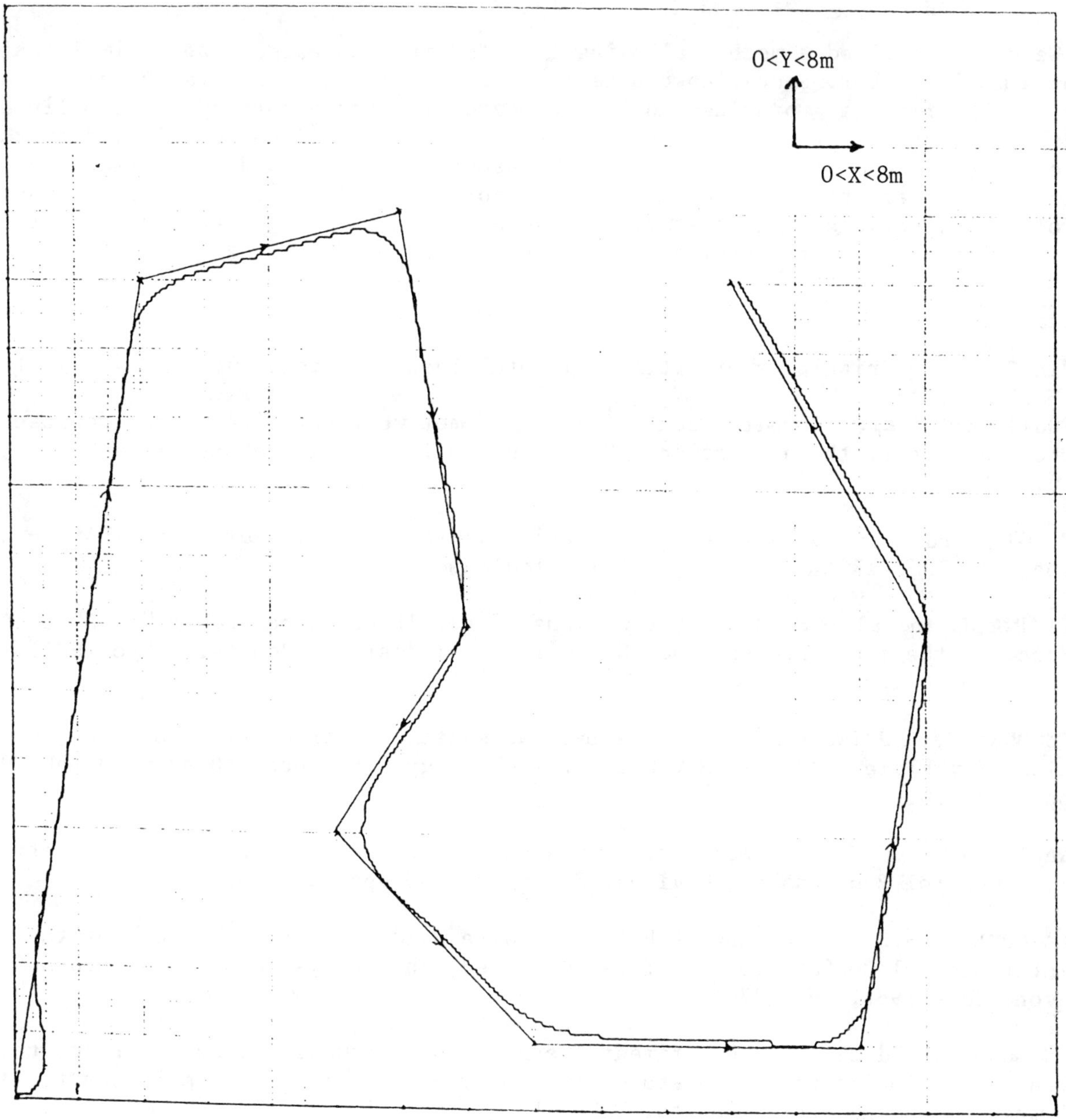

FIGURE 6 : EXAMPLE OF REFERENCE AND FOLLOWED TRAJECTORIES
Small steps on the followed trajectory are quantification effects from the DAC which drive the analog plotter.

VII - CONCLUSION

We have described a method allowing to determine and execute an optimal trajectory for a mobile robot. Execution can take into account initial data as well as acquired data. The first are provided to the on-board microprocessors by a radio-linked computer, the second are acquired by sensors during travel and interpreted by the same computer. The different functions (ultrasonic telemeter and I.R. triangulation) are implemented around INTEL 8085 microprocessors or will be implemented (guidance control). Applications of this study include essentially the guidance of industrial vehicles without a priori physically realized trajectory.

BIBLIOGRAPHY

(1) NILSSON J. - Principles of Artificial Intelligence - Tioga Publishing Co., 1980

(2) FUJII S. and al.- Computer control of a locomotive robot with visual feedback - Proc. of the 11 th International Symposium on Industrial Robots. Tokyo 1981 , pp. 219-226

(3) BRIOT M and al.- Le système de perception du robot mobile autonome HILARE - 2ème congrès AFCET/IRIA, 12-14 sept. 1979, Toulouse.

(4) IIJIMA J. and al.- Elementary functions of a self contained robot "YAMABICO 3.1." Proc. of the 11th International Symposium on Industrial Robots, Tokyo 1981, pp. 211-218

(5) TSUMURA T., FUJIWARA N. - An experimental system for processing movement information of vehicle - 28th IEEE Vehicular Technology Conf. Proc. Denver, March 1978, pp. 163-168.

(6) NAKAMURA T, FUKUI I. - Navigation of a robot vehicle by slit pattern detection. IFAC Control Problems and Devices, Budapest 1980, pp. 233-239.

(7) LARCOMBE M.H.E. - Tracking stability of wire guided vehicles - Proc. of the Ist International Conference on Automated Guided Vehicle Systems - Stratford-upon-Avon, June 1981, pp. 137-144.

(8) TSUMURA T. and al. - An experimental system for automatic guidance of roboted vehicle following the route stored in memory - Proc. of the 11th International Symposium on Industrial Robots, Tokyo 1981, pp. 187-193.

Some Problems of Mechanics and Control for Pneumatic Industrial Robots

A. Yu. Ishlinsky
Institute for Problems of Mechanics
USSR Academy of Sciences
Moscow, USSR

F.L. Chernousko
Institute for Problems of Mechanics
USSR Academy of Sciences

V.G. Gradetsky
Institute for Problems of Mechanics USSR
Academy of Sciences

Abstract

Theoretical and experimental results obtained in the Institute for Problems of Mechanics, USSR Academy of Science is discussed. Some basic problems of mechanics, control and sensors for Soviet industrial robots are considered. Kinematic and dynamic properties of the robots are studied including mechanical stiffness, accuracy and transient behaviour of manipulators. Optimal minimum-time regimes of robots are calculated and realized experimentally. Some types of sensors for industrial robots are discussed. Operation of the system is described which consists of Soviet pneumatic robot Tsiklon, Japanese video sensor Fuji Electric and microprocessors for information processing and control.

INTRODUCTION

Since a manipulating robot is a controlled mechanical system with one or more manipulators and control computer designed for manipulation of various objects (workpieces, components, processing tools etc.), achievements of many sciences are involved in design of robots. Mechanics plays a vital, often leading, role in the effort. In particular, pneumomechanics is used in designing of the Soviet industrial robots Tsiklon, Ritm, PR-10 and others. Many electromechanical robots, such as Soviet robots Universal-5, RPM-2S, have pneumatic drivers of the gripper motions. Theoretical and experimental results obtained for industrial robots Tsiklon, Universal, Ritm, can be extended for other similar models of manipulators.

In the recent years the Department of Control Systems and the Robotics Laboratory in the Institute for Problems of Mechanics, USSR Academy of Sciences, have been carrying out research on mechanics and optimal control of manipulators, and on development of sensors enabling robots to percieve their environment. This paper describes some results in this field obtained in the Institute.

Problems of kinematics and dynamics of manipulators are discussed. Some results on dynamics of manipulators are presented, especially for manipulators with flexible links.

Some results on optimization of motion of manipulators are presented.

Problems of measurement in robotics are discussed and some pneumatic sensors developed in the Institute are described.

An informational diagnostic robot is described which has been designed and is based on pneumatic industrial robot Tsiklon and video detector Fuji Electric. This robot controlled by microprocessor has a wide range of possible applications.

KINEMATIC ASPECTS

From the kinematic point of view, an industrial manipulator is a multi-component machine with components linked by cylindrical hinges. The mechanical diagram of a manipulator contains, as a rule, translation and rotation kinematic pairs. This considerably affects the classification. The number of branches in a mechanical diagram is usually associated with the number of manipulators of the robot, and the number of kinematic pairs in the branches is equal to the number of moving and guiding degrees of freedom for the respective manipulators. The geometry of the manipulator's operating zone is also determined by the number, type, relative position of the kinematic pairs and relative sizes of its links.

The kinematic pairs are responsible for the robot degrees of freedom in different coordinate systems: orthogonal (Soviet robot Sprut-1), cylindrical (Soviet robots Tsiklon, Universal-3, MP - 5), spherical (Soviet robot RS - 4), combined type (MP - 1).

Description of the manipulator kinematics can be aided by results of the research on kinematics of rigid body systems obtained earlier in the research on gyroscope systems and gimbal suspensions. A systematic description of kinematics of such objects including geometry of gimbal suspensions and theory of finite rotations of rigid bodies is given in [1]. General results on kinematics of rigid body systems obtained in books [1, 2] are useful in problems of kinematics of robots.

SOME PROBLEMS OF DYNAMICS OF ROBOTS

In study of dynamics of manipulators use is commonly made of a mechanical model of rigid bodies (rods) linked by hinges. This rigid model is often sufficient as a first approximation. In a number of cases, however, the flexibility of manipulator greatly affects the accuracy of its operations and hence must be taken into consideration in a design of a control system. Such real mechanical properties of manipulators as elasticity of its links, flexibility and free play of its hinged joints and transient behaviour of its pneumatic system must be included in mechanical models of manipulator dynamics. We considered several mechanical models of manipulators having these properties:

a) a manipulator consists of elastic rods, and the mass of a manipulator is small compared with a mass of a load;

b) a manipulator consists of elastic rods, and the mass of a manipulator is big compared with a mass of a load;

c) a manipulator consists of rigid bodies connected by flexible hinged joints;

d) a manipulator consists of rigid bodies, and transient behaviour of pneumatic system is taken into consideration.

The first model a) was considered in the papers [3, 4] and corresponds to manipulators with big linear dimensions. Eigenfrequencies and eigenvectors of free elastic oscillations of a manipulator with a load were calculated for different configurations of manipulators and loads. Some results for a manipulator shown in Fig. 1 are presented in Fig. 2. Here we consider a manipulator which consists of two links 1, 2 of equal length l. The load is a point mass m. Three eigenfrequencies are given by formulas

$$\omega_1^2 = aq\left[3(q-p)\cos^2\alpha + 3q\cos\alpha + 3p + q(p+1)\right]^{-1}$$
$$\omega_{2,3}^2 = 2a(3\cos\alpha + p + 4 \pm P)^{-1} \tag{1}$$

Other eigenfrequences are much higher than ω_1, ω_2, ω_3. The following notations are used in (1)

$$a = \frac{3E_1I_1}{ml^3}, \quad p = \frac{E_1I_1}{E_2I_2}, \quad q = \frac{C_1}{E_2I_2}$$
$$P = \left[4(p+3)\cos^2\alpha + 6(p+4)\cos\alpha + 13 + 4p + p^2\right]^{1/2} \tag{2}$$

The links 1, 2 of the manipulator are assumed to be elastic rods with circular cross-sections. In formulas (2), E_1 and E_2 denote the moduluses of elasticity for the material of the links, I_1 and I_2 are the moments of inertia of their cross-sections with respect to their diameters, C_1 is a torsional stiffness of the first link, α is an angle between the links, see Fig. 1. Note that the torsional stiffness of the second link does not appear in formulas (1), (2) because the load is assumed to be a point mass.

Free oscillations with the lowest frequency ω_1 are perpendicular to the plane where both links of the manipulator lie. The oscillations with frequencies ω_2, ω_3 take place in the plane of the manipulator and are perpendicular to each other. The angle β between the second link and the direction of the oscillations of the load with the frequency ω_2 is given by

$$\beta = \text{sign}(\sin\alpha) \arccos\left[(P+2+p+3\cos\alpha+2\cos^2\alpha)/(2P)\right]^{1/2} \qquad (3)$$

Fig. 2 shows the non-dimensional eigenfrequencies of elastic oscillations $\Omega_i^2 = \omega_i^2/a$ as functions of the angle α for the fixed values of the parameters $p = 1.5$, $q = 1.125$. When $\alpha \to 0$ or $\alpha \to \pi$ then $\Omega_3 \to \infty$. Similar calculations of free oscillations were performed also for other configurations of manipulators.

In the paper [4] the general approach was developed for the analysis of non-stationary elastic oscillations of a manipulator (for the first model a)). This approach is based on the asymptotic theory of oscillations. The motion of the rigid manipulator is assumed to be known: the angles of rotation of its hinges are some given functions of time. We found an asymptotic solution of the equations of motion for the elastic manipulator. This solution uses eigenfrequencies and eigenvectors of free oscillations and the given motion of the rigid manipulator. We heve obtained explicit formulas for all principal mechanical parameters of the system: elastic displacements, velocities, forces and torques acting in the structure of the manipulator. The equations of motion of the elastic manipulator were integrated also numerically by a computer. The results of this straightforward simulation have been compared with asymptotic analytical solutions and coincide with high accuracy. Some results are presented in Fig. 3

where the component u_x of the elastic displacement of the load m (parallel to the axis x) and the torque M_2 (see Fig. 1) are shown as functions of time. Here the manipulator performs certain three-dimensional motion. Note that the displacement consists mostly of oscillations with frequencies ω_1, ω_2 while the torque contains also oscillations with higher frequency ω_3. This fact can be easily explained if one recalls that the torque is obtained from the displacement by means of differentiation.

Similar theoretical analysis and computer simulation were performed also for other kinematic schemes and mechanical models b), c) of manipulators mentioned above. Elastic non-stationary oscillations were calculated for different motion of manipulators.

We analysed elastic flexibility of industrial robots also by experimental methods. For the Soviet pneumatic robot Tsiklon shown in Fig. 4 we measured static displacements due to static loads. The measuring technique included loading of the manipulator by constant forces applied at different characteristic points, fixing of different joints of the manipulator and measurement of elastic displacements at different places, see Fig. 4. Some characteristic experimental results are shown in Fig. 5; here F is a force (in kilograms) and δ is a displacement (in millimeters). All experimental diagrams "force-displacement" are linear with high accuracy. The analysis of obtained results showed that the main contribution in the total flexibility of the manipulator (approximately 50%) is due to the flexibility of the vertical axis of rotation; the second largest contribution is due to the joint of the horizontal progressive motion of the hand. The measuring technique described above was applied also to other Soviet pneumatic robots (Ritm et al). This technique permitted to determine principal origins of flexibility which have influence on the accuracy of manipulators. The results of measurements were used for identification of elastic parameters in the mechanical models of manipulators (for the model c)).

We have carried out experimental analysis of elastic oscillations of the Soviet industrial robot Universal. Dynamic characteristics were obtained by means of accelerometers which measured accelerations along three orthogonal directions at different points of the manipulator while the robot performed some typical motions (rotations, vertical motions, progressive motions of the hand). Accelerations were amplified, recorded on the magnetic tape and then processed by a computer. Algorithms and programs of data processing included obtaining spectral characteristics and correlation functions. As a result of this experimental research we studied dynamic processes in pneumatic and electromechanic joints of the robot. As an example, spectral characteristic of the hand of the robot Universal is shown in Fig. 6; the length of the hand here is maximal. Eigenfrequencies of the hand are $\omega_1 = 14$ Hz, $\omega_2 = 20$ Hz$_1$, $\omega_3 = 31$ Hz. If the length of the hand is equal to 20% of the maximal length, then the eigenfrequencies become approximately 1.4 times greater.

OPTIMIZATION OF REGIMES AND PARAMETERS OF INDUSTRIAL ROBOTS

We used methods of optimal control for obtaining optimal regimes of working operations of robots. We applied Pontryagin's maximum principle [5], analytical methods of optimal control developed in the book [6] and the numerical method of successive approximations described in the survey paper [7].

Some time-optimal regimes were calculated for a manipulator with rigid links shown in Fig. 1. Here we consider horizontal two-dimensional motions of a manipulator with two rotational degrees of freedom; a load is a point mass. The control torques M_1, M_2 of the motors acting at the two rotational axes of the manipulator are assumed to be bounded by a constant M : $|M_1(t)| \leqslant M$, $|M_2(t)| \leqslant M$. The problem is to find the functions $M_1(t)$, $M_2(t)$ and the corresponding motion of the manipulator for the time-optimal regime. Initial and terminal positions of the load are fixed, time T of the operation is to be minimized. Some results obtained numerically by means of the method [7] are shown in the Fig. 7 . Here the optimal trajectories of the load and respective minimal times T for several initial and terminal positions are given. Note that optimal trajectories depend on the configuration of the manipulator: unbroken lines in the

Fig. 7 correspond to configurations with $\alpha < 0$, dash lines correspond to $\alpha > 0$, and arrows show the direction of motion. Time-optimal trajectories differ notably from straight lines.

Time-optimal regimes were obtained also for two-dimensional horizontal motions of the industrial robot Universal. This robot has three degrees of freedom for horizontal motions: rotation of the platform about the vertical axis O, rotation of the hand about the vertical axis O_1 and change of the length O_1O_2 of the hand, see Fig. 8. As the point load O_2 has two degrees of freedom, therefore the position of the load does not determine unique configuration of the manipulator. Experimental measurements were carried out which showed that all three motions are almost independent: time of each motion depends (in a nonlinear way) only on the angle or the distance of this motion and does not depend on other motions. The obtained experimental data were used in calculating time-optimal regimes and corresponding initial and terminal positions of the manipulator for given initial and terminal positions of the load. One example of optimization is shown in the Fig. 8. Here the prescribed transfer of the load can be obtained by one motion, namely by rotation of the platform about the axis O by the angle of 90°. This regime shown in the left part of Fig. 8 requires 3.6 seconds. The optimal regime shown in the right part of Fig. 8 uses all three motions of the manipulator and requires only 2.3 seconds. Time-optimal regimes calculated numerically were realized experimentally for the robot Universal. The difference between theoretical and experimental durations of optimal regimes does not exceed 3%.

We considered also some problems of optimization of parameters for pneumatic robots. Transient behaviour for rotational and vertical motions was studied. Dynamics of vertical motion of the robot Tsiklon is described by the following nonlinear system of differential equations

$$m\ddot{y} = mg - (p-p_0)S, \quad \dot{m}_0 = -kp\,\varphi(p/p_0) \tag{4}$$

Here m is a mass of the moving part of the robot, g is the gravity acceleration, y is a vertical displacement, p is a pressure of the air in the chamber, p_o is the atmosphere pressure, m_o is a mass of the air in the chamber, φ is a given function, k is a constant. Typical vertical motions obtained by means of numerical integration of the system (4) are shown in Fig. 9. The oscillations here are due to elasticity of the air in the chamber. Optimal values of parameters of the robot Tsiklon were found which correspond to the best quality of its transient behaviour. In particular, optimal value of the parameter k was calculated; this parameter is connected with the diameter of the orifice through which the air passes from the chamber into the atmosphere.

MEASUREMENT OF MECHANICAL PARAMETERS OF THE MOTION OF ROBOTS

Improvement of functional possibilities of the modern robotics systems largely depends on their measuring systems or sensor parameters. Various physical measuring concepts: mechanical, electrical, acoustical, pneumatic, electromagnetic, optical, are applied to robotics. Combination of higher accuracy and reliability is particularly efficient in measurement of motions, distances near location zone, linear sizes, availability of parts by means of pneumatic proximity sensors (Fig. 10). The number of sensors depends on a specific robotics problem and the automation process (assembly, punching, shaping, grinding, welding etc.). In automation of the punching or molding process a robot has to be equipped with one pneumatic grip sensor and two pneumatic sensors of orientation in the press molding (Fig. 10 a, b). The sensors shown in Fig. 10 were developed in the Institute for Problems of Mechanics, USSR Academy of Sciences. These sensors consist of turbulent (Fig. 10 a, 1 - 5) and laminar (Fig. 10 a, 7) resistors, "flop-nozzle" elements (Fig. 10 b, 9), fluid capacities, functional modules and fluid channels (Fig. 10 a, pressure P_{B1}, P_{B2}). The mode of operation of the pneumatic detector is the following: when the object of manipulation (Fig. 10 a, 8) is near to the nozzle, the pressure goes up in the output channels (P_{B1}, P_{B2}). When the object (Fig. 10 b, 10) is in the gripper of the robot, the nozzle of the sensor is near to it and the the pressure rises. The relationship between the relative pressure P_{B1}/P_o and distance x/d_o is shown in Fig. 10 c. The experimental dependence between distance x/d and designing parameters y/L is shown in Fig. 10 d;б; P_{B1} - output pressure, P_o - supply pressure, x - distance, d - nozzele diameter.

The results and computer methods for calculation pneumatic robot sensors are summarized in [8 - 10]. The system of sensors for a robot should have parameters corresponding to each particular case of applications. The first stage in design of the detector - information system of a robot is connected with processing primary data of measurement by reading certain sensors or monitoring their parameters (sensitivity, selectivity, gain etc.). The subsequent stages should be devoted to formation of control algorithms with regard to the summarized primary measurement data.

INFORMATION-DYAGNOSTIC ROBOT BASED ON INDUSTRIAL ROBOT TSIKLON AND VIDEO-DETECTOR FUJI ELECTRIC

The designed system was made in the Institute on the base of two-level controls (Fig. 11). The system include the lower cycle programming control level and the upper decision making control level. The upper control level chooses a certain program motion of the lower level depending on the result of pattern recognition microprocessor video-sensor system. The upper decision making control level is Fuji Electric control system [11] and the lower cycle programming control level is control console of the industrial robot Tsiklon. Two manipulators of Tsiklon change their program motion due to upper level controls. The microprocessor is connecting with program cycle system by special interface unit . The objects to be inspected adopted by robot gripper and transportated to the control position of the video-detector (Fig. 11). The robot gripper takes the object to be inspected and transfer it to the control position of video detector (Fig. 11). The detecting signal, which is the result of on-line pattern recognition and decision making procedure, changes the program motion of the robot. The system includes the feedback loop. Photo shows one version of the system with industrial robot Tsiklon. This version is intended for inspection of motor car cylinder unit.

CONCLUSIONS

We presented here theoretical and experimental results on roboitcs devoted to mechanics, control and sensors for industrial robots. Some of these results are applicable to pneumatic as well as to electromechanical and other types of robots. The informational diagnostic robot described above widens possibilities of pneumatic robots and allows to use them for inspection and sorting.

The authors are grateful to L.D.Akulenko, N.N.Bolotnic, D.M.Gorinevsky, A.A.Gukasyan, A.A.Kaplunov, M.Yu.Rachkov, V.B.Veshnikov who took part in the research described in this paper.

BIBLIOGRAPHY

1. Ishlinsky A.Yu. Orientation, Gyroscopes and Inertial Navigation. Moscow, Nauka, 1976.
2. Ishlinsky A.Yu. Mechanics of Relative Motion and Forces of Inertia. Moscow, Nauka, 1981.
3. Akulenko L.D., Mikhailov S.A. and Chernousko F.L. Simulation of Dynamics of a Manipulator with Elastic Links. Izvestiya of the USSR Academy of Sciences, Mechanics of Solids, N 3, 1981.
4. Chernousko F.L. Dynamics of Controlled Motions of Elastic Manipulator. Izvestija of the USSR Academy of Sciences, Engineering Cybernetics, N 5, 1981.
5. Pontryagin L.S., Boltyansky V.G., Gamkrelidze R.V. and Mishchenko Ye.F. Mathematical Theory of Optimal Processes. Moscow, Nauka, 1976; Wiley-Interscience, 1962.
6. Chernousko F.L., Akulenko L.D. and Sokolov B.N. Control of Oscillations. Moscow, Nauka, 1980.
7. Chernousko F.L. and Lyubushin A.A. Method of Successive Approximations for Solution of Optimal Control Problems. Optimal Control Applications and Methods, vol. 3, N 2, 1982.
8. Gradersky V.G., Paroy A.A. Industrial Robot with Fluently Braking Motion of the Pneumatic Driver. Vestnik Mashinistroieniya, N3, 1981.
9. Gradetsky V.G. Analysis of Pneumatic Sensors of Industrial Robots. Robotic Systems in the Industry, Papers of the 2nd National Conf. of Industrial Robots, Minsk, May, 1981.
10. Bejasov I., Gradetsky V., Tsvetkov G., Nenov C. Fluid Elements and Systems, Sofia, Technika, 1976.
11. Nitta Y. Visual Identification and Sorting with TV-camera Applied to Automated Inspection Apparatus. Proc. of the 10th ISIR, March, 1980, Italy.

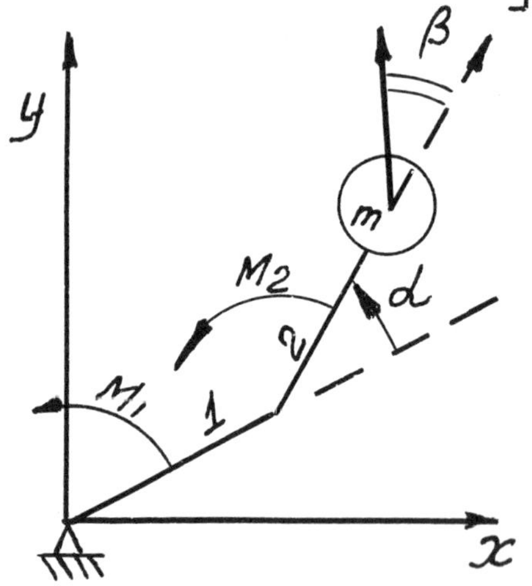

Fig. 1

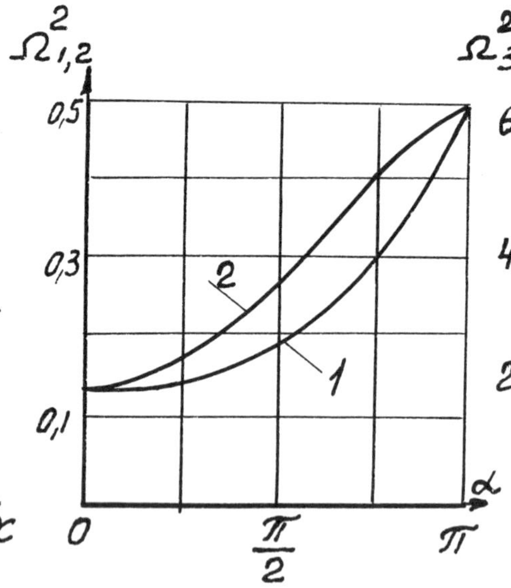

Fig. 2 a

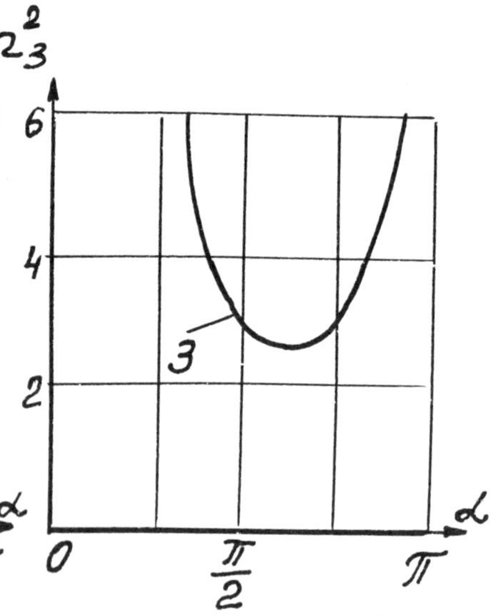

Fig. 2 б

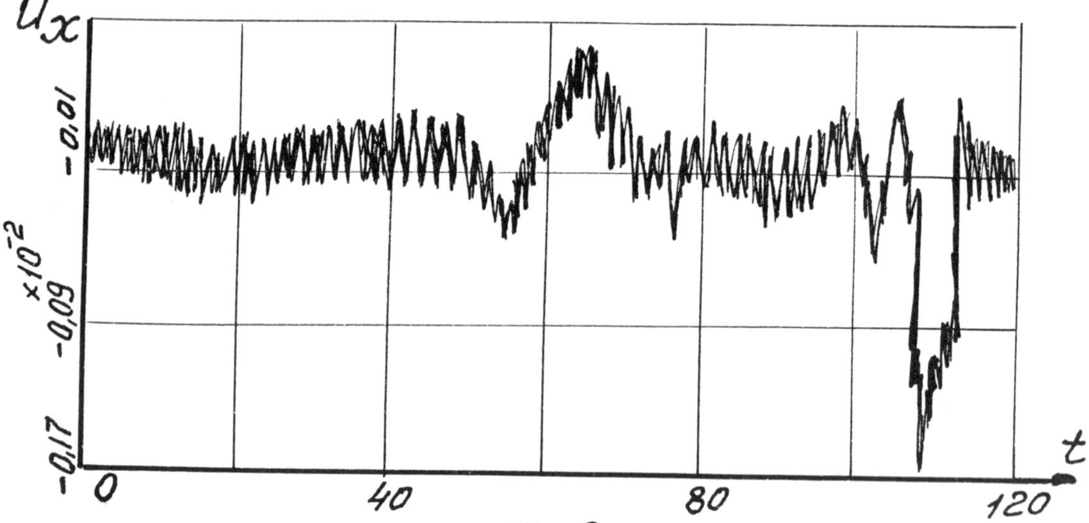

Fig. 3 a

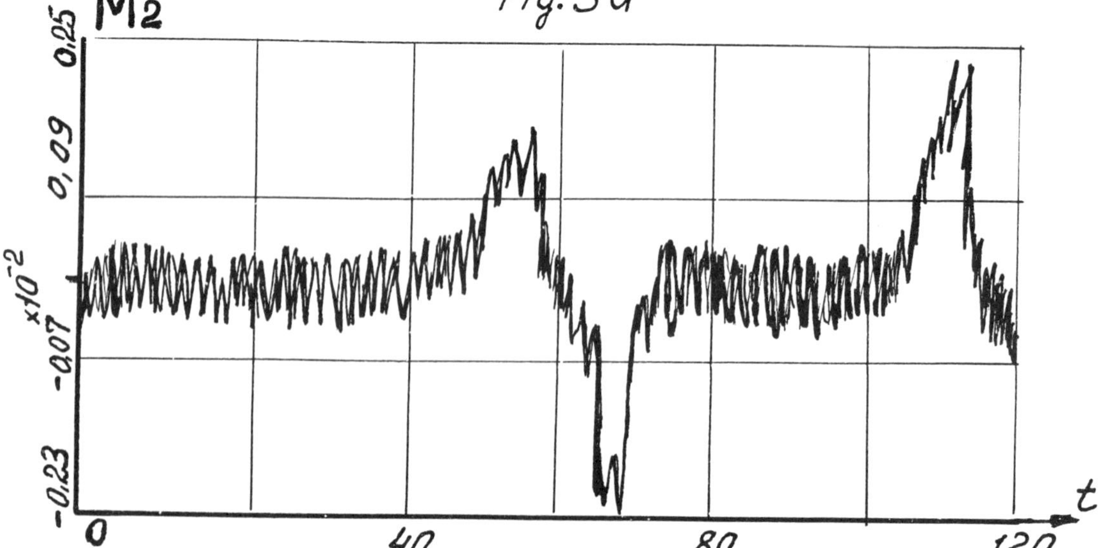

Fig. 3 б

13-80

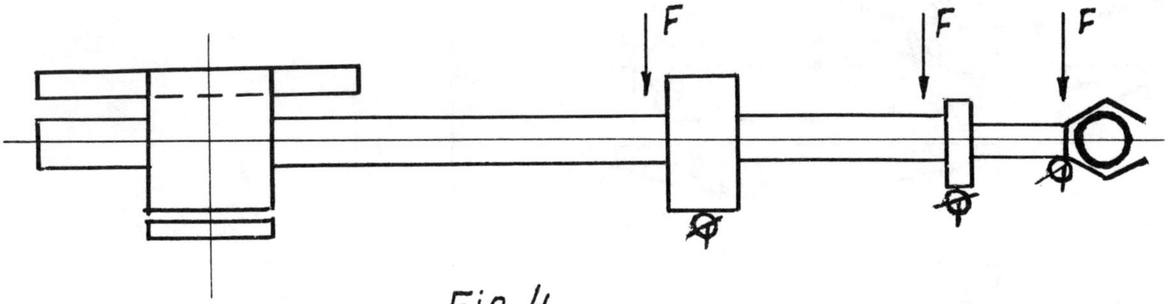

Fig. 4

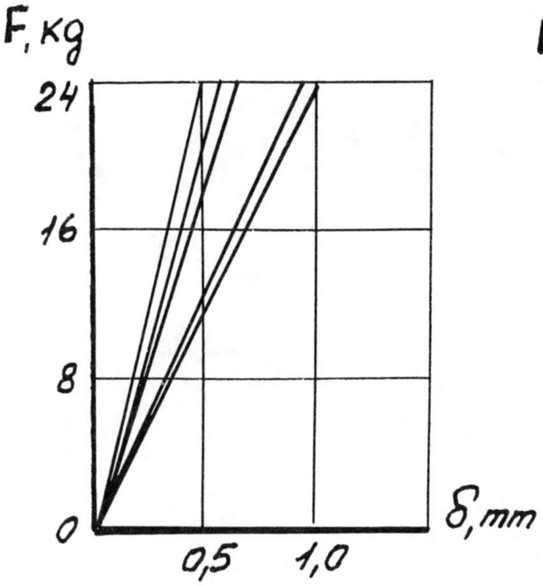

Fig. 5 a

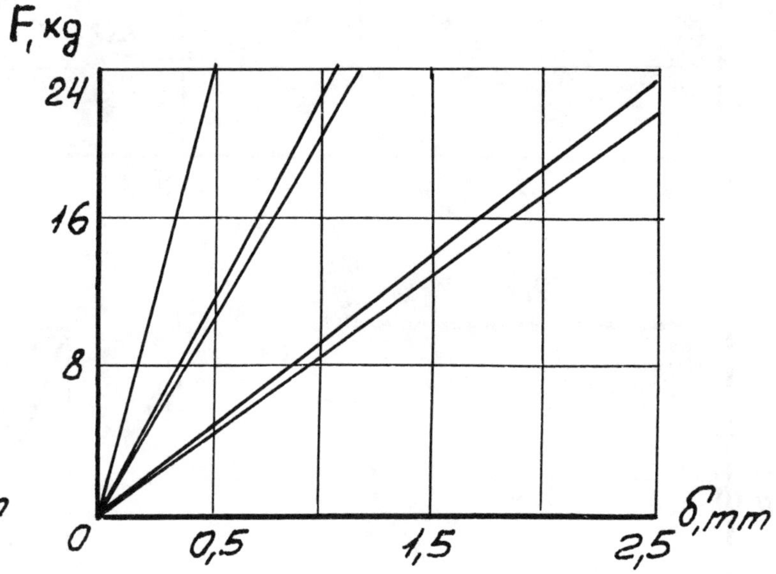

Fig. 5 в

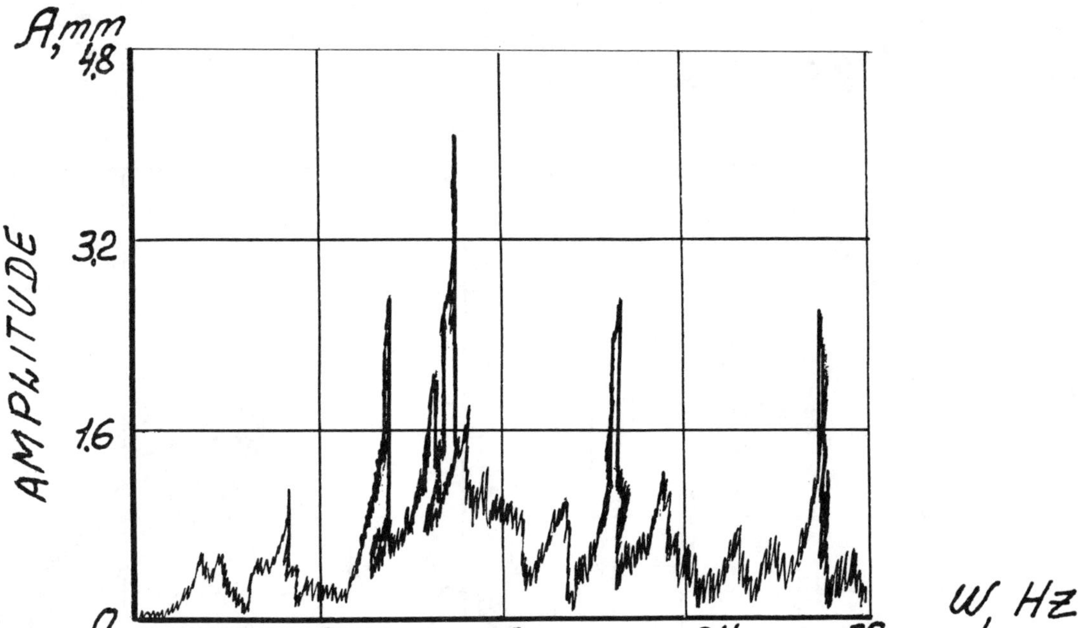

Fig. 6

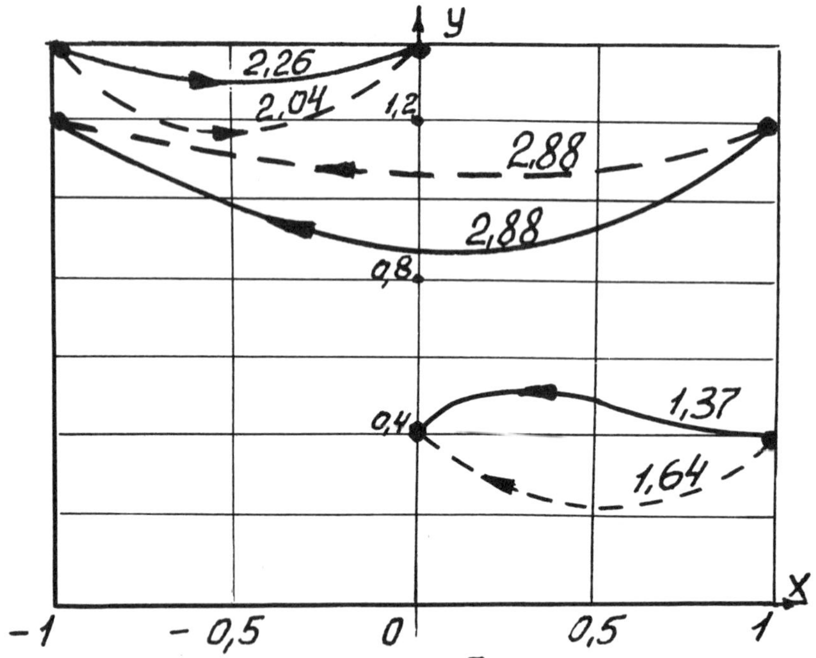

Fig.7a

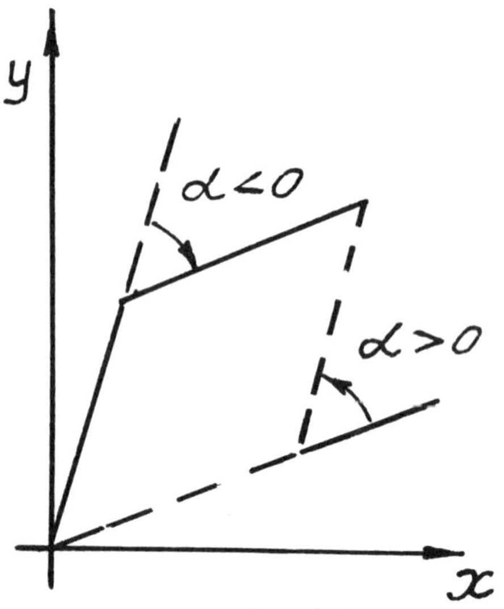

Fig.7в

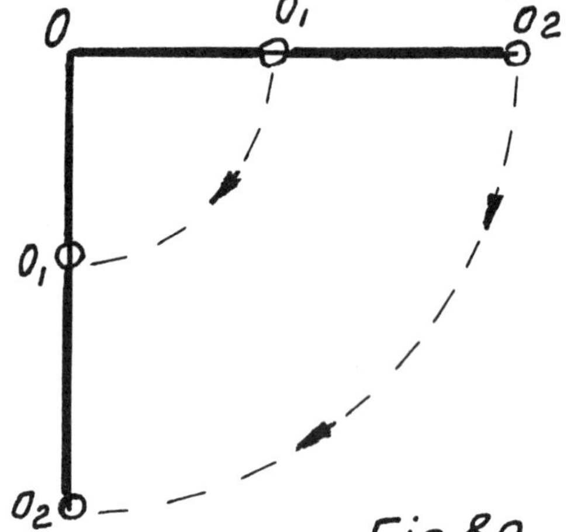

Fig.8a

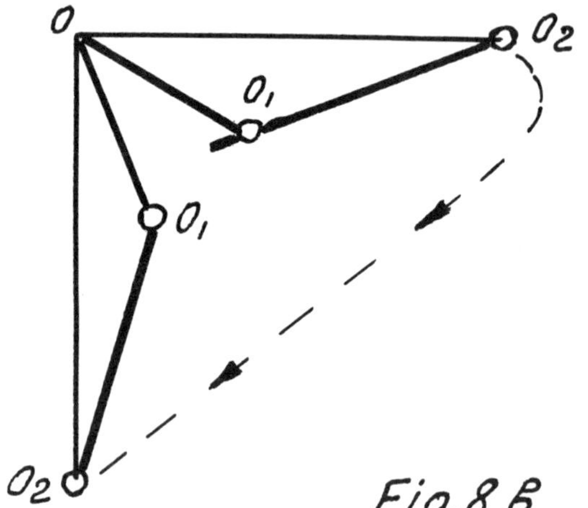

Fig.8в

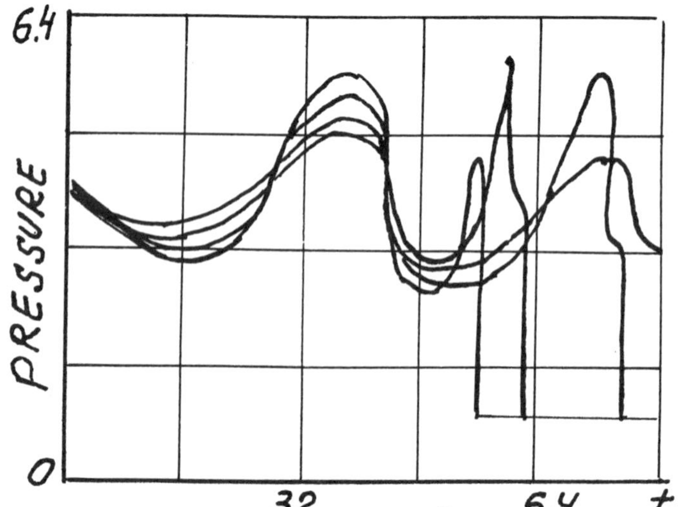

Fig. 9a

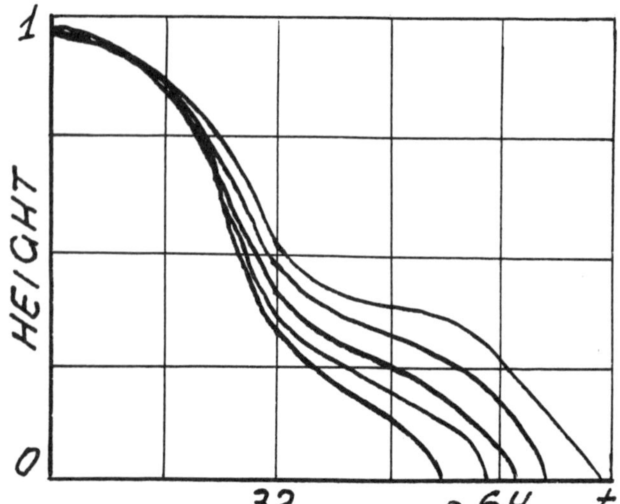

Fig. 9в

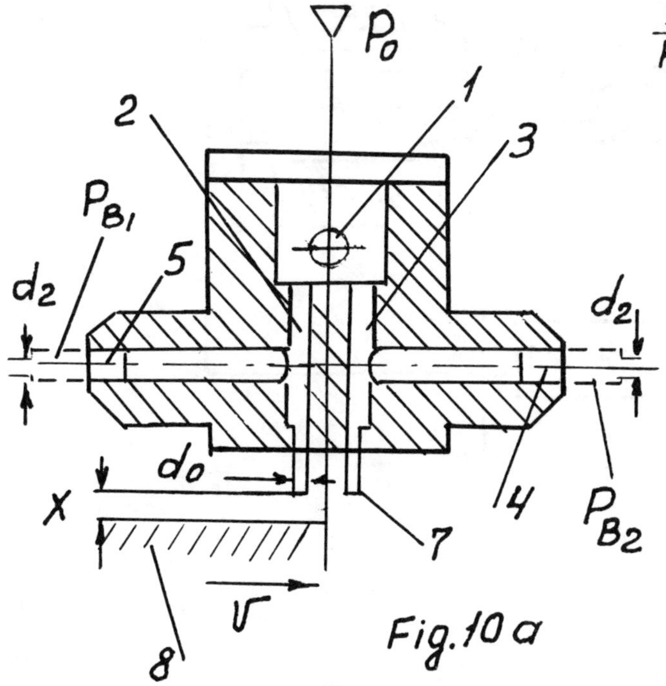

Fig.10 a

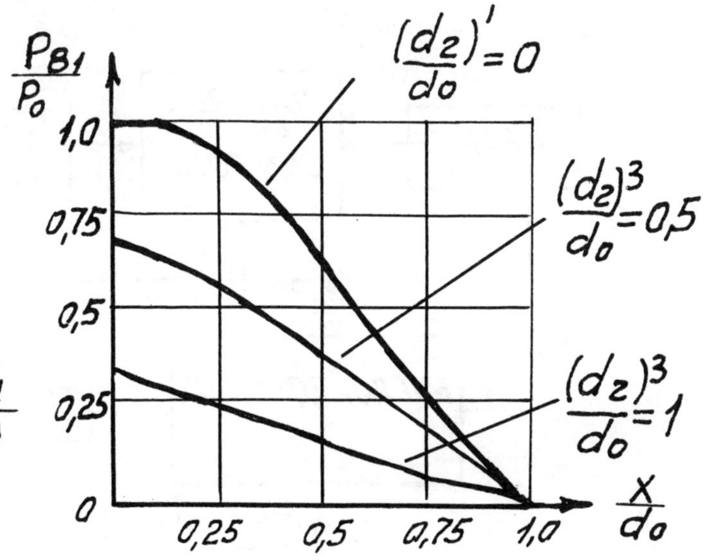

Fig.10c

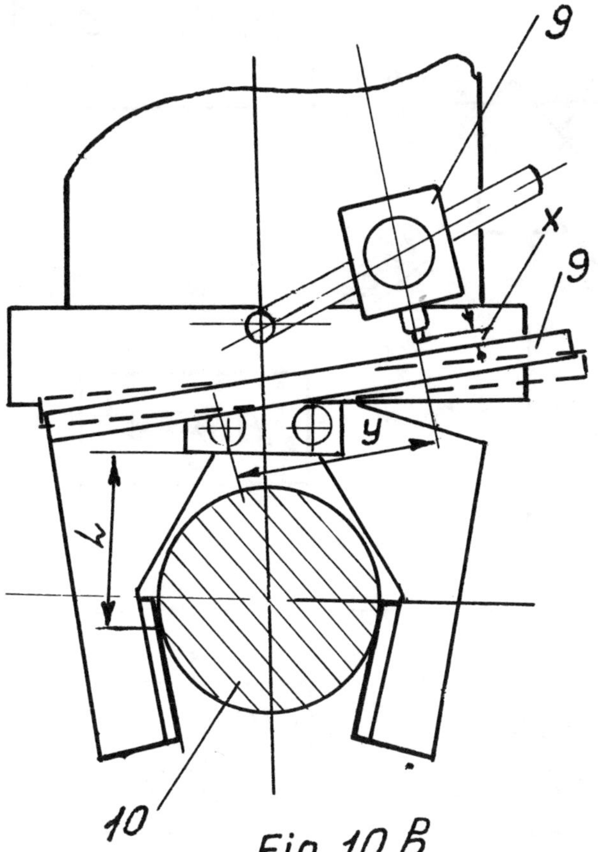

Fig. 10 B

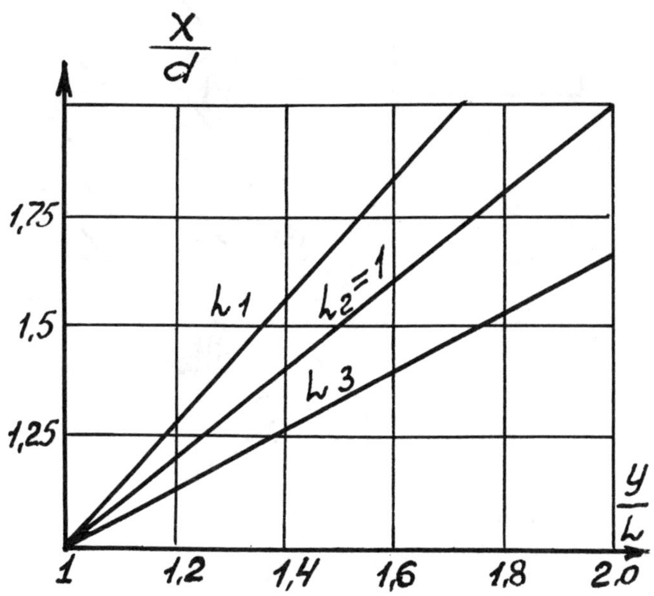

Fig.10 d

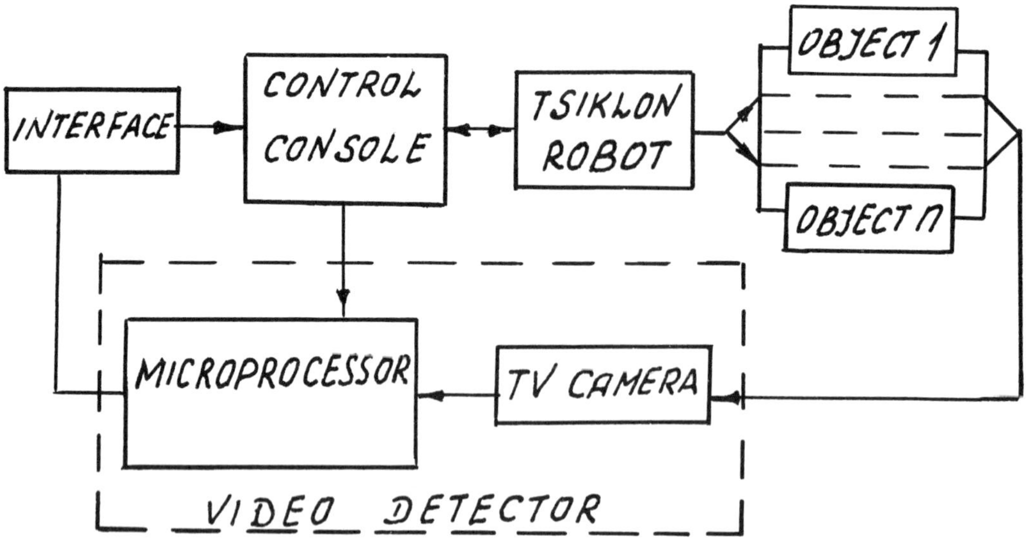

Fig. 11a

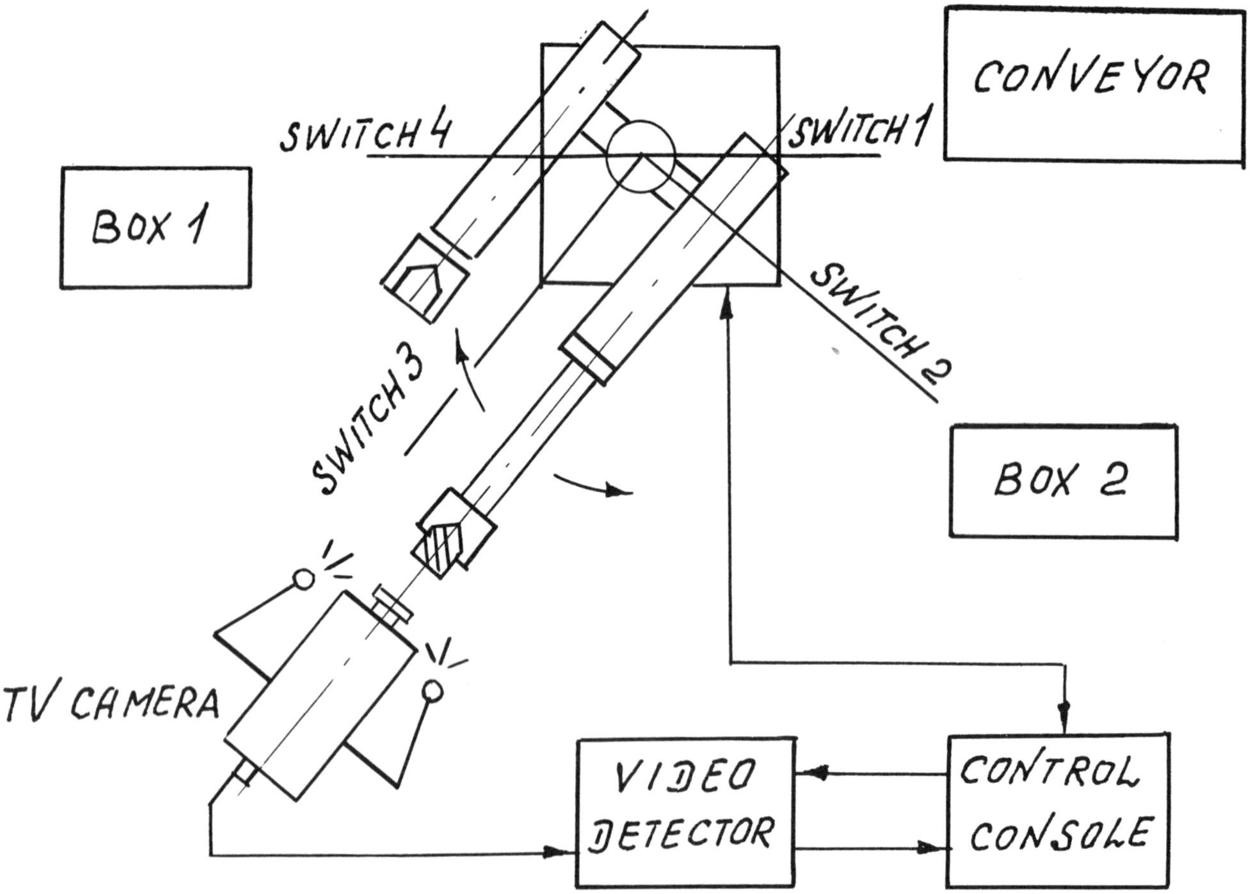

Fig. 11b

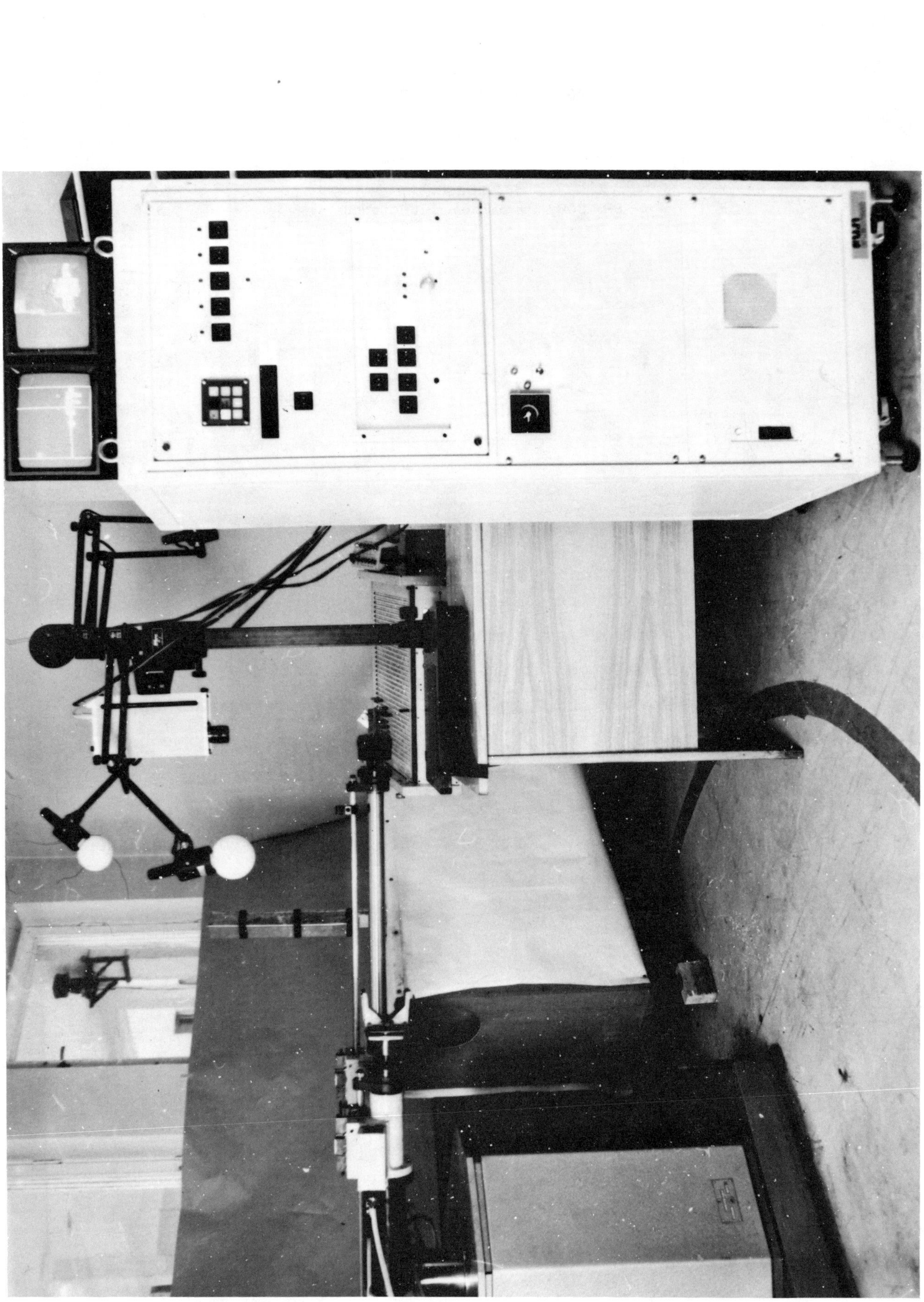

Illustration 1

Self-Adaptive Control of High Performing Translation Module

A. Bourjault
ERA 906—Microsystemes et Robotique
Besancon, France

P. Andre
ERA 906—Microsystemes et Robotique

R. Sepaser
ERA 906—Microsystemes et Robotique

Abstract

This paper deals with the study and realization of an electro-hydraulic translation module control system, intended for fast and precise displacements of loads with an a-priori unknown mass, and able to vary after a certain number of work-cycles. The system being highly non-linear, the implementation of an analogical control system would require a sharp adjustment of its parameters; that is why we have devised a numerical control system, as a medium of a spectral strategy, splitting up the motion in two parts: a fast displacement, aiming at a wrong target, followed by a precise positioning, on a short distance. In order to compensate for the slow derivative (supply pressure variation, dry frictions) a self-adaptive algorithm is used.

INTRODUCTION

Little or middle mass produced articles assembly (85 % of the industrial production) requires automation appliances offering a high adaptability-degree. Therefore, a few years ago, we started the study of a programmable motricity-module, able to constitute manipulators or assembly automatons with universal vocation for producing small homogeneous lots of diverse articles.

The fundamental characteristics of these sets are :
- modularity, concerning both motricity organ and mechanical structure, materially juxtaposable and functionally co-ordinable to others.
- programmability : program planning realizable by learning in real utilization conditions and working up through a non-especially qualified hand.
- self-adaptability, so as to take into account modifications of robot environment (inertia variations, transverse accelerations).

In this context, the prototype of an electro-hydraulic programmable translation module has been conceived and realized in collaboration between Sormel Inc. and Laboratoire d'Automatique de Besançon.

DESCRIPTION OF PROGRAMMABLE TRANSLATION MODULE

We focused our study on 0.5×10^{-3}m stroke module. The performances, fixed by the schedule of conditions were :
- accuracy : repositioning hysteresis inferior to .1 millimeter
- motions : minimum speed limit 1.5 meter per second and no overtravel
- orders to be given in absolute value and numerically
- transported mass : maximum value : 15 kg.

The constitution of this translation module is the following (See Fig.1) :
- a linear differential jack (especially produced by Integral Hydraulic), with a 500 mm useful stroke ; piston diameter : 14 mm and stem diameter : 10 mm (area ratio between both sides of piston is approximately 0.5) ; working under a 100 bar pressure, it can provide a 700 N force. This type of jack has the advantage of being controlled by a two-way valve, which is much easier to perform than a four-way valve, necessary to govern a double-acting symmetrical jack.
- a servovalve governed by a D.C. motor without moving contact. The servovalve position-sensor conceived and realized in our laboratory, is an original opto-electronic sensor without any friction and with low inertia. It provides analogical information, proportional to the position of a spot of light moving on a sensitive surface (letters patent outstanding). So fitted out, the servovalve characteristics are : positioning precision superior to 3°/oo, time response with low amplitude (few hundredths of radians) 4 ms, no overtravel, critical damping obtained by corrective networks.

Besides, it seemed necessary to fix a superior limit to accelerations and decelerations so as to eliminate the risks of losing any of the transfered articles. The limit chosen is : 57 m/s^2, the nominal value being 56,25 m/s^2.

DISADVANTAGES OF ANALOGICAL CONTROL SYSTEM

At first, we have experimented a PD servosystem, to guarantee the translation positioning of the jack (see Fig.2). But the real system being strongly non-linear, its dynamic characteristics depend upon the wanted stop abscissa X_c (modifications of compressed oil volume) and upon the transfered load M ; therefore, obtaining stable servo-control without overtravel imposes to adapt (in a rather sharp way) the derivative action gain to X_c and M. Such an ability is difficult to confer to an analogical control-system. For an isolated module and which is transfering a purely inertial load, the a-priori adaptation can be eventualy considered. Then it would be sufficient to precise, while programming, the transfered-mass value for each elementary translations which constitute the work cycle. But in the case of a robot composed of several modules, each one materializing its degrees of freedom in translation or rotation, the a-priori adaptation is in fact impossible ; the load, generally variable during motion (mechanical system with varying geometry) is rarely exclusively inert.

Those considerations brought us to opt for the realization of a digital control-system (based on a monochip microcomputer of the 8048 family). The numerical techniques actually allow the utilization of control-algorithms, more varied, then more efficient than the classical linear regulators', namely the self-adaptive algorithms which take into account the system's dynamic parameters variations (X_c, M, dry frictions), and adapt control parameters. That point is important, because the working conditions are likely to be repeated in short term (a few tons of successive cycles).

DIGITAL CONTROL SYSTEM

Global strategy of control

In order to find a solution to the dilemma speed and accuracy, we split the dificulty by splitting the action into 2 main stages :
- displacement stage, during which obtained positioning is near to wanted one (a wrong target is aimed at), it abscissa being $X_f (X_f \leqslant X_c)$
- a positioning stage, on a short distance (approximately 2 mm).

Besides, in order to minimize abruptness in accelerations and decelerations, we have chosen a control allowing acceleration control. That's why displacement stage is composed of 3 phases :
- a moving-off phase, with constant accelerations : 5.5,6 g on 20 mm.
- a cruising speed phase, at maximum speed (servovalve opened at the utmost)
- a braking phase, with constant deceleration : - 5.5, - 6 g on 20 mm.

Let's study now moving-off, braking and positioning phases.

Study of moving-off

The law of servovalve opening (with constant accelerations) is read :

(1) $\theta = \alpha\, t$ with :

$$\begin{cases} \alpha = \dfrac{S\,\gamma_o}{k}\sqrt{\dfrac{2\,S}{S\,P_u - 2\,M\,|\gamma_o|}} \\[2ex] P_u = P_a - 2\,\dfrac{f}{S} \qquad (\ f : \text{dry frictions} \ ;\ P_a : \text{supply pressure}) \\[2ex] M = M_o + M_u \qquad (M_u : \text{transfered mass} \ ;\ M_o : \text{mass without load}) \end{cases}$$

We can thus obtain constant acceleration by opening servovalve according to a linear time law. Reciprocally, if servovalve is opened according to known law, measurement of piston speed at given time allows to determine displaced mass. (See Appendix). From this observation, we have worked out a self-adaptive control method, which is composed of 2 parts :

1 st part : elaboration of in-situ program, through emulation procedure : a law of opening is imposed to servovalve, corresponding to working unloated. After a reference time t_r, we measure speed v_e reached by the module, and we deduct mass M from it. Then we can calculate the theoretical law (1) of servovalve opening, so as to obtain piston motion with constant acceleration $\gamma_o = 56.25$ m/s^2 (The calculations required for this phase are made off-line through emulation system).
Besides, the reference input of the automatic valve-control system being coded with 8 bits, it lies between -127 and +127 quanta (1 quantum = 1.18 10^{-3} Radian). Therefore, this reference input signal is worked out with an 8- bit counter which is increased or decreased (depending on the case) up to ± 127, at the rate of a clock of period T_o :

$$(2) \quad T_o = \frac{|\theta_{Max}|}{127\,|\alpha|} = \frac{0.15\ k}{127\ S\ |\gamma_o|}\sqrt{\frac{S\ P_u - 2\ M\ |\gamma_o|}{2\ S}}$$

2 nd Part : the value of M found thus is approximate since P_u and P_o may vary, because of the possible modifications of supply pressure P_a and dry frictions f. In order to compensate for these variations, we have implemented an algorithm which adapts temporal law of servovalve opening to real working conditions. To moving-off (i+1), we apply :

$$(3) \quad \theta = \alpha^{i+1} t \text{ with : } \alpha^{i+1} = \frac{\alpha^i}{\mu^i}\,,\ \mu^i = \frac{v_e^i}{v_r}\,.$$

which corresponds to recursive equation upon acceleration γ :

$$(4) \quad \gamma^{i+1} = \gamma_o\sqrt{\frac{S\ P_u - 2\ M\ |\gamma^{i+1}|}{S\ P_u - 2\ M\ |\gamma^i|}} \qquad \text{that converges on } \gamma_o\ .$$

It corresponds also to the recurrence upon the clock-period which increases θ :

$$(5) \quad T^{i+1} = T^i\ \mu^i = T^i\ \frac{v_e^i}{v_r}$$

Then, the self-adaptation of the temporal law of servovalve opening can be obtained through modification of period T, allowing a very accurate modulation of the law and realizing a steady control of the servovalve, contrary to what would happen if T was kept constant and increases in opening-angle were modulated. In practice, to each moving-off i, one measures speed v_e^i attained at the end of a reference time t_r and one calculates the ratio $\mu^i = v_e^i/v_r$ and then T^{i+1}.

Study of braking

In this phase, an open loop control cannot be used as before because the braking-length is critical whereas moving-off length is not.

Let v_d and θ_d be respectively the piston speed and the servovalve opening-angle, at the start of braking. Since a slowing down with constant deceleration is imposed, piston-speed v must follow a parabolic law, to abscissa X_f, abscissa of the end of braking :

$$v = v_d\sqrt{1 - \frac{X}{X_f}} \qquad \text{with} \qquad X_f = \frac{v_d^2}{2\gamma_o}$$

Hence the control-law :

$$(6) \quad \theta = \theta_d \sqrt{1 - \frac{X}{X_f}}$$

This relation may be considered as the equation of a guide-curve giving a one-to-one mapping between the position of the mobile and the relative opening of the servovalve.

During the cruising speed phase, or when it does not exist, at the end of moving-off, speed v_d reached is measured. The value of $2\ M\ |\gamma_o|$ is deduced from the value of period T^i really used during moving-off, and from ideal period T_o :

$$(7) \quad 2\ M\ |\gamma_o| = S\ P_o\ (1 - (\frac{T^i}{T_o})^2) + 2\ M_o\ |\gamma_o|$$

As v_o and $2\ M\ |\gamma_o|$ are known, the opening of servovalve at the start of braking phase θ_d can be calculated :

$$(8) \quad \theta_d = \frac{S\ v_o}{k} \sqrt{\frac{2\ S}{S\ P_u + 2\ M\ |\gamma_o|}}$$

that allows to use relation (6), θ being increased (or decreased) at the rate of period T_1.

However, the variations of P_a or f may disturb the working. To cope with that disadvantage, we compare X continuously (i.e. with a period much inferior $T_o\ T_1$) to theoretical value X_o deduced from equation (6) ; as the case may be θ is decreased faster or not. In practice, to avoid the calculation of X through equation (6), a guide-curve has been memorized for a given value of θ_d^r ; an appropriate transformation allows to calculate the corresponding value for any value of θ_d.

Study of positioning

As soon as braking phase is finished ($\theta = 0$), the control system performs another program that realizes numerical servocontrol according to an equation equivalent to :

$$\theta = A\ e + B\ \dot{e} \qquad \text{with : } e = X_c - X$$

A = 15 rd/m

B = function of X_c and M, lies between 0.05 rd.s/m (piston in, no load) and 0.17 rd.s/m (piston out, M = 15 kg).

The adaptation of the servocontrol to X_c and M is realized through modification of period T_2, by using a table of precalculated values. Because of the strategy of control which consists of bringing the mobile to stopping close to final position by way of using a controlled deceleration and then by a commutation on an automatic servocontrol program, the amplitudes of the steps are 6 to 8 quanta to the utmost; an overtravel of 1/8, that is 12,5 %, or a great overdamping can be allowed, since the stroke will be low, then of short duration. In that case, the accuracy required on B, then on T_2, can be indifferent (to 20 %). M and X_c may be quantized with 16 levels, and it is possible to build up a table with 256 values of T_2.

RESULTS

The control system described before has been established on a mono-chip 8048 computer. The tests performed on a prototype have allowed to verify that all the performances of the schedule were reached ; the repositioning after a forward and return stroke is better than 0.1 mm, the duration of a cycle being less than 1 s. ; the positioning servosystem is stable for any value of X_c as M = 16 kg. In the absence of information about M and P_a, the self-adaptability gives the expected results, if there is no overlapping between moving-off and deceleration phases (which supposes a length inferior to 4 cm between two successive orders). We are working now on getting some improvements on that point.

CONCLUSIONS

The obtained results will allow an industrialization of the module, these works having been carried on with SORMEL Inc. .

Such performances have been reached because, on the one hand during a cycle we have been beyond the dilemma accuracy-stability (splitting up of the motion into two stages : displacement and positioning), on the other hand on a few cycles, we have introduced a procedure of self-adaptation, which allows to get out of knowing certain parameters and to compensate for slow variations of the plant ; these procedures allow to give repetitive works to non-performing (then low-cost) robots.

BIBLIOGRAPHY

Taco J. Viersma, Analysis, Synthesis and Desing of Hydraulic Servosystems and Pipelines, Elsevier Ed., 1980.

Harris C.J. Billings S.A., Self-Tuning and Adaptive Control, Peter Peregrinus Ltd., New-York, 1981

D. Simon, Contribution à l'étude de modules de motricité hydrauliques pour robots d'assemblage, Th. Doct - Ing., Besançon, 1980

APPENDIX : Calculation of the transfered mass during moving-off phase.

The law of opening of the servovalve when acceleration γ of the piston is constant is :

$$\theta(t) = \frac{S\gamma}{k} \sqrt{\frac{2\,S}{S\,P_u - 2\,M\,|\gamma|}}\; t$$

with : $M = M_o + M_u$

M_o = mass of travelling module, unloaded running

M_u = useful mass

One can use the law of opening for servovalve working unloaded (γ_o is the wanted acceleration) as reference law :

$$\theta(t) = \frac{S\,\gamma_o}{k} \sqrt{\frac{2\,S}{S\,P_u - 2\,M_o\,|\gamma_o|}}\; t$$

This law, applied to working unloaded, leads to reference velocity v_r, after a time t_r ; when applied to working loaded, after the same time t_r, it leads to real velocity v_e :

$$v_r = \gamma_o t_r$$

$$v_e = \gamma \ t_r$$

Hence :

$$\Theta(t_r) = \frac{S \gamma_o}{k} \sqrt{\frac{2 S}{S P_u - 2 M_o |\gamma_o|}} \ t_r = \frac{S \gamma}{k} \sqrt{\frac{2 S}{S P_u - 2 M |\gamma|}} \ t_r$$

That is :

$$\gamma = \gamma_o \quad \frac{S P_u - 2 M |\gamma_o|}{S P_u - 2 M_o |\gamma_o|}$$

Assuming : $\mu = \dfrac{v_e}{v_r}$, we can find :

$$M = \frac{S P_u - \mu^2 S P_o}{2 |\gamma|} \quad , \text{with } P_u = P_o + 2 \frac{M_o |\gamma_o|}{S}$$

$$\gamma = \frac{v_e}{t_r}$$

<u>Conclusion</u> : the measure of v_e (real velocity after a time t_r) allows to calculate the total transfered mass (S, P_u and P_o are characteristics of the module).

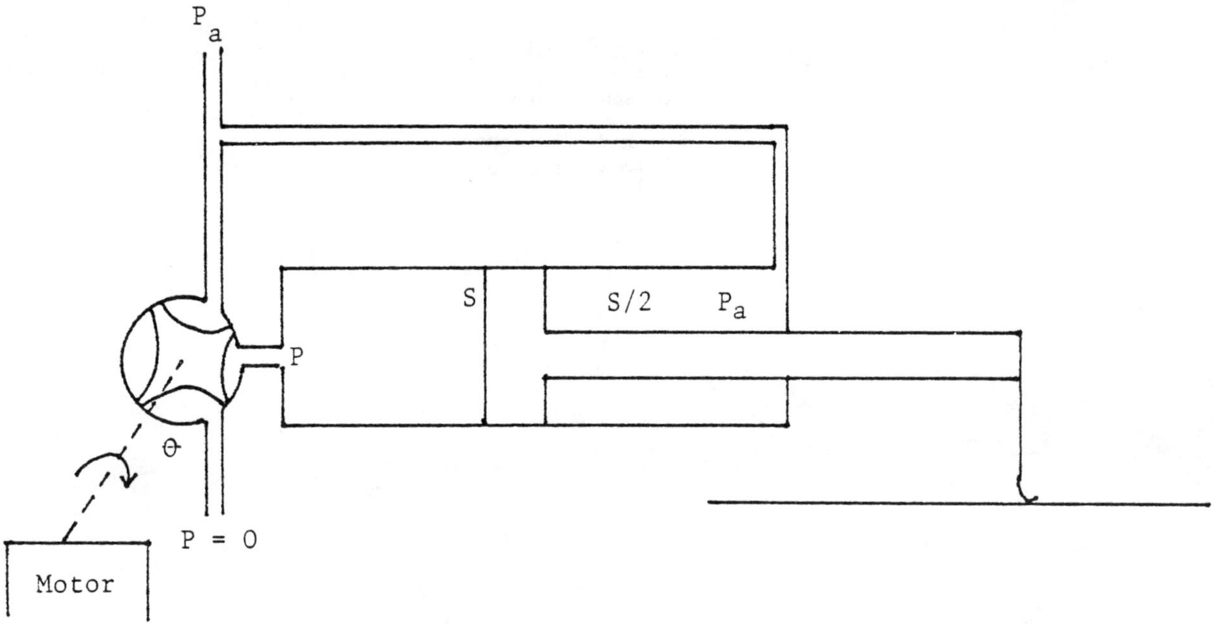

Figure 1 : Servovalve and piston.

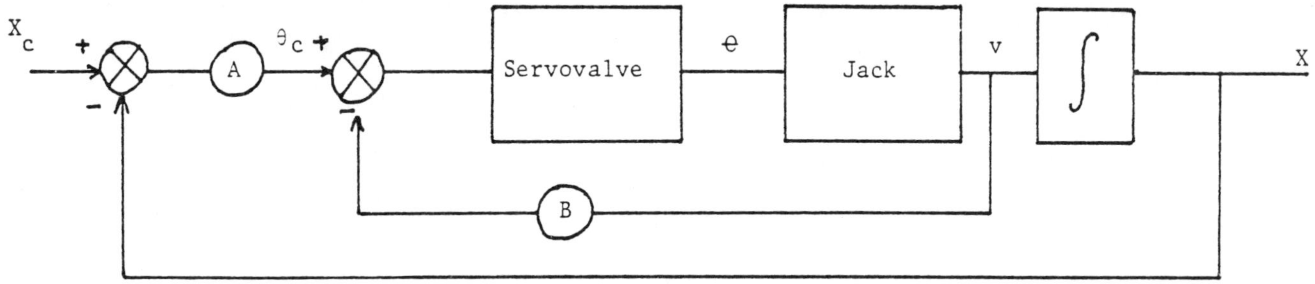

Figure 2 : Analogical servocontrol (P.D.).

Robotics Applied to Rapid Excavation

Reinaldo J. Chavez
U.S. Army—Meradcom
Fort Belvoir, Virginia

Charles A. Amazeen
U.S. Army—Meradcom

Abstract

This paper describes the system design and modifications associated with the implementation of a hierarchical 8 bit microprocessor control structure to the excavation cycle of a J.I. Case Model 35 backhoe. The design criteria and software development associated with using four absolute position optical encoders to determine the state of the system is discussed, as well as the feedback--control software development associated with driving proportionally controlled electro-hydraulic valves. The interface requirements and the structure and types of data and control lines used are described. Preliminary results from the prototype demonstration tests are discussed.

INTRODUCTION

Recognizing the US Army Corps of Engineers has a need for applying industrial robotic technology to high risk and labor intensive tasks performed in the environment of a modern battlefield, this laboratory started a small investigative project in a specific robotic application area - robotics applied to rapid excavation (e.g., field fortifications). The excavator used was a modified J.I. Case Model 35 Backhoe mounted on a highly mobile Mercedes Unimog truck.

The conditions of the modern battlefield make this task extremely hazardous to the human operators due to the advent of highly accurate, technologically sophisticated weapons systems and the possibility of nuclear, biological and/or chemical contamination of the battlefield environment.

The goal, then, of the project was to eliminate or reduce, through automation, as much of the manual digging operation of the backhoe as time and funding permitted. Also, while in the process of doing this, designing a microcomputer based control system (including the software algorithms) that, in addition to digging, could be applied to other tasks utilizing the modified backhoe (e.g., a heavy lift manipulator).

With this goal in mind, what will be presented in this report will be the preliminary results associated with controlling the backhoe arm through a series of sequential motions. The complete cycle controlling program will be only discussed from a conceptual viewpoint while the more primitive state driving and state sensing program are explained in detail.

HIERARCHICAL CONTROL SYSTEM

The microprocessor control system consists of three separate central processing units (CPU's) of the 6502 type. There are two levels of control associated with this system. The first, or primitive level, consists of those programs which determine the state or current position of the backhoe arm (bucket, dipper, boom and swing); and then, the necessary voltages that drive the proportionally controlled electro hydraulic servo valves. (See figure 1)

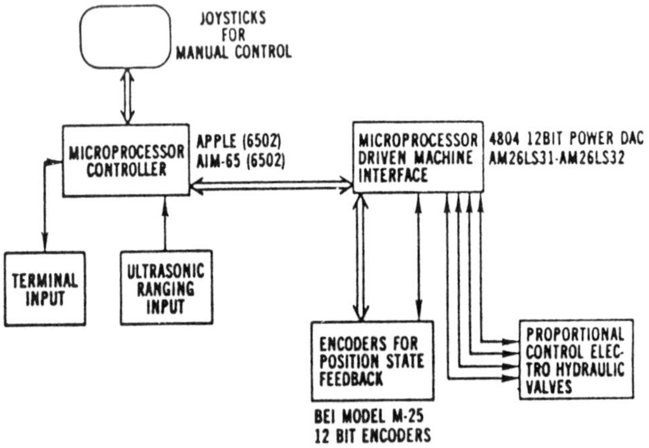

Figure 1
Control Hierarchy

The higher level consists of a single overall system control program which interacts with the human operator in a real time basis using a "menu" driven approach. From the menu, one of three operation modes can be chosen. The backhoe can either be entirely controlled by the operator using joysticks (that are also used for an interrupt controlled emergency override), or the operator can choose a hole pre-defined in the control program, or the operator can input the dimensions of a hole to be excavated. These operation modes can be found in the program as subroutines or, as called in PASCAL*, procedures. The microprocessors selected were two Rockwell AIM 65's and one Apple II system. This equipment was used due to availability and budgetary constraints.

* The higher level control program is written in UCSD PASCAL and utilizes on line floppy disks that store the segmented procedures.

SYSTEM CONTROL PROGRAM

The system control program is the software which will operate the excavator while running in the Apple II and interfacing by an RS 232 serial connection to the subordinate AIM 65 processors which will maintain the servo loops.

The high functions necessary to operate in the hole template mode have been run using graphic simulation to observe results and have operated successfully in that environment. This simulation is not presented due to the restrictions as to the length of this paper. The total controller (to include the template mode) has not been completed in detailed code at this time; it exists as the top stage of a top down development. This stage identifies functions to be performed, shows the organization and control flow, and provides the basis for writing the detailed software. The structure of this present skeleton is patterned on PASCAL as its implementation language and the detailed code of the controller will be written in PASCAL except for a few assembly language I/O drivers. The initial draft of detailed code is expected to be ready for evolutionary testing in the fall of 1982 after completing checkout of the four servo loop subsystems under control of AIM 65.1 and AIM 65.2.

STATE DETERMINATION

In the state sensing portion of the state determination program, the state vector is uniquely defined by the four angles α, β, γ, and ϕ. These angles are determined by the digital (10-bit) output of the optical encoders mounted on the bucket, dipper, boom, and swing assembly of the backhoe. (See figure 1A)

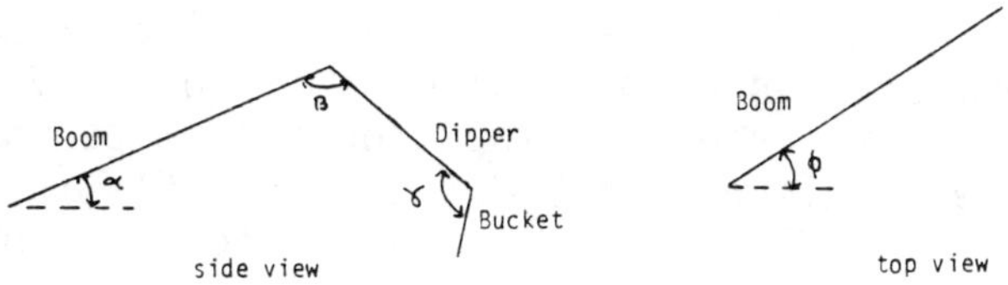

Angle	Encoder#
ϕ	1
α	2
β	3
γ	4

Figure 1A
State Determination

The state determination program developed uses the 6502 CPU Assembly Language instruction set to control both the data bus (AM26LS31 - AM26LS32) and the encoder select enable lines.[1]

Basically, the state determination program manages or strobes the optical encoder's 10-bit word onto the data bus, such that only one 10-bit word is on the bus line per decoded encoder enable. This sequence prevents any type of uncertainty on the data bus from occurring. It is, of course, apparent that the RS422 tri-state buses (uses AM26LS31 - AM26LS32) is enabled or in the transmit mode continuously. By continuously enabling this parallel bus, each encoder enabled onto the bus for transmission of data to take place. The main advantage is one of hardware reduction, namely, one RS422 parallel bus services as many devices as are strobed.

The enabling sequence is program controlled by the state determination program[2] using 8-bit words written onto the output registers of the AIM 65's Versatile Interface Adapter (VIA). The 6522 Versatile Interface Adapter (VIA) is an enhanced version of the 6520 Peripheral Interface Adapter (PIA)[3]. The main difference between the VIA and the PIA is that the VIA provides all the functions of the PIA plus 2-bit counter/timers and an 8-bit shift register. The shift register is a particularly useful addition since it allows for fast transfer of serial data between the two AIM 65's without a special bus interface. This area will be described in more detail in the next section.

There are only 2-bits used of the 8-bit word sent to the output registers of the VIA for the encoder enable. The actual optical encoder is generated by a demultiplexed/decoded high output generated from a SN74156/SN7404 encoder select circuit.

The AIM 65 applications connector which ties in directly to the applications VIA, provides pins PB3 and PB2 for the coded signal that is used by the demultiplexer/decoder to output lows to the HEX inverters. As previously stated, PB3 and PB2 are 2-bits of an 8-bit word sent under program control to the VIA. The actual assignment of the control words bits as either input or output ports of the VIA is done utilizing the memory mapped registers provided to the AIM's (CPU) Central Processing Unit by the applications VIA. See the state determination program for the actual HEX address.

Once the 10-bit optical encoder word is selected into the VIA pins PA0-PA7 and PA0-PB2, then a transfer of their values is done using the accumulation command LDA from the HEX address that has the memory mapped I/O port. At this point, trivial calculations can be made to determine that state vector or the state of the backhoe. The complete flow chart showing the state determination program as well as both a basic driven encoder select program and an assembly language encoder select program is shown at Fig. 2.

1. See Reference (2) pp. 25-30
2. See Reference (6) pp. 5-14-6-5
3. See Reference (6) pp. 5-14-6-5

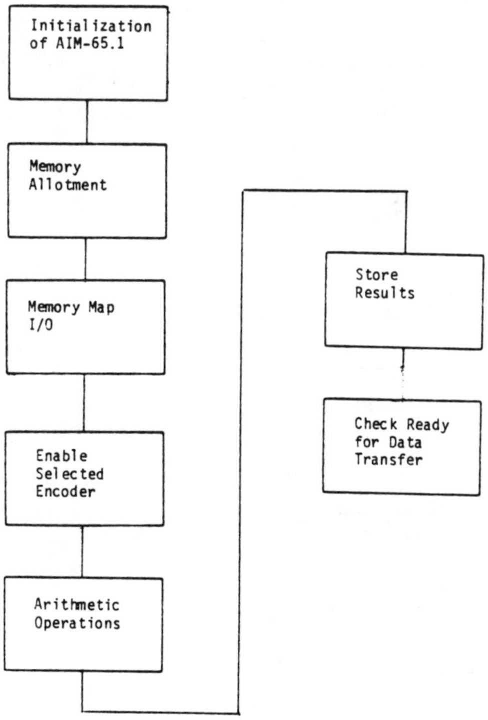

Figure 2
State Sensing/Determination Program

```
K * = 300
/43
0300 A9 LDA #00
0302 8D STA A003
0300 A9 LDA #FC
0307 8D STA A002
030A AD LDA #04
030C 8D STA A000
030F AD LDA A001
0312 85 STA 30
0214 AD LDA A000
0317 18 CLC
0318 38 SEC
0319 E9 SBC #04
031B 85 STA 31
031D A9 LDA #00
031F 8D STA A000
0322 AD LDA A001
0325 85 STA 32
0327 AD LDA A000
032A 18 CLC
0328 38 SEC
032C E9 SBC #08
032E 85 STA 33
0330 A9 LDA #10
0331 8D STA A000
0335 AD LDA A001
0338 85 STA 34
033A AD LDA A000
0330 18 CLC
033E 38 SEC
033F E9 SBC #10
0341 85 STA 35
0343 A9 LDA #30
0345 8D STA A000
0348 AD LDA A001
0328 85 STA #36
0340 AD LDA A000
0350 18 CLC
0351 38 SEC
0352 E9 SBC #30
0354 85 STA 37
0356 00 BRK
0357 00 BRK
0360 00 BRK
```

Figure 2
Assembly Language Encoder Select Program

Once the state vector calculations are made[4], if required, then the Matrix Transfer program is used. Specifically, what is denoted by Matrix is the unique set of four 10-bit words which define a point in configuration space as a location point for the final bucket position. The transfer portion is initiated upon either a request from the Apple microprocessor or the other AIM 65 microprocessor. The requests from the Apple microprocessor are done under separate program control and are periodic system status updates. The actual request is received from the Apple by the AIM 65.1 utilizing a RS232C port controlled by two 6850 Asynchronous Communications Interface Adaptors (ACIA)[5]. The request from the other AIM 65 is received and enabled utilizing the shift registers of the VIA's and the control lines CB1,2 and PB4,5.

The RS232C Protocol[6] used is essentially one of first determining the Data Terminal (DTR) device and the Data Set (DSR) device. In this application, the DSR device is the Apple microprocessor and the DTR device, the AIM 65 microprocessor. The driving rationale for this decision being one of minimizing the hardware modifications to either microprocessor.

Once the DTR and DSR have been selected, the standard sequence for communications is initiated. (See figure 3)

4. See Reference (3) pp. 251-252
5. See Reference (4) pp. 4-2-4-11
6. See Reference (5) pp. 2-4-2-9

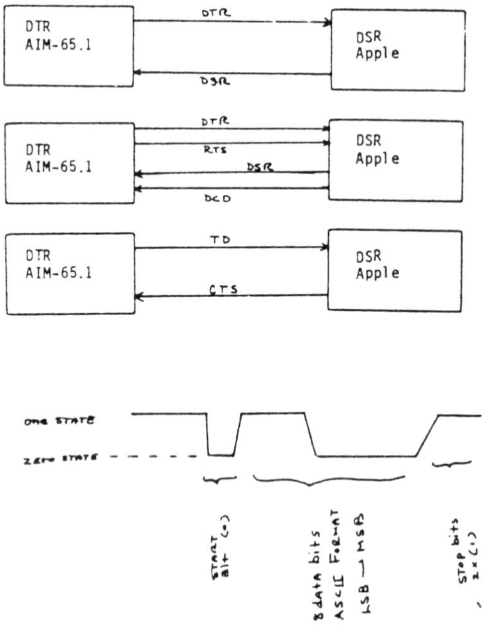

Figure# 3
RS232C Protocols

The communication rate has been selected as 9600 baud with 8-bit words, two stop bits, one start bit, and no parity check. The framing and bit width all being taken care of by the on board ACIA's.

The configuration of the start, data, and stop bits are shown in figure 3. The bit width calculation is based upon 11-bits transmitted per frame, e.g.

Baud Rate Calculation

CPS = Characters per second

9600 baud = Y (CPS) X 11 bits

Y(CPS) = 9600 ÷ 11

Bit Width = 11 ÷ 9600

The basic operation of the RS232C data link is controlled through the programmable ACIA's on both the AIM 65 interface board and the Apple RS232C interface board. The registers of the ACIA on the AIM 65 have been inserted in memory locations HEX 1000 through HEX 1008. The HEX address 1000 is the assigned base addresses and all the internal registers are accessed relative to this base address. The software developed to initiate the AIM 65 ACIA module first executes a reset operation, then writes the baud rate, word length, and number of stop and start bits into the R6551 ACIA control register. Also, the internal clock rate is selected by loading into the ACIA control register bit four a logic one.

A simple driver program is shown at figure 4. Note that an ASCII HEX 34 is sent to the transmit register of the ACIA via the STA 1000 command after first loading the value immediately into the accumulator using the LDA #34 command.

At this point, the data is sent to the Apple's ACIA and is received in register HEX COAD for later use by the Main Matrix Program in the Apple microprocessor.

Now as to the data transfer between the state sensing AIM 65 and the state driving AIM 65, the difference in data transfer between the AIM's and the Apple is due to the use of the shift registers on the AIM 65's. The angle change or relative position is sent out in a 96-bit data stream consisting of four blocks of 24-bits of information. Each block contains the 10-bit word that is in a one-to-one correspondence with the absolute position and another 8-bits used for addressing and sign. The serial driver is shown using one memory location in Figure 5. Note that the 8-bit value is loaded from location :0400 to the shift register and is outputted on CB2 with CB1 containing the timing information of the shift pulses. There is also a dummy load to location :A00A for initialization of the shift register.

```
K * = 400
/12
0400 A9 LDA #00
0402 8D STA 1001
0405 8D STA 1000
0408 A9 LDA #1E
040A 8D STA 1003
040D A9 LDA #09
040F 8D STA 1002
0412 A9 LDA #34
0414 8D STA 1000
0417 4C JMP 0412
041A 00 BRK
041B 00 BRK
```

Figure 4
ACIA Driver Program

```
K * = OFDC
/15
OFDC A9 LDA #00
OFDE 8D STA A00B
OFEI A9 LDA #14
OFE3 8D STA A00B
OFE6 A9 LDA #20
OFE8 8D STA A008
OFEB A9 LDA #00
OFED 8D STA A009
OFFO AD LDA 0401
OFF3 8D STA A00A
OFF6 EA NOP
OFF7 EA NOP
OFF8 EA NOP
OFF9 00 BRK
OFFA 00 BRK
```

Figure 5
Serial Driver Program

With the completion of the data transfer between the two AIM 65's, the feedback loop is completed, which is necessary for the system program to operate.

STATE DRIVING

Since the AIM 65.2 microcomputer Controller of the electrohydraulic valves has no position sensors connected directly to it, all position data must be received from AIM 65.1 so that appropriate voltages can be sent to the power digital to analog converters which, in turn, control the electrohydraulic valves. This AIM 65.1/65.2 interface consists of a four-line serial data port that is used to transmit serial data between each of the AIM 65's 8-bit shift registers, where two lines from AIM 65.2 (out of four available) are hardwired to the corresponding two lines on the AIM 65.1. One line pair is used to transfer the control clock pulses while the other pair is used to transfer the serial data (encoder address and associated angular difference data). The handshake control of the data transfer is done using two lines (PB6 and PB7 on AIM 65.2 to PB4 and PB5 on AIM 65.1, respect). The serial data consists of a 96-bit data stream with the encoder address sent in one 8-bit field and the associated angular difference data sent in two 8-bit fields (contiguous to the encoder address).

Note that each 8-bit field as it is received is stored in consecutive memory locations so that the AIM 65.2 control program can, when data is needed, sequentially pick-out of memory the encoder address and angular difference data.

The reason for off loading this function to another microcomputer was the resulting increase in control speed due to parallel processing.

After receiving the serial data, the control program written in the BASIC programming language resident on the AIM 65, then "PEEKS" to the memory locations holding the first encoder address (eventually, the 2nd, 3rd, 4th) and associated angular difference data and converts these binary values to decimal equivalents. The results are decoded and checked for errors. If no errors are found, the program proceeds to find the power digital to analog converter (PDAC) output voltage versus angular difference equation associated with each link.* From this equation, the PDAC voltage is found which corresponds to that angular difference condition. This value is then converted to a decimal number based on the resolution of the PDAC (11 effective bits @ ±24 volt power transistor supply voltage or $24/(2^{11}-1)$ = 11.7 mV per bit change) and with the encoder/link address, outputted to the parallel output port of the AIM 65.2. Since this output port actually consists of two 8-bit ports, Port A (PA) and Port B (PB) - as well as the four serial communications lines, the eight least significant bits of the voltage value are outputted on Port A, while the remaining three most significant bits of the voltage value are outputted to Port B along with the sign bit and the 2-bits used for the encoder/link address (the remaining 2-bits on Port B are used for handshake control of the serial data transfer).

Following the output of address and data (including sign bit) to Ports A & B is a strobe of several milliseconds duration that "enables" the address decoder which, in turn, sends a strobe to the addressed PDAC.

* An empirical equation found by measuring the response of each link to an applied voltage.

Now, if errors were found in the decoded address and data, the program would shutdown the backhoe, print a message stating the reason for shutdown, and query the user as to what to do next. The program has two responses to the user's reply - terminate program or begin again by requesting more data from AIM 65.1.

This process is continued until the angular differences, given for each link, approach a tolerance value* close to zero, corresponding to zero volts on the output voltage - angular difference equation. If a link does over-shoot the final position, the angular difference data received from AIM 65.1 will be decoded as a negative value which corresponds to a negative PDAC output voltage. Consequently, the direction of link motion will be reversed towards the correct final position. A block diagram of the output port of the AIM 65.2 and all that is connected to it is shown in Figure 6A and Figure 6B respectively.

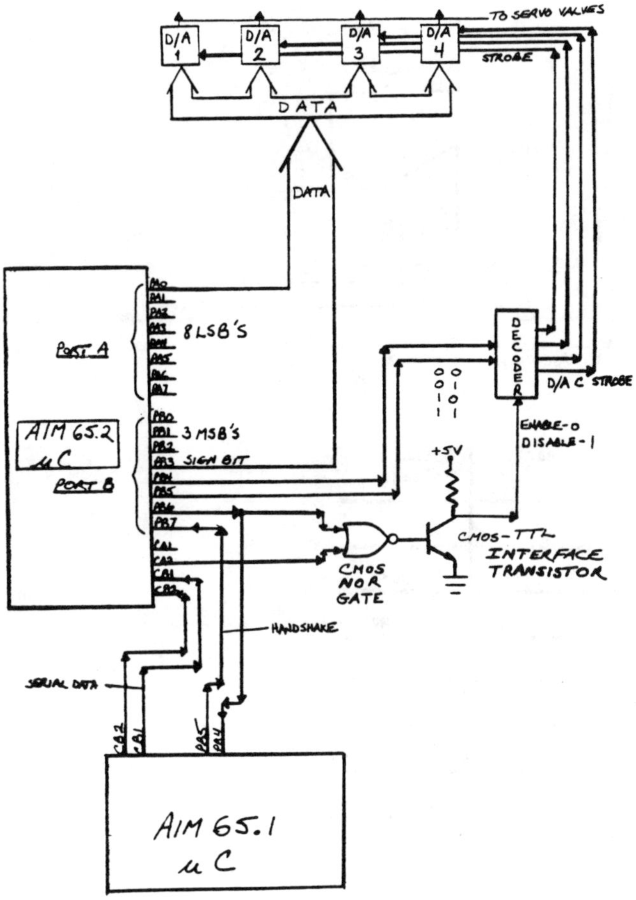

Figure 6A
AIM 65.2 Output Port

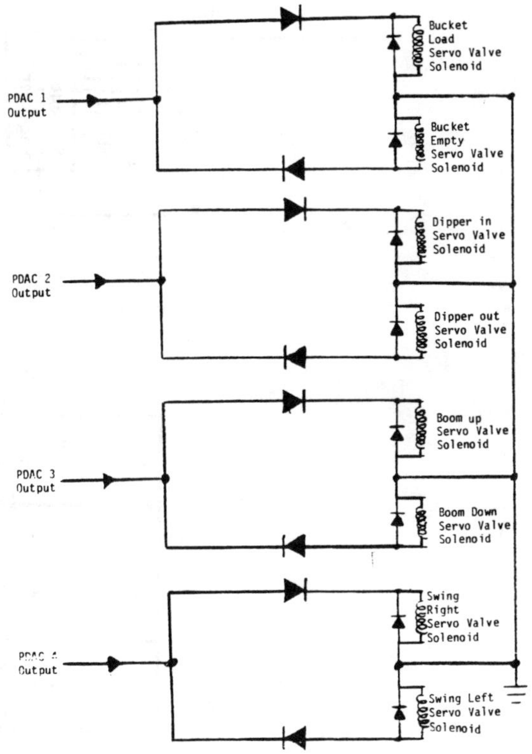

Figure 6B
PDAC Output-to-
Servovalve Solenoid

* An empirical value based on the response of each link.

The power digital-to-analog converter consists of a 12-bit storage register with strobed inputs a 12-bit digital-to-analog converter, and a power output stage (see Figure 7). The maximum drive voltage the servo valves will see from the PDAC's is 15 volts at 100% connect time.* The resolution of the PDAC's given a supply voltage of 24 volts to the power stage and an effective word length of 11-bits, is 11.7 MV. Maximum drive current the PDAC can continously supply to a load (in this case, a 27 ohm solenoid) is ± 1.2A.

The electrohydraulic valve that replaced the case valve on the case backhoe was the Monsun-Tison HV03 proportionally controlled valve with an attached Monsun-Tison EHC3 electrohydraulic converter, optional joysticks were added. With these valves a nearly linear output hydraulic pressure input vs input drive current can be attained.

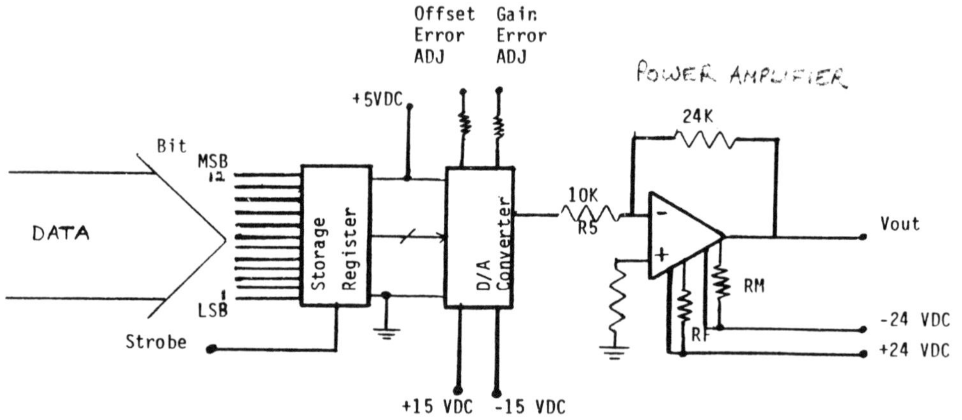

Input Code (offset binary) SB	MSB'S	LSB'S	Output Voltage
1	111	11111111	+24.00V
1	111	11111110	+23.988V
1	000	00000000	+11.7 MV
0	111	11111111	0.000V
0	111	11111110	-11.7 MV
0	000	00000000 1	-23.976V
0	000	00000000	-23.988V

Figure 7
PDAC AND INPUT CODE TABLE

* A higher voltage can be used (up to 24 volts) if the voltage source is pulsed (50% duty cycle). This reduces hysteresis that is developed in the solenoids of the EHC3. Also, with the higher voltage, an increase in backhoe arm "power" can be realized.

HOLE STATE DETERMINATION

During the initial stages of work with the automated backhoe, the overall system controller will be "blind" as to the hole state. The backhoe will be digging in soft earth so the controller will, at this point, assume that the bucket is removing a cubic matrix square with each bucketfull of earth and, when finished with the complete digging cycle, assume the desired hole was dug to the specification inputted by the user. But, after successful demonstration of the complete system controller, a sensor(s) will be added to supply this information to the controller. A sensor that was determined could supply this information (hole state and bucket digging efficiency) is a rangefinder.

Using a rangefinder, the controller would be capable of making a hole survey (if rangefinder was properly mounted) by comparing the three-dimensional hole matrix that it generated in its control program to the results of the rangefinder's SCAN. Also, the controller could determine how much material the bucket was removing (bucket digging efficiency) by curling the bucket to a pre-determined position then using a rangefinder (again, properly mounted) to indicate the level of material in the bucket by examining the distance between the sensor and the material in the bucket.

In evaluating the results of tests conducted on a specific rangefinder (Polaroid's Ultrasonic Ranging System), the conclusion was the rangefinder performed as stated in the manufacturer's technical literature (most importantly, the 7 degree beam width) but, the control program would have to take into account the environmental conditions that affect the rangefinder's performance (specifically, extremes in temperature) and the transducer of the ranging system would have to be weather-proofed from, among others, dust, exhaust gases, wind, rain, freezing rain and snow.

RESPONSE OF SYSTEM

This section is included for completeness only since results are preliminary and a full report is expected in the near future.

A preliminary analysis has been made of the dynamic response of the electro-hydraulic servo system using the partial fraction expansion. The sum of the first and second order terms are used to solve the third order response system of the electrohydraulic valves. The results of the analysis have been constructed.[7] A worst case as well as desired case response is shown at figure 8. Note that the physical parameters are shown for a typical valve.[8]

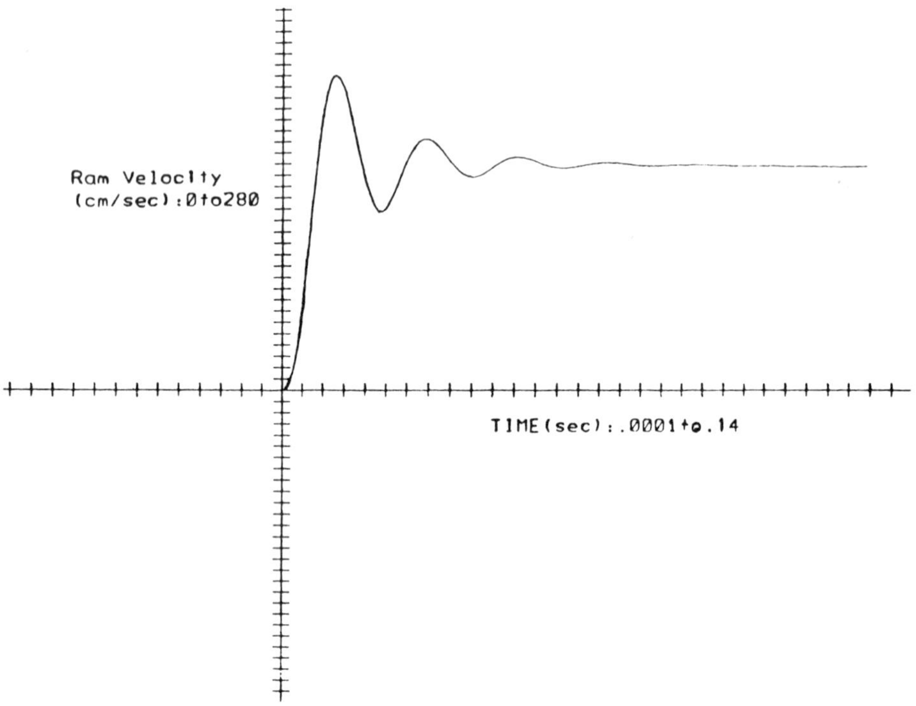

Figure 8
Servo Response

7. See Reference (1) pp. 131-132
8. See Reference (1) pp. 135-136

CONCLUSION

Currently the state determination, state sensing, and data transfer (96-bit stream) programs have been designed, developed, and tested on the prototype demonstrator. All hardware modifications have been completed.

In the future, it is expected that a complete open loop/closed loop servo response with a full system will be analyzed to determine deviations from expected value due to the mechanical uncertainties in the system and the maximum drive rate of the hydraulic system.

Additionally, it is anticipated that considerable work needs to be done in interfacing the main control program and controller so that considerations such as:

(1) Reference point motion

(2) Inertial load changes

(3) Real time response

(4) Multiple processing

are included in the prototype demonstration.

ACKNOWLEDGEMENTS

The authors would like to thank the members of the CS/CI Laboratory's Robotics/ A.I. Group. In particular Charles Clark, Rick Dupont, Harry Keller, and Don Kelley. Special thanks to Marlene Brumbaugh for her patience in putting together the final copy.

REFERENCES

(1) Clark, R.N., "Introduction to Automatic Control Systems", Wiley, 1962.

(2) Osborne, A., "6502 Assembly Language Programming", McGraw-Hill, 1980.

(3) Wakerly, J.F., "Microcomputer Architecture and Programming", Wiley, 1981.

(4) Manual "Model 7710 ACIA", California Computer Systems, 1981.

(5) Manual "Data Communications Interface-4051", Tektronix, Inc., 1981.

(6) Manual "R6500 Microcomputer System Hardware", Rockwell International, 1980

Microcomputer Control of Free Ranging Robots

Dr. R. A. Cooke
Trent Polytechnic
Nottingham, United Kingdom

Abstract

A truly mobile robot must be capable of detecting and avoiding obstacles in its path. This paper begins with an introduction to mobile systems, and a discussion of the particular advantages of autonomous free-ranging robots. An ultrasonic ranging system is described which enables the robot to avoid obstacles. Software and hardware aspects of this system are discussed. A system of ultrasonic mapping is described whereby the robot's surroundings in simplified form are memory-mapped for possible display on a V.D.U. The paper concludes with some possible industrial applications.

1) INTRODUCTION

Mobile robots which rely on path following, using buried wire (inductive loop) techniques, have the inherent disadvantage in that should an obstacle block the path the robot must either stop, to prevent damage to itself or the obstacle, or merely attempt to push the obstacle aside.

The latter course of action is not desirable and therefore the robot is immobilized until the obstacle is moved. The former action necessitates the use of some form of obstacle location. In the simplest case this could be some form of skirt surrounding the robot into which is built a series of microswitches. When any one of these is actuated, by the skirt touching an obstacle, the robots drive motors are switched off.

The disadvantage of this technique is that some contact must be made with the obstacle. Furthermore the robot must stop before damage is done. In turn this implies that the speed of the robot, in areas where obstacles are likely, is severely restricted unless some form of contactless obstacle location is employed. The nature of the obstacle itself may also preclude contact, for example if the obstacle is fragile, very hot or even radioactive.

A truly mobile robot must be able, not only to detect an obstacle before contact, for the above reasons, but must also be able to modify its path according to the number and size of the obstacles.

The disadvantage of such a system is the added cost and comlexity necessary to achieve this end, although against this is balanced the uninterupted, although modified, path of the robot. The total cost of the prototype system, not including drive motors and robot body, was apportioned as a percentage of the total robot cost and calculations show an increase in cost of some 3%.

2) DESIGN CONSIDERATIONS

If an industrial free ranging robot is to be considered then it should combine the advantages of both the path-following and the path-modifying modes.

A feasible system could utilise the path-following mode until an obstacle was sensed whereupon this mode would be overidden by the autonomous path-modifying mode.

When the robot had passed around the obstacle and regained its original path then the path-following mode would regain control.

Any mobile robot must possess;

a) a means of propulsion (usually elecric motors) and associated drive mechanisms (gearboxes, clutches, tracks or

wheels)
b) a power source (batteries)

A free ranging robot, in addition, must possess;

c) an obstacle sensor and interfaces
d) some decision making logic (microcomputer and software)

This paper is concerned with items c) and d).

2.1) The obstacle sensor and interface

If a contactless system is to be employed then the choice lies between optical, ultrasonic and microwave (radar) devices. A modified commercially available ultrasonic ranging kit was eventually chosen for the prototype.(reference 1). The reasons for this are;

a) the robot may be working in conditions of low or zero ambient light which precludes optical devices, or at least renders them ineffective unless some form of laser ranging is used. This may not be possible for reasons of worker safety. A form of television imaging could have been used but was rejected on the grounds of cost.

b) the working environment may be very noisy from an R.F. (radio frequency) point of view. This would be the case in workshops where welding or R.F. heating was done. In addition heavy electrical machinery would contribute to R.F. noise. This detracts from using some form of microwave sensor.

The author realised that ultrasonic noise may also be present but considered that this could be filtered out with relative ease.

The system operates as follows;

a) the output from the ultrasonic sensor is modified and, in the latest version, is compared with the pre-programmed desired distance, from the obstacle, at which the robot must begin path modification.

b) if the actual distance to the obstacle is equal to or less than the pre-programmed distance the robots drive is stopped and the ultrasonic scanner turned sideways. The robot turns left or right by the appropriate drive motor being energised.

c) both motors are now driven so that the robot is moving parallel to, but at a programmed distance from, the front of the obstacle At regular intervals, temporally or spacially, the scanner turns to face the obstacle in order to verify its presence. When the corner of the obstacle is located the robot makes a further turn. During this part of the system operation the distance travelled has been recorded in the microcomputer user memory (RAM).

The robot is now travelling down the side of the obstacle, moving approimately in the direction of the goal but displaced from the original path. When the next corner is reached, a third turn is made.

d) the robot is now moving back towards its original path line and is comparing the distance it travels with that recorded during c).

e) a final turn back onto the original path is made when the distance travelled during d) is the same as that recorded during c).

In its original form the ultrasonic ranging kit provides a three digit L.E.D. display showing the distance from the transducer to the obstacle expressed in tens, units and decimal parts of feet. The range of the system is up to 35ft. with a resolution of 0.1 ft. (1.2").

The kit was modified, in order to present compatible data to the microcomputer, in the original version by enabling a NOR gate when the 'tens of feet' digit was being displayed, the tens, units and decimal L.E.D. drives being multiplexed. In this system used originally, the NOR gate produced a 'high' output whenever the robot approached from a distance greater than 10.0 ft. to 9.9 ft. The output was sampled by the microcomputer and when 'high' caused program modification.

This system was only used in a feasibility study since the main disadvantage is the 'hard-wired' inflexibility of the distance at which obstacle avoidance takes place.

The present systemuses a software approach whereby the ' units of feet' signal to the L.E.D. drive is sampled at intervals and compared with a distance, stored in memory, at which evasive action must take place.

Fig. 1 shows the interface between the microcomputer and the ultrasonic sensor.

The system allows the distance between obstacle and robot, at which the robot will turn, to be software variable between 1.0 and 9.0 feet. This range is considered suitable for most industrial situations envisaged.

2.2) The microcomputer and software

A 6800 microprocessor-based development kit was used with 1.0 Kbytes of on-board user RAM(reference 2). This kit provides a 6821 user PIA (peripheral Interface Adapter) which contains 2X8bit ports. Thus 16 lines are available to the user to be used as input or output lines. The actúal configuration of port lines is shown in Fig.2.

Port A is configured as byte input with b0 through b3 giving binary representation of the obstacle distance up to 9 feet, b4 being cleared if the obstacle distance is 9.9 feet or less, set if greater than 10 feet.

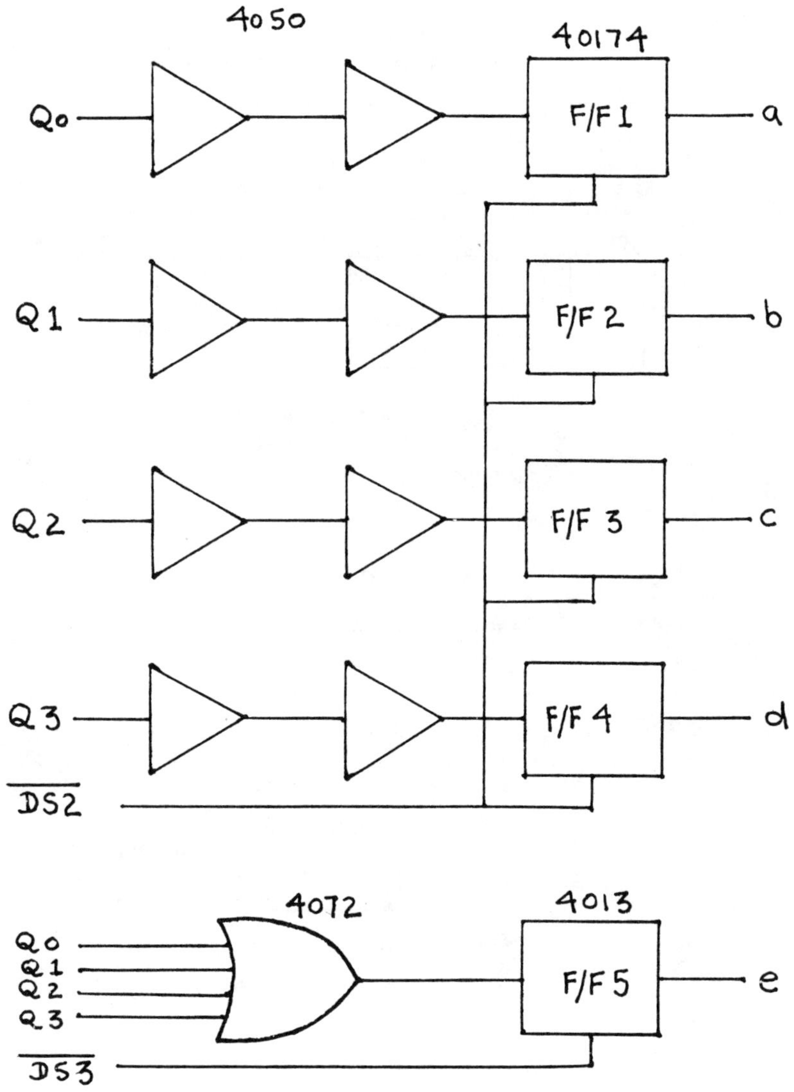

FIG. 1.

The turn status is given by b5, which is set by the output from a binary counter when a clockwise or anti-clockwise turn has been completed. The present system allows 90degree turns only although work is underway to allow greater flexibility in the choice of turning angle.

Port B is configured as byte output with b0 and b1 being

the control bits to the stepper motor on which the ultrasonic sensor is mounted. This feature allows the sensor to remain facing the obstacle when the robot body has turned.

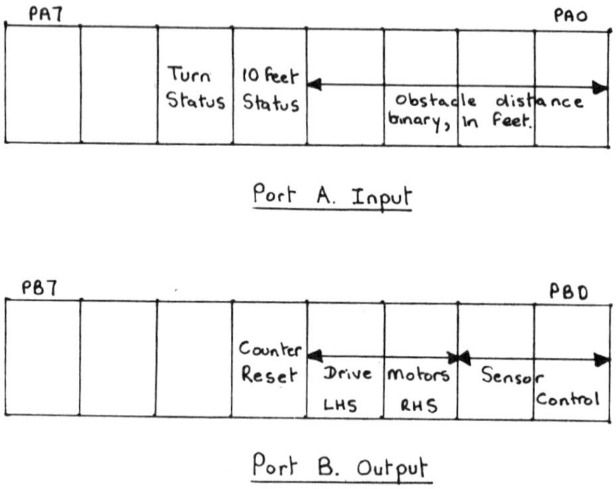

FIG. 2.

The right and left tractor drives are controlled via b2 and b3 respectively whilst b4 is the ' counter reset' bit which clears the counter associated with turning left or right. This is shown in Fig. 3.

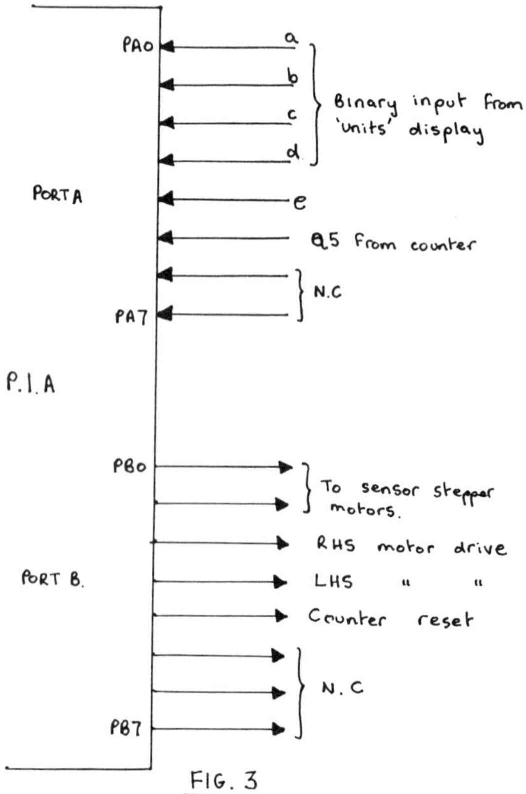

FIG. 3

The software, which in the system under development was stored on cassette and loaded each time as required, would, in an industrial robot, be contained in ROM.

The software was written in M6800 assembly language (reference 3) and the object program obtained via the polytechnic mainframe cross assembler. The current system software operates as follows;

1) the robot travels forward (both left and right tractor drives ON) in normal mode. At this time the robot could be following the path of a buried guidance wire. Port A is periodically scanned and the distance to any obstacle in the path of the robot compared to a pre-set distance held in the robots microcomputer RAM.

2) when these distances are equal the left and right tractor drive outputs from Port B go 'low' and the robot stops.

3) the sensor control bits are activated, in sequence, so that the stepper motor turns the ultrasonic scanner through 90 degrees anticlockwise and DIST 1, the distance to the nearest obstacle on the left is recorded. The scanner is now turned 180 degrees.

4) the distance to any obstacle on the right, DIST 2, is now recorded. If DIST 1 is less than DIST 2, the robot will execute a 90 degree turn clockwise. (left motor drive ON, right motor drive OFF, turn status recorded). The sensor now rotates anticlockwise (sensor control sequenee activated, counter reset after rotation).

5) both drive motors are turned on for 5 seconds, the counter incrementing in this mode and then both drives are turned off. The sensor is now rotated anticlockwise to test for the corner of the obstacle. If the obstacle is still present then the 5 second drive sequence is repeated.

6) when, as a result of the above, the sensor no longer detects the obstacle, the robot performs an anticlockwise turn of 90 degrees and the scanner rotates clockwise.

7) the robot is now moving up the side of the obstacle and scanning forward. After 5 seconds the scanner rotates anti clockwise to reaffirm the presence of the obstacle, this being repeated, as above, until the next corner is located.

8) as the robot moves along the back of the obstacle the counter is now decremented until its contents are zero. At this point the robot executes a clockwise turn, regaining its original forward path.

If the robot were following a guidance loop in the normal mode then extra software would be required so that immediately before the last clockwise turn it would execute a 'search' for the buried guidance cable. This has not been included in the current work.

The disadvantage of the method described is that when moving along the front and back of the obstacle (steps 5 and 8) distances moved are measured in terms of the number of revolutions of the driving wheels. If any slippage occurs therefore between drive wheel and ground then errors will result. A tracked vehicle would be better from this point of view although this would not obviate the problem.

The overall system was tested in the laboratory and performed according to expectations. The obstacles used in the first tests were rectangular but when cylindrical objects were introduced the ultrasonic sensor interpreted these as rectangular with side dimensions less than the diameter of the cylinder. Since the 'sampling' time (see step 5) was somewhat arbitrarily set at 5 seconds this did not cause problems. If the temporal resolution was made shorter however such that sampling took place every second (this would mean that the robot would travel a shorter distance around the obstacle) then collisions may occur with cylindrical objects.

The actual rate at which sampling takes place is determined of course by the tractor velocity and the resolution desired.

The system works well for single obstacle avoidance although if a second obstacle was encountered during the circumnavigation of the first, extra sofwtware would be but would be in the form of a sub-routine added to the original program. The basic single obstacle software occupied most of the 1Kbyte user RAM available on the microcomputer development kit.

3) AN ULTRASONIC MAPPING TECHNIQUE

Whilst working on the problems associated with obstacle avoidance using an ultrasonic scanner it became apparent that an alternative approach could be adopted.

In the system already described the ultrasonic sensor moves only when the robot turns, so that the sensor still faces the obstacle. The software facilitates only 'forward' and 'sideways' looking modes.

One alternative to this, which is under current investigation at Trent, is to stop the robot when the obstacle-robot distance is equal to that held in memory and then allow the ultrasonic sensor to scan in the horizontal plane to locate the edges of the obstacle. The angles through which the scanner has moved, from its normal 'forward' position to each edge of the obstacle, are recorded and the decision to execute a clockwise or anticlockwise turn of the robot is then based on the minimum scan angle. This is illustrated in Fig.4.

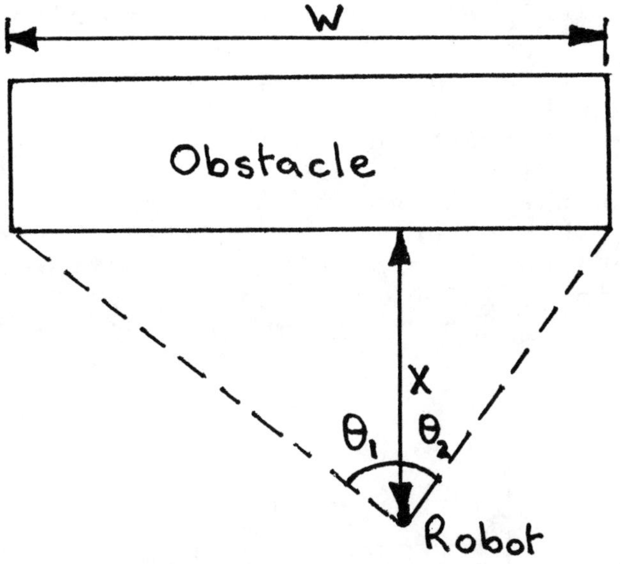

$$W = X[\tan\theta_1 + \tan\theta_2] \quad \theta_1 > \theta_2$$

FIG. 4.

One important feature of this technique is that using a simple algorithm, the width of the obstacle and the position of the robot relative to it are established. This data may then be used when the robot is travelling around the obstacle instead of temporal sampling.

Additionally the sensor can locate further obstacles to each side of the one originally located and , allocating a section of RAM for this information, a map of the robots environment is established. As the robot moves around the obstacle this information can be updated.

As the above mentioned work was proceeding the author realised that it may have implications reaching beyond that of the autonomous mobile robot. If the robot were connected in the simplest case via an umbilical to a VDU then a memory mapped display would show to an operator the position of the robot and the obstacles surrounding it, within the range of the scanner. Although the resolution of such a system would not approach that of an image derived from a television camera there are occasions when the latter is not feasible (e.g. extremely dusty or smoke filled regions or in conditions of darkness)

Work is now proceeding to produce on a VDU an image in the form of blocks representing a plan view of the obstacles, the current position of the robot also being shown.

In the first instance, as the robot approaches the obstacles the image will appear as a series of lines of arbitrary thickness. Having chosen the optimum path the robot proceeds around the first obstacle thus changing the image seen on the VDU as the contents of the section of memory reserved for the VDU are constantly updated. The original information is enhanced by providing data on the depth of the obstacles. This is illustrated in Fig. 5.

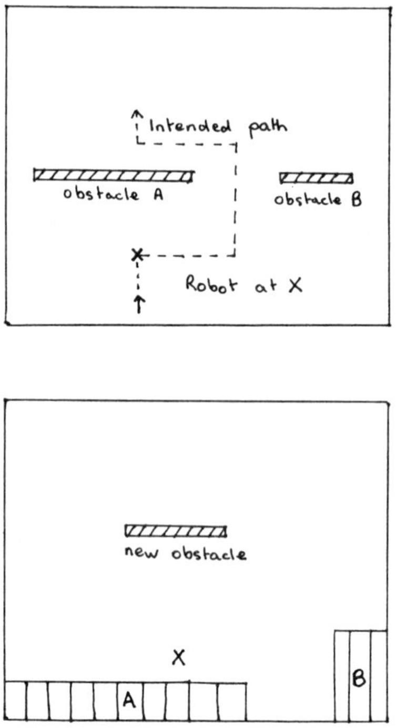

FIG. 5

The operators view would be that the robot was stationary and that the obstacles were moving. This method obviates the need to scroll the screen when the robot approached the screen periphery.

4) INDUSTRIAL APPLICATIONS

Mobile robots guided by induction techniques are not uncommon. The system described in 2) would enhance their mobility.

In situations where guidance wires are not possible the

system would work in an autonomous mode proving that the memory was expanded from the 1Kbyte prototype. Here the robots normal route would be programmed into memory and the 'obstacle avoidance' would exist possibly as an interrupt sub-routine to the 'normal route' program.

If the robot were to be used in a telechiric capacity (reference 4) then under conditions of limited visibility the ultrasonic mapping technique would be useful. One application could be in firefighting whereby the robot would enter a smoke filled area either ot provide remotely an ultrasonic map of obstacles encountered or, if it were being steered by the operator, the VDU map would enable the obstacles to be avoided. In the latter case the robot could carry fire extinguishers to be deployed on instructions from the operator.

Ultrasonic techniques at present do not provide sufficient resolution or discrimination to be used in tasks requiring detailed vision. However this paper has attempted to show that they can provide a cost-effective alternative to traditional vision systems for some low-resolution applications.

BIBLIOGRAPHY

1) Polaroid Ultrasonic Ranging Kit Manual
Polaroid 1980

2) Microcomputer experimentation using the Motorola MEK 6800D2
L. A. Leventhal
Prentice Hall 1981

3) The M68 XAS cross assembler guide.
Trent Polytechnic Computer Services 1982

4) Robotics and Telechirics.
M. Thring
Industrial and Commercial Training
June 1980

Interpolation for 2-Dimensional Contouring Control

Bong Jin Lee
Precision Machinery & Technology Center, KAIST
Seoul, Korea

Tae Seok Nho
Precision Machinery & Technology Center, KAIST

Abstract

The interpolator is usually built-in hardware (logic circuitry), and the interpolator, fabricated in a single LSI chip, is recently used in most contouring control systems, making them more compact. However, the LSI interpolator has many technical difficulties and a very high cost in its fabrication. To solve these problems, a method of interpolating by software was explored and a program which, executed by INTEL's 8085 microprocessor and can distribute the input pulses of up to 4.0 [kpps] for the linear interpolation and 3.0, [Kpps] for the circular interpolation was developed. An algorithm and the flow chart of the interpolation program was used to reduce the execution time.

INTRODUCTION

The control of the feed mechanism of an NC machine tool or an industrial robot is to be classified into 2 categories: Positioning Control (Point-to-point Control) and Contouring Control (Continuous Path Control). In the former, the positioning accuracy at the end point only is important, while in the latter, the path and the speed of movement as well as the positioning accuracy at the end point are of consequence.

The generation of position reference signals in contouring control is usually implemented by interpolation. Interpolation is to be defined as: Calculation of the coordinate values of the points on the given path using the coordinate values of the starting point and the end point only. Interpolation is to be classified into several kinds according to the shape of the given path, and they are called the linear interpolation, the circular interpolation, the parabolic interpolation, and so on. However, the linear interpolation and the circular interpolation are frequently employed.

As for the method of interpolation, the MIT method, the DDA method, the Algebraic Operation method and the Least Error method are developed up to present, and the principles or the block diagrams of their circuits can be found in the references [2]~[8]. But the circuitry is so complicated that the size of the hardware will become too large (equivalent to about 200 SSI's)if it is to be implemented using the general-purpose SSI's like the TTL, etc. In the developed countries, the interpolator is fabricated in a single LSI chip to make the system more compact. However, the LSI interpolator not only has many technical difficulties but also requires very high cost, in its fabrication.

To solve these problems, we tried to find the method of interpolating by software. As the MIT method, the DDA method and the Least Error method are found very difficult to be implemented by software, the Algebraic Operation method is adopted. This method was originally developed in Tokyo Institute of Technology, and its principle of operation is described in the references [2]~[4]. However, as this method requires the hardware for calculation of very high speed, and it is also very difficult to design an interpolation program according to the algorithm introduced in the references [2]~[4], we tried to reduce the execution time, and developed a better algorithm which can easily be handled by the assembly language and consequently reduce the execution time. According to this new algorithm INTEL's 8085 microprocessor can distribute the input pulses of up to 4.0 [Kpps] for the linear interpolation and 3.0 [Kpps] for the circular interpolation. In the following, the algorithm and the flow chart of the interpolation program are described.

LINEAR INTERPOLATION

Consider the case shown in Fig. 1-(a), where the machine is moved along the line segment $\overline{AB}$. To simplify the equations, the line segment $\overline{AB}$ is translated so that the point A may coincide with the origin 0 as shown in Fig. 1-(b).

Let the coordinate of the point B after translation be (Xe,Ye). Then, the equation of the line segment $\overline{AB}$ is given as $y = (Ye/Xe)x$,and the condition for the moving point P (x_i,y_j) to be located in the upper part of the line segment $\overline{AB}$ is given as $y_j > (Ye/Xe)x_i$, or $Xey_j - Yex_i > 0$. So, if the determinant $D_{i,j}$ is let as

$$D_{i,j} \triangleq Xey_j - Yex_i, \qquad (1)$$

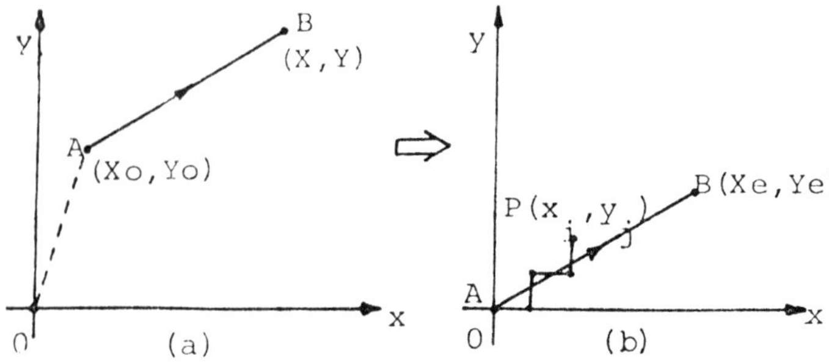

Fig. 1. Translation of coordinate for linear interpolation.

then, the following conclusions can be obtained:

1. The point $P(x_i, y_j)$ is located in the upper part of the line segment $\overline{AB}$ for positive $D_{i,j}$.
2. The point $P(x_i, y_j)$ is located on the line segment $\overline{AB}$ when $D_{i,j} = 0$.
3. The point $P(x_i, y_j)$ is located in the lower part of the line segment $\overline{AB}$ for negative $D_{i,j}$.

Using the above facts, the machine can be moved along the line segment $\overline{AB}$ within the error of 1 pulse if it is moved according to the following rules:

1. The machine is moved by 1 [pulse] in the positive direction of x-axis when $D_{i,j} \geqq 0$.
2. The machine is moved by 1 [pulse] in the positive direction of y-axis when $D_{i,j} < 0$.

The result of linear interpolation according to this principle is shown in Fig. 2, where Xe = 6 and Ye = 5.

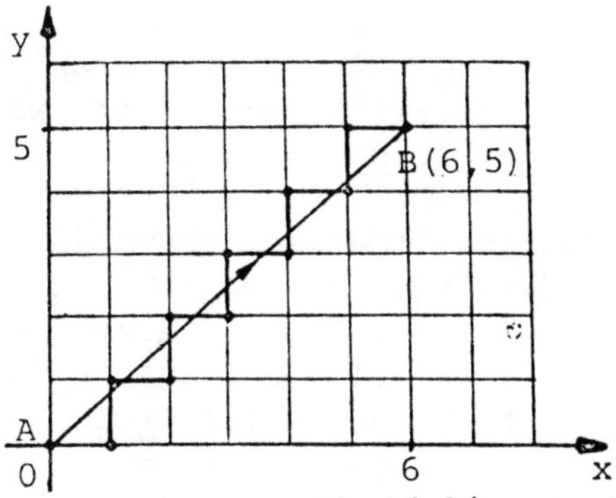

Fig. 2. The result of linear interpolation.

To execute the above process, it is necessary to calculate the value of $D_{i,j}$ and to determine the sign (polarity) of $D_{i,j}$ at each point. According to the equation (1), the value of $D_{i,j}$ seems to be easily obtained by subtracting Ye multiplied by x_i from Xe multiplied by y_j. However, this calculation is very difficult for the 8-bit microprocessor. The reason is that Xe, Ye, x_i and y_j are all 3 byte (24bit) data, bacause the coordinate value of 3 bytes is necessary to cover the stroke of the machine sufficiently in the system in which the least command increment is 1 [μ] . Hence, another mothod of calculation should be found.

After a careful examination, the interpolation process according to the rules mentioned above can be summarized as follows:

1. The initial values of x_i, y_j and $D_{i,j}$ are all zero: $X_0 = 0$, $y_0 = 0$ and $D_{0,0} = 0$.
2. The new values of x_i and $D_{i,j}$ after each movement when $D_{i,j} \geqq 0$ are given as follows:

$$x_{i+1} = x_i + 1, \qquad (2)$$
$$D_{i+1,j} = D_{i,j} - Ye. \qquad (3)$$

3. The new values of y_i and $D_{i,j}$ after each movement when $D_{i,j} < 0$ are given as follows:

$$y_{j+1} = y_j + 1, \qquad (4)$$
$$D_{i,j+1} = D_{i,j} + Xe. \qquad (5)$$

The Eqs. (2), (3), (4) and (5) show that the new values of x_i, y_j and $D_{i,j}$ after each movement can easily be obtained, and the flow chart of the linear interpolation subprogram according to the above algorithm is shown in Fig. 3.

The sequence of process concerned with Fig. 3 is as follows. Initially, the microcomputer is given the data which are necessary for interpolation. And then, if the interpolation command pulse is applied to the RST 7.5 interrupt pin, the interpolation subprogram is initiated. At the beginning, x-axis or y-axis is selected according to the sign of the determinant $D_{i,j}$ at the current position. Then, it is checked whether there are any spare distance to move in the direction of the axis selected, and if there are any spare distance, the machine is moved along that axis and vice versa. And then, the new value of the determinant $D_{i,j}$ after that movement is calculated. Finally, it is checked whether the machine arrived at the final point after that movement. If the machine did not yet arrived at the final point, the process returns to the main program waiting for the next interpolation command pulse, and if the machine arrived at the final point, the process returns to the main program after executing the process for the end of interpolation.

CIRCULAR INTERPOLATION

There can be 2 kinds of circular interpolation according to the direction. However, the circular interpolation in CCW direction only is to be described in this paper, since the circular interpolation in CW direction can be converted to that in CCW direction by taking the mirror image of the given path with respect to the line y = x.

Consider the case shown in Fig. 4-(a), where the machine is moved along the circular arc $\widehat{AB}$. To simplify the equations, the circular arc $\widehat{AB}$ is translated so that the center C of the circular arc $\widehat{AB}$ may coincide with the origin 0 as shown in Fig. 4-(b). Let the coordinates of the points A and B after translation be (Xs,Ys) and (Xe,Ye) respectively. Then, the equation of the circular arc $\widehat{AB}$ is given as $x_i^2 + y_j^2 = Xe^2 + Ye^2$, and the condition for the moving point $P(x_i,y_j)$ to be located in the outer part of the circular arc $\widehat{AB}$ is $x_i^2 + y_j^2 > Xe^2 + Ye^2$, or $x_i^2 + y_j^2 - (Xe^2 + Ye^2) > 0$. So, if the determinant $D_{i,j}$ is let as

$$D_{i,j} \triangleq x_i^2 + y_j^2 - (Xe^2 + Ye^2), \qquad (6)$$

then, the following conclusions can be obtained.

1. The point $P(x_i,y_j)$ is located in the outer part of the arc for positive $D_{i,j}$.
2. The point $P(x_i,y_j)$ is located on the circular arc $\widehat{AB}$ when $D_{i,j} = 0$.
3. The point $P(x_i,y_j)$ is located in the inner part of the arc for negative $D_{i,j}$.

Using the above facts, the machine can be moved along the circular arc $\widehat{AB}$ within the error of 1 [pulse] if it is moved according to the following rules;

1. The machine is moved by 1 [pulse] in the negative direction of x-axis when the determinant $D_{i,j} \geq 0$.
2. The machine is moved by 1 [pulse] in the positive direction of y-axis when the determinant $D_{i,j} < 0$.

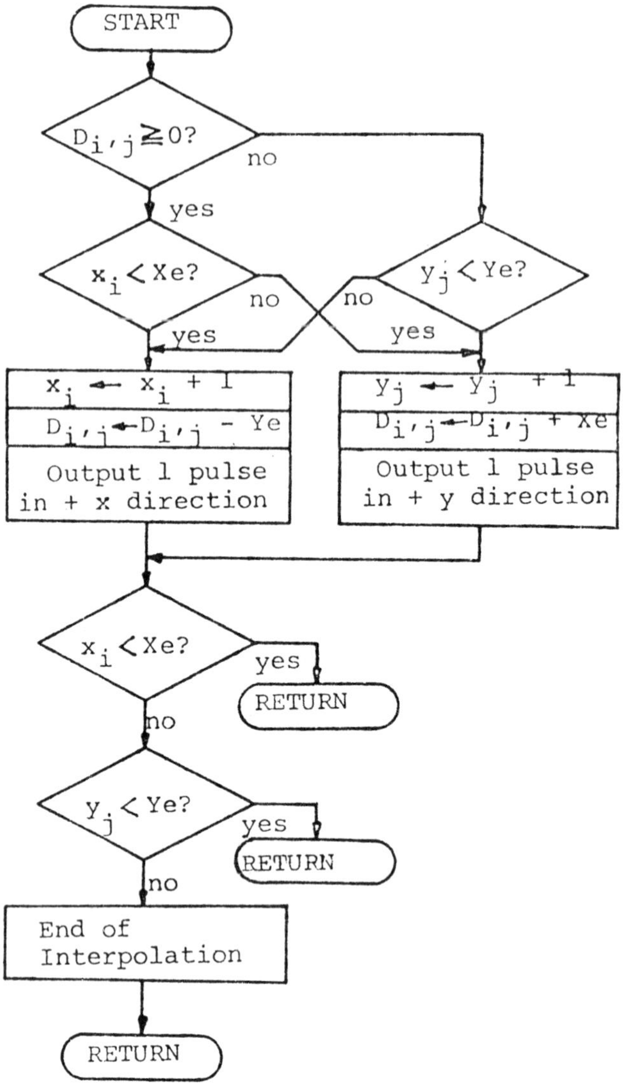

Fig. 3. The flow chart of linear interpolation subprogram.

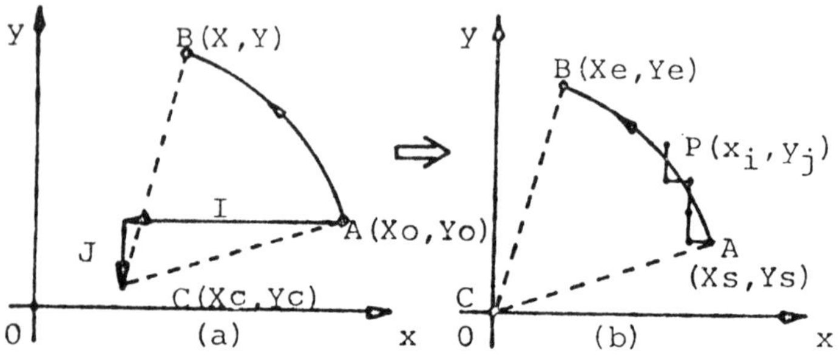

Fig. 4. Translation of coordinate for circular interpolation.

The result of the circular interpolation according to these rules is shown in Fig. 4, where $Xs = 6, Ys = 0$, $Xe = 0$ and $Ye = 6$.

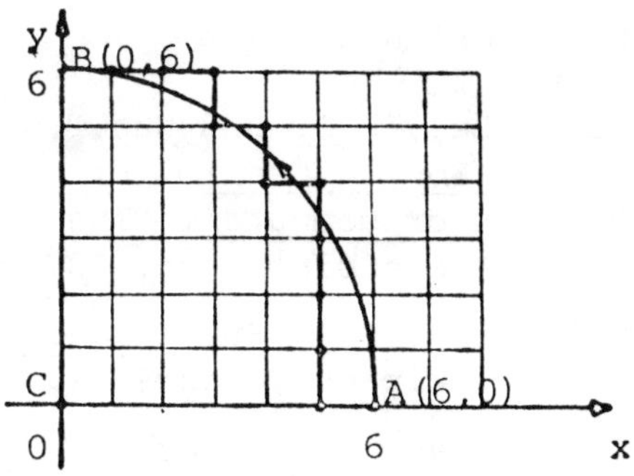

Fig. 5. The result of circular interpolation.

To execute the above process, it is necessary to calculate the value of $D_{i,j}$ and to determine the sign (polarity) of $D_{i,j}$ at each point. The calculation of $D_{i,j}$ according to the Eq. (6) is very difficult and time-consuming task for the microprocessor, and we should find another way of caculation. After a careful examination, the interpolation process according to the rules mentioned above can be summarized as follows:

1. The initial values of x_i, y_j and $D_{i,j}$ are all known.
2. x_i decreases by 1 after each movement when $D_{i,j} \geqq 0$.
3. y_j increases by 1 after each movement when $D_{i,j} < 0$.

Hence, the following conclusions can be obtained:

1. The initial values of x_i, y_j and $D_{i,j}$ are given as $x_o = Xs$, $y_o = Ys$ and $D_{o,o} = 0$.
2. The new values of x_i and $D_{i,j}$ after each movement when $D_{i,j} \geqq 0$ are given as

$$x_{i+1} = x_i - 1 \qquad (7)$$

$$D_{i+1,j} = D_{i,j} - (2x_{i+1} + 1) \qquad (8)$$

3. The new values of y_j and $D_{i,j}$ after each movement when $D_{i,j} < 0$ are given as

$$y_{j+1} = y_j + 1, \qquad (9)$$

$$D_{i,j+1} = D_{i,j} + (2y_j + 1). \qquad (10)$$

The Eqs. (7), (8), (9) and (10) show that the new values of x_i, y_j and $D_{i,j}$ after each movement can easily be obtained and the flow chart of the circular interpolation subprogram according to the above algorithm is shown in Fig. 6. The sequence of process concerned with Fig. 6 is the same as that concerned with Fig. 3.

CONCLUSION

We tried the method of software interpolation, and succeeded in developing an algorithm which can easily be handled by the assembly language. The interpolation program designed according to the algorithm was tested with INTEL's 8085 microprocessor, and it was confirmed that the program could distribute the input pulses of up to 4.0 [Kpps] for the linear interpolation and 3.0 [Kpps] for the circular interpolation. These pulse rates correspond to 40 [mm/s] and 30 [mm/s] respectively, for the system in which the least command increment is 10 [μ], and are thought to be available only in ordinary machining except thread cutting. However, the situation becomes worse for the system in which the least command increment is 1 [μ].

The consideration of these facts shows that several ten times or more of speed-up is necessary, which is at present impossible with the 8-bit general-purpose microprocessor. If a 24 bit microprocessor specially designed using the bit-slice microprocessor design technique is to be used instead of the 8-bit microprocessor, the input pulses of up to 250 [Kpps] can be distributed successfully, which speed is comparable to that of the LSI interpolator.

BIBLIOGRAPHY

[1] Lee, B.J. and Nho,T.S. The Positioning Control of Robot Using Microcomputer. The 11th ISIR Proceedings, pp. 237-244, 1981.

[2] Lee, B.J. Suchi Je-eo. Hyeonmun Chulpansa, Seoul, 1978.

[3] Inaba, S. Yasashii NC Yomihon. Nippon Noritsu Kyokai, Tokyo,1977.

[4] Yamagishi, S. NC Kosaku Kikai. Nikkan Kogyo Shinbunsha, Tokyo, 1975.

[5] Pressman, R.S. and Williams, J.E. Numerical Control & Computer Aided Manufacturing. John Wiley and Sons, New York, 1977.

[6] Koren, Y., Shani, A. and Ben-Uri, J. Numerical Control of a Lathe. IEEE Trans. Ind. Appl., IGA-6, No. 2, pp. 175-179, 1970.

[7] Koren, Y. Interpolation for a Computer Numerical Control Systems. IEEE Trans. Comput., C-25, No. 1, pp.32-

[8] Danielsson, P.E. Incremental Curve Generation. IEEE Trans. Comput., C-19, No. 9, pp. 783-793, 1970.

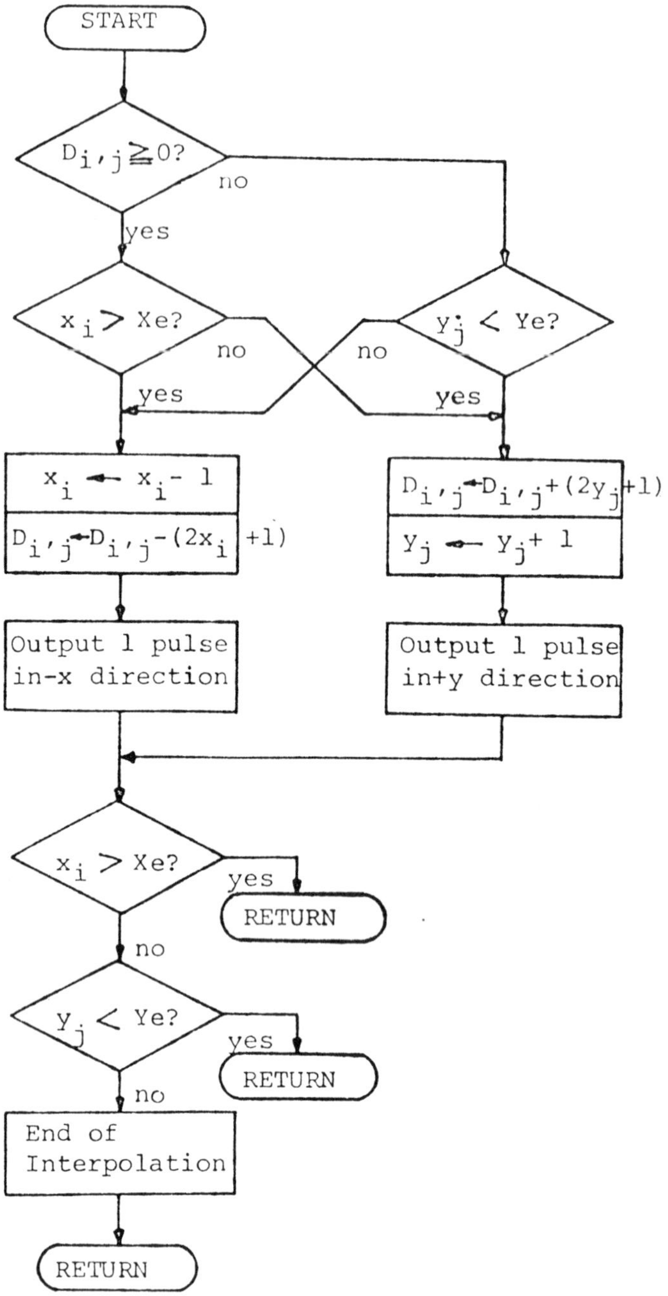

Fig. 6. The flow chart of circular interpolation subprogram.

Microprocessor-Based Controller for the Intelligent Robot System

Ching-Fang Lin
University of Wisconsin-Madison
Madison, Wisconsin

Abstract

Sophisticated and accurate control algorithms for robot control in industrial applications are a challenge in robotics research. This paper describes the use of special purpose processors in an advanced control scheme for the control of an intelligent robot system. Modern control with a nonlinear system inverse, system identification, and state estimation employing the Kalman filter, are used for the robot arm applications. The proposed control scheme can be used to reduce oscillations during fine motions of the robot arm and accommodate changes in loading in fine assembly operations.

INTRODUCTION

This paper describes the use of special purpose processors in a hierarchical microprocessor based control structure for an advanced robot control scheme using Kalman filters for system identification and state estimation. Much work is needed on fast, highly accurate control of manipulative mechanisms because current robots exhibit considerable low frequency vibration during high speed operations which may cause difficulty in assembly of small components. The described control scheme can be used to reduce oscillations during fine motion operation and can accommodate changes in loading so that flexible fine assembly operations are physically possible. Sophisticated and accurate control algorithms for robot arm motions in industrial applications are a challenge in the robotics research.

Some of the difficulties which have to be recognized when using this approach are that the dynamic equations for the robot model are strongly nonlinear and highly coupled. This makes the solution difficult to solve on-line with today's computer limits on execution speed. The control scheme shown was derived with these limits in mind. Significant gains in manipulator performance require the consideration of structural dynamic models, sophisticated control techniques, and the exploitation of advanced computer technology.

Our proposed system consists of a general purpose processor, which is either a 16-bit or a 32-bit microprocessor, as a supervisory machine to the intelligent robot system. This host machine should be able to perform orientation matrix calculation, trajectory transformation, path planning, coordinate system transformation, communication with other processors, gross motion control and fine motion control. A high-speed single chip special purpose processor should be attached to the host machine to do most of the intensive computations in the real-time environment such as real-time interpolation between set points and computation of the correct torque for each joint of a robot arm. [1]

MICROPROCESSOR-BASED SYSTEMS

In present day computer-controlled manipulators, conventional servomechanism techniques are being used. However, the motion dynamics of an 'n' degree-of-freedom manipulator is inherently nonlinear and can only be described by a set of 'n' highly coupled nonlinear second order ordinary differential equations. The nonlinearities arise from inertial loading, coupling between neighboring joints, and gravitational loading of the links. Furthermore, the dynamic parameters of a manipulator vary with the position of the joint variables which are themselves related by complex trigonometric transformations. The servomechanism approach models the varying dynamics of a robot arm inadequately and neglects the coupling effect of the joints. As a result, the robot arm moves only at slow speed with considerable vibration. [1]

Fig. 3 of ref. 2 is a multi-microprocessor structure of a Hitachi robot controller [2,3]. It has a master microprocessor (an HMCS6800) that supervises the whole system, processes data through teaching, does calculations, and communicates with other microprocessors. There may be a simple mathematic processing unit, sine table or other form of arithmetic units that is tied to the master microprocessor that speeds up the computation time. The lower level microprocessor, also an HMCS6800, is responsible for the servo control of joints. The upper and lower level microprocessors are linked by a dual port of RAM.

For a more complicated system that requires visual recognition, such as a robot deriveter [4], an extra 16-bit microprocessor is needed for the visual sensor that gives 2^{16} bits resolution. For other microprocessors like the master microprocessor and the arm processor, an 8-bit system is enough. However, the trend is to switch the 8-bit systems to 16-bit systems or even 32-bit microprocessors such as the Motorola MC68000 [5,6].

Another reference for conventional robots is the current model of the PUMA robot arm control scheme as shown in Fig. 1 of ref. 1. The controller consists of an LSI 11/02 and six 6503 microprocessors each with a joint encoder, digital-to-analog convertor (DAC), and a current amplifier. The control is hierarchically structured. At the top is the LSI 11/02 microcomputer which serves as a supervisory computer. At the lower level are the six 6503 microprocessors, one for each degree of freedom. The two major functions of the LSI 11/02 are the on-line user interaction with subtask scheduling from the user's VAL commands (VAL is the high level language developed by Unimation for its PUMA robot [7]) and the subtask coordination with the six 6503 microprocessors to carry out the command. On-line interaction with the user includes parsing, interpreting and decoding the VAL commands, and reporting appropriate error messages to the user. Once VAL is decoded, various internal routines are called in the EPROM of LSI 11/02 which includes coordinate system transformations, trajectory planning, acknowledgement from 6503s that motion has completed, and lookahead to perform continuous path interpolation. [1].

At a lower level, the 6503 is an integral part of the joint controller which controls each axis of motion. It communicates with the LSI 11/02 through a Unimation designed interface board. The microprocessors compute the error signal and send it to the analog servo board to compensate the error. The main function of the 6503s include:

1) Receiving and acknowledging set points from the LSI 11/02 and performing interpolation between the current joint value and the desired joint value.

2) Reading the register value that stores the incremental values from the encoder of each axis.

3) Updating the error signal from joints.

4) Converting the error signal to voltage using the DAC and sending the voltage to the analog servo board to move the joints.

The PUMA model is more sophisticated than the Hitachi model, but the main disadvantage for both models is that the feedback gains are constants and prespecified. They cannot update the feedback gains under varying payloads. Since a robot is a highly nonlinear system, the inertial loading, the coupling between joints, and the gravity are all position dependent; and at high speeds the inertial loading terms can change drastically. Therefore, the system does not perform well under varying speeds and payloads. In fact, the PUMA arm moves with noticeable vibration at reduced speed, i.e., the fine motion. [1,8] One solution to this is the use of digital control with feedforward components computed by a special purpose processor. [1]

To be able to handle sophisticated control techniques like adaptive control and inverse plant using Newton-Euler formulations in robotics, we need the ability to implement computation intensive algorithms in the form of special purpose computer architecture or even special VLSI chips. [8]

The processor, termed as the arithmetic processor (ARP), is suitable for real time control where sophisticated control strategies require very large numbers of high precision arithmetical operations to be performed for every input/output transaction. In particular, ARP is intended for real time control of a robot arm. It should function as an attached processor of a general purpose microprocessor. [8]

Conceptually, the ARP lies between the Floating Point System AP120B, which is a high performance numerically oriented attached processor, and the Intel 8087, which is a single chip numerically oriented attached processor in the Intel 8087 family components. The ARP must work with floating point numbers. It should be much simpler than the AP120B and should differ from the 8087 by having its own on-chip program memory, and input/output buffers to facilitate real time application. [1,8]

The common denominator among the special purpose processors are:

1. A computation environment in which the majority of operations are arithmetic.

2. The use of multiple microprocessors to support high operation rates.

3. High bandwidth of memory to supply operands. [8]

For a multi-robot system working in cooperative tasks, a centralized control may be inefficient and unreliable. A better solution may be a decentralized control with distributed local controllers, one for each robot. This system will focus on the coordination schemes for the supervisory computer, the control architecture of the local controller, and the control implementation technique in the local controller. [8]

The following is a brief discussion of the programming method of current robotic systems and the proposed robotic system.

Robotic control systems are constructed by a programming algorithm, i.e., they are controlled by calculations based on an algorithm. [2] Programs are divided into tasks and subroutines. Tasks are supervised by the monitor program according to interrupts from the robot and the control console. Subroutines perform specific operations. The main tasks are teaching, manual modification, automatic execution of processes, and pass checks. The controller calculates data for processing, then drives the robot to the desired point, checks conditions, determines appropriate actions, and continues until all passes have been completed. [2,9]

Conventional robots use a playback type of operation by guiding the robot through its motion first by a manual operation called teaching pendant. For example, in Hitachi or PUMA robots, there is usually a hand held control box or terminal that controls manual operation and teaching. A human operator can instruct the robot by means of the hand held control box. Dials or high level language programming exists on these operating consoles and the operator can set the console or send commands to indicate data such as the size of objects or the set points. The data will then be stored in the memory of the robot. [2,3,10]

However, it is only practical when all the details of motion are known. In an application environment involving decision making based on sensed information and synchronization with other equipment or robots, the programming cannot be made only by teaching but, requires some form of written programming. Also, when programmed by teaching, complex operations like assembly and insertion of screws can be tedious and difficult. Teaching doesn't provide mechanisms for making adjustments while at the same time robots apply complicated branch and exceptional logic that are hard to teach. [8] Therefore, an off-line programming scheme that programs the robot through the use of remotely generated point coordinate data, function data, and logic data is needed. [7] It can be accomplished through the use of information from CAD/CAM programs that need accuracy, aligning reference frames, and the usage of high level programming languages.

The off-line programming can be classified into three categories: [3]

1. Teaching command - for manual operation of robot and for input and modification of location data. Commands are inputs to the controller which has 4 motions:

    a. independent motion of joints.

    b. motion of arm and wrist w.r.t. ground.

    c. motion of arm w.r.t. wrist.

    d. motion of arm and wrist w.r.t. shifted cartesian coordinates.

2. Program control - used for the description of motion sequence of robot and input through control console.

    a. motion - several motion commands such as the linear interpolation and simultaneous control of six axes.

    b. vision interaction and triggering (generation of location data based on visual information).

    c. program flow control - synchronization of multiple robot motions.

    d. arithmetic logic - calculation for location data.

3. Monitor command - selection of task control.

    a. editing and modification of robot motion program.

    b. select teaching and execution mode.

    c. storing and loading of program and location data.

    d. displaying programs and location data on terminal.

    e. recovery from errors.

The following modes are possible in the editing mode; modification, teaching and execution, partial execution, step execution, and alternate processing. These modes help in the debugging of a program. [3]

Of all the above functions multi-robot coordination is particularly distinguished. By the simple combination of visual information, input commands, and conditional branch commands such complicated operations can be easily programmed. Multi-robot coordination has been made simple due to the development of a language structure in which motion flow involving multiple robots can be written in a single program. Through translation procedures, the source program is automatically divided into programs for individual robots, and the synchronization commands between the robots are also automatically inserted as required in each program. [3]

In order to develop an intelligent robotic system which can be readily used by a broad spectrum of users, suitable higher level languages must be developed. They must include the ability to utilize and process the information provided by a wide variety of sensors (e.g., the use of interactive graphics coupled with information about the parts being assembled can be coupled to a 2-D visual recognition system to provide near real time 3-D information and a simple display of the part may be coupled to an interactive graphic programming method). At a still higher level, there are questions about application of artificial intelligent (AI) and knowledge-based system to robotics task specifications. [1,8] It is very important that user-oriented ways be found for the specification of the task robot. After the specification of the task robot is performed, the results must be translated automatically into the appropriate robotic system.

The following are some high level languages: [8]

WAVE: provides the programmer with the ability to specify point to point robot motion.

AL: coordinates motion between 2 robots, upon sensor, trajectory calculation, and compile time planning value (AI like language).

VAL: truly high level language.

Programming in these languages are similar to FORTRAN or PASCAL. However, there is the obvious difference that one is always programming w.r.t. a revolving geometric situation.

In theory, these languages solve many problems perceived in programming by teaching, however, new problems arise:

1. Obstacle avoidance and trajectory calculation. Robots must be aware of and avoid all objects in its workspace. This is particularly difficult in a multi-arm system.

2. Complexity. This requires a high level of skill from the programmer.

3. Multiplicity of devices. The device communication becomes complicated.

To solve these problems we introduce the following ideas:

a. coupling CAD geometric databases to robot programming languages.

b. use of computer graphics in robot programming.

c. examination of a distributed operating system to manage the multiplicity of devices which must work together in a robot-based manufacturing cell. [8]

It is possible to extend CAD to include programming by graphical techniques or by robot programming as well. Information contained in the database has the potential to simplify robot language or form the base for a graphical programming language, (e.g., once the location is known, relative position of holes, etc. can be determined from the database). In fact, the point of interest can be determined beforehand at the time of compilation thus reducing the execution time required. If the appropriate points are coupled by labelling them in the database, procedural programming may be simplified and the base of graphic programming can be formed. For example, in object avoidance, an algorithm can refer to data in the geometric modelling database. [8]

The above approaches can improve robot language capabilities. The study of coupling between the CAD geometric modelling databases and robot programming may provide useful feedback on CAD systems themselves that in fact may improve the simulation design of new robotic systems.

MODERN STATE ESTIMATION, IDENTIFICATION, AND CONTROL TECHNIQUES WITH A NONLINEAR SYSTEM INVERSE FOR ROBOT ARM APPLICATIONS

The proposed control scheme as shown in Figure 1 will be presented here in three forms. All three forms are similar except for the method used for system identification. In this figure the entire control system is composed of six distinct portions: The trajectory command generator, regulator, inverse plant, robot arm, identification, and estimation. First, the desired trajectory, calculated a priori, is generated real time, on line by the trajectory command generator. Next, the regulation block serves as the trajectory regulator, i.e., the trajectory perturbation controller, with the trajectory feedback information as the input and the trajectory perturbation as the output. In the next step, the inverse plant determines the control torques based on the desired trajectory and the trajectory perturbation from the trajectory regulator. The control torques are then applied to the robot arm. Next, the resulting motion is measured in terms of the position and velocity of each joint. The motion measurement together with the control torques constitute the input to the identification block. The system is identified using one of the three methods to be discussed. Finally, Kalman filters are used to estimate the state at the present time in the presence of plant and measurement noise. This estimated state is then used in the calculation of the next control torque for the robot arm.

The trajectory that the robot will be commanded to follow is defined by the Trajectory Command Generator. The control system transforms the desired trajectory and robot estimated state into an appropriate control U that forces the robot to generate a trajectory that approaches the command trajectory, even in the presence of process noise w and measurement noise γ. The desired trajectory $(v^*,\dot{v}^*,\ddot{v}^*)$, where v^*, $\dot{v}^*$, and $\ddot{v}^*$ are the desired position, velocity, and acceleration of each joint respectively, is an optimal trajectory generated a priori. The states, $X = (v,\dot{v})$, are related to the only measurable variables available. These are the position and angular velocity of each joint as measured from position resolvers and tachometer feedback loops respectively. The equations of motion for the robot arm are of the form:

$$\dot{X} = f(X,U) + w \tag{1}$$

$$Z = h(X) + \gamma$$

The state equations and the equations of motion are strongly nonlinear and highly coupled. In this equation $h(X) = X$. The inverse plant represents an inverse of the robot response to control inputs, and the combined relationship of the input to the inverse plant to the robot response is, ideally, an identity. Therefore, the relationship can be used to advantage by closing the feedback loop in front of the inverse plant (the author adopted this idea from References 9 and 10), as shown in Fig. 1. As a result, because of the invariance of the feedback control response the control loop can be made to behave in a linear fashion over the entire operational range of the robot. The design of the regulator is much simpler by closing the feedback loop in front of rather than after the inverse plant (where the total control U is composed of the nominal control U^* generated by the inverse plant with the desired trajectory as the input, and the perturbation control δU). The latter approach is commonly used today. Applications of the concept have shown that the total complexity of the control system can be reduced. Furthermore a better balance between the open-loop (a priori) and closed-loop (a posteriori) control is a consequence of the method.

Inverse Plant (Newton-Euler). The purpose of the inverse plant is to determine what control torque or force input to each of the links in the robot will cause the robot to follow the desired trajectory. Therefore the input to the inverse plant is the desired trajectory (position, velocity, and acceleration) of the end effector and the output of the inverse plant is the control torque or force which is required to produce the desired trajectory. For robotic arm problems the inverse plant is usually modeled with the Lagrange Euler (L-E) or Newton Euler (N-E) equations. Both equations give dynamic models of kinematic mechanisms, a robot is an open loop kinematic chain, and can be applied to this problem. The L-E equations involve a larger number of additions and multiplications than the N-E method for a given number of links, therefore the N-E method will be used due to its increased efficiency.

The N-E method is divided into two parts. The first part identifies the kinematic relationships of each of the links to each other and finds the velocity and acceleration of each link starting from the base working toward the end effector. The second part of the N-E method starts with the forces and torques exerted on the end effector from the environment and calculates the forces and torques exerted on each link starting from the end effector working toward the base. From this information the input torque (rotational joint) or input force (translational joint) can be found that will cause the robot to generate the desired trajectory.

Inverse Plant Calculation Using Newton-Euler Formulation. The inverse plant problem is defined as:

Given $\{v_1, v_2, \ldots v_n\}$, $\{\dot{v}_1, \dot{v}_2, \ldots \dot{v}_n\}$
$\{\ddot{v}_1, \ddot{v}_2, \ldots \ddot{v}_n\}$.

Find: $\{\tau_1, \tau_2, \ldots \tau_n\}$

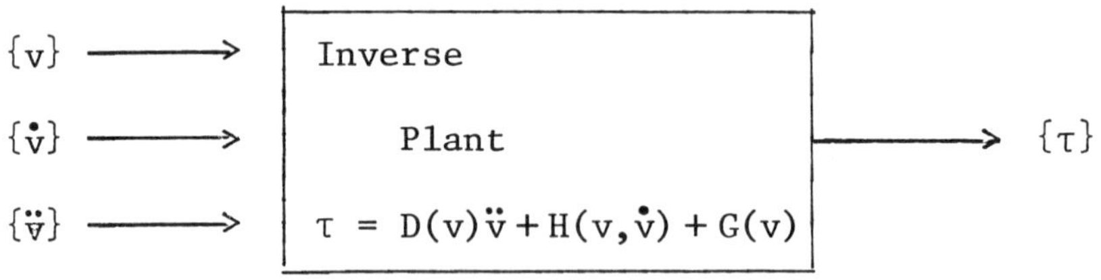

Kinematic Model of a Robot Arm. Any robot arm can be modeled as a series of rigid links connected together by either revolute or prismatic joints. One end is free while the other end is attached to a supporting base. Motion of the arm is provided by the relative motion of the links.

Each joint has one degree of freedom, so that for an N degree of freedom robot arm there will be N+1 links (one is the base) and N joints. Figure 2 shows a simplified model of a robot arm. For the model shown, the generalized coordinates are v_1, v_2, ..., v_N where v_i is the relative position/orientation between adjacent links. The general expression for the equation of motion of a robot arm can be written as:

$$\{\tau\} = [D(v)]\{\ddot{v}\} + \{H(v,\dot{v})\} + \{G(v)\} \qquad (2)$$

where: $[D(v)]$ = Inertia matrix, dependent upon position, $(N \times N)$

$\{H(v,\dot{v})\}$ = Coriolis and centripetal vector matrix, dependent on position and velocity, (N×1)

$\{G(v)\}$ = Gravity vector matrix, dependent on position, (N×1)

The equation above consists on N nonlinear coupled differential equations and is difficult to solve.

1. System Identification Using Generalized Equation of Motion for Robots

The first method of system identification uses the general equation of motion (2) to estimate the state of the robot before measurements are made and then uses the estimation block to calculate the estimated state after the measurements (Fig. 3). This identification procedure uses a Continuous Dynamics, Discrete Measurements Extended Kalman Filter (Nonlinear Kalman Filter). The linearized matrix $A \equiv \frac{\partial f}{\partial x}$, which is needed in computing the error covariance matrix P, can be identified from the state obtained from Eq. (2) and the control torque without actually solving $\frac{\partial f}{\partial x}$. This method is very difficult if not impossible to implement due to the difficulty in solving the equation of motion to determine the state. The equations of motion are typically six highly coupled nonlinear differential equations. This is a very difficult problem to solve and could not be done in real time with today's computer technology and therefore is not recommended.

2. System Identification Using Plant Model

The second method [11] of system identification uses the plant model to identify the system matrices [A] and [B] which can be used to find ϕ_{k-1} and Γ_{k-1} for the discrete system model. The discrete model is then used in a Linear Discrete Kalman Filter for updating the state estimation after the measurements are made. The state can also be estimated by using just the [A] and [B] matrices by using a Linear Continuous Dynamics, Discrete Measurements Kalman Filter. With this method the ϕ_{k-1} and Γ_{k-1} matrices never need to be calculated. The [A] and [B] matrices are found as a result of identifying the acceleration of each joint given the input torque and measured values for position and velocity of each joint. We can rewrite the equation of motion (2) as:

$$\{\tau\} = [C]\ddot{v} + \{\overline{D}\} \tag{3}$$

where $\{D\}$ is an $n \times 1$ vector and $[C] = [\{C\}_1,\ldots,\{C\}_n]$ is an $n \times n$ matrix with $\{C\}_i$, $i = 1,\ldots,n$ as an $n \times 1$ vector. [C] and $\{\overline{D}\}$ need to be found for $v(t_i)$ and $\dot{v}(t_i)$ as follows:

For:

$$v = v(t_i) \qquad \ddot{v} = \begin{Bmatrix} 0 \\ 0 \\ \vdots \\ 0 \end{Bmatrix}$$

$$\dot{v} = \dot{v}(t_i)$$

By the use of N-E formulation we can calculate $\{\tau\}_0$ associated with the assumed acceleration. Substituting $\{\tau\}_0$ and $\ddot{v}$ into the revised equation of motion (3), we have

$$\{\overline{D}\} = \{\tau\}_0 \tag{4}$$

Furthermore, for

$$v = v(t_i) \qquad \ddot{v} = \begin{Bmatrix} 0 \\ 0 \\ \vdots \\ 1 \\ \vdots \\ 0 \end{Bmatrix} \leftarrow \text{ith row}$$

$$\dot{v} = \dot{v}(t_i)$$

using the same procedure we used to find $\{\tau\}_i$ $i = 1, 2, \ldots, N$ and substituting into the revised equation of motion, we have:

$$\{C\}_i = \{\tau\}_i - \{\tau\}_0. \tag{5}$$

Having $[C]$ and $\{\overline{D}\}$ for $v = v(t_i)$ and $\dot{v} = \dot{v}(t_i)$ we can write:

$$\{\ddot{v}\} = [C]^{-1}(\{\tau\} - \{\tau\}_0) \tag{6}$$

Thus, $\ddot{v}$ can be calculated if τ, $v(t_i)$, and $\dot{v}(t_i)$ are specified. The advantage of the above equation for a complicated system is clear. For any particular value of v and $\dot{v}$, we can write the above equation in the form:

$$\begin{Bmatrix} \dot{v} \\ \ddot{v} \end{Bmatrix} = \underbrace{\begin{bmatrix} 0 & I \\ 0 & 0 \end{bmatrix}}_{\equiv [A]} \begin{Bmatrix} v \\ \dot{v} \end{Bmatrix} + \underbrace{\begin{bmatrix} 0 \\ [C]^{-1} \end{bmatrix}}_{\equiv [B]} (\{\tau\} - \{\tau\}_0) \tag{7}$$

Since $[C]^{-1}$ and τ_0 are functions of v and $\dot{v}$, solutions to the above equation are only available by numerical procedures. This method also provides an estimate of the acceleration for each joint; this is then added to the estimated state for position and velocity from the state estimation. This is necessary because the acceleration is not available from the measurement and can not be used in the Kalman filter estimation of the state. See Fig. 4 for a block diagram of the control configuration.

3. System Identification Using a Kalman Filter to Identify System Matrices

The third method of system identification uses a linear Kalman filter to estimate the system discrete matrices Γ_{k-1} and ϕ_{k-1} to be used in the estimation block of the control scheme to determine the state of the robot joints after measurement and noise. The Kalman filter for the estimation of the system discrete matrices is developed as follows. Given the state equations:

$$\begin{aligned} X_k &= \phi_{k-1}X_{k-1} + \Gamma_{k-1}u_{k-1} + B_{k-1}w_{k-1} \\ Z_k &= H_kX_k + v_k \end{aligned} \tag{8}$$

Note H_k is equal to the identity matrix I. Defining $\overline{X}_k^T \equiv [\phi_{k-1} \; \Gamma_{k-1}]$ we can rewrite the state equations as:

$$X_k = \overline{X}_k^T \begin{bmatrix} X_{k-1} \\ u_{k-1} \end{bmatrix} + B_{k-1}w_{k-1} \tag{9}$$

$$Z_k = X_k + v_k$$

Using $X_k = Z_k - v_k$, we have

$$Z_k = \bar{X}_k^T \begin{bmatrix} X_{k-1} \\ u_{k-1} \end{bmatrix} + \bar{v}_k \quad (10)$$

where

$$\bar{v}_k = B_{k-1} w_{k-1} + v_k$$

$$E[\bar{v}_k] = 0 \quad (11)$$

$$E[\bar{v}_k \; \bar{v}_k^T] = B_{k-1} Q_{k-1} B_{k-1}^T + R_k = \bar{R}_k$$

Defining m as the number of joint controls (degrees of freedom) and n = 2m then for the ith component of Eq. (10)

$$Z_{k,i} = \bar{X}_{k,i}^T \begin{bmatrix} X_{k-1} \\ u_{k-1} \end{bmatrix} + \bar{v}_{k,i} \quad (12)$$

where the measurement $Z_{k,i}$ is a scalar quantity and $\bar{X}_{k,i}^T = [\phi_{k-1,i}, \Gamma_{k-1,i}]$ with $\phi_{k-1,i}$ and $\Gamma_{k-1,i}$ are ith row of ϕ_{k-1} and Γ_{k-1} respectively. Defining $\bar{H}_k \equiv [X_{k-1} \; u_{k-1}]$ Eq. (12) can be rewritten as:

$$Z_{k,i} = \bar{H}_k \bar{X}_{k,i} + \bar{v}_{k,i} \quad (13)$$

Note that since X_{k-1} is not available we must use $\hat{X}_{k-1}^+$ in the $\bar{H}_k$ matrix. The above equation can be considered as a measurement equation for the system parameters. Now assuming that the dynamic equation for the system parameters are

$$\bar{X}_{k,i} = \bar{X}_{k-1,i} + \bar{w}_{k-1,i}$$

$$E[\bar{w}_{k-1,i} \; \bar{w}_{k-1,i}^T] = \bar{Q}_{k-1,i} \quad (14)$$

and letting errors in the assumption reside in the noise term $\bar{w}_{k-1,i}$ we can use a linear Kalman filter to estimate the system parameters. The Kalman filter estimates of the system before the measurement are:

$$\hat{\bar{X}}_{k,i}^- = \hat{\bar{X}}_{k-1,i}^+$$

$$\bar{P}_{k,i}^- = \bar{P}_{k-1,i}^+ + \bar{Q}_{k-1,i} \quad (15)$$

and finally the Kalman filter estimates of the system after the measurement are:

$$\bar{K}_{k,i} = \bar{P}_{k,i}^- \bar{H}_{k,i}^T (\bar{H}_{k,i} \bar{P}_{k,i}^- \bar{H}_{k,i}^T + \bar{R}_{k,i})^{-1}$$

$$\bar{P}_{k,i}^+ = (I - \bar{K}_{k,i} \bar{H}_{k,i}) \bar{P}_{k,i}^- \quad (16)$$

$$\hat{\bar{X}}_{k,i}^+ = \hat{\bar{X}}_{k,i}^- + \bar{K}_{k,i} (Z_{k,i} - \bar{H}_{k,i} \hat{\bar{X}}_{k,i}^-)$$

Therefore from the definition of $\hat{X}_{k,i}$ the desired identification of the discrete matrices $\hat{\phi}_{k-1,i}$ and $\hat{\Gamma}_{k-1,i}$ can be found. A block diagram of the control scheme used in this method is shown in Fig. 5. This method should give the most accurate results with a minimum of execution time required between samples.

ACKNOWLEDGEMENT

This paper represents excerpts from various projects. The author would like to thank the following graduate students for their assistance in the following projects:

James R. Roll - Control of Robotic Devices.

Connell Yuk and Shyh-Hao Kang - Microprocessor-Based Controller for the Robot Arm.

Indera Nurhadi - Inverse Plant Calculation Using Newton-Euler Formulation.

Special thanks to James R. Roll for his excellent input and help in completing this paper.

REFERENCES

1. C. S. G. Lee, T. N. Mudge, J. L. Turney. "Hierarchical Control Structure using special purpose processor for the control of robot arms." Pattern Recognition and Image Processing. IEEE Computer Society. June 13-17, 1982, Las Vegas, Nevada.

2. N. Miyake, S. Ando, R. Tokuno. "Automation of gas cutting and multipass welding process for pressure vessel nozzles utilizing in industrial robot." Hitachi Review, Vol. 30 (1981), pp. 217-222.

3. M. Kohno, H. Horino, M. Isobe. "Intelligent assembly robot." Hitachi Review, Vol. 30 (1981), pp. 211-216.

4. J. M. Vranish. "The robotic deriveter." Robotic Today/Winter 81-82, pp. 45-47.

5. Report on Industrial Robots in Japan. Available from Technical Database Corp., La Canada, CA 91011.

6. H. M. Morris. "Where do Robots fit in Industrial Control." Control Engineering, Feb. 1982, pp. 58-64.

7. R. L. Tarvin. "An off-line Programming Approach." Robotic Today, Summer 1981. pp. 30-35.

8. University of Michigan Center for Robotics and Integrated Manufacturing, "Research." pp. 12-54.

9. "Solving the basics can lead to better robot design." The Industrial Robot, Dec. 1981.

10. A. Ray. "Updating the control system of a Trallfa Industrial Robot." The Industrial Robot, Sept. 1981, pp. 166-169.

11. George Meyer and Luigi Cicolani. "Application of Nonlinear Systems Inverses to Automatic Flight Control Design-System Concepts and Flight Evaluations." pp. 10-1 to 10-29, AGARD-AG-251, Theory and Applications of Optimal Control in Aerospace Systems, July 1981.

12. William R. Wehrend, Jr. and George Meyer. "Flight Tests of the Total Automatic Flight Control System (TAFCOS) Concept on a DHC-6 Twin Otter Aircraft." NASA TP-1513, 1980.

13. S. H. Koozekanani, R. B. McGhee and H. T. Change. "A Recursive Free-Body Approach to Computer Simulation of Human Postural Dynamics," to appear in IFAC Symposium on Control Aspects of Prosthetics and Orthotics.

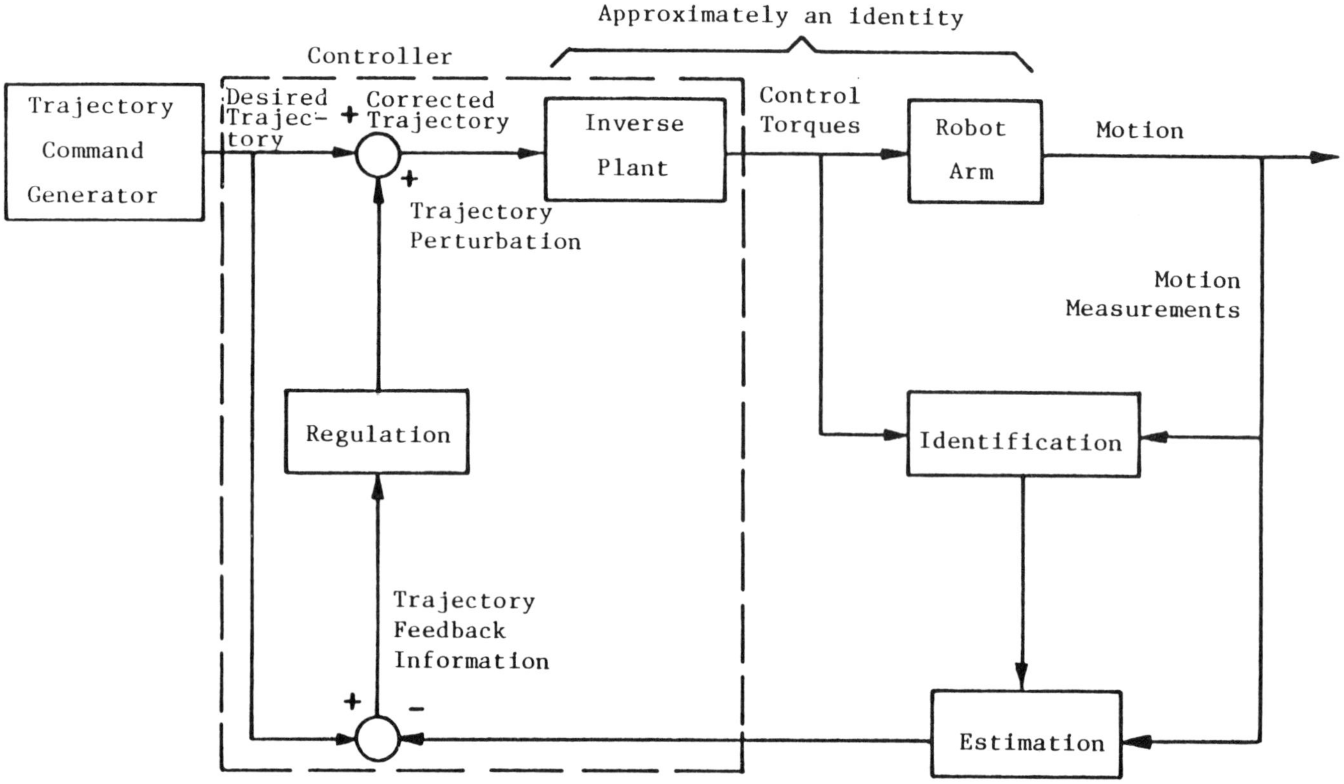

Fig. 1 Block Diagram of Modern State Estimation, Identification and Control Techniques with a Nonlinear System Inverse for Robot Arm Applications.

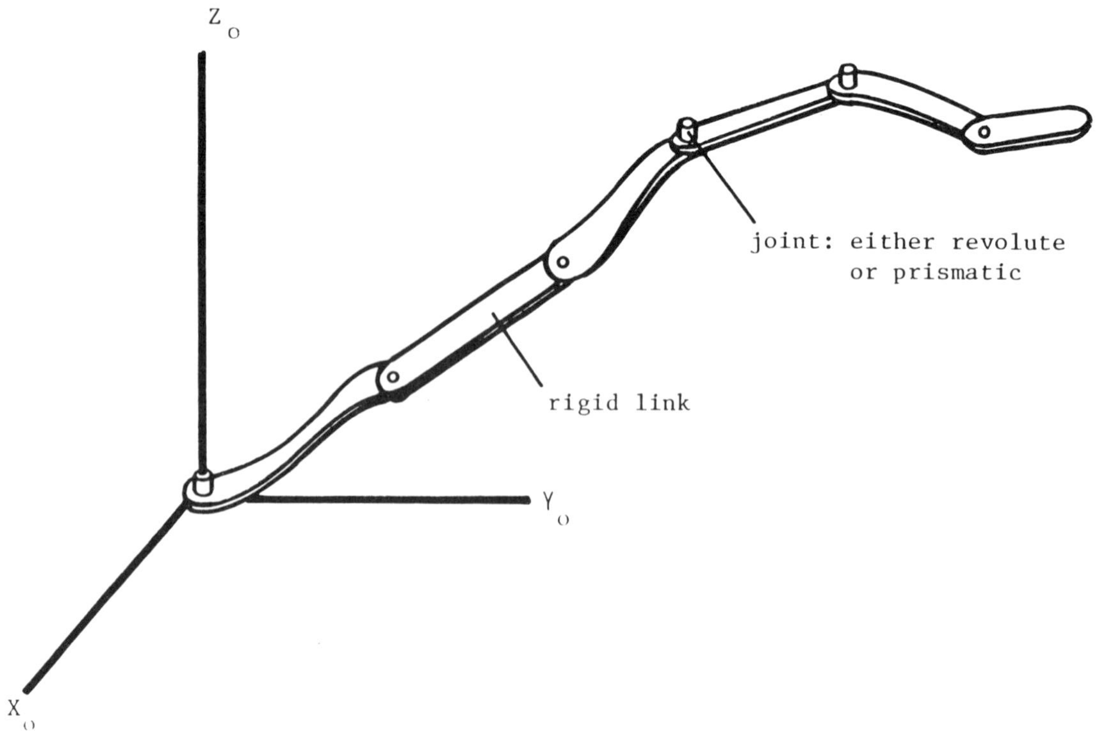

Fig. 2 A Model of a Spatial Robot Arm

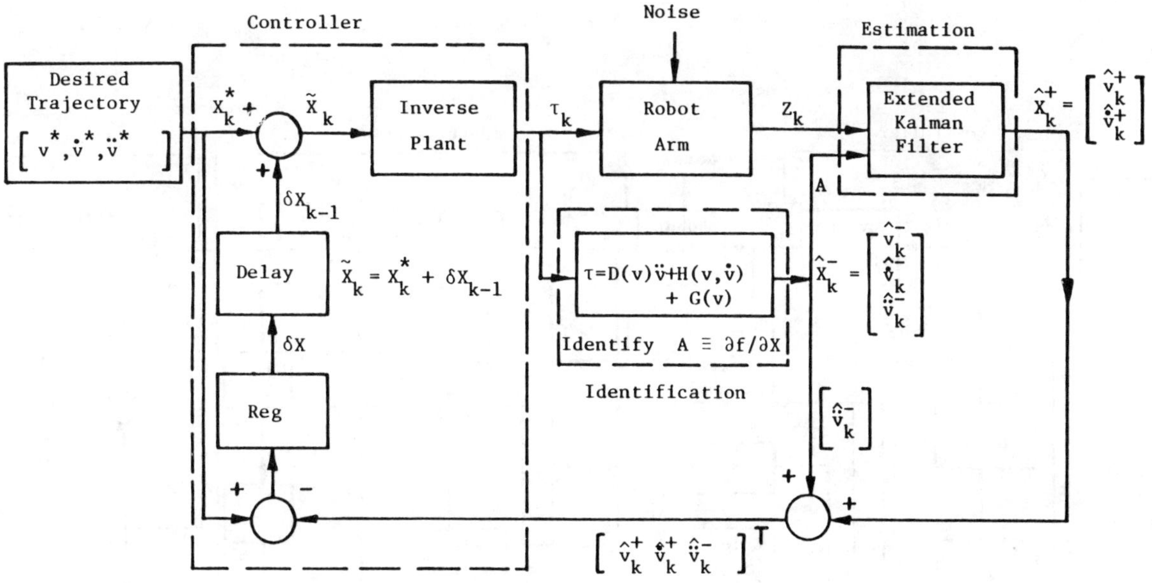

Fig. 3 Block Diagram of Control System Using General Equation of Motion to Identify the System State Before Measurements.

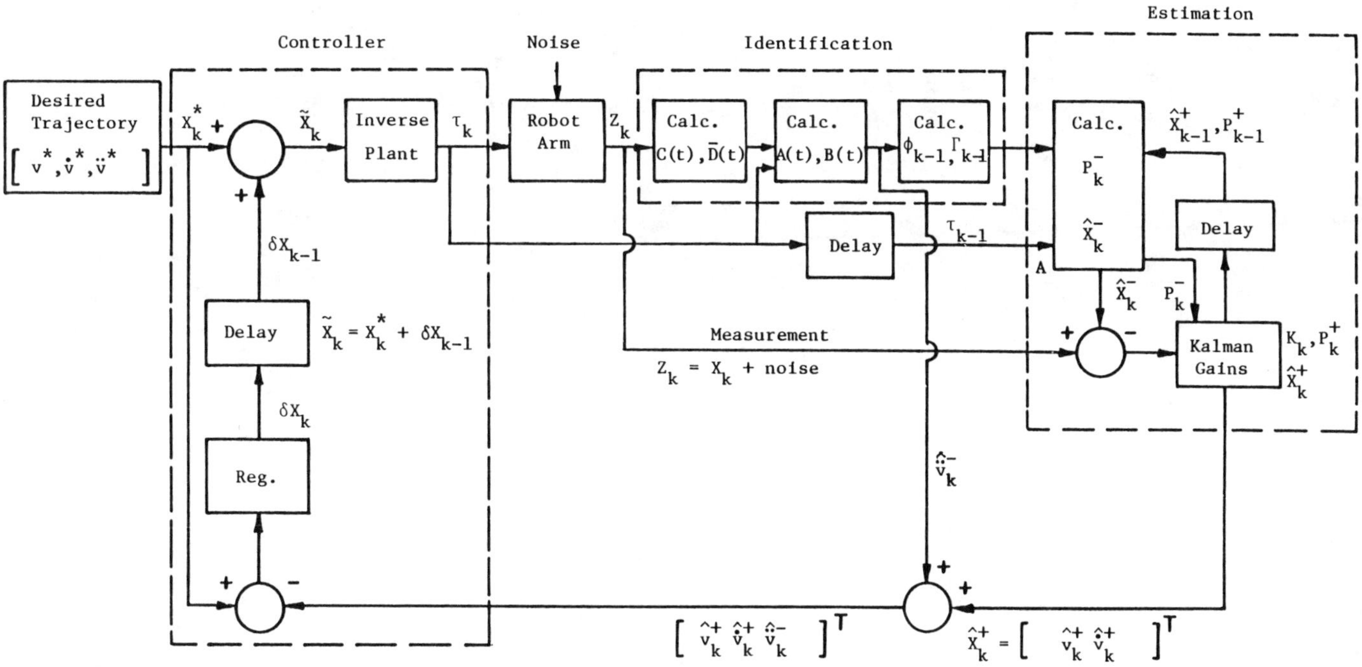

Fig. 4 Block Diagram of Control System with Identification from Plant Model

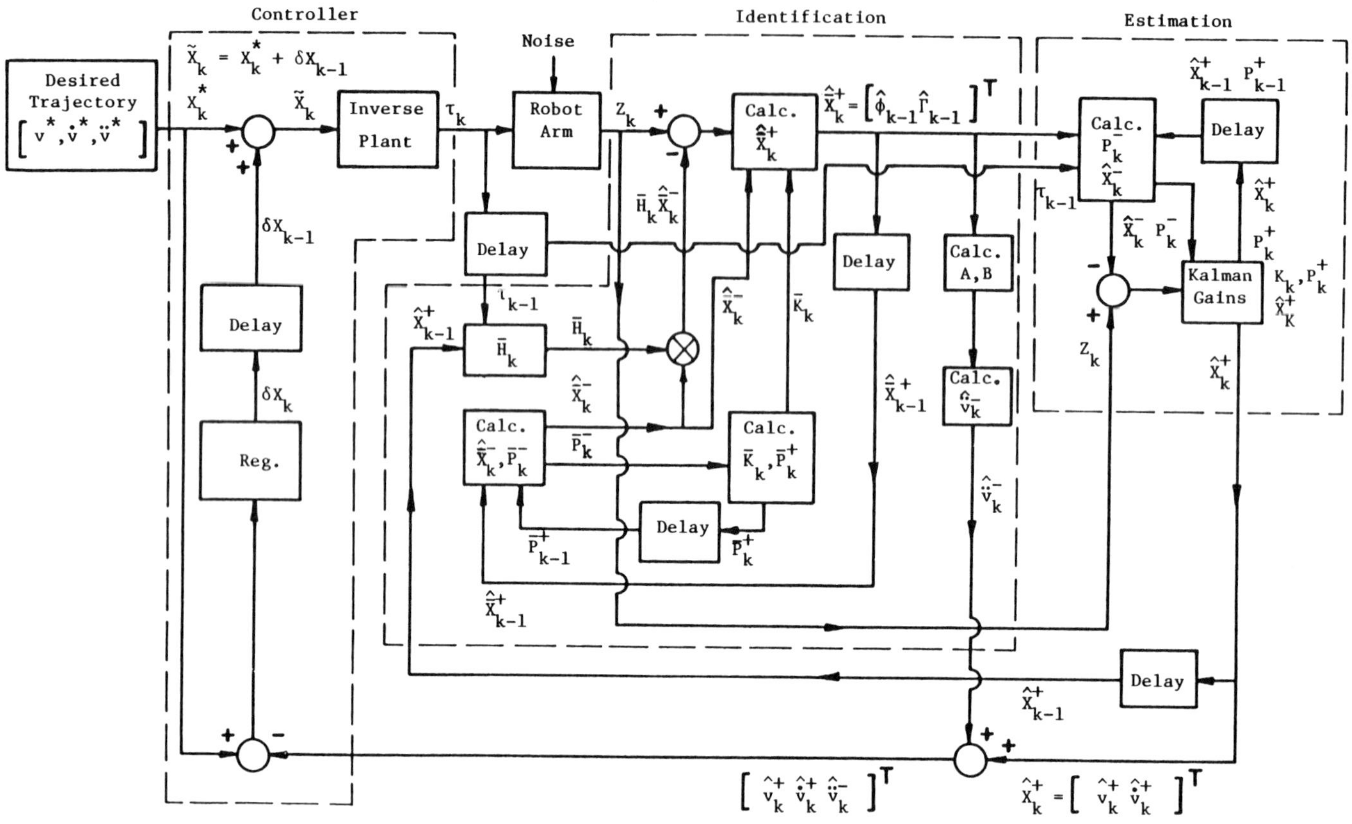

Fig. 5 Block Diagram of Control System Using Kalman Filter for System Identification

Robotic Nutplate Sealing Cell for F-16 Air Surface Panels

Peter E. McCormick
General Dynamics
Fort Worth, Texas

Abstract

Air frames require sealant between parts in order to make them fuel and air-tight. The F-16 fighter plane has many interchangeable panels which are fastened with screws and "sealed dome" nutplates. These nutplates are sealed with a polysulfide sealant which is applied to the nutplate flange and surrounding part with a filet sealant gun. The sealant robot implementation schedule, present method, and the various concepts being pursued to apply sealant robotically are discussed.

INTRODUCTION

Approximately 48 percent of General Dynamics' factory labor is spent in subassembly and assembly operations. These operations consist mostly of manual labor without the benefit of automated systems. General Dynamics' management reviewed these shop processes in detail and determined that several were good candidates for automation. Three specific operations were reviewed for final selection. Those operations were:

- Manual part finishing operations - Deburring, removal of cutting marks, and elimination of sharp edges.
- Manual part indexing for automatic riveting.
- Manual nutplate sealing applications.

We found that these specific operations shared a common trait. They were labor-intensive, and to date, no aerospace firm had automated any of these systems for production utilization. As we continued our candidate program selection process, we developed data for each of the manual processes and reviewed the respective benefits associated with automating each. The current manual nutplate sealing process stood out as our most promising candidate for automation.

Current F-16 part production requires a substantial application of FMS-1044 sealant. The sealant is applied to fasteners, such as standard nutplates, to establish a complete fuel tight structure. Current application of FMS-1044 sealant is highly labor-intensive, requiring one operator as much as eight hours to complete a fuel cell panel. Cleaning operations also add to the task, since currently there is no system to dispense sealant precisely. The manual motions required to accomplish these tasks are highly repetitive and result in operator fatigue.

The objective of the Robotic Nutplate Sealing Project is to automate the current manual labor-intensive method of applying FMS-1044 sealant to nutplate fasteners by establishing a nutplate sealing cell. Internal monies are funding this entire program. It is hoped to have an on-line system by early 1984.

General Dynamics' primary goal for technology development is increased productivity - with the ultimate goal of a completely modernized, automated factory. When successfully completed, this program will contribute significantly to that goal by making possible more extensive and efficient utilization of flexible automation, i.e., robotic systems.

The proposed nutplate sealing cell will contribute to this automated factory in both specific and general terms. Specifically, it will greatly reduce or completely eliminate a tedious, labor-intensive job that is performed industry-wide. The panels enumerated in this report are not the only sealed panels on the F-16 and, since all aircraft have sealed areas, the technology gained in this cell can be directly transferred to many other applications. Indirect benefits will involve advancing the state-of-the-art in the areas of off-line programming, relocatable contour sequences, and new end effector techniques. All these factors will contribute to our prime goal - a better product at a lower cost.

In summary, the sealing cell concept offers the distinct advantages of significant labor savings, introduction of a new robotics technology for the aerospace industry, and future savings in sealant material usage.

IMPLEMENTATION PLAN FOR ROBOTIC NUTPLATE SEALING

Parameters for the nutplate sealing cell will be based on the parts in Groups 1 and 2, as in Figure 1. Group 1 parts have more nutplates, thus take longer to seal than Group 2 or 3 parts. These panels will be clamped manually into a repeatable position on a table for processing by a robot. This robotic cell will be located in the existing manual sealing area. A section of this area will be designated for robotic implementation such that, although it is easily accessible, it will be conducive to parallel manual operation in order to avoid production delays during robotic implementation. This avoidance of production delays will be paramount in the application plan throughout the project.

Before introducing the cell to the production environment, the robot will be interfaced with a sealant delivery system, and part holding will be designed and prototyped in the Robotics Laboratory. This will minimize development time spent on the production floor.

When the system is fully prototyped and debugged, the cell will be integrated into the work environment, and off-line programming from an engineering data base will be used. As experience and production data increase, more parts will be integrated into the cell to further increase productivity.

Successful implementation will depend upon meeting several signficant criteria as shown in Figure 2.

TIMELY CAPITAL ACQUISITION

Each time General Dynamics begins planning a new robotics cell, it becomes increasingly apparent that a sound plan for acquiring necessary capital equipment (robot, positioners, etc.) is vital to timely completion of the program.

In view of this, the necessary steps were taken to include related capital in the 1983 budget. The implementation schedule shown in Figure 3 projects the time required for capital acquisition. Included in this schedule are the periods necessary for official specification writing, bid collecting involving applicable vendors, and shipping and receiving lead-time. We understand that, regardless of the quality of our design or integrity of our development work, an automatic sealing cell is difficult without a robot. Therefore, expeditious acquisition of this robot in time to meet scheduling constraints will be a prime consideration.

EQUIPMENT MODIFICATION

Since the first robot was installed at General Dynamics, we have found it necessary to modify the existing robot controllers through user input/output and peripheral equipment addition, and add an entire host computing network to the robot cells. The robot hardware has been modified in various ways, from adding cables and valves to designing an innovative new faceplate capable of automatically changing end effectors during processing.

This experience in equipment modifications lends itself towards a broad base for understanding the changes necessary to implement the nutplate sealing cell. The process, along with the sealant delivery system, will need modification. The area will require modification in order to allow parallel processing (manual and robotic) during implementation. Even the job functions of the workers will have to be modified through training programs and hands-on experience with the robot.

GROUP I PARTS:

Part No.	Part Name	WBS	IND	T/C	Effectivity	Hrs.
16[illegible]1-801	Skin Assy.	1132	1	758	All	4
16B5303-804	Skin Assy.	1132	1	758	All	6
16B5[illegible]-803	Skin Assy.	1132	1	758	All	6
16B331-A13	Skin Assy.	0143	1	759	All	6
16B6529-A77	Frame Assy.	0143	1	759	All	4
16B6529-A99	Frame Assy.	0143	1	759	All	4

GROUP II PARTS:

Part No.	Part Name	WBS	IND	T/C	Effectivity	Hrs.
16B410-A118	Upper Duct Assy.	0128	1	757	1	2
16B410-A119	Upper Duct Assy.	0128	1	757	2	2
16B1270-39	Bulkhead Assy.	1125	1	757	1	1
16B1410-811	Beam Assy. R.H.L.	0126	1	757	S	1
16B1410-817	Beam Assy. R.H.L.	0126	1	757	S	1
16B1706-809	Beam Assy.	0126	1	757	S	1
16B1706-811	Beam Assy.	0126	1	757	S	1
16B1706-817	Beam Assy.	0126	1	757	2	1
16B2270-A14	Bulkhead Assy. FS 243	1122	1	757	2	1.5
16B2270-A15	Bulkhead Assy. FS 243	1122	1	757	2	1.5
16B2410-819	Beam Assy. RH	0126	1	757	S	1
16B2410-835	Beam Assy. RH	0126	1	757	S	1
16B2410-839	Beam Assy. RH	0126	1	757	S	1
16B5233-3	Bulkhead Assy.	0134	1	758	All	1
16B5302-3	Skin Assy.	1132	1	758	All	1
16B5302-5	Skin Assy.	1132	1	758	All	1
16B5507-3	Cover Assy.	1131	1	758	All	.25

GROUP III PARTS:

Part No.	Part Name	WBS	IND	T/C	Effectivity	Hrs.
16B1250-3	Bulkhead Assy.	1125	1	757	1	.5
16B1250-805	Bulkhead Assy.	1125	1	757	S	.5
16B1560-3	Cover Assy.	1125	1	757	1	.5
16B5113-A801	Web Assy. LH	0134	1	758	2	.5
16B5113-A802	Web Assy. RH	0134	1	758	2	.5

Part No.	Part Name	WBS	IND	T/C	Effectivity	Hrs.
16B5123-803	Web Assy.	0134	1	758	1	.5
16B5123-811	Web Assy.	0134	1	758	1	.5
16B5123-813	Web Assy.	0134	1	758	2	.5
16D705-801	Tube Assy.	1131	1	758	All	.25
16P106-801	Union Assy.	1131	1	758	All	.25
16P111-6	Conn. Assy.	1131	1	758	E	.5
16P111-801	Conn. Assy.	1131	1	757	2	.5
16P111-802	Conn. Assy.	1131	1	757	2	.5
16P112-1	Housing Assy.	1131	1	758	[illegible]	.25
16P112-3	Housing Assy.	1131	1	758	2	.25
16P118-3	Union Assy.	1131	1	758	All	.25
16P119-1	Elbow Assy.	1131	1	758	All	.25
16P120-1	Union Assy.	1131	1	758	All	.25
16P122-1	Union Assy.	1131	1	758	All	.25
16P122-3	Union Assy.	1131	1	758	2	.25
16P123-1	Bellmouth Assy.	1131	1	758	All	.25
16P129-3	Tube Assy.	1131	1	758	All	.25
16P129-801	Tube Assy.	1131	1	758	All	.25
16P130-1	Tube Assy.	1131	1	758	All	.25
16P130-3	Tube Assy.	1131	1	758	All	.25
16P134-3	Fitting Assy.	1131	1	758	All	.25
16P150-801	Tube Assy.	1131	1	758	All	.25
16P158-3	Elbow Assy.	1122	1	757	2	.25
16P161-3	Union Assy.	1125	1	757	2	.25
16P168-3	Bellmouth Assy.	1125	1	757	2	.25
16P212-1	Fitting Assy.	1132	1	758	All	.25
16P216-1	Nipple Assy.	1132	1	758	All	.25
16P216-3	Nipple Assy.	1122	1	757	2	.25
16P231-1	Fitting Assy.	1131	1	758	All	.25
16P232-1	Nipple Assy.	1132	1	758	All	.25
16P233-1	Fitting Assy.	1125	1	757	All	.25
16W173-3	Retainer Assy.	1162	2	761	All	.5

Figure 1 Candidate Parts List

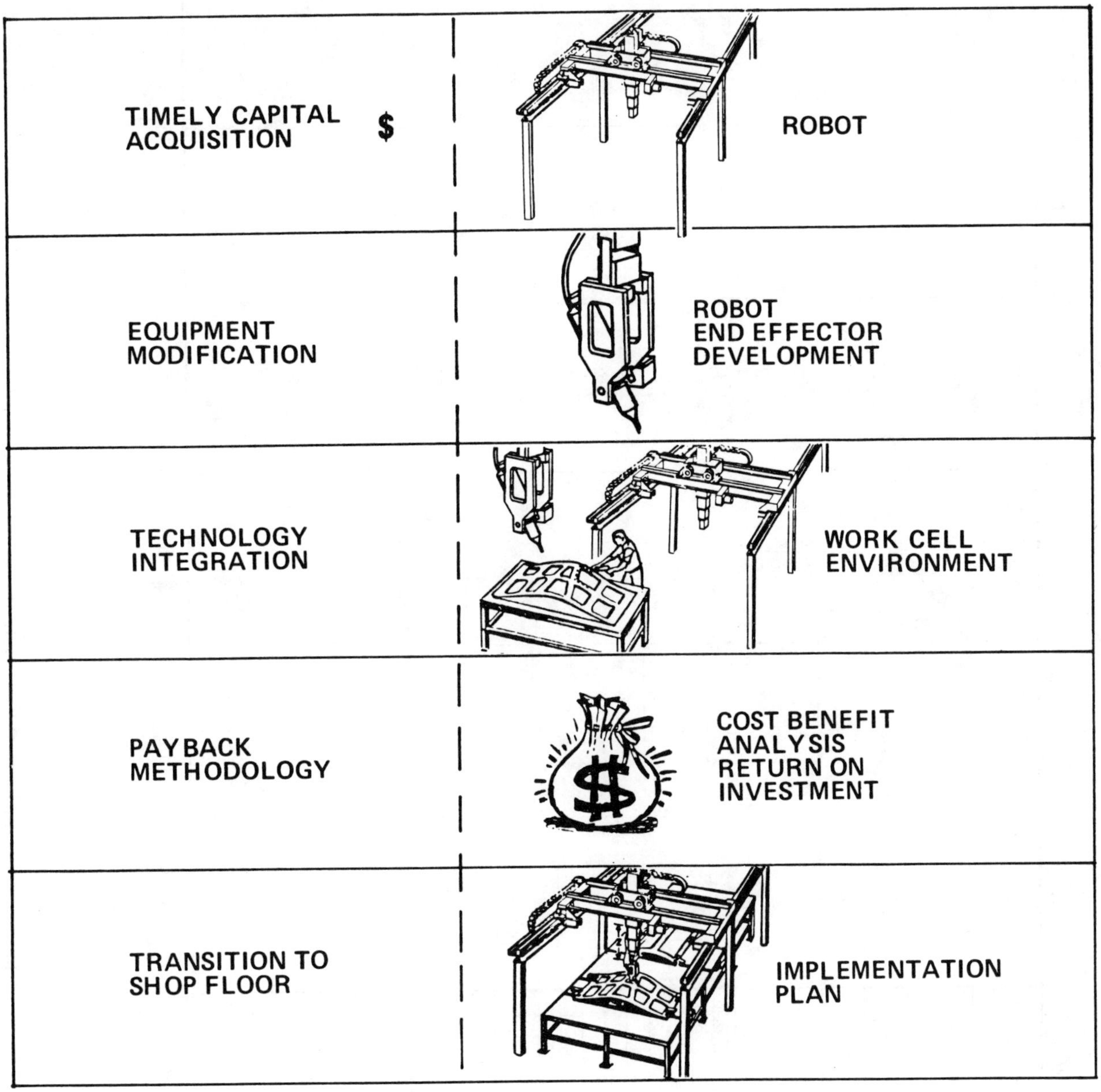

Figure 2 Significant Implementation Criteria

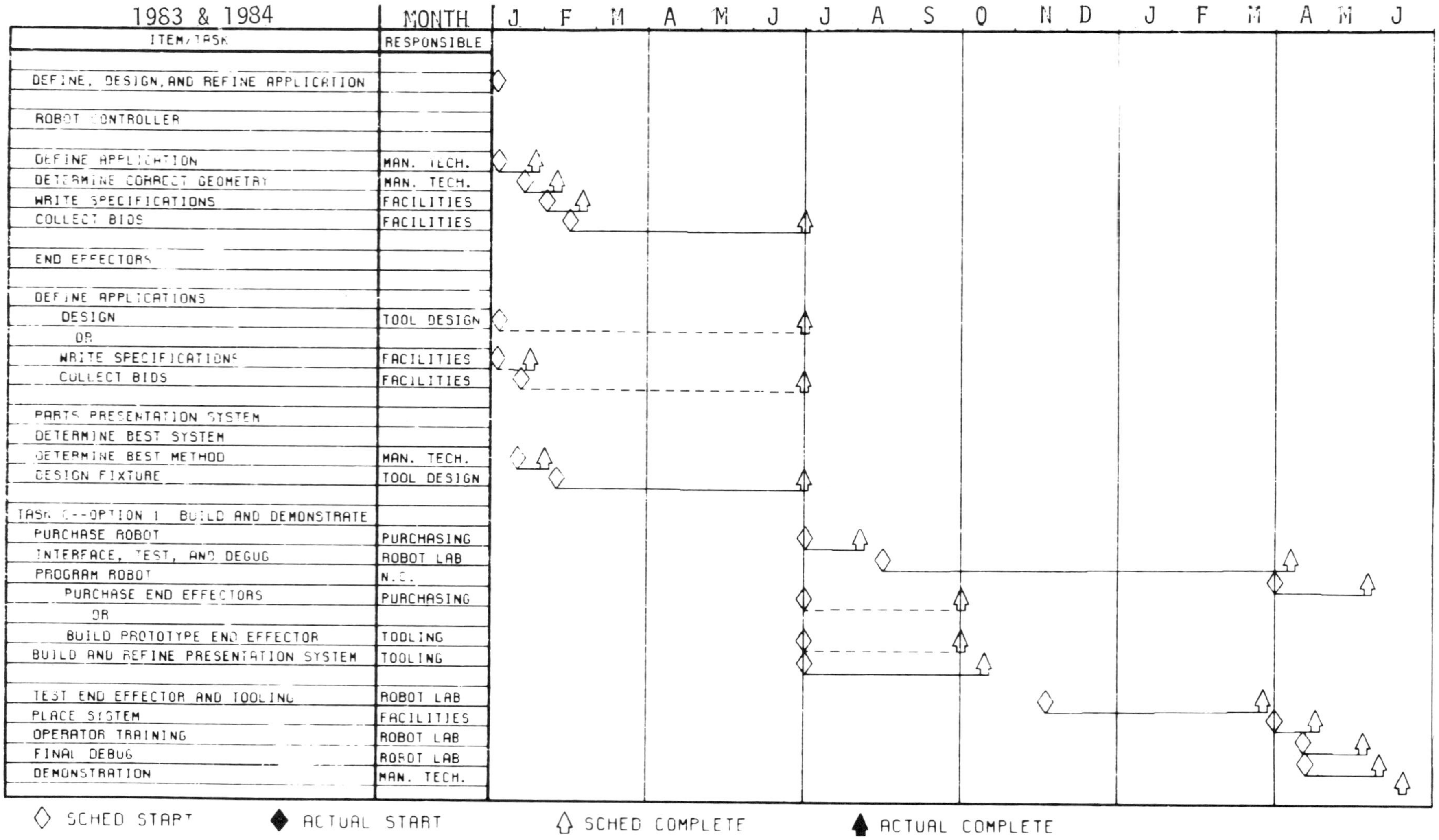

Figure 3 Implementation Schedule

TECHNOLOGY INTEGRATION

Regardless of the uniqueness of a concept, increased productivity cannot be achieved unless the automated cell is placed into production. Our main objective is to produce F-16's and to produce each one better, faster, and at lower cost. Thus, our robotic program has always had production enhancement as a beneficial cost driver. All our previous lab efforts have resulted in production-operative robotic cells that are now used to process F-16 parts.

METHOD OF PAYBACK ANALYSIS

General Dynamics, through the Technology Modernization Program, developed a computer-based system to determine return on investment and payback criteria at all levels of capital investment. The system is dependent only on a baseline-level input, as supplied by program management, and provides an excellent financial guidepost.

In addition to the initial cost-benefit analysis (which will use standard corporate financial formula for calculating return on investment and payback period), a finite cost-benefit tracking system will be used.

TRANSITION TO THE SHOP FLOOR

The problems associated with making computer/robotic technology a functional tool for general use in the workplace must be addressed. The tool should not require a technical expert to use, operate, or maintain and should have the skill and intelligence required to perform intricate tasks, logical thinking, and complex procedures.

PRESENT MANUAL METHOD

Many of the sheet metal assemblies and parts used in F-16 integral fuel tanks must be fully sealed. A number of these parts also incorporate interchangeable/replaceable panels that are attached with screws and "sealed-dome" type nutplates. These nutplates are sealed with a polysulfide sealant that is applied to the nutplate flange and surrounding part (Figure 4). Manufacture of these parts is a highly labor-intensive process; some parts require eight hours to complete.

Sixty different parts are worked in this particular area. The parts can also be broken into three groups according to processing time as follows:

Classification	Description	Count	Avg. # Fasteners	Avg. Man-Hours
GROUP 1	Large	6 Parts	400	6
GROUP 2	Medium	16 Parts	150	1.5
GROUP 3	Small	38 Parts	10	0.35

Group 1 parts have the highest potential per part cost savings because of higher fastener counts. A typical Group 1 part is shown in Figure 5. These are the largest parts processed in this task center, so an automated system developed for Group 1 parts would have an adequate working envelope to handle Group 2 or 3 parts.

Presently, Group 1 parts are brought into the task center in small batches (lots of two to four parts) with the nutplates attached. The parts are then cleaned, an adhesion promoter is applied to the areas to be sealed, sealant is applied individually to the nutplates, and the sealant is cured. These steps are performed

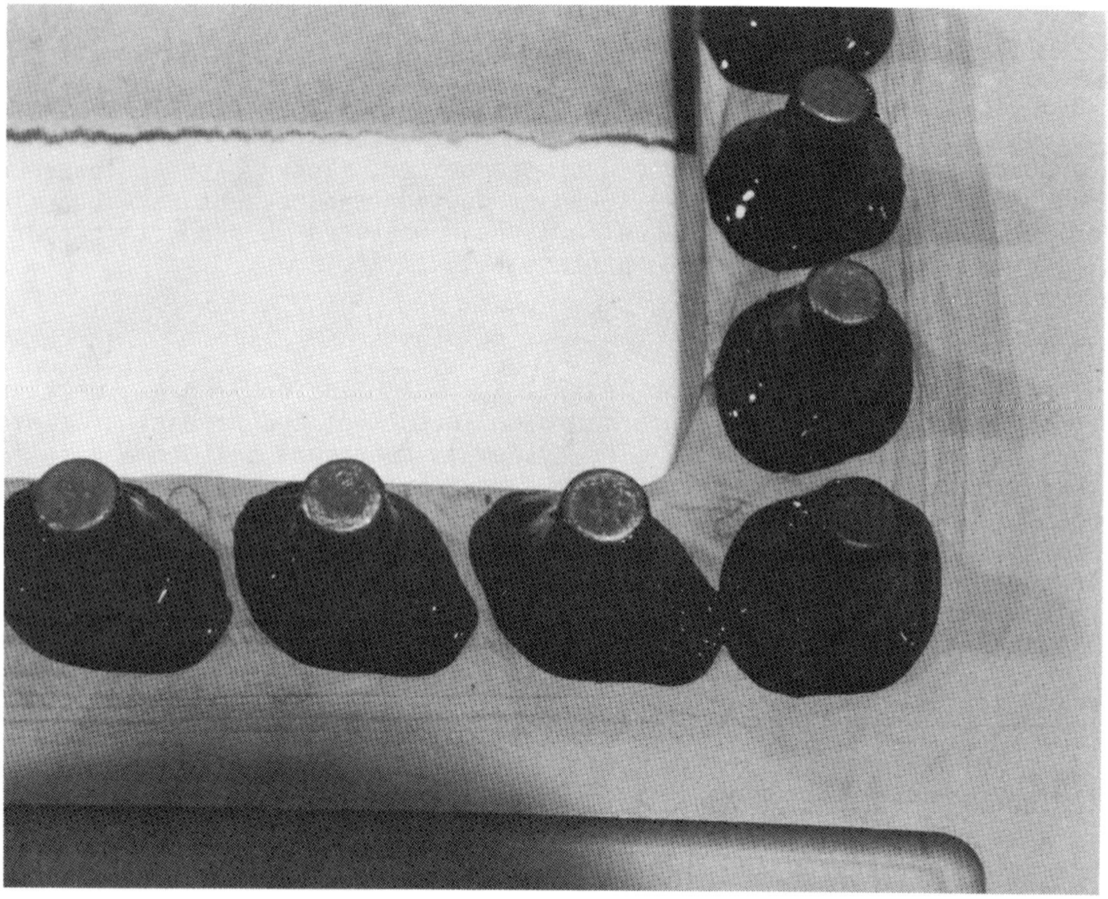

Figure 4 Typical Sealed Nut Plates

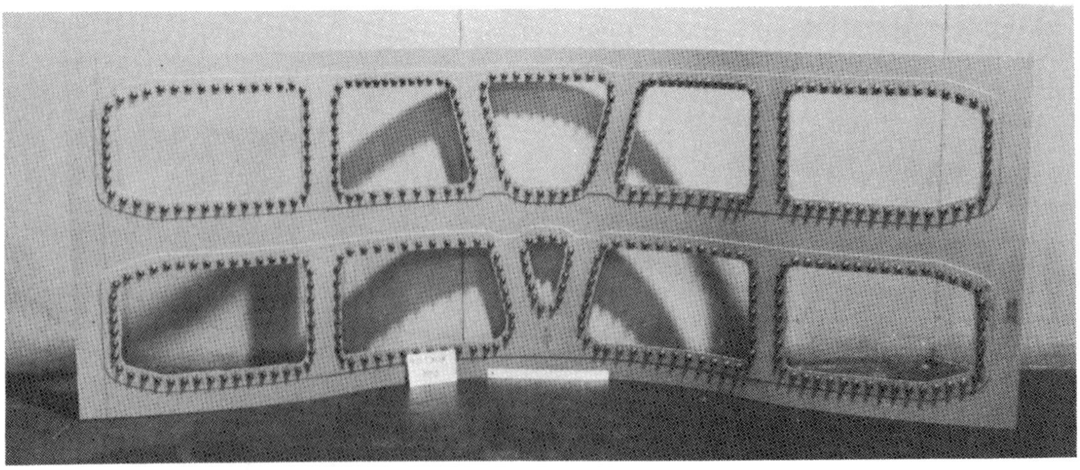

Figure 5 Typical Group 1 Parts

sequentially for each part, and the parts are inspected before leaving the area. Each step is detailed below:

1. Cleaning - The nutplates and surrounding parts are cleansed with solvent (MIL-C-38736) and wiped dry with cheesecloth.

2. Adhesion Promoter (P 6140) - The promoter, a thin liquid, is applied with a brush to the surfaces to be sealed. It must air dry for at least 15 minutes, but not more than 24 hours, before application of sealant.

3. Sealant (FMS-1044-5 B2) - The sealant used on the nutplates is a highly viscous compound and has the consistency of a paste. It is a two-part material that is mixed prior to use. After mixing, it may be stored at -20° F for 21 days (maximum) and thawed (6 or 7 minutes) when ready to use. The working life of this particular compound is two hours but will vary slightly with changes in temperature and humidity. The recommended placement and amount of sealant is specified in Figure 6. Sealant is applied with an air-powered filleting gun (SEMCO Model No. 250-06), shown in Figure 7. The techniques used vary somewhat from worker to worker. Generally, the procedure is to work around the nutplate with the applicator tip, starting at the flange and working up. Considerable "working" of the sealant with the applicator tip is done to produce a good seal with a smooth, even amount of sealant. The workers are very meticulous and the quality of the seal is very good.

4. Cure - The cure time under standard conditions (77° F and 50% relative humidity) is approximately 37 hours, so curing is usually accelerated by placing the part in an oven for a period of time.

Manual sealing, though effective, is time-consuming and tedious. Excessive amounts of sealant are often used because flow from the gun cannot be controlled. This results in excessive cleaning operations.

A photograph of the present nutplate sealing area appears in Figure 8. At the present time, an average of four workers are directly involved in nutplate sealing.

Cleanliness of the area is important because the sealant must stay uncontaminated to cure properly. The sealing area is off to one side of the main assembly line and is underneath a mezzanine to minimize dust and dirt. In fact, after the sealing operation is completed and before sealant cure, the workers use MIL-C-38736 solvent to clean the part of excess sealant (from dripping and spills) and any accumulated dust.

As stated earlier, the sealant used is mixed in mass quantities at another location, put into cartridges, and refrigerated. The life of the sealant while frozen is limited, and the room temperature working life is very short. In addition, for each cartridge that is filled, a sample of sealant is kept for records and testing. This sealant sample is applied to a test fixture, tagged, and observed. If the sample fails (dries out, fails to cure, etc.) within 30 days, all parts sealed with that sealant are called back and reworked. This is a rare occurrence, but the possible hazards dictate this caution.

The implementation of a robotic nutplate sealant cell will not solve all the problems inherent in such an operation. However, once an efficient cycle of operation is programmed, many advantages will surface, such as reduced sealant waste and elimination of boredom in processing. The implementation will have to survive

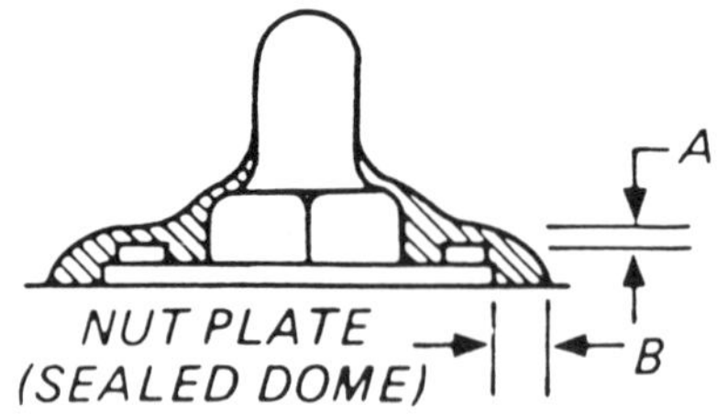

DIM	THICKNESS(IN.)	
	MIN	MAX
A	.05	.09
B	.18	.22

RECOMMENDED DIMENSIONS FOR FILLETS

Figure 6 Sealant Application Specifications

Figure 7 Air Powered Filleting Gun

in a true factory environment; General Dynamics will use factory performance and other crucial factors as priority criteria in the cell design.

Since the nutplate sealing cell will be designed to seal fuel areas of F-16 air surface panels on the factory production floor, the robot cell will be designed and implemented in compliance with existing quality control specifications. During the design stage and market search, return on investment (ROI) and payback period will be justified under corporate standard accounting practices.

ROBOT METHOD

Design of the nutplate sealing application, detailed in this section, will be subject to the following guidelines:

- Specifications for the robot system will be generated after compilation of: 1) an industry survey identifying available hardware, and 2) design criteria identifying specific needs.

- The end effectors will be designed relying on data involving sealant properties, military specifications on sealant application, and engineering data on the nutplate panels.

- Tooling will be designed based on robot work envelope, part size, and plant lay-out.

General Dynamics proposes to automatically apply sealant to each nutplate using an industrial robot to position a sealant delivery system. An operator will be required to load and unload parts and to prepare the parts for sealant application (clean and apply promoter). The primary development effort will center on the sealant applicator (end effector) and on techniques for consistently and reliably applying sealant to ensure high quality.

The automated system will consist of four major subsystems:

1. Robot and related controller, including hardware/software interface.

2. Part presentation system, with associated loading/unloading station.

3. End effector (sealant applicator), with supporting hardware.

4. Sealant delivery system.

ROBOT

Two very similar robotic system concepts are being considered for the sealing task. The main difference in the two is the part presentation methodology. The two methods are a gantry robot (Figure 9) and an inverted jointed arm robot (Figure 10). The two robots are very similar since the gantry robot might also be a jointed arm system. Both robots will be overhead for processing the part. However, since the inverted jointed arm will be fixed in space, the part presentation system, described in the next section, would need to be more elaborate for this type of robot. Both systems would need six axes of motion (three position and three orientation), excluding end effector movement.

The weight requirement of either system will be approximately 30 pounds (13.5 kg) at the tool tip for the sealant delivery system and the end effector housing.

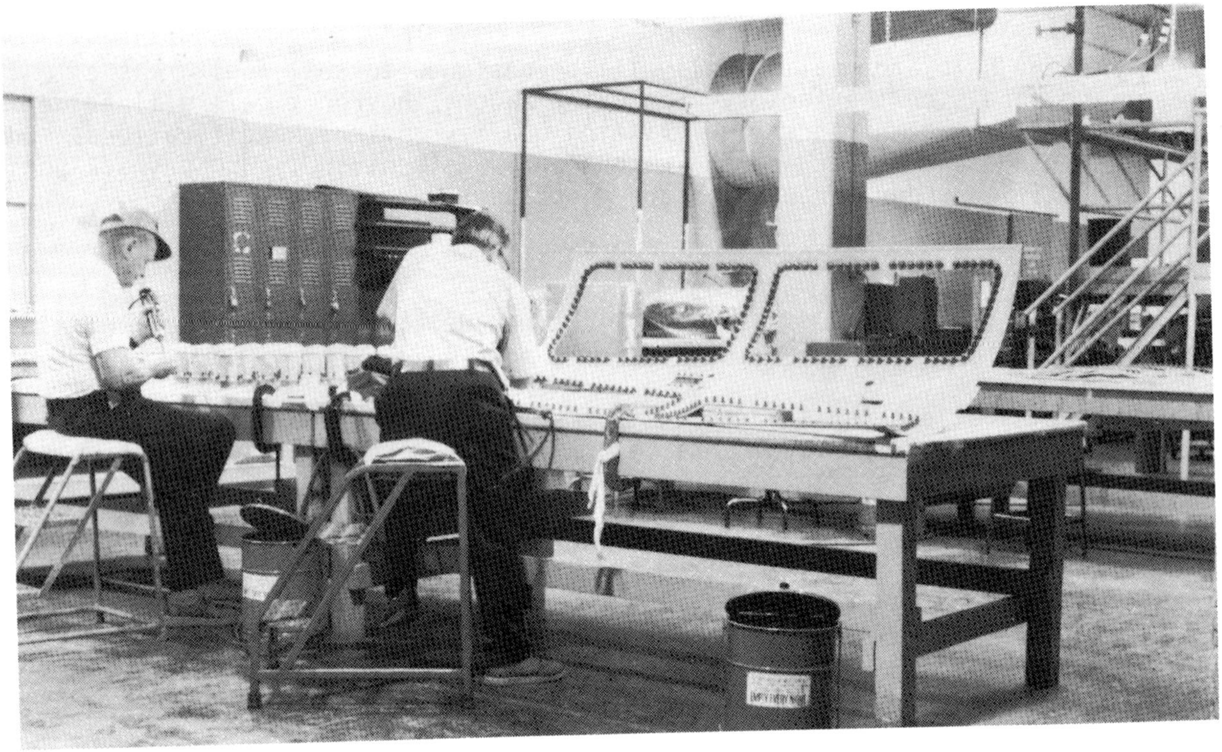

Figure 8 Present Sealing Area

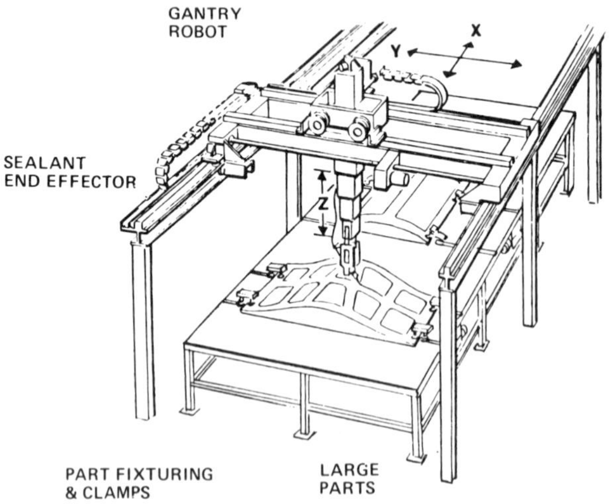

FEATURES:

- o Continuous Robot Sealing
- o Productivity Improvement 3:1 Over Manual Method
- o Large Table for Multiple Work Stations
- o Reduction or Elimination of Excess Sealant Usage

Figure 9 Gantry Robot Concept

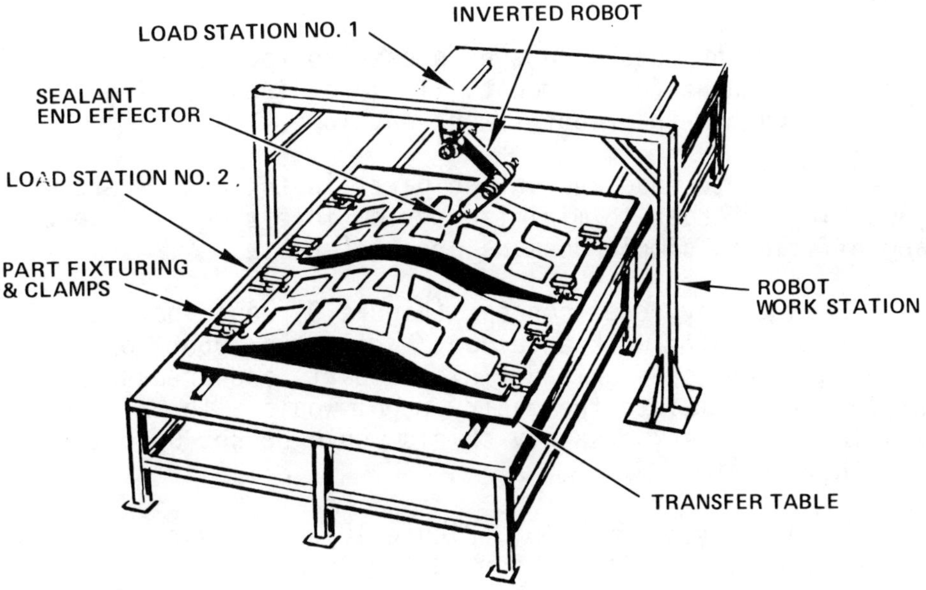

FEATURES:

- o Continuous Robot Sealing
- o Productivity Improvement 3:1 Over Manual Method
- o Additional Table Space for Extra Tasks
- o Overhead Robot Saves Floor Space
- o Reduction or Elimination of Excess Sealant Usage
- o Efficient Load/Unload Concept

Figure 10 Inverted Jointed Arm Robot Concept

The necessary work envelope including either the gantry or the transfer table to allow an operator area for preprocessing is about 240 cubic feet. However, the area the robot must physically access at any given time is much smaller. The weight requirements of the cell are low enough to leave the power source an independent decision-based on vendor availability of hardware.

PART PRESENTATION SYSTEM

The parts to be processed must be presented to the robot in precisely the same location that was taught to the robot. This requires tooling that can accurately locate the parts within the robot's work envelope.

It is desirable to load and unload parts in the area where the robot is not currently working in order to minimize system idle time. To keep the system productive and efficient, simple designs are preferable.

The overhead gantry robot concept (Figure 9) would have a work station at each end of the gantry robot's work envelope. While the robot applies sealant at one station, the operator could unload finished parts and prepare new parts for processing at the other. Since the gantry robot would alternate between both stations for the sealing operation, a safety shield would be set up between the active robot station and the operator.

The inverted jointed arm robot (Figure 10) could work over a positioned transfer table. The concept is similar to the gantry robot concept. The major difference lies in having the part move under the robot rather than the robot move to the part. The movement is relative. The operator prepares a part at one end of the table while the part at the other end of the table is positioned under the robot arm. After the robot processes the part, the table is indexed so the recently prepared part can be processed, the finished part unloaded, and new part prepared for processing. Since this transfer table takes the place of the gantry, this tooling would be developed in-house, and the initial capital investment would be lower. Still, the work envelope would not be as extensive as the gantry system nor as versatile for different applications.

SEALANT APPLICATION END EFFECTOR

The end effector must accurately place sealant around the nutplate, forming a perfect seal. There are two general approaches to this task:

1. The first method is to simulate as closely as possible the manual method, with the robot manipulating a standard-type sealant gun. Several sophisticated controllers can be programmed to move robots through contoured path routines that can be automatically relocated to other positions and orientations. A filleting gun arrangement similar to the one used in the manual operation (Figure 7) could be mounted on the robot arm. This gun would have a precisely controlled sealant flow to avoid excess sealant problems. Once the contour routine is programmed with the elliptical and circular application paths, the robot could be rapidly programmed by positioning the end effectors over each nutplate and calling the circular subroutine. Preliminary tests have shown that this method produces unpredictable results due to varying sealant flow qualities.

2. Another method, such as using a cup-shaped applicator, not only may prove to be faster, but may produce higher quality parts than copying the "manual" application method now used. General Dynamics used a crude sealant end

effector which is basically nothing more than a manual sealant gun mounted to the robot arm for testing various sealant nosepieces (Figure 11). A T-3 robot was used to manipulate the end effector over the nutplates (Figure 12). The nosepiece on the left (Figure 11) is filled with sealant while flush with the part then slowly lifted away and swirled to minimize sealant stringing. The right nosepiece relies upon the robot's wrist roll to smear sealant around the nutplate as sealant is forced out of the nosepiece.

SEALANT DELIVERY SYSTEM

Currently, cartridges containing sealant are loaded into a cannister that pneumatically forces sealant out of the gun when triggered. Sealant flow control is important in order to maintain high quality. Again, two basic delivery system concepts are possible:

1. The present filleting gun concept could be used to control the amount of sealant flowing out of the tip. Flow can be controlled by adjusting the feed rate of the plunger inside the cartridge with either a precise volume of air or a motor drive with feedback. A motor drive is probably more effective because its position can be precisely controlled. Air is compressible and thus cannot be depended upon to deliver precise amounts of sealant. Also, with a motor drive the cartridge cannister does not need to be airtight, allowing the system itself to automatically replace an expended cartridge. The motor could easily retract the pistion plunger to simplify cartridge changing, and a modified automatic dispensing storage freezer could hold the cartridges until needed.

2. Initial studies show that mixing the sealant compound and the accelerator at the tool point is expensive and quality control is difficult. However, the advantage of mixing at the tool point is that no cartridge changing device is needed. Further tool point mixing possibilities will be pursued early in the study to determine whether this or the cartridge dispensing system is the more reliable and efficient.

Robotic science is at a point where this automatic sealant application system will be cost-effective and require a minimal amount of development work before it is put into production. Our major effort will involve integration of these various systems and the development of a sealant end effector. A joint effort between Industrial Engineering, Facilities Planning, Factory Planning, and Manufacturing Technology will results in an integrated plan involving these four design areas.

OFF-LINE PROGRAMMING

The robotic sealant cell will be equipped with off-line programming capabilities and inventory control logging in order to minimize reprogramming time when engineering changes are made. The system will be equipped with a post-processor capable of interpreting Machine Control Language Commands (Figure 13) for future integration into General Dynamics' factory control system.

SUMMARY

Sealing F-16 nutplates robotically will be accomplished at General Dynamics within a year. Preliminary research work has shown that robots are capable of applying sealant twice as fast as a person. Off-line programming will allow this flexible robotics cell to apply sealant on a multitude of parts. This project is of major importance to the aerospace industry and the completely automated factory of the future.

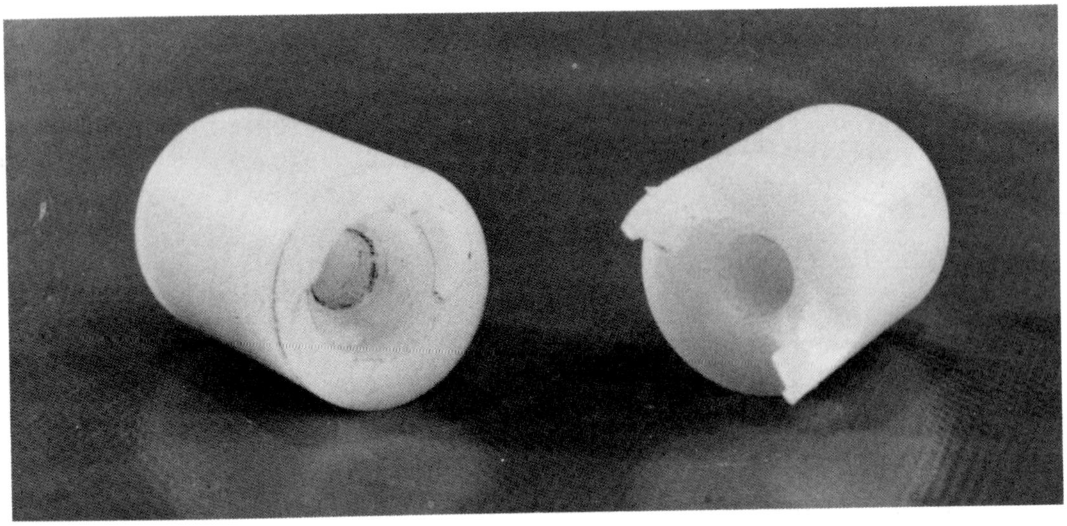

Figure 11 Two Variations of the Cup Shaped Applicator

Figure 12 Prototype Sealant End Effector

```
$$
$$
$$              MCL PROGRAM FOR SEALING NUTPLATES
$$
$$  DEFINES PART GEOMETRY
$$

    P1 = POINT/1,4,0
    P2 = POINT/1,6,0                      Defines a pattern of nut-
    P3 = POINT/1,8,0                      plates to be sealed.
    P4 = POINT/1,10,0

    PAT 1 = PATTTERN/LINEAR, P1, P2, P3, P4

$$
$$  P5 = POINT/5,75,0
    P6 = POINT/7,75,0                     Defines another pattern of
    P7 = POINT/9.75,0                     nutplate to be sealed

    PAT 2 = PATTERN/LINEAR, P1, P2, P3

$$
$$                                        Other Patterns to be
                                          defined.

$$
$$  PVEC = VECTOR/0,1,0
    PROLL = VECTOR/1,0,0

$$
$$  APPROACH POINT TO FIXTURE B
$$
    USEFRM/FIXB
    BAPP = POINT/0,0,0                    Defines a point & orienta-
    BVECT = VECTOR/1,0,0                  tion for approaching fixture
    BROLL = VECTOR/1,0,0                  B.

$$
$$
$$  STRING/STATE
    INTEGER/ANSWER, ATTACH, PASS          Declaring string integer &
    REAL/AMT                              real type variable


$$
$$
    MULTAX/[illegible]                    Declares all future motion
$$                                        commands will be 6-axel
$$  ASK OPERATOR TO ATTACH SEALANT
$$  GUN
$$  LET PASS TO ATTACH SEALANT GUN
$$  WEHN PASS
$$  SEND CONSOL, TO ATTACH SEALANT
$$  GUN
$$
    ALT ANSWER = NO
    SEND/CONSOL, ENTER YES WHEN SEALANT GUN IS ATTACHED WHILE (ANSWER =
    EA. NO)
    RECEIVE/CONSOL, ANSWER                Loops until operator
    END OF/WHILE                          enters yes.
    ENDOF/WHEN
```

```
$$
$$  DECIDES WHICH FICTURE IS READY FOR ROBTO
$$  WHEN (PASS  EA.)
    WHILE (STATE WQ. N READY)
    SEND/CONSOL, ENTER Ø IF NEITHER FIXTURE IS READY
                 ENTER 7 IF FIXTURES A IS READY
                 ENTER 2 IF FIXTURE B IS READY
    RECEIVE/CONSOL, ANSWER
    WHEN (ANSWER = EA  1 OR 2) LET STATE = READY
    END OF/WHEN
    END OF/WHILE
    END OF/WHEN

    LOOPS  UNTIL ONE OF THE FIXUTRES, EITHER A OR B HAS A PART WHICH IS
    READY FOR SEALING.

$$
$$  CHECK SEALANT LEVEL
$$
    SEND/SLEVEL, CHECK
    RECEIVE/SLEVER, ANT
    WHILE (AMT LT. .20)  SEND/CONSOL, FILL SEALANT CONTAINER
                         RECEIVE/SLEVEL, AMT
    END OF/WHILE

    CHECKS TO SEE FI THERE IS ENOUGH SEALANT.  IF NOT, LOOPS UNTIL SEALANT
    REACHES A CERTAIN LEVEL SEND/SLEVEL,CHECK INSTRUCTS THE CONTROL SYSTEM
    TO CHECK THE AMOUNT OF SEALANT.  RECEIVE/SLEVEL, AMT RETURNS THE
    AMOUNT OF SEALNAT CURRENTLY AVAILABLE.

$$
$$  ALIGN SEALANT GUN WITH ACTIVE FIXTURE
$$  AND BEGINS SEALING

    DEVICE/ROBOT
    WORK PT/G POINT, GUECT, GROLL
    WHEN (ANSWER .EQ. 1)
         GOTO/AAPP, AVECT, ROLL, AROLL
         CYCLE/SEAL
         TLAXIS/PVEC, ROLL, PROLL
         GOTO/PAT 1
         GOTO/PAT 2

       CYCLE/NO MORE
       LET ANSWER = 2

    ELSEN

    WHEN (ANSWER = EQ. 2)
         GOTO/BAPP, BVECT, ROLL, BROLL
         CYCLE/SEAL
         TLAXIS/PVEC, ROLL, PROLL
         GOTO/PAT 1
         GOTO/PAT 2

         CYCLE/NO MORE
         LET ANSWER = 1

    END OF/WHEN
    PASS = PASS = 1
```

Notes

Device/robot indicates that all future motion will be done by the robot.

WORKPT defines a point & orientation for the sealant gun which is to be alinged with a point & orientation for a nutplate specified in the GOTO command.

If answer 1 = 1 then fixture A is active; otherwise fixture B is active.

Cycle/Seal readies the sealant flow

GOTO/PAT 1 and TLAXIS specify the orientation and position of sealant points to be aligned with the sealant guns.

Cycle/no more unreadies the sealant flow let anser = 2 indicates that fixture B is the next active fixture.

The same commands are used when the value for answer is 2 except the value for answer is changed to 1 at the end in order to alternate active figures.

Figure 13   Sample MCL Program

Teaching Robots By Example

M. Selfridge
The University of Connecticut
Storrs, Connecticut

A. Levas
The University of Connecticut

Abstract

Planning ability provides robots with the flexibility to cope with complex and uncertain manufacturing environments, but requires that robots be able to learn new plans. Robots should acquire new plans by observing examples of those plans. A robot manipulator system learns the plan for assembling an electric switch by "watching" a teacher assemble the switch. The success of this system suggests that teaching robots new plans by example may indeed prove a useful technique for robots in the future.

I. Introduction

Robots are increasingly being used in situations requiring them to exhibit flexible behavior. In such situations point-to-point motion sequences are inadequate [1]. Instead, robots will probably use planning mechanisms, coupled with sensors of various kinds, to carry out their tasks [1]. Such mechanisms allow robots to perform in situations where object position cannot be exactly specified in advance and in which unanticipated events can occur. However, training robots which use plans requires teaching them new plans. This paper describes one method, "showing" the robot an example of the desired behavior and having it infer the underlying plan.

To see why teaching plans to a robot is important, consider a possible robot application, that of a shop specializing in short production runs producing things from fairly standard parts. Such a shop cannot use special purpose equipment, since runs are short, therefore using robots may be the only way to automate. They would use planning mechanisms, for the reasons outlined above. However, short production runs require frequent retraining, and the robot must acquire new plans frequently. Teaching new plans by example appears to be an effective solution to this problem.

The approach described here is motivated by informal observations of humans in similar situations. It appears that people often learn new plans by observation. That is, people watch someone else doing something, and then they use their general planning knowledge and knowledge of the world to infer the plan underlying what they saw. They can then carry out the same general action in different situations. This suggests that robots could learn new plans the same way. It also suggests that general planning knowledge and knowledge of the world are prerequisites for learning by example. That is, a robot in a short production run environment will have to know general rules about manipulating objects, and will have to know the positions and orientations of those objects if it is to learn by example. It will also, like the person, require learning mechanisms which use this knowledge.

The rest of this paper will describe a system which learns a new plan by example. This discription will encompass the system's hardware, its world knowledge and planning knowledge, and its learning mechanisms. A sample of the system's performance will then be presented, and finally conclusions and suggestions for future research.

II. The System

The system has several hardware and software components. The robot manipulator is a five degree of freedom Microbot Minimover-5, having a parallelogram-type hand with a touch sensor. The arm is controlled by an Apple II microcomputer which is in turn connected to a DEC VAX-11/780 via a serial line. All Apple software is written in BASIC, while all VAX software is written in LISP. The system can operate in one of two modes. In top-down mode, planning programs running on the VAX send the Apple manipulator movement commands which the Apple executes. In bottom-up mode, the user uses the Apple keyboard to "manually" move the manipulator, and to designate certain manipulator positions. The Apple enables the user to control the manipulator through the keyboard, and when points are designated sends them to the plan learning program on the VAX. It is this capacity to report the coordinates of the designated points which enables the system to "see" the user's example.

Overall control of the system rests with an executive program on the VAX. This program accepts task specifications from a user and calls a standard planning program to carry them out (see [5] for an introduction to robot planning programs). If the planner has the knowledge needed to achieve the task, it does so, calling the Apple to move the manipulator as appropriate. When the task is achieved, the planner returns control to the executive. If, however, the planner has no plan to achieve the task, it returns control immediately to the executive with an error flag set. The executive notes the error flag and calls the learning program to acquire a plan for that task. This initializes the Apple to accept manual control of the manipulator from its keyboard and to return designated points. As the user designates points, the learning program attempts to infer the plan underlying the example. When learning has concluded, control is again returned to the executive which is again ready to accept task specifications from the user.

While the system does not now have visual input, it has been given the kind of knowledge available from a high-performance visual system. It knows that the workspace contains several objects, and it knows their locations, shapes, sizes, and orientations. For simplicity, the system treats the post hole and contact slot as objects made of empty space. The chart below shows the knowledge the system has (simplified for exposition).

	size			location				
object	x	y	z	x	y	z	orientation	type
post	2	.5	.5	8	8	0	Vertical	solid
contact	2	.3	.6	4	4	0	Vertical	solid
base	1	3	1	7	6	0	Horizontal	solid
hole	.7	.7	.7	7	7	.3	Perpendicular	space
slot	.8	.5	.7	7	5	.3	Perpendicular	space

The system also knows about four primitive manipulator motions which can be initiated by the planning program. All plans must ultimately translate into sequences of these primitives. The following table gives these:

GRASP -- close the gripper until tight using the touch sensor.

OPEN-FINGERS -- open the gripper to maximum width.

ROTATE-WRIST -- pitch or roll the wrist to a designated orientation.

MOVE -- move the fingertips to a designated point.

For the example to be presented in this paper, the system has nine planning rules. Each rule has two parts, a goal part and a plan part. Their interpretation within the planning system is "to achieve <goal-part>, achieve the components of <plan-part> in order." These rules are general in the sense that they have nothing to do with making particular electric switches, but are rules about picking things up and putting things into holes in other things. They are not intended to be fully general planning rules, but merely to be adequate in this example. Simplified versions of these rules are presented below:

GOAL: AT-GRASP-PT X
PLAN: "MOVE hand to a point within X's Y and Z boundaries and centered within its X boundary."

GOAL: AT-OVER-PT X
PLAN: "MOVE hand to a point within X's X and Y boundaries and greater than its upper Z boundary."

GOAL: AT-HORIZONTAL-HAND-PT
PLAN: "ROTATE-WRIST until the hand is horizontal."

GOAL: AT-INTERSECTED-PT X Y
PLAN: "MOVE hand to a point such that X is within Y."

GOAL: FINGERS-AROUND X
PLAN: 1) AT-OVER-PT X
2) OPEN-FINGERS
3) AT-GRASP-PT X

GOAL: HOLDING X
PLAN: 1) FINGERS-AROUND X
2) GRASP

GOAL: OVER X Y
PLAN: 1) HOLDING X
2) AT-OVER-PT X

GOAL: VERTICAL X
PLAN: 1) HOLDING X
2) AT-HORIZONTAL-HAND-PT

GOAL: INSERT X Y
PLAN: 1) OVER X Y
2) VERTICAL X
3) AT-INTERSECTED-PT X Y

The heart of the system is the plan inference module. Based on work in learning, inductive inference, and goal inference [2,3,4], this module examines the manipulator positions designated by the user and tries to infer the plan behind those positions. It does so using the following bottom-up interpretation of planning rules: "if you see the user achieving the subgoals in <plan-part>, then infer that he is achieving <goal-part>." The plan inference process is organized around a short-term memory (STM) onto which the designated manipulator positions are placed as soon as received from the Apple. Every time something changes in the STM, the plan inference module searches the planning rules to see if the current STM configuration matches the plan-part of any rule. If it does, the matching configuration is removed from the STM and replaced with the goal-part of that rule. This constitutes a change in the STM, of course, and so the rules are examined again. The process continues until the user stops designating points and no more rules' plan-parts match the STM. At this point the STM contains a series of sub-goals which constitute the plan inferred to be behind the example.

III. Learning to Make a Switch

This section shows the system learning the plan for assembling the electric switch. In this example "DP" stands for "designated point", the point designated by the user for the attention of the system.

```
% lisp loadplanner

Franz Lisp, Opus 36

-->(runrobot)

ENTER GOAL: MAKE SWITCH1
SEARCHING PLANS...NO PLANS FOUND TO MAKE SWITCH1

ENTER TRAINING MODE (Y/N)? Y
ENABLING MANUAL CONTROL

DP: (8 8 0.2 -90 0 0 1)     GOAL INFERRED: HOLDING POST

DP: (7 7  5  -90 0 0 1)     GOAL INFERRED: AT-OVER-PT HOLE
                            GOAL INFERRED: OVER POST HOLE

DP: (7 7  5    0 0 0 1)     GOAL INFERRED: AT-HORIZONTAL-HAND-PT
                            GOAL INFERRED: VERTICAL POST

DP: (7 7 1.2   0 0 0 1)     GOAL INFERRED: AT-INTERSECTED-PT POST HOLE
                            GOAL INFERRED: INSERT POST HOLE

DP: (4 4 0.2 -90 0 0 1)     GOAL INFERRED: HOLDING CONTACT

DP: (7 5  5  -90 0 0 1)     GOAL INFERRED: AT-OVER-PT SLOT
                            GOAL INFERRED: OVER CONTACT SLOT

DP: (7 5  5    0 0 0 1)     GOAL INFERRED: AT-HORIZONTAL-HAND-PT
                            GOAL INFERRED: VERTICAL CONTACT

DP: (7 5 1.2   0 0 0 1)     GOAL INFERRED: AT-INTERSECTED-PT CONTACT SLOT
                            GOAL INFERRED: INSERT CONTACT SLOT
```

```
SEQUENCE CONCLUDED

BUILDING NEW PLANNING RULE:
GOAL: MAKE SWITCH1
PLAN: 1) INSERT POST HOLE
      2) INSERT CONTACT SLOT

-->(runrobot)

ENTER GOAL: switch1
SEARCHING PLAN LIST...PLAN RETRIEVED

[...planning program proceeds to execute the new learned
 plan which accesses the previously known planning rules
 and assembles the switch...]

-->(exit)
%
```

IV. Conclusions

Future development of this system will proceed in several directions. First, a top-down vision system is being added to the system to allow it to actually see the objects in the workspace. Second, the system is being given the ability to verify its inferences with the user using voice input and output. Third, the capabilities of the system are being expanded to include implementation on a full-size industrial robot system, and capabilities such as product testing as well as manufacture.

The successful performance of the system described in this paper suggests that using planning systems and plan inference can be an important part of teaching robots new actions in real-world situations. In particular, important part of teaching robots new actions in real-world situations, and that teaching robots new plans by example is a practical technique.

BIBLIOGRAPHY

[1] Birk, J. and Kelly, R. (1980) Proceedings of the Workshop on the Research Needed to Advance the state of Knowledge in Robotics, printed by the NSF and the Univ. or R.I.

[2] Angluin, D. and Smith, C. (1982) A Survey of Inductive Inference: Theory and Methods. Technical Report #250, Yale University, New Haven, CT

[3] Soloway, E. and Riseman, E.M. (1977) Levels of Pattern Description in Learning. Proc. 5th Int. Conf on Art. Int., August, 1977, Cambridge, MA

[4] Wilensky, R. (1978) Understanding Goal-Based Stories. Ph.D. Thesis, Technical Report #140, Yale University, New Haven, Ct.

[5] Winston, P. (1977). Artificial Intelligence. Addison-Wesley Publishing Company, Reading, Mass.

Non-Adaptive and Adaptive Control of Industrial Robots

M. Vukobratovic
Mihailo Pupin Institute
Beograd, Yugoslavia

D. Stokic
Mihailo Pupin Institute

N. Kircanski
Mihailo Pupin Institute

Abstract

The control synthesis for the robotic systems in which parameters are partially unknown is considered. Proposed are synthesis of robust, non-adaptive, decentralized control which has to stabilize the robot for all allowable variations of the parameters. If the robust non-adaptive control cannot withstand all expected variations of parameters, a synthesis of indirect adaptive control is proposed. The non-adaptive and adaptive control syntheses are illustrated by simulation of an industrial robot with unknown payload mass.

INTRODUCTION

The most general task in robotic practice can be regarded as a task of practical stabilization of the robot around the given nominal trajectory. This control task has been already considered in our previous papers [1-3]. However, in these previous papers we have assumed that the parameters of the mechanism are precisely defined in advance. Thus, we have synthesized nominal programmed control using well-defined centralized model of the system. Then, the model of deviation from the nominal trajectory and control is observed and decentralized control is synthesized. However, in this we have assumed that the nominal control perfectly compensate the coupling among the subsystems on the nominal level. This means that all parameters of the system are well known in advance. In many cases in robotic practice the assumption that the parameters of the system are precisely known, is valid. However, in general case this need not to be true, since some parameters of the robot (e.g. payload, clearences, frictions) might be uncertain and variable. On the other hand, the synthesis of nominal centralized programmed control demands off-line computation of the complete model of the robotic system, what might be unconvenient in some cases. Thus, in such cases we prefer to avoid the synthesis of nominal centralized control and to synthesize control only on the subsystem level [4].

In this paper we shall consider stabilization of the robotic system with variable parameters of the mechanism. This problem has been studied by several authors [5, 6]. Mostly, they tried to approach the problem by adaptive control synthesis, but none of them analyzed the practical stability of the complete nonlinear model of the system. So nobody analyze the real necessity to introduce adaptive control, the implementation of which might be complex and expensive.

Our aim is to synthesize simple non-adaptive and robust decentralized control which can whithstand all expected variations of parameters. However, if it is impossible to find such simple robust control it is necessary to introduce adaptive control. In the case of parameter variation centralized nominal control is not usefull. We consider system as a set of subsystems, each associated to one degree of freedom of the mechanism. For each subsystem we synthesize local controller which stabilizes decoupled (free) subsystem [7]. Then we directly examine practical stability of the global (coupled) system for the given set of allowable parametric values. Thus, we can find the local controllers which can stabilize the global system for given set of allowable parameter values. If, the set of allowable parameters values is given, it might appear that we cannot find unique local regulators which accomodate the system for all values of parameters. In that case we have to introduce global control or adaptive control must be introduced.

MATHEMATICAL MODEL AND CONTROL TASK

Let us consider the robotic system S consisting of the mechanical part of the system S^M and actuators S^i. The system has n degrees of freedom each powered by one actuator S^i. The model of the actuator is given by linear time-invariant system:

$$S^i: \quad \dot{x}^i = A^i x^i + b^i u^i + f^i P_i, \qquad \forall i \varepsilon I = \{i: i=1,2,\dots,n\} \quad (1)$$

where $x^i \varepsilon R^{n_i}$ is the vector of state coordinates of S^i, n_i is the order of the model S^i, $A^i \varepsilon R^{n_i x n_i}$ is system matrix, $b^i \varepsilon R^{n_i}$ is input distribution vector, $f^i \varepsilon R^{n_i}$ is the load distribution vector, $u^i \varepsilon R^1$ is the input to actuator S^i, P_i is the driving torque (load) acting upon the i-th d.o.f. (actuator). The model of the mechanical part of the system is given in a general form [4]

$$S^M: \quad P = H(q, \theta)\ddot{q} + h(q, \dot{q}, \theta) \tag{2}$$

where $q(t)\varepsilon R^n$ is the vector of internal mechanism coordinates $\theta\varepsilon R^{n_p}$ is the vector of mechanism parameters, $P(t)\varepsilon R^n$ the vector of driving torques (forces); $H(q, \theta)$ is a full-rank positive definite inertial matrix-valued function, $h(q, \dot{q}, \theta)$ is a vector function due to gravity, centrifugal and Coriolis´ effects. It is also assumed that in each x^i two state coordinates coincide with q^i and $\dot{q}^i$. Thus, the state vector of the overall system S is given by $x=(x^1,\ldots,x^n)^T$, the order of the system S is $N=\sum_{i=1}^{n} n_i$. The following control task is imposed. The nominal trajectory of the state coordinates of the system S is given $x^o(t)=(x^{o1T}(t), x^{o2T}(t),\ldots$ $\ldots,x^{onT}(t))^T$, $\forall t\varepsilon T$, $T=\{t:t\varepsilon(0, \tau)\}$, τ - given time period. The control should be synthesized ensuring practical stability of the system around the nominal trajectory, so that $\forall x(0)\varepsilon X^I$ and $\forall\theta\varepsilon\Theta$ implies $x(t, x(0))\varepsilon X^t(t)$, $\forall t\varepsilon T$, where $X^I CR^N$ is given finite region of allowable initial conditions, and $X^t(t)CR^N$, $\forall t\varepsilon T$, is given finite region of allowable system states at arbitrary moment $t\varepsilon T$. Here, it is assumed that $x^o(0)\varepsilon X^I$ and $x^o(t)\varepsilon X^t(t)$, $\forall t\varepsilon T$. Here, θ denotes the set of allowable values of parameters, $\theta\varepsilon R^{n_p}$. In the scope of decentralized control synthesis it is necessary to perform the distribution of the model of complete system into the local subsystems and the coupling. The distribution may be done in various ways. Since the coupling P_i is a function of the state coordinates x^i of subsystem S^i, it is possible to extract some terms from the model of the mechanical system part (2) and associate them to free subsystems. Diagonal elements of matrix $H(q, \theta)$ can be presented in the form

$$H_{ii}(q, \theta) = \bar{H}_{ii}(\theta) + \varepsilon_H(q, \theta) \tag{3}$$

where $\bar{H}_{ii}(\theta)>0$, $\varepsilon_H=H_{ii}-\bar{H}_{ii}(\theta)$.

The i-th element of vector $h(q, \dot{q}, \theta)$ can be written as

$$h_i(q, \dot{q}, \theta) = \bar{h}_i(\theta)q_i+\varepsilon_h(q, \dot{q}, \varepsilon) \tag{4}$$

where $\varepsilon_h=h_i-\bar{h}_i q_i$. Subsystems models S^i can be written in the form

$$S_i: \quad \dot{x}^i=\bar{A}^i(\theta)x^i+\bar{b}^i(\theta)N(u^i)+\bar{f}^i(\theta)\Delta P_i \tag{5}$$

where $\bar{A}^i=\bar{C}^i(\theta)(A^i+f^i\bar{h}(\theta)T_I^i)$, $\bar{b}^i=\bar{C}^i(\theta)b^i$, $\bar{f}^i=\bar{C}^i(\theta)f^i$, $\bar{C}^i(\theta)=(I_{ni}-$ $-f^i\bar{H}_{ii}(\theta)T^i)^{-1}$, $\Delta P_i=P_i-\bar{H}_{ii}T_i\dot{x}^i-\bar{h}_i T_1 x^i$; $T_I^i\varepsilon R^{1xn_i}$, $T^i\dot{x}^i=\ddot{q}^i$, $T_1^i x^i=q^i$. The introduction of $\bar{H}_{ii}(\theta)$ in the subsystem model S_i is very common: each actuator is considered with inertiality (moment of inertia or equivalent mass) which includes not only actuator´s own inertiality, but also the inertiality of the mechanical part of system, which is moved by the considered actuator. The introduction of $\bar{h}_i$ in the subsystem model S^i means that the actuator model is considered with gravitational load, which is a function of q_i.

NON-ADAPTIVE CONTROL SYNTHESIS

In order to synthesize decentralized control for practical stabilization of the robotic system S, we shall observe approxiamtive decentralized model of the system, i.e. we shall observe system S as a set of decoupled subsystems:

$$S^i: \quad \dot{x}^i = \bar{A}^i(\theta)\dot{x}^i + \bar{b}^i(\theta)u^i \qquad (6)$$

which are coupled through $f^i(\theta)P_i$. The control task will be solved as follows. First the local controllers are synthesized for free (decoupled) subsystems and then practical stability of the global system (1), (2) is checked. However, matrices of subsystems (6) depend on the unknown parameters $\theta\varepsilon\Theta$. First we shall synthesize simple linear control for one chosen value of the parameter vector θ and then we shall examine whether such control is robust enough to whithstand all allowable parameter variations, i.e. whether such control can stabilize global system S for all assumed values $\theta\varepsilon\Theta$. Let us consider free subsystems (6) for the chosen parameters values $\theta=\Theta_d$:

$$S^i: \quad \dot{x}^i = \tilde{A}^i x^i + \tilde{b}^i u^i \qquad (7)$$

where $\tilde{A}^i=\bar{A}^i(\theta_d)$ and $\tilde{b}^i=\bar{b}^i(\theta_d)$. Since subsystems (7) are linear, time--constant systems the synthesis of local controllers which have to stabilize (7) can be performed in following way. Let us assume that there exists programmed control $\bar{u}^{oi}(t)$ satisfying:

$$\dot{x}^{oi}(t) = \tilde{A}^i x^{oi}(t) + \tilde{b}^i \bar{u}^{oi}(t), \qquad \forall t\varepsilon T, \quad \forall i\varepsilon I \qquad (8)$$

The model of deviation of free subsystem (7) from nominal trajectory $x^{oi}(t)$ and nominal programmed control $\bar{u}^{oi}(t)$ is given by:

$$S^i: \quad \Delta\dot{x}^i = \tilde{A}^i \Delta x^i + \tilde{b}^i \Delta u^i, \qquad \forall t\varepsilon T, \quad \forall i\varepsilon I \qquad (9)$$

where $\Delta x^i(t)=x^i(t)-x^{oi}(t)$, $\Delta u^i(t)=u^i(t)-u^{oi}(t)$. The stabilization of the subsystem (9) around nominal trajectory $x^{oi}(t)$ and nominal control $\bar{u}^{oi}(t)$ can be achieved by minimization of the criterion:

$$\bar{J}(\Delta x^i(0)) = \int_0^\infty (\Delta u^i r_i \Delta u^i + \Delta x^{iT} Q_i \Delta x^i)\exp(2\beta t)dt \qquad (10)$$

where $\beta>0$ is prescribed degree of exponential stability of free subsystem (9). Here, we observe criterion over the infinite time interval although the practical stability on the finite time interval is desired. The control minimizing the criterion (10) under the constraint (9) is given by

$$\Delta u_i^L(t) = -r_i^{-1}\tilde{b}^{iT} K_i \Delta x^i(t), \qquad \forall t\varepsilon T, \quad \forall i\varepsilon I \qquad (11)$$

where $K_i\varepsilon R^{n_i x n_i}$ is the positive definite matrix, which is the solution of the corresponding algebraic Ricatti equation. Now, the control

$$u^i(t) = \bar{u}^{oi}(t) + \Delta u_i^L(t) = \bar{u}^{oi}(t) - r_i^{-1}\tilde{b}^{iT} K_i \Delta x^i(t), \quad \forall i\varepsilon I \qquad (12)$$

should stabilize free subsystem (7) around the nominal trajectory $x^{oi}(t)$. Obviously, such control is suboptimal with respect to criterion (10).

As the next step in control synthesis we have to examine whether the global system S (1), (2) (when coupling and parameter variations are taken into account) is practically stable. This analysis can be carried-on using aggregation-decomposition method as has been presented in [4]. This analysis can be used to establish iterative procedure for the choice of weighting matrices Q_i and r_i in (10) and the choice of parameter values Θ_d in (7), so that the synthesized local controllers (12) can ensure practical stability around nominal trajectory $x^o(t)$ and for all allowable parameter variations $\theta\varepsilon\Theta$.

If the influence of coupling among subsystems and influence of parameter variations are too strong it might happen that it is impossible to find out unique local controllers (12) which are robust enough to whithstand nonlinear coupling and all parameter variations. In this case it is necessary to introduce either global control which should compensate destabilizing influence of coupling and parameter variations, or adaptive control which should change the local gains in accordance to parameter variations. As we have already recognized in our previous papers [1 - 3], coupling among d.o.f. of the robotic systems is represented by load P_i which can be directly measured. Thus, if we introduce force transducers we can establish force feedback as a global control. We have proposed the global control in the form [4, 7]:

$$\Delta u_i^G(t) = -K_i^G \, \text{sign}(\Delta u_i^L) P_i \qquad (13)$$

where $K_i^G \varepsilon R^1$ is the global gain. This global control law is chosen in order to directly compensate the influence of coupling and parameter variations. Such control is thus robust to parameter variations. However, the implementation of force feedback imposes introduction of force transducers increasing the price of the control system. Direct implementation of force feedback involves also technical problems regarding the presence of noise on the transducers. The proposed control scheme is presented in Fig. 1.

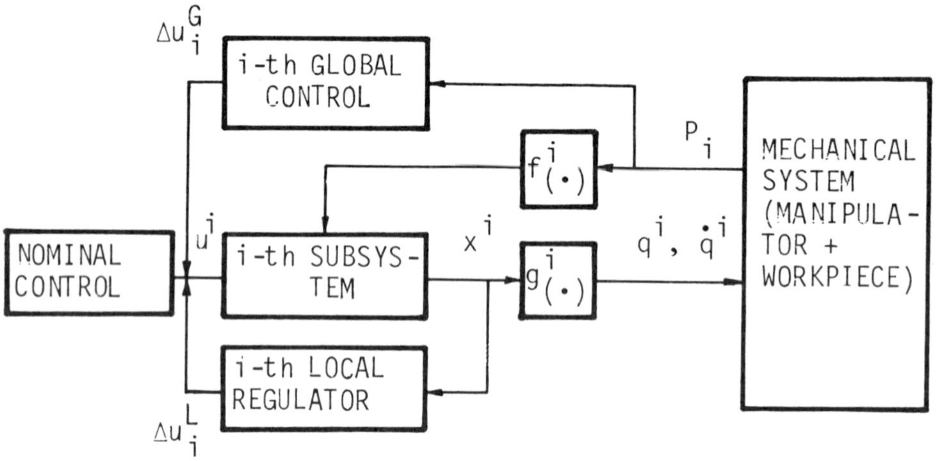

Fig. 1. Non-adaptive control scheme

ADAPTIVE CONTROL SYNTHESIS

If the proposed non-adaptive control cannot accomodate all expectable parameter variations it is necessary to introduce adaptive control which should change local control gains in accordance with parameter variations. However, since parameter variations are not known in advance it is necessary to estimate parameters θ. Unfortunately, many estimation procedures (see, for example, [8, 9]) cannot be directly applied to robotic systems because they need the complete system model as a part of control algorithm. This would seriously burden the computational efforts and make the real-time implementation difficult, if not impossible. In order to overcome these difficulties we had to use some specific properties of manipulator systems. Instead of calculation of the complete system model we can use a "part" of systems dynamics which is related to the estimated parameters (payload mass and inertias). Denoting by n the index of the link which corresponds to end-effector and by n_p the number of payload dynamic parameters, one obtains [10]

$$\Delta P\Big|_{q=q_o(t)} = \sum_{\ell=1}^{n_p} (H_{n\ell}(q_o)\ddot{q}_o + h_{n\ell}(q_o, \dot{q}_o))\Delta\theta_\ell \tag{14}$$

where $q_o(t)$ represents the nominal trajectory and $\Delta\theta_\ell$ the dynamic parameters of payload. It should be noted the equ. (14) permits the calculation of the deviation of the generalized force vector which enables the system to follow the nominal trajectory in presence of finite parameter variations. As shown in paper [10] the number of floating-point multiplications/additions for the calculation of $H_{n\ell}(q)$ and $h_{n\ell}(q, \dot{q})$ is nearly n times less than that for dynamic model matrices $H(q, \theta)$ and $h(q, \dot{q}, \theta)$. This is of fundamental importance for real-time implementation.

The basic ideas for the presented estimation technique can be found in [11]. These results are also used in the paper [5] on the centralized manipulator model which is too complex for real-time application. In order to pose the estimation algorithm, consider the inequality

$$[\Delta P(t_k) - S_n(\xi(t_k))\Delta\theta] < \varepsilon(t_k) \tag{15}$$

where $\xi = [q, \dot{q}]$, $S_n(\xi) = H_n(\xi)T^n\xi + h_n(\xi)$ is vector defined in (14), $\Delta\theta$ is the variation of parameter θ from its nominal value, and $\varepsilon(t) > 0$ is a given scalar function. Expression (15) represents an infinite set of inequalities as it can be written for $\forall t_k \varepsilon T$. It is therefore necessary to discretize the time interval T: $T = \{t_k\}$, with $k=1,\ldots,[\tau/\Delta t]$, Δt - time--interval of discretization. The estimation $\Delta\theta_e^i$ of the real value of $\Delta\theta^i$ can be calculated by the recursive algorithm

$$\Delta\theta_e(t_{k+1}) = \begin{cases} \Delta\theta_e(t_k), \ (15) \text{ - fulfilled} \\ \Delta\theta_e(t_k) + P_{r_\theta}[\|S_n(\xi_k)\|^{-2}\cdot S_n^T(\xi_k)(\Delta P_k - S_n(\xi_k)\Delta\theta_e(t_k))] \end{cases} \tag{16}$$

where $P_{r_\theta}[\cdot]$ is the operation of projection to the parameter space Θ. In the simplest case the operator P_{r_θ} can be substituted by a scalar multiplicative constant a_i. Obviously, the rate of change of the estimation vector ("quasistationarity condition") can be decreased by decreasing the value of a_i. On the other hand, that increases the adaptation time. A compromise value should be chosen by the designer. As described in the previous section, the synthesis of local regulators (8 - 12) is exactly valid when θ is constant. When θ is variable over time $t\varepsilon T$, the procedure can be regarded as correct only when the rate of change $\theta(t)$ is much less than that corresponding to the system dynamics. This is the quasistationarity condition which can be fulfilled by decreasing the constant a_i of the estimation algorithm (15). In that case the synthesis of local regulators (8 - 12) should be repeated for all $\theta\varepsilon\Theta$, where Θ is the domain of all possible parameter vectors. Fortunately, in the case of robotic systems, only one parameter could be consider as variable (payload mass). Now, it is easy to form the function $D_i(\theta)$ (D_i - matrix consisting of local feedback coeffitients of the i-th subsystem) by repeating the optimization procedure for several $\theta\varepsilon(\theta_{min}, \theta_{max})$. The general block scheme of the algorithm is given in Fig. 2. The efficiency of the dynamic estimator and the complete adaptive algorithm may be best recognized through numerical example presented below.

EXAMPLE

The synthesis of the proposed non-adaptive and adaptive control will be illustrated by a hydraulic 6 d.o.f. manipulation system, presented in

Fig. 3. Three d.o.f. form the minimal configuration, while the remaining three represent the gripper. Manipulator parameters (member lengths, masses, moments of inertia) are given in Table 1.

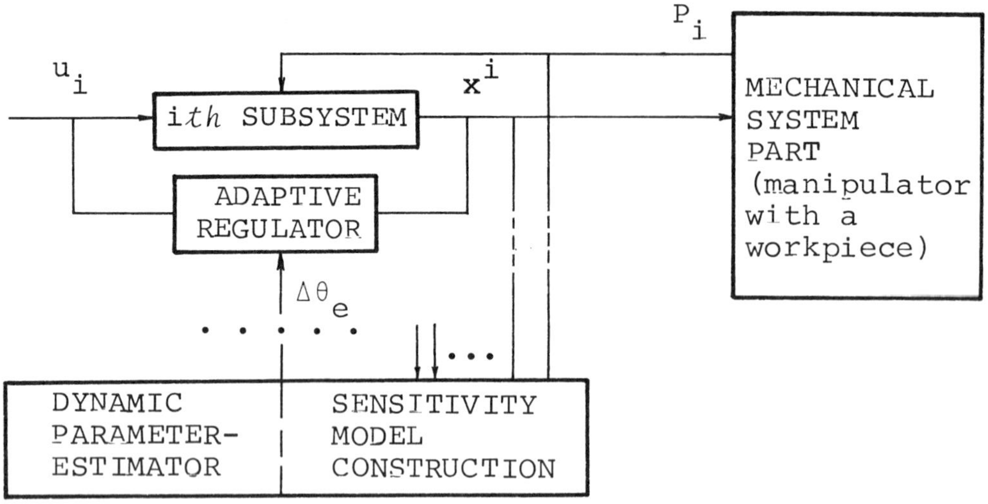

Fig. 2. Adaptive Control Scheme

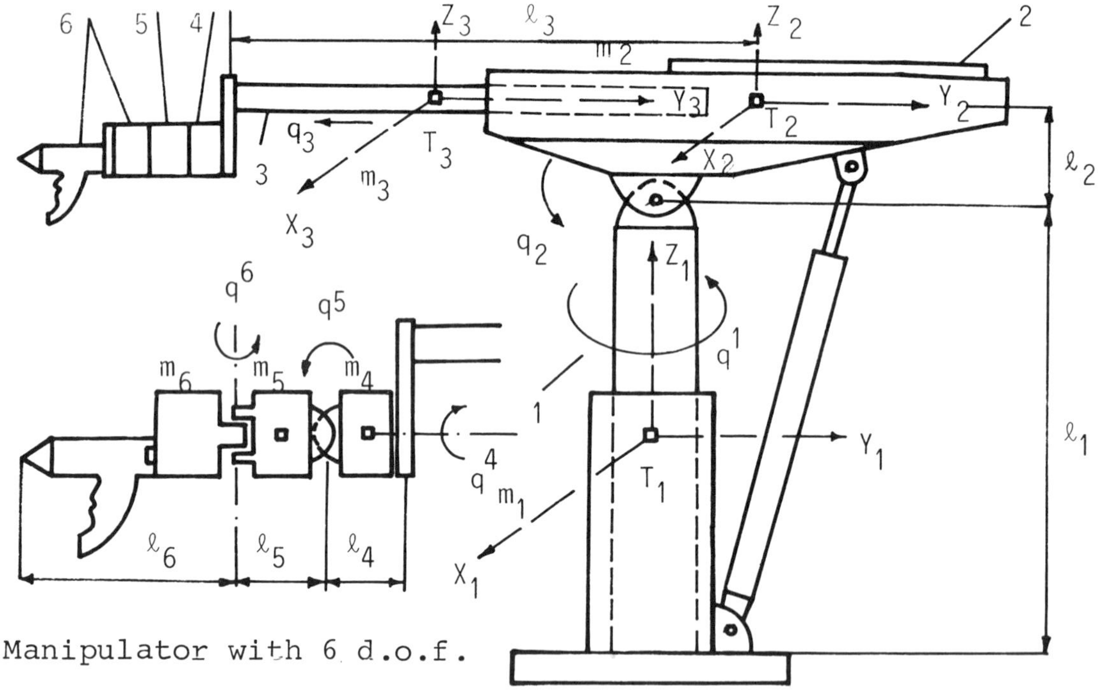

Fig. 3. Manipulator with 6 d.o.f.

Member	1	2	3	4,5	6
Mass m_i (kg)	-	27.8	22.34	2.33	0.33
Length ℓ_i (m)	1.2	0.142	1.14	0.14	0.26
J_{ix} (kgm^2)	-	2.98	1.21	0.004	.0007
J_{iy} (kgm^2)	-	-	-	.004	.0009
J_{iz} (kgm^2)	0.322	3.701	1.21	0.004	.0009

TABLE 1. Parameters of the mechanism

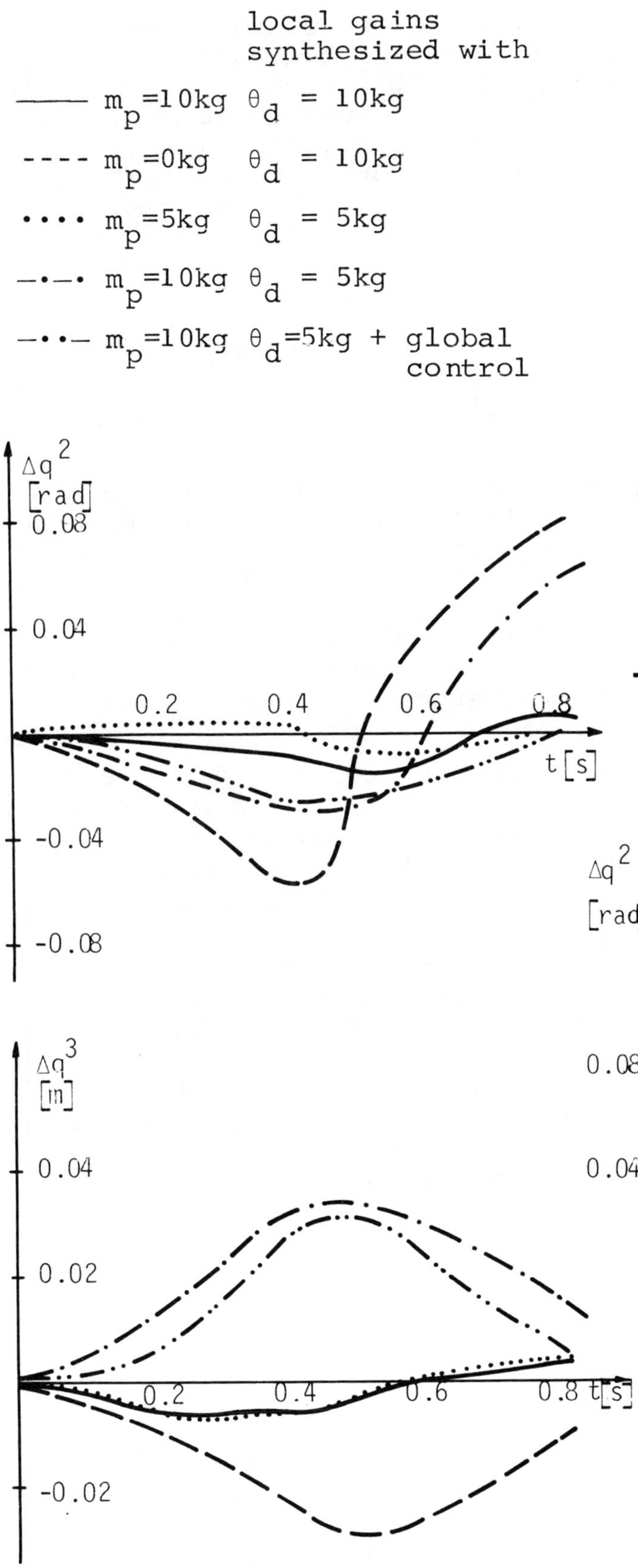

Fig. 4. Simulation results - non--adaptive control

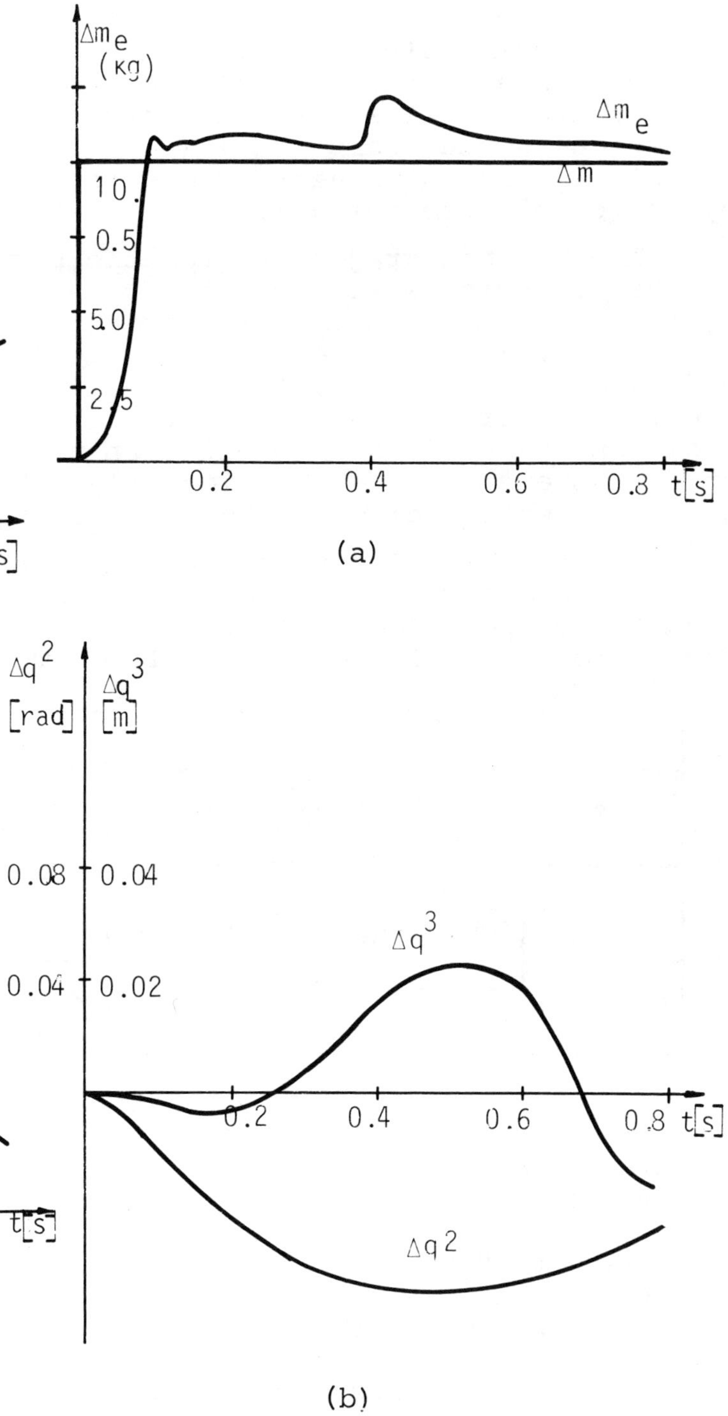

Fig. 5. Simulation results-adaptive control

Hydraulic cylinders, controlled by hydraulic servovalves, are modeled by models of order $n_i=3$. The state vector of the hydraulic actuator is $x^i=(\ell^i, \dot{\ell}^i, p^i)$, where ℓ^i is the piston travel, $\dot{\ell}^i$ is the piston velocity, p^i is the oil pressure in the cylinder. The first three manipulator degrees of freedom are driven by means of hydraulic cylinders while the three d.o.f. of gripper are powered by wane actuators having rotational motion. In this second case, the subsystem S^i state vector is taken as $x^i=(q_i, \dot{q}_i, p^i)$, where q_i, $\dot{q}_i$ are the corresponding angle and angular velocity. The actuator inputs are presented by servovalve coil currents, which are constrained by amplitude. As the first two d.o.f. are rotational and linear actuators are used to drive them, the connection between the actuator model state coordinates x^i and the state coordinates of the mechanical system part model (i.e. the corresponding angles and angular velocities) is not linear. These nonlinear relations are shown in Fig. 1 by vectorial function $g(\cdot)$.

The nominal trajectory for the particular manipulation task is given in the following way: the tip of the manipulator should move along straight line from the point A (coordinates of which with respect to the absolute coordinate system are $(0., 1.14, 1.342)^T$) to the point B (with coordinates $(-0.31, 1.14, 1.65)^T$) while the internal angles of the gripper should be kept constant ($q^4=0.$, $q^5 = -1.57$rad, $q^6=0.$). The triangular velocity profile of the manipulation tip is accepted and the time duration of the movement is $\tau=0.8$s. The only unknown parameter is the payload mass $\theta=m_p$ which can get values belonging to the region $\Theta=\{m_p: m_p \varepsilon (0, 10\text{kg})\}$.

First we shall try to synthesize non-adaptive control which could whithstand all variations of the payload, in the assumed range. We have synthesized local controllers (11), (12) for the maximal allowed mass of the payload in (7) i.e. for $\theta_d=10$kg. The local gains are given in Table 2.

	SUBSYSTEM	1	2	3	4	5	6
$\theta_d=0$	posit.feed. vel.feed.	423.45 90.50	161.77 25.40	154.35 2.30	35.11 0.54	35.09 0.52	35.00 0.35
$\theta_d=5$kg	posit.feed. vel.feed.	745.10 159.74	212.89 44.03	154.38 2.65	36.12 2.21	36.10 2.18	35.28 0.87
$\theta_d=10$kg	posit.feed. vel.feed.	1090.77 234.01	349.16 74.40	154.42 3.01	37.86 4.03	37.83 4.00	35.57 1.37

TABLE 2. Local gains (for $\beta=5$)

We have simulated the dynamics of the manipulation system S (from Fig. 3) during tracking of nominal trajectory with the synthesized local controllers. If the payload is $m_p=10$kg the tracking is satisfactory, but in the case when $m_p=0$ the tracking is poor, since local gains are too high for the manipulation system without payload. The results of simulation are shown in Fig. 4, where $\Delta q^i(t)$ denotes deviations of the corresponding coordinate from its nominal value. We have tried to stabilize the system with local controllers synthesized for the mass of the payload in (7) $\theta_d=5$kg. The results of tracking the nominal trajectory (Fig. 4) with these local gains (Table 2) shows good tracking for $m_p=5$kg and for $m_p=0$kg, but for $m_p=10$kg these gains are too low. These means that we cannot find unique local controllers (12) which are able to whithstand all assumed parameter variations. Thus, we have to introduce global control (13) in the form of force feedback. Simulation with local controllers (with θ_p=5kg) and with global force-feedback (13) shows that the tracking with such control is better but it is not still quite satisfactory.

In order to stabilize manipulation system S for all possible payloads we have synthesized adaptive control algorithm as it has been presented before. The local gains in (11) have been synthesized for various payload masses (Table 2.). Figs. 5 a) and b) show the simulation results of parameter estimation and tracking of desired nominal trajectory with adaptive control law. It can be seen that we now achieve satisfactory tracking of nominal trajectory for arbitrary payload mass in a given range. However, such control is much more complex for implementation than the simple robust nonadaptive control law presented above.

REFERENCES

[1] Vukobratović K.M., Stokić M.D., "Simplified Control Procedure of Non-linear Large-Scale Mechanical Systems" (in Russian), Avtomatika and Telemechanika, No. 11, Also in English, Autom. and Remote Control, Vol. 39, No. 11, 1973.

[2] Vukobratović K.M., Stokić M.D., "Contribution to the Decoupled Control of Large-Scale Mechanical Systems", Automatica, Jan., No. 1. 1980.

[3] Vukobratović K.M., Stokić M.D., "A Procedure for Interactive Dynamic Control Synthesis of Manipulators", Trans. on Systems, Man and Cybernetics, Sept/Oct. issue, 1982.

[4] Vukobratović K.M., Stokić M.D., Control of Manipulation Robots: Theory and Application, Monograph, Springer-Verlag, 1982.

[5] Timofeyev A.V., Ekalo V.J., "Stability and Stabilization of Programmed Motion of Robots-Manipulators", (in Russian), Avtomatika and Telemechanika, No. 10, pp. 148-156, 1976.

[6] Dubowsky S., DesForges T.D., "The Application of Model-References Adaptive Control to Robotic Manipulators", Trans. of the ASME, Journal of Dynamic Systems, Measurement and Control, Vol. 101, No. 3, pp. 193-200, 1979.

[7] Vukobratović K.M., Stokić M.D., "One Engineering Concept of Dynamic Control of Manipulators", Journal of Dynamic Systems, Measurement and Control, Trans. of the ASME, Vol. 103., No. 2., pp. 108-118, 1981.

[8] Guinzy, N.J., A.P.Sage, "System Identification in Large-Scale Systems with Hierarchical Structures", Comp. Elect. Engng, 1,1, 1973.

[9] Chemouil, P., Katebi M.R., Sastry D., Singh M.G., "Maximum a Posteriori Parameter Estimation in Large-Scale Systems", Automatica, Vol. 17, No. 6, 1981.

[10] Vukobratović K.M., Kirćanski N., "Computer Assisted Sensitivity Model Generation in Manipulation Robots Dynamics", Mechanism and Machine Theory, No. 3, 1982.

[11] Feldbaum, A.A., "Theory of Dual Control", Autom, & Remote Control, No. 9, pp. 1240-1249, 1960.

The Positioning Accuracy of Industrial Robots

Lee-Chi Yu
Beijing Machine Tool Research Institute
Beijing, China

XiaoZu Zhang
Shenyang Industrial College
of Machine Tool Company
Peoples Republic of China

Abstract

Along it's travel path, and at certain key points, a gripper with the object being gripped is investigated on "play-back" positions and orientations with respect to those of "teach-in". Based on a thorough analysis of their differences, tolerant positioning accuracy and necessary conditions are discussed. This investigation provides a sufficient basis of formulating reasonable operating rules and control algorithms for different operating procedures of industrial robots having no information fed-back from the object to be gripped.

As we know, positioning accuracy is comprehensively resulting from many factors, including errors of both linear and angular movements. It is not adequate that only the position error is mentioned since position errors and orientation errors are both involved with the gripper and the gripped object. When many different movements are required, it is imperative to select correct positioning accuracy for each trajectory. However, this accuracy are not always reached by a particular working robot due to some reason or others.Therefore, analysis work should be done of the diviation from the teach-in value so that the actual positioning accuracy may be improved.

DETERMINATION OF POSITIONING ACCURACY

Let gripper and the gripped object be an integral rigid body. The position and orientation of this bady in space are determined by 6 parameters, i,e. three degrees of freedom in position and three degrees of freedom in orientation. Assume that we have a cube with the side length of h before us. The position of the cube may be described by vectors leading from the original of the reference coordinate to certain representative points; the orientation of the cube may be described by a three dimension coordinate as well as by an orientational vector, $\bar{a}$. As shown in Fig.1. $\bar{R}=\bar{P}+\overline{CR}$ the length of $\overline{CR}$ heing $\frac{\sqrt{3}}{2}h$ with the direction opposite to $\bar{a}$. the relationship between $\bar{p}$ and $\bar{a}$ is as follows:

$$\bar{R} = \bar{P} - \frac{\sqrt{3}}{2} h \bar{a} \qquad \text{--------(1)}$$

According to the algorithm proposed by R.H. McEntire[1], the orientation vector should be expressed as

$$\bar{a} = d(\bar{a}_1 + \bar{a}_2 + \bar{a}_3) \qquad \text{----------------(2)}$$

and

$$\bar{a}_1 = [\lambda_1, \mu_1, \gamma_1]^T$$

$$\bar{a}_2 = [\lambda_2, \mu_2, \gamma_2]^T$$

$$\bar{a}_3 = [\lambda_3, \mu_3, \gamma_3]^T$$

where

$$\lambda_1 = \frac{1}{\sqrt{1+E_1^2+U_1^2}}, \quad \mu_1 = E_1\lambda_1, \quad \nu_1 = U_1\lambda_1,$$
$$\lambda_2 = U_2\mu_2, \quad \mu_2 = \frac{1}{\sqrt{1+E_2^2+U_2^2}}, \quad \nu_2 = E_2\mu_2 \qquad (2)$$
$$\lambda_3 = E_3\nu_3, \quad \mu_3 = U_3\nu_3, \quad \nu_3 = \frac{1}{\sqrt{1+E_3^2+U_3^2}}$$

E_1, E_2, E_3 —— determined values from the measurement

U_1, U_2, U_3 —— values only related to E_1, E_2, E_3 .

The position vector $\bar{P}$ (refer to Fig. 1)

The point C is the three perpendicular faces of the cube will intersect at the corner. The equation for each plane is related to its associated unit vector. Let $\bar{P}=[Xc,Yc,Zc]^T$, the equations for the X,Y,Z face are:

$$\begin{aligned} \lambda_1 X + \mu_1 Y + \nu_1 Z &= \lambda_1 X_1 + \mu_1 D \\ \lambda_2 X + \mu_2 Y + \nu_2 Z &= \mu_2 Y_1 + \nu_2 D \\ \lambda_3 X + \mu_3 Y + \nu_3 Z &= \nu_3 Z_1 + \lambda_3 D \end{aligned} \qquad \text{-------------- (3)}$$

A simultaneous solution of equation (3). The solution at the point C(Xc,Yc,Zc) is :

$$\begin{aligned} X_c &= \Delta_x / \Delta_0 \\ Y_c &= \Delta_y / \Delta_0 \\ Z_c &= \Delta_z / \Delta_0 \end{aligned} \qquad \text{------------------------ (4)}$$

Where

$$\Delta_0 = \begin{vmatrix} \lambda_1 & \mu_1 & \nu_1 \\ \lambda_2 & \mu_2 & \nu_2 \\ \lambda_3 & \mu_3 & \nu_3 \end{vmatrix} \qquad \Delta_x = \begin{vmatrix} C_1 & \mu_1 & \nu_1 \\ C_2 & \mu_2 & \nu_2 \\ C_3 & \mu_3 & \nu_3 \end{vmatrix}$$

$$\Delta_y = \begin{vmatrix} \lambda_1 & C_1 & \nu_1 \\ \lambda_2 & C_2 & \nu_2 \\ \lambda_3 & C_3 & \nu_3 \end{vmatrix} \qquad \Delta_z = \begin{vmatrix} \lambda_1 & \mu_1 & C_1 \\ \lambda_2 & \mu_2 & C_2 \\ \lambda_3 & \mu_3 & C_3 \end{vmatrix}$$

X_1, Y_1, Z_1, D —— determined values from the measurement.

The position error vector $\bar{\rho}$, orientation error vector $\bar{A}$ and angular error θ may be described in the following (refer to Fig. 2 and 3)

$$\bar{\rho} = \bar{R}_t - \bar{R}_r = [x_t - x_r,\ y_t - y_r,\ z_t - z_r]^T \qquad (5)$$

$$|\bar{\rho}| = \sqrt{(x_t - x_r)^2 + (y_t - y_r)^2 + (z_t - z_r)^2} \qquad (6)$$

$$\bar{A} = \bar{a}_t - \bar{a}_r = [\lambda_t - \lambda_r,\ \mu_t - \mu_r,\ \nu_t - \nu_r]^T \qquad (7)$$

$$|\bar{A}| = \sqrt{(\lambda_t - \lambda_r)^2 + (\mu_t - \mu_r)^2 + (\nu_t - \nu_r)^2} \qquad (8)$$

$$\theta = 2\sin\frac{|\bar{A}|}{2} \qquad (9)$$

$$\text{or } \theta = \cos^{-1}(\lambda_t\lambda_r + \mu_t\mu_r + \nu_t\nu_r)$$

Where $\bar{R}_t, \bar{a}_t$ — "taught" position vector and orientation vector

$\bar{R}_r, \bar{a}_r$ — position vector and orientation vector of the average repeated position

Only in taking into consideration both position and orientation accuracies can the positioning accuracy be fully determined. Of course, the repetitive accuracy may be also obtained through large quantity of repetitive measurements, which are the basis in selecting accepted positioning accuracy.

CONSIDERATIONS IN THE SELECTION OF ACCEPTED ACCURACIES

Following factors should be counted in when we select accepted accuracies:

1. Weight, shape and dimension of the to-be-gripped;

2. Positioning mode of motions or rotations;

3. Repetitive positioning accuracy under special or nominal working conditions;

4. Main constraints for basic trajectories.

Table 1. Accepted positioning accuracies(mm)

Grade Accuracy	load capacity (kgf)		
	0.1-0.63	1-6.3	10-63
high	±0.008-0.063	±0.01-0.08	±0.0125-0.10
moderate	±0.08-0.63	±0.1-0.8	±0.125-1.0
ordinal	±0.8-6.3	±1 - 8	±1.25 - 10

In table 1 accepted positioning accuracies are listed which are common to general jobs.

One more factor we should consider is the speed, which, correlating with the repetitive accuracy, may be expressed in speed redundant coefficient K_T [2].

$$K_T = \frac{1+\beta K_v^2}{2K_v}$$

the appoximation equation of which is:

$$K_T^* = \zeta \frac{\sqrt{S\Delta}}{\sqrt[3]{G}\,V} \qquad \text{----------(10)}$$

where G - load capacity

V - speed

S - stroke

Δ - positioning accuracy

ζ - positioning mode coefficient

(ζ =60-120)

Let Δ_1 be the measured value for repetitive positioning accuracies under the working conditions of G_1, V_1; Δ_2 the accepted value for positioning accuracies under G_2, V_2. The ideal value (i.e all "teach-in" errors eliminated) must meet the requirement

$$\frac{V_2}{V_1} \cdot \sqrt[3]{\frac{G_2}{G_1}} \leq \sqrt{\frac{\Delta_2}{\Delta_1}} \qquad \text{------------(11)}$$

When a robot must perform different operating predures requirement of the following should be met:

$$\frac{\xi \sqrt{S_i \Delta_i}}{\sqrt[3]{G_i}\, V_i} \geq K_T^{*} \qquad \text{------------(12)}$$

Where the subscript i denotes different trajectory point number, for instance, if there should be two basic points in each trajectory, then for a gripper to be used for five different objects, we have i=1,2,······, 10.

ELIMINATION OF POSITIONING ERRORS

Errors obtained from equations (5)······(9) are basically those out of the robotic system, most of which may be eliminated, while the relative errors obtained from repetitive positioning measurements are involved with some unknown factors and the positioning mode itself.

According to Singer theorem, the displacement of a free rigid body from one place to another may be calculated by moving the selected poles linearly and rotating them about the axis they pass through. Referring to the measured errors, the gripper may be streamlined to the desired or "teach-in" state by translating the original point for a $|\bar{\rho}|$ along the error vector $\bar{\rho}$ and rotating for a θ about the vector $\bar{n}$ passing through the original point. $\bar{n}$ is determined by the equation

$$\bar{n} = \bar{a}_r \times \bar{a}_t \qquad \text{-------------(13)}$$

We may conclude that the prerequisite for eliminating positioning errors is that the actuating mechanism should possess 6 DOF. If we use 4 X 4 matrices, T_{Nt} and T_{Nr}, to represent, respectively, the gripper states at the time of "teach-in" and the repetitive "play-back", then their relationship may be expressed as

$$T_{Nt} = T_{Nr} \cdot B\,(\bar{n}, \theta, \rho) \qquad \text{-----------(14)}$$

and B is the transformation matrix, depending on $\bar{n}, \theta, \rho$

$$B\,(\bar{n}, \theta, \rho) \qquad \text{-------------(15)}$$

Obviously, if B is the unit matrix, then the gripper is in the goal state, the linearized relationship equation of meating this requirement for each compoent member of the actuation mechanism is as follows

$$\Delta T_N \approx \sum_{i=1}^{N} U_{Ni} \cdot \Delta q_i \qquad \text{-------------(16)}$$

Where Δq_i is the slight increment of the general coordinate of the ith component member; U_{Ni} is actuation mechanism--4 X 4 order matrix, depending on the state; i, the series number of actuating component member (i=1,2...N); N, joints of the actuation mechanism.

It is a question of reverse resolution for us to compensate the positioning errors with their corresponding component. Generally speaking, the question of reverse resolution would be involved with some more complicated algorithms (e.g. multi-joint arms). However, in our case, what we are asked to do is to adjust the positioning procedures since we have already had moving trajectories before us. It is not necessary to plan, a second time, gripper moving trajectories and its articulation. In fact, we may set up a series of coefficient matrices, U_{Ni}, beforehand, which my alleviate the amount of "teach-in" work and lessen the " blindness" in the process of adjustment, so that satifactory result might be obtained.

Reference

(1) R.H. McEntire: "Three dimensional accuracy measurement methods for robots" "The Industrial Robot" Vol 3 № 3 1976

(2) Yu Li-Ji (i,e, Yu Lee-Chi) "Estimation method of Robot Movement Parameters" "Automation in Mechanical Engineering" 1980. № 1.

(3) Yu Lee-Chi and Chang Hsiao-tsu: "An Approach to Control Hierarchy and Control Algorithm for Industrial Robot" Proceedings of 11th ISIR: 1981 PP825-836

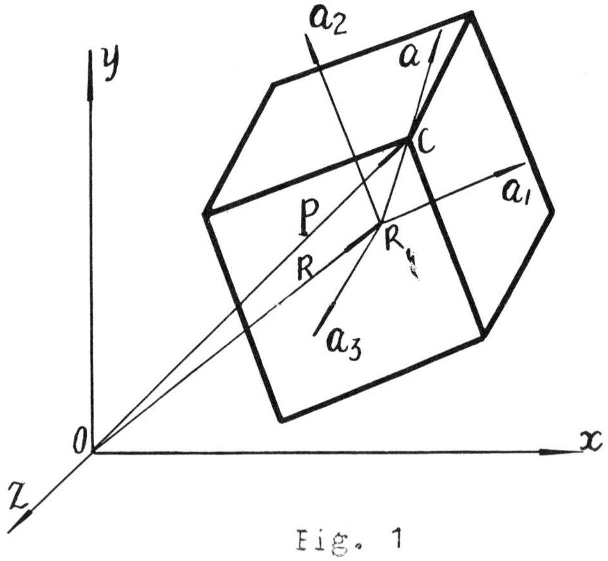

Fig. 1

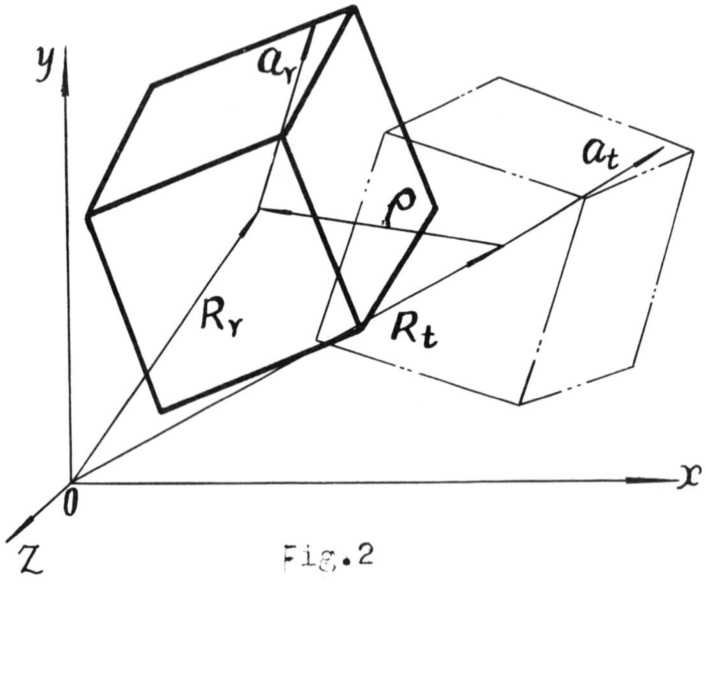

Fig.2

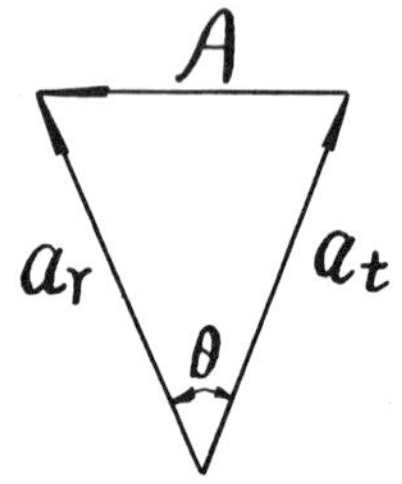

Fig.3

CHAPTER 14

ADVANCED VISION AND TACTILE SENSORS

Problems of Vision Directed Robots in an Unstructured Parts Handling Environment

Chuck Raymond
University of Minnesota
Minneapolis, Minnesota

Max Donath
University of Minnesota

W. Richard Olson
Honeywell, Inc.
Minneapolis, Minnesota

Abstract

Typical tasks undertaken by an integrated robot manipulator and vision system fall into the category of picking up unoriented parts and loading them on to machines. The programming of vision-controlled robots will be introduced followed by a discussion of the problems which exist when implementation on commercially available systems is attempted. These include, (1) touching (and overlapping) parts, (2) reliable thresholds for developing binary images, (3) scene complexity, (4) recognition time and (5) safety. A number of solutions to these problems have been explored.

INTRODUCTION

As technology moves into the competitive 80's, manufacturing organizations are increasingly looking to flexible automation as a major factor in obtaining a competitive advantage. The motivation for vision directed robots comes from the desire to utilize the flexibility inherent in programmable robots which is often lost due to the need for custom designed tooling (hard automation) necessary for parts presentation. The use of vision allows the designer to incorporate inspection into the task being automated which is not unlike the way people (consciously or unconsciously) inspect parts as they are performing manual operations.

The revoltuion in VLSI technology promises to continue the drop of computational cost associated with computer vision systems. Existing industrial vision systems can provide an easily reprogrammable parts presentation system for a large class of parts. This paper provides an introduction to the use of vision directed robots followed by a review of the problems and limitations involved in using such systems.

Vision-directed assembly robots are composed of three parts: the machine vision system, the computer controlled manipulator and the programming system software. The machine-vision system accepts information in the form of a video image, extracts specific data about the scene (e.g. the presence or absence of a part, the parts location and orientation, etc.) and passes that information to the robot controller. The vision processor can use several techniques to analyze the picture information. These can broadly be classified into two groups: systems which analyze grayscale information and those which reduce their images to binary pictures before an analysis is done. Gray scale processors have the ability to deal with complex images which require picture shading information in order to understand the scene. In order to reduce the computational requirements of the analysis algorithms, binary vision systems intentionally throw out much of the gray scale information, reducing their image to a scene which consists of pure black or white regions. Binary vision systems can further be classified into those machines which execute connectivity analysis algorithms, those which use special window functions and those which use template matching to extract information.

Connectivity analysis and feature extraction are general purpose techniques which are used to locate and classify certain patterns (objects) in a binary image. They can recognize the occurrence of a pretaught pattern in a scene, as long as patterns do not touch, without the need for specially programmed algorithms.

Systems which use window functions require the vision programmer to decide where in the image it would be useful to know if a pixel or group of pixels is active and the significance of their being on and off. This is often not as difficult as it would seem but it does require that custom software be written for each part being recognized. Because their recognition algorithms are tailor-made for each application, these systems can usually do recognition much faster than the first method and can often be configured to handle touching parts.

In their simplest form template matching systems compare a pretaught pattern of pixels (i.e. a template) to a blob on the image. If the pixels match, the system declares that an occurrence of the part has occurred; if no match is found, the template is moved around on the image until one is found or all possibilities are exhausted. These techniques of image analysis are not exclusive, many systems can utilize several in combination to quickly analyze fairly complex images.

The robot controller makes decisions based on the vision-supplied information, e.g. is this part within reach? or, is this the right type of part?, and takes actions based on the decisions made, i.e. don't reach for parts that are too far, or pick up the part if it is the one expected. Robot motions can be classified into three types: simple slewing, joint interpolated, and straight line motion. In the simple slewing mode, the controller commands all the joints to move in order to position and orient the gripper, each joint then moves independently to its goal. Since each joint will have different displacements to rotate or translate, each will arrive at its goal position at different times. The robot motion will be jerky and the path that the end effector takes on its way to the goal is difficult to predict intuitively. Joint interpolated motion solves some of these problems and is available on many industrial robots today. In joint interpolated motion the controller moves each joint in such a way that, as the robot goes from its start to its goal position, each joint starts and stops at the same time. Robot motion is then smooth and the path more predictable. To move the robot end effector along straight lines from one arbitrary position and orientation to another, the robot controller must do a series of complex calculations in real time. The path between the gripper positions is then predictable, a requirement for many tasks such as welding or inserting one object into another. Two of the disadvantages of straight line motion are (1) some of the individual joints may have to move at high velocity in order for the end effector to move at constant speed and (2) it is not always possible to move from one position and orientation to another due to kinematic constraints of the robot. Despite these limitations, the ability to execute straight line motion makes the generation of application programs much easier.

The system software which supports the production of user application programs is the most important part of the robot control system. If the robot is too difficult to program, it will not be practical to use the machine for small batch runs even if it is mechanically well suited for the job. Robots are currently programmed using several techniques, of which the two most important are teach playback and interpreted control language. The teach playback method requires the operator to move the robot to each position in the robots task and record these positions. The robot is then able to play back fixed sequences of movements. The teach playback method of programming does not lend itself well to applications which require decision making and editing. The interpreted robot control language makes available the full power of a dedicated computer to control not only the robot, but accessory equipment such as conveyors and bowl feeders. The control language technique stores only endpoints, destinations to which the robots gripper is required to move. The controller generates a path between these points as the program is executed. By providing a framework into which decision making algorithms can be embedded, a language allows the robot to make decisions based on changes in its environment. A well designed language can facilitate the creation of new paths based on sensory information.

A CASE STUDY

The operation of a vision directed robot is best explained by examining a specific robot system executing a particular task. Consider the problem of acquiring unoriented parts from trays using the Puma-Univision system from Unimation. The vision processor used in the system is the Machine Intelligence Corp. VS-100. It uses a D.E.C. LSI-11 16 bit minicomputer to execute connectivity and feature extraction algorithms. The VS-100 is used as a peripheral by the robot controller, it responds to commands it receives over a dedicated parallel communication link. The PUMA robot controller is a multi-processing system using six

6502 microprocessors and an LSI-11 minicomputer to interpret the robot control language. The power of the controller lies in the fact that it allows the user to write simple programs which:

- o Execute straight line motions with controlled accelerations
- o Represent motions of the robot in convenient terms such as cartesian coordinates instead of joint angles
- o Control external equipment such as conveyors or bowl feeders
- o Embed logic algorithms into the control framework allowing the robot to change its actions based on its environment

The following is an example of a program segment, written in Unimation's robot control language VAL, which picks up parts using vision information. The simple instructions which the user writes represent a rather complex series of actions the robot system must execute in order to complete the instruction. The degree to which the VAL language hides these complexities from the user is a good measure of the utility of the system. This program segment does not include any error checking although this is done in practice to ensure that the locations to which the robot is going to move do not cause interference with other objects in the work space.

```
1  OPEN

2  MOVE PLACE 1

3  PICTURE

4  MOVE PLACE TWO

5  LOCATE BLOCK, 3

6  MOVE TO CAMERA: BLOCK: ABOVE

7  MOVES TO CAMERA: BLOCK

8  CLOSE

9  DRAW 0, 0, 100

10 MOVE MACHINE PLACE

11 OPEN

12 END
```

The system execution will be described as the robot controller interprets each step of the VAL program. All the locations in the program (i.e. PLACE 1, PLACE TWO) are just convenient names for positions and orientations of the gripper in the robot's workspace which have been taught using the teach pendent (or derived from a world model data base) except for the location BLOCK which is vision derived.

1 OPEN: The first step is straightforward. It commands the robot controller to open the gripper.

2 MOVE PLACE 1: The second step causes the robot to move along a joint interpolated path to position the end effector at the location which has been named "PLACE 1". For this particular case a more precisely defined path is not needed thus reducing the amont of computation required.

3 PICTURE: The third instruction causes the robot controller to send a "take picture" command to the vision system. A rough outline of the sequence of events which follow is shown in Fig. 1. The vision system digitizes the image from the video camera and stores it as a matrix (256 x 256) of pixels (picture elements). Binary vision refers to the technique of just storing one of two possibilities, light or dark, as the value for each pixel. Each pixel is bright if the brightness of the corresponding position on the gray scale image is above the threshold level; otherwise it is set to dark. The binary image is much easier to analyze than the original gray scale, since object boundries are implicitly defined as places where the pixel values change. The vision processor executes a series of procedures in order to reduce the nature of the scene.

The matrix of binary pixels is used as input to a connectivity analysis process which builds a table of information for each object in the scene. Objects, referred to as blobs, are simply defined as groups of bright pixels which touch. The table consists of a record for each object which contains the area and the X and Y coordinates of the centroid of the object. If the area of the blob is less than a certain minimum, a record for the blob is not entered into the table and it is ignored during all further processing. This reduces the amount of time spent analyzing small groups of pixels which are obviously not of interest but are hard to remove due to noise in the picture.

The table of blob locations and sizes along with the pixel buffer then becomes input to the feature extraction process. The feature extraction process computes a set of orientation independent features about each object in the blob table. The features the VS-100 can calculate are total area, number of holes, hole area, perimeter, length of the major and minor axis of an equivalent ellipse (one having the same second moment), length of the longest possible distance from the centroid to the blob edge, length of the shortest possible distance from the centroid to a blob edge, the roundness or compactness (area/(2* *perimeter)), the overall length, and the overall width. The features calculated are added to the record for each blob in the blob table.

This is now used as input for the classification process which attempts to identify the blob as an occurrence of one of the previously stored prototypes. The blobs are classified on the basis of how well their features match the prototype feature set. Fig. 2 shows how the classification process works for a blob being classified using two features. Notice that the clusters represent the combinations of the two features (a two dimensional feature vector) which correspond to acceptable occurrences of the desired patterns. If the coordinates of the blob's features lie within one of these boundaries it is classified as belonging to one of these prototypes, otherwise it remains an unknown blob. This often used classification rule is called the nearest neighbor algorithm, since blobs are classified as belonging to a prototype if their feature coordinates lie nearest to that prototype (as long as it is within a certain predetermined distance). The nearest neighbor algorithm is not always the best classifying rule but its shortcomings are offset by its ease of calculation. In experiments on industrial parts, C.A. Rosen, et al (ROSEN 1979) have found this classifier and set of features to be successful. Occasionally, however, the two dimensional projections of different parts will result in blobs so similar that they cannot be discriminated with this feature set and the nearest neighbor algorithm. Two alternatives are possible, either use more complex classification algorithms and/or add an identifying feature onto the part which will be thresholded as part of the binary blob. At this point the prototype with which each blob matches most favorably is entered in the table for that blob. In order to manipulate the object, the parts orientation must be determined. The orientation of the blob is the angle which some key feature of the blob makes with respect to the camera coordinate system. The key feature is selected by the programmer and stored in the prototype information table for each type of blob. Fig. 3 shows the orientation angle of a blob which is an occurrence of prototype "wrench" whose orientation feature has been specified as the major axis of the blob.

There now exists a table which contains information about all the blobs in the scene. Specifically, the table contains what prototype class the blob best represents, its location and orientation and a set of features which has been calculated about the blob. The vision system now sends a processing complete message to the robot controller.

4 MOVE PLACE TWO: While all this vision processing has been going on (about 1 sec on the VS-100) the robot controller has been moving the gripper to the position labeled "PLACE TWO" in order to be ready to pick up the next part. Because of the systems ability to move the robot in parallel with processing the vision information, the vision processing time does not necessarily contribute to the overall cycle time of the system. The controller registers the fact that the picture processing is done, so that no waiting is necessary when the locate command is reached.

5 LOCATE BLOCK, 3: When the vision processing is done, the controller sends the locate command to the vision processor. The vision processor searches its blob table for any blobs which match the prototype named "BLOCK". If a match is found, the coordinates and the orientation angle of this blob are sent to the robot controller; otherwise a (no "BLOCK" in view) message is sent back. If no part was found, the controller jumps back to program step 3 and retakes a picture. When a part is found, the controller uses the coordinates to build a location named "BLOCK". In general, the position of an object in space is described by six variables, three translations (X,Y,Z), and three rotations (roll, pitch, yaw). All of these must be specified in order to drive a six axis robot to a unique position. (On certain robots, such as the PUMA series, uniqueness is only achieved by specifying additional constraints on the manipulator configuration.) The vision system can only ascertain three of the six variables, two displacements and one rotation. The robot controller must then assume default values for the other three components to arrive at a unique position and orientation (i.e. location). In the VAL language, as well as other languages, locations can be thought of as a plan for calculating the position and orientation in space of an object relative to another (the origin) analogous to a vector. The origin for a location can be described relative to another location by chaining them together, i.e. making the position and orientation of one location the origin of the next. Each location in space can have a coordinate system associated with it which facilitates locating other objects relative to it. The location "BLOCK" was defined in the camera coordinate system. Therefore, the origin of "BLOCK" is the origin of the camera coordinate system.

6 MOVE TO CAMERA: BLOCK: ABOVE: This command causes the robot controller to move the robot end effector to the location designated by the chained location "TO CAMERA: BLOCK: ABOVE". To evaluate the chained (or compound) position/orientation the location string is analyzed from left to right determining the relative positions in space. The first term (TO CAMERA in this case) always originates at the origin of the coordinate frame of the robot. The location "TO CAMERA" thus describes the location of the origin of the camera coordinate system relative to the robots coordinate frame origin (see Fig. 4). "TO CAMERA" is now used as the origin for location "BLOCK", which points to the centroid of the part which was previously determined by the vision system. The location "ABOVE" is described with respect to the centroid of the part and refers to a location above and behind the part. Since the position and orientation of location "ABOVE" is the last in the chain, it becomes the destination of the robot end effector. Although not apparent to the user, the VAL language uses homogeneous coordinate frame transformations to implement this technique of location representation (PAUL 1981). Since "ABOVE" is defined with respect to "BLOCK", a change in the location "BLOCK" will not affect the location of the end effector with respect to the "BLOCK" a very desirable property.

7 MOVES TO CAMERA: BLOCK: GRIP: The S in MOVES instructs the controller to move the end effector along a straight linear path to the effective location designated by the chained string. The location "GRIP" has its origin at the centroid of the object and points to the position and orientation where the robot should grip the object. It is often the case that a part should not be gripped at its centroid. Chained locations give the programmer the option of specifying any gripping configuration relative to the centroid of the part.

8 CLOSE: The "CLOSE" instruction closes the pneumatic gripper.

9 DRAW 0,0,100: The instruction causes the robot controller to move the end effector 100 millimeters in the positive Z direction in the base coordinate system of the robot with no change in the X and Y directions.

The robot now has acquired a part without the use of hard automation parts presentation equipment. Note that the part was randomly located in the field of the camera on a surface parallel to the image plane of the camera. No mention has been made, however, about how the prototypes feature sets used by the vision system were obtained.

VISION SYSTEM TRAINING

In menu based systems such as the VS-100 vision system, the user is shielded from the complexities of the vision processing algorithms. The system uses general connectivity, feature extraction and classification procedures which, for the most part, remain unchanged no matter what types of parts are processed. The user does not have access to these algorithms nor are tools provided to allow the user to write his own algorithms. This philosophy is in contrast to systems which do provide the user with software development capability with which to write special purpose algorithms for each application such as those available from Automatix Corp. and many others. The benefit of menu philosophy is that the VS-100 is relatively easy to configure to a specific application. The user is freed from the task of software development. Instead he or she just selects a series of software parameters and switches which facilitate specific applications. The drawback of this approach is a loss of flexibility. Often special purpose algorithms can be conceived which will handle a particular type of scene that may be too complex for the general purpose processes to handle, (e.g. the touching parts problem). Although custom software can be obtained from the vendor, it detracts from the desired flexibility of a turnkey vision system.

The vision system is taught to recognize new patterns by repeatedly analyzing images of the new prototype. Since the vision is based on a 2D projection, a different prototype must be stored for each stable position of the object that results in a different pattern. A set of features and their variability for the pattern is calculated and stored as information defining the prototype. A new prototype is automatically compared to all others previously stored in order to determine which features are sufficient to discriminate between the prototypes. This facilitates the determination of the minimum number of features needed and thus minimizes the recognition time.

PROBLEMS AND LIMITATIONS ENCOUNTERED

2D VISION

Machine vision systems today, do not incorporate range data. The images which appear before the camera are all assumed to a plane which remains fixed with respect to the robot. The research community has proposed several techniques to incorporate range data into the sensing system, among these are:

- The use of two cameras in a stereo vision arrangement (Yakimovsky and Cunningham 1976).
- The use of automatic focusing equipment to derive range data optically using one camera.
- The use of sonar detectors to obtain range data nonoptically.
- The use of laser range detectors (Nitzan, Brain, Duda 1977).
- Structured illumination with perspective transformations and mapping (Vanderbrug 1979).

Unfortunately, none of these techniques are available in commercial systems today. Practical solutions for the practicing automation engineer then consist of ways to circumvent the limitations of the system by clever structuring of the problem.

This amounts to building the robot's workspace in such a way that parts can never be piled on one another while being viewed by the camera. The easiest way to deal with this problem is to make sure that all parts presented to the vision system lie on a plane. For a large class of parts the best way to ensure they are in a plane is by storing them using trays or racks of parts. The parts can be randomly oriented in the plane eliminating the need for egg carton type separators.

Some parts by nature or tradition are presented to the work cell piled in bins. Two approaches to acquiring these types of parts have been used. Mechanical methods can be used to get the parts onto a plane for vision processing. Honeywell has developed a general purpose bowl feeder which will feed a large class of parts onto a conveyor belt without the need for custom designed escapements. General purpose vision is then used to acquire the parts from the belt. Object Recognition Systems, Inc. has demonstrated a system which can direct the robot to pick parts directly from a bin by using the robot as a range data sensor. The system views the top of the pile and uses customer designed vision algorithms (tailor-made for each type of part) to find the planer coordinates of a part in the pile. It then directs the robot to a position over the part known to be on top of the pile. The robot then moves in a straight line down until sensors in the gripper detect contact with the part. The robot gripper closes on the part, but its orientation in the gripper is not well defined (remember we have only measured 4 of the 6 variables which determine the position of an object) and may require regripping before it can be manipulated. Techniques integrating vision and gripper designs (particularly the vacuum type gripper) have been studied at the University of Rhode Island (BIRK 19XX).

BINARY VISION

In binary vision thresholding is crucial. The applications engineer must be sure that proper thresholding is taking place if the results of binary image analysis are to be relied upon. When a vision system thresholds a gray scale image into a binary one in order to reduce processing complexity, much of the information contained in the gray scale image is purposely discarded. The consequence is that no amount of vision processing can recover the information which is lost due to incorrect thresholding. An image is properly thresholded if all objects which the system is required to recognize result in blobs in the binary image which are acceptable matches of the associated prototype. This implies that thresholding must be consistent from training prototypes to actual application. Fig. 5 shows how the gray scale brightness of a pixel is mapped (thresholded) into a binary value depending on the value of the threshold. Images with high contrast will be much less sensitive to changes in thresholding. The high contrast required for reliable thresholding can usually be obtained by the use of a number of clever lighting and filtering techniques known as structured lighting. In a lab environment high contrast is easily achieved with the use of backlighting. In production backlighting is often not desirable due to dirty environment, expense or physical constraints. If the parts have distinguishing spectral characteristics, colored lighting and filters can be used to enhance image contrast. However, the spectral characteristics of the camera being employed will have to be taken into consideration (many solid state cameras are somewhat sensitive to the infra red). Bright lights or strobes can be used to drown out variations in existing light conditions. Fig. 6 shows a lighting technique used in the GM CONSIGHT (HOLLAND 1979) system for obtaining high contrast in a dirty environment.,

In some cases, lighting conditions may vary to the extent that fixed thresholding cannot be used to obtain proper images. In these cases, binary vision systems which allow the user to program the algorithms may use adaptive thresholding to obtain reliable binary images. The concept is based on placing a reference target (an object or visual pattern) in the field of view at all times. When an image is analyzed, it is checked for the existence of the reference target. If it is not in view, the system recognizes that the image was not properly thresholded, and automatically changes the threshold until the reference target reappears. Some scenes may be too complex to analyze using binary vision. In cases where the parts are variable in brightness and/or have identifying surface variations such as three dimensional contours, gray scale vision may be the only recourse. Instead of letting the thresholding process define objects on the basis of binary light intensity, gray scale processors define object boundries using more general techniques referred to as edge algorithms (BALLARD and BROWN 1982). Gray scale vision algorithms are much more complex and therefore require more processor power, memory and time.

TOUCHING PARTS

The SRI algorithms based vision systems recognize parts by calculating features about regions which are simply defined as groups of like-colored touching pixels. Problems arise when a region is composed of two prototype regions. Fig. 7 illustrates the problem. Classification is done based upon the composite global features of a region which is formed by the combination of two prototype regions. Here combining regions can be defined as those which touch or occlude each other. Techniques which may have potential to handle combined regions by doing recognition based on local features such as corners and holes have been developed in the research community (BOLLES and CAIN 1982) but are not yet available on commercial systems. The alternatives then involve either structuring the problem so the parts cannot touch or thresholding the image so that only regions internal to the part appear as blobs on the binary image. Fig. 8 shows a system acquiring parts from trays using the latter technique. While the vision system cannot classify touching parts, it can return the location of unrecognized regions in its view. This information can be used to direct the robot to bump the parts so that they move apart. While this makes for interesting demonstrations, the increased cycle times and the unpredictability of the bumping operation may preclude its use in a production environment.

SCENE COMPLEXITY

Scenes which are properly thresholded and do not contain touching parts may still be too complex for a vision system to analyze. With a typical image array of 256 x 256 pixels, a large number of objects may potentially be in the field of view. During the connectivity phase of processing, a table entry is made for each object in the image. If the number of objects becomes too large, there may not be enough memory for the table. This limitation restricts the VS-100, for example, to dealing with approx. 15 objects without exhausting memory. There are two ways of limiting the number of objects considered by the vision system. First is the minimum blob size parameter mentioned earlier. The other is the window function built into many vision system. The window function allows the user to command the system to ignore all blobs outside a programmable window on the image, see Fig. 9. This normally is an excellent way to break up the scene into smaller simpler problems which can be handled separately. Unfortunately, this technique breaks down for the Univision system. While the vision system is capable of changing its window in response to commands from an external computer, the interface software in the robot controller does not provide a way to send these commands to the vision system. The user may then have to structure the problem in such a way as to limit the number of parts in view. In an application where the robot was presented with a tray which could contain up to 27 parts, the approach used was to only view a third of the tray at a time. The robot was required to move the tray twice in order to view and pick up all the parts. While this increased the cycle time, it utilized the flexibility of the robot to simplify the vision processing task. Alternatively, separate cameras or improved interface software allowing movable windows could provide the same capability.

Vision systems typically have a maximum number of prototype feature sets that can be stored (e.g. nine for the VS-100). In many applications, the limited number of prototypes can be a serious limitation. Consider a sorting task where the robot is to sort a large variety of circuit boards as they are presented on a conveyor belt. In general, the problem with adding more prototypes is that the recognition time increases as a function of the number of objects in view times the number of prototypes. This is further complicated by the fact that more than one prototype may be needed to recognize an object if a variety of stable resting positions exist for the part.

RECOGNITION TIME

With current technology, the recognition time for machine vision systems using the general feature extraction approach, as discussed earlier, is roughly in the range of .25 to 1 second depending on the complexity of the application. This implies several things about the nature of the vision and robot system. First, the vision information is used to drive the robot in an open loop manner. The system takes a picture, calculates a new desired location, and then moves to that location. Open loop operation implies that the pickup accuracy of the system is limited by (1) the linearity of the mapping between vision coordinates and actual robot positions and (2) to the size of a pixel projected into the pickup plane. Closed loop operation would involve using the vision system to measure the location of the robot and the goal location, finding the difference between the two and using that error to continuously drive the robot to the goal location. Vision sampling times being what they are, this type of approach would cause an unacceptable reduction in the bandwidth of the manipulator. Another implication of open loop operation is that the parts must not move in an unpredictable fashion during the time between the taking of a picture and the end of robot motion based on that information. One conclusion that can be drawn is that people should be excluded from the workspace for reliable operation.

BINARY VISION AND SAFETY

Industrial robots, especially those with vision, are relatively new to industry; they attract attention. People want to get close to see what's going on. Eventually they get in the way and injury may very well result. Because binary vision systems ignore objects which are not thresholded into the binary image, they cannot be relied upon to recognize foreign objects in the field of view. Vision systems cannot be relied upon to recognize people which may invade the robots workspace. External systems such as mechanical or light fences and, or watchdog sensory system, such as those being developed at the National Bureau of Standards (KILMER 1982) which stop the robot if people intrude, are a requirement for robotic applications where absolute security cannot be provided.

CONCLUSION

We have found the following set of simple rules can be useful guidelines for introducing automation engineers to applications involving machine vision for parts acquisition. These serve as guidelines for systems commercially available to date and do not in any way preclude the development of more powerful systems with greater flexibility.

1. The parts should lie in a plane with a limited number of stable positions.

2. The parts should be at rest or moving in a predictable way.

3. The parts should have high contrast features for reliable thresholding.

4. The features should be related to the orientation of the part.

5. The number of parts in view at once should be limited.

6. The number of different objects which can be recognized is limited.

7. The parts must not be touching unless the region thresholded into the binary image is internal to the part.

8. The vision system must not be relied upon to recognize foreign objects in the workspace.

9. People should be excluded from the robot workspace.

BIBLIOGRAPHY

Ballard, D. H. and Brown, C.M.: Computer Vision. Prentice Hall 1982

Birk, J.R. et al: "An Orienting Robot for Feeding Work Pieces Stored in Bins", IEEE TRANS on Systems Man and Cybernetics SMC-11, No. 2, pp 151-160 (1981)

Bolles, R.C. and Cain, R.S.: "Recognizing and Locating Partially Visible Objects: The Local-Feature-Focus Method". Journal of Robotics Research. Vol. 1, No. 3, pp 57-82 (1982)

Duda, R.O. and Hart, P.E.: Pattern Classification and Scene Analysis. Wiley (1973)

Fu, King-sun (Editor): Computer Vol. 15, No. 12 (1982)

Gonzalez, R.C. and Safabakhsh, R.: "Computer VIsion Techniques for Industrial Applications and Robot Control:. Computer Vol. 15, No. 12 (1982)

Holland S.W. et al: Consight-1, Dodd, G.G. and Rassol, L. (EDS), Computer Vision and Sensor Based Robots, Plenum Press (1979) pp81-100.

Kilmer, R.D.: "Safety Sensor Systems for Industrial Robots". Proceedings: Robots VI Conference (1982)

Nitzan, D., Brain, A.E. and Duda, R.O.: "The Measurement and Use of Registered Reflectance and Range Data in Scene Analysis". Proceedings: IEEE, Vol. 65, No. 2 (1977)

Rosen, C. and Nitzan, D. et al: Machine Intelligence Research Applied to Industrial Automation. 1st through 8th Reports, NSF Grant APR75-13074, SRI International, Menlo Park, CA. (1973-1979)

Paul, R.P.: Robot Manipulators: Mathematics, Programming, and Control The MIT Press (1981)

Vanderbrug, G.J.: "A Vision System For Real Time Control of Robots" Proceedings: 9th International Symposium on Industrial Robots (1979)

Yakimovsky, Y. and Cunningham, R.: "A System for Extracting Three-Dimensionsl Measurements from a Stereo Pair of TV Cameras". Computer Graphics and Image Processing 7, pp 195-210 (1978)

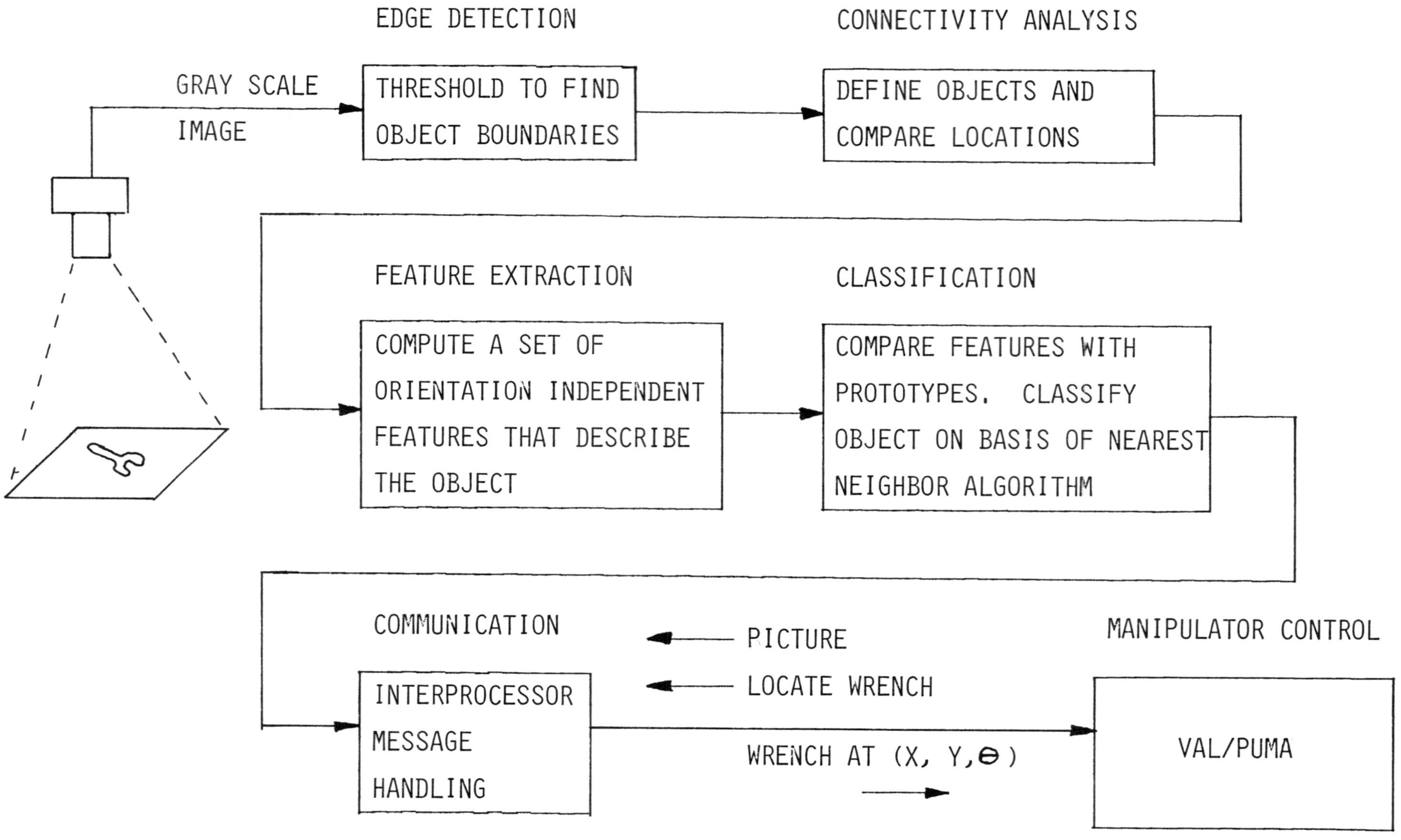

FIGURE 1. FUNCTIONAL BLOCK DIAGRAM OF UNIVISION SYSTEM

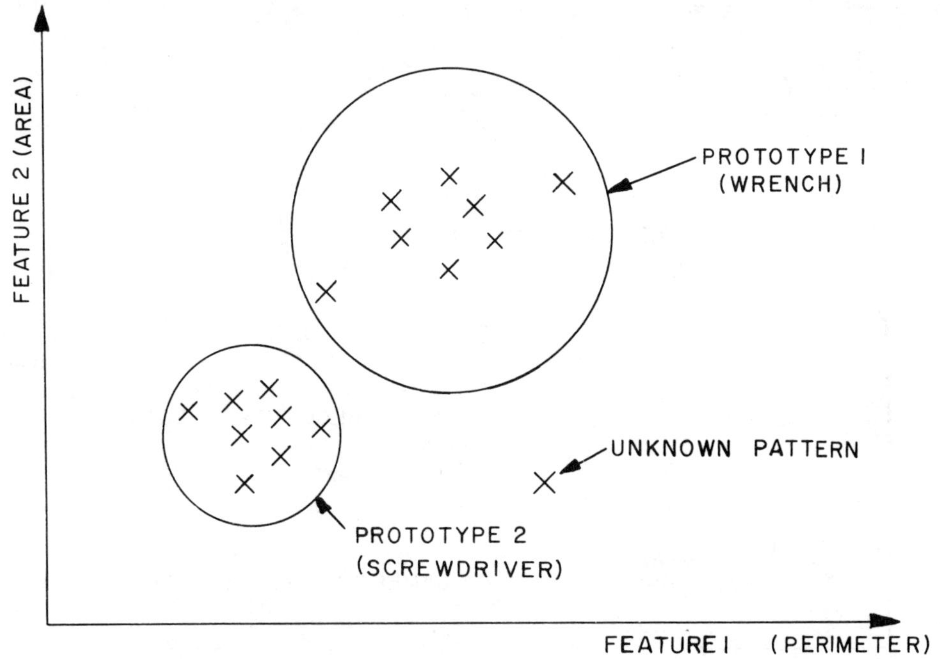

FIG. 2 CLASSIFYING PATTERNS ON THE BASIS OF TWO FEATURES

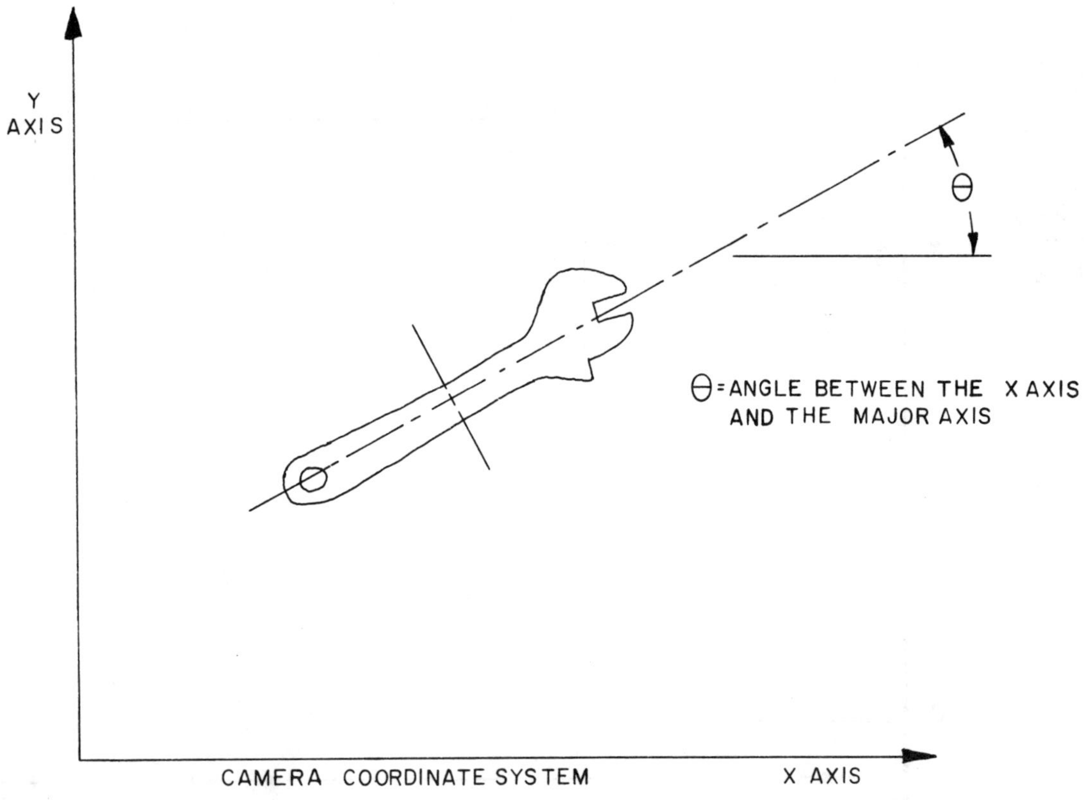

FIG. 3 ORIENTATION ANGLE

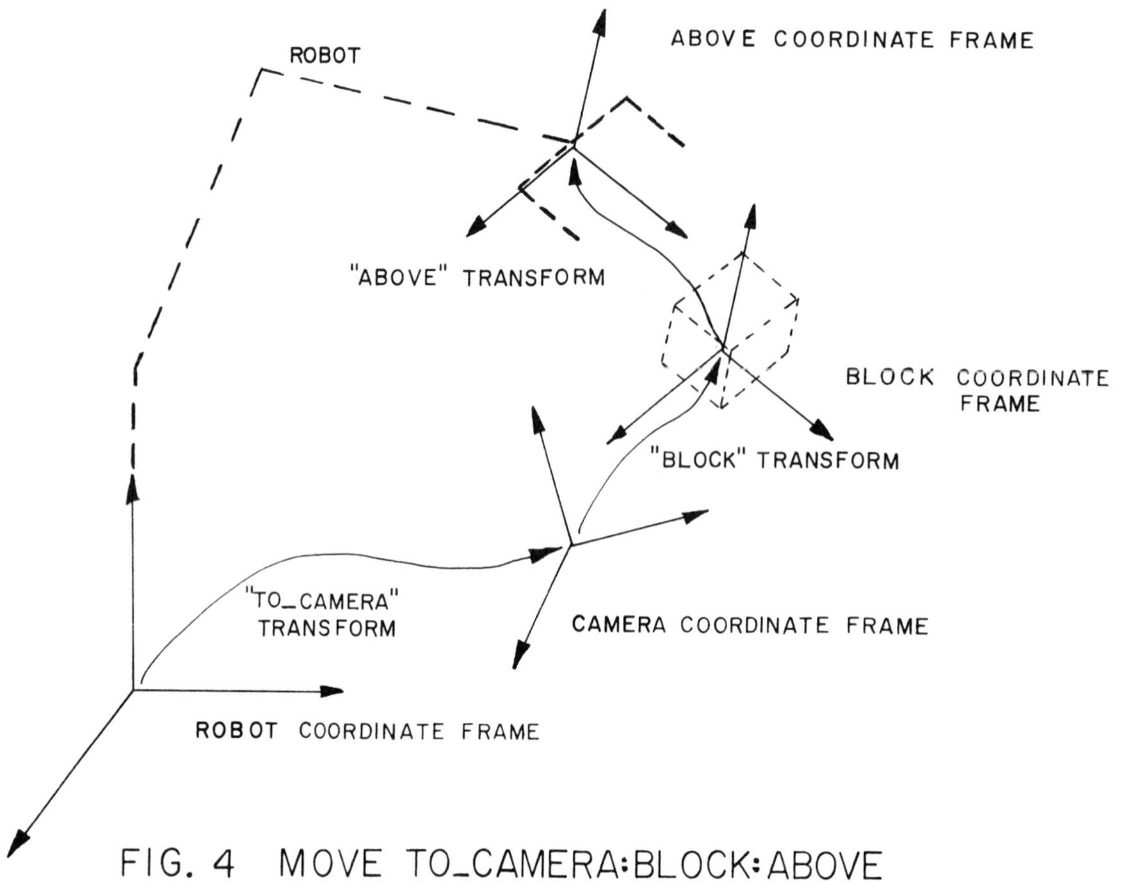

FIG. 4 MOVE TO_CAMERA:BLOCK:ABOVE

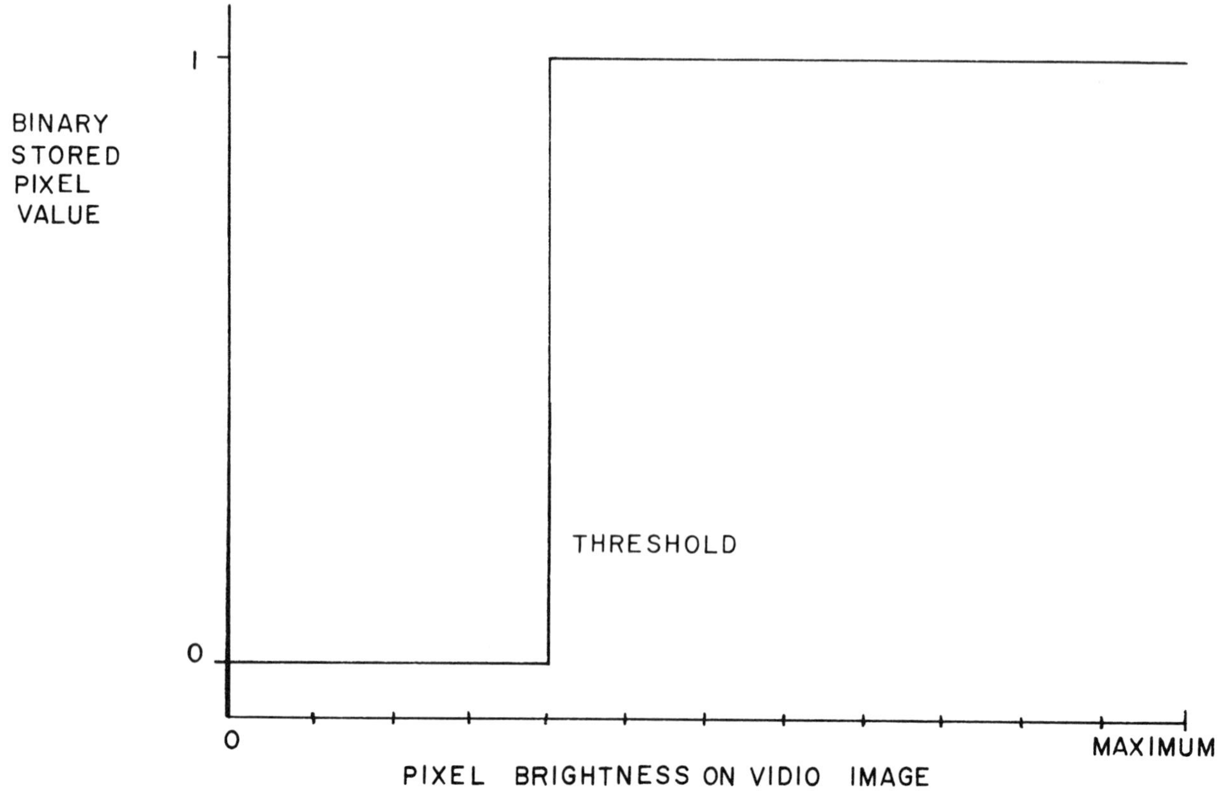

FIG. 5 OBTAINING BINARY IMAGES

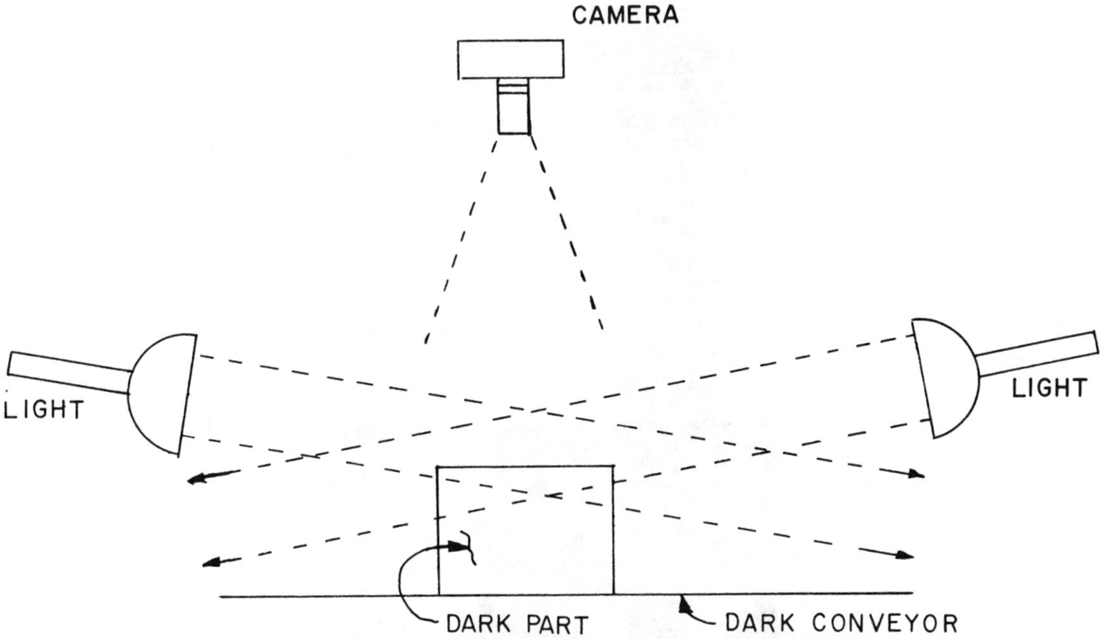

FIG. 6 STRUCTURED LIGHTING UTILIZING THE PHYSICAL DIMENSIONS OF OBJECTS TO OBTAIN HIGH CONTRAST IMAGES

FIG. 7 CLASSIFICATION BASED ON REGIONAL FEATURES IS LIMITED TO NONTOUCHING OBJECTS

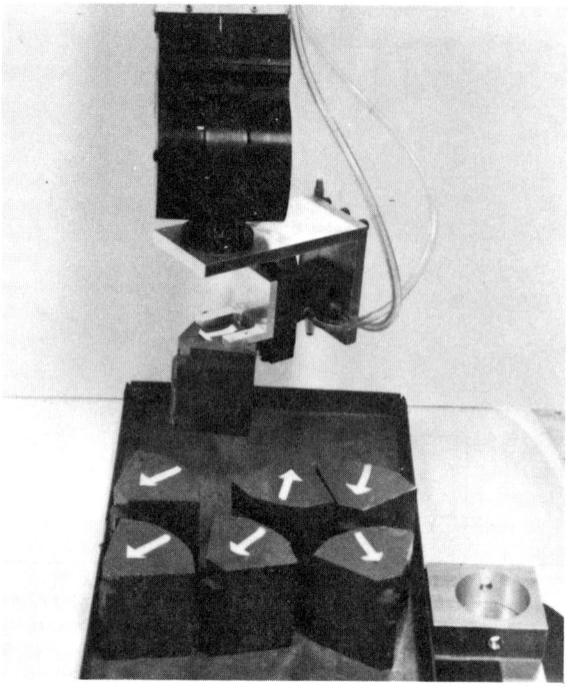

FIGURE 8 - ACQUIRING TRAYS OF PARTS

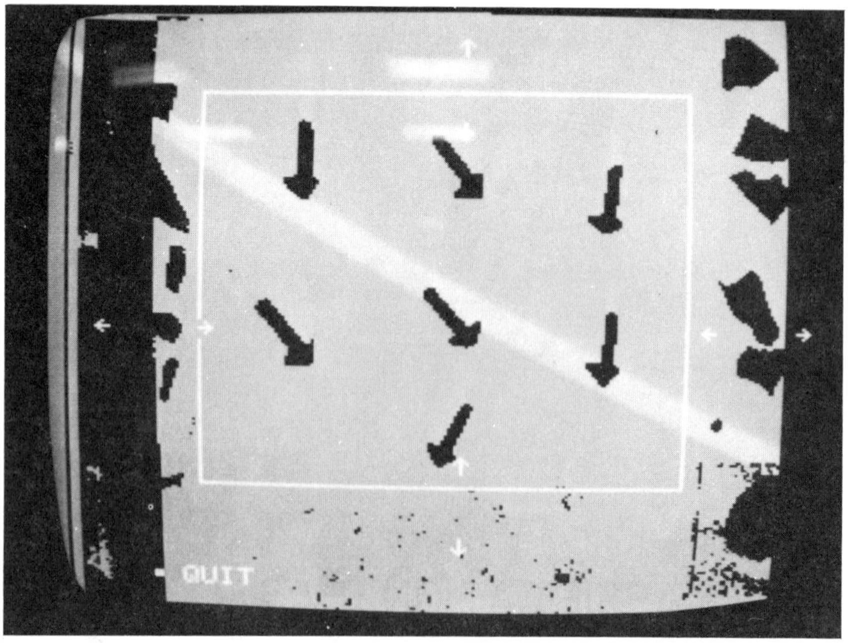

FIGURE 9 - BINARY IMAGE WITH WINDOW

Development of a Machine-Vision-Based Robotic Arc-Welding System

Jan Kremers
SRI International
Menlo Park, California

Carl Blahnik
SRI International

Alfred Brain
SRI International

Ronald Cain
SRI International

Jeff DeCurtins
SRI International

Jose Meseguer
SRI International

Norman Peppers
SRI International

Abstract

Complete automation of arc welding will require a sensing system that can determine not only the location of the weld joint (for path guidance), but also properties such as fitup condition and geometry for use in selection and real-time modification of welding procedures and parameters. This paper describes the development of a laboratory prototype, machine-vision guidance system for robotic arc welding. Acquisition and analysis problems are discussed and techniques for their successful solution are described.

I INTRODUCTION

The work reported in this paper was supported by a group of eight Industrial Affiliates, with additional funding from the U.S. Navy under NAVSEA Contract No. N00024-82-C-5320.

Research into adaptive-control mechanisms for arc welding has identified several factors that must be controlled to produce a high-quality weld. Among these are surface finish, penetration depth, bead width and shape, and the cross-sectional area and shape of the fusion and heat-affected zones [1,2,3]. The following parameters can be adjusted to influence these factors (see Figure 1):

* Arc voltage and current or speed of wire feed

* Torch travel Speed

* Torch standoff distance

* Torch position with respect to the joint line

* Torch attitude.

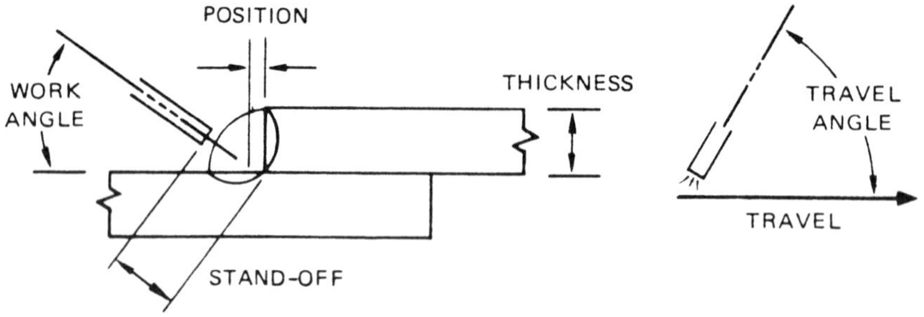

FIGURE 1 TORCH ATTITUDE DEFINITION FOR LAP WELD

Real-time sensing of some or all of the above factors (e.g., penetration depth) and control of all these parameters may be necessary in some critical applications, such as the fabrication of nuclear-reactor fuel rods, piping and pressure vessels [4,5]. For many less critical but economically important applications, however, it is sufficient to sense only joint location, geometry, and fitup. For example, in gas-metal arc welding (GMAW) of mild-steel plate it has been found that penetration depth depends quasi-linearly on the angle of torch travel and exponentially on the standoff distance (see Figure 1)[6]. Two of the applications for which this level of control is adequate include shipbuilding and fabrication of heavy earthmoving equipment.

Position sensors currently available for arc welding fall into three general categories:

* Mechanical or contact devices, including electromechanical probes and various forms of guide rollers and pins.

* Eddy-current sensors.

* Systems employing weld parameter analysis, i.e., real-time measurement and analysis of arc voltage or current waveforms, combined with controlled perturbation of welding torch position [7].

All of these devices or systems suffer from a number of disadvantages, among which are the following:

* None is capable of providing the simultaneous measurement of three-dimensional joint location, geometry, and fitup required for automatic adjustment of welding parameters and procedures or of torch motion, attitude, and position.

* Mechanical, electromechanical and eddy-current sensors must operate either in contact with or in close proximity to the workpiece. This requirement entails mechanical wear or increased probability of damage, with consequent loss of accuracy.

* Mechanical, electromechanical and eddy-current sensors must be mechanically reconfigured for different types of joints. Eddy-current and weld parameter sensors may be applicable only to certain types (e.g., fillet joints, or beveled butt joints on thick materials).

* Weld parameter sensors may show wide variations in performance with different welding processes, metal transfer modes (in the case of GMAW), material thicknesses, and joint types. This variability limits the flexibility of application and could make initial setup difficult and time-consuming.

We believe that a satisfactory position-and-geometry sensing system for general use in arc welding must meet the following design criteria:

* The sensing system must be process-independent, that is, it must operate in conjunction with any welding process without requiring any special alteration or adjustment of welding parameters or techniques.

* The sensing system must require no special workpiece preparation (e.g., painting or cleaning) beyond that needed for conventional manual or mechanized welding.

* The sensor must provide three-dimensional information on joint location, spatial trajectory, dimensions, and geometry within tolerances appropriate to the welding process and application. This means that the spatial resolution of the sensor may vary between 0.005 inches for some gas-tungsten arc (GTA) applications to 0.030 inches GMAW and flux-cored arc welding (FCAW).

* The sensing system must operate in real time during welding, although a two-pass approach (i.e., a preweld inspection pass followed by "blind" welding) may be satisfactory for small, well-fixtured parts.

* The sensor package must be small enough to cause minimal spatial intrusion, so as to duplicate to the extent possible the flexibility of manual welding.

* The sensing device and control system must be at least as mechanically and electrically reliable as conventional welding apparatus.

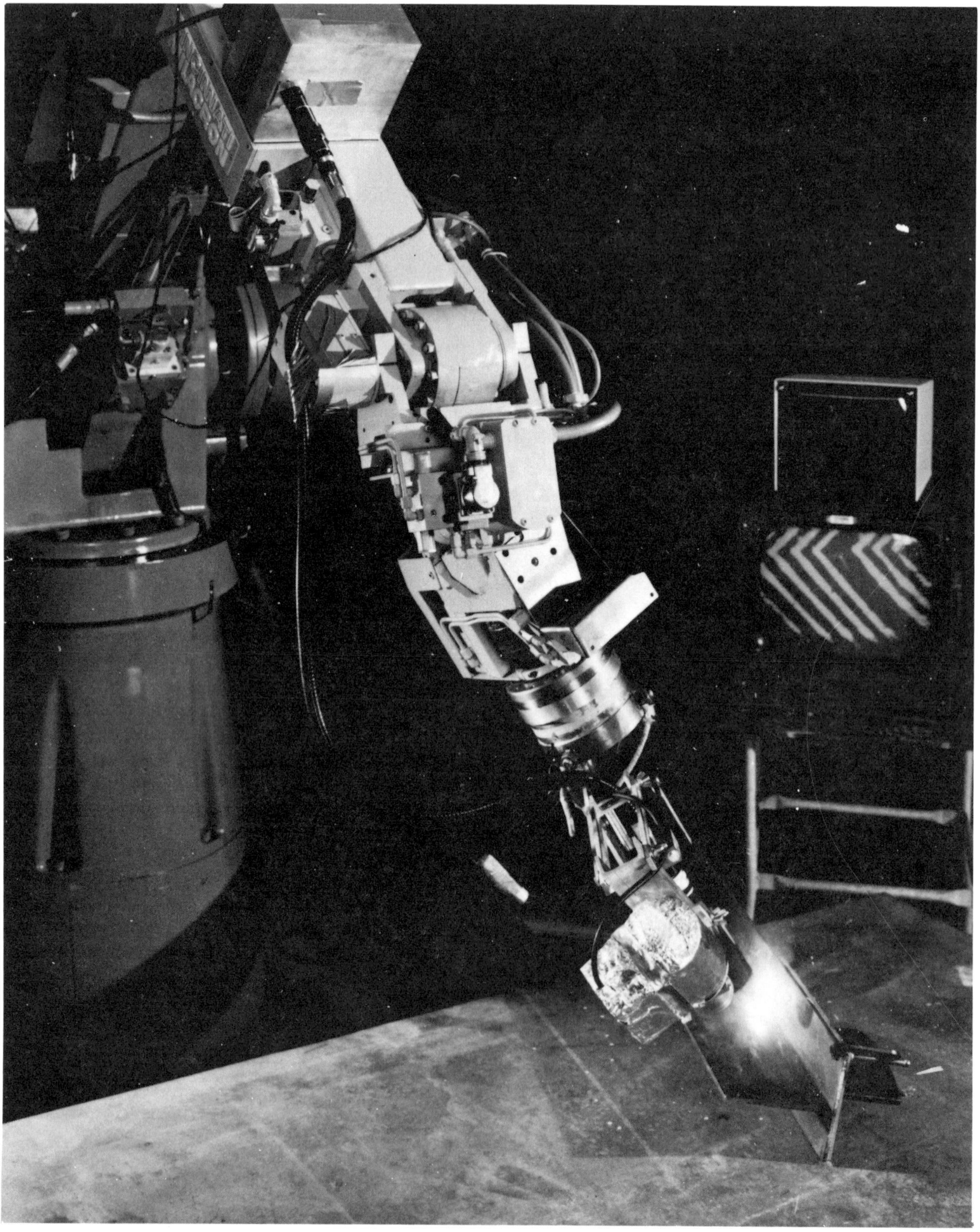

FIGURE 2 EXPERIMENTAL VISUALLY GUIDED ROBOT ARC-WELDING SYSTEM

* The system must be easy to set up and program. It must be possible for a welding engineer or knowledgeable plant worker to reconfigure the system for a new workpiece without prior extensive experience in computer programming or sensor technology.

SRI is developing a real-time machine-vision-based sensing system that we believe can satisfy most, if not all, of these design goals. This work is based on the results of several previous efforts involving the use of machine vision for manipulator servo control [8,9] and for rapid extraction of three-dimensional shape and geometry information from simple binary imagery and structured illumination [10 thru 14]. Our approach to this problem and the results of our work are described below.

II VISUAL GUIDANCE OF AN ARC-WELDING ROBOT

Our developmental real-time, visually guided arc-welding system (shown in Figure 2) is implemented on a Cincinnati-Milacron T3 robot. The image-processing software for this system was originally implemented on a DEC PDP-11/34 computer, and has been recently revised and reimplemented on a VAX-11/750 system. A DEC LSI-11/23 is used as a "smart" interface between the image processing computer and the Cincinnati-Milacron robot controller. During our initial development program, a modified General Electric Model TN2500 CID camera was used as the image acquisition device. This has recently been replaced by a new sensor of our own design.

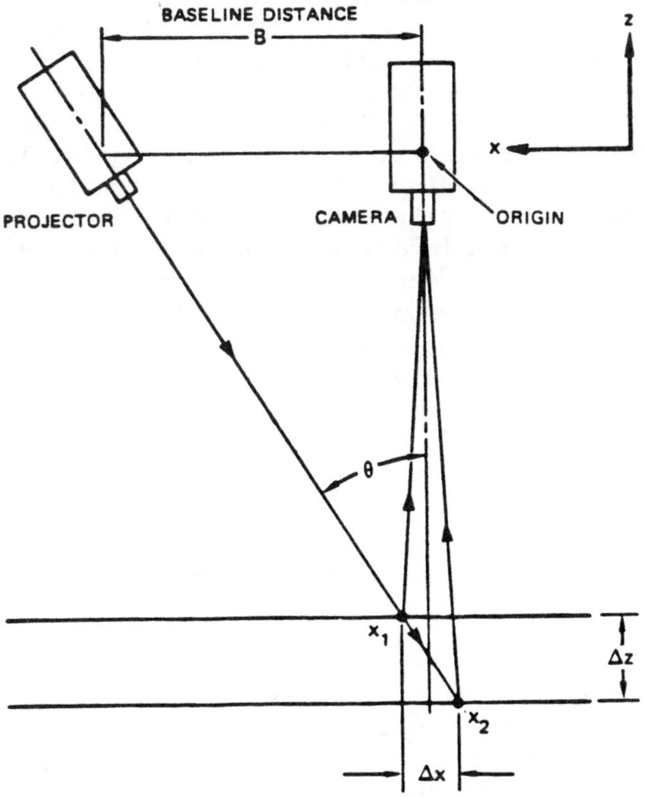

FIGURE 3 DEPTH MEASUREMENT WITH CAMERA AND PROJECTOR

The following sections provide an overview of the system's operation and a more detailed description of some of its major components.

A. Overview

Our basic approach to the use of machine vision for control of arc welding involves the real-time acquisition, of a series of three-dimensional images of the workpieces immediately ahead of the welding arc. This is accomplished by means of an optical range sensor carried on the robot end-effector along with the welding gun. These images are assembled and analyzed to construct a three-dimensional model of the joint area in computer memory. In this model, the workpiece surfaces are represented either as planes or as localized clusters of nonplanar, three-dimensional data points. The latter representation is used in the case of tack welds and other small, unusually shaped volumes.

This model of the joint area is analyzed to extract the location of the joint line, the geometric disposition of the joint surfaces, and the width of the joint gap. The orientation of the joint is determined from the known orientation of the sensor, preprogrammed workpiece data and the observed joint geometry.

The process of locating the weld joint is controlled by procedures and data that are specified by the user during system setup. These data include an approximate description of the weld joint, expressed as a collection of planar surfaces and tolerance limits on joint location and fitup. Utilization of the same planar surface model by both the data analysis and programming subsystems is intended to make the system easy to program--either by a human operator using a conventional graphics device, or, eventually, directly from a CAD/CAM data base. The user must also furnish the system with welding parameter values, including those for voltage, current, torch standoff, and torch attitude, together with descriptions of the welding procedures to be used for various joint geometries and orientations as well as to accommodate different fitup conditions.

The observed geometry and spatial trajectory of the joint line, together with the preprogrammed torch attitude and welding procedure information, are used to compute a group of position and orientation setpoints for the robot end-effector that carries the welding torch and sensor. This information is sent to the robot controller, which performs Cartesian interpolation between adjacent points and generates motion commands to the robot. To ensure proper synchronization of welding-parameter changes and robot motion, welding-parameter information, supplied by the user, is also sent to the robot controller for output to the welding apparatus.

The following sections describe the development and operation of the various components of this system in greater detail.

B. Image Acquisition

1. Basic Imaging Approach

The basic technique employed by our system for acquisition of three-dimensional images is that of optical triangulation, or "structured illumination." This technique, which has been used extensively for problems of this type, involves projecting a beam or sheet of light onto the target from one viewpoint and observing the location and shape of the reflected pattern with a camera situated at another viewpoint. An image of a larger area of the target surface may be obtained by projecting more than one light sheet, or by means of a mechanically scanned light beam. If the perspective transform of the camera (a 4 x 4 matrix describing the camera position, attitude, focal length, and scale factors) and its relative distance from the projector are known, then the distance between the camera and an illuminated point on the target can be computed by using the location of this point in the camera image. The optical geometry of this technique is shown in Figure 3. The mathematics employed, as well as a description of the calibration procedures for this apparatus, may be found in [15].

2. Real-Time Imaging

Obtaining an image of a weld joint in the presence of the welding arc poses three main problems:

* The high-intensity light emitted by the welding arc adversely affects the contrast of the image obtained by the camera.

* Flying particles of luminescent spatter from the welding arc and puddle produce streaks in the TV image. These streaks introduce noise into the image of the projected light pattern and, in some cases, may lead to erroneous interpretation of the image.

* Smoke emitted from the arc can obscure the scene, either because of its density or by scattering light from the arc directly into the camera lens.

We have devised and implemented an imaging technique that overcomes these problems. Our approach is based on three principal elements:

* Placement of the camera so as to view the workpiece a short distance (between 2 and 4 inches) ahead of the welding arc; in addition, equipping the camera lens with a suitable lens shade to prevent direct entry of arc-emitted light.

* Use of a compact, high-intensity light source in the projector.

* Shortening the exposure time of the camera to the smallest possible interval necessary to obtain a reliable image. This has the effect of reducing spark images from streaks to discrete points that may be easily removed by the image processing software. In addition, this technique enhances image contrast by minimizing the effect of spurious illumination from the welding arc.

3. Implementation

For our initial implementation of this sensing technique, we used a small slide projector that cast a structured light pattern onto the workpiece at a distance of approximately 3 inches ahead of the welding arc. The light source used for this projector was a compact, short-arc, xenon flash lamp that was operated at a pulse energy of approximately 100 watt-seconds. Our camera was a General Electric model TN2500 CID device equipped with a rotating mechanical shutter of our own design. Use of this shutter reduced the camera exposure time from 30 ms to 1 ms. Because of its size, the camera-shutter unit was mounted on the robot's forearm and a coherent fiber optic bundle was used to transmit the image from the lens mounted on the end effector. Figure 4 shows this system as it was mounted on the robot.

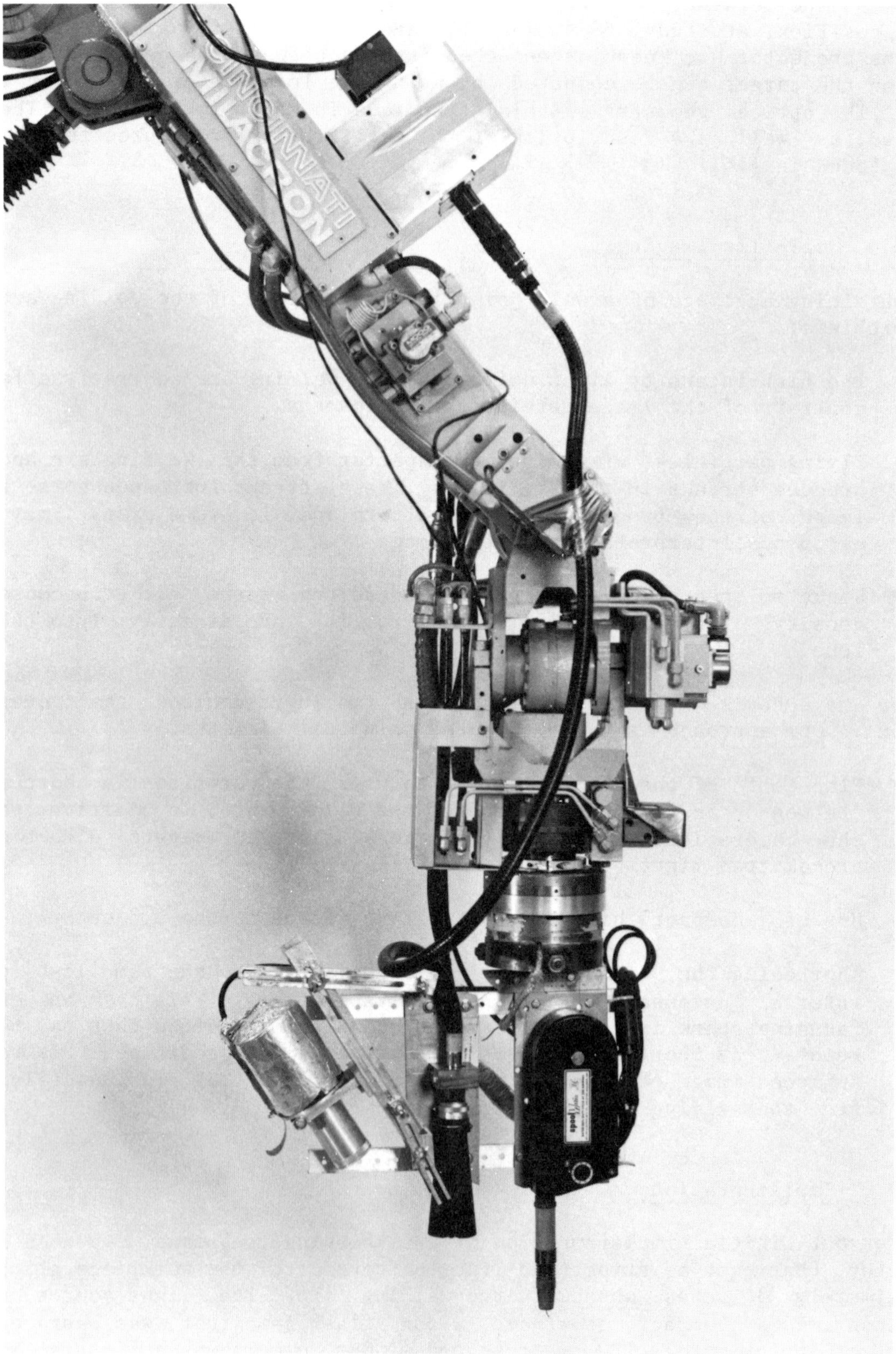

FIGURE 4 GENERAL ARRANGEMENT OF THE VISUAL-GUIDANCE SYSTEM

The light structure consisted of a coded pattern of parallel lines. The area covered was approximately 4 inches square and the average distance between adjacent lines was approximately 0.25 inches. The use of a pattern having a relatively large spatial extent enabled us to analyze a long section of the weld joint from a single picture. It was believed that this technique would simplify the processing of image data obtained from such complex geometries as interior box corners [14]. The use of a coded pattern was necessary to allow us to identify each pattern line unambiguously on any surface in the image. Figure 5 shows a typical real-time image of a fillet joint.

4. Results

This system was successfully used to obtain images to control the robot in welding low-carbon, hot-rolled steel plate. The results obtained with this apparatus confirmed the feasibility of our techniques for real-time imaging, image processing and analysis, and robot control.

A major problem developed, however, in using this approach to acquire images from fillet joints on ground or wire-brushed aluminum or steel surfaces. The images obtained in these cases were found to contain numerous extraneous images of the light pattern. These were caused by multiple reflections of the incident light between the two workpieces. No camera-projector geometry or processing algorithm was found that could reliably eliminate these reflections or produce a correct interpretation of the image.

To overcome this problem we have developed a new sensing system that is based on similar principles, but utilizes a different technique for generating the light pattern and observing its reflection from the workpiece. The new device employs only a single light sheet instead of the multiple bar pattern used previously. This apparatus is also more compact and capable of providing much better spatial resolution (plus or minus 0.008 inches) than could be obtained before. Since we are presently seeking patent coverage for this device, we cannot publish any additional details at this time.

C. Image Analysis

The data points output by the image acquisition system are assembled by the image-processing software into a three-dimensional model of the joint area. In this model, each workpiece surface is represented as one or more flat planes. Curved surfaces appear as piecewise polyhedral approximations.

Each of these planes is represented by an equation of the form

$$Ax + By + Cz + D = 0 \quad , \qquad (1)$$

where the coefficients A, B, and C define a unit vector that is normal to the plane (hence specifying its orientation in space) and D is the distance from the plane to the coordinate origin.

The following sections describe the procedures for constructing this model and for using it in joint line location and in measurement of fitup and geometry. These procedures are currently implemented on a DEC VAX-11/750 computer in the C language.

1. Model Construction

The objective of the model constructor is to determine the equations of the planes "seen" by the imaging system as it is moved along the weld joint. This procedure operates as follows:

(1) Three-dimensional (x-y-z) data from a group of succeeding images are collected into a "bundle." The size of this bundle is determined by the accuracy and speed requirements of the particular welding application. Figure 6 shows a typical data bundle for a straight fillet joint.

(2) A "seed plane" is fitted to this bundle by arbitrarily selecting three data points arranged in a triangular pattern. The surface normal for this seed plane is computed by using two sides of this triangle as vectors and calculating their cross product. This yields coefficients A, B, and C in Equation (1) above. Figure 7 shows a such a seed plane for the data of Figure 6.

(3) The remaining points in the triangle formed by the three initial points are tested to see whether they lie on this seed plane. The test is conducted by verifying that each point satisfies Equation (1) within a specified tolerance.

(4) Failure of this test for a sufficient number of points indicates that the three corners of the initial seed plane do not lie on the same surface. If such is the case, the original seed plane is subdivided into 4 smaller planes and Steps 2 and 3 are repeated. This partitioning process continues recursively until the triangular seed planes become too small or until the test specified in Step 3 is satisfied.

(5) When a suitable seed plane is found, it is extended outward by testing all the surrounding data points until the edge of the bundle is reached, or until a group of points is found that no longer satisfies the equation for the discovered plane (i.e., the edge of the surface is reached). Note that this expansion process may subsume adjacent seed planes that were created in previous subdivision steps. Figure 8 shows the progress of this expansion step. The search for coplanar points is greatly facilitated by the fact that the imaging system "scans" in a direction roughly perpendicular to the joint line.

(6) This process continues incrementally as sufficient data are available from the scanner to form bundles of the required size. Figure 9 shows the completed fillet joint model.

The output of this procedure is a set of lists that includes all of the extracted planar surfaces discovered, ordered by their top-to-bottom and left-to-right appearance in the bundle, (i.e. plane lists) explicit lists of the data points that form the boundary of each surface (i.e., edge point lists), and a list of regions containing data points that do not fit any planar surface. These "unexplained" regions typically indicate irregularities, such as tack welds.

2. Joint Line Location

The true location of the joint line is determined by comparing the model created from the real-time image data with a joint prototype created by the user. This joint prototype defines the weld joint by specifying the geometric and positional relationships among the various surfaces and edges that comprise the joint and its surroundings. The joint prototype also contains tolerance information for each of the specified values. For example, as currently implemented, a simple fillet joint is defined as a pair of surfaces whose normals point towards each other and intersect at an angle of 90 degrees, plus or minus 5 degrees. A lap joint is defined as a set of parallel surfaces separated by a specified distance (i.e., the plate thickness plus an allowance for warpage and imperfect fitup). Additional conditions might be specified in the joint prototype to accommodate the possible appearance of tack welds or the size and shape of any edge preparation or root gap.

The model-prototype matching procedure analyzes each bundle of data after it has been processed by the model constructor. This procedure operates by searching the plane lists generated by the model constructor, looking for a set of surfaces that satisfies the conditions specified by the model. This search is aided by the use of prior knowledge of the joint line location, obtained either from previous pictures or from information specifing the vantage point at startup. The output of this procedure is a simple data structure specifying the planes that comprise the joint, together with pointers to the appropriate edge point lists.

Creation of the joint prototype presently involves writing a subroutine that embodies the requisite searching and testing procedures. We are now investigating methods for automating the production of these prototypes by means of a graphics system or a CAD/CAM data base.

D. Guiding the Robot During Welding

Derivation of guidance and control information for the robot and welding apparatus involves combining the data furnished by the joint locator, described in the previous section, with welding procedure information obtained from a preprogrammed welding-procedure data base. In general, this guidance procedure operates as follows:

(1) The surface normal parameters supplied by the model constructor and joint locator, together with geometric data from the joint prototype, are used to define a weld joint coordinate system. This system has its origin fixed on the joint line and may be considered to travel along this line as the weld is made. The purpose of this system is to provide a consistent, standardized coordinate frame for specification of torch attitude and motion parameters. The definition of this coordinate system is part of the joint prototype and specifies the vector relationships among the axes of the joint system, the joint surface normals, and, if necessary, the joint line trajectory. Figure 10 illustrates a typical system for a fillet joint. In this figure, vectors xj, yj, and zj represent the axes of the joint coordinate system, T is the torch vector, d is the standoff distance, c is a cross-joint offset, and w and t are the work and travel angles, respectively.

(2) The edge point lists output by the joint locator are used to compute the joint trajectory and fitup. This process may involve fitting a simple curve (e.g., straight line or circle) to these data for smoothing and interpolation. The fitup measurements are analyzed, in consultation with the welding procedure data base, to determine the procedure (e.g., "weaving" or "stringering") needed to achieve proper joint fill and bead shape. In the future, these measurements will also be used to control welding parameters such as voltage and current.

(3) The welding procedure data base is consulted to determine the correct torch attitude angles with respect to the joint coordinate system. These angles are then transformed from joint coordinates into the robot orientation angles needed to position and orient the torch properly at each point on the joint line.

(4) Cartesian and robot joint angle interpolations are performed between adjacent points to assure smooth robot motion. The resulting data points and welding control commands are then output to the robot.

The procedure described above operates on each data bundle as it is output by the joint locator. After all joint points in a bundle have been processed, control is returned to the model constructor so that it can continue assembling and analyzing image data.

III CONCLUSIONS

The system described in this paper is exhibiting increasingly better performance as we develop improved image acquisition hardware and faster, more powerful, image-processing algorithms. Our original system, which used a DEC PDP-11/34 computer for image processing, was capable of welding at a maximum speed of approximately 15 inches per minute, with an average spatial resolution of 0.125 inches on simple weld joints. In our current implementation, the 11/34 has been replaced by a DEC VAX-11/750, and the original flash-lamp-based image acquisition system by an improved device of our own design. This new system is capable of welding more complicated joints at speeds of up to 30 inches per minute with a spatial resolution of 0.012 inches. We are currently investigating a "pipeline processing" technique to replace the sequential technique described above. Implementation of this approach is expected to further enhance the system's performance.

While our robotic arc-welding system is somewhat more complex than many of the commercial systems now available [16,17], we believe that this or a comparable approach is necessary if the system is to be truly flexible and easily programmed to perform a variety of welding tasks.

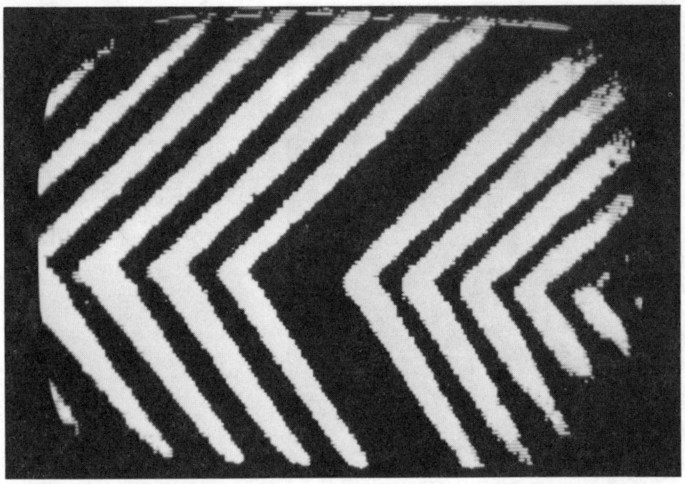

FIGURE 5 REAL–TIME FILLET JOINT IMAGE

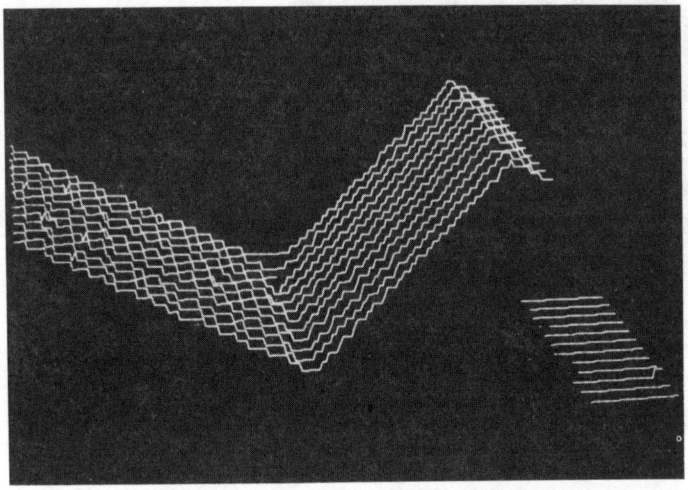

FIGURE 6 RAW DATA FOR FILLET JOINT, PERSPECTIVE VIEW

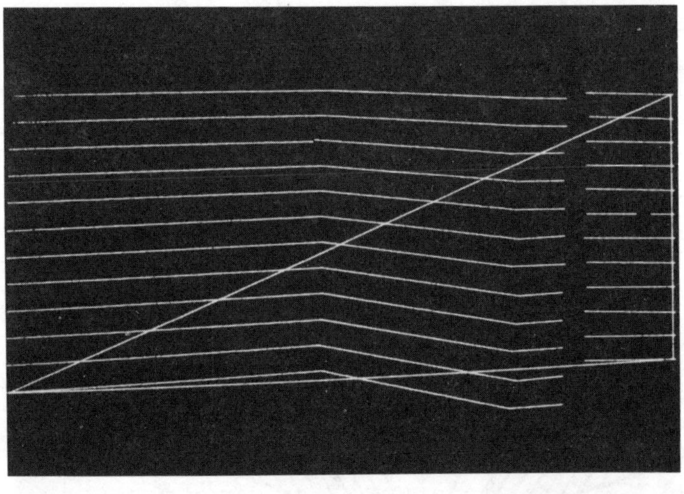

FIGURE 7 SEED PLANES FOR FILLET JOINT

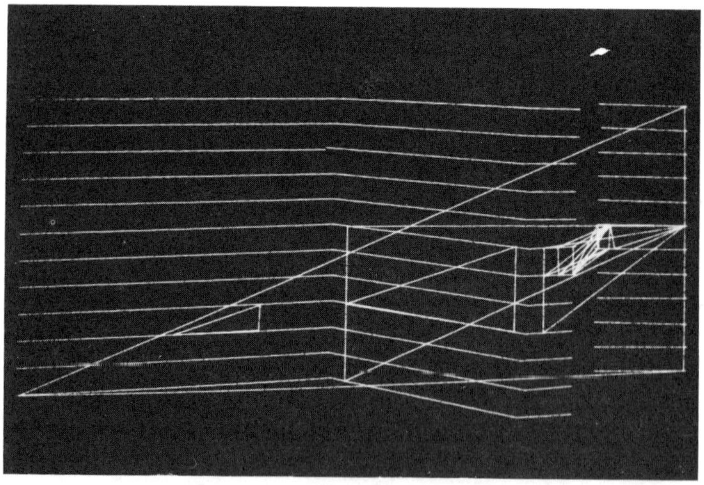

FIGURE 8 SEED PLANE EXPANSION FOR FILLET JOINT

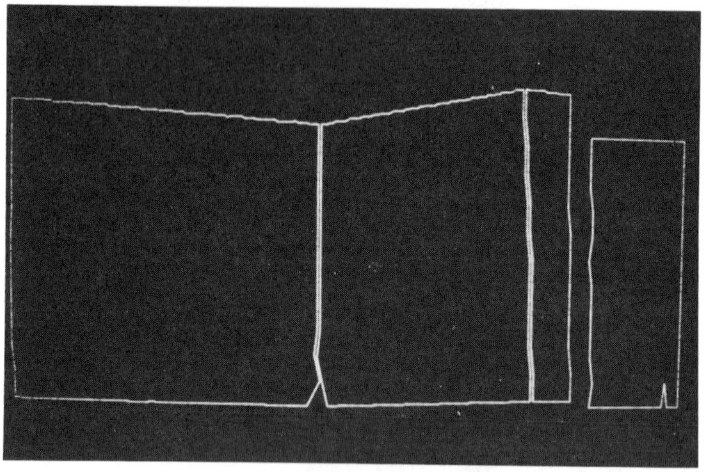

FIGURE 9 PLANAR MODEL FOR FILLET JOINT

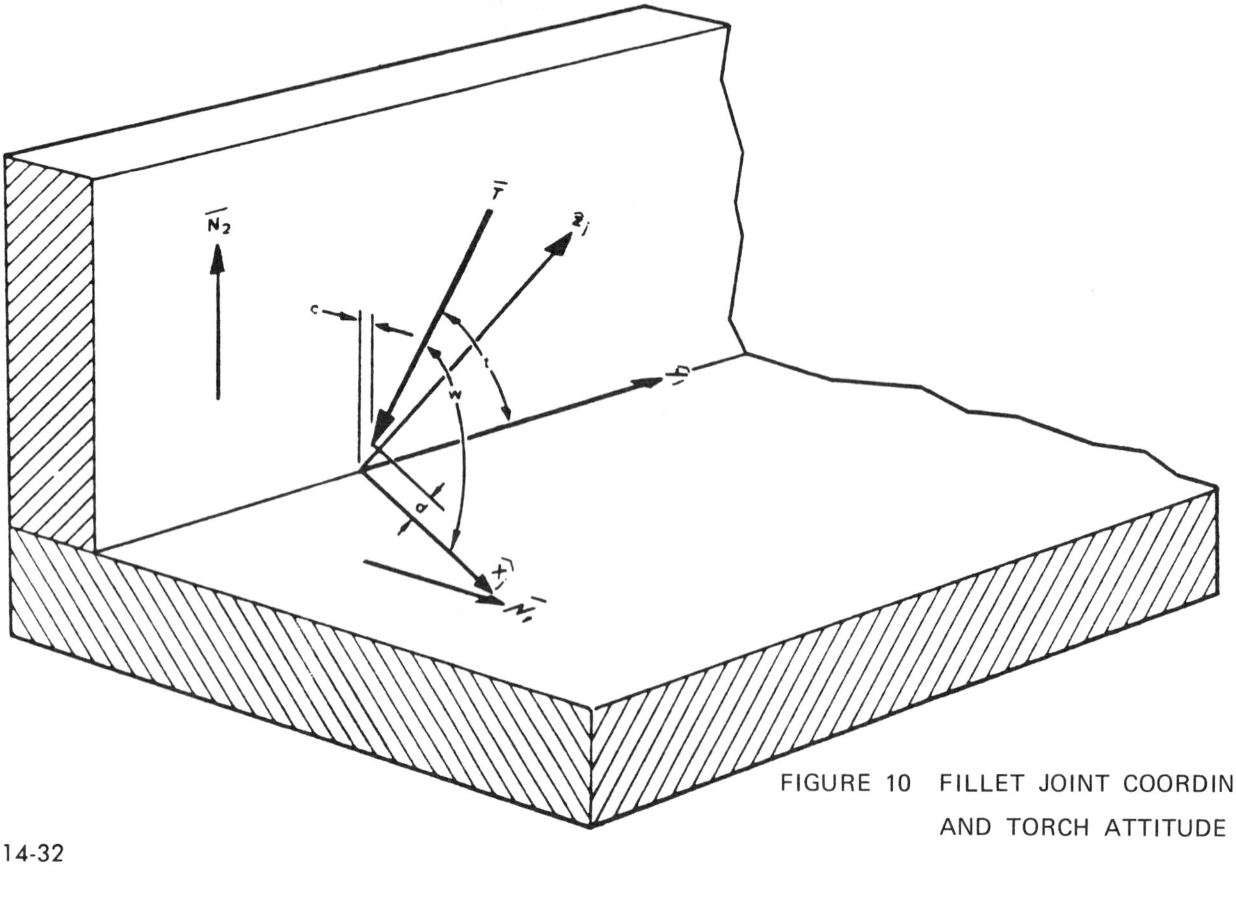

FIGURE 10 FILLET JOINT COORDINATES AND TORCH ATTITUDE ANGLES

REFERENCES

1. R. W. Cain, "Weld Quality Meter for Adaptive Control," Proc. IEEE Southwest Conf., pp. 1-6, Louisville, Kentucky (May 1973).

2. Welding Handbook, 7th. ed., Vol. 1, (American Welding Society, Miami, Florida, 1976).

3. C. F. Tseng and W. F. Savage, "The Effect of Arc Oscillation," The Welding Journal, Vol. 50, pp. 777-786 (November 1971).

4. A. P. Bennett and C. J. Smith, "Improving the Consistency of Weld Penetration by Feedback Controls," Proc., Fabrication and Reliability of Welded Process Plant, London, England (1976).

5. G. E. Cook, "Feedback Control of Process Variables in Arc Welding," Proc. JACC, Paper FA7-B, San Francisco, California (August 1980).

6. H. B. Cary, Modern Welding Technology, (Prentice-Hall, Inc., Englewood Cliffs, New Jersey, 1979).

7. The Merrick Corp., promotional literature, (1981).

8. G. J. Agin, "Servoing With Visual Feedback," Technical Note 149, Artificial Intelligence Center, SRI International, Menlo Park, California, (July 1977).

9. G. J. Vanderbrug, J. S. Albus, and E. Barkmeyer, "A Vision System for Real Time Control of Robots," Proc. 9th Int. Symp. on Industrial Robots, Washington, D.C. (March 1979).

10. G. J. Agin, "Representation and Description of Curved Objects," Stanford AIM Memo 173, Stanford University, Stanford, California, (October 1972).

11. R. J. Poppelstone et al., "Forming Models of Plane- and Cylinder- Faceted Bodies From Light Stripes," 4th International Joint Conference on Artificial Intelligence, pp. 664-668 (1975).

12. P. M. Will and K. S. Pennington, "Grid Coding: A Preprocessing Technique for Robot and Machine Vision," 2nd International Joint Conference on Artificial Intelligence, pp. 66-70 (1971).

13. Y. Shirai, "Recognition of Polyhedrons With a Range Finder," Bulletin of the Electrotechnical Laboratory, 35(3), pp. 96-102 (1971).

14. A. Yoshiaki et al., "Automatic Control of Arc Welding (Report VI), "Recognition for Intersection of Two or Three Planes," Trans. JWRI, Vol. 6, No. 1 (1977).

15. R. C. Bolles, J. H. Kremers, and R. A. Cain, "A Simple Sensor to Gather Three-Dimensional Data." Technical Note 249, SRI International, Menlo Park, California (July 1981).

16. P. Villers, "One Vision System Handles Parts Assembly and Weld Seam Tracking", Sensor Review, Vol. 2, No. 3 (July 1982).

17. Robotic Vision Systems, Inc., promotional literature (1982).

Piezoelectric Polymers: New Sensor Materials for Robotic Applications

P. Dario
Centro "E. Piaggio"
Pisa, Italy

C. Domenici
Centro "E. Piaggio"

R. Bardelli
Centro "E. Piaggio"

D. De Rossi
Centro "E. Piaggio"

P.C. Pinotti
Centro "E. Piaggio"

Abstract

Recently discovered piezoelectric polymers are considered potential materials for robotic sensors and actuators. The preparation process of polyvinylidene fluoride (PVF_2), the most known and active among piezoelectric polymers, is presented, and considerations on the design of polymer devices are given. Ruggedness, light weight, easy conformability to complex surfaces, potential low cost, pyroelectric sensitivity, and large bandwidth are some of the properties of PVF_2 film. Several piezo and pyroelectric polymer contact, touch, tactile and proximity sensors for robotics, as well as special actuators suitable for micromanipulation, have been developed and are described in this paper.

1. INTRODUCTION

Piezoelectricity, the capability of some crystalline materials to change their dimensions when subjected to an electric field or, conversely, to produce an electric signal when mechanically deformed, is a well known property of polar natural crystals,such as quartz and Rochelle salt,and of polycrystalline ceramics, such as barium titanate.

Piezoelectric activity has been also recognized in natural and artificial polymers, but only after the discovery of a very strong piezoelectric effect in polyvinylidene fluoride (PVF_2 or PVDF), reported by Kaway (1) in 1969, this matter received considerable attention.

Since then, piezoelectric properties of polymers have represented an attractive field of investigation for polymer physicists interested in the comprehension of the basic mechanism of this effect, as well as for engineers, who tried to take advantage of the inherent characteristics of this class of materials for several applications in different fields. In fact these materials, which couple a strong piezoelectric activity (comparable with that of quartz) with availability in large areas and in thin film (as thin as 6 microns) and thick sheet (1-2 mm) forms, ruggedness, light weight, easy workability and potential low cost, represent a valuable choice for the design of several piezoelectric devices.

The inherent broadband operation of piezoelectric polymers allows to develop mechanical transducers with sensing and actuating capabilities ranging from quasi static to ultrasonic applications. Owing to the particular symmetry of their crystalline structure, many of the piezoelectric polymers (and PVF_2 among them) also possess a relevant pyroelectric activity; therefore, a change in temperature of these materials originates an electric signal. This effect, which is concomitant with the piezoelectric effect, can be sometimes regarded as a drawback causing undesired artifacts when mechanical signals must be detected. Nevertheless, appropriate design can largely reduce the influence of this effect, when undesired, while pyroelectric properties can be usefully exploited in the construction of several pyroelectric devices.

Practical application of piezoelectric polymers in commercial devices has not obtained yet the widespread diffusion that might be expected considering their attractive features. Audio frequency transducers (microphones, headphones, loudspeakers), hydrophones, ultrasonic transducers (for non destructive test or medical imaging, arrays), actuators (display devices, fans, light shutters), pyroelectric sensors (infrared detectors, intrusion detectors, vidicons) and biomedical sensors and actuators represent the areas on which substantial research effort has been centered and where a few devices have already reached maturity for commercial diffusion. Extensive reviews of applications of piezoelectric polymers have been presented by, among the others, Broadhurst et al. (2), Sessler (3) and De Rossi et al. (4).

Uncertanty in the commercial availability of piezoelectric PVF_2, inadequate knowledge of the stability characteristics of the piezoelectric properties of polymer materials and difficulties in defining proper technological processes to use such materials in the construction of practical devices (a problem frequently encountered when dealing with new materials) are the major reasons which prevented so far a wider diffusion of piezo and pyroelectric polymer devices. However, most of the aforesaid problems have probably been overcome.

Our laboratory has been studying preparation and application of piezoelectric polymers for a few years.

Based on our experience, we think that several properties of piezoelectric polymers are potentially useful for the construction of sensors and actuators for robotics.

In this paper, after a brief description of the preparation and characterization methods of piezoelectric polymers, the applications of these materials to the field of robotics are discussed, with particular emphasis to the ones which have been developed or are presently studied in our laboratory.

2. PREPARATION, CHARACTERIZATION AND USE OF PIEZOELECTRIC POLYMERS

Piezoelectric properties have been discovered and investigated in numerous natural (wax, wood, bone, cellulose, silk, collagen, DNA, RNA, etc.) and synthetic (polyvinylchloride, polyacrylonitrile, polyvinylfluoride, polyvinylidenefluoride, polymethylmethacrylate, nylon 11, etc.) polymers (5).

However, the piezoelectric activity of PVF_2 is much higher than that of the other polymers and therefore this material is the preferred choice for the development of piezo and pyroelectric polymer devices. In this paper we shall focus our attention onto the properties and applications of PVF_2, but also features and advantages of other polymeric materials will be discussed, when appropriate.

PVF_2 is a semicrystalline polymer which can be obtained, in the form of powder, pellets, film, sheets and tubes, from different commercial sources.

The crystalline phase of commercial PVF_2 is usually arranged to yield a centrosymmetric unit cell, thus this raw material is not piezoelectric. The usual procedure adopted to transform the non active crystalline phase, known as α, in the noncentrosymmetric β phase, which can be rendered piezo and pyroelectric, is to stretch uniaxially or biaxially the commercial film.

Uniaxial stretching of PVF_2 film is performed in our laboratory in a purposely designed apparatus, under controlled conditions of temperature, draw ratio and velocity (6). After an annealing treatment which greatly improves the dimensional stability of the stretched film, the desired pattern of electrodes is deposited on the sample, usually by conventional vacuum metallization techniques. Aluminum, nickel-chromium and chromium-gold are commonly used as electrode materials.

The poling process, consisting in applying a strong electric field (usually higher than 500 KV/cm) between the electrodes, provides a permanent orientation of the molecular dipoles within the film.

Electric field, temperature and time are important poling parameters, which contribute to determine the final piezoelectric properties of the PVF_2 sample and their stability (7)(8).

The described preparation process, synthetized in Fig.1, is only an example of the possible ones and actually several different techniques can be adopted to transform the α phase PVF_2 into the β phase PVF_2 (for instance by casting from different solutions (9)), to stretch the polymer film (for instance up to draw ratios of 7:1, to enhance mechanical and electric properties (10)) or to pole the sample at high field (for example by corona discharge (11)).

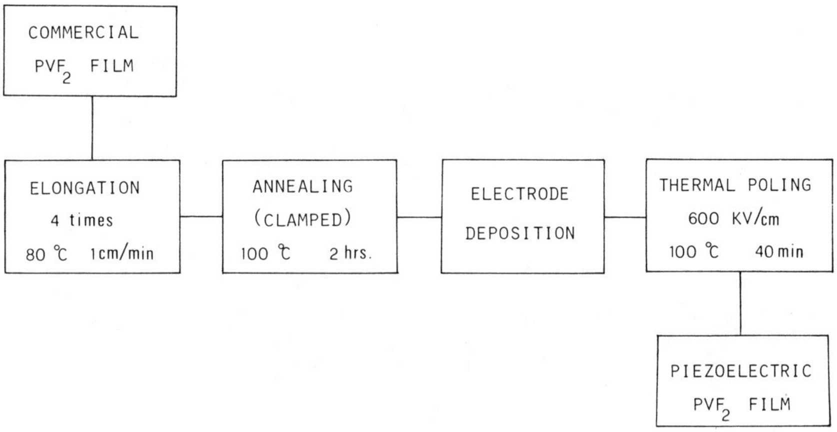

Figure 1 Standard preparation process of piezoelectric PVF_2 film

The preparation process varies from polymer to polymer, depending on its structural morphology: for example such polymers as PVF and PVC do not need to be oriented. Appropriate techniques are also adopted to render piezoelectric other polymeric materials, such as copolymers, blends or composites between polymers and ceramics, which have been developed to obtain particular mechanical properties or to try to reduce the drawbacks or overcome the limitations of presently available piezoelectric polymers.

The measurement of different piezo and pyroelectric constants and of mechanical and dielectric properties not only validate various preparation processes, but it also represents an essential information for the designer of useful devices.

In our laboratory we devoted particular attention to set up appropriate procedures and instrumentation to perform accurate measurements of different piezoelectric parameters.

These parameters, together with other provided by commercial suppliers or published in the literature for polymers and other piezo and pyroelectric materials, are reported in Table 1.

As already mentioned, piezoelectric polymers are featured by several properties which are in principle attractive for the construction of sensors and, with major limitations, actuators useful for robotic applications.

TYPICAL CONSTANTS OF SOME PIEZOELECTRIC AND PYROELECTRIC MATERIALS

Material	Density $10^3 Kg/m^3$	Young's modulus $10^9 N/m^2$	Relative dielectric constant	Acoustical impedance 10^6 Rayl	Piezoelectric constants 10^{-12} C/N	Pyroelectric coefficient $10^{-5} C/m^2 \cdot °K$
PVF_2	1.78	2.5	12	3.5	$d_{31} = 21$ $d_{32} = 2.3$ $d_{33} = -31$	- 4
PVC	1.4	2.7	5.3		$d_{31} = 1.5$	0.1
Copolymer PVF_2 + PTFE			2		$d_h = 12.1$	2.7
PZT + Polymer	2.8	2.5	150		$d_{33} \approx 200$	7
Quartz	2.65	80	4.5	15.3	$d_{11} = 2.3$	0
PZT-5H	7.5	50	3400	34.2	$d_{31} = -274$ $d_{33} = 593$ $d_h = 45$	≃20
TGS	1.7		30		d = 50	40

TABLE 1

Piezoelectric polymers are sensitive (PVF_2 is more sensitive than quartz and even more than ceramics, at least as a transformer of mechanical into electric energy) and can be so thin to be easily conformed to complex surfaces; they can be produced in a continuous roll process and in large areas, as it is not possible either with quartz or ceramics.

Patternsof electrodes can be defined on polymer films by depositing different metals through masks which can be produced by conventional or chemical machining, as well as by etching, with appropriate photolitographic techniques, the continuous metallization previously deposited on the polymer film.

The film itself can be cut with different tools, ranging from a pair of scissors to sophisticated laser techniques.

Some techniques which are typical of polymer technology, such as injection moulding and thermoforming, can be also used to obtain piezoelectric polymer transducers in a fast and cheap way (12).

Working modes of piezoelectric materials are established by the configuration of the matrix of their piezoelectric coefficients (d_{ij}, g_{ij}), which determine the link between mechanical input and electric output (or viceversa) for the transducer.

For PVF_2, the piezoelectric coefficients d_{31}, d_{32} and d_{33} govern membrane mode (d_{31}, d_{32}) and thickness mode (d_{33}) operations. The capability of a thin PVF_2 film to behave as a membrane sensor, a peculiar feature of piezoelectric polymers, greatly enhances its sensitivity and gives the user a large number of design options.

In fact, a thin polymer film is not stiff enough, usually, to mantain a predetermined shape, but it must be supported. This means that the performance of the sensor is determined mainly by the characteristics of the supporting structure and that higher sensitivity to a stress rather than to another or suitable values of resonant frequency (just to mention some) can be obtained by appropriate design.

As it is well known, piezoelectric materials do not allow static measurements because their internal resistance, though very high, is not infinite and the electric charge produced by a mechanical stress is lost with a time constant that is determined essentially by the values of both the resistivity and the dielectric constant of the polymer, as well as, when the transducer is connected to a measuring instrument, by the input impedance of such instrument.

Usually, the inherent time constant of a PVF_2 sensor is about 100 s and low frequency cut off can be kept as low as 0.1 Hz by using commercially available high input impedance amplifiers. The measurement of quasi static signals is therefore possible with piezoelectric polymer sensors.

Thickness mode, resonant frequency of a piezoelectric polymer sensor is determined by film thickness and by sound velocity within the polymer. The resonant frequency on thin film PVF_2 sensors can be as high as several hundreds MHz, and in fact ultrasonic applications of piezoelectric polymers are intensively studied (also in our laboratory) and are believed to possess exciting perspectives.

Of course, lower frequency resonances are easily observed when the polymer is used in some configurations. However, broadband operation still features piezoelectric polymer transducers.

Excellent linearity and very low hysteresis are positive characteristics of PVF_2, making this material superior, at this regard, to presently available conductive elastomers proposed and used in several touch sensors for robots.

Temperature dependence and aging characteristics of piezo and pyroelectric properties have been studied in polymers, as an important issue to decide practical usability of this class of materials. Data provided by commercial suppliers and results published in a recent paper (13) demonstrate that the piezoelectric properties of PVF_2, after an initial decay which is proportional to the working temperature, remain stable for years. There is no substantial decrease of activity up to a working temperature of about 70°C; moreover remarkably high activity (about 50% of the initial value at a temperature of 110°C) is mantained at higher temperatures. This feature, together with the ruggedness and inertness to chemical agents of PVF_2 (which is widely used for the construction of pumps and pipes for chemical industry, owing to its solvent resistance), renders this material particularly suitable for use in hostile environments such as the ones where robots often operate.

Finally, it is appropriate to spend a few words on the issue of the commercial availability of piezoelectric polymer materials.

Kureha, which has represented for a long time the only commercial source of piezoelectric PVF_2, suspended providing this material to foreign customers. After a few years period of extremely difficult availability, piezoelectric PVF_2 is now sold in a few different forms by Pennwalt (King of Prussia, Pa, USA), by Thorn EMI (Hayes, Middlesex, UK) and by Solvay & Cie (Bruxelles, Belgium).

Because of present small scale production, the price of piezoelectric PVF_2 is still much higher than it could be in principle.

In our laboratory we have developed the technology to prepare, characterize and use piezoelectric polymers, starting from raw material. This facility enables us to alter properly some of the original properties of the material and to produce also "special" piezoelectric materials tailored for specific applications.

3. ROBOTIC APPLICATIONS OF PIEZO AND PYROELECTRIC POLYMERS

It is now widely recognized that the most significant characteristic of the next generation of robots will be the availability of advanced sensory capabilities. These future robots are expected to reduce the gap between the present sophysticated control capability, allowed by the outstanding progress of microelectronic technology, and the sometime primitive level of the state of the art in current robotic transducer technology.

Piezo and pyroelectric polymers can be used to develop several devices able to enhance, in different ways, the ability of present robots to interact with the outer world.

Although some robotic applications of PVF_2 are being presently considered in a few laboratories in different countries, very little has been published on this specific subject.

Without any ambition of being exhaustive, in this paper we intend to review the few proposed applications of piezo and pyroelectric polymers to the field of robotics, to discuss the ones presently under development in our laboratory and to present further possible robotic applications that appear particularly promising.

The matter is classified, with reference to areas of application, as follows:

A. Contact sensors. This is probably the less sophysticated approach to the complex issue of tactile sensing. However, piezoelectric polymers can offer simple and effective solutions to some problems of object detection.

An example of a PVF_2 contact sensor has been described by Toda (14). A PVF_2 bimorph, has been constructed by bonding together two 9 μm thin PVF_2 films, as illustrated in Fig. 2. In the proposed configuration one film is driven by an ac voltage while the other, connected to a suitable amplifier, detects the consequent movement of the whole structure. The amplitude and the phase of the output signal are alterated when the bimorph touches an object.

According to Toda, a PVF_2 "soft-touch" contact sensor, 7.5 mm long and 6 mm wide, can detect the position of nearby objects with an accuracy of about 0.01 mm over a range of a few millimeters.

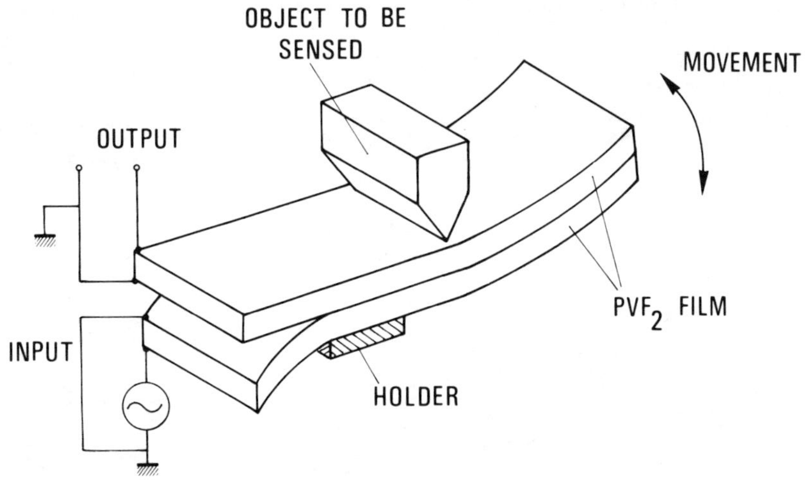

Figure 2 PVF_2 bimorph "soft touch" contact sensor

The stiffness of the described transducer is very low , so that this contact sensing device should be used only in particular conditions, where air flow or inertial effects are negligible, to avoid possible spurious signals. Manipulation of delicate objects in clean environments, in which high sensitivity and "soft" touch are recommended, represents the ideal field of use of a PVF_2 contact sensor.

B. Touch sensors. We use this term to differentiate a class of transducers that can provide "touch sensing over an area within which there is spatial resolution" from the transducers which, following the definition given by Harmon (15) for tactile sensing, are able to provide "continuously variable touch sensing".

An example of a touch sensing device using piezoelectric polymers is represented by a platform for recognizing the position and orientation of objects which has been developed in our laboratory.

The proposed apparatus is intended for use in assembly operation, where randomly oriented objects have to be picked up by a robot for successive operations.

In the present configuration, schematically depicted in Fig. 3, a pattern of 256 circular piezoelectric sensors has been defined on a single PVF_2 sheet by means of a rather unconventional technique which allows a significant reduction of constructive complexity.

Details of the constructive features of this touch sensing platform have been given in a recent paper (16).

The object to be recognized must be pressed against the sensing platform (or, in a different working configuration, the platform must be pressed against the object).

The shape of the object is reconstructed by scanning electronically the 256 PVF_2 sensors and identifying by 1 the pressed sensors and by 0 the non-pressed sensors.

A suitable algorithm has been implemented on the 4K ROM memory of a μP control unit to calculate and display the center of area (which can be identified with the center of mass for planar, homogeneous objects) and the orientation of a principal axis of inertia.

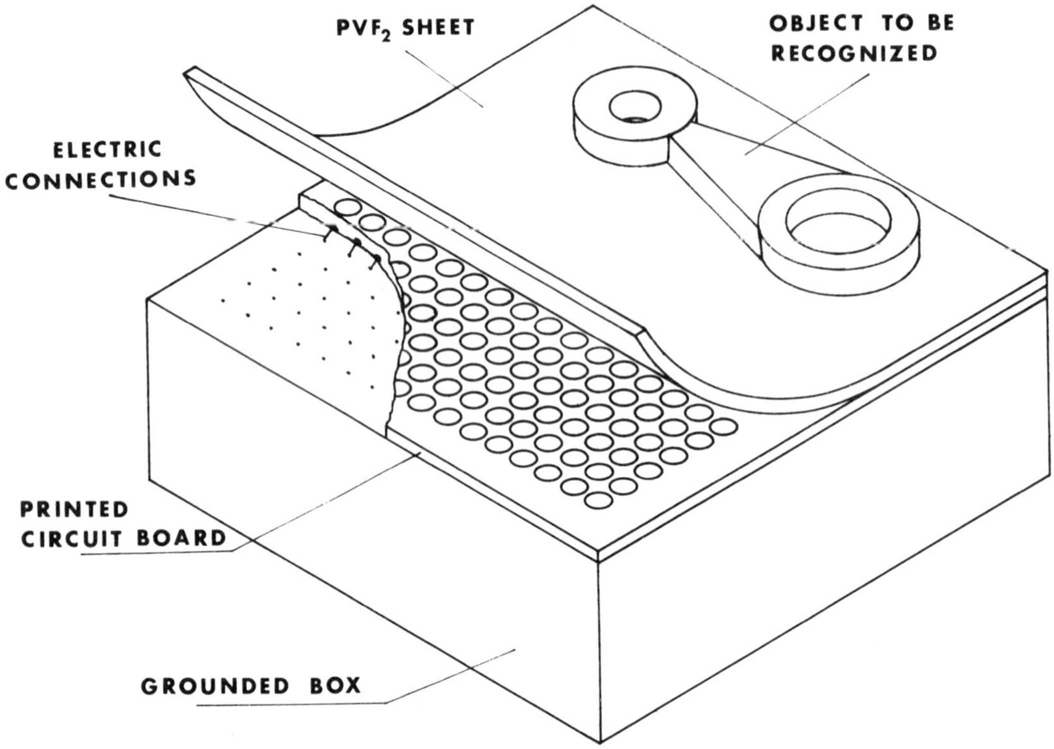

Figure 3 Structure of a PVF_2 matrix sensing platform. The multiplexing and preamplifying unit is contained inside the grounded box

The whole reading and calculating process can take no longer than 0.1 s, while errors in the calculation of the center of mass and of the orientation angle are of the order of 1 mm and 3°, respectively.

The pressure P that should be exerted on each sensor to produce a detectable threshold voltage V is given by the following equation:

$$P = -\frac{V\,\varepsilon_{33}\,\varepsilon_0}{d_{33}t} = -\frac{V}{g_{33}\,t}$$

Where: d_{33} = piezoelectric strain constant

$g_{33}=d_{33}/\varepsilon_{33}\cdot\varepsilon_0$ = piezoelectric stress constant

ε_{33} = relative dielectric constant

ε_0 = dielectric constant of the vacuum

t = thickness of the film

For the PVF_2 sheet used for the preparation of the matrix sensor (final thickness t = 100 μm, ε_{33}= 12, ε_0= $8.85 \cdot 10^{-12}$ F/m, d_{33}=-30pC/N, g_{33}= =0.188 V·m/N) a pressure of about $3.5 \cdot 10^2$ N/m^2 is needed to produce a voltage V = 10 mV which can be easily distinguished from the electric noise produced during the scanning of the matrix.

The input impedance of the adopted voltage amplifiers is finite (usually about $10^{13}\Omega$), therefore the open circuit condition assumed to calculate the previous equation is no longer valid and a higher pressure is actually necessary to produce the requested threshold voltage.

Experiments performed on the proposed device confirmed its merits of ruggedness, sensitivity, potential low cost which render the matrix platform suitable to work in an industrial environment. At this regard, the possible, easy and fast substitution of the PVF_2 matrix can be regarded as a valuable advantage.

Problems in the reliability of the electronic unit, partly due to the inherently difficult handling of the signals produced by high impedance sources, and practical complications, introduced by the necessity of pressing the object against the sensing matrix resulted to be the main drawbacks of this system.

Work is in progress both to improve the electronic unit (by means of accurate shielding and, more significantly, by adopting a single amplifier for each sensor; this latter modification would also allow the continuous reading of the output of each sensor, possibly moving the proposed device into the category of tactile sensors) and to eliminate the need for pressing the object to recognize, with the final goal of rendering this touch sensing platform competitive with existing visual recognition systems.

C. Tactile sensors. Tactile sensitivity is rather diffusely recognized as an urgent need for robots of the next generation, concomitantly with vision (17). The convincement that tactile sensors should be skin-like is at least as much diffused.

According to Harmon (18), a skin-like glove sensor would be perhaps the most desirable solution for robot tactile sensing.

Notwithstanding several, often attractive solutions to this problem have been proposed by numerous authors, we believe that piezoelectric polymers can be an answer to the need of a skin-like sensor for robotics and prostheses. Preliminary considerations and results on the research activity presently performed in this field in our laboratory, are presented in a paper which also appears in these Proceedings (19).

Here we shall only mention that a flexible film of piezo and pyroelectric PVF_2, with properly distributed sensitive areas and conformed to the different parts of a robot can actually reproduce the behaviour of the human skin, whose piezo and pyroelectricity have been already reported (20)(21).

Pressure sensitivity, a thermal sensitivity which even allows to distinguish different materials by touch, and compliance (achieved either by covering the sensing polymer surface with a suitable rubber or by applying the sensing film over a rubber compliant substrate) are properties obtainable with PVF_2 film sensors. The addition of a different class of piezoelectric polymers, such as poly(γ-benzyl-glutamate) and poly(γ-methyl L-glutamate), which are sensitive to shear stresses (22)(23), could add to the tactile polymer sensor also slip detection capability.

The feasibility of the direct coupling between piezoelectric polymer film and an integrated circuit has been already demonstrated (24). This technology seems particularly promising for robotics, where more and more sophisticated peripheral data-processing is recognized as an important need. The major inherent limit of piezoelectric polymers is represented by their lack of static response, that, even if a feature also of human skin (but this is still a matter of debate among physiologists) could be a serious drawback in robotic applications. Different solutions have been considered to solve this problem, but one of the most satisfactory has been proposed and experimentally demonstrated by Schoemberg and co-workers at the University of Utah Research Institute (25).

In the configuration depicted in Fig. 4 a PVF_2 film is driven to emit through a layer of a suitable elastomeric material ultrasonic waves which are detected by a second PVF_2 film.

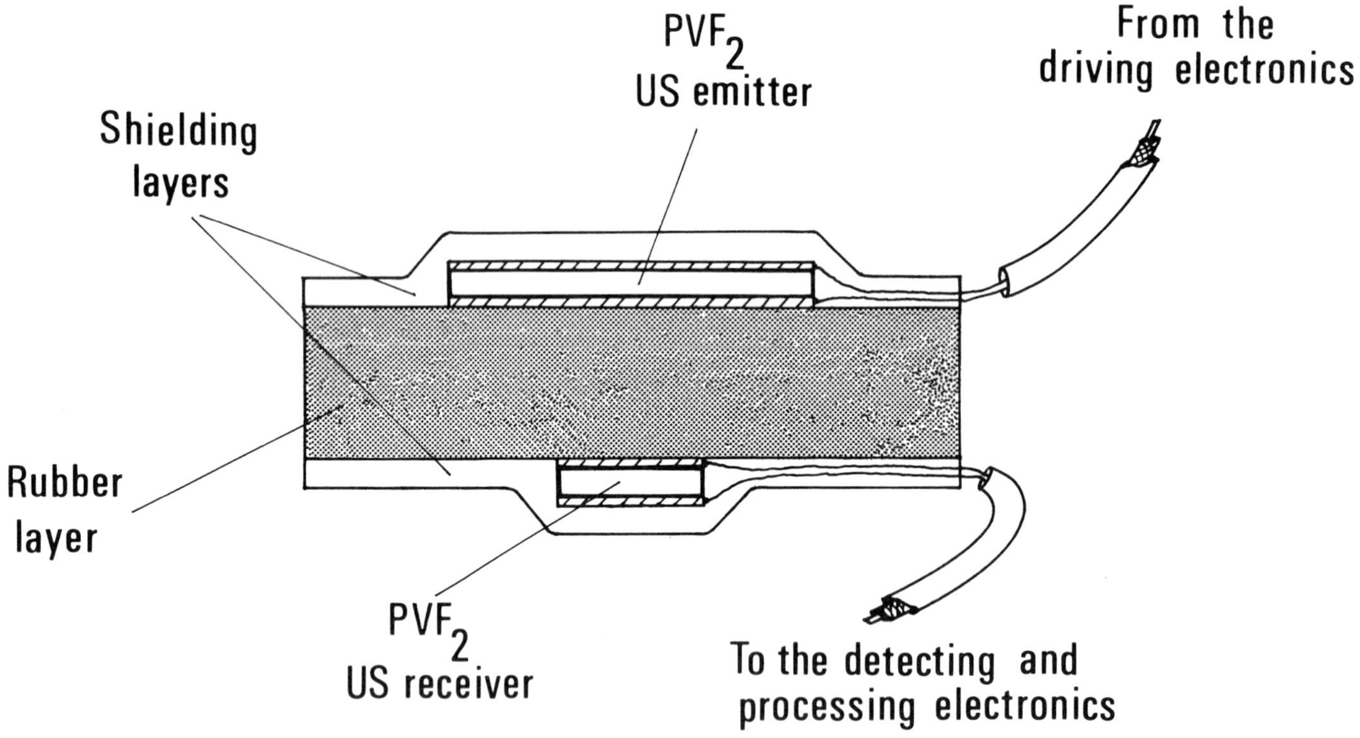

Figure 4 Scheme of a PVF_2 tactile transducer

The time of flight of the ultrasonic waves through the compliant material layer is related to the thickness of such a layer and varies when the transducer is pressed and deformed. The duration of the time of flight is electronically transformed in an analogic signal which, after calibration, provides a direct measurement of the static pressure exerted on the transducer.

With the proposed technique the influence of the pyroelectric effect is completely eliminated and this often represents an advantageous feature. On the other hand, hysteresis introduced by the elastomeric material and increased electronic complexity are the main limitations of this kind of PVF_2 tactile transducer.

D. Proximity sensors. Proximity sensing is precious for the robot to avoid obstacles, find objects and position its hands. Although optical devices are mostly used for this class of sensors, ultrasonic transducers can represent an interesting alternative choice.

As already mentioned, PVF_2 has inherently large bandwidth ($Q \simeq 10$) and very low acoustical impedance, comparable with those of water and human tissues. The consequent good impedance matching leads to very efficient energy transfer between PVF_2 and the transmitting medium, usually water.

The low value of the electromechanical coupling factor (K_T) and the high value of the dielectric losses represent severe limitations to the extensive use of piezoelectric polymers as ultrasonic transducers, especially as transmitters. Nevertheless, several attractive applications of piezoelectric polymer ultrasonic transducer have been devised and intensive research is carried on to modify the preparation process of existing polymers or to develop new polymeric materials more competitive with piezoelectric ceramics (26).

A very efficient configuration which makes use of curved PVF_2 film transducers to transmit and receive ultrasonic waves in air has been also devised by Schoemberg and co-workers to provide prosthetic hands with sensory capabilities (25). A PVF_2 US transducer of this kind is positioned on a finger and used both to detect the presence of objects and to locate the position of the finger itself with reference to a second PVF_2 transducer mounted on the wrist. The same arrangement can be analogously used in robotic hands and possibly spatial resolution further improved by means of arrays of US transducers.

E. Pyroelectric sensors. Infrared transducers are commonly used for intrusion detection systems for military services and home surveillance.

Pyroelectric polymers, such as PVF_2, possess a number of properties which make them potentially competitive with existing pyroelectric materials (ZnO, CdS, TGS, etc.) not only for the construction of pyroelectric detectors, but also for the development of targets for more sophisticated instruments such as calibrated radiometers (27) and infrared vidicon camera tubes (28). PVF_2 pyroelectric sensors are featured by high detectivity (typically 10^9 $f^{1/2}$cm $Hz^{1/2}/W$ for a 1 cm^2 area device, where f is the frequency of the pulsed thermal signal, in Hz), broad spectral response, very high Curie point and high responsivity (about 2μA/W).

The effectiveness of PVF_2 pyroelectric sensors as intrusion detectors, already demonstrated by experimental and commercial devices, can find useful applications in the robotic field, where improved capability to detect the presence of human operators in the field of action of the robot is strongly requested. Purposely designed transducers or, with appropriate techniques and configurations, even the same active areas of the possible PVF_2 robotic skin can be used as pyroelectric sensors.

Non contact, temperature variation measurements, performed by PVF_2 pyroelectric sensors, can be also useful to control robot operation when objects having a variable temperature (for instance heat forged pieces) must be handled.

F. Actuators. Piezoelectric materials deform when an electric field is applied. The amount of this displacement is usually very small both for ceramics and for polymers. For example, the value of the strain S_1 for a sample of piezoelectric material working in extensional mode and free of moving, and submitted to the electric field E_3, is given by the equation:

$$S_1 = d_{31} E_3$$

Introducing the mean numerical values of d_{31} for PVF_2 and PZT and supposing to apply electric fields compatible with the need to avoid the depolarization of the two piezoelectric materials, the values of S_1 are in the order of 10^{-4} and 10^{-5} for PVF_2 and PZT, respectively. The use of bimorph structures, consisting of two (or more) PVF_2 films bonded together, and electrically connected in such a way that one layer contracts while the other expands, largely increases the amplitude of bending motion (29)(30).

Because of the motion amplitude magnification ratio is proportional to the ratio of bimorph lenght to its thickness, a magnification of a few thousand times can be obtained for a 6 μm thin PVF_2 film, while for a 0.5 mm thick ceramic bimorph this amplitude magnification is limited to about a hundred times (31).

Moreover, as observed by Linvill (32), the maximum mechanical output power per unit volume for PVF_2 is 5 times larger than for PZT ceramic, as the maximum electric field which can be applied to PVF_2 without depolarization is about 300 KV/cm, against 3 KV/cm for PZT.

PVF_2 bimorphs, featured by low power consumption, small volumes and simple structures,have been used in several devices for driving very light weight structures (33)(34).

Although the force that can be exerted by this class of PVF_2 actuators, as well as the one achievable with other kinds of PVF_2 actuator structures such as the cylindrical configuration proposed by Dameron and Linvill (35), is very small, nevertheless the feasibility of PVF_2 actuators for robotic use has been recently demonstrated by Umetani and Suzuki (36). According to the authors, PVF_2 micromanipulators for handling very delicate and light weight objects can be of interest for operation in cellular biology, bacterial culture, integrating circuit manufacturing, and micro-mechanical manipulation.

Besides performing mechanical work, piezoelectric manipulators can be equipped with an electronic feedback circuit to provide also tactile sensitivity (37).

A micro-stabber which deflects about 80 μm under a voltage of 300 volts and senses a $1.5 \cdot 10^{-6}$N force, a thruster which displaces about $\pm$ 140 μm under the same voltage, and a composed micromanipulator which is able to position small objects within an area of 330 μm x 500 μm again under the same driving voltage of 300 volts have been described by the same authors, but they only represent an example of the possible applications of piezoelectric polymer actuators.

4. CONCLUSION

Piezoelectric polymers, copolymers and composites, are featured by several unique properties. However, their relatively recent discovery and difficult availability, prevented so far their extensive use in practical devices.

In the field of robotics, the applications of the most known and efficient piezo and pyroelectric polymer, PVF_2, as sensor and/or actuator material are still in their infancy. Although limited by a few drawbacks, an increasing use of piezoelectric polymers for the development of rugged, cheap and sensitive robotic sensors and efficient actuators for micromanipulation is expected.

New attractive fields, such the possible combination of piezoelectric polymers and integrated circuits to develop "smart" sensors begin to be investigated.

It is reasonable to think that this class of polymeric materials could contribute to overcome the present contraddictory situation in which robots endowed with a highly advanced microelectronic "brain" are condemned to a poor "life" by their inadequate sight, hearing and tactile sensing.

BIBLIOGRAPHY

1. H. Kawai: "The piezoelectricity of poly(vinylidene fluoride)". Jpn. J. Appl. Phys.,8, pp. 975-976, 1969.

2. M.G. Broadhurst, S. Edelman, G.T. Davis: "Piezoelectric and pyroelectric application of plastics". Organic Coatings and Plastics Chemistry, 42, pp. 241-245, 1980.

3. G.M. Sessler: "Piezoelectricity in polyvinylidene fluoride". J. Acoust. Soc. Am.,70(6), pp. 1596-1608, 1981.

4. D. De Rossi, P.M. Galletti, P. Dario, P.D. Richardson: "The electro-mechanical connection: piezoelectric polymers in smart artificial organs". Accepted for publication on "asaio journal".

5. R. Hayakawa and Y. Wada : "Piezoelectricity and related properties of polymer films". Advances in Polymer Science, 11, pp. 1-55, 1973.

6. F. Pedrini, R. Francesconi, R. Bedini: "Microprocessor based control system for preparation of piezoelectric polymer film". Proceedings of the ISMM Eighteenth International Symposium on Mini and Microcomputers and Their Application, pp. 111-114, Davos, March 2-5, 1982.

7. N. Murayama, K. Nakamura , H. Obara and M. Segawa: "The strong piezoe-

lectricity in PVDF".Ultrasonics, 14, pp. 15-23, 1976.

8. A. De Reggi: personal communication.

9. K. Tashiro, H. Todakoro, M. Kobayashi: "Structure and piezoelectricity of poly(vinylidene fluoride)". Ferroelectrics, 32, pp. 167-175, 1981.

10.E.L. Nix, L. Holt, J.C. McGrath: "Highly drawn poly(vinylidene fluoride) with enhanced mechanical and electrical properties". Ferroelectrics, 32, pp. 103-114, 1981.

11.D.K. Das-Gupta and K. Doughty: "Piezoelectricity in uniaxially stretched and corona poled polyvinylidene fluoride". J. Phys. D.: Appl. Phys., 13, pp. 95-105, 1980.

12.F. Micheron and C. Lemonon: "Moulded piezoelectric transducers using polar polymers". J. Acoust. Soc. Am., 64(6), pp. 1720-1721, 1978.

13.D. De Rossi, A.S. De Reggi, M.G. Broadhurst, S.C. Roth, G.T. Davis: "Method of evaluating the thermal stability of the pyroelectric properties of polyvinylidene fluoride: effects of poling temperature and field". J. Appl. Phys., 53(10), pp. 6520-6525, 1982.

14.M. Toda: "A PVF_2 piezoelectric bimorph device for sensing presence and position of other objects". IEEE Trans. Electron. Devices, Vol. ED-26, N°5, pp.815-817, 1979.

15.L.D.Harmon: "Touch sensing technology: a review". Society of Manufacturing Engineers Technical Report N° MSR80-03, 1980.

16.P.Dario, R.Bardelli, D.De Rossi, L.R.Wang, P.C.Pinotti: "Touch-sensitive polymer skin uses piezoelectric properties to recognize orientation of objects", Sensor Review, Vol.2, N°4, pp. 194-198, 1982.

17.S.S.M.Wang and P.M.Will: "Sensors for computer controlled mechanical assembly". The Industrial Robot, Vol.5, N°1, pp.9-18, 1978.

18.L.D.Harmon: "Automated tactile sensing". Society of Manufacturing Engineers Technical Report N° MSR82-02, 1982.

19.R.Bardelli, P.Dario, D.De Rossi, P.C.Pinotti: "PVF_2 skin-like tactile sensors for robots and prostheses". 13th Symposium on Industrial Robots and Robots 7 Conference and Exposition, Chicago, April 19-21,1983.

20.S.Lang: "Pyroelectricity: occurrence in biological materials and possible physiological implications". Ferroelectrics, Vol.34, pp.3-9, 1981.

21.H.Athenstaedt, H.Claussen, D.Schaper: "Epidermis of human skin: pyroelectric and piezoelectric sensor layer". Science, Vol.216, pp.1018--1029, 1982.

22.M.Date, S.Takashita, E.Fukada: "Temperature variation of piezoelectric moduli in oriented poly(γ-methyl L-glutamate)". J.Polym. Sci., Vol.8, pp.61-70, 1970.

23.T.Furukawa and E.Fukada:" Piezoelectric relaxation in poly(γ-benzyl--glutamate).J.Polym.Sci.,Vol.14,pp.1979-2110, 1976.

24. R.G.Swartz and J.D.Plummer: "Integrated silicon-PVF_2 acoustic transducer arrays". IEEE Trans. Electron Devices, Vol.ED-26, N°12, pp.1921-1931, 1979.

25. A.A.Schoemberg, C.D.Baker, D.M.Sullivan, H.E.Booth: "Ultrasonic sensory feedback transducer development". University of Utah Research Institute, Progress Report TR4-536-008, 1982.

26. A.Nannini, C.Domenici, P.Dario, D.De Rossi: "Valutazione dell'influenza di alcune procedure di preparazione sulle proprietà ultrasoniche del polivinilidene fluoruro e di alcuni suoi derivati". Symposium on "Le immagini e le tecniche acustiche nella diagnostica medica, nelle prove non distruttive e nella sismologia", Rome, May 13-14, 1982.

27. G.W.Day, C.A.Hamilton, P.M.Gruzensky, R.J.Phelan, Jr.: "Performance and characteristics of polyvinylidene fluoride pyroelectric detectors". Ferroelectrics, Vol. 10, pp.99-102, 1976.

28. J.E.Jacobs and S.A.Remily: "Optimized PVF_2 targets for infrared vidicon camera tubes". Infrared Physics, Vol.19, pp.1-7, 1979.

29. M.Toda and S.Osaka: "Electromotional device using PVF_2 multilayer bimorph". Trans.IECE of Japan, Vol.E 61, N°7, pp.507-512, 1978.

30. M.Toda: "Design of piezoelectric polymer motional devices with various structures". Trans.IECE of Japan, Vol.E 61,N°7,pp.513-518,1978.

31. M.Toda: "Voltage-induced large amplitude bending device-PVF_2 bimorph its properties and applications". Ferroelectrics, Vol.32,pp.127-133,1981

32. J.G.Linvill: "PVF_2-Models, measurements device ideas: "Technical Report 4834-3, Stanford Electronics Laboratory, 1978.

33. M.Toda and S.Osaka: "Application of PVF_2 bimorph cantilever elements to display devices". Proc.of the S.I.D., Vol.19-2,pp.35-41, 1978.

34. M.Toda and S.Osaka: "Vibrational fan using the piezoelectric polymer PVF_2". Proc. IEEE, Vol. 67, N° 8, pp. 1171-1173, 1979.

35. D.H.Dameron and J.G.Linvill: "Cylindrical PVF_2 electromechanical transducers". Sensors and Actuators, Vol.2, pp.73-84, 1981/82.

36. Y.Umetani and H.Suzuki: "Piezo-electric micro-manipulator in multi--degrees of freedom with tactile sensibility". Proc.10th ISIR, pp. 571-579, 1980.

37. Y.Umetani: "Principle of a piezo-electric micro-manipulator with tactile sensibility", Proc.8th ISIR, pp. 406-413, 1978.

Robot Sensing for a Hierarchical Control System

M. Shneier
National Bureau of Standards
Washington, District of Columbia

E. Kent
National Bureau of Standards

J. Albus
National Bureau of Standards

P. Mansbach
National Bureau of Standards

M. Nashman
National Bureau of Standards

L. Palombo
National Bureau of Standrds

W. Rutkowski
National Bureau of Standards

T. Wheatley
National Bureau of Standards

Abstract

An hierarchical sensory system is tailored to the special needs of a robot manipulator. It uses geometric models of objects and knowledge of the robot's position to generate expectations about the environment. The expectations are matched with sensory input from a variety of sources, including visual, tactile, proximity, and force and torque sensors. In the implementation, the visual sensor uses structured light techniques to calculate three-dimensional properties of objects. The low levels of the sensory hierarchy deal with less-processes information than higher levels, and are expected to perform simpler tasks. At each level, the sensor input is matched with the expectations derived from the models, and the differences update the understanding of the environment. Each level in the sensory-processing hierarchy communicates information to corresponding levels in the hierarchical control system that drives the manipulator.

1. Introduction

The problems of sensory processing in the robot domain are somewhat specialized in comparison with the problems in a totally unstructured world. The task of sensory processing is to supply information about the positions, identities, and motions of objects relative to the robot. The information must be kept up to date at all times, and must facilitate the performance of the robot's task. It usually involves three-dimensional rather than two-dimensional processing, and is derived from more than one sensor.

There are a number of constraints on the domain, however, that make the task tractable. The objects to be manipulated are usually prespecified, and are described in detail by geometric models. The layout of the workspace is also known, and can be modified to ease the sensing task. Special lighting can be employed, such as planes or grids of light, to simplify the extraction of three-dimensional information using a visual sensor. There is also a continuity in the environment that can be exploited. Successive sensory inputs are usually closely spaced in time, and much of the processing can be carried over from one input to the next. Indeed, to calculate some of the required information, such as velocities and accelerations, a succession of inputs is essential. Information from the robot control system can also be used to generate predictions about subsequent or current viewing conditions.

This paper describes an approach to constructing a sensory-processing system for a robot manipulator that exploits the constraints on the environment. It interacts with a multilevel robot control system developed at the National Bureau of Standards (Albus *et al.*, 1981), and allows sensory-interactive servoing in real time. The next section describes the design of the system and how it interacts with the control system. The following section contains a short description of the current implementation.

2. System Design

The sensory system is designed to provide the control system with the information it requests in as timely a manner as possible. The control system is organized hierarchically, and different levels of the hierarchy require different information from the sensory processing system. Low levels of control usually need only simple information from the sensory system, while higher levels, which plan actions over significant time spans, might need much more complex information. As a result, the sensory system is also naturally organized hierarchically, with sensory input being more extensively processed as it passes up from level to level. The control system can interrogate each level of the sensory hierarchy, and extract information at various stages of processing.

For example, for low level control operations, such as visual servoing, the control system needs to know little more than the range to the closest point in the field of view of the camera. Higher level control operations, such as the planning of action sequences, require the identification of objects and surfaces by name, implying the use of object models and recognition strategies.

The requests at each level of control are made at vastly different time scales. The low levels require information as rapidly as possible, to ensure that the servoing is smooth and that the robot does not run into unexpected objects. The higher level processing, however, occurs at a much slower rate. This allows time for more detailed processing, necessary to supply the more complex information. With suitable sensors, and with separate processors in the hierarchy running asynchronously, it is possible to supply all levels of control with information that is suited to their needs.

The sensory processing system describes all data in terms of its shape, extent, and location. It assigns names to the objects that it recognizes, and uses information stored with the object models to enhance their descriptions. When it cannot recognize an object, it is nevertheless able to describe its shape, and predict its appearance at later times. The system works both bottom-up and top-down. Raw data is sensed at the lowest levels, and is described at various levels of detail as it progresses up the hierarchy. Model data is input from the top, and filters down to influence the processing of the sensed data. The output of each level of the hierarchy is presented to the control system in parallel; it is also passed up to the next level of sensory processing. Information is thus made available to the control system as soon as it is available. Figure 1 shows the interactions between the control system and the sensory system for a number of levels.

Except at the lowest levels, the sensory processing system is mainly concerned with building a description (or model) of the workspace and the objects in it. It is from this model that it answers queries put to it by the control system, and not from the raw data alone. This allows information to be collected over a period of time, and enables knowledge from the object models to influence the responses. The system computes expectations about future sensory inputs, and compares these with the actual data. Discrepancies are used to update the model of the workspace and keep it current. The model is built using all available sensory input, including visual touch and proximity sensors, and readings of joint angles and forces.

Internal knowledge about the workspace is also available. This takes several forms. Models are available of all the objects expected to appear in the work area. The models are geometrically complete, three-dimensional descriptions of the objects, and include information about surface markings and reflectances. More general knowledge is available about the work area itself, such as the existence of trays and vises, and the locations of light sources. There is a third kind of knowledge, collected by the system itself, which concerns the current context of the sensory system. This includes the location and motion of the observer, obtained from the control system, and which sources of illumination are currently in use.

The workspace model is constructed in a hierarchical manner, to conform with the levels in the sensory system. Low levels describe the moment-to-moment events in the world, and have virtually no memory. Higher levels describe the objects in the world and they ways in which they interact with the manipulator over time. The general form of the hierarchy is shown in Figure 2. Level 0 describes the raw input data (e.g., image intensities). Level I describes parts of objects (e.g. connected components). Level II describes whole objects, and level III describes the whole scene.

Level 0 is concerned simply with describing the raw input from each sensor. For example, intensity, color, and range information may be available on a pixel by pixel basis from a visual sensor. No attempt is made at this level to correlate the outputs from different sensors.

Level I deals with properties of aggregates of data points. At this level connected components are identified and useful spatial and temporal features, such as corners and velocities, are computed. There is some attempt at this level to match components from different sensory modalities. Both this level and the previous one have essentially no input from higher levels of the hierarchy. They operate largely to describe the input in a context-independent way. It is possible to focus the attention of the system, however, and direct it to compute only features of a given kind. For example, straight lines or corner features can be found without knowing about the contents of the scene. But, if knowledge of the scene is available, it can be used to verify that corners of a particular orientation exist, or to restrict the areas in which to look for them. No overt use of context-dependent information is made at these levels, however.

Level II is where the information derived from the data (bottom up) meets the information derived from the models and the workspace description (top down). This level attempts to attach names to parts of the input data, or assign them to known categories. Groups of the features extracted at the previous level may represent objects known to the system. Successful recognition permits access to the knowledge stored with the object models, which is used in turn to improve the representation of the object and to resolve ambiguities elsewhere in the scene. Unseen features of the objects can also be inferred, as well as information such as how to pick them up. If an object cannot be named, its description is passed up to the next level, and its shape and motion parameters are used to predict its appearance at future times. For known objects, it is possible to predict their appearances precisely if the viewing position is known relative to the objects. For objects that have only been seen in a few views, it is not possible to predict the appearance of unseen parts, but it is possible to predict the appearance of the visible parts as the viewpoint changes.

Level III describes the scene as a whole, as opposed to describing the individual objects in it. Relationships between objects are inferred, such as "A is on B", or "A and B move together", even if level II only supplies information about the positions or motion parameters of the individual objects. If objects have been classified at level II, context-dependent information will be available to assist the interpretation. If not, it is still possible to describe spatial and temporal relationships between unidentified objects and other parts of the scene.

There are two main processes comprising each level above level 0. These are illustrated in Figure 3. First, an interpretative process calculates properties of the incoming sensor data, as processed by the preceding levels. Second, a predictive process operates on the workspace model to decide what those properties are expected to be. The interpretation of the scene is based on the data from lower levels, the past history of processing, on general knowledge about interpreting the world, and on expectations generated by the predictive process. The predictive process is based on data from the control system about any actions of the manipulator that affect the sensor outputs, and on inputs from higher levels of the sensory hierarchy.

The top down component of the hierarchy starts with three-dimensional geometric models, obtained from a knowledge base of object types and expected scene contents. This information is transformed at successive levels into expectations about the appearances of objects or surfaces and features that are appropriate to the incoming data with which it is to be matched. The transformation takes into account the kind of sensor, the relationship between the sensor and the object, the illumination conditions, and other objects that may cause occlusions. The models are stored as lists of surfaces, edges, and vertices, with properties. Also available are reflectance information and descriptions of surface markings, and any other desirable information.

At the topmost level, where three-dimensional models are matched with three-dimensional sensor input, the only transformation required is a hidden-surface elimination, based on the viewpoint from which the object was sensed. This can be obtained from the known position of the sensor (read from the control system if not fixed) and the expected orientation of the object. Usually, the relative positions of sensor and object are not known exactly. This means that some error must be allowed in the appearance of the predicted object. To account for this, a "generic view" of the object is computed (Koenderink and van Doorn, 1979; Brooks, 1981). A generic view is one in which small perturbations of viewing position do not alter the topology of the viewed object (e.g., no surfaces appear or disappear). It is possible to subdivide the space around an object into regions in such a way that any viewpoint within a region can be represented by a single generic view. These views can be stored with the object models in the knowledge base, and used as the basis for constructing the expected appearances of objects.

The next level down is concerned with two-dimensional projections of the objects, and the predictive process must generate expectations about the locations of features, for example, corners in an image. This can easily be done by projecting the hidden surface view onto a plane and computing the desired features from that, taking into account the kind of illumination in use.

At each level in the hierarchy, the predictive process must be able to accept information from the interpretative process about objects that are not part of the expected model received from higher levels. These objects can enter into the expectancy generating process only in that transformations based on changes in orientation may be applied to them as well as to the known objects.

Figure 4 shows an example of how the interpretative and predictive components of the sensory system interact. Here the predictive process forms an expectation that a corner feature will appear at a particular position and orientation, with an expected apex angle. The interpretation process compares this expectation with the sensor data, and, if the difference between the two is small, the predicted feature is relocated to the position most consistent with the sensor data. That is, the shape of the corner is obtained from the prediction, which does not suffer from noise problems, while the position of the corner is obtained from the real world. The ideal feature, relocated to the best fit position, is passed up to the next level. This follows from the assumption that the prediction provides a better description of the feature, while the sensor data provides the correct position. The error in the match is passed to the predictive process, which updates its predictions for the next scene.

At the next level, a similar situation arises. Predictions about objects are compared with collections of features extracted from the input data. The best fit of features from the predicted object and features from the data is used to determine the ideal object to be passed to the next level. The error is again passed to the prediction process for this level and used to correct errors in the orientation and position of the object in the description of the workspace.

The interpretation algorithms may discover that there is no match with the predicted features. This indicates an incorrect assumption on the part of the prediction, an unknown object, or bad data. In this case, the features or objects are simply described in the same terms as the recognized features, but are passed to the next level without being matched.

All levels of the system run asynchronously, although low levels may be synchronized to particular sensors. This means that a processor may not always be ready to accept the output from a lower level, or that a higher level processor might not receive its input when it needs it. In the former case, the output from the lower level may simply be discarded, under the assumption that more current input will be available when the higher level requests it. Otherwise, the output may be buffered. In the latter case, new expectancies are still being generated by the predictive process, even though there may be no data to match them with. In this situation, the predictions are passed forward as the best guess about the environment. When new data become available, any drift in this data is corrected. The workspace model is constantly being updated by the sensor data, so that it always represents the current best guess about the state of the environment. If the model includes velocities and accelerations, the guesses will remain accurate for longer periods.

Another feature of the sensory hierarchy is the ability to focus attention, at each level, on a small number of features. Every level performs a default set of operations, causing it to attend to certain types of features and areas of interest. To optimize the matching process, there are algorithms that identify the most discriminable features in the predicted scene. Attention is focussed on areas and features most likely to disambiguate the scene. These features will often be sufficient to analyze the whole scene without further processing. The robot control system may also request that certain features be attended to, for instance to enable it to servo in on a particular surface of an object.

3. Implementation

The implementation is illustrated in Figure 5, in which rectangular boxes represent interfaces to other systems or to sensors, and ellipses represent processors. The arrows on the paths between the components indicate the primary direction of data flow, but all pathways are bi-directional. In general, commands proceed down the network, and data proceed upward, being further interpreted at each level.

At the lowest level, there are tactile, force and torque, and proximity sensors, and a structured light vision system (Albus et al., 1982). The robot arm is equiped with a camera, mounted on its wrist, which moves as the arm moves. Also attached to the robot wrist are two light sources. These are photographic flash tubes that can be selectively triggered under computer control. The duration of either flash can also be controlled, allowing variations in the integrated brightness of the illumination. One of the light sources provides floodlighting of the whole region visible to the camera, while the other projects a pair of light planes into the camera's field of view. A calibration allows the distance to be calculated for all points illuminated by the planes of light. Another calibration for the floodlit images allows displacements to be calculated given the distance obtained from the planes of light.

The hardware that controls the sensors is currently concerned largely with the visual sensor. It allows switching between the light sources, setting the light intensities, and adjusting a threshold used to convert the input from the camera to a binary image. The image is compressed by run-length encoding, and is passed up to the first vision processor.

The first stage of vision processing formats the data for higher stages, and also peforms important measurements to provide fast responses to the control system. The processor is responsible for setting the threshold and intensity values for the hardware, and instructing it about what light source to employ. When the pair of planes of light are used as the illumination, this processor uses triangulation to calculate the range to the closest point in the scene (or in a given window). The range to the point can be read directly by the robot controller, and used to update its trajectory (The details are given in Albus et al., 1982). This is the fastest feedback loop for visual processing, although other sensors can supply information at much higher rates.

From the first stage vision processor, the image data is passed to a second processor that performs one of two tasks. The first task, and the one that is usually performed, is a connected components analysis of the image. Components are extracted and various properties, such as moments, principal axes, and boundaries are measured. The components are stored in a tree structure and are passed in this form to the next processor.

The alternative mode for the processor is entered when analysis indicates that the pictures being received are deteriorating in quality. An automatic procedure is entered to adjust the exposure parameters used by the first stage of vision. A new table of such values is generated using the results of analyzing several pictures. The table is sent to first stage vision, where it is used to establish the intensity and threshold values for taking pictures.

The tree structure of components is input to the next processor, which is responsible for feature extraction. This processor receives commands from higher levels concerning the features to be extracted (such as corners, holes, etc.), and the regions of the image in which to extract them. A set of structures describing the features is generated, and passed up to the next processor.

On the bottom right of Figure 5, the rectangular box represents the hardware associated with all the non visual sensors. Information from these sensors is received and processed by a processor, which outputs its results to the control system and to the routines that match input with predictions. In general, the force and torque, proximity, and touch sensors operate at much higher rates than the visual sensor, and currently there is little processing performed on their output.

It is at the next stage of processing that knowledge about the workspace is first used directly. The workspace knowledge is derived from the database of parts and a geometric model of the last-sensed state of the work area. Initially, the subset of parts that is expected in the current task is extracted from the database and converted into a form suitable for sensory processing. Knowledge of the task allows two levels of sensory models to be constructed. The higher level is a three-dimensional description of each object, based on surfaces, edges, and vertices. The lower level contains generic descriptions of the expected two-dimensional appearances of each object during the sequence of actions that defines the task. This decomposition is similar to that used in ACRONYM (Brooks, 1981).

The two-dimensional workspace model information is used indirectly by the previously-mentioned processor that extracts features, in that it determines the commands that it receives about what features to look for and where they should be found. The next level of processing, however, refers directly to the workspace model. This processor is divided into three sections, for three different tasks. The first task handles the interactions between the two-dimensional workspace model and the vision system. The second handles the interactions between the other sensors (proximity, force and torque, and touch) and the workspace model. The third task is to integrate the information from the different sensory sources and format it for higher levels.

The first task (visual interpretation) takes features from the input image and features from the two-dimensional image predicted by the model and tries to reconcile them. Its main functions are to enhance the quality of the detected features by comparing them with predicted information, and to adjust the predicted position information using the image data. For example, a corner may be detected at a certain location (x,y), with the angle between its edges measured as θ degrees. The model might have predicted the position to be (x',y') and the angle to be θ' degrees. The output of the process would be a corner with position (x,y), but with angle θ'.

Ideally, the actions of the task responsible for dealing with non-visual sensors are analogous to those for vision. The nature of the sensors, however, precludes this in general. For example, tactile sensing can only confirm or deny the workspace model expectations. It is not possible to decide how large the error is if the sensor does not respond when it is expected to. For proximity sensing, an acceptable range of values is predicted by the model and compared with the actual data. The task outputs either the original sensor data, if it is in range, or an error signal.

The third task is responsible for reconciling information from the multiple sources. Where features from vision and other sensors concern similar information, a check is made to ensure that the information is consistent. For example, if a proximity sensor measures a range, z, for a feature, and the visual sensor measures a range, z', for the same feature, the ranges must lie within some tolerance of each other. If they do, a new range is calculated from the two measurements. If not, a precedence scheme is used, based on the regions of highest accuracy of each sensor. Thus, at large distances, the visual sensor has precedence, but as the range decreases, more and more emphasis is placed on the responses from the proximity sensor. When data cannot be reconciled, error messages are sent to the control system and the world model.

The workspace model uses the normal and error outputs to update its predictions, while the control system uses them to adjust its actions. For the control system, it is in many cases possible to supply three-dimensional feature locations based on range information obtained using the structured light.

The output of the two-dimensional reconciliation task serves as the input to the final stage of the hierarchy currently defined. This processor is responsible for three-dimensional recognition, using three-dimensional workspace model information and the combined features extracted from all the sensors. This stage, too, uses predictions derived from the workspace model to identify and name features and to recognize objects. The set of measured features and relations between them are matched with features derived from the model. Currently, a least squares fitting method is used to bring the points into correspondence after matching. Once this is done, the model can supply any features that are missing or invisible, as well as other information, such as legal grip points, that is not obtainable using the sensors. The match also supplies the three-dimensional positions of all points whose positions were not previously known. The output of this stage allows the control system to move to objects or surfaces by name instead of by location. The output is also used by the workspace model to update its expectations about the future positions and appearances of the objects.

The system hardware used to implement the design consists of a number of microprocessors (currently, Intel 8086/8087 processors). These run asynchronously in parallel, with little overhead other than handshaking to pass data. Communication takes place over dedicated point-to-point data paths that are handled as buffered DMA transfers. A low bandwidth broadcast bus ties all the processors together for occasional system messages and initialization.

Control in the system is usually top-down, with data and error signals passing upwards. In most cases, only small amounts of data pass between the sensory processing system and the control system, although fairly large amounts of data pass between processors in the sensory hierarchy. The processors operate asynchronously, and this requires some mechanism for identifying what information is current. A system-wide clock is provided for this purpose, and data is tagged with its time of acquisition. Thus each processor can ensure that it is working with the most current data available to it.

4. Discussion and conclusions

This paper has described a system for extracting sensory information from the world in such a way as to allow a robot control system to respond quickly and meaningfully to changing conditions. The sensory system interacts with the robot controller at successively higher levels of abstraction. These levels are matched by the levels of planning in the control system, and allow information to pass as quickly as it is needed between the two systems.

The existence of a workspace model allows the sensory system to respond to control queries even when it has not received current sensory input. This is because the world model constantly attempts to update its predictions about the environment based on past sensory input and the robot's current position and velocity. Thus, a response predicted by the world model can be used by the control system if the sensory processing cannot respond fast enough. So long as this does not happen too often, there should be little appreciable difference in performance, especially considering that lower levels of the sensory system are providing more immediate range and feature information at much higher rates.

The system described above is designed to provide information to the control system about the names and locations of objects and features on request. It achieves its goals by decomposing problems into specialized subtasks. Each subtask can respond rapidly using its own special expertise, and the control system can be supplied with intermediate results during the processing. The result is a smooth interaction between sensing and control, and the ability to perform complex actions in the real world using a number of sensors to provide feedback on a moment-to-moment basis.

Bibliography

1. J. S. Albus, A. J. Barbera, R. N. Nagel, Theory and practice of hierarchical control. In 23rd IEEE Computer Society International Conference, September 1981, pp 18-39.

2. J. S. Albus, E. Kent, M. Nashman, P. Mansbach, L. Palombo, and M. Shneier, A 6-D vision system. Proc. SPIE Technical Symposium, Crystal City, Virginia, May 1982.

3. R. A. Brooks, Symbolic reasoning among 3-D models and 2-D images. Artificial Intelligence 17 1981, 285-348.

4. J. J. Koenderink and A. J. van Doorn, The internal representation of solid shape with respect to vision. Biol. Cybernetics 32, 1979, 211-216.

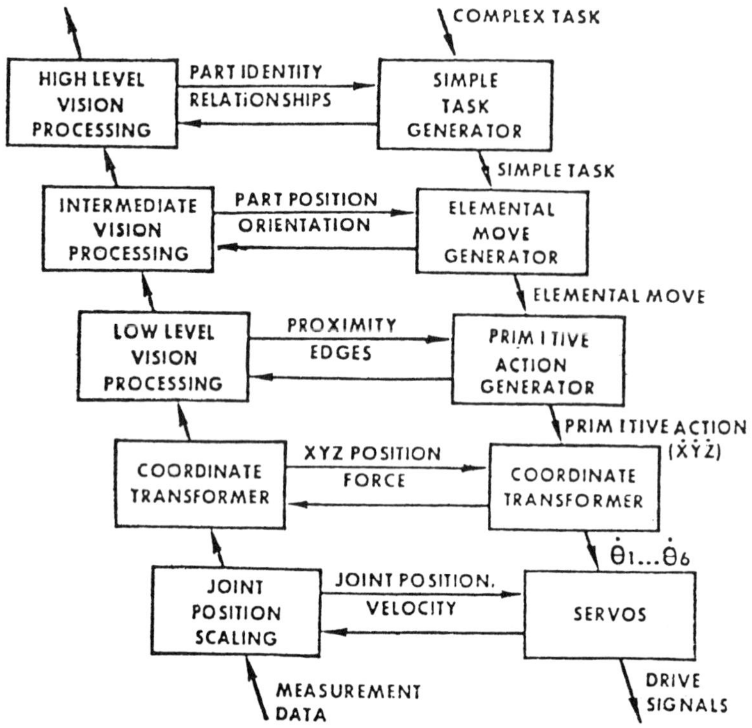

Figure 1. The hierarchical sensory-interactive robot control system. The modules on the left perform sensory processing, while those on the right carry out the control task.

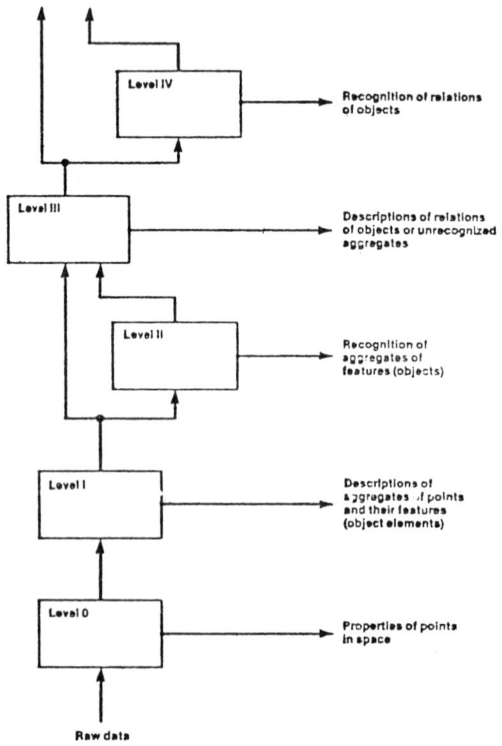

Figure 2. The levels in the sensory processing hierarchy, and the information output at each level.

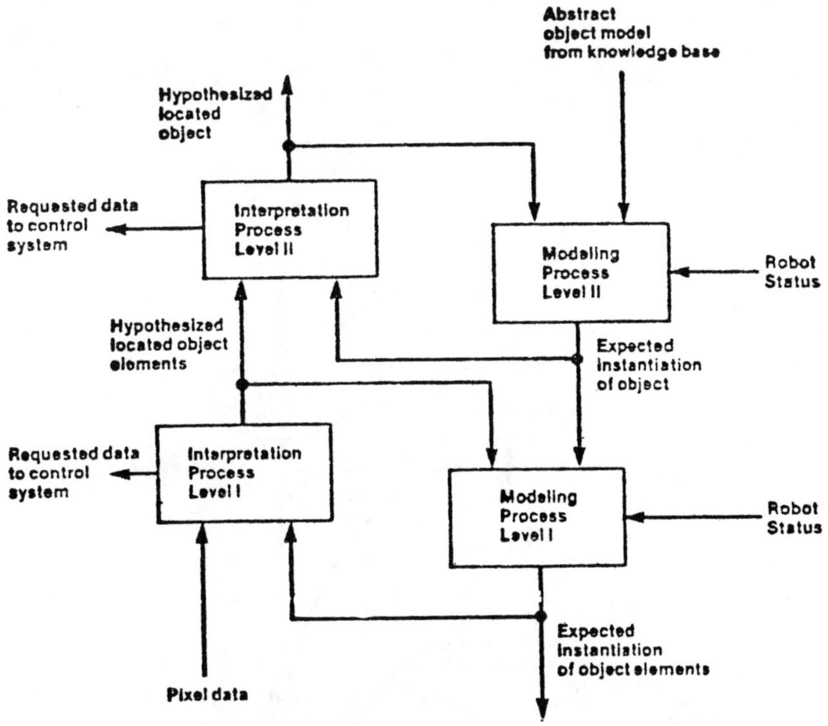

Figure 3. The two processes operating at each level. At the left is the interpretative process, and at the right the predictive process.

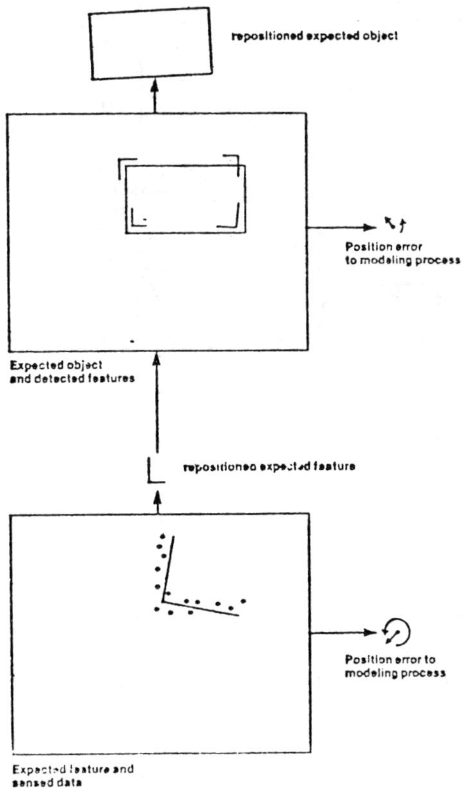

Figure 4. An example of how the interpretative and predictive processes work together to interpret a scene.

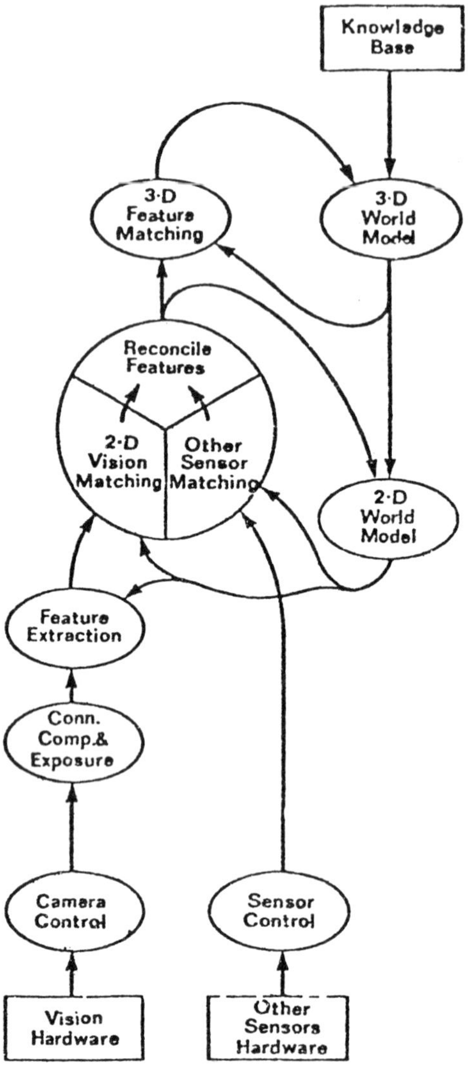

Figure 5. The sensory-processing system. Rectangular boxes represent interfaces to sensors or to other systems. Ellipses represent processors. The arrows point in the direction of principal data flow.

Programming a Vision System

G. Cividini
Polytechnic of Milano
Milano, Italy

G. Villa
Polytechnic of Milano

M. Gini
University of Minnesota
Minneapolis, Minnesota

Abstract

We present LIVIA, a language for writing application programs for industrial vision systems. LIVIA is a Pascal-like language with specific instructions for vision. The objects are identified by the values of their invariant features. Operations on the objects can be done using the elementary set theory. The recognition of objects in a data base and the selection of any of their features is easily programmed.

INTRODUCTION

The importance of providing robots with more sophisticated vision sensors is widely recognized. Visual feedback for manipulation and assembly operations is highly desirable to increase reliability and to reduce the constraints on the positioning of parts. Visual inspection of manufactured products to determine if they meet their standards plays an important role in industrial automation [7][12][13].

The availability of industrial vision systems and their integration into the manufacturing process indicate a growing interest by industry in automatic vision [4][5].

Flexibility, programmability, and reliability are the key issues for vision systems, although cost and processing time are still important. Portability will become more and more important as soon as software packages become available for various applications.

There is an increasing need for systems that can be easily modified to adapt to new situations, such as changes in lighting conditions, or different classes of parts, or different inspection procedures, or new ways of specifying models.

What is needed is a language in which one can specify the control structures that relate the various functions to be performed on the image as well as a language in which one can specify models of objects. A language and a programming environment specifically designed for industrial vision systems will dramatically reduce the software costs, opening new application areas.

For many aspects we can foresee the same development for vision software we have seen in languages for controlling robots. From simple systems based on "teaching by training", sophisticated programming languages have been developed that make it easier to program robots for complex tasks [8]. Unfortunately not too many languages are available for vision. Recent proposals do not seem to make much progress in this area [9].

The declining cost of computation equipment together with the increasing cost of labor are rapidly pushing the designers of manufacturing systems into new directions. Various functions that were previously separated are integrated into a single distributed system.

A modular system for industrial automation, MODIAC, has been designed in Italy by academic and industrial groups as part of a national effort to advance programmable automation. The field of application of MODIAC ranges from the control of continuous processes to manufacturing processes. As part of the MODIAC system we have designed a general purpose vision system for industrial applications, GYPSY [9].

In this paper we present a language, LIVIA, designed for programming the GYPSY system. We give an overview of the system and its logical architecture. Then we describe in more detail the features of LIVIA. Examples of programs follow.

A VISION SYSTEM FOR INDUSTRIAL APPLICATIONS

The main task we had in mind in developing GYPSY (a Growing but still Young Portable SYstem) was to develop a flexible system that can be ready for effective factory use in a short time, but that allows the easy incorporation of extensions.

We have set up a general framework and some tools, that allow us to configure the system, to verify the module interfaces, and to write application programs. We have first developed a collection of modules to form a skeleton system. We have incorporated modules to handle several applications and we are in the process of adding more.

GYPSY is programmable at different levels:

- user level. The final user of the system is not interested in the internal aspects; he wants a simple way to communicate what he wants to do. A programming language, LIVIA, is available for writing application programs. A simple interactive system based on menus is available for the teaching phase;

- application level. Application dependent packages can be written for specific fields of applications as arc welding, or manipulation of particular types of objects. For instance, we are writing a package for the computation of features of 3D objects obtained by projecting a taper light beam [6];

- system level. Any of the basic components of the system can be modified either by changing some code or by adding new functions. Since the system is programmed in high level language using well-defined interfaces it is easy to put modules together to form a system. A simple but powerful operating system can be used to link the various modules and to check the consistency of the interfaces.

The main use of GYPSY is currently for identification of objects to be assembled by a robot. Object images are analyzed during the teaching phase to acquire their features and to build a model. The model is a set of features precisely describing the object. More complex models based on graphs can be constructed for objects with many holes allowing use of the maximal clique program [3].

In the operation phase the images are analyzed to compare their features with the models. A large set of characteristics is computed, scenes with many objects are analyzed, objects inside objects are detected. Even capsized parts can be recognized.

The present implementation of GYPSY is based on the use of a GE TN-2200 solid state TV camera with 128 x 128 pixels and 256 grey levels. The camera is connected through a DMA interface to a DEC MINC-11 computer with 28K words of memory and a LSI 11/02 CPU. A video terminal DEC VT 105 is used for interaction with the user, and a color monitor ISC 8001 provides a graphic display of the image and the results.

The programs are written in OMSI Pascal, except the camera acquisition routine which is in MACRO-11. RT-11 is the standard operating system. A PDP-11/34 can be used for the development of programs. The implementation of GYPSY on MODIAC will be done as soon as the complete hardware and the basic software will become available. We do not expect any difficulties because of the way in which GYPSY has been designed.

LOGICAL ORGANIZATION OF GYPSY

From a logical point of view GYPSY is divided into two levels, a low level process and a high level process. The low level process, connected to the camera, performs the basic computations on the image. The high level process executes application programs and communicates with higher level processes. This organization has been designed to take advantage of the multiprocessing capabilities of the system that will be used for its final implementation.

```
------------------------------------------------------------------------------
                                supervisor
------------------------------------------------------------------------------
|                                                                            |
| LIVIA       editor    interpreter of  teaching   application   linker     |
| compiler              intermediate     module     dependent                |
|                       code                        modules                  |
|                                                                            |
|                       intermediate                                         |
|                       code                                                 |
|                                                                            |
------------------------------------------------------------------------------
                            communication handler
------------------------------------------------------------------------------
|                                                                            |
| image          run       threshold   features     recognition  application |
| acquisition    length                extraction                dependent   |
|                code                                            modules     |
|                                                                            |
| image             grey levels        display      library                  |
| preprocessing     histogram          routines     management               |
------------------------------------------------------------------------------
```

Fig. 1 - The logical organization of GYPSY

The lower level process includes threshold selection, binarization, run-length coding, sequential extraction of features, recognition, teaching by showing, management of a library of models, and other modules for image preprocessing [1][2][14][15].

Other modules contain screen display routines which are dependent on the specific hardware. Modules that depend on specific applications can be inserted here. For instance we have one module using the maximal cliques program that we use to recognize objects not completely in the field of view [3].

The user can interact through a terminal on which a menu of commands is available to start various programs. An operating system allows him to reconfigure the system by adding or changing any of the modules. System level programming takes place mainly at this level.

Permanent storage space for programs and libraries of models is available on floppy disks.

The higher level process is mainly used for programming the low level process. It includes the compiler for the LIVIA language, the interpreter of the intermediate code, editor and general purpose software.

The supervisor allows the user to select how to operate the system. Standard modes of operations are teaching-by-showing, developing LIVIA programs, executing LIVIA programs, and creating a new collection of modules on the low level process. The consistency of the module interfaces is checked at this point. As result of the linking process a table is generated containing the addresses and the names of the routines, so that the LIVIA compiler can generate the appropriate references.

Application dependent programs can be added. If they require specific modules on the low level system the user is responsible for adding the appropriate routines. A second terminal is available for entering LIVIA programs.

Other processes can communicate with the vision machine both directly by feeding commands through the communication handler or through the high level process by sending commands in LIVIA to be compiled and interpreted or by sending commands already compiled to be interpreted.

The communication with the robot mainly takes place by sending directly commands to the communication handler or, for more complex operations, through LIVIA programs. In many cases the robot asks the vision system directly to take a picture and to return the identity, position, and orientation of the object in view. The responsibility of transforming the coordinates of the vision system into the coordinates of the robot is given to the robot program [10].

LIVIA: THE USER PROGRAMMING LANGUAGE

We have designed a language, LIVIA, for programming GYPSY. LIVIA is a Pascal-like language with special instructions for vision. We have decided to base most of the operations on the elementary set theory.

LIVIA has two data types, numbers and sets. Numbers can be integer or real. Sets represent groups of objects. Each object is described by a set of feature values. A list of those features is given in Appendix. Some features are position dependent, some are position independent. The last ones characterize the objects and constitute what we call an object model.

The model is usually constructed with a teaching-by-showing process by showing various times the same object in different positions and orientations. Data from CAD data bases could be used to obtain the model. Various libraries of models can be stored on disks. Each library of models can contain any number of elements.

The abstract model that we are using for sets is illustrated in Fig. 2.

```
-----------------------------------------------------------------
|                                                               |
|    memory of blobs               memory of models             |
|                                                               |
|     ---------------------------------------------------       |
|     | blob name       nearest model        distance   |       |
|     |                                                 |       |
|     ---------------------------------------------------       |
-----------------------------------------------------------------
```

Fig. 2 - The abstract model of the set

Each set is composed by three component parts. The memory of blobs contains all the objects identified in the image with their feature values. The memory of models contains all the models that should be used. The correspondence table is constructed connecting the blobs with the models. The correspondence is generated by a classification algorithm. Each blob is connected to the nearest model. The value of the distance considered acceptable for recognition can be set by the user.

The function DISTANCE returns the distance of the indicated blob from the corresponding model. NEAREST returns a set constituted by the model nearest to the indicated blob. Using them the user can write different classification algorithms.

Each time a picture is taken the system constructs a set of the blobs seen by the camera. We call this set TEL. We may construct a set of all the elements that can be found in any of the available libraries of models. We call this set MEM. All the other sets are created from those two sets.

Using the set operations we can intersect the sets TEL and MEM, or find any subset with specific features, as a certain number of holes or a certain range of perimeter.

Operations available on sets include elementary set operations (union, intersection, and set difference), selection of elements with specific features, and extraction of single elements from any set. Those single elements are considered as sets with a single element, so that the same operations can be applied to them.

The operation INTER computes the intersection of two sets, i.e. the set of elements that are members of both the sets. The correspondence table is used to figure out what objects are in both the sets. The operation UNION computes the union of two sets, i.e. the set of elements that are members of either one of the two sets. DIFF computes the set difference, i.e. the set of elements that are members of the first set and not members of the second. Those operations can be freely intermixed in the same instruction.

Operators are available to select objects with specific features. Selections can be applied to any set to construct subsets of elements satisfying the desired properties.

The user has the option of selecting elements using either the feature values of the models or those of the blobs. Different selectors should be used according to the desired values. Since it is possible to have in the set only blobs or only models, and since the user should not be aware of that, there is a priority scheme. Each of the selectors is applied to its specific class (blobs or models) before and then, if its class is empty, to the other class.

We will explain this mechanism by an example. Let SCREWS be a set containing the blobs A and B, and the models C, D and E. B and C are connected in the correspondence table. The set is illustrated in Fig. 3, assuming that only the Area feature is used.

```
-----------------------------------------------------------------
|     A    Area = 31                  C    Area = 22             |
|     B    Area = 19                  D    Area = 10             |
|                                     E    Area = 71             |
|                                                                |
|                                                                |
|     ----------------------------------------------------       |
|     | blob name        nearest model          distance |       |
|     | B                 C                        ....  |       |
|     ----------------------------------------------------       |
|                                                                |
-----------------------------------------------------------------
```

Fig. 3 - An example of a set

The instruction MEDIUMSCREW := SCREWS * AREA < 20 inserts into the set MEDIUMSCREW the blob B and the models C and D. If we want to use the values of the models for the selection we should write the instruction MEDIUMSCREW := SCREWS * AREAMD < 20. In this case the set contains only the model D. After that the selectors AREA and AREAMD will return the same results.

The control structures available in LIVIA are the same as in Pascal (do..while, repeat..until, if..then, if..then..else, go to). I/O is performed exactly as in Pascal. Procedures are not yet available.

There are a few specific instructions for vision operations.

TAKEAPICTURE takes a picture, computes all the features of the objects and construct the set TEL. The previous content of TEL and of any set derived from it are cleared.

TAKEMEM (name) constructs the set MEM inserting into it all the models present in the named library. The previous content of MEM is cleared.

SETAREA (val) sets the minimum area (in pixels) that will be considered by the system. In this way the user can decide to do not consider objects that are too small or that can be little spots.

SETTHRESHOLD (val) sets the threshold used for binarization.

SETDISTANCE (val) sets the distance that is considered acceptable by the classification algorithm.

The various elements of the set can be selected by using INIT to start at the beginning of the set, and NEXT to go to the next element in the set. The predicate ENDSET tells whether the set has been completely examined. ELEM returns the current element of the set.

Once a particular element of the set has been identified it is possible to extract each one of its feature values by means of extraction operators. The extraction operators are the same as the selectors. When they are used as extractors they return a numeric value.

EXAMPLES OF LIVIA PROGRAMS

Some examples of short programs are illustrated here.

The first problem is to select in the library GROUP#2 the objects that have area greater than 200 pixels and more than one hole. NAMEINT is an extractor which returns the integer identifying the object, while NAMEEXT returns its symbolic name.

```
PROGRAM WRITE;
VAR NUM: NAME;
    SET: AA;
BEGIN
    TAKEMEM ("GROUP#2");
    AA:= MEM * AREA>200 * NHOLES>=1;
    PRINT ("NAMES OF MODELS WITH AREA>200 AND NHOLES>1 ");
    PRINTELN ;
    INIT (AA);
    NEXT (AA);
    WHILE NOT ENDSET (AA) DO BEGIN
        NAME:= NAMEINT (ELEM (AA));
        PRINT (NAMEEXT (NAME)); PRINTELN ;^?
        NEXT (AA);
        END;
END.
```

The next program takes a picture, selects objects with specific features, and recognizes them using a library. The intersection between TEL and MEM contains the recognized objects. We use then two different extractors for the same feature, since we want to compare the value of the blob with the value of the model. AREA returns the area of the object and AREAMD is the area from the corresponding model.

```
PROGRAM SELECT;
VAR NUM: NAME, AR, ARMD;
    SET: BB;
BEGIN
    TAKEAPICTURE;
    TAKEMEM ("GROUP#1");
    BB:= TEL INTER MEM;
    INIT (BB);
    NEXT (BB);
```

```
    PRINT ("RECOGNIZED OBJECTS"); PRINTELN (1);
    PRINT ("NAMES    AREA OBJECTS SEEN    AREA MODELS");
    PRINTELN ;
    WHILE NOT ENDSET (BB) DO BEGIN
        NAME:= NAMEINT (ELEM (BB));
        PRINT (NAMEEXT (NAME));
        AR:= AREA (ELEM (BB));
        ARMD:= AREAMD (ELEM (BB));
        PRINT ("         ", AR, "        ", ARMD);
        PRINTELN ;
        NEXT (BB);
        END;
END.
```

The last program is used to inspect parts. Objects not recognized and without holes are to be discarded. Their position is printed so that a robot could be used to remove them. The set UNKNOWN is constituted by the objects that are not recognized and that have no holes. In fact the intersection between TEL and MEM is the set of recognized objects. The difference between TEL and the set of recognized objects constitutes the set of non recognized objects. Among this set the objects that do not have holes are then selected using the selector HOLES. The coordinates of their centers of gravity are then printed.

```
PROGRAM INSPECTION;
VAR NUM: COORDX, COORDY;
    SET: UNKNOWN;
BEGIN
    TAKEMEM ("KNOWN");
    SETAREA (200);
    TAKEAPICTURE;
    UNKNOWN:= (TEL DIFF (TEL INTER MEM)) * HOLES = 0;
    INIT (UNKNOWN);
    NEXT (UNKNOWN);
    WHILE NOT ENDSET (UNKNOWN) DO BEGIN
        COORDX := XBAR (ELEM (UNKNOWN));
        COORDY := YBAR (ELEM (UNKNOWN));
        PRINT ("DEFECTIVE PART IN ", COORDX, " ",COORDY);
        NEXT (UNKNOWN);
        END;
END.
```

CONCLUSIONS AND ACKNOWLEDGMENTS

We have presented LIVIA, a high level language for programming industrial vision systems. We have given examples of use.

This work was partially supported by the Progetto Finalizzato per l' Informatica, Sottoprogetto P3, Obiettivo MODIAC of the National Council of Research of Italy.

BIBLIOGRAPHY

[1] G. J. Agin, Computer vision systems for industrial inspection and assembly, Computer Magaz., pp 21-31, May 1980.

[2] A. K. Agrawala, A. V. Kulkarni, A sequential approach to the extraction of shape features, Computer Graphics and Image Processing, N. 6, pp 538-557, 1977.

[3] R. C. Bolles, Robust feature matching through maximal cliques, SPIE, Vol 182, Imaging Applications for Automated Industrial Inspection and Assembly, pp 140-149, 1979.

[4] B. Carlisle, The PUMA/VS-100 robot vision system, Proc. First International Conference on Robot Vision and Sensory Control, Stratford-upon-Avon, UK, 1981.

[5] R. T. Chin, Machine vision for discrete part handling in industry: a survey, IEEE 1982 Industrial Applications of Machine Vision, pp 26-32, 1982.

[6] Dai Wei, M. Gini, The use of taper light beam for object recognition, in Robot Vision, IFS Ltd., forthcoming.

[7] G. G. Dodd, L. Rossol, Computer vision and sensor based robots, Plenum Press, New York, 1979.

[8] G. Gini, M. Gini, Interactive development of object handling programs, Computer Languages, Vol. 7, pp 1-10, 1982.

[9] G. Gini, M. Gini, A general Purpose Vision Sensor, Proc. 26th SPIE Technical Conference, San Diego, Ca, 1982.

[10] G. Gini, M. Gini, The integration of manipulation and vision for assembly and quality control, International Journal of Production Research, to appear.

[11] B. Gudmundsson, An interactive high level language system for picture processing, Computer Graphics and Image Processing, N. 18, pp 392-403, 1982.

[12] R. P. Kruger, W. B. Thompson, A technical and economic assessment of computer vision for industrial inspection and robotic assembly, Proc. IEEE, Vol. 69, pp 1524-1538, December 1981.

[13] J. L. Mundy, J. F. Jarvis, Automatic visual inspection, in Applications of Pattern Recognition, K.S. Fu Ed., CRS Press, New York, 1980.

[14] A. Rosenfeld, A. Kak, Digital Picture Processing, Academic Press, New York, 1976.

[15] J. S. Weska, R. N. Nagel, A. Rosenfeld, A threshold selection technique, IEEE Trans. on Computers, Vol. C-23, pp 1322-1326, 1974.

APPENDIX

Position independent features computed both for the blobs and for the models. For each of them there are two selectors available, one used when the feature value of the blob is desired, the second for the feature value of the model:

Perim: the perimeter of the blob, computed with the distance method

Area: the area of the blob (in pixels)

DMin: the minimum distance from the centroid to the perimeter

DMax: the maximum distance from the centroid to the perimeter

Holes: the number of holes

HArea: the total area of all the holes (in pixels)

AvHArea: the average area of the holes (in pixels)

AvHDist: the average distance of holes from the blob centroid

Compact: the ratio of the square of the perimeter to the area of the blob

FrHDist: the distance from the farthest hole to the blob centroid

NrHDist: the distance from the nearest hole to the blob centroid

HoHoDist: the distance between the centroids of the farthest hole and the nearest hole

MinMaxDif: the difference between MaxAng and MinAng. It may be used to test whether the object is capsized

FrHNrHDif: the difference between FrHAng and NrHAng. It may be used to test whether the object is capsized

Features available only in the models:

NameInt: the integer identifier of the model

Nameext: the string of characters associated with each model. It represents its external name

Position dependent features computed for the blobs:

Name: the identifier of the blob

MomX: the first moment with respect to X

MomY: the first moment with respect to Y

XBar: the X coordinate of the blob centroid

YBar: the Y coordinate of the blob centroid

Dad: the identifier of the blob which contains the blob examined

Sons: the number of blobs interior to this blob

FrHName: the name of the hole farthest from the blob centroid

NrHName: the name of the hole nearest to the blob centroid

MinAng: the angle between the X axis and the line passing through the blob centroid and the point on the blob perimeter closest to the centroid

MaxAng: the angle between the X axis and the line passing through the blob centroid and the point on the blob perimeter farthest from the centroid

FrHAng: the angle between the X axis and the line passing through the blob centroid and the hole centroid (FrHName) farthest from the blob centroid

NrHAng: the angle between the X axis and the line passing through the blob centroid and the hole centroid (NrHName) nearest to the blob centroid

XMin: the X coordinate of the point on the blob perimeter closest to the blob centroid

YMin: the Y coordinate of the same point

XMax: the X coordinate of the point on the blob perimeter farthest from the blob centroid

YMax: the Y coordinate of the same point

Compliant Part Mating and Minimum Energy Chamfer Design

Michael P. Hennessey
3M Company
St. Paul, Minnesota

Abstract

Theoretical models of compliantly supported rigid pegs entering compliant holes and minimum energy chamfers help to further understand the mechanics of assembly. Key features of the "peg-in-hole" model include large angle solutions and rotational hole compliance. Different assembly phases are identified and qualitative features of the insertion force versus depth plot are determined. Minimum energy chamfers have been designed according to either minimum insertion work (energy), or equivalent minimum frictional work (energy). An experiment attempted to verify the existence of these chamfers.

1.0 COMPLIANTLY SUPPORTED RIGID PEG ENTERING A COMPLIANT HOLE

1.1 INTRODUCTION

A field of study related to industrial automation is part mating. Part mating is primarily concerned with the mechanics of assembly, studying the forces and displacements of the parts involved so as to recommend ways of making the assembly easier. In many assembly situations (e.g., putting a bearing into a housing) there are two parts involved: one is fixed (the "hole") and another part (the "peg"), while being supported, is assembled to the first part as shown schematically in Figure 1.1. The subsequent analysis will presume this type of assembly.

In general, there are two classifications of part mating: Rigid and Compliant. In rigid part mating the parts involved do not store a substantial amount of elastic energy during the mating process, whereas in compliant part mating one or both of the parts store a substantial amount of elastic energy during the mating process. Much work has been done in rigid part mating and some areas of rigid part mating have been recently extended to compliant part mating.[2,3] This paper concentrates on compliant part mating.

Compliant part mating is primarily concerned with the mechanics of assembly of parts that cannot be considered rigid compared to the support. Recently, part mating theory has been extended to handle restricted cases that involve compliant hole walls (e.g., no friction, small angles).[4,5] An even more general model which takes into account combined lateral and rotational hole compliance will be developed below.[4,6] This model has been used to analyze a single "peg-in-hole" assembly.

In the analysis that follows, various simplifying assumptions will be made. These assumptions are: (1) quasi-static 2 dimensional model, (2) Hookean springs for compliant elements, (3) rectangular peg and hole with chamfer, and (4) the peg and the hole are supported compliantly at their compliance centers. A compliance center is a point where a fictitious support acts to provide independent lateral and rotational support. The compliance center of the peg is located along its center axis and the symmetrical compliance centers of the hole are positioned so the sides of the hole do not interfere with the assembly. Figure 1.2 illustrates the initial configuration of the peg and the hole.

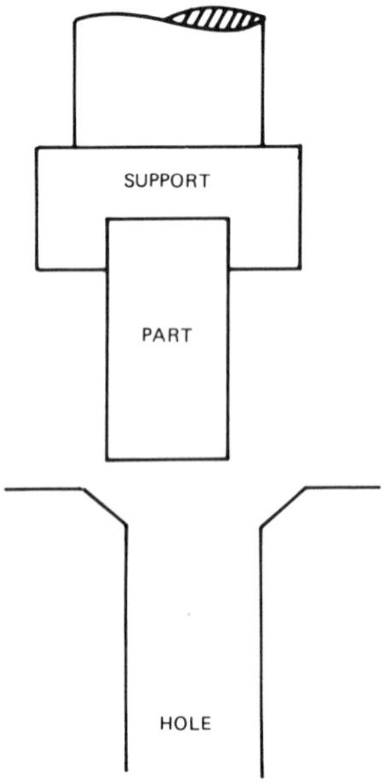

Figure 1.1 Initial configuration of peg and hole.

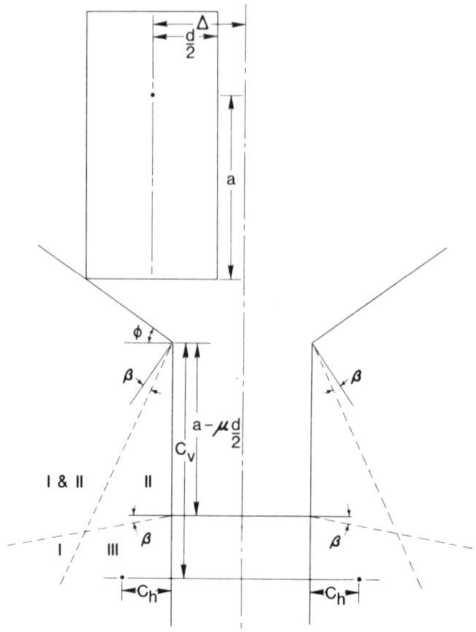

Figure 1.2 Initial configuration of peg and hole.

The general quasi-static phases of successful assembly to be considered are (a) chamfer crossing, followed by (b) one-point contact, (c) two-point contact, (d) resumption of one-point contact, and (e) line contact. The assembly process is continuous except for the transition from chamfer crossing to one-point contact. This is because the normal force changes direction abruptly when the corner of the peg meets the corner of the chamfer.

1.2 LATERAL AND ROTATIONAL COMPLIANCE HOLE

1.2.1 Introduction

In the most general case a model must incorporate both lateral and rotational hole wall compliance. Initially the peg and hole are positioned as shown in Figure 1.2 (μ – friction coefficient, $\tan \beta = \mu$) and during the assembly the hole walls deform outward, enlarging the hole. By Chasle's Theorem, this deformation may be treated as a translation ($\delta x_1, \delta x_2$) and a rotation ($\delta\theta_1, \delta\theta_2$) of the hole walls taken in either order. The assembly phases to be analyzed are (1) chamfer crossing, (2) one-point contact, and (3) two-point contact shown in Figures 1.3 - 1.5 respectively. The assembly equations which involve contact forces and displacements may be derived by imposing (1) equilibrium requirements, (2) force deformation relations, and (3) geometric compatibility requirements during each phase of the assembly.[7] These equations will now be listed for each assembly phase.

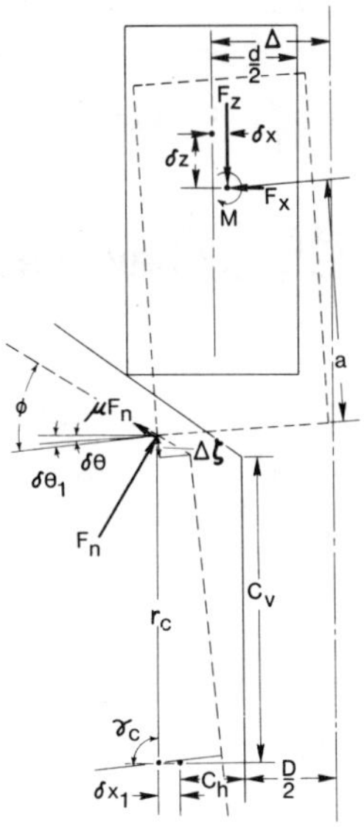

Figure 1.3 Chamfer crossing.

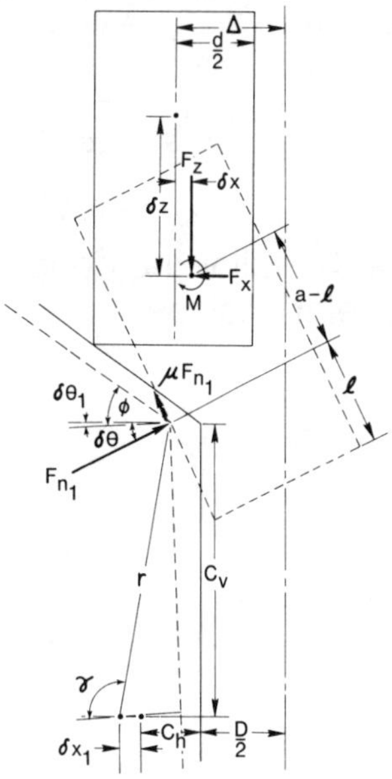

Figure 1.4 One-point contact.

1.2.2 Chamfer Crossing

The following equations may be derived from Figure 1.3:

Equilibrium Requirements

$$
\begin{aligned}
F_x &= F_n[\sin(\phi - \delta\theta_1) - \mu \cos(\phi - \delta\theta_1)] \\
F_z &= F_n[\cos(\phi - \delta\theta_1) + \mu \sin(\phi - \delta\theta_1)] \\
M &= F_n\{a[\sin(\phi + \delta\theta - \delta\theta_1) - \mu \cos(\phi + \delta\theta - \delta\theta_1)] \\
&\quad - \frac{d}{2}[\cos(\phi + \delta\theta - \delta\theta_1) + \mu \sin(\phi + \delta\theta - \delta\theta_1)]\}
\end{aligned}
\tag{1.1}
$$

Force-Deformation Relations

$$F_x = K_x \,\delta x$$

$$M = K_\theta \delta\theta$$

$$K_{x_1}\delta x_1 = F_n[\sin(\phi - \delta\theta_1) - \mu\cos(\phi - \delta\theta_1)]$$

$$K_{\theta_1}\delta\theta_1 = F_n\left[(\Delta\zeta + C_V)(\sin\phi - \mu\cos\phi) + \left(\frac{\Delta\zeta}{\tan\phi} - C_h\right)(\cos\phi + \mu\sin\phi)\right] \tag{1.2}$$

Geometric Compatibility Requirements

$$\frac{\Delta z}{\tan\phi} = \delta x + \delta x_1 + a\sin\delta\theta + r_c[\cos(\gamma_c - \delta\theta_1) - \cos\gamma_c] + d\sin^2\frac{\delta\theta}{2}$$

$$a + \Delta z = \delta z + a\cos\delta\theta + \frac{d}{2}\sin\delta\theta + r_c[\sin(\gamma_c - \delta\theta_1) - \sin\gamma_c] \tag{1.3}$$

For convenience the following dependent insertion variables ($\Delta\zeta, r_c, \gamma_c$) have been introduced by the following relationships

$$\Delta z + \Delta\zeta = \left(\Delta - \frac{CD}{2}\right)\tan\phi \tag{1.4}$$

$$r_c^2 = (C_V + \Delta\zeta)^2 + \left(\frac{\Delta\zeta}{\tan\phi} - C_h\right)^2 \tag{1.5}$$

$$\cot\gamma_c = \frac{\frac{\Delta\zeta}{\tan\phi} - C_h}{C_V + \Delta\zeta} \tag{1.6}$$

Chamfer crossing begins when the insertion distance $\Delta z = 0$, and ends when $\Delta z = (\Delta - \frac{CD}{2})\tan\phi$; here C is the clearance ratio. Also to avoid the other type of one-point contact, the compliance center of the peg must be located a distance at least $\frac{d}{2\tan(\phi + \delta\theta - \delta\theta_1 - \beta)}$ from the end of the peg where $\delta\theta > \delta\theta_1 \geq 0$.

1.2.3 One-Point Contact

Immediately following chamfer crossing is the brief "discontinuity phase" followed by one-point contact. In a similiar manner (see Section 1.2.2) the following equations may be derived with the aid of Figure 1.4.

Equilibrium Requirements

$$F_x = F_{n_1}(\cos\delta\theta - \mu\sin\delta\theta)$$

$$F_z = F_{n_1}(\sin\delta\theta + \mu\cos\delta\theta) \tag{1.7}$$

$$M = F_{n_1}\left[(a - \ell) - \frac{\mu d}{2}\right]$$

Force-Deformation Relations

$$F_x = K_x\,\delta x$$

$$M = K_\theta\,\delta\theta$$

$$K_{x_1}\,\delta x_1 = F_{n_1}(\cos\delta\theta - \mu\sin\delta\theta)$$

$$K_{\theta_1}\,\delta\theta_1 = F_{n_1}\{C_V[\cos(\delta\theta - \delta\theta_1) - \mu\sin(\delta\theta - \delta\theta_1)]$$

$$- C_h[\sin(\delta\theta - \delta\theta_1) + \mu\cos(\delta\theta - \delta\theta_1)]\} \tag{1.8}$$

Geometric Compatibility Requirements

$$\Delta - \frac{CD}{2} = \delta x + \delta x_1 + (a - \ell)\sin\delta\theta$$

$$+ r[\cos(\gamma - \delta\theta_1) - \cos\gamma] + d\sin^2\frac{\delta\theta}{2}$$

$$a + \left(\Delta - \frac{CD}{2}\right)\tan\phi = \delta z + (a - \ell)\cos\delta\theta \tag{1.9}$$

$$+ \frac{d}{2}\sin\delta\theta + r[\sin(\gamma - \delta\theta_1) - \sin\gamma]$$

Here ℓ is the insertion distance and one-point contact begins when $\ell = 0$. Also, the angle of the peg with respect to the left hole wall ($\delta\theta - \delta\theta_1$) must be nonnegative and since $\delta\theta \geq 0$ Equations 1.7 and 1.8 imply that a $\geq \frac{d}{2}$. Again, the distance between the compliance center of the hole and the contact point (r) is dependent, so is the angle γ. The required relations are

$$r^2 = c_v^2 + c_h^2 \tag{1.10}$$

and

$$\cot\gamma = \frac{-c_h}{c_v} \tag{1.11}$$

Equations (1.5) and (1.6), of course, reduce to Eq. (1.10) and (1.11) when

$$\Delta z = \left(\Delta - \frac{CD}{2}\right) \tan \phi$$

One-point contact (two-point contact) ends and two-point contact (resumption of one-point contact) begins when

$$\ell \sin \delta\theta + d \cos \delta\theta = D + \delta x_1 + r[\cos(\gamma - \delta\theta_1) - \cos \gamma] \quad (1.12)$$

Manipulation of the above equation with the one-point contact equations yields the following quadratic in $(a - \ell)$

$$\sin \delta\theta (a - \ell)^2 + B(\delta\theta, \delta\theta_1)(a - \ell) + C(\delta\theta, \delta\theta_1) = 0 \quad (1.13)$$

where $B(\delta\theta, \delta\theta_1)$, $C(\delta\theta, \delta\theta_1)$ are complicated expressions. Since Equation 1.13 will yield two real roots or two complex roots for $(a - \ell)$, a resumption of one-point contact is possible. This has been verified by computer runs. Also, the largest value of ℓ for which one-point contact is possible is $a - \frac{\mu d}{2}$.

1.2.4 Two-Point Contact

In general, one-point contact is followed by two-point contact (see Figure 1.5). Similarly the following equations may be derived.

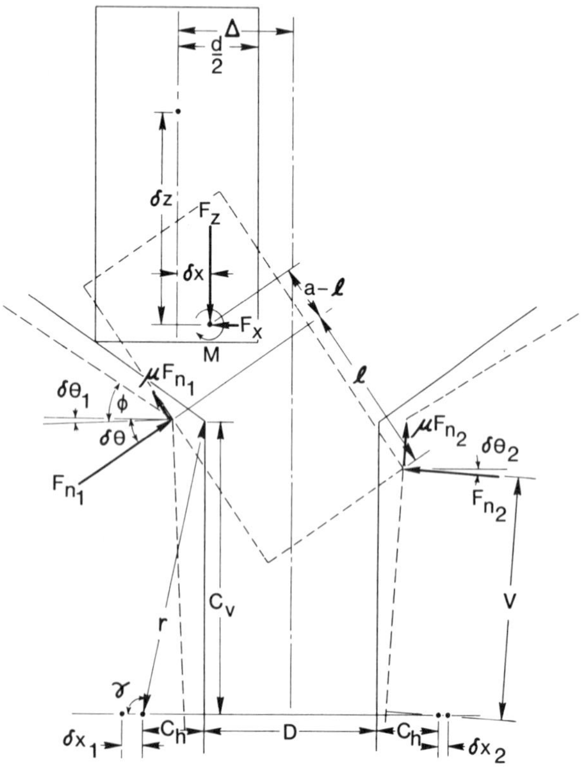

Figure 1.5 Two-point contact.

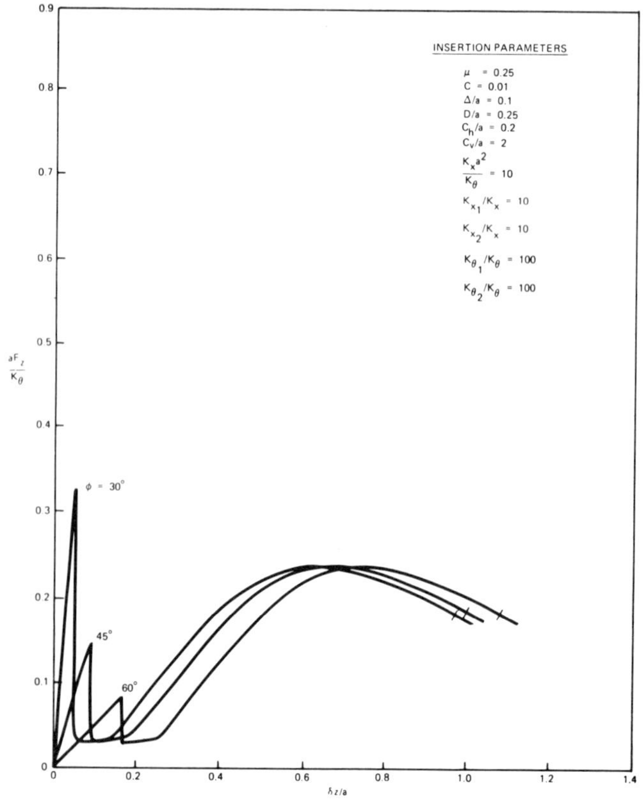

Figure 1.6 Effect of ϕ on F_z versus δz.

Equilibrium Requirements

$$F_x = F_{n_1}(\cos\delta\theta - \mu\sin\delta\theta) - F_{n_2}(\cos\delta\theta_2 - \mu\sin\delta\theta_2)$$

$$F_z = F_{n_1}(\sin\delta\theta + \mu\cos\delta\theta) + F_{n_2}(\sin\delta\theta_2 + \mu\cos\delta\theta_2) \tag{1.14}$$

$$M = F_{n_1}\left[(a - \ell) - \frac{\mu d}{2}\right] - F_{n_2}\{a[\cos(\delta\theta + \delta\theta_2) - \mu\sin(\delta\theta + \delta\theta_2)]$$

$$- \frac{d}{2}[\sin(\delta\theta + \delta\theta_2) + \mu\cos(\delta\theta + \delta\theta_2)]\}$$

Force-Deformation Relations

$$F_x = K_x\delta x$$

$$M = K_\theta\delta\theta$$

$$K_{x_1}\delta x_1 = F_{n_1}(\cos\delta\theta - \mu\sin\delta\theta)$$

$$K_{\theta_1}\delta\theta_1 = F_{n_1}\{C_V[\cos(\delta\theta - \delta\theta_1) - \mu\sin(\delta\theta - \delta\theta_1)]$$

$$- C_h[\sin(\delta\theta - \delta\theta_1) + \mu\cos(\delta\theta - \delta\theta_1)]\} \tag{1.15}$$

$$K_{x_2}\delta x_2 = F_{n_2}(\cos\delta\theta_2 - \mu\sin\delta\theta_2)$$

$$K_{\theta_2}\delta\theta_2 = F_{n_2}(V - \mu C_h)$$

Geometric Compatibility Requirements

$$\Delta - \frac{CD}{2} = \delta x + \delta x_1 + (a - \ell)\sin\delta\theta$$

$$+ r[\cos(\gamma - \delta\theta_1) - \cos\gamma] + d\sin^2\frac{\delta\theta}{2}$$

$$a + \left(\Delta - \frac{CD}{2}\right)\tan\phi = \delta z + (a - \ell)\cos\delta\theta + \frac{d}{2}\sin\delta\theta \tag{1.16}$$

$$+ r[\sin(\gamma - \delta\theta_1) - \sin\gamma]$$

$$\ell\sin\delta\theta + d\cos\delta\theta = D + \delta x_1 + \delta x_2 + r[\cos(\gamma - \delta\theta_1) - \cos\gamma]$$

$$+ 2C_h\sin^2\frac{\delta\theta_2}{2} + V\sin\delta\theta_2$$

For convenience V has been introduced by the relation

$$V \cos \delta\theta_2 = -\ell \cos \delta\theta + d \sin \delta\theta + C_h(\sin \delta\theta_1 - \sin \delta\theta_2) + C_V \cos \delta\theta_1 \tag{1.17}$$

1.2.5 Results and Discussion

Once the assembly equations have been solved for the entire assembly process (using a generalized Newton-Raphson method)[8], the effect of various insertion parameters on the solution can be determined. Of primary interest is the effect of the insertion parameters on the "insertion force versus depth" plot (i.e., F_z versus δz).

Each of the assembly phases may be identified in the force versus depth plot (e.g. see Figure 1.6, $a = 1, K_\theta = 1$). The chamfer crossing region is seen to be very linear, followed by a discontinuity where one-point contact begins. The force during one-point contact is also reasonably linear, and in this instance almost constant. After one-point contact the insertion force rises sharply during two-point contact to a maximum and then gradually declines to where one-point contact resumes. This resumption of one-point contact is typically of short duration as the peg snaps back to line contact. The insertion force in this region also tends to be very linear. Finally, line contact occurs and is represented by the end point of the curve.

In Figure 1.6 the effect of ϕ on F_z versus δz is shown for several ϕ. As the chamfer becomes flatter (ϕ small) the insertion force during chamfer crossing rises and vice versa. During one-point contact and two-point contact the solution is primarily shifted as ϕ varies.

2.0 MINIMUM ENERGY CHAMFER DESIGN

2.1 Introduction — Chamfer Design in General

From Section 1.2 (Figure 1.6) and previous work in Part Mating it is apparent that the slope and shape of the chamfer can greatly affect the insertion forces which arise during chamfer crossing. Knowledge of how these forces depend on the chamfer's slope and shape is essential if better chamfers are to be designed which enable the parts involved to be assembled more easily.

Design criteria have included the following: (1) minimum peak force, (2) constant force, (3) minimum vertical work/energy, and (4) minimum frictional work/energy during chamfer crossing. In general, the chamfers obtained by applying any one of these criteria also depend on the following three factors: friction, geometry (e.g. of peg), and compliance (of peg support and hole). Specification of these factors and one of the design criteria then determines the desired chamfer shape provided it exists. For example, chamfers have been designed where (1) the peg is supported laterally, (2) supported by rotational compliance, and (3) for doorlatch tongue shapes where there are two sources of friction.[1]

2.2 MINIMUM ENERGY CHAMFER DESIGN — Rotational Peg Support Case

2.2.1 Introduction

In this paper the emphasis will be on the design of minimum energy chamfers where the peg is supported by rotational compliance. This is the original minimum energy chamfer problem proposed by D.E. Whitney and partially solved for small angles. In the present formulation, the small angle assumption will be dropped. Thus far, the problem has not been solved completely; however, major results such as the derivation of the optimal slope and wedging chamfers have been obtained.

In certain instances the peg will be supported with rotational compliance (K_θ) as shown in Figure 2.1. Here the chamfer is an arbitrary shape y(x) and the peg is represented by a line segment of length ℓ. Initially the end of the peg is in contact with the top of the chamfer (0, y_o) and inclined at an angle θ_o to the vertical. During the assembly to be considered, the peg's support point moves vertically downward along the $\bar{y}$-axis, while the end of the peg slides along the chamfer with friction. Chamfer crossing (assembly) ends when the contact point is at the bottom of the chamfer (x_o, 0).

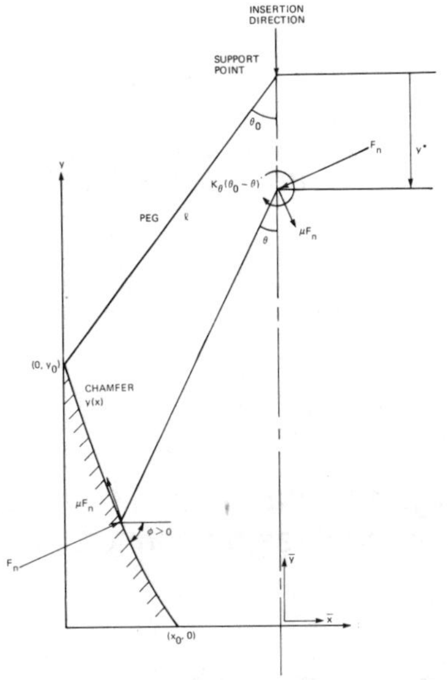

Figure 2.1 Chamfer crossing - rotational peg support.

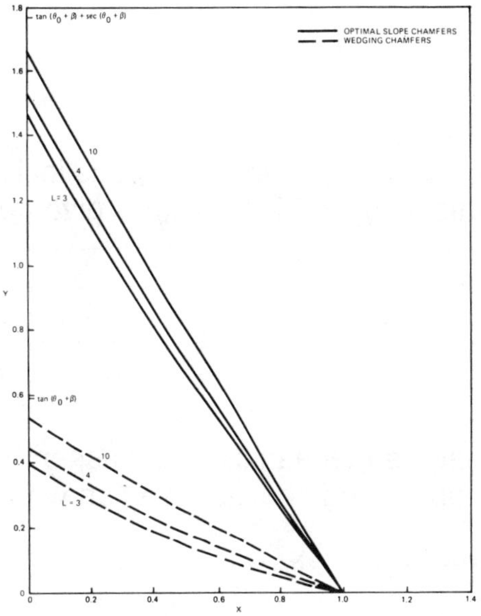

Figure 2.2 Optimal and wedging chamfers for various L, $\mu = 0.2, \theta_o = 20°$

With the aid of the free-body diagram of the peg provided in Figure 2.1, the mechanics of the assembly can be analyzed. The normal contact force (F_n) can be solved for by balancing moments at the support point,

$$F_n = \frac{(K_\theta/\ell)(\theta_o - \theta)}{\sin(\phi - \theta) - \mu\cos(\phi - \theta)} \tag{2.1}$$

where

$$\ell \sin\theta = \ell \sin\theta_o - x$$

$$\tan\phi = -y' \tag{2.2}$$

The normal force may then be expressed in terms of x, y′. However, it is more convenient to do the analysis in the $\bar{x}$, $\bar{y}$ coordinates and then transform back to the x, y coordinates later on. Therefore, F_n is given by:

$$F_n = \frac{K_\theta\left[\theta_o + \sin^{-1}\left(\frac{\bar{x}}{\ell}\right)\right]\sqrt{1 + \bar{y}'^2}}{(\bar{x} - \mu\sqrt{\ell^2 - \bar{x}^2}) - \bar{y}'(\sqrt{\ell^2 - \bar{x}^2} + \mu\bar{x})} \tag{2.3}$$

where

$$\bar{x} = x - \ell\sin\theta_o$$

$$\bar{y} = y \tag{2.4}$$

Since the normal force is nonnegative,

$$\bar{y}' \leq \frac{\bar{x} - \mu\sqrt{\ell^2 - \bar{x}^2}}{\sqrt{\ell^2 - \bar{x}^2} + \mu\,\bar{x}} \tag{2.5}$$

provided the initial offset angle (θ_o) is not too large:

$$\cot\theta_o > \mu \tag{2.6}$$

For chamfers with larger slopes or pegs inclined at larger initial offset angles, wedging will occur. Also, the vertical contact force (F_y) is given by:

$$F_y = \frac{K_\theta\left[\theta_o + \sin^{-1}\left(\frac{\bar{x}}{\ell}\right)\right](1 - \mu\,\bar{y}')}{(\bar{x} - \mu\sqrt{\ell^2 - \bar{x}^2}) - \bar{y}'(\sqrt{\ell^2 - \bar{x}^2} + \mu\,\bar{x})} \tag{2.7}$$

Now that the mechanics of the assembly have been analyzed for a general chamfer shape, minimum frictional and vertical insertion work criteria will be used to design chamfers.

2.2.2 Frictional Work Criterion

A natural criterion for designing chamfers involves finding a chamfer which minimizes the frictional work (dW_μ) given by the product of the frictional force (μF_n) and an increment in the distance (ds) through which the end of the peg moves anti-parallel to this frictional force while in contact with the chamfer, i.e.

$$dW_\mu = \mu F_n\, ds \tag{2.8}$$

where s is the arclength along the chamfer. The total frictional work (W_μ) is then given by:

$$W_\mu = \mu K_\theta \int_{\bar{x}} \frac{\left[\theta_o + \sin^{-1}\left(\frac{\bar{x}}{\ell}\right)\right](1 + \bar{y}'^2)\, d\bar{x}}{(\bar{x} - \mu\sqrt{\ell^2 - \bar{x}^2}) - \bar{y}'(\sqrt{\ell^2 - \bar{x}^2} + \mu\bar{x})} \equiv \int_{\bar{x}} I_\mu(x, \bar{y}')\, d\bar{x} \tag{2.9}$$

where appropriate substitutions have been made. The frictional work is seen to depend on the chamfer's slope ($\bar{y}'$) but not its shape ($\bar{y}$). It may also be shown that chamfers designed according to the minimum frictional work criterion are the same as those designed according to minimum vertical work criterion ($\min \int F_y\, dy^*$)!

2.2.3 Calculus of Variations Analysis

The Calculus of Vatiations may be used to find the chamfer shape $\bar{y}(x)$ (or y(x)) such that the frictional work is minimized.[9] Legendre's necessary condition[10] for a minimum is satisfied ($I_{\mu_{\bar{y}'\bar{y}'}} \geq 0$) and Euler's equation reduces to an equation which may be solved for $\bar{y}'$ using the quadratic formula to yield:

$$\bar{y}' = \frac{\bar{x} - \mu\sqrt{\ell^2-\bar{x}^2}}{\sqrt{\ell^2-\bar{x}^2} + \mu\bar{x}} - \sqrt{\frac{\left[\theta_o + \sin^{-1}\left(\frac{\bar{x}}{\ell}\right)\right]\left[1 + \left(\frac{\bar{x} - \mu\sqrt{\ell^2-\bar{x}^2}}{\sqrt{\ell^2-\bar{x}^2} + \mu\bar{x}}\right)^2\right]}{\left[\theta_o + \sin^{-1}\left(\frac{\bar{x}}{\ell}\right)\right] + c\left(\sqrt{\ell^2-\bar{x}^2} + \mu\bar{x}\right)}} \tag{2.10}$$

where c is an integration constant. It may be argued that $c \geq 0$ and the $-$ sign must be used on the discriminate. Since I_μ does not depend on $\bar{y}$ it follows that the optimal slope chamfer will result when $c = 0$. Also, for large ℓ the rotational peg support case reduces to the lateral peg support case.[1] The slope is bounded since (use $c = 0, \infty$);

$$\frac{\bar{x} - \mu\sqrt{\ell^2-\bar{x}^2}}{\sqrt{\ell^2-\bar{x}^2} + \mu\bar{x}} - \sqrt{1 + \left(\frac{\bar{x} - \mu\sqrt{\ell^2-\bar{x}^2}}{\sqrt{\ell^2-\bar{x}^2} + \mu\bar{x}}\right)^2} \leq \bar{y}' \leq \frac{\bar{x} - \mu\sqrt{\ell^2-\bar{x}^2}}{\sqrt{\ell^2-\bar{x}^2} + \mu\bar{x}} \tag{2.11}$$

By using one of the boundary conditions, it is observed that the $c = 0, \infty$ chamfers serve as an envelope for the rest of the chamfer shapes. This bound has to do with the fact that an optimal slope exists which depends on x and that wedging has been avoided.

Equation (2.10) cannot be integrated easily (if at all) by elementary methods; however, when $c = 0, \infty$ direct integration (pseudo-elliptic form) is possible. For $\ell \sin\theta_o \geq X_o$ solutions will be obtained. These solutions correspond to the case where the peg's support point does not interfere with the chamfer (see Figure 2.1) during assembly. The solutions are (in original coordinates):

$$Y^{c=\infty} \equiv Y^W = \sqrt{L^2-\bar{X}(1)^2} - \sqrt{L^2-\bar{X}^2} + \frac{\mu L}{\sqrt{1+\mu^2}} \ln\left[\frac{(\sqrt{1+\mu^2}\sqrt{L^2-\bar{X}^2} + \mu L)}{(\sqrt{1+\mu^2}\sqrt{L^2-\bar{X}(1)^2} + \mu L)} \cdot \frac{(L + \sqrt{1+\mu^2}\,\bar{X}(1))}{(L + \sqrt{1+\mu^2}\,\bar{X})}\right] \tag{2.12}$$

where

$$Y^W = y^W/x_o, \quad X = x/x_o, \quad L = \ell/x_o,$$

$$\bar{X}(X) = X - L\sin\theta_o \tag{2.13}$$

$$\mu > 0, \quad \cot\theta_o > \mu, \quad L\sin\theta_o > 1$$

and

$$Y^{c=0} \equiv Y^O = Y^W + \frac{L}{\sqrt{1+\mu^2}}\left[\sin^{-1}(\bar{X}(1)/L) - \sin^{-1}(\bar{X}/L)\right] + \frac{\mu L}{\sqrt{1+\mu^2}} \ln\left[\frac{\left(\sqrt{L^2-\bar{X}^2} - \mu\bar{X}\right)(L^2-(1+\mu^2)\,\bar{X}(1)^2)}{\left(\sqrt{L^2-\bar{X}(1)^2} - \mu\bar{X}(1)\right)(L^2-(1+\mu^2)\,\bar{X}^2)}\right] \tag{2.14}$$

2.2.4 Results

Figure 2.2 shows some representative optimal slope and wedging chamfers; both of which are slightly concave. If the aspect ratio ($S = y_o/x_o$) is too small the peg will wedge into the chamfer. If the aspect ratio is between the aspect ratio of the wedging chamfer and the optimal slope chamfer a numerical solution (see Equation 2.10) may be found for some $C: 0 \leq C \leq \infty$ And finally, for aspect ratios larger than the aspect ratio of the optimal slope chamfer the minimum energy chamfer reduces to the optimal slope chamfer. Also as L increases all chamfers become steeper and at $L \to \infty$ the wedging and optimal slope chamfers are straight line chamfers with aspect ratios $\tan(\theta_o + \beta)$ and $\tan(\theta_o+\beta)+\sec(\theta_o+\beta)$ respectively, where $\tan\beta = \mu$.

2.3 MINIMUM ENERGY CHAMFER EXPERIMENT

2.3.1 Introduction

An experiment was conducted which attempted to support the existence of the minimum energy chamfers developed in Section 2.2. Although it is not possible to verify experimentally that a chamfer is a minimum energy chamfer (since it must be compared to an infinite number of chamfers), some insight can be gained by comparing it to a finite number of chamfers — such as a family of straight line chamfers. In the experiment only three chamfers were used: one optimal slope chamfer and two straight line chamfers. One of the straight line chamfers was flatter (chamfer #1, S = .6) than the optimal slope chamfer and the other was steeper (chamfer #3, S = 3.75) than the optimal slope chamfer (chamber #2, S = 1.40, $\mu = 0.15$, L = 3, θ_O = 15°). According to the theory developed, the flattest chamfer will be nonoptimal primarily because the insertion forces are too large, and the steepest chamfer will be nonoptimal because the insertion forces must act over a very large distance. Also, since minimum vertical work chamfers are the same as minimum frictional work chamfers, either criterion may be implemented experimentally. Since the vertical work criterion is simpler the experiment was based on this criterion.

2.3.2 Experimental Results

In Figure 2.3 the experimental apparatus is shown and Figure 2.4 plots the insertion force data for each of the chamfers. Table 2.1 compares the experimental results with the theory predicted. Good agreement was achieved (experimental error < 10%) so that chamfers with aspect ratios larger or smaller than the aspect ratio of the optimal slope chamfer will be nonoptional with respect to insertion energy.

Figure 2.3 Experimental apparatus — close-up of peg, chamfer, and Draper force sensor.

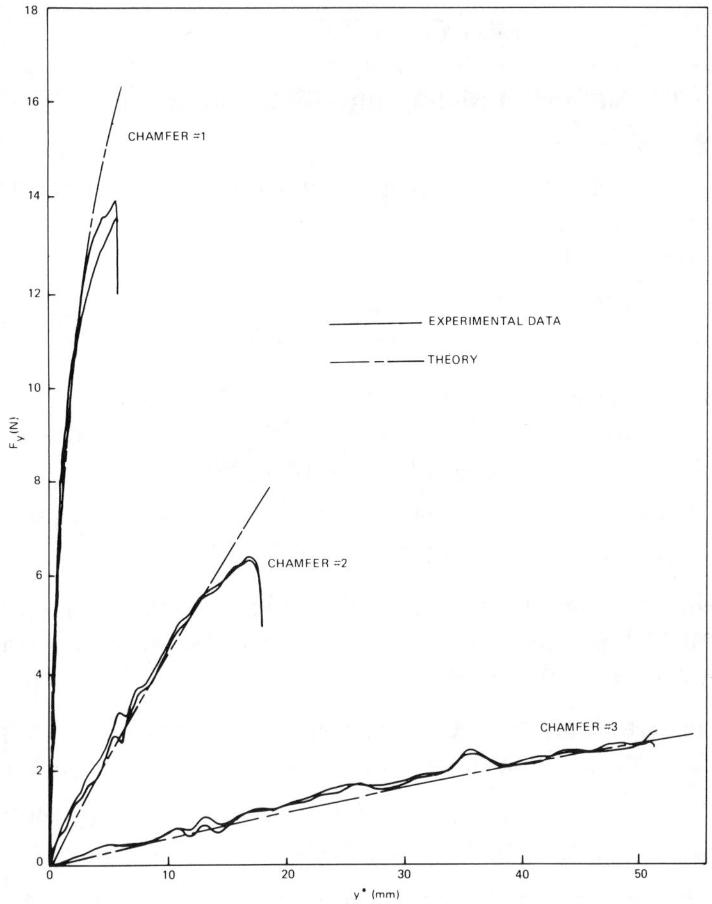

Figure 2.4 Insertion force versus depth — all chamfers.

Table 2.1 Comparison of Theoretical Energies with Experimental Energies

Chamfer	Theoretical Energy For $\mu = 0.15$ (Nmm)	Experimental Energy For $\mu = 0.15$ (Nmm)	Theoretical % Energy More Than Optimal	Experimental % Energy More Than Optimal
1	77.41	73.31	22.2	18.1
2	63.35	62.07	0	0
3	74.85	80.26	18.2	29.3

3. SUMMARY

A theoretical model of a compliantly supported rigid peg entering a compliant hole was used to understand the mechanics of an idealized assembly in terms of the assembly phases, the friction, geometry and compliance. This led to the classification of different assembly phases and an examination of the "insertion force versus depth plot".

The sensitivity of this plot during chamfer crossing to the slope of the chamfer has led to the area of chamfer design. Chamfers were designed subject to minimum energy criteria and an optional slope chamfer was determined. An experiment was conducted which supported the existence of minimum energy chamfers.

LIST OF REFERENCES

1. Hennessey, M.P., Compliant Part Mating and Minimum Energy Chamfer Design, Master's Thesis, MIT, September 1982.
2. Whitney, D.E. Rigid Part Mating, C.S. Draper Laboratory Report No. P-1447, Session No. 8, November 1981, presented at the 2nd Annual Seminar on Advanced Assembly Research, Cambridge, MA, November 1981.
3. Whitney, D.E., et al., Part Mating Theory for Compliant Parts, First Report, C.S. Draper Laboratory Report No. R-1407, September 1, 1979 to August 31, 1980.
4. Gustavson, R.E. and Hennessey, M.P., Compliant Part Mating, C.S. Draper Laboratory Report No. P-1447, Session No. 9, November 1981, presented at the 2nd Annual Seminar on Advanced Assembly Research, Cambridge, MA 1981.
5. Nevins, J.L., et al., Exploratory Research in Industrial Modular Assembly, C.S. Draper Laboratory Report No. R-1218, September 1, 1977 to August 30, 1978.
6. Whitney, D.E., et al., Part Mating Theory for Compliant Parts, C.S. Draper Laboratory Report No. P-1390, November 1981, presented at the 9th Conference on Production Research and Technology, Ann Arbor, MI, November 1981.
7. Crandall, Stephen H., Dahl, Norman C., and Lardner, Thomas J., An Introduction to the Mechanics of Solids, second edition, Chapter 2, McGraw-Hill, New York, 1972.
8. Whitney, D.E., Optimum Step Size Control for Newton-Raphson Solution of Nonlinear Vector Equations, IEEE Trans. Auto. Control, Vol. AC-14, No. 5, October 1969.
9. Sokolnikoff, I.S. and Redheffer, R.M., Mathematics of Physics and Modern Engineering, second edition, Chapter 5, Section 15, McGraw-Hill, New York, 1966.
10. Courant and Hilbert: Methods of Mathematical Physics, Vol. , pp. 214-216, Interscience Publishers, Inc., New York, 1953.
11. Dwight, Herbert B., Tables of Integrals and Other Mathematical Data, fourth edition, p. 49, MacMillian, New York, 1961.
12. Gradshteyn, I.S., and Ryzhik, I.M., Table of Integrals, Series, and Products-Corrected and Enlarged Edition, pp. 157, 160, Academic Press, New York, 1980.
13. Arai, T. and Kinoshita, N., The Part Mating Forces that Arise when using a Worktable with Compliance, Assembly Automation, August 1981, pp. 204-210.

ACKNOWLEDGEMENTS

This material is based upon work performed at the Charles Stark Draper Laboratory and supported by the National Science Foundation under Grant No. DAR79-10341. Any opinions, findings, and conclusions or recommendations expressed in this publication are those of the authors and do not necessarily reflect the views of the National Science Foundation.

This paper is based on the author's Master of Science thesis from the Department of Mechanical Engineering, MIT, September 1982.

Low Cost Absolute Position Data for Robots

Seymour Lanton
ILC Data Device Corporation
Bohemia, New York
Bart Sakasai
ILC Data Device Corporation

Abstract

Position sensing is one of the key components in a robot feedback system. Absolute position encoding is desirable in many applications. Unfortunatley, absolute position encoding with a resolution of 10 bits (1024 counts) or more is very expensive. The following discussion is on a rugged, resolver-based, low-cost absolute encoding system. This resolver system is centered on a new state-of-the-art hybrid microelectronic resolver-to-digital converter. A single turn system with a 1:1 correspondence to the robot movement as well as a dual transducer system where very high resolution can be attained through a multi-turn concept are discussed.

INTRODUCTION

A reliable low cost absolute position encoder system for robot application is needed. Absolute position data has been available for some time, however at prohibitive prices. A reliable resolver based position encoding system is now available through a development in low cost resolver-to-digital conversion electronics. This new industrial grade converter is based on a custom LSI chip, innovative circuit layout and cost saving packaging. Although the converter is new, the circuit design is of a proven design which has been in use in aerospace and military applications for many years.

BACKGROUND

Resolver

The resolver is a rugged electromechanical position transducer which is part of the synchro family of transducers. It is basically a variable transformer which has its output as a function of the angular shaft position (Figures 1 and 2). It is similar to an electric motor in its rugged construction. The stator consists of windings which are 90° apart, located on a slotted cylindrical lamination and the rotor is made up of a shaft, lamination, windings, slip rings and brushes. Brushless resolvers which couple the primary excitation voltage through a rotary transformer rather than through slip rings and brushes are available. The brushless resolvers reduce friction error from brushes, eliminate slip ring wear and brush bounce, and are safe for use in hazardous environments. For intrinsically safe requirements such as in paint spraying applications, Factory Mutual approved resolvers are also available.

The resolution of a resolver is theoretically infinite while its accuracy is typically about 7 minutes of arc for a size 11 brushless. Resolvers of this class are about $80 each (100 pieces) while a resolver with about 3 minute accuracy is about $90 each (100 pieces). The small size of a resolver (e.g., size 11 equals 1.1 inch diameter X 1.6 inch length) makes it ideal for placement into size limited locations such as in a robot arm.

Resolver-to-Digital Conversion Electronics

Since the resolver is an analog position sensor, a resolver-to-digital converter is required to convert the analog data into a digital format capable of interfacing with a microprocessor. Until recently, a converter with a 14 bit (16,384 counts per revolution) resolution was in the range of $280 each (100 pieces), however a low cost hybrid microelectronic converter designated the SDC-19100 is now available for about $120 each (100 pieces) which now makes the absolute 14 bit encoding systems truly affordable (Figure 3).

These new converters are of a type II tracking loop design (Figure 4), which allows for a real time data output. Basically, the converter operates as follows: It receives the modulated sine/cosine outputs of a resolver which is typically 11.8V line-to-line with a frequency of 400Hz or 2.5KHz. The shaft position information which rides on the carrier frequency is processed into a sin (θ - Ø) format where θ is the shaft position and Ø is the internally generated angle. This difference signal is then demodulated and fed into a circuitry which produces clock pulses that are counted by the up and down counter. The output of the counter is continually updated and always available. These converters are resistant to reference signal variation and electrical noise because of the ratiometric amplitude comparison technique used.

The features of the converter include (Figure 4) a 3-state 8 bit byte output for microprocessor interfacing, velocity output (VEL) scaled typically to ±5 volts, direction output (U), count output (CB), maximum tracking rates of 19,200, 4800 and 1200 RPM for 10 bit (1024 counts), 12 bit (4096 counts) and 14 bit (16,384 counts) units respectively, accuracies of ±21, ±8.5 and ±5.3 minutes, 1 LSB repeatability which is equivalent to 0.35° (10 bit unit) 0.088° (12 bit unit) and 0.022° (14 bit unit), and a transparent latch which allows for uninterrupted tracking even while an inhibit is applied for asynchronous data extraction. Since the heart of the converter is a specially designed LSI, the parts count is low and consequently the mean time between failure (MTBF) is high.

The unique layout design of the converter allows for a single substrate design for all 24 variations. The input voltage and frequency variations are controlled solely by selecting the correct electrical connections within the microcircuit and therefore the component count remains the same for all models.

To maintain the high reliability of a resolver based position encoding system, the active devices within the hybrid microelectronic are sealed with a junction coating. In addition, the outer cover is sealed with an electronic grade RTU compound. A 48 hour burn-in at the maximum operating temperature to weed out infant mortality further assures a reliable system.

APPLICATION

Absolute Position - Single Turn

For a single turn absolute position encoding, the resolver can be attached to a shaft with a 1:1 correspondence to the required axis. This is possible since a robot axis rarely rotates beyond 360° (Figure 4). A 14 bit (16,384 count) system has a unique digital code associated with every 0.022° of arc, thus the data output is absolute. An incremental system can be replaced with this low cost absolute system depending on the requirement. For example, if a 512 pulses per revolution incremental device is being used at a rate of 32 revolutions to attain full scale displacement of the axis, then a 14 bit (16,384 counts) resolver based single turn absolute system can attain the same resolution (Figure 5).

In many applications only the column, shoulder and elbow movements require a resolution in the range of 16,384 counts (14 bits) while the wrist movements can be satisfied with only 4096 counts (12 bits) resolution.

Absolute Position - Multi-Turn

For application requiring very high resolutions, an absolute multi-turn position encoding system can be implemented through a combination of two position transducers. One low resolution transducer is used as a "coarse" indicator while the second higher resolution transducer is used as a "fine" indicator. In order to implement a truly absolute system both transducers must be of the absolute type (Figures 6 and 7). In order to maintain the high reliability of a resolver system it is recommended that a dual resolver system be used. However, depending on the application a potentiometer or a low resolution absolute optical encoder may be used as a coarse indicator.

As an example, if the full scale movement requires a minimum position sensing resolution of 163,840 counts at a full scale displacement of 300° of arc then a dual transducer system is required. The resolution for this example is about 0.0018° of arc. Any combination of a 14 bit resolver system and a 10 bit resolver system (Note: resolver-to-digital converters generally have a minimum resolution of 10 bits), a potentiometer with a 4 bit A/D or a 4 bit absolute optical encoder can be used.

Support Functions

Since the converter contains a 3-state latch it can interface directly with an 8 bit microprocessor or by enabling both 8 bit bytes interface with a 16 bit microprocessor. The velocity output which is a by-product of the type II tracking loop design, although limited to ±5 volts scaling, is an added bonus for most applications. The tracking rates of the converters are fast enough for almost all robotic applications. The 20 revolutions per second tracking rate for a 14 bit (16,384 counts) converter is equivalent to 327,680 counts per second. The 1 LSB repeatability of the SDC-19100 converter is an important criteria, especially for the playback or the teaching type robot applications, since the movements fed into it by the operator can be repeated accurately with the resolver system.

CONCLUSION

For applications requiring absolute position encoding, a low cost system based on the rugged resolver and state-of-the-art microelectronic technology is currently available. This system allows for placement of the shaft position sensor in size limited areas and can be implemented as a single turn or a dual transducer multi-turn absolute system. Although potentiometers or absolute optical encoders can be used for the coarse position input, it is recommended that a resolver be used, especially for harsh environments.

References:

1. ILC Data Device Corporation, "Synchro Conversion Handbook"

2. Muirhead Vactric, "Synchro Engineering Handbook"

3. B. Sakasai, "Shaft Encoders", Measurements and Control, September 1982, p. 178-185.

4. S. Muth, "Consider Basic Parameters When Choosing S/D Converters", EDN, June 23, 1982.

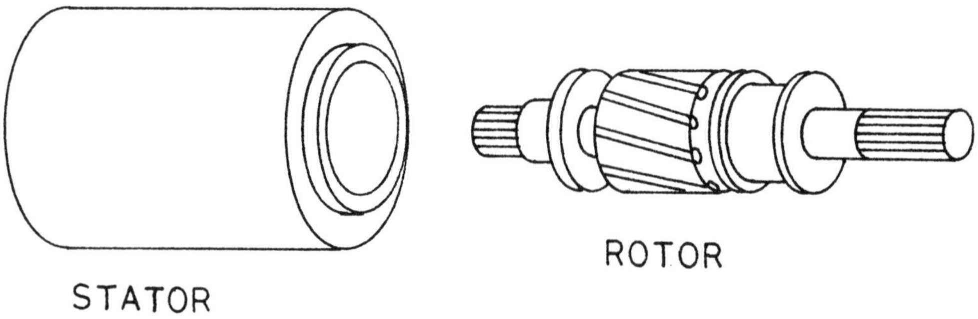

Figure 1 - The brushless resolver couples the primary excitation voltage through a rotary transformer rather than through slip rings and brushes thereby eliminating friction error, slip ring wear and brush bounce.

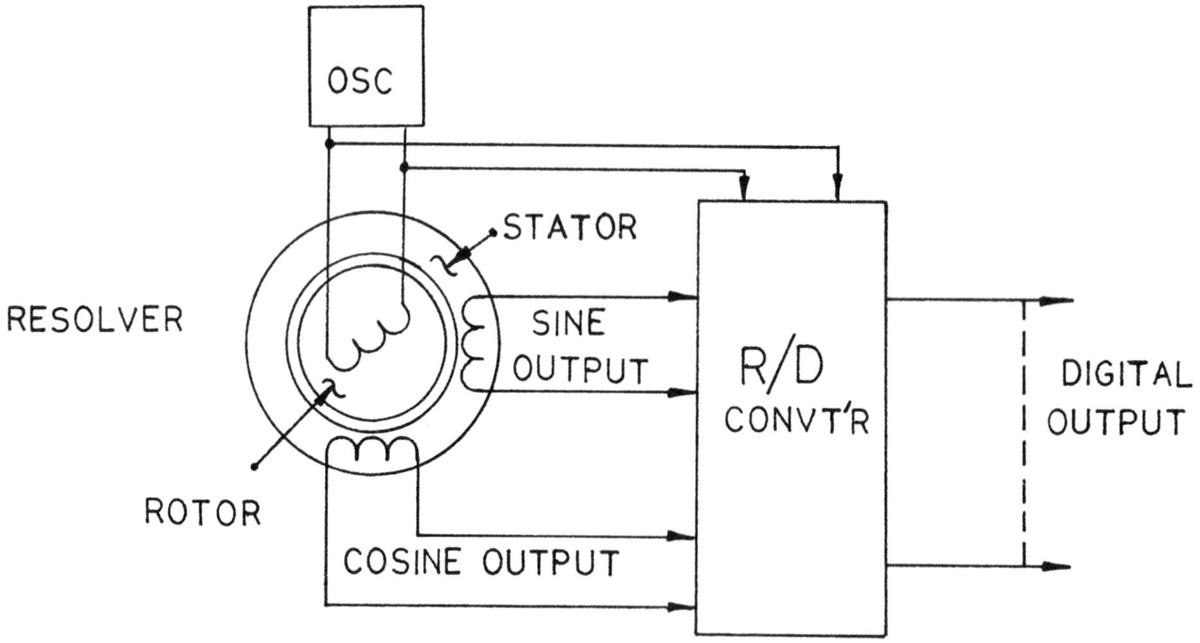

Figure 2 - The rotor of the resolver is excited by a reference source which is typically 26V at 400Hz or 2.5 KHz. The output provided by the resolver stator is 11.8V at the reference frequency. In order to allow for the implementation of a cost effective square wave reference source, the 26V/11.8V resolver can be driven at 4.4V.

Figure 3 - The custom LSI chip, innovative substrate layout and cost saving packaging of this new converter allows for the implementation of a rugged low cost absolute position encoder system.

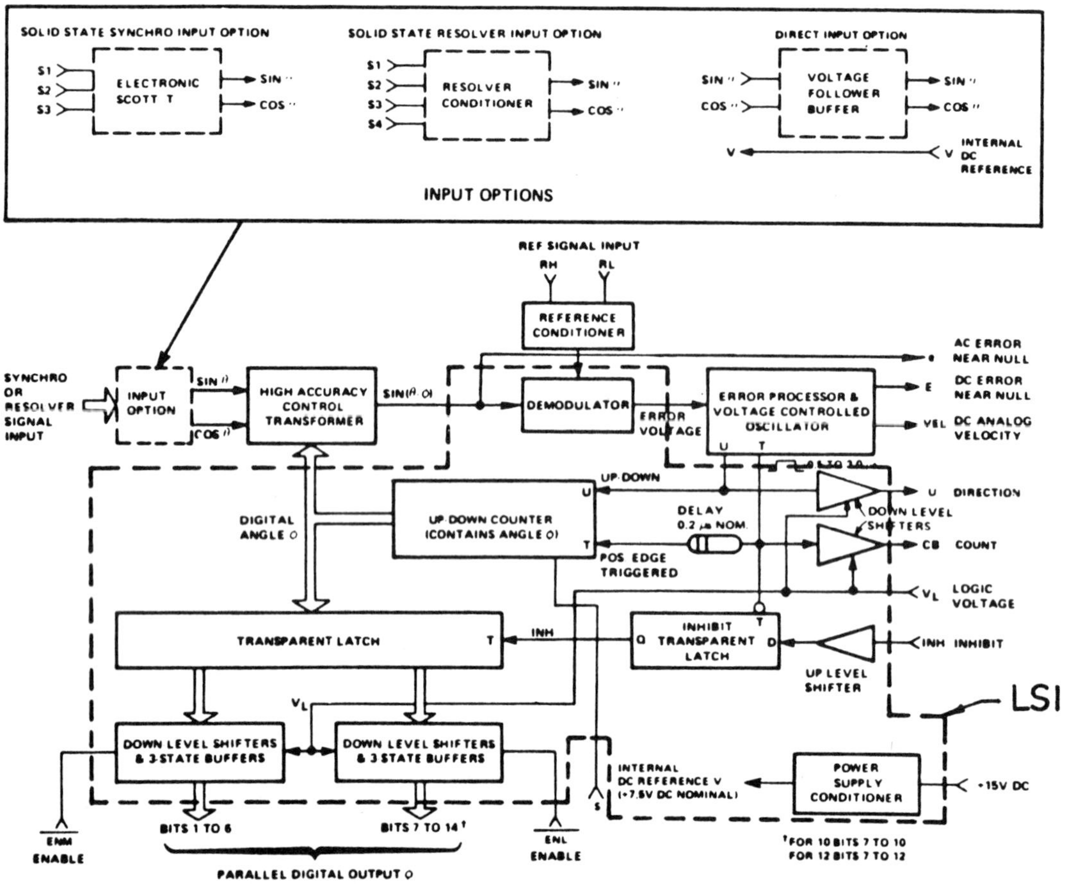

Figure 4 - As shown in the above diagram, most of the functions have been incorporated into the custom LSI chip. Since the component count has been greatly reduced with the use of the LSI chip, the Mean-Time-Between-Failure (MTBF) increased. Input options allow for resolver as well as synchro type inputs.

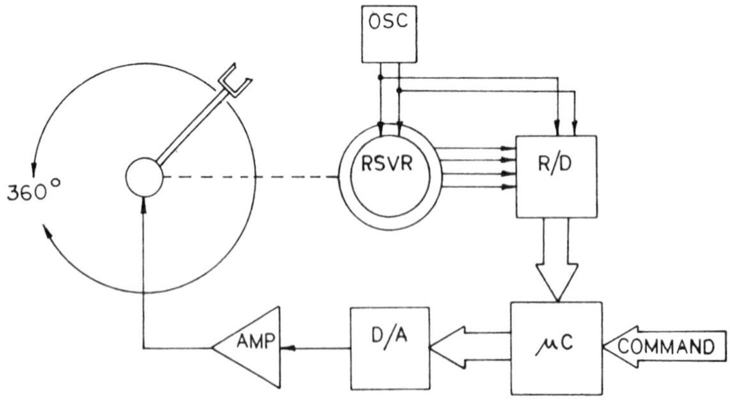

Figure 5 - Since the robot axis movement rarely exceeds 360°, a 1:1 correspondence between the robot movement and the resolver (RSVR) rotation yields an absolute encoding system. Since the low cost resolver-to-digital (R/D) converters are available in resolutions up to 14 bits (16,384 counts), an incremental system based on a 1024 count X 16 revolution can readily be converted into an absolute system.

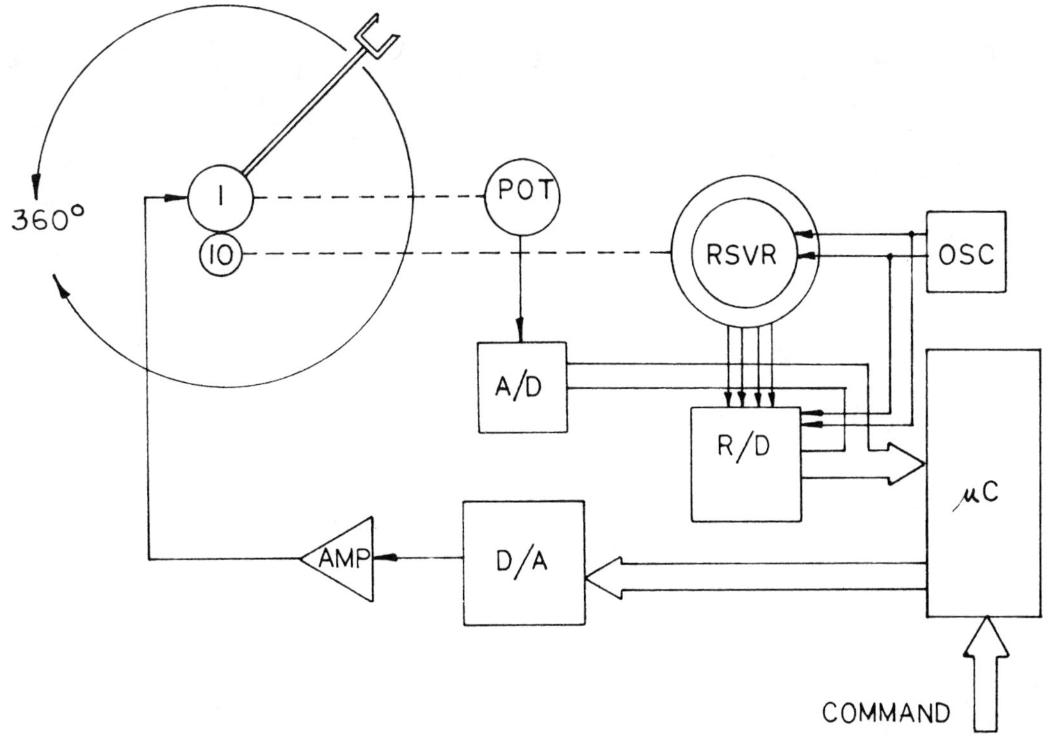

Figure 6 - For absolute multi-turn applications, a dual transducer system can be used. The above example depicts a 163,840 count per full scale robot movement system. The coarse position indicator which is in 1:1 correspondence with the robot movement identifies the sector in which the robot axis lie and the 10:1 rotation of the resolver, breaks down each 1/10 of the coarse movement into 16,384 counts. For harsh environments, it is recommended that the coarse indicator be replaced with a resolver.

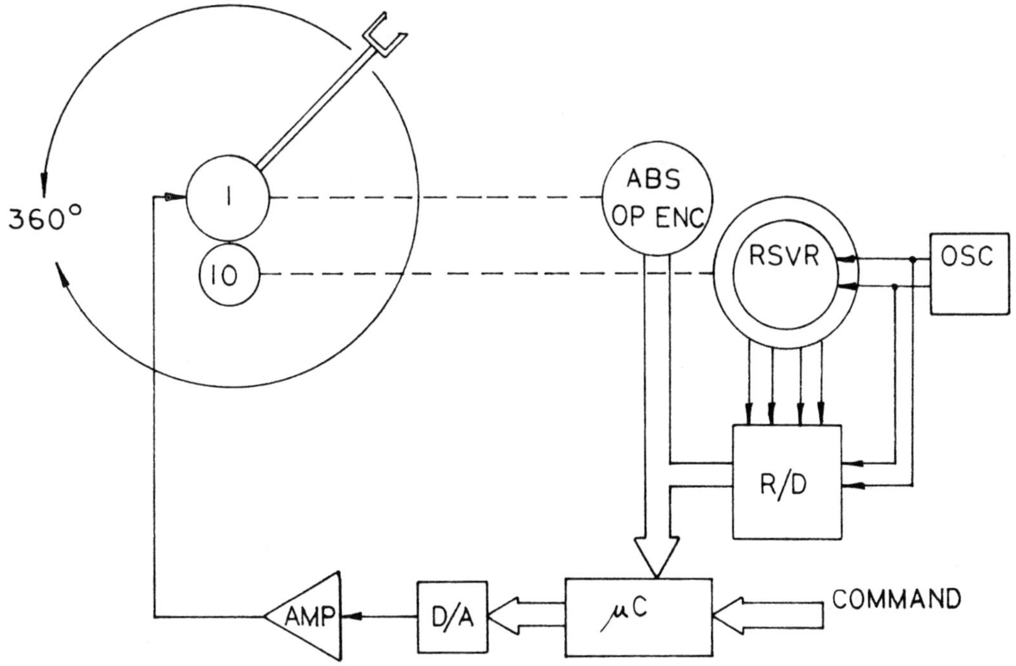

Figure 7 - The use of a low resolution absolute optical encoder eliminates the requirement for a separate converter since the self-contained optical encoder provides digital data. However, for rugged requirements the resolver should also be used to provide coarse as well as fine position information.

CHAPTER 15

ROBOTICS PROGRAMS IN EDUCATIONAL COMPLEXES

Training the Key to Success in the Use of Robots

David M. Osborne
ASEA Inc.
Troy, Michigan

Abstract

Proper training, before and after a robot installation, is an important factor in the economic use of robots. The main areas in which training affects the use of robots are as follows: (1) removing the robot's false mystique, (2) properly utilizing the robots potential, and (3) maintaining the robot for greatest productive life. The paper examines the timing, curriculum, personnel selection, and goals of each of these areas, and outlines a program that could be implemented by the user to train his personnel in the use of robots.

ROBOT INSTALLATION, THE NEED FOR TRAINING

The training needed to properly support a robot installation is different than the training needed for other equipment in the industrial realm. The differences can be atributed to several factors, among which are: their newness, the misperception of robots, and the flexible nature of their use. Robot training also differs from other training in it's potential for giving a return on investment, when properly implemented.

In a relatively short time, a host of companies have entered the robot supply business. Although some of these companies have been in business for a long time, the robot business is new to them. Different customers are approached with the robot product, and new ways of doing business are instituted. No matter how much development the product has undergone, it is new, and confusing, to the user. There are no past examples for the potential user to fall back on. A new mill may be presented to the market as an improvement of an existing product. The customer has some basis for comparison, and a certain previous knowledge of the features being presented, but robots are new even to those that sell them, and few people have any experience.

In some instances, the company that is selling the robot is not new to the customer. Machine tool builders with a long reputation and other makers of industrial products are also marketing robots. The robot product is still sufficiently new that it presents new problems to the seller and the customer alike. The entire manufacturing business and indeed the entire western culture has much to learn about robots and due to the constraints of economics, very little time to learn it.

Robots are also new in their technology. Robots use the very latest of computer and sensory technology. Even very simple robot moves can be the result of complex computer analysis. A customer's ability to understand new equipment and to properly use it can be greatly strained when any high technological process is involved, but especially so if the technology is used in a fundamentally new way as with robots. Many company personnel that will make decisions about robots, received their formal education after the technology that has allowed the creation of the modern robot was developed. Some continuing education is needed.

Perhaps more confusing to the potential customer that the youth of the robot market or the high technology it uses, is the ability of the robot to enter traditionally human facets of manufacturing. Jobs that once could only be accomplished by human labor, can now be simply programmed into the robot and commanded at will. The programmable feature of robots is often misunderstood and for it to be properly utilized a clear picture of what can be accomplished with it must be gained. Robots are misperceived.

Because so much fiction has preceeded the use of real robots, many that would use them do not because they cannot separate the facts about robots from the images that have grown up in the fiction. Even the worker that may have only limited contact with a robot after it is installed, may consider the robots potential, a threat. Perhaps worse for the manufacturing industry is the erroneous view that robots are the universal solution to the current problems of faultering productivity. Training is the least expensive way to educate potential users to the real abilities of the modern robot. Robots are much too expensive to buy one just to see what it can do. Like any feasibility study, robot education can be seen as an investment. With proper training all the people that work on the decision to use robots can most adequately do there jobs and save money on the larger scale.

WHO MUST BE TRAINED ABOUT ROBOTS

For the best robot installation, people must be trained from areas of the company that may seem far from the actual use of the robot. Although each case is different the people we generally expec to see trained include:

- Process Engineers
- Tooling Engineers
- Material Handling Engineers
- Plant Engineers
- Engineering Staff Designing the Product
- Maintenance Staff
- Production Operators
- Production Supervision
- Personnel Staff
- Purchasing Staff
- Plant Personnel working near the robot that may become uneasy at its presence

While it is obvious to most that the people that will actually implement the robot and that will daily use it as a part of their work must receive training, it is a relatively new concept that those that will design the product upon which the robot will work must become familiar with its capabilities.

There is no back log of trained personnel which can repair and install a robot. A company may have maintenance staff that is acquainted with the technology used in the robots, but the actual implementation of the robot as a total of many parts is so new as to preclude the existence of a large number of trained people that can work on them confidently. Plant personnel already skilled in the technology of computers and other robot components, can learn the necessary robot skills quickly but must be specifically trained in the particular features and needs for the equipment to be used.

It is somewhat unusal to consider training people from the personnel and the purchasing departments just because a new type of equipment is being used, but the unique features of robots will sometimes necessitate decisions and scheduling on their part that will be both easier and more successful if they are properly trained.

THE LEVELS OF TRAINING

While we have considered that many different type of people and with different levels and types of education must be trained, we must consider that this training can be very simple and inexpensive if that is what is called for. Proper training would not take any more time or effort than is necessary to do the job.

For purposes of the program presented here, we will consider five levels of training:

1. Extensive Technical Training
2. Simple Technical Training
3. Operation Training
4. Features/Economies Training
5. Awareness Training

Extensive technical training would include the most intricate details of the working of the robot. The details for the customer to properly maintain the robot without the help of the vendor. This training might not be as necessary if the user is to have a service contract with the vendor, but must still be considered for the manufacturing engineering staff that will rely on the robot's ability to accomplish a task. Simple technical training might include the development of skills such as the precussory repair of the robot and its ability to interact with other equipment.

Operational training would include such features as the programming of the robot and the day to day operation the robot will see when in use. Education on the features and the economies of robot operation could include the ability of various robots to accomplish the different tasks of the user and the relative merits the robot has for the economic operation of the plant.

Awareness training is a simple explanation of the facts about robots that will allow everyone to properly judge a robot on its real features.

WHO NEEDS WHAT TRAINING

The following graph shows the training that might be expected for each of the groups mentioned.

	1	2	3	4	5
Process Engineers		✓	✓	✓	
Tooling Engineers			✓	✓	
Material Handling Engineers			✓	✓	
Plant Engineers				✓	
Engineering Staff Designing the Product				✓	
Maintenance Staff	✓		✓		
Production Operators			✓		
Production Supervision				✓	

Personnel Staff					✓
Purchasing Staff					✓
Plant Personnel working near the robot that may become uneasy at its presence					✓

THE GOALS OF TRAINING

Just as the training needs of different personnel vary with their contact with the robot, the goals of that training vary. The major goals of training include: removing the robots false mistique, maximizing the robots potential, caring for the robot equipment properly, and selecting the robot that will be of greatest use to the user.

When robot awareness training is effective, the student should no longer harbor false notions about the use or misuse of robots. Robots are not some monster created to hurt workers, neither are they able to mow lawns. While there is little direct benefit for the customer in providing awareness training for their workers, many potential problems can be avoided if such training is provided. At the very least the current popular interest in robots can create a positive feeling toward the users company as one of innovation and with enough concern for it's workers to at least educate them in the use and features of new equipment.

Education that is to provide the manufacturing staff with the ability to work with robots as a part of the production process has goals that are far more reaching. There should be material that will allow the engineer to calculate the maximum return possible for a robot installation, and to select both the robot and the surrounding equipment to accomplish this end. The student should be able to avoid installations that would make unrealistic demands on the equipment and which might jeopardize the large operation if the robot component does not respond exactly as expected. We would expect a student that would be able to recognize other possible installations that would be good candidates for robot use.

Maintenance education should give the student the ability to repair a robot quickly and the ability to quickly diagnose a problem once it develops. The routine maintenance that must be applied should be covered along with the symptoms of neglect that would alert the maintenance personnel to possible dangers. When special repair or fault tracing equipment is needed of available, its use should also be a part of this type of class.

A CURRICULUM TO ACHIEVE THE GOALS

Much of the training that is needed by the customer can be provided by the robot vendor. Most of the larger robot suppliers have regualr training that can be used to meet our goals. The full repertoire of training that we would expect fo fill all our needs is listed below:

Electrical Maintenance

Mechanical Maintenance

Scheduled Maintenance

Robot Use Engineering

Non Technical Features

Demonstrations and Literature
for Awareness Training

Some of the individual classes might be combined to allow several aspects of robotic technology to be taught at once. The table shows how the classes could be used by each of the personnel groups.

	Electrical Maintenance	Mechanical Maintenance	Scheduled Maintenance	Robot Use Engineering	Non Technical Features	Demonstrations and Literature for Awareness Training	Robot Programming
Process Engineers				✓	✓		✓
Tooling Engineers				✓	✓		✓
Material Handling Engineers				✓	✓		✓
Plant Engineers			✓	✓	✓		✓
Engineering Staff Designing the Product					✓		✓
Maintenance Staff	✓	✓	✓				✓
Production Operators					✓		✓
Production Operators					✓		✓
Production Supervision			✓		✓	✓	
Personnel Staff						✓	
Purchasing Staff						✓	
Plant Personnel working near the robot that may become uneasy at its presence						✓	

TIMING

An important aspect of proper training is the correct timing of the training. Although the robot might be ordered and the engineering started long before the installation, the training for operators should not preceed the actual use of the robot by more than a few months. Most robots are complicated enough that a long time without a physical robot to work with might allow the trained personnel to forget their skills before they are of any use to the company. It is also of little use to train personnel after their work with a robot is passed. (Although there might very well be another installation to follow that could take advantage of the late training) the time line graph which follows shows how each of the classes offered would fit with the order and use of a particular robot.

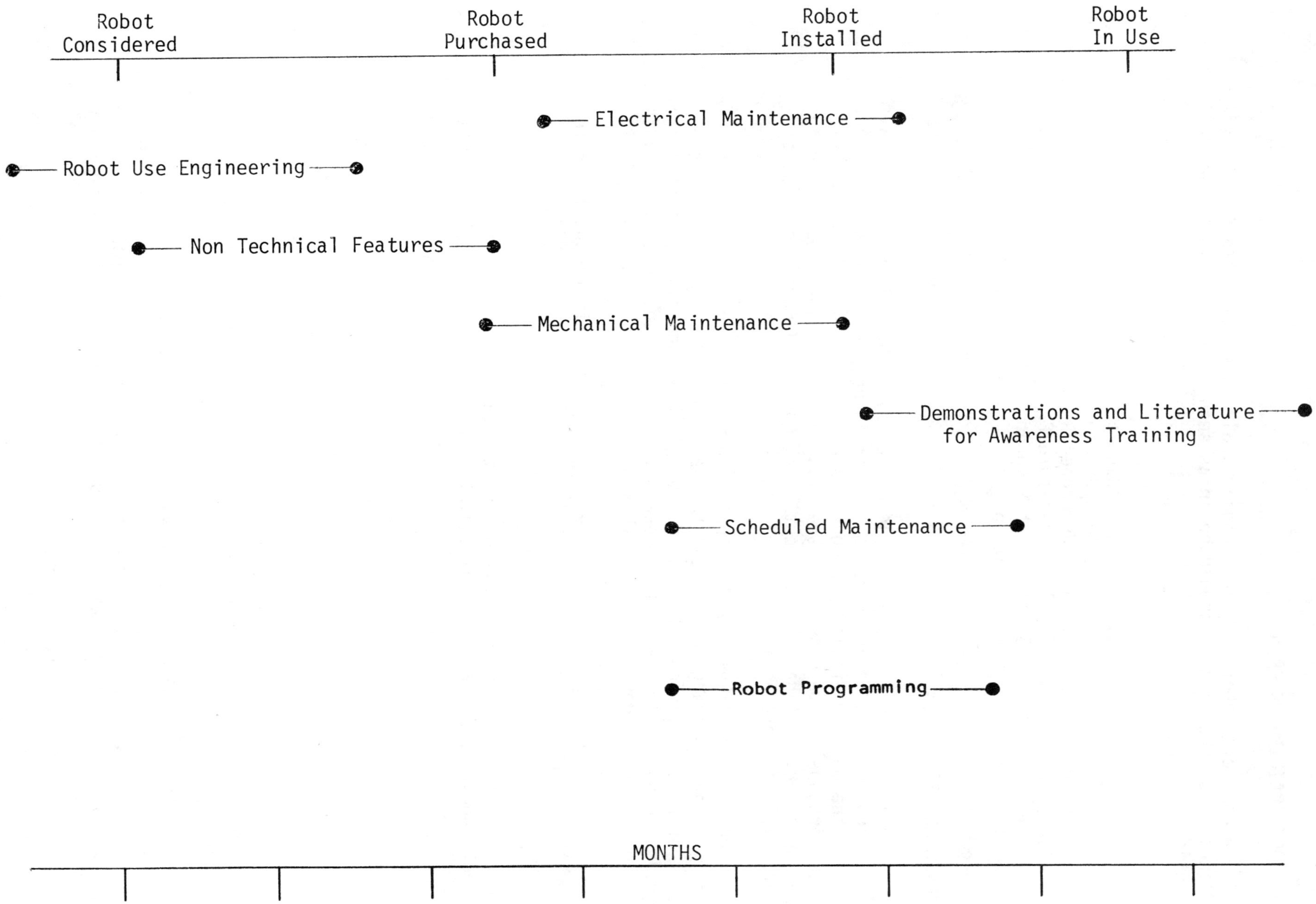
Robot Considered
Robot Purchased
Robot Installed
Robot In Use
Electrical Maintenance
Robot Use Engineering
Non Technical Features
Mechanical Maintenance
Demonstrations and Literature for Awareness Training
Scheduled Maintenance
Robot Programming
MONTHS

A ROBOT TRAINING PROGRAM

A complete robot training program would have the following six steps. Planning for the program should begin as soon as a robot installation is considered.

1. Set Time Table
2. Plan Curriculum/Set Dates with vendors/ allocate space
3. Chose Personnel/Notify
4. Plan Literature/Print
5. Implement Program
6. Follow Up

PERSONNEL SELECTION

While the user may have little choice over who will work with the robot and who will be assigned to take training, care should be taken to select the most beneficial personnel whenever possible. We consider the best personnel to be those that are: willing, able, in-line, not ready for transfer, that have the correct background, and have the greatest potential for interaction with the robots.

No employer likes a worker that is unwilling, but we find that there are many hard working individuals that are not willing to take on the task of robot training. Some by virtue of their age of educational background do not feel they will succeed with the training. Others feel that by keeping the older safe skills they will remain on secure ground with their status as a tradesman or professional. Whatever the reason, it is best to ask each of the prospective training candidates about their feelings for the training.

We must also consider the very real situation that some of our good workers are not suited for some of the training. The use of computers in robots and of some very special equipment demands for some of the training level of intellect not found in every worker. While alsmost everyone could be enriched by the training, if only a limited number of people may be trained (this is the usual case) then we should consider the best and most capable candidates for our training.

Care should be taken that we also observe the rules of seniority within the plant to select the person who has earned the right to attend training. Although we might take a person with less seniority as a candidate for training it is good to make it obvious that the decision was made on some educational merit rather than the capricious decision of management.

How long will the student be working with the robot? If the time is to be a short one, then it might be best to train another at the onset of the training, rather than be required to train another person at a later date. Just as important, will the student be working with the robot on a regular basis? If the amount of contact is small, then perhaps the selection should fall to someone else.

COSTS

Costs will of course vary but a few rules of thumb can help to approximate the outside costs. Robot classes offered by a vendor, about his equipment, will usually cost about $ 1000.00/student week, when given at the vendor's facility. The cost to have an instructor from the vendor come to the plant to offer a course using the users equipment, are about $1000.00/Instructor day. Many vendors give free classes with the sale of each robot. The cost to the user of time spent by personnel during training must be compared to the cost of improper robot utilization and lost production by unnecessary dowtime.

How Should the University Teach Robotics?

Barry Irvin Soroka
University of Southern California
Los Angeles, California

Abstract

A university robotics curriculum cuts across the traditional disciplines of Computer Science, Electrical, Industrial, and Mechanical Engineering. An Introduction to Robotics provides an overview of current techniques suitable for both managers and for those preparing for careers in research and development. Many topics are important: kinematics, dynamics, control theory, sensors language design, artificial intelligence, economics, and applications. Follow-up courses in Robot Software and Analytical Methods are tailored for the two cultures from which students seem to come. The paper presents speculation on laboratory facilities needed for robotics education.

NEEDS AND PROVIDERS

No quantitative study or prediction has yet been made of America's current and future manpower needs in robotics. Nonetheless, certain needs have become obvious since the beginning of the Robot Revolution:

- personnel to maintain robots and robot systems,
- applications engineers,
- researchers,
- development teams,
- college teachers,
- large-scale retraining of current personnel, and
- general public education.

To fulfill these needs, there exist a number of providers of manpower and retraining:

- self-education,
- short courses,
- technical schools,
- 2-year colleges,
- 4-year colleges,
- universities,
- professional organizations (such as RI/SME).

Universities cannot fill all these needs, but they play an important role in producing R&D staff and in teaching robotics to both new students and old.

ROBOTICS AND THE UNIVERSITY

Historically there has been considerable interaction at the research level between universities and the robot industry. Stanford and MIT (and their associated local research arms) have been responsible for major breakthroughs in the following areas:

- robot design and kinematics,
- force sensing and control,
- compliance,
- simulation of robot systems,
- computer vision, and
- robot programming languages.

These research achievements have been translated into the industrial robots sold today, and the graduates of university robotics programs work for many robot companies.

Now there is a pressing need for education in robotics, and students both young and old are asking the university for service. Most entering engineering students display a macabre interest in robotics, fueled by current media hype. The university should expose these students to enough real robotics facts so that students gain realistic views of the state of the art and the current problems of robotics. On the retraining side, universities have an obligation to help update the engineering skills of industrial personnel in their communities. If local industry stagnates, then so does the university.

Universities must upgrade their personnel and facilities to meet the challenge of robotics education. Few current faculty have training or experience in robotics, and only a handful of PhDs are produced each year with a background in robotics. Texts are scarce, and the sources of robotics information are scattered among many conferences and journals. Professors must retrain themselves before they can teach robotics to their students. Industry can play a vital role in robotics education by lending personnel and experience to local universities.

Robotics is inherently interdisciplinary, cutting across the boundaries of traditional engineering departments. Robotics is sexy, and many departments claim that robotics belongs to them: Mechanical Engineering, Industrial Engineering, Electrical Engineering, Computer Science, and others. (A similar situation existed when Computer Science was claimed by both EE and Mathematics.)

At USC, the Electrical Engineering Department took the lead in establishing a robotics curriculum by hiring a professor with the relevant background. Since that time, other departments have joined in a cooperative effort, now guided by a Steering Committee on Robotics Education (SCORE). SCORE includes members of all the academic departments of the School of Engineering, meets monthly, and reports directly to the Dean. Among the duties of the Steering Committee are:

- coordinating robotics courses across departments;
- acquiring and maintaining laboratory facilities for robotics education;
- informing the faculty of seminars and conferences;
- maintaining liaison with local industries regarding their educational needs.

CURRICULUM

USC now offers four specific courses in robotics:

EE 545	Introduction to Robotics;
EE 546L	Basic Robotics Laboratory;
EE/CS 547	Software Methods in Robotics; and
ME/EE 548	Analytical Methods in Robotics.

These courses are described in more detail below. They are supplemented by numerous related courses throughout the School of Engineering in fields such as:

- mechanics;
- control theory;
- CAD/CAM;
- engineering economics;
- computer vision; and
- artificial intelligence.

After completing formal courses, promising students are encouraged to continue work in robotics by means of seminars and individual research.

We have decided to wait a while before seriously considering the implementation of a degree program in robotics. For the moment, students are best advised to take their degrees in a traditional department, with a specialization in robotics.

EE 545: INTRODUCTION TO ROBOTICS

This course provides a survey of the state of the art and open problems in robotics. It is aimed at two groups of students: (1) those who will continue into advanced courses (R&D), and (2) those who need only an overview of the field (managers). The sole prerequisite is graduate standing in Engineering at USC. By the end of EE 545, students have a common vocabulary and shared experience which enable them to evaluate meaningfully the claims of vendors and publicists. Students enter with many misconceptions about what robots can and cannot do; they must be untaught most of the impressions they have picked up from the public media.

EE 545 has now been taught twice, to a total of about 140 students, more than half of them full-time employees of local industry, observing the lectures over USC's Educational TV Network. The TV Network provides talk-back capabilities (so that remote students may interrupt the lecturer with questions) and a courier service (for handouts and homeworks). Initially a lab was included as part of EE 545, but students at distant TV sites were unable to trek to campus with much frequency, and the lab has been spun off as EE 546L.

The observed mix of students has been:

- 40% Electrical Engineering;
- 40% Computer Science;
- 20% Mechanical Engineering, Industrial Engineering, and others.

We expect the distribution of students in EE 545 to equalize across the departments of the School of Engineering as its existence becomes more generally known. There is an interesting phenomenon, which I call "the two cultures," which makes teaching such a course a real challenge: about half the students come from math-intensive backgrounds, in control theory or mechanics; the other half come from computer backgrounds, in operating systems or programming languages. It is very easy to find exercises which are trivial for one group and impossible for the other.

The outline of EE 545 is shown in Figure 1. This is a view of robotics in-the-large, not restricted by the blinders of any single academic department or subfield. One topic, "The Factory," deserves special emphasis. The average student has never set foot in a factory, and it is important for the robotics instructor to convey the flavor of the modern manufacturing matrix.

There is currently no single satisfactory introductory text in robotics, so the following texts are used for EE 545:

(1) Richard P Paul, Robot manipulators: mathematics, programming, and control (Cambridge MA: MIT Press, 1981). This is basically a math book. Paul is heavy on kinematics, as we would expect of a Mechanical Engineer. His chapter on Control presumes considerable background in control theory. His discussion of programming is aimed more at mathematical issues than at robot programming languages. Applications are unmentioned.

(2) Joseph F Engelberger, Robotics in practice: management and applications of industrial robots (AMACOM, 1980). You can't claim to know robotics if you haven't read this book. I require that students read about one chapter per week, and they are tested on one application selected at random from Engelberger's catalog. This book effectively conveys the atmosphere of modern industrial robotics.

(3) Shahid Mujtaba & Ron Goldman, AL user's manual (Third Edition, Stanford University, Computer Science Department, Report STAN-CS-81-889, December 1981). I chose the AL robot programming language of Stanford to illustrate the capabilities and pitfalls of modern robot programming. It presents a history of robot languages and a set of graded examples of robot programs. To understand it, students need familiarity with any one computer programming language, and this familiarity is now universal.

(4) Other topics are covered by xeroxed handouts of survey articles (vision, geometric modelling) or other journal materials.

Lacking is an anatomy text for robots, which would describe the design and construction of robots from a practical point of view. Few manufacturers are willing to share their proprietary expertise, so we may have to wait a while before Soul of a New Robot appears.

EE 545 has two exams and several homeworks. The exams test familiarity with terms and concepts, and the homeworks require that students delve deeply into several of the topics covered. Last year's course required the following homeworks: (1) solve a robot arm; (2) annotate a control diagram; (3) select an application, design a multi-arm workstation, and program the application using AL. Because most of the students are remote, it isn't feasible to require their use of on-campus computers for actual programming problems. Were this not the case, I would assign homeworks involving USC's extensive library of robot simulators and robot design aids.

EE 546L: BASIC ROBOTICS LABORATORY

This course provides hands-on experience in programming small robots in USC's Student Robotics Laboratory. EE 545 is a co-requisite for the lab. We currently have two work-stations, each consisting of a Microbot MiniMover5 robot, an Apple-2 computer, and two floppy disks. Twenty students per term take the lab, utilizing the lab 10 hours per week total. Students work in pairs, with lab sessions two hours in length. Exercises include: simple applications (stacking, palletizing); complex programming problems (handwriting); user interface programs; arm solutions; synchronization of multiple arms; implementation of straight-line trajectories; and advanced projects in hardware, software, and mechanisms. Programming languages used are BASIC and 6502 assembler. Various language and robot manuals are used as texts.

For an introductory lab such as EE 546L, a standard industrial robot is not appropriate, altho advanced students would benefit strongly from working on such equipment. A large arm is not cost-effective for novice use, and it presents a significant safety hazard.

A number of companies now market small robots for educational purposes:

Colne (Armdroid);
Intelligent Artefacts;
Microbot;
Mitsubishi;
Rhino;
Systems Control (Smart-Arms).

These robots are inadequate for the vast bulk of experiments from which students would benefit.

An ideal educational robot would permit experiments in the following areas:

* Servo. Students should be able to tweak gains in order to get a feel for the physical meaning of the control parameters about which they are being taught.

* Trajectory. Students should be able to experiment with different sorts of trajectories, including straight-line, coordinated joint, and following a moving belt.

* Dynamics. There should be a good model of the dynamics of the arm so that students can experiment with different control algorithms, ranging from naive linearized models to schemes which include gravity and coupling.

* Programming. Students should be able to perform human factors experiments in robot programming, using alternative input devices (teach-box, menu, voice) and alternative languages. It should be possible to link the robot system to on-campus CAD/CAM and vision systems.

* Applications. The robot should be sturdy and repeatable enough to permit prototype applications programming.

We offer the following specifications of a robot suitable for educational use:

Cost	about $7000;
Reach	about 0.75-m;
Repeatability	0.5-mm;
Control	closed loop;
Computer	68000 or equivalent.

Unfortunately, there are no commercial offerings which match these specifications.

USC plans to upgrade the Student Robotics Laboratory to 10 workstations, each configured as follows:

robot arm;
local computer;
gripper sensors;
camera & binary digitizer;
back-lit table.

This system will permit a wide range of experiments, including practical exercises in computer vision.

The workstations will be unified by the following shared equipment:

a conveyor belt running around the lab;
a network enabling the shared use
of a printer and hard disk;
a host of binary sensors
(light beams and microswitches).

We plan to expand the laboratory to its final configuration in a series of smaller steps, hopefully with help from the industries to whom our students go.

EE/CS 547: SOFTWARE METHODS IN ROBOTICS

This is an advanced course with the following prerequisites:

introduction to robotics;
compilers;
operating systems.

Topics include:

robot programming languages;
robot architectures and operating systems;
human factors in robot interfaces;
geometric modelling and simulation;
collision avoidance and grasping;
planning robot tasks.

Exercises will include extensive software projects, including simulations and robot programming languages.

This course was prototyped as a Special Topics seminar in the fall of 1982, with ten alumni of the original EE 545. The class studied and compared existing commercial robot programming languages, including AL, AML, HELP, RAIL, and VAL.

ME/EE 548: ANALYTICAL METHODS IN ROBOTICS

This is an advanced course with the following prerequisites:

introduction to robotics;
control theory.

Topics include:

homogeneous transformations;
formal description of robot manipulators;
kinematic equations and their solution;
differential relationships;
trajectories;
control;
static forces;
compliance.

Exercises will involve computer manipulation of kinematic and dynamic equations, simulation of control systems, and experiments with actual arms (pending equipment).

DIFFICULTIES

Let me sketch a few of the problems which arise when trying to teach robotics in the university setting. First, the academic department is an extremely powerful beast, and interdisciplinary activities, such as robotics, pose a significant challenge to its power. Few professors can teach a robotics course on their own; most courses require the cooperative efforts of several faculty from several departments. Unfortunately, team-teaching is difficult to implement: who will get the tuition dollars? how will partial teaching credits be allocated to professors who team-teach a course? These are not insuperable problems, but they are non-routine.

Another problem stems from the immense current popularity of robotics. Professors need extensive administrative help in handling the deluge of inquiries which arise when word gets out that College X has a course in robotics. These inquiries include the following:

questions from prospective students;
demands for free advice;
proposed speaking engagements;
requests for interviews and news stories;
openings of meaningful exchange.

Professors are not generally trained to handle this influx of contacts, and it is important that the college take steps to free professors from an unwanted public relations role.

THE UNIVERSITY-INDUSTRY PARTNERSHIP

There are some things which universities cannot teach: reading books can't make an applications engineer, only practical experience can. So universities will not be producing applications engineers experienced in predicting, detecting, and fixing the subtle implementation bugs which make industrial robotics so challenging. Similarly, the goal of a university research project in robotics is not the production of a practical working industrial process. Because successful robotics requires both theory and practice, it is likely that the best robotics personnel would be produced by a an intelligent mixture of college courses and factory-floor experience.

Robotics offers a significant opportunity for university-industry cooperation, a two-way street. Today's college students are tomorrow's industry employees. Do you want them to be trained in robotics? From what schools do your employees come? See what can be done to upgrade robotics education at the colleges which are your manpower suppliers.

Let the university know that you are willing to share your expertise by sending lecturers to campus and by inviting classes to visit your factory. If you're a vendor of robots or other manufacturing equipment, then donate equipment, and make sure the college knows how to put it to good use. Together, university and industry can help each other survive the Robot Revolution.

FIGURE 1.

Outline of USC's EE 545, Introduction to Robotics.

INTRODUCTION
- What is a robot?
- Why have robots?
- The factory.
- Statistics.
- Precursors of robots.
- Parts of a robot.

EFFECTORS
- The geometry of 3-space.
- Kinematics, arm solutions, and trajectories.
- Actuators, drives, and instrumentation.

LOW-LEVEL CONTROL
- Servo and control.
- Architectures of robot control systems.
- Low-level teaching.
- Dynamics.
- Force sensing and control.

SENSORS
- Vision.
- Proximity.
- Tactile.

HIGH-LEVEL CONTROL
- How to make a robot programming language.
- A survey of robot programming languages.
- The AL language for robot programming.

PLANNING & ARTIFICIAL INTELLIGENCE
- Geometric modelling.
- Simulation.
- Planning collision-free paths.
- Representing and planning assemblies.

CONCLUSION
- Special topics.
    - Locomotion.
    - Hands & grasping.
- Applications & economics.
- Expected future developments.

The Use of Intelligent Chips in Educational Robot Systems

Bernard J. Schroer
The University of Alabama in Huntsville
Huntsville, Alabama

Gary L. Workman
The University of Alabama in Huntsville

Larry L. Fullerton
The University of Alabama in Huntsville

Abstract

In recent years, there has been an explosion in the use of microcomputers and small intelligent special purpose chips. One application of these intelligent chips is to perform special purpose tasks within the robot system. This paper presents the use of these chips in a simulated manufacturing work station. This work station consists of a parts bin with a vision and image analyzer, a machining station with an X-Y table and drill press, a fixture station, an assembly station, and a materials handling station consisting of a MiniMover-5 robot. All the stations are controlled by a microcomputer. This paper discusses the operation of the system, the computer interfaces of the vision system, the machines, the robot, the user, interfaces, and programming aspects of the various intelligent chips.

INTRODUCTION

The University of Alabama in Huntsville has initiated a program to update its capabilities and its curriculum in manufacturing engineering. As part of this program, we are developing a simulated cellular manufacturing facility as an educational tool for engineering students. The original products of this facility are university paperweights. This paper illustrates our technological approach to developing a manufacturing cell and outlines some potential problems in designing a system similar to the one described here. As indicated in the description of the system, we chose to use intelligent subsystems wherever possible.

CELL DESCRIPTION

A simulated manufacturing work cell has been fabricated for manufacturing small $2\frac{1}{2}$'' x $2\frac{1}{2}$'' university paper weights. The cell consists of the following work stations:

Station A: Parts bin using video camera
Station B: Part re-positioning fixture
Station C: X-Y table and drill press
Station D: Steel university logo $1\frac{1}{2}$'' diameter
Station E: Magnetic fixture
Station F: Glue sponge
Station G: Part assembly

The physical layout of the cell is given in Figure 1. Several photographs of the cell are given in Figures 2 and 3. The manufacturing flow is as follows. A blank is placed anywhere in the parts bin at Station A. The vision system locates the part and sends the location to the microprocessor. The microcomputer then instructs the robot to pick up the blank and transfer it to Station C. Should the blank be on its side the robot will transfer the blank to Station B where the blank is properly oriented.

At Station C the blank is placed on the X-Y table. The table moves the blank under the drill press and the drilling of four holes is simulated. The robot then moves the blank to Station G.

The robot picks up the magnetic fixture at Station E and uses the fixture to pick up the steel university logo at Station D. The logo is brushed against the glue sponge at Station F and then placed on the blank at Station G.

It should be noted that Station C was originally designed by Dr. John W. Nazemetz and his students at Oklahoma State University. His engineering drawings were used to fabricate the X-Y table and to modify the drill press.[1,2]

SYSTEM SCHEMATIC

The system schematic of the manufacturing cell is given in Figure 4. This schematic illustrates the computer interfaces between the work stations. The cell consists of the following electronic hardware:

1. Apple II Plus microcomputer with 48K RAM, one floppy disk drive, MX-80 printer, and TV monitor.

2. MiniMover-5 robot with ARMBASIC firmware.

3. Sanyo video camera with Dithitizer firmware and video monitor.

4. Three CY512 stepper motor controllers.

5. Two Phillips stepper motors on X-Y table and one Hallmark stepper motor on the drill press.

Intelligent Positioning Stepper Motor Controller[4]

The CY512 is a programmable peripheral controller chip designed to control stepper motors using an instruction sequence that may be stored internally in a program buffer. The CY512 architecture may be partitioned into the following functional subsystems: (See Figure 5.)

- input data subsystem;
- output data subsystem;
- program parameter storage;
- mode flags and pins;
- program storage buffer;
- instruction selection, decoding, and control mechanisms;
- position register.

The input data subsystem accepts commands and the output data subsystem holds the output control signals to the stepper drive circuitry and includes the associated direction and pulse timing lines.

The program parameter storage subsystem is used to store the step rate parameters, ramp rate parameter, and to maintain a 16-bit position register. The position register is incremented (or decremented) when stepping in the clockwise (or CCW) direction. The position register is used when absolute position commands are specified. The 16-bit step counter is used when relative commands are employed. The contents of the position register change with every step, while the step counter register contents remain unchanged until a specific command is used to change them.

The mode flags and mode select pins are used during command execution to perform the appropriate action or to interpret data or input signals correctly. The CY512 contains a program buffer that allows the user to store a sequence of instructions that can be executed upon command. This provides all of the benefits of stored program execution that have made computers such powerful tools.

The instruction decoding and control subsystem performs the actual execution of commands. The CY512 contains a 16-bit position register that can be read to determine the current location. The CY512 will accept relative and absolute position commands; however, the position register always indicates absolute position.

The CY512 kit consists of a printed circuit board containing all the logic to operate a stepper motor. (See Figure 6). The kit includes switches for CY512 inputs, buffers with LED indicators on CY512 output signals, abort/terminate logic for closed loop motor operation, unipolar driver circuit for the stepper motor, and associated peripheral parts such as crystal, socket, capacitors, etc. There are three edge connectors which make the various signals available to the system.

MiniMover-5 Manipulator[3]

The MiniMover-5 is a five-jointed mechanical arm with lifting capacity of 8 ounces when fully extended. Each joint is conrolled by a stepper motor. The end of the hand can be positioned anywhere within a partial sphere with a radius of 17.5 inches. The hand has a maximum opening of 3 inches. A built-in sensor permits adaptive control so that the presence of an object as well as the size of the object may be determined as the hand closes.

The MiniMover-5 ARMBASIC software consists of the following six commands:

```
PRINT "@STEP", D,J1,J2,J3,J4,J5,J6

PRINT "@CLOSE", D

PRINT "@SET", D

PRINT "@RESET"

PRINT "@READ"

        INPUT V1,V2,V3,V4,V5,V6

PRINT "ARM", N
```

The STEP command causes all six stepper motors to move simultaneously. The arguments are:

D = is the delay between the pulses to the motors (=0.66+0.0127D) in

J1-J6 = number of steps that each of the motors is to move.

J1 clockwise base swivel
J2 downward shoulder bend
J3 upward elbow bend
J4 upward right wrist
J5 upward left wrist
J6 open hand

The CLOSE command causes the hand to close until the grip switch indicates that the fingers have closed on an object or have touched each other.

The SET command places the manipulator into the manual mode. The manipulator can then be controlled by the Apple II keyboard.

The RESET command zeros the contents of the internal position registers and turns off the current to all the six motors.

The READ command allows the six internal position registers to be read into variables within the users program (V1-V6).

The ARM command selects the arm to which any subsequent ARMBASIC commands will be addressed, thus permitting alternate control of two arms (N = 1 or 2).

The BASIC pick and place program to pick up the blank from Station A and place it on the X-Y table is given in Table I. The programming aspects of using the robot becomes very easy when using this set of instructions. The primary drawback is speed. Interpretive BASIC is slow, but for simulative procedures it is acceptable.

```
20 DIM P(6),R(6),D(6),P1(200,2)
30 S = 175
40 TH - 55
49 GOTO 100
50 D$ = CHR$ (4)
60 PRINT D$;"PR#3"
65 PRINT D$;"BLOAD DSCAN 4.2.OBJ
    "
70 GOTO 10
100 REM DATA
110 DATA 1450,2000,-691,82,82,1
    100
120 DATA 1450,2194,-691,82,82,1
    100
125 DATA 1000,1644,-468,61,61,7
    63
130 DATA -292,1644,-468,61,61,7
    63
140 DATA -292,1760,-540,61,617
    60
150 DATA -292,1760,-540,61,61,9
    80
165 DATA 0,1000,-300,0,0,0
170 DATA 0,0,0,0,0,0
200 GOTO 2000
1000 REM
1005 PRINT "@RESET"
1007 RESTORE
1009 A = X
1010 FOR I = 1 TO 9
1015 READ D(1),D(2),D(3),D(4),D(5),D(6)
1017 IF i > 2 AND I < 6 THEN D(6) = Z
1020 FOR J = 1 TO 6
1030 P(J) = (R(J) - D(J)) * - 1
1040 NEXT J
1045 PRINT S,P(1),P(2),P(3),P(4),P(5),P(6)
1047 IF I = 3 THEN GOSUB 1200:P(6) = 0
1050 PRINT "@STEP",S,P(1),P(2),P(3),P(4),P(5),P(6)
1060 PRINT "@READ"
1070 INPUT R(1),R(2),R(3),R(4),R(5),R(6)
1080 NEXT I
1090 RETURN
1200 PRINT "@CLOSE"
1210 PRINT "@READ"
1220 INPUT R(1),R(2),R(3),R(4),R(5),R(6)
1230 Z = R(6)
1240 RETURN
2000 REM VIDEO DRIVER
2001 RESTORE
2002 POKE 49280 + (16 * 7),TH
2004 CALL 7424
2005 X = 0
2007 FOR A = 1 TO 120
2008 P1(A,2) = 0
2009 NEXT A
2010 FOR B = 0 TO 127 STEP 2
2020 FOR A = 8192 TO 8958 STEP 512
```

```
2030 P = PEEK (A + B)
2040 IF P = 128 THEN GOTO 2070
2050 IF P = 255 THEN POKE (A + B),247
2060 GOTO 2100
2070 POKE (A + B),255
2080 X = X + 1
2090 P1(X,1) = A:P1(X,2) = B
2100 NEXT A
2110 NEXT B
2120 FOR A = 1 TO X
2130 IF P1(A,2) > 57 AND P1(A,2) <  67 THEN GOSUB 1000
2140 NEXT A
2150 GOTO 2000
10098 A = X + 1
```

PROGRESS TO DATE

The project has been underway for five months. The mechanical portions of the project required the least amount of effort. The major effort has been on the interfaces between the computer and the subsystems. The programming of the robot was relatively simple since the ARMBASIC firmware was purchased. The major effort has been on (1) the image processing software, (2) the stepper motor software, and (3) the communication software between the microcomputer and the peripherals.

During the development of the cell the following observations have been made:

- The actual programming of the robot is relatively simple using the ARMBASIC software. A simple pick and place routine can be written after several hours of reading the manuals.

- The time required to learn to program the stepper controllers was considerably more than originally anticipated.

- The image processing software required to locate the blank at Station A had to be developed in phases. The first phase which has been completed is the location of a blank positioned parallel to the horizontal plane.

- If the position and attitude are known prior to the grab at Station A, the algorithm is simple detection. However, with either attitude or position unknown, the grabbing algorithm becomes very complex.

- The determination of the absolute position of the arm and the stepper motors since the system is open loop. The solution was to have a known reset position from which all future positions are calculated. This resulted in oversizing the stepper so that no slippage would occur.

- The stepper controller is a position or movement controller and not a motion controller. The stepper can be brought from zero to a constant velocity for a defined number of revolutions so that position can be exactly programmed. However, in a real world situation the stepper controller cannot perform a smooth motion involving more than one axis.

- Video memory mapping using the Apple II is cumbersome to use because of the organizational structure of the microcomputer.

FUTURE PLANS

Once operational the cell will be used in the undergraduate and graduate manufacturing and production courses in the Industrial and Systems Engineering Department. Already a number of students have expressed an interest in both using the cell and in assisting in the further refinements of the cell.

The following refinements to the cell have been identified:

- Development of a fixture to hold down the blank on the X-Y table (Station C).
- Construction of a material handling system for transferring blanks to Station A and for transferring finished paperweights from Station G. Fischer Technique modeling will be used.
- Development of algorithm to locate blank at Station A when blank is placed randomly.

BIBLIOGRAPHY

1. A Microprocessor Controlled Physical Simulator: An Aid to Manufacturing Education, D. D. Ashcraft, Master's Thesis, Oklahoma State University, 1981.

2. Adding and Implementing a Vertical Drill Axis to an Existing X-Y Table, D. L. Tabb, Special Problems INDEN 5350, Oklahoma State University, 1981.

3. Robotics Reference and Applications Manual, Microbot Inc., 1981.

4. CY512 Intelligent Positioning Stepper Motor Controller, Cybernetic Micro Systems, 1981.

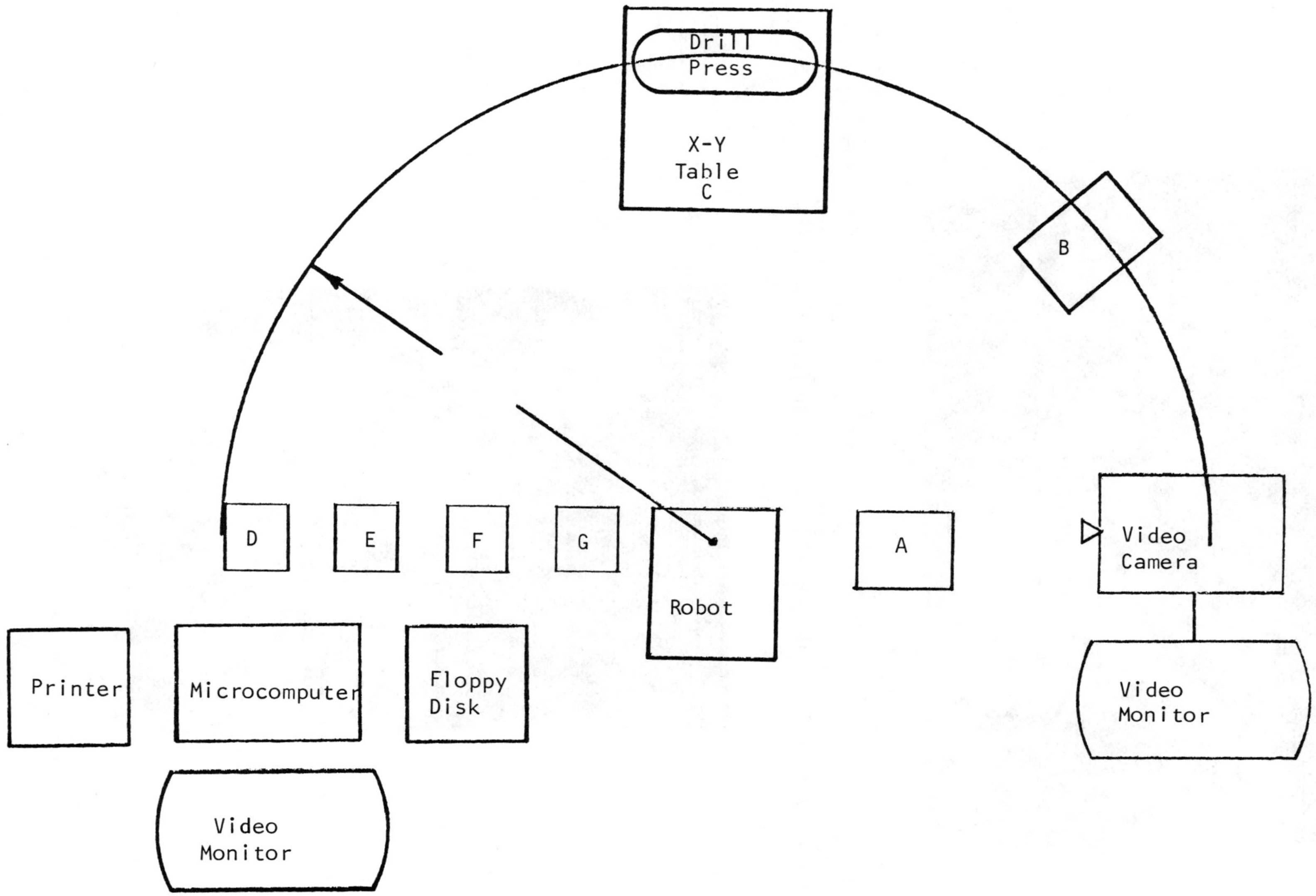

Figure 1. Physical Layout of Cell

Figure 2 Overall View of the Simulated Manufacturing Cell

Figure 3 View of Robot Transferring Object from X-Y Table

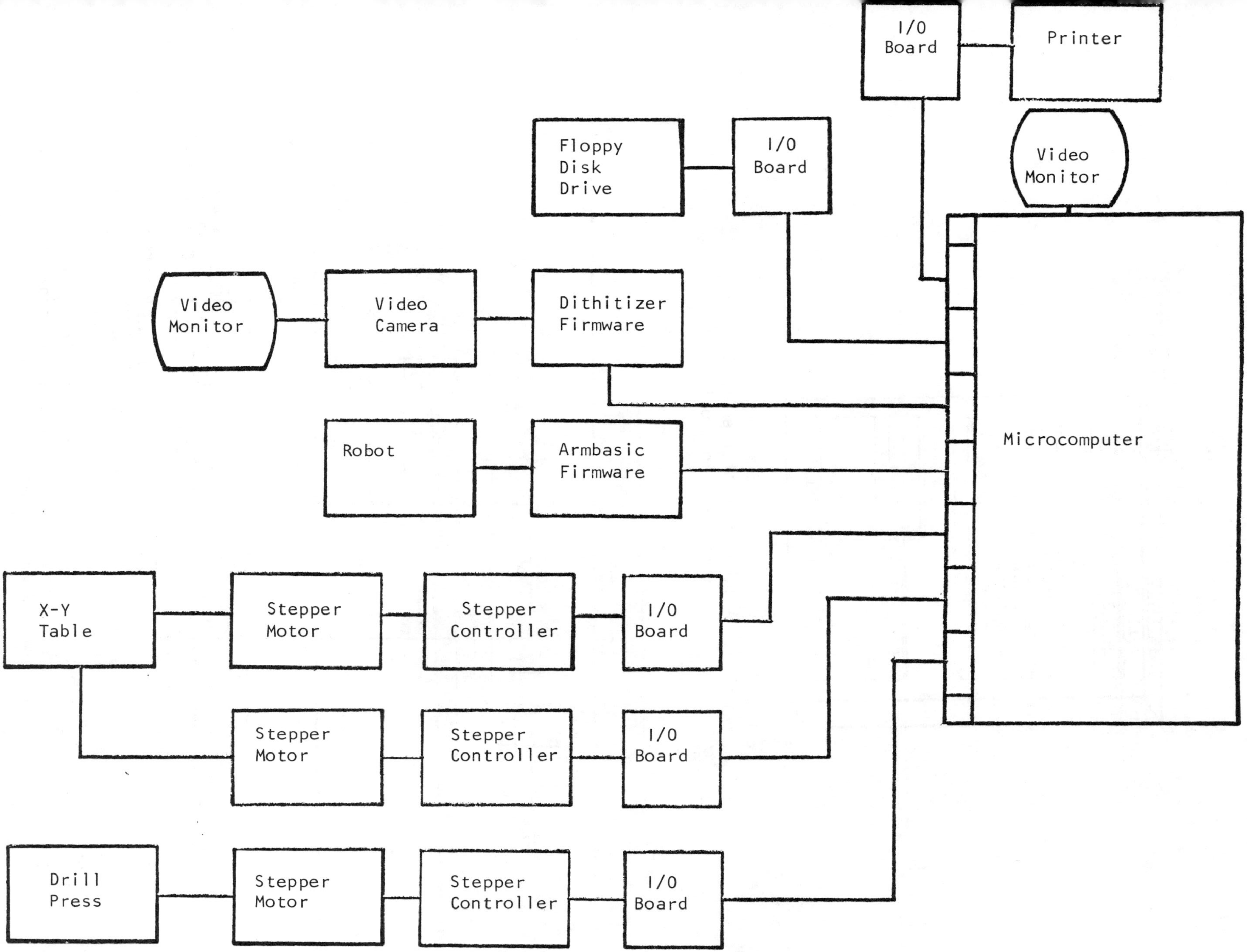

Figure 4. System Schematic of Cell

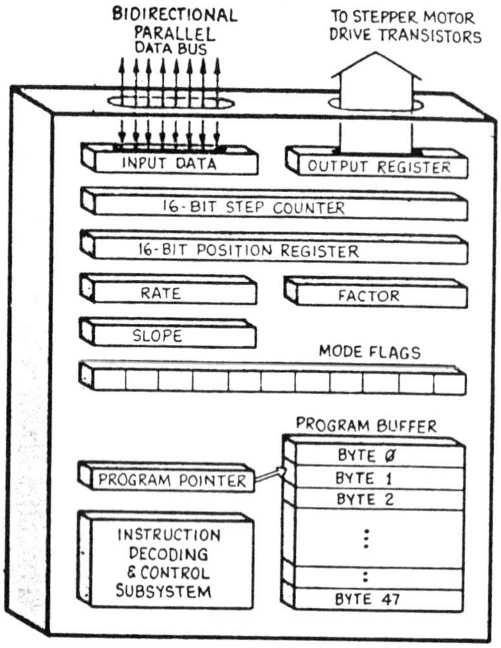

Figure 5 Architecture of the CY512 Stepper Motor Controller

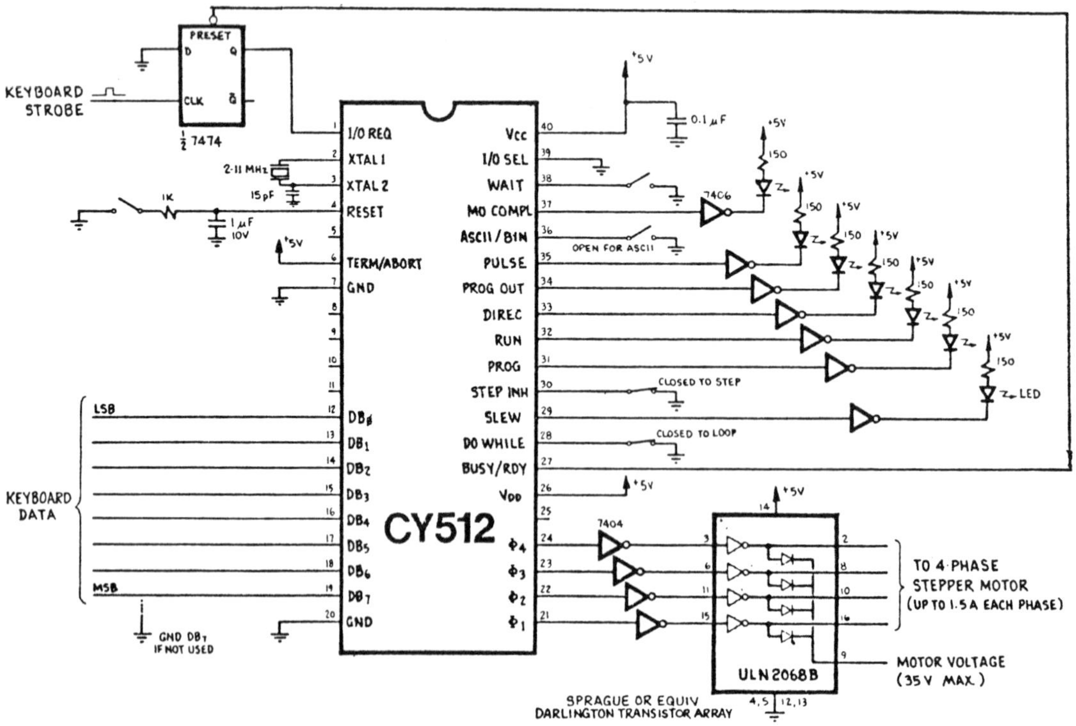

Figure 6 The CY512 Stepper Motor Controller Subsystem Showing all the Hardware Required for Use

Education in Robotics in Poland

Adam Morecki
Technical University of Warsaw
Warszawa, Poland

Andrzej Kaczmarczyk
Technical University of Warsaw

Tadeusz Strzelecki
Technical University of Warsaw

Abstract

Some selected problems of university, technical university, and industrial education in the field of robotics are presented. The present system of education based on separate educational studies on mechanical, electronical control, humanistic, sociological, praxiological, and others is not suitable for an interdisciplinary subject as robotics. The possible changes of this system will be discussed. The most important questions concern how to organize university education: from beginning on graduate level or post graduate study, to the ratio of designers to researchers. The necessity of continuing post graduate education for industrial engineers is emphasized.

INTRODUCTION

Further development of automation, specially flexible manufacturing systems, depends to some degrees on the advanced design and application of industrial manipulators and robots. The effective use of these devices is possible on grounds of knowledge of their construction and properties. It can be anticipated that the future of robotics considerably depends on the educational process.

In the last few years in mechanical engineering schools in Poland the subject "Industrial Robots" is tought on the fourth year. The main goal is to educatc a new generation of technicians, familiar with new type of flexible machines.

At present the most important questions are: how to organize the university education at various levels: elementary, undergraduate, graduate and post graduate. What should be the proportions between the designers and researche workers etc. The necessity to continue the post diploma education, mostly for industrial engineers, provided by Universities or Technical Associations of Mechanical or Electrical Engineers is underlined. The third direction - the training provided by the Makers of robots, especially for the users is described.

This paper presents some ideas concerning the above mentioned problems.

TYPES OF EDUCATION

The education in the area of manipulators and robots or robotics in Poland can be classified and systematized according to the following criteria:

1. Level of education,
2. Required scope of knowledge,
3. Use of specialists.

The three above mentioned criteria will be briefly discussed.

Level of education

The education can be divided into five categories, according to educational institutions which may be:

- technical high schools, mechanically or electronically oriented,
- technical Universities with mechanical, electrical, electronical and control faculties on undergraduate, graduate and post graduate levels,
- various technical associations focused on the improvement of professional level of engineers,
- factories specialized in production of robots and manipulators,
- possibilities of an individual.

Scope of knowledge

Directions can be distinguished as follows:

- technical high schools:
  - → general exploitation and service of robots in industry /mechanical system/,
  - → general exploitation and service of robots in industry /control

system/,
→ special manufacturing process;

- at Universities:
  → mechanics of robots,
  → control of robots,
  → sensing and machine intelligence,
  → design and synthesis of robots and man-machine systems.

Use of specialists

- reorganization of production under industrial conditions,
- economical, ergonomical and social aspects of robotization.

THE CURRICULUM FOR TECHNICAL HIGH SCHOOLS

The Course "Industrial robots" is tought on the fourth year of some technical schools. In about 100 hours four main topics are discussed, namely:

- industrial robot as an instrument,
- mechanisms and drives,
- transducers and limiters of displacements for robots and manipulative objects,
- control systems of robots.

In the first topic the system interpretation of different kinds of human work is discussed. Starting from description of work without instruments, through automation of different tasks, robotized systems are explained. Economical, technical and social aspects of robots belonging to 1, 1.5, 2, 2.5 and 3rd generation are discussed.

Within the scope of second item the problem of the number of degrees of freedom, kinematical diagrams and working volume are described. Next, the mechanical, pneumatic and magnetic grippers are discussed. At the end some selected problems of hydraulic, pneumatic and electric actuators are considered.

Third item includes different types of limiters such as: potentiometers, resolvers, digital measurements of position, force and strain transducers, optical transducers and correlators.

The last fourth item concerns the goal and the function of the control system. Different control systems like sequential, open and closed loop systems, trajectory programming are discussed. Teaching process and pattern recognition systems are also presented [1]. Taking into account the background of the pupils most of the above mentioned problems are presented in the descriptive form. Only as some selected problems the elements of mathematics are applied, specially to describe dynamical requirements for the control systems. To present the amplitude and phase diagrams of first, second and higher order systems to discuss its stability and to analyse the equation of motion of servo-mechanism the operator calculus is very useful. Of course, the theory of linear systems considered on the University level is rather simple but at technical high schools it is difficult to understand and apply from methodological and conceptual point of view.

The collected experience is small now to generalize.

SELECTED CURRICULAE FOR TECHNICAL UNIVERSITIES

Starting from the academic year 1980/81 robotics is regularly lectured at the Faculty of Electronics of the Technical University in Wrocław. The lectures are delivered by the employees of the Institute for Technical Cybernetics. For the students specializing in Automatic Control two courses are obligatory, namely:

- "Basis of functioning of automatic subject" /VII semester, 2 h in a week/. The main topics are as follows: pattern recognition, proofs of theorems and solving of problems, modelling of complex systems, automatic take up of decisions.
- "Basis of industrial robots design" /VIII semester, 2 h of lectures plus 2 h of seminar/ lectured by A.Kaczmarczyk.

The content of this subject was as follows:

1. Basic terms and systematics,
2. Kinematic scheme and main mechanical parameters,
3. Constructional solution of robot mechanics,
4. Devices of robots,
5. Mobile robots and their trajectories,
6. Grippers and technological heads,
7. Control systems, characteristics and systematics,
8. Examples of control systems,
9. Developed programming systems and hierarhic and groups control,
10. External transducers and sensors of robots,
11. The characteristics of industrial robots and their testing,
12. Production and market of industrial robots,
13. Applications in synthetic form.

A special subject "Robotization of industrial process" is lectured on 9th semester /4 h of lectures, 2 h of seminar, 2 h of designing and 2 h of laboratory exercises in the future/.

The lectures based on the systematics are presented in Fig.1 in form of a map. The attention was drawn to the not uniformed problem of terminology and classification. The same problem is concerned the classification of control systems.

In the end each student was asked to answer in written form a set of problems like: drive system, positioning, control systems PTP or CP type, intelligent robot, kinematic diagram of five basic types of robots, systems of coordinates, grippers, transducers and so on.

At the Technical University of Warsaw for many years a small group of students, specially in Applied Mechanics, took their M.Sc. degree in robotics, under supervision of A.Morecki and his co-workers.

Prior to writing dissertation it is necessary to attend some courses, specially: mechanics of machines, automatic control and dynamic measurements. These three disciplines give knowledge of the structure, kinematics and dynamics of open kinematic chains with many degrees of freedom, which are the fundamental part of each manipulator or robot. In the frame of automatic control the basic problems of linear and nonlinear systems, modelling methods and algorithms are presented. In dynamical measurements the principles of mathematical description of an object, transducer and instrumental path are all taken into account. The devices for measuring such quantities as: translational and rotational displacements, velocities, acceleration, force, torque, pressure, flow and temperature are discussed. Data processing, transmission and recording are important part of the subject and are illustrated by separate laboratory demonstrations. In particular: bridge circuits, amplifiers, filters, cable and FM/FM radiotelemetry, A/P, P/A converters, magnetic tape recorders, large-scale data systems are discussed during the lectures. Besides of obligatory topics with prerequisites /theoretical mechanics, Elementary Computer Sciences adn mathematics/ and postrequisites /theory of construction of machines and advanced control of systems/ a few elective courses are lectured on higher semesters such as: Bionics of Motion, Robotics, Advanced Foundations of Experiment, Machine Systems etc. [2].

They must be included prior to semester projects and diploma work, which may be supervised by specialists in robotics. Typical curriculum for the subject "Manipulators" /VIII semester/ are listed below:

1. Basic terms and systematics /3 h plus 1 h/,
2. Properties of skeleton-muscle-nerve system and design of manipulators /6 h plus 2 h/,
3. Rehabilitative manipulators /4 h plus 3 h/,
4. Structure, kinematic, dynamics of technical manipulators /3 h plus 2 h/,
5. Control of motion of technical manipulators /2 h plus 1 h/,
6. Engineering solutions of technical manipulators /2 h plus 1 h/,
7. Manipulators for underwater and space activity /1 h plus 1 h/,
8. Walking machines /2 h plus 1 h/,
9. Introduction to industrial robots /3 h plus 1 h/,
10. Review of basic types of industrial robots /4 h plus 2 h/.

The lectures are followed by exercises in laboratory and under factory conditions. The total time devoted to the subject is 30 h plus 15 h /exercises/. [3,4].

Depending on the taken courses and completed projects some of the students are better prepared to designing, research work /mechanics or control/ or application as the others.

At the Technical University of Warsaw a post-diploma study in the area of "Digital Control of Machines and Industrial Robots" was organized recently.

The aim of this type of training was to acquaint the designers and industrial engineers with the problem of:

- design, application and exploitation of the automated systems incorporating industrial robots,
- design and building of robots and special mechanical devices for automation of technological processes.

The course was available for engineers with at least two-year industrial practice.

The participants were dispatched from industry to the University for 9 one-week sessions separated from each other with one month breakes. The total number of hours /lectures, exercises and laboratories/ was 366 which means 41 h for each session. Home work occupied about 15 hours a month. The participants finished the course with positive results obtained a diploma from the University.

The curriculum for the course was as follows:

- selected problems of mathematics,
- introduction to digital control,
- control systems and microprocessor systems,
- laboratory of digital control,
- mechanization and sutomation of technological processes,
- industrial manipulators and robots,
- laboratory of industrial manipulators and robots,
- the principles of computer programming system,
- computer systems for the control of technological systems,
- laboratory of application of computer systems,
- programming of control systems of mechanical devices,
- selected problems of theory of servo-mechanisms.

TEACHING PROVIDED BY ENGINEERING ASSOCIATIONS

Some additional training of engineers is provided by special Center established in 1948 by the Society of the Polish Mechanical Engineers and Technicians /SIMP/. During the 35 years of teaching activity over 10 000 courses were organized, and were attended by about 450 000 participants, who extended their personal knowledge or sometimes altered their professional profiles.

In the mentioned period of time over 450 conferences and technical symposia were organized for about 300 000 participants [5,6]. Besides a special post-diploma studies in technology and machine building were organized in the form of controlled selfeducation and the studies in design of factories for the Machine Building Industry.

In the last years a special division of SIMP for Industrial Robots and Manipulators was established /Chairman A.Morecki/. The responsibility for teaching and additional training of engineers was taken by T.Strzelecki. The main activity in this area is provided in the form of public lectures and courses. An important factor, which makes this activity more attractive for industrial engineers is the possibility to obtain the title of a specialist which gives them official priviledge to perform expertizes for industrial companies.

Another type of education provided by the Association of Engineers and Technicians is a so-called course of the 3rd order, intended for the industrial workers in the area of Industrial Robots RIMP-1000 -"Flexible Production Automation through RIMP-1000 robots application".

The main goal of this type of training is to transfer a certain portion of theoretical and practical information, which consists a basis for modernization of production process in Factory, and to supervise a proper exploitation.

The participants of the course were recruited from the workers of dactories, graduated from high schools with mechanical, electronical or industrial automatic orientation. The educational process is divided in two parts. The first theoretical part is lectured at the Society. The second, practical part, takes place in the laboratory where RIM-1000 robots and peripheral devices were installed.

The number of participants is limited to 10 persons. The lectures for mechanical and electronic technicians are delivered separately. The practical exercises are conducted in three-person groups. The total number of hours in the course is 62. The final examinations are taken at the end of the course.

EDUCATION OF USERS OF ROBOTS

This type of education also called the company or factory education was provided in the last years by various makers or users of robots produced in Poland.

We will present the course organized by MERA-PIAP Company in Warsaw. The course was intended for the users of PR-02 and IRb industrial robots [7].

Three types of training were selected:

i. Training of robot operators,
ii. Utilization of robots,
iii. Service training.

Each type is briefly discussed below.

Type /i/ of training is the basic one, and is included in two other types. It was intended for qualified workers thirty hours course included 15 hours of practical exercises. The curriculum consisted of the basic information on building or robots, programming and service /75% of teaching time/, inspection and preservation.

Type /ii/ of training is intended for engineers /less for the technicians/. The course lasts 60 hours and includes 24 hours of practical exercises.

Though the course was mainly focused on training an operator, it included also extensive and general information on programming and applications /60% of time/.

Type /iii/ of training is forseen for the technicians. Besides of the training in operation of the robots /67 h/ the course included 15 h of practical exercises, some information on programming and service /75% of time/.

PROPOSED IMPROVEMENT OF THE PROCESS OF EDUCATION

Basing on the experience related to education collected in the last years some conclusions concerning the present state of the subject and some proposals for the future can be made.

It seems that three basic types of education should be continued:

/i/ at the high school and university level - basic education,
/ii/ at the level of courses organized by engineering societes-improvement of knowledge or reorientation of professional profile,
/iii/ at the level of the courses organized by the users and makers - necessary information for everyday practice, service and exploitation.

The University level training can be performed either at the faculties or departments for mechanical or electrical engineering or in the form of interdisciplinary studies, because the object - robot - is a system, which consists of five basic subsystems: mechanical part, drive, control sensing and measuring systems.

Depending on the level of robotization in the country the education can be realized from the first course or on graduate level and post graduate level. A proper proportions between the number of designers, researchers and future users of robots should be preserved.

The training provided by different Engineering Associations should be limited to the extension of knowledge, or reorientation of professional profile. The form of short, a few week courses, is very useful in practice.

What concerns the third type of training, different forms applied by makers or users should be accepted. A special attention should be given to practical exercises performed under real factory conditions.

It is important to underline here that all three described types of education are necessary. Neither of them can be substituted by another one. In our opinion it will be useful to coordinate the whole process of education in the future.

The periodic exchange of experiences at the conferences of symposia of robotics should be very useful.

Table 1 presents possible types of education in robotics and means for coordination. Some basic and typical courses are shown. Depending on the type and scope of education different sets of courses are organized.

CONCLUSION

The current system of professional training in the field of robotics in Poland was presented.

Some imperfections of the current education imply that necessary future changes have to be considered.

BIBLIOGRAPHY

[1] Niederliński, A. *Industrial Robots* /in Polish/. High School and Pedagogic Publisher, Warszawa, 1981.

[2] Morecki, A. and Oderfled, J. *Education in TMM at the Technical University of Warsaw*. Proceedings, Intern. Symp. on Education in Theory of Machines and Robotics, SEMEMATRO'82, Blagoevgrad, Bulgaria, Sept. 20-22,1982.

[3] Morecki, A. *Bionic Manipulators*. /in Polish/, PWN-Polish Scientific Publisher, Warszawa, 1976.

[4] Morecki, A., Ekiel, J., Fidelus, K. *Cybernetic Systems of Limb Movements in Man, Animals and Robots*. Ellis Horwood Ltd. Publishers, Chichester, Holsted Press: a division of John Wiley and Sons. New York, Chichester, Brisbane, Toronto, 1983.

[5] Scientific Session on occasion of XXX years of SIMP activity in improvement of technical specialists lectures - addresses- - conclusions. Leszno-Rydzyna, 28-29.10.1978 /in Polish/.

[6] Proceedings. Scientific Conf. on "Improvement of the work conditions through application of industrial robots". Warszawa, 19-20.05.1980.

[7] Buć, J. and Morecki, A. *Study and Application of the Industrial Robots in Poland.* Proceedings of the 9th International Symp. on Industrial Robots, March 13-15, 1979, washington, D.C., USA.

Table 1

Type of education / Range of education	Technical High Schools mechanical or electrical type	Technical Universities or colleges of engineering	Association of Mechanical or Electrical Engineers & Technicians	Makers and users of industrial robots
Encyclopaedic	A L	A L E	A	A P
Research or Design in Robotics	—	A P L (E) S	—	—
Building of Robots	—	A P L E S	—	S P
Application of Robots	A P E	(E)	(E)	(E)
Equipment Testing Training	A L P	A L P L	(E) (E)	(E) (E) (E)

A - lectures
L - laboratory
S - seminar
E - exploitation
P - project
(E)- elective subject

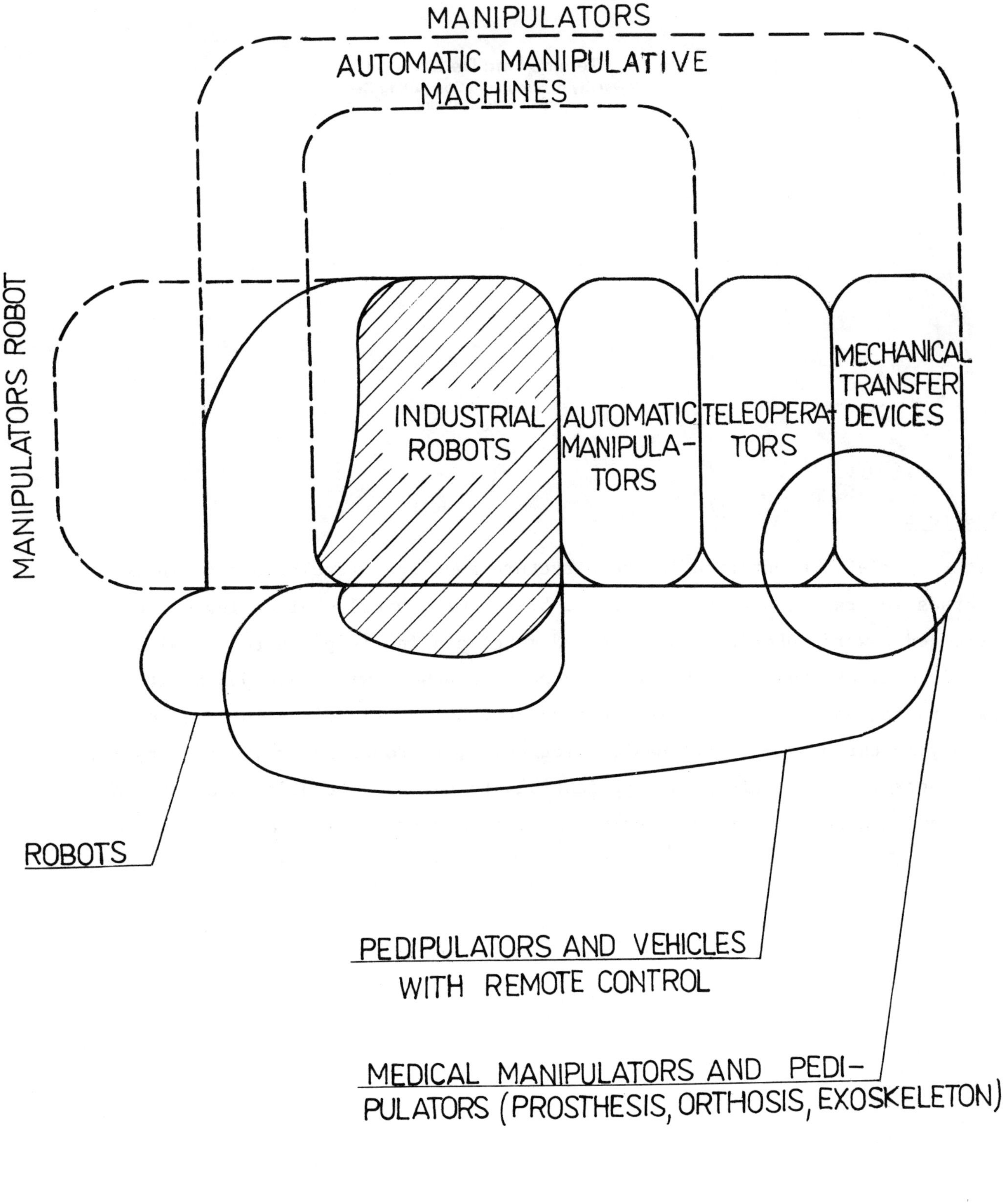

Fig. 1

Systematics of the problems presented to the students

A Engineering Curriculum Oriented to Robotics and Electronic Control

Geraldo Lino De Campos
University of Sao Paulo Computing Center
Sao Paulo, Brasil

Abstract

After a brief presentation of the structure and legal regulations of engineering courses in Brazil, a curriculum for a 5 year mechanical engineering degree oriented toward robotics and automated manufacturing is presented. This curriculum, excluding the basic and general courses, will have 55% of its credits in basic mechanical engineering; the balance is distributed almost evenly in three areas: (1) basic electronics, microprocessors, and software; (2) control and robotics; (3) applications to the manufacturing process. A continuing education program derived from this curriculum is also described.

SOME HIGHLIGHTS ON ENGINEERING EDUCATION IN BRASIL

The strutucture of undergraduate engineering courses in Brazil obeys strict regulations of the Federal Government, that establishes a fair percentage of the curriculum, wich should comprise basic courses, courses of general interest and specific courses for the intended degree. Each school is required to complete the required credits with courses oriented toward specialization in one field or to attend regional peculiarities.

In the case of mechanical engineering, the mandatory courses should cover the following topics:

BASIC COURSES	SPECIFIC COURSES
Mathematics	Applied mechanics
Physics	Applied thermodynamics
Chemistry	Mechanic materials
Mechanics	Mechanic systems
Data processing	Thermal systems
Drawing	Fluidmechanical systems
Electricity	Manufacturing processes
Strength of materials	

These courses should comprise at least 2940 hours of work, leaving 660 hours for the courses fixed by each school. Obviously , these numbers can be exceeded if the school so decides. As a rule, schools require 4000 to 4800 hours for granting a degree. The curriculum is designed for a 5 years program.

Although some schools allow the student to freely choose electives, most - including the Polytechnic School of the University of São Paulo - offer closed packages of electives, leading to a kind of especialization which will determine the legal attributions of the engineer. Whithin this framework, the evolution of curricula must be planned well in advance; the evolutionary way - start offering new courses - that works very well in American Universities can not be used.

BRIEF DESCRIPTION OF THE PROPOSED CURRICULUM

The University of São Paulo has about 36000 undergraduate and 12000 graduate students, with a faculty of over 4700. Most graduate students are part-time, however. The school of engineering, called Polytechnic School, has 3750 undergraduate and 1600 graduate students, with a faculty of 422. The school grants degrees in civil, mechanical, electrical, mining, chemical and naval engineering.

In 1981, the Director of the Polytechic School, reconizing the need for an update in the Mechanical Engineering curriculum,initiated a study of the kind of knowledge a mechanical engineer shall need in the next few years. The main conclusions will be detailed in the next section. Based on these conclusions, a draft curriculum has been proposed. This paper is based on this proposal.

In 1982 a commitee was alpointed for studying the draft curriculum and propose the establishment of a package of electives, leading to a new specialization in the mechanical engineering degree. Some bureaucratic restrictions precluded the immediate implementation of the specialization, but the commitee recognized the urgent need for knowledge in this area, and decided instead to introduce immediatelly a few courses for all the mechanical engineering students.

An important result of the commitee work was the establishment of a continuing education program, starting in early 1983, and described later in this paper.

KEY POINTS IN A MODERN MECHANICAL ENGINEERING CURRICULUM

There are three closely related features in the modern concept of mechanical engineering:

- CAD, CAM and CIM
- Robots as a key element for flexible automation
- Microprocessors as machine elements

Let us review briefly the reasons and consequences of these concepts.

The robots shines in the scene of modern manufacturing, but for economic success it must be backed by a host of other technics and devices. Its usage just for a fixed repetitive task is questionable; a dedicated mechanism will do the same, perhaps at a significantly lower cost. The key work for robot justification is flexibility. If this flexibility is not exploited properly, there is no reason for the robot.

The effective exploitation of this flexibility requires a significative change in the whole manufacturing process, entraining areas outside mechanical engineering: management, production scheduling and control, relations with the labour force, etc., and almost all phases of product conception, design and development.

The natural way of designing a product intended to be automatically manufactured is by automatic means; this justifies the CAD, CAM and CIM part of the curriculum. Its advantages are well known and beyound the scope of this paper.

The proposed curriculum includes the following courses in this area:

COURSE	HOURS
Industrial Design	45
Programming Systems	60
Computer-aided design and manufacturing	60
Machine tools II	90
Numerically controlled machine tools	90
TOTAL	345

Industrial design covers the steps involved in the design of an industrial product, with emphasis on the use of computer-aided means. Computer-aided design and manufacturing studies these means. Programming systems covers the main topics of software, wich are an essential prerequisite to understand computers and design and production systems. It is also important for the microprocessors courses. Machine tools II extends the standard course in machine tools, with emphasis on automation.Numerically controlled machine tools is self-explanatory.

Surprisingly, there is comparatively little specific knowledge about robot design and construction. Apart from the electronic parts, it is a quite standard mechanism, using principles teached in the usual mechanical engineering courses. The control of the movements of a robot is a standard control problem. Although there are courses in control, new ones were proposed, essentialy to give an integrated approach combining analog and digital methods.

Robot applications are more difficult to teach; first, because it is a very new field and still unsettled; and second, because it is not a technic in itself, but just a new alternative - although of a new kind - to be considered in the design of a manufacturing process, specially justifiable when the company is seen as a live and evolving entity. It is also important to study sensors and instruments to receive information from the external world in order to provide inputs for controling the robot.

The courses in this group are the following:

COURSE	HOURS
Electromechanical control I	60
Electromechanical control II	60
Electromechanical control III	60
Industrial Robots	60
Industrial Instrumentation	30
Graduation project	60
TOTAL	330

Electromechanical control I and II should cover control systems, stability, controlability, optimal control, the Pontryagin principle and applications, and gradient methods, with a unifield approach for continuous and discrete systems. Electromechanical control III should cover special applications, such as usage of dynamic programming to control problems, and peculiarities of hydraulic and pneumatic systems used in robot motion control. Industrial robots covers the main applications of robots - materials handing, machine loading, painting, soldering - with emphasis in the relationship to other automatic equipments and economic justification. Industrial instrumentation covers the sensors used with robots, including applications of computer vision. The graduation project is a one semester activity conducted individually by each student, that is expected to integrate the learnings of several courses into a project; as a rule the design and construction of a simple piece of equipment. We have no examples in mechanical engineering, but in the electronic engineering program a typical equipment designed and constructed has the complexity of a video terminal, with shows that quite complex equipment can be designed and produced by the students. The students are really motivated by this activity, and work much more than the formally assigned 60 hours to the course, in the case of more ambitious projects.

Microprocessors are becoming one very important component in industrial products; the robot itself depends heavily on it. In consumer goods, its usage should increase exponentially, pioneered by the automative industry. At time of this writing, one brazilian-made car already uses a microprocessor for controlling its automatic transmission. The usage of microprocessors in devices until recently entirely mechanical demands new skills from the mechanical engineer, specially the ability to understand the fundamentals of electronics, and to dialog with electronic engineers.

The following courses were included to fulfill these needs:

COURSE	HOURS
Switching circuits	60
Digital systems I	60
Digital systems II	60
Digital systems laboratory	60
Microprocessors applied to mechanical engineering	60
Microprocessors laboratory	60
TOTAL	360

Switching circuits, digital systems I and digital systems II form a set of courses that cover Boolean algebra, Karnaugh maps, combinatorial and sequential circuits, numerical representations, binary arithmetic, codes, circuit design, races and hazards, basic components of digital circuits: gates, registers, multiplexers and demultiplexers, memories (RAMs, ROMs, PROMs, EPROMs), interface circuits, A/D and D/A converters. The laboratory work is concentrated in the digital systems laboratory; the ideal solution would be to have laboratory work in each one of these courses, but this would take too much time for a mechanical engineering program. Microprocessors applied to mechanical engineering studies one microprocessor, development systems, serial and paralel interfaces, interfaces for controlling power devices, and data acquisition and control systems. The laboratory begins with familiarization with microprocessors, followed by the design and implementation of simple interfaces, as seen in the theory course. If possible, this course should be integrated with the graduation project, developping the microprocessor and interfaces that will eventually be needed.

The curriculum includes, obviously, the other usual mechanical engineering courses, that are of no concern here. In short, about 20% of credits are in basic and general courses, 55% in basic mechanical engineering courses, and 25% in courses of this specific program.

THE CONTINUING EDUCATION PROGRAM

The continuing education program was designed to be covered in one year, divided in four 10 week terms. Each course comprises 30 lecture hours, to be given off-hours. The first term comprises three courses, and the remaining ones two courses, with a third night reserved for laboratory or related work.

The courses and their distribution are the following:

1st term	2nd term	3rd term	4th term
Applications of computers to mechanical eng.			
Introduction to industrial Robots	CAD, CAM and CIM	Numerical control machines	Process control
Electronics for mechanical engineers	Introduction to digital systems	Electro-mechanical devices	Microprocessors and applications

Each row in the above table, except the first, represents a different line of study, and can be taken independently. The first course is a broad overview of the whole program, and of some topics not covered, like usage of software packages for structural analysis, production scheduling and control, etc.

Several of these courses were already examined; let's see the rationale behind the others.

In the second row, process control was included for being a rapidily expanding field, as a direct result of a government effort to reduce imports. CIM and Robots are introduced as means to achieve flexible automation and improve quality, and not as an end in itself.

The third row intends to give a general view of digital electronics and fundamentals of computers to practising engineers, many of them never exposed before to computers, even in their undergraduate courses. The microprocessors and applications program include laboratory work, where the student is expected to make a simple interface and code a small program to control it. It might seem quite ambitious, but experience in other continuing education programs has shown that even people with no background in engineering are able to do this kind of work. The idea here is not to shape a design engineer, but to give the necessary background to establish a dialog between mechanical and electronic engineers, working together in the development of new products.

CONCLUSIONS

The key issue in robot usage is productivity, and this can not be achieved by the robot alone. So, an undergraduated curriculum must present a well balanced mix of basic mechanical engineering, control theory, design aids, and a working knowledge of (micro)computers and electronics, both as machine elements and for the understanding of robots and other flexible production machinery.

The curricula proposed - for an undergraduate program and for a continuing education program - will certainly be improved with time. It is very difficult to foresee the evolutionary trends, and anticipate the kind of knowledge our students will need. Neverthless, an immediate action should be taken by universities in those countries that can not afford tobe left behind in the flexible manufacturing revolution.

ACKNOWLEDGMENTS

The author thanks the other members of the commitee, Profs. Valter F. A. Alves, Ettore Bresciani Filho, Luiz N. F. França, Edith Ranzini, and Eitaro Yamane.

Robot Education for Engineers in Manufacturing

Gary C. Hammond
GMI Engineering & Management Institute
Flint, Michigan

Abstract

The robotic facilities at GMI Engineering & Management Institute have been developed with several objectives in mind. The primary objective is to educate both undergraduate engineering students and practicing engineers through CE programs. The philosophy not only covers the fundamentals in classroom experience, but also provides considerable hands on experience through the use of the eight industrial robots in the lab. Another objective provides applied research for industry in cooperation with plant personnel and through the use of student projects in conjunction with the robots. This presentation outlines our teaching and laboratory experience with the undergraduate course and various CE programs as well as describe some interesting applied research projects.

Introduction

GMI Engineering & Management Institute of Flint, Michigan is an accredited 5-year cooperative engineering and management college granting degrees in Electrical, Industrial, and Mechanical Engineering and in Industrial Administration. Present enrollment is 2,079 engineering and 343 industrial administration students. Besides providing educational opportunities for the co-op student body, GMI also has an extensive Continuing Education (CE) program. Because of GMI's close ties with industry, courses developed for our engineering and CE programs are designed to meet the current needs of practicing engineers and managers.

With GMI being located within the heart of the nation's industrial belt and with so many industries (such as the automobile industry) getting extensively involved in the use of robots, it made sense for GMI to provide educational opportunities in the field of robotics. Since robotics is an extension of automated manufacturing, the task of developing the course was given to Professor Jack Lane of the Mechanical Engineering Department whose expertise was in the field of automated manufacturing.

Development of "Introduction to Robotics" Course

After several years of working with industry in the area of robotics, serving on several GM and RI robotic committees, and based upon his knowledge obtained in automated manufacturing (of which he was both a professor and consultant), Professor Jack Lane developed and conducted a pilot Robotic course at GMI Engineering & Management Institute commencing July, 1981. At approximately the same time Professor Gary Hammond, whose expertise was in instrumentation, control systems, and sensors, teamed up with Professor Lane to help set up a robotic lab. A good working relationship soon developed and much was accomplished toward the development of the robotic course.

Because of GMI's close association with industry (strictly GM in the past, but now with over 100 companies) and because of the plant experience that our co-op students have, the decision was made to design the robotics course to be of direct benefit to practicing engineers. The robotics course, ME 475, is a three credit course consisting of two hours of lecture and two hours of lab per week. For GMI's 12 week semester, this provides 22 hours of lecture and the same amount of lab time. Lecture includes such topics as robot classifications, components, tooling, matching the robot to the workplace, planning and implementation, reliability and maintenance, economic considerations, safety, and future developments. See Figure 1 for the course outline.

The course description is given as follows:

ME 475, Robotics

Industrial robots are studied to determine their design variations, methods of programming, operating characteristics, and flexibility for manufacturing operations. A systematic approach to their successful implementation is related to various industrial applications along with tooling and interfacing considerations. New developments in robotics are related to their application for the computer aided manufacturing systems of the future.

WEEK 1	Session 1	Course Induction, Robot Definition
	Session 2	Robot Classification (type of: 1) Program Control & 2) Path Generated, 3) Design Variations, 4) Functional Type, 5) Technology Type, 6) Method or Programming
WEEK 2	Session 3	Robot Classification (cont'd)
	Session 4	Generic Programming
WEEK 3	Session 5	Robot Components (Mechanical Unit, Power Source, Control System & Programming, & Tooling)
	Session 6	Robot Components (cont'd)
WEEK 4	Session 7	Hydraulic, Pneumatic Systems and Servo Valves
	Session 8	Positional and Velocity Sensors (Encoders, Resolvers, Tachometers)
WEEK 5	Session 9	Description of Analog & Digital Control Systems
	Session 10	Robot Operating Characteristics (Operating Speeds, Repeatability, Load Capacity, Work Envelope, Memory, Programming, Limitations, Mounting Restrictions)
WEEK 6	Session 11	Robot Operating Characteristics (cont'd)
	Session 12	End Effectors (Tools & Grippers)
WEEK 7	Session 13	EXAM
	Session 14	Matching Robots to Workplace (Workplace Configurations, Floor Layout, Material Handling Requirements, Backup, Accessibility for Maintenance and Operators, and Facilities)
WEEK 8	Session 15	Matching Robots to Workplace (cont'd)
	Session 16	Reliability and Maintenance
WEEK 9	Session 17	Robot Planning and Implementation
	Session 18	Advantages of Robots
WEEK 10	Session 19	Economic Considerations
	Session 20	Safety
WEEK 11	Session 21	Future Developments
	Session 22	EXAM
WEEK 12	Session 23	Review
		FINAL EXAMINATION*

* The final exam is optional and can be used to replace a lower six week exam grade.

Figure 1, ME 475 "Introduction to Robotics" - Outline

Although GMI had been given a 6CH version I Cincinnati Milacron on an allied transfer from one of the GM divisions, it was not operational for the first semester. However, this did not deter students from performing meaningful lab projects. Considerable time was spent in preparing tooling (including arc welding, resistance welding, and grippers) along with investigating and preparing for interfacing, programming and maintenance of the robot. It was a valuable lesson that a robot is not simply acquired, powered up and instantly operational. As a matter of fact, it can be a very frustrating experience as one tries, almost in vain, to coordinate all of those, such as the electricians, plumbers, hydraulic people, etc., who are involved in installing the robot. (It took over 12 weeks from the time the first robot arrived in the lab until it was made operational). Much of this delay was due to lack of maintenance personnel at GMI and to the problem of scheduling around contractors (GMI was undergoing extensive renovation at the time, which is not uncommon to most organizations undergoing change).

At the present, practical experience is obtained during the lab time as the students learn to program each of the following eight types of industrial robots available in the 5100 square foot robotic lab; an ASEA (IRb60, electrical), 3 Cincinnati Milacrons (6CH, version 1, hydraulic; HT3, version 3, hydraulic; 726, version 4, electrical), Copperweld (CR-50, pneumatic), General Electric (P5, electrical), GMF (AO, electrical), and a PRAB (4200, hydraulic). See Figure 2 for a floor plan of the Robotic Lab, and Figure 3 for overall pictorial views of the lab facility. This facility gives the student experience on both servo and non-servo types as well as on electrical, hydraulic, and pneumatic robots. It also lets them use robots with 4, 5 and 6 degrees of freedom, with load capabilities of a few pounds to several hundred pounds, with those providing a high degree of repeatability ($\pm$ .002) and with some that are not as repeatable ($\pm$ .050). One of the robots even has a tracking option.

The students are each given a semester project (some are individual while others are team projects) which involve some designing and building of tooling, interfacing and programming. The laboratory period is divided into two segments. The first hour of the lab period is devoted to instructing students on the operation of the various robots including start up, programming, and interfacing procedures. The second hour of this lab period is given to the students to work on their projects or to practice programming the various robots.

The ME 475, "Introduction to Robotics", course is presently offered to Junior level mechanical engineering students and to Senior level electrical and industrial engineering students. The enrollment this past and present semester has been limited to 50 students with four two hour laboratory periods. This means approximately 12 students per lab, thus providing each student with sufficient time for hands-on experience. The mechanical engineering students receive courses in manufacturing processes and instrumentation that prepare them to take the course at a junior level. It is probable that by the time electrical and industrial engineering students become seniors they will have received a corresponding background from their industrial experiences.

Laboratory/Research Projects

As stated previously, this robotic course was designed to prepare students with a working knowledge of robots and their relationships to the workplace. Upon completion of the course, many of the students are given work assignments during the co-op phase of their education that are directly related to robotic activities.

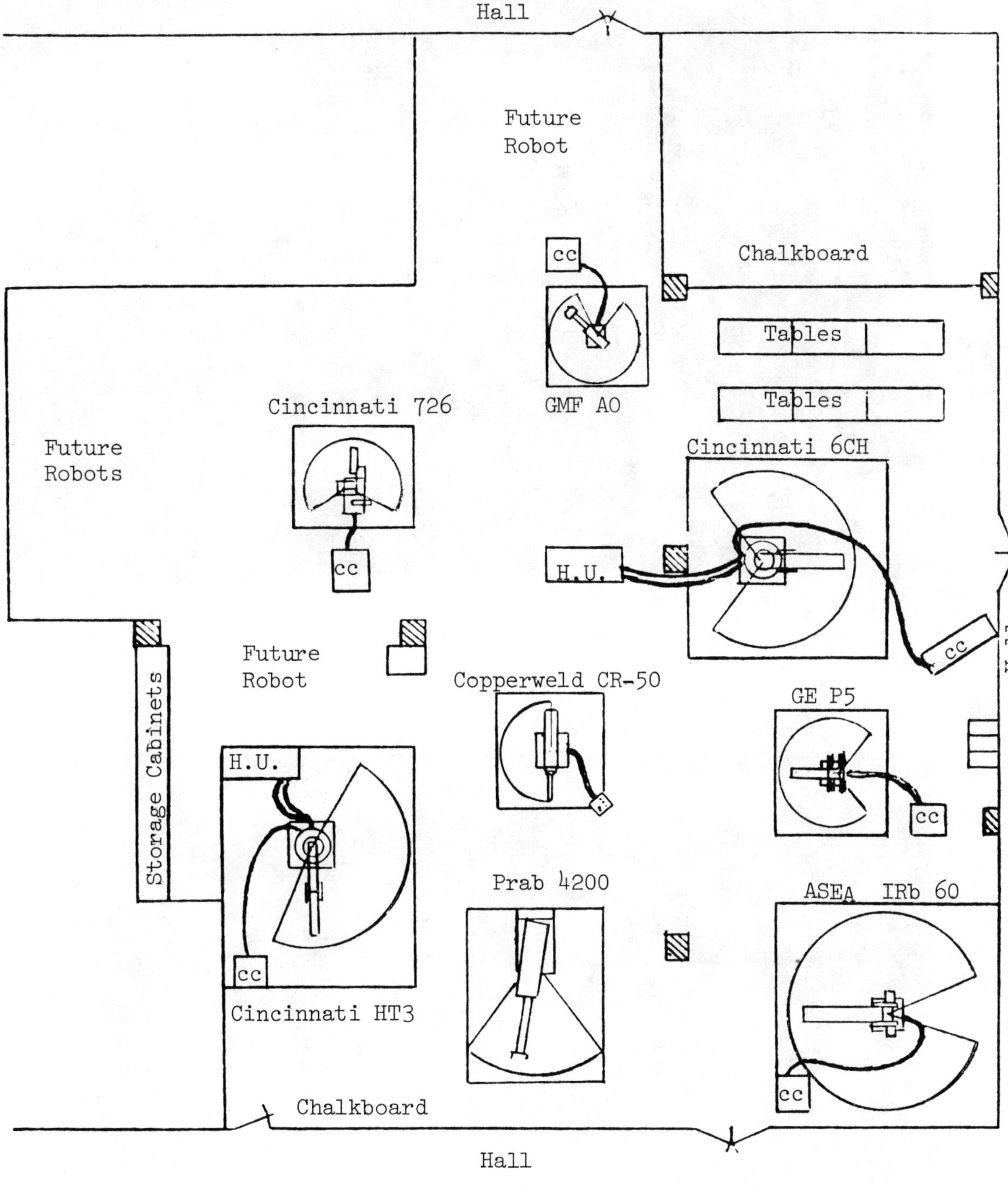

Key

H.U. - Hydraulic Unit
cc - Control Console
SCALE: 1 cm = 5 ft.

FIGURE 2 - GMI Robotic Laboratory

Figure 3 - Overall Views of GMI Robotic Laboratory

Others are assigned a thesis project, also required for graduation, that involves some aspect of robotics. Typical thesis topics usually fall into two basic areas. One type is where the plant has an existing robot and the student is assigned the project of making it operational. This usually consists of designing tooling, interfacing and programming to perform the desired function. The other type of thesis project is where the student has to investigate the plant for possible robotic applications and to determine which ones are most feasible from all aspects, especially from an economic justification. This type of project usually is completed with the recommendation that a particular robot (model, including options, etc.) be purchased and be applied to a particular operation.

To prepare the student for their work experiences, our robotics course must have, in conjunction with the pertinent lecture material, realistic laboratory experience that will prepare them for industry demands. Two key elements permit this realistic laboratory experience. First is the availability of the eight industrial robots similar to if not exactly the same as those that will be encountered in industry. Second is GMI's close relationship to industry which permits the opportunity to immediately determine future needs as related to possible robotic applications. We are often contacted by various industries and plants asking us to investigate the feasibility of applying robots to certain applications, designing some tooling, or some other aspect of applying robots to a particular situation. Thus, we provide applied research for industry in cooperation with plant personnel through the use of student projects. This provides for meaningful laboratory experiences, not just "busy work". An additional benefit is that our CE program participants have the opportunity to see and work with functional robots performing real tasks.

Some of the current industrial projects involve inserting valves in engine heads (see Figure 4), inserting spark plugs, inserting valve lifters, mounting wheels, measuring forces required to insert valves in heads with and without remote center compliance, and investigating some quick change tooling for a local tool supplier. Other projects involve resistance welding (see Figure 5), arc welding, machine load and unloading, palletizing, and simple assembly. These projects were selected to demonstrate common robotic applications.

New Courses in the Offering

Besides the introductory course entitled, "Introduction to Robotics", there will be two more courses offered in Robotics. "Robotic-Applications" which deals with applications, sensors, and interfacing in more detail will be offered to the students starting Fall semester of '83. The course will be taught in a similar manner as the first course with a two hour lecture and two hour lab session per week for a total of three credits. The prerequisites for this course will be senior level status and ME 475 "Introduction to Robotics". The other course, "Robotic Interfacing and Communication", soon to be offered by the electrical engineering department, will include such topics as: analog and digital electronics, controls, vision, speech and networks.

Continuing Education Robotic Programs

Presently there are two continuing engineering education courses being taught at GMI's facilities. The first one is a three day robotic course offered by GMI's CE department and is entitled, "Robotics in Manufacturing". This program consists of 14 hours of lecture on basics such as classifications, systems, components, advantages, justification, planning and implementation, operating characteristics,

Figure 4 - Valve Insertion Using Robot

Figure 5 - Resistance Welding Using Robot

matching to the workplace, end effectors, programming, and future developments. These lectures are given by Professors Lane and Hammond. The remaining ten hours are devoted to lab time with four two hour blocks given to general instruction as to operating and programming procedures for the various robots. This is accomplished by breaking the students up into four groups, each being assigned to one of four faculty members. During each of the two hour lab periods, 2 of the eight robots are studied. After each two hour lab period, the students are reassigned another lab instructor and study another two of the robots. At the completion of the four lab periods, every student will have seen each of the eight robots operated (start up procedures, programming, operating characteristics, etc.,) as well as learning such details as load capabilities, repeatability, special features on the various models, cost, etc. For the remaining two hour lab period, the students can pick a robot of particular interest to them or be assigned one. They are encouraged to go through the start up procedure and to program their assigned robot to perform some task. This often requires some simple interfacing to complete their task. See Figure 6 for schedule of typical three day programs.

Student evaluations have proven the program to be excellent in meeting their needs. They particularly enjoy the opportunity to see and operate so many different types of robots, a very positive feature of our program. It is not uncommon that immediately following the program, several participants will contact the various robot suppliers for further details and quotes on particular models. This program is open to any engineer or technician from any company. However, the enrollment is limited to 30 so that each student will have an opportunity to participate in the lab periods.

Another robotic course is offered by the GM Educational and Training group and is limited to GM personnel. They in turn subcontract GMI for the lab and basic lecture portion of their course, this part being very similar to that previously described. However, this course is a five day program with the remaining two days being devoted to GM personnel from various GM divisions giving talks on specific robotic applications and a local GM plant tour showing some robots in production activities.

These two programs run approximately every other week with between 25 to 30 participants per program. Assuming the same continuing rate for the remaining three months of the first year of operation, it will mean approximately 500 to 600 CE program participants and 170 to 200 co-op engineering students will have gone through one of the programs offered by either GMI or the GM Educational & Training Department. Although it was anticipated that most of the CE program participants would be from the immediate community, (Michigan, Ohio, Indiana, Illinois and Ontario) a surprising number come from as far as California, Florida and Quebec.

Future CE Programs

As the new robotic courses are developed for our regular engineering students, it is planned that corresponding continuing engineering education courses will be offered to the public.

Hour \ Day	Monday	Tuesday	Wednesday
8-9	Introduction and Robot Definition	Robot Components	Robot Planning and Implementation
9-9:50	Robot Classification Systems	Robot Components (cont'd)	Control Systems Analog & Digital
9:50 to 10:10	BREAK		
10:10 to 11:00	Classification Systems (cont'd)	End Effectors (Tools & Grippers)	Robot Operating Characteristics
11-12:00	General Programming	Matching Robots to the Workplace	Advantages of Robots
12:00	LUNCH		
1-2:00	First Lab 4 Groups	Third Lab 4 Groups	Robot Justification
2-2:50	Each Group Assigned to 2 Robots	Each Group Assigned to Third 2 Robots	Future Robot Developments
2:50 to 3:10	BREAK		
3:10 to 4:00	Second Lab 4 Groups	Fourth Lab 4 Groups	Free Lab 4 Groups
4:00 to 5:00	Each Assigned to Second 2 Robots	Each Assigned to Last 2 Robots	Program and Operate Selected Robot

Figure 6 - "ROBOTICS IN MANUFACTURING" Schedule

CHAPTER 16

KINEMATICS

A Geometric Approach in Solving the Inverse Kinematics of PUMA Robots

C.S.G. Lee
University of Michigan
Ann Arbor, Michigan

M. Ziegler
University of Michigan

Abstract

A geometric approach to derive a consistent joint solution of a six-link PUMA robot is presented. An algorithm established the link coordinate systems for a robot. Based on the link coordinate systems and human arm geometry, various arm configurations of a PUMA robot can be identified with the help of three configuration indicators. These indicators enable one to find a solution from the possible four solutions for the first three joints, and a solution from the possible two solutions for the last three joints. The method can be extended to any robot arm with rotary joints.

1. Introduction

An industrial robot is a general purpose manipulator having several rigid links connected in series by revolute or prismatic joints. One end of the chain is attached to a supporting base while the other end is free and attached with a tool to manipulate objects or perform assembly tasks. The motion of the joints result in relative motion of the links. Mechanically a robot is composed of an arm (or main frame) and a wrist subassembly plus a tool. It is designed to reach a workpiece located within its work-volume. The arm subassembly consists of three degree of freedom movement. The combination of the movements will position the wrist unit at the workpiece. The wrist subassembly unit usually consists of three rotary motions. The combination of these motions will orient the tool according to the configuration of the object to ease pickup. Hence for a six-link robot, the arm subassembly is the positioning mechanism, while the wrist subassembly is the orientation mechanism.

Robot arm kinematics deals with the study of the geometry of motion of a robot arm with respect to a fixed reference coordinate system as a function of time without regard to the forces/moments that cause the motion. Thus it deals with the spatial configuration of the robot as a function of time, in particular the relations between the joint-variable space and the position and orientation of a robot arm. The kinematics problem usually consists of two subproblems - the direct and inverse kinematics problems

The direct kinematics problem is to find the position and orientation of the end effector of a manipulator with respect to a reference coordinate system, given the joint angle vector $\vartheta = (\vartheta_1, \vartheta_2, \vartheta_3, \vartheta_4, \vartheta_5, \vartheta_6)^t$ of the robot arm. The inverse kinematics problem (or arm solution) is to calculate the joint angle vector ϑ given the position and orientation of the end effector with respect to the reference coordinate system. Since the independent variables in a robot arm are the joint angles and a task is generally stated in terms of the base or world coordinate system, the inverse kinematics solution is used more frequently.

2. Links, Joints, and Coordinate Transformation

A mechanical manipulator consists of a sequence of rigid links connected by either revolute or prismatic joints. Each joint-link pair constitutes one degree of freedom. Hence for an n-degree-of-freedom manipulator, there are n joint-link pairs with link 0 attached to a supporting base where an inertial coordinate frame is usually established for this dynamic system, and the last link is attached with a tool. In order to describe the translational and rotational relationship between adjacent links, a Denavit-Hartenberg matrix representation for each link is used [1] and shown in Fig. 1. From Fig. 1, an orthonormal coordinate frame system ($\mathbf{x}_i, \mathbf{y}_i, \mathbf{z}_i$) is assigned to link i, where the $\mathbf{z}_i$ axis passes through the axis of motion of joint $i+1$, and the $\mathbf{x}_i$ axis is normal to the $\mathbf{z}_{i-1}$ axis pointing away from it, while the $\mathbf{y}_i$ axis completes the right hand rule. With this orthonormal coordinate frame, link i is characterized by two parameters: a_i, the common normal distance between the $\mathbf{z}_{i-1}$ and $\mathbf{z}_i$ axes, and α_i, the twist angle measured between the $\mathbf{z}_{i-1}$ and $\mathbf{z}_i$ axes in a plane perpendicular to a_i. Joint i which connects link $i-1$ to link i is characterized by a distance parameter d_i measured between the $\mathbf{x}_{i-1}$ and $\mathbf{x}_i$ axes and a revolute joint variable ϑ_i which is the joint angle between the normals and measured in a plane normal to the joint axis. If joint i is prismatic, then it is characterized by an angle parameter ϑ_i and a joint variable d_i.

Once the link coordinate systems have been established for each link, a homogeneous transformation matrix, $\mathbf{A}^i_{i-1}$, can easily be developed relating the ith coordinate frame to the $i-1^{th}$ coordinate frame. Using the $\mathbf{A}^i_{i-1}$ matrix, one can relate a point $\mathbf{p}_i$ at rest in link i and expressed in homogeneous coordinates with respect to the ith coordinate system to the $i-1^{th}$ coordinate system established at link $(i-1)$ by:

$$\mathbf{p}_{i-1} = \mathbf{A}^i_{i-1}\ \mathbf{p}_i \qquad (1)$$

where $\mathbf{p}_{i-1} = (x_{i-1}, y_{i-1}, z_{i-1}, 1)^t$; $\mathbf{p}_i = (x_i, y_i, z_i, 1)^t$;

$$\mathbf{A}^i_{i-1} = \begin{bmatrix} \cos\vartheta_i & -\cos\alpha_i \sin\vartheta_i & \sin\alpha_i \sin\vartheta_i & a_i \cos\vartheta_i \\ \sin\vartheta_i & \cos\alpha_i \cos\vartheta_i & -\sin\alpha_i \cos\vartheta_i & a_i \sin\vartheta_i \\ 0 & \sin\alpha_i & \cos\alpha_i & d_i \\ 0 & 0 & 0 & 1 \end{bmatrix} \text{; for a \textit{rotary joint}} \qquad (2)$$

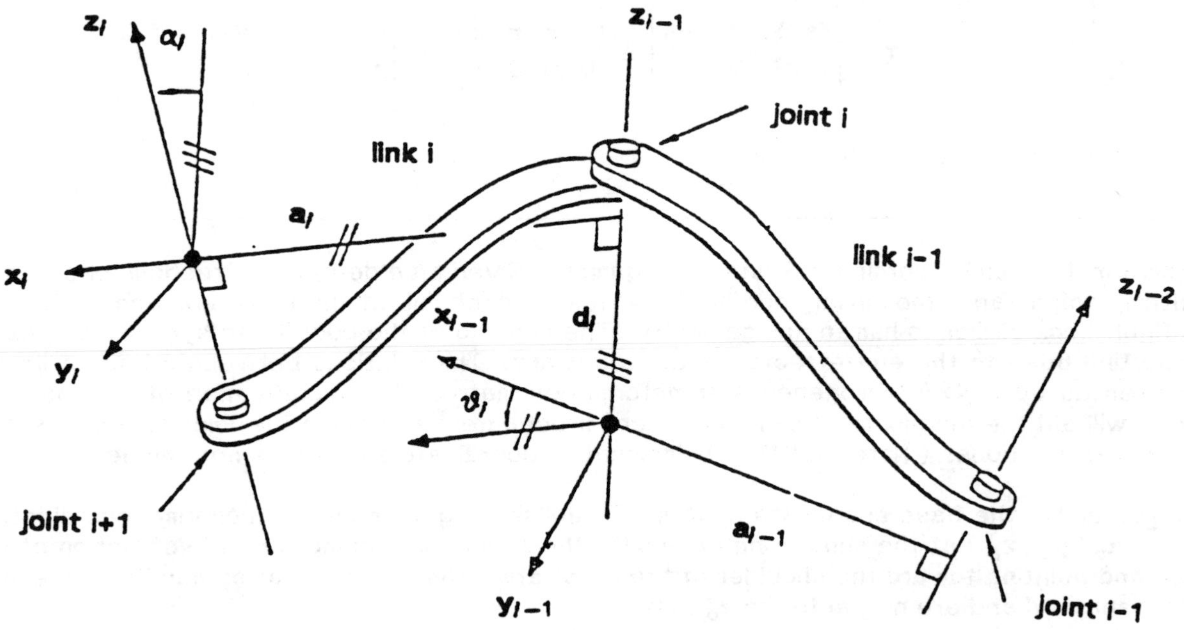

Figure 1 Link Coordinate System and Its Parameters

$$A^i_{i-1} = \begin{bmatrix} \cos\vartheta_i & -\cos\alpha_i \sin\vartheta_i & \sin\alpha_i \sin\vartheta_i & 0 \\ \sin\vartheta_i & \cos\alpha_i \cos\vartheta_i & -\sin\alpha_i \cos\vartheta_i & 0 \\ 0 & \sin\alpha_i & \cos\alpha_i & d_i \\ 0 & 0 & 0 & 1 \end{bmatrix} \quad ; \text{ for a } \textit{prismatic joint} \tag{3}$$

With the basic rules for establishing an orthonormal coordinate system for each link and the geometric interpretation of the joint and link parameters, a procedure for establishing *consistent* orthonormal coordinate systems for a robot is outlined in Algorithm 1. An example of applying this algorithm to a PUMA robot arm is given in Fig. 2.

The six A^i_{i-1} transformation matrices for a PUMA robot have been found based on the coordinate systems established in Fig. 2. These A^i_{i-1} matrices are listed in Fig. 3.

3. Kinematic Equations for Manipulators

The homogeneous matrix T^i_o which specifies the position and orientation of the ith coordinate frame with respect to the base coordinate system is the chain product of successive coordinate transformation matrices of A^i_{i-1}, expressed as:

$$T^i_o = A^1_o A^2_1 \cdots A^i_{i-1} = \prod_{j=1}^{i} A^j_{j-1} = \begin{bmatrix} x_i & y_i & z_i & p_i \\ 0 & 0 & 0 & 1 \end{bmatrix} = \begin{bmatrix} R^i_o & p^i_o \\ 0 & 1 \end{bmatrix} \quad ; \text{ for } i = 1,2 \ldots n \tag{4}$$

Specifically for $i = 6$, we obtain the T matrix, $T = A^6_o$, which specifies the position and orientation of the end-point of a manipulator with respect to the base coordinate system. This T matrix is used so frequently in robot arm kinematics that it is called the "arm matrix". Consider the **T** matrix to be of the form:

$$\mathbf{T} = \begin{bmatrix} \mathbf{x}_6 & \mathbf{y}_6 & \mathbf{z}_6 & \mathbf{p}_6 \\ 0 & 0 & 0 & 1 \end{bmatrix} = \begin{bmatrix} \mathbf{n} & \mathbf{s} & \mathbf{a} & \mathbf{p} \\ 0 & 0 & 0 & 1 \end{bmatrix} = \begin{bmatrix} n_x & s_x & a_x & p_x \\ n_y & s_y & a_y & p_y \\ n_z & s_z & a_z & p_z \\ 0 & 0 & 0 & 1 \end{bmatrix} \tag{5}$$

where (see Fig. 4):

Algorithm 1. (Link Coordinate System Assignment) Given an n degree-of-freedom robot arm, this algorithm assigns an orthonormal coordinate system to each link of the robot arm according to the arm configurations similar to human arm geometry. The labeling of the coordinate systems begins from the supporting base to the end-effector of the robot arm. The relations between adjacent links can be represented by a 4×4 homogeneous transformation matrix. The significance of this assignment is that it will aid the development of a consistent procedure for deriving the joint solution as discussed in the later sections. (Note that the assignment of coordinate systems is not unique.)

D1. [Establish the base coordinate system] Establish a right-handed orthonormal coordinate system ($\mathbf{x}_o$, $\mathbf{y}_o$, $\mathbf{z}_o$) at the supporting base with the $\mathbf{z}_o$ axis lies along the axis of motion of joint one, and pointing toward the shoulder of the robot arm. The $\mathbf{x}_o$ and $\mathbf{y}_o$ axes can be conveniently established and are normal to the $\mathbf{z}_o$ axis.

D2. [Initialize and Loop] For each i, i=1,...,n, perform steps D3 to D6.

D3. [Establish joint axis] Align the $\mathbf{z}_i$ with the axis of motion (rotary or sliding) of joint i+1. For robots having left/right arm configurations, the $\mathbf{z}_1$ and $\mathbf{z}_2$ axes are pointing away from the shoulder and the "trunk" of the robot arm.

D4. [Establish the origin of the ith coordinate system] Locate the origin of the ith coordinate system at the intersection of the $\mathbf{z}_i$ and $\mathbf{z}_{i-1}$ axes or at the intersection of common normal between the $\mathbf{z}_i$ and $\mathbf{z}_{i-1}$ axes and the $\mathbf{z}_i$ axis.

D5. [Establish $\mathbf{x}_i$ axis] Establish $\mathbf{x}_i = \pm (\mathbf{z}_{i-1} \times \mathbf{z}_i) \; / \; || \mathbf{z}_{i-1} \times \mathbf{z}_i ||$ or along the common normal between the $\mathbf{z}_{i-1}$ and $\mathbf{z}_i$ axes when they are parallel.

D6. [Establish $\mathbf{y}_i$ axis] Assign $\mathbf{y}_i = + (\mathbf{z}_i \times \mathbf{x}_i) \; / \; || \mathbf{z}_i \times \mathbf{x}_i ||$ to complete the right-handed coordinate system. (Extend the $\mathbf{z}_i$ and the $\mathbf{x}_i$ axes if necessary for steps D8-D11).

D7. [Find Joint and Link Parameters] For each i, i=1,..,n, perform steps D8 to D11.

D8. [Find d_i] d_i is the distance from the origin of the $i-1^{th}$ coordinate system to the intersection of the $\mathbf{z}_{i-1}$ axis and the $\mathbf{x}_i$ axis along the $\mathbf{z}_{i-1}$ axis. It is the joint variable if joint i is prismatic.

D9. [Find a_i] a_i is the distance from the intersection of the $\mathbf{z}_{i-1}$ axis and the $\mathbf{x}_i$ axis to the origin of the ith coordinate system along the $\mathbf{x}_i$ axis.

D10. [Find ϑ_i] ϑ_i is the angle of rotation from the $\mathbf{x}_{i-1}$ axis to the $\mathbf{x}_i$ axis about the $\mathbf{z}_{i-1}$ axis. It is the joint variable if joint i is rotary.

D11. [Find α_i] α_i is the angle of rotation from the $\mathbf{z}_{i-1}$ axis to the $\mathbf{z}_i$ axis about the $\mathbf{x}_i$ axis.

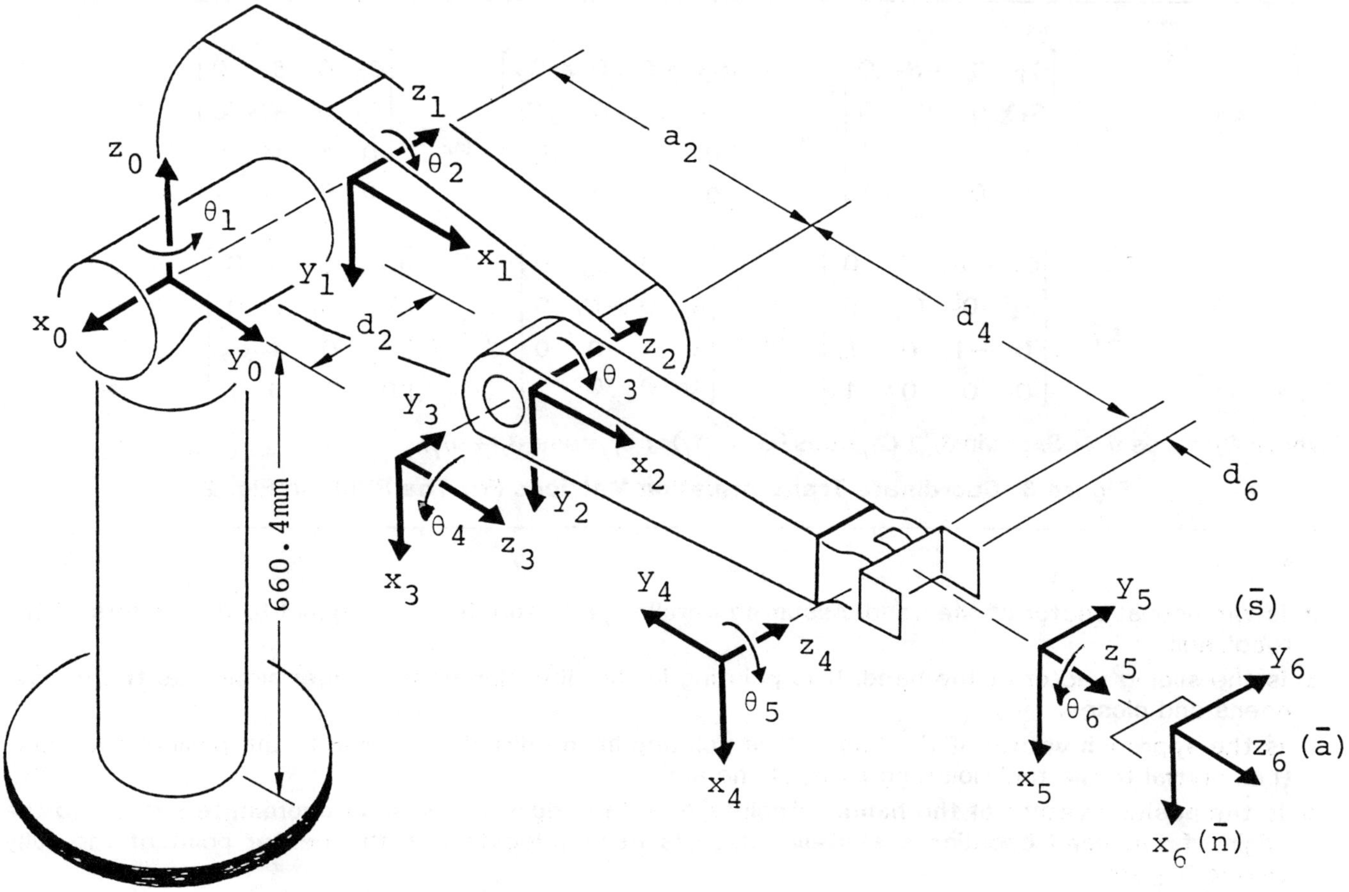

PUMA Robot Arm Link Coordinate Parameters					
Joint i	**ϑ_i**	**α_i**	**a_i**	**d_i**	**Range**
1	90	-90	0	0	-160 to +160
2	0	0	432 mm	149.5 mm	-225 to +45
3	90	90	0	0	-45 to +225
4	0	-90	0	432.0	-110 to +170
5	0	90	0	0	-100 to +100
6	0	0	0	56.5	-266 to +266

Figure 2 Link Coordinate Systems For A PUMA Robot

$$\mathbf{A}_o^1 = \begin{bmatrix} C_1 & 0 & -S_1 & 0 \\ S_1 & 0 & C_1 & 0 \\ 0 & -1 & 0 & 0 \\ 0 & 0 & 0 & 1 \end{bmatrix} \quad \mathbf{A}_1^2 = \begin{bmatrix} C_2 & -S_2 & 0 & a_2C_2 \\ S_2 & C_2 & 0 & a_2S_2 \\ 0 & 0 & 1 & d_2 \\ 0 & 0 & 0 & 1 \end{bmatrix} \quad \mathbf{A}_2^3 = \begin{bmatrix} C_3 & 0 & S_3 & 0 \\ S_3 & 0 & -C_3 & 0 \\ 0 & 1 & 0 & 0 \\ 0 & 0 & 0 & 1 \end{bmatrix}$$

$$\mathbf{A}_3^4 = \begin{bmatrix} C_4 & 0 & -S_4 & 0 \\ S_4 & 0 & C_4 & 0 \\ 0 & -1 & 0 & d_4 \\ 0 & 0 & 0 & 1 \end{bmatrix} \quad \mathbf{A}_4^5 = \begin{bmatrix} C_5 & 0 & S_5 & 0 \\ S_5 & 0 & -C_5 & 0 \\ 0 & 1 & 0 & 0 \\ 0 & 0 & 0 & 1 \end{bmatrix} \quad \mathbf{A}_5^6 = \begin{bmatrix} C_6 & -S_6 & 0 & 0 \\ S_6 & C_6 & 0 & 0 \\ 0 & 0 & 1 & d_6 \\ 0 & 0 & 0 & 1 \end{bmatrix}$$

where $C_i \equiv \cos\vartheta_i$; $S_i \equiv \sin\vartheta_i$; $C_{ij} \equiv \cos(\vartheta_i + \vartheta_j)$; $S_{ij} \equiv \sin(\vartheta_i + \vartheta_j)$.

Figure 3 Coordinate Transformation Matrices For The PUMA in Fig. 2

n is the normal vector of the hand. Assuming parallel-jaw hand, it is orthogonal to the fingers of the robot arm.

s is the sliding vector of the hand. It is pointing in the direction of the finger motion as the gripper opens and closes.

a is the approach vector of the hand. It is pointing in the direction normal to the palm of the hand. (i.e., normal to the tool mounting plate of the arm.)

p is the position vector of the hand. It points from the origin of the base coordinate system to the origin of the hand coordinate system, which is usually located at the center point of the fully closed fingers.

The arm matrix T for the PUMA robot arm shown in Fig. 2 is found to be:

$$\begin{aligned} n_x &= C_1[C_{23}(C_4C_5C_6 - S_4S_6) - S_{23}S_5C_6] - S_1[S_4C_5C_6 + C_4S_6] \\ n_y &= S_1[C_{23}(C_4C_5C_6 - S_4S_6) - S_{23}S_5C_6] + C_1[S_4C_5C_6 + C_4S_6] \\ n_z &= -S_{23}[C_4C_5C_6 - S_4S_6] - C_{23}S_5C_6 \end{aligned} \quad (6)$$

$$\begin{aligned} s_x &= C_1[-C_{23}(C_4C_5S_6 + S_4C_6) + S_{23}S_5S_6] - S_1[-S_4C_5S_6 + C_4C_6] \\ s_y &= S_1[-C_{23}(C_4C_5S_6 + S_4C_6) + S_{23}S_5S_6] + C_1[-S_4C_5S_6 + C_4C_6] \\ s_z &= S_{23}(C_4C_5S_6 + S_4C_6) + C_{23}S_5S_6 \end{aligned} \quad (7)$$

$$\begin{aligned} a_x &= C_1(C_{23}C_4S_5 + S_{23}C_5) - S_1S_4S_5 \\ a_y &= S_1(C_{23}C_4S_5 + S_{23}C_5) + C_1S_4S_5 \\ a_z &= -S_{23}C_4S_5 + C_{23}C_5 \end{aligned} \quad (8)$$

$$\begin{aligned} p_x &= C_1[d_6(C_{23}C_4S_5 + S_{23}C_5) + S_{23}d_4 + a_2C_2] - S_1(d_6S_4S_5 + d_2) \\ p_y &= S_1[d_6(C_{23}C_4S_5 + S_{23}C_5) + S_{23}d_4 + a_2C_2] + C_1(d_6S_4S_5 + d_2) \\ p_z &= d_6(C_{23}C_5 - S_{23}C_4S_5) + C_{23}d_4 - a_2S_2 \end{aligned} \quad (9)$$

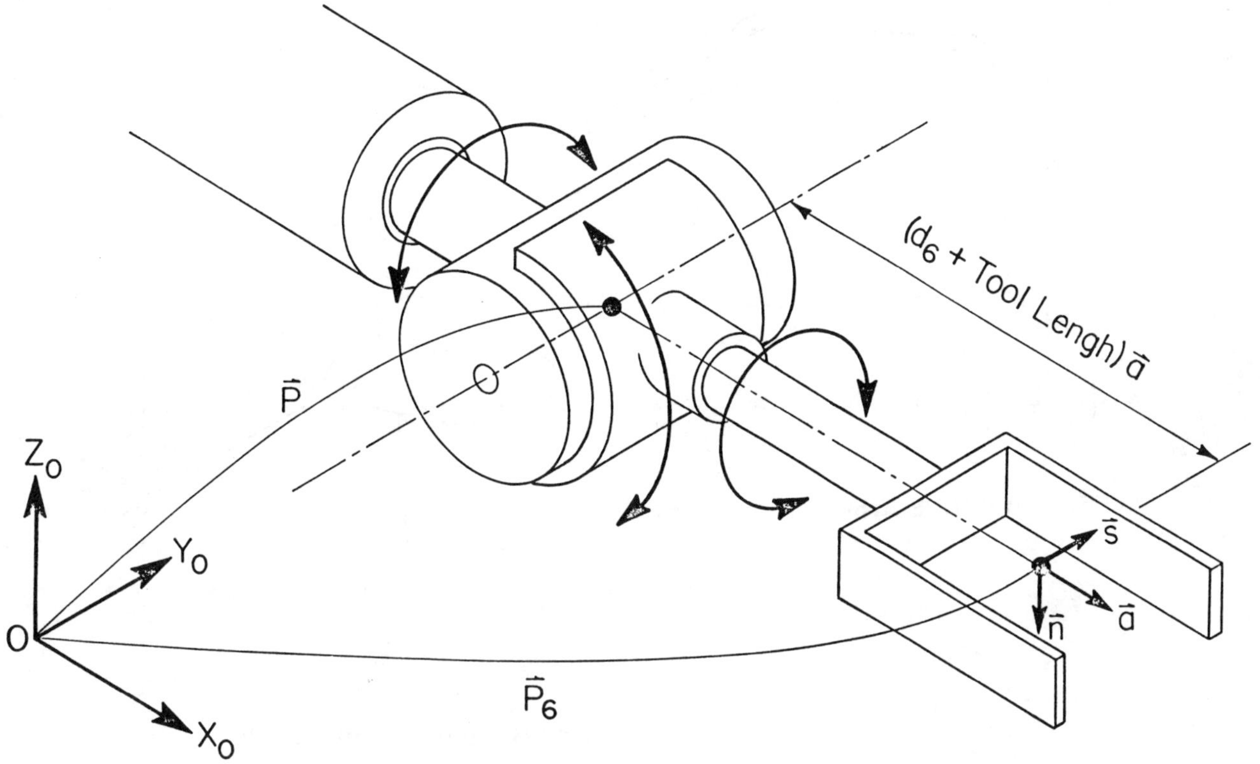

Figure 4 Hand Coordinate System and [n , s , a]

4. The Inverse Kinematics Solution of a PUMA Robot Arm

This section presents a geometric approach to derive a consistent joint angle solution of a PUMA robot given the arm matrix. The approach calls for the definition of various possible arm configurations based on the link coordinate systems and human arm geometry. These arm configurations are then expressed in an exact mathematical way to allow the construction of arm configuration indicators and their corresponding decision equations. In general, there are four possible solutions to the first three joints and two possible solutions to the last three joints for a six-link PUMA robot. Associated with the joint solution are three indicators, two with the solution of the first three joints and one with the last three joints. The first two indicators allow one to determine one solution from the possible four solutions for the first three joints. Similarly, the third indicator allows one to select a solution from the possible two solutions for the last three joints. These arm configuration indicators are prespecified by a user for finding the inverse solution. As a verification of the joint solution, the arm configuration indicators can be determined from the decision equations which are functions of the joint angles. This approach can be generalized to solve the inverse kinematics problem of most present day industrial robots with rotary joints.

4.1. Definition of Various Arm Configurations

For the PUMA robot arm shown in Fig. 2 (and other rotary robot arms), various arm configurations are defined according to human arm geometry and the link coordinate systems which are established using Algorithm 1 as: (Fig. 5)

- RIGHT (shoulder) ARM: Positive ϑ_2 moves the wrist in the *positive* z_o direction while joint 3

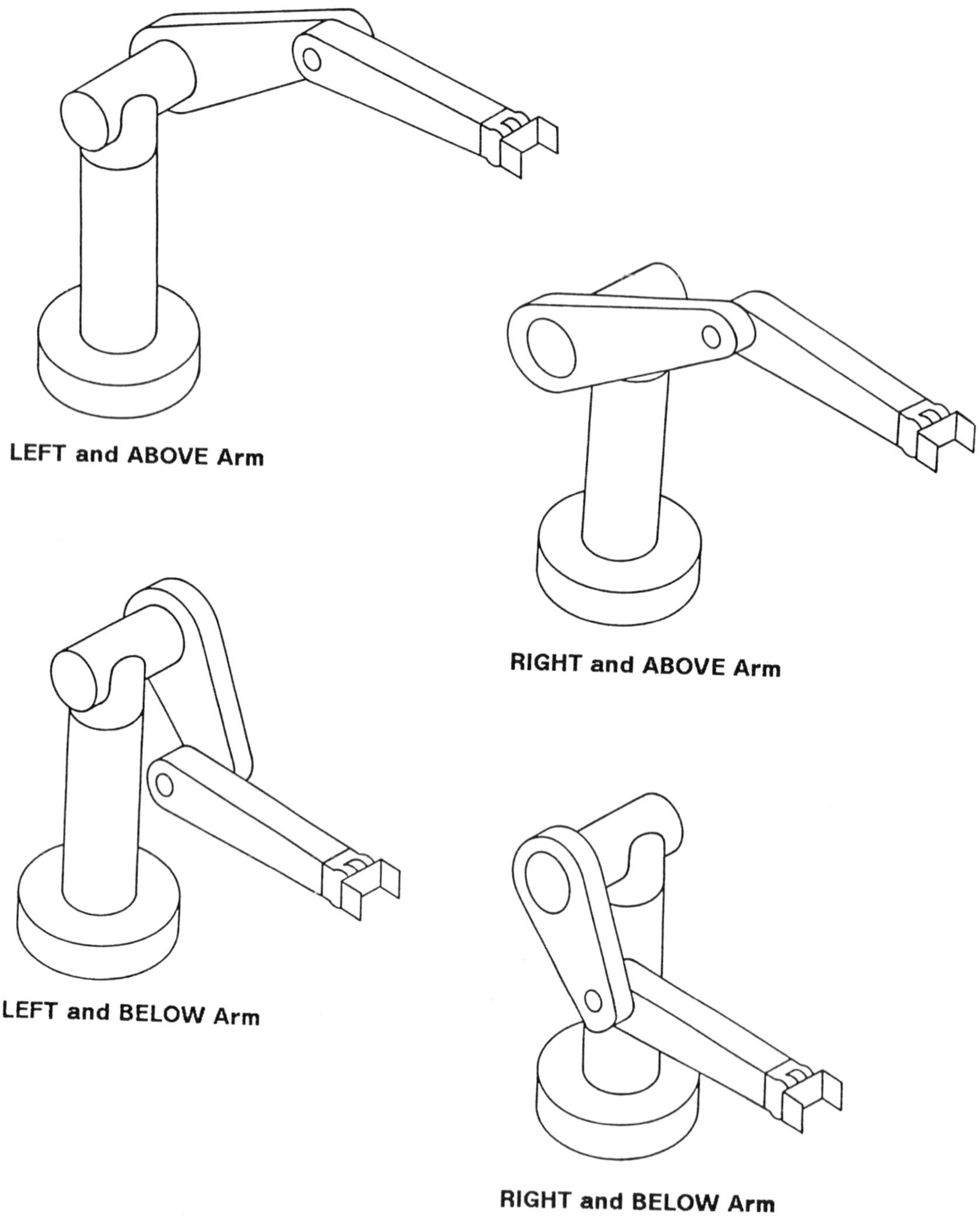

Figure 5. Definition of Various Arm Configurations

is not activated.

- LEFT (shoulder) ARM: Positive ϑ_2 moves the wrist in the *negative* $\mathbf{z}_o$ direction while joint 3 is not activated.

- ABOVE ARM (elbow above wrist): Position of the wrist of the $\begin{Bmatrix} RIGHT \\ LEFT \end{Bmatrix}$ arm with respect to the shoulder coordinate system has $\begin{Bmatrix} negative \\ positive \end{Bmatrix}$ coordinate value along the $\mathbf{y}_2$-axis.
- BELOW ARM (elbow below wrist): Position of the wrist of the $\begin{Bmatrix} RIGHT \\ LEFT \end{Bmatrix}$ arm with respect to the shoulder coordinate system has $\begin{Bmatrix} positive \\ negative \end{Bmatrix}$ coordinate value along the $\mathbf{y}_2$-axis.
- NOFLIP WRIST: The sliding unit vector of the hand coordinate system **s** and the $\mathbf{y}_5$ unit vector have a positive dot product or **n** is equal to $\mathbf{y}_5$.
- FLIP WRIST: The sliding unit vector of the hand coordinate system **s** and the $\mathbf{y}_5$ unit vector have a negative dot product or **n** is equal to $-\mathbf{y}_5$.

(Note that the definition of the arm configurations with respect to the link coordinate systems may have to be slightly modified if one uses different link coordinate systems.)

With respect to the above definition for various arm configurations, two arm configuration *indicators* (ARM and ELBOW) are defined for each arm configuration. These two indicators are combined to give one solution out of the possible four joint solutions for the first three joints. A third indicator (WRIST) gives one of the two possible joint solutions for the last three joints. These indicators are defined as:

$$ARM = \begin{cases} +1 & ; \ RIGHT\ arm \\ -1 & ; \ LEFT\ arm \end{cases} \tag{10}$$

$$ELBOW = \begin{cases} +1 & ; \ ABOVE\ arm \\ -1 & ; \ BELOW\ arm \end{cases} \tag{11}$$

$$WRIST = \begin{cases} +1 & ; \ NOFLIP \text{ for } the\ wrist\ soluiton \\ -1 & ; \ FLIP\ the\ wrist\ solution \end{cases} \tag{12}$$

The signed values of these indicators are prespecified by a user for finding the inverse kinematics solution. These indicators can also be set from the knowledge of the joint angles of the robot arm using the corresponding decision equations. We shall later give the decision equations which will determine these indicator values. The decision equations can be used as a verification of the inverse kinematics solution.

4.2. Arm Solution for the First Three Joints of A PUMA Robot Arm

From the kinematics diagram of a PUMA robot arm as in Fig. 2, we define a position vector **p** which points from the origin of the shoulder coordinate system $(\mathbf{x}_o, \mathbf{y}_o, \mathbf{z}_o)$ to the point where the last three joint axes intersect as (see Fig. 4):

$$\mathbf{p} = \mathbf{p}_6 - d_6\mathbf{a} = (p_x, p_y, p_z)^t \tag{13}$$

and finding the position vector of $\mathbf{T}_o^4$, we have:

$$\begin{bmatrix} p_x \\ p_y \\ p_z \end{bmatrix} = \begin{bmatrix} C_1 \left(a_2C_2 + d_4S_{23}\right) - d_2S_1 \\ S_1 \left(a_2C_2 + d_4S_{23}\right) + d_2C_1 \\ d_4C_{23} - a_2S_2 \end{bmatrix} \quad (14)$$

Joint One Solution. If we project the position vector **p** onto the $x_o - y_o$ plane as in Fig. 6, we obtain the following equations for solving ϑ_1:

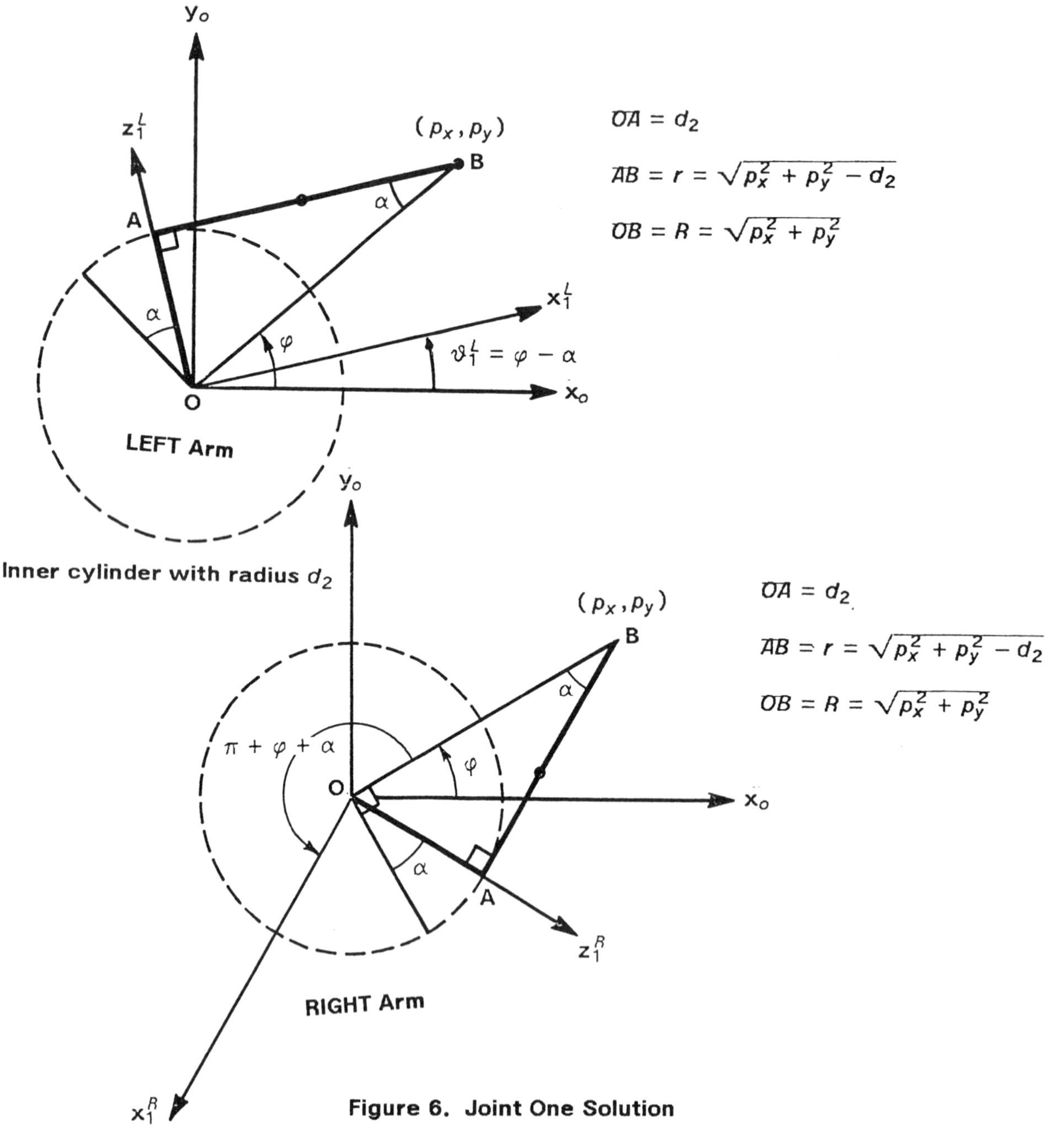

Figure 6. Joint One Solution

$$\vartheta_1^L = \varphi - \alpha \;;\; \vartheta_1^R = \pi + \varphi + \alpha \;;\; r = \sqrt{p_x^2+p_y^2-d_2^2} \;;\; R = \sqrt{p_x^2+p_y^2} \tag{15}$$

$$\sin\varphi = \frac{p_y}{R} \;;\; \cos\varphi = \frac{p_x}{R} \;;\; \sin\alpha = \frac{d_2}{R} \;;\; \cos\alpha = \frac{r}{R} \tag{16}$$

where the superscript indicates the LEFT/RIGHT arm configuration. From Eqs. 15-16, we obtain:

$$\sin\vartheta_1^L = \sin(\varphi - \alpha) = \sin\varphi\cos\alpha - \cos\varphi\sin\alpha = \frac{p_y r - p_x d_2}{R^2} \tag{17}$$

$$\cos\vartheta_1^L = \cos(\varphi - \alpha) = \cos\varphi\cos\alpha + \sin\varphi\sin\alpha = \frac{p_x r + p_y d_2}{R^2} \tag{18}$$

$$\sin\vartheta_1^R = \sin(\pi + \varphi + \alpha) = \frac{-p_y r - p_x d_2}{R^2} \;;\; \cos\vartheta_1^R = \cos(\pi + \varphi + \alpha) = \frac{-p_x r + p_y d_2}{R^2} \tag{19}$$

Combining Eqs. 17-19, we obtain the sine and cosine functions of ϑ_1 respectively:

$$\sin\vartheta_1 = \frac{-\,ARM \cdot p_y\sqrt{p_x^2 + p_y^2 - d_2^2} - p_x d_2}{p_x^2 + p_y^2} \;;\; \cos\vartheta_1 = \frac{-\,ARM \cdot p_x\sqrt{p_x^2 + p_y^2 - d_2^2} + p_y d_2}{p_x^2 + p_y^2} \tag{20}$$

where positive square root is taken in these equations and ARM is defined as in Eq. 10. In order to evaluate ϑ_1 for $-\pi \le \vartheta_1 \le \pi$, an arc tangent function, $atan2(\frac{y}{x})$, which returns $\tan^{-1}(\frac{y}{x})$ adjusted to the proper quadrant will be used. It is defined as:

$$\vartheta = atan2(\frac{y}{x}) = \begin{cases} 0^\circ \le \vartheta \le 90^\circ & ;\ \text{for } +x \text{ and } +y \\ 90^\circ \le \vartheta \le 180^\circ & ;\ \text{for } -x \text{ and } +y \\ -180^\circ \le \vartheta \le -90^\circ & ;\ \text{for } -x \text{ and } -y \\ -90^\circ \le \vartheta \le 0^\circ & ;\ \text{for } +x \text{ and } -y \end{cases} \tag{21}$$

From Eq. 20, ϑ_1 is found to be:

$$\vartheta_1 = atan2\left[\frac{\sin\vartheta_1}{\cos\vartheta_1}\right] = atan2\left[\frac{-\,ARM \cdot p_y\sqrt{p_x^2 + p_y^2 - d_2^2} - p_x d_2}{-\,ARM \cdot p_x\sqrt{p_x^2 + p_y^2 - d_2^2} + p_y d_2}\right] \;;\; -\pi \le \vartheta_1 \le \pi \tag{22}$$

Joint Two Solution. To find joint 2, we project the position vector **p** onto the $x_1 - y_1$ plane as shown in Fig. 7. From Fig. 7, we have four different arm configurations. Each arm configuration corresponds to different values of joint two as:

Arm Configurations	ϑ_2	ARM	ELBOW	ARM · ELBOW
LEFT and ABOVE arm	$\alpha - \beta$	-1	+1	-1
LEFT and BELOW arm	$\alpha + \beta$	-1	-1	+1
RIGHT and ABOVE arm	$\alpha + \beta$	+1	+1	+1
RIGHT and BELOW arm	$\alpha - \beta$	+1	-1	-1

Table 1. Various Arm Configurations for Joint Two

where $0^\circ \le \alpha \le 360^\circ$ and $0^\circ \le \beta \le 90^\circ$.

From the above table, we can express ϑ_2 in one equation as

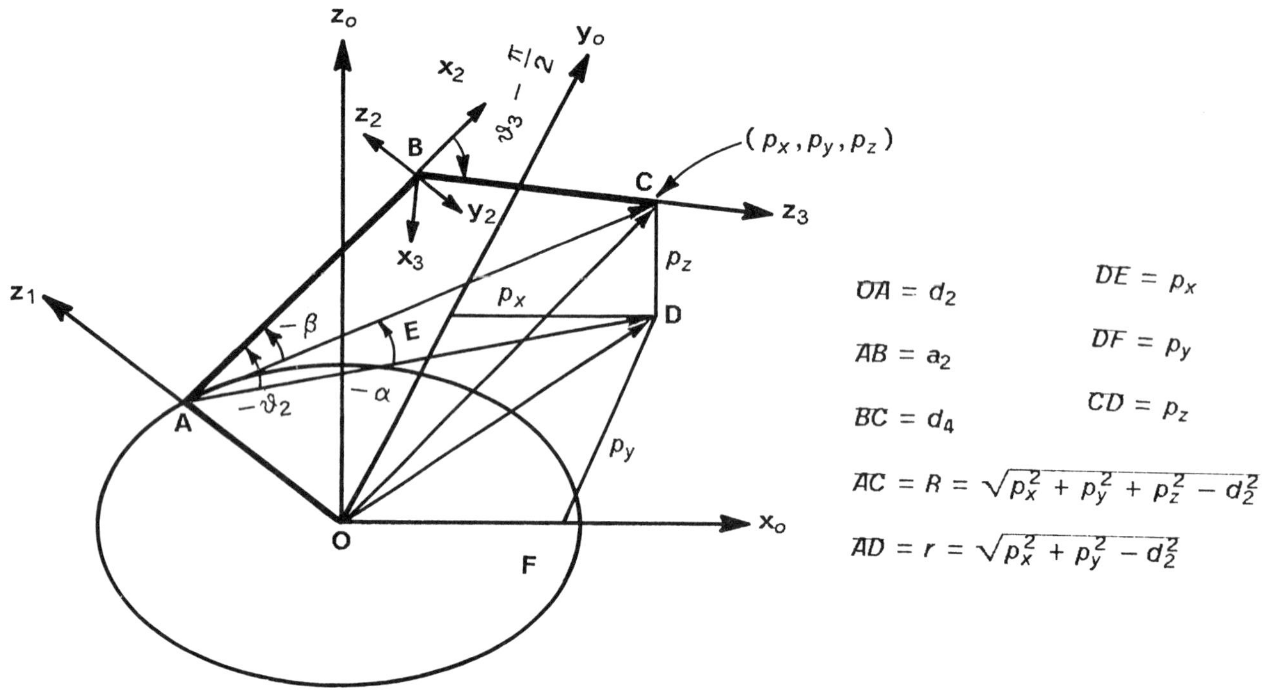

Figure 7. Joint Two Solution

$$\vartheta_2 = \alpha + (ARM \cdot ELBOW)\,\beta = \alpha + K \cdot \beta \tag{23}$$

where the combined arm configuration indicator $K = ARM \cdot ELBOW$ will give an appropriate signed value and the "dot" represents a multiplication operation on the indicators. From Fig. 7, we obtain:

$$R = \sqrt{p_x^2 + p_y^2 + p_z^2 - d_2^2} \quad ; \quad r = \sqrt{p_x^2 + p_y^2 - d_2^2} \tag{24}$$

$$\sin\alpha = -\frac{p_z}{R} = -\frac{p_z}{\sqrt{p_x^2 + p_y^2 + p_z^2 - d_2^2}} \quad ; \quad \cos\alpha = -\frac{ARM \cdot r}{R} = -\frac{ARM \cdot \sqrt{p_x^2 + p_y^2 - d_2^2}}{\sqrt{p_x^2 + p_y^2 + p_z^2 - d_2^2}} \tag{25}$$

$$\cos\beta = \frac{a_2^2 + R^2 - d_4^2}{2a_2R} = \frac{p_x^2 + p_y^2 + p_z^2 + a_2^2 - d_2^2 - d_4^2}{2a_2\sqrt{p_x^2 + p_y^2 + p_z^2 - d_2^2}} \quad ; \quad \sin\beta = \sqrt{1 - \cos^2\beta} \tag{26}$$

From Eqs. 23, 25 and 26, we can find the sine function of ϑ_2

$$\begin{aligned}\sin\vartheta_2 &= \sin(\alpha + K\cdot\beta) = \sin\alpha\cos(K\cdot\beta) + \cos\alpha\sin(K\cdot\beta) = \sin\alpha\cos\beta + K\cos\alpha\sin\beta \\ &= \sin\alpha\cos\beta + (ARM \cdot ELBOW)\cos\alpha\sin\beta\end{aligned} \tag{27}$$

Similarly, we can find the cosine function of ϑ_2:

$$\cos\vartheta_2 = \cos(\alpha + K\cdot\beta) = \cos\alpha\cos\beta - (ARM \cdot ELBOW)\sin\alpha\sin\beta \tag{28}$$

From Eqs. 27 and 28, we obtain the solution for ϑ_2:

$$\vartheta_2 = atan2\left[\frac{\sin\vartheta_2}{\cos\vartheta_2}\right] \quad ; \quad -\pi \le \vartheta_2 \le \pi \tag{29}$$

Joint Three Solution. For joint 3, we project the position vector **p** onto the x_2-y_2 plane as shown in Fig. 8. From Fig. 8, we again have four different arm configurations. Each arm configuration corresponds to different values of joint three as:

Arm Configurations	$(p_2^4)_y$	$\cos\vartheta_3$	ARM	ELBOW	ARM · ELBOW
LEFT and ABOVE arm	≥ 0	≤ 0	-1	+1	-1
LEFT and BELOW arm	≤ 0	≥ 0	-1	-1	+1
RIGHT and ABOVE arm	≤ 0	≥ 0	+1	+1	+1
RIGHT and BELOW arm	≥ 0	≤ 0	+1	-1	-1

Table 2. Various Arm Configurations for Joint Three

where $(p_2^4)_y$ is the y-component of the position vector from the origin of (x_2, y_2, z_2) to the point where the last three joint axes intersect.

From Fig. 8, we obtain the following equations which lead us to finding solution for ϑ_3:

$$R = \sqrt{p_x^2 + p_y^2 + p_z^2 - d_2^2} \quad ; \quad \vartheta_3 = \pi - \varphi + \frac{\pi}{2} = \frac{3\pi}{2} - \varphi \tag{30}$$

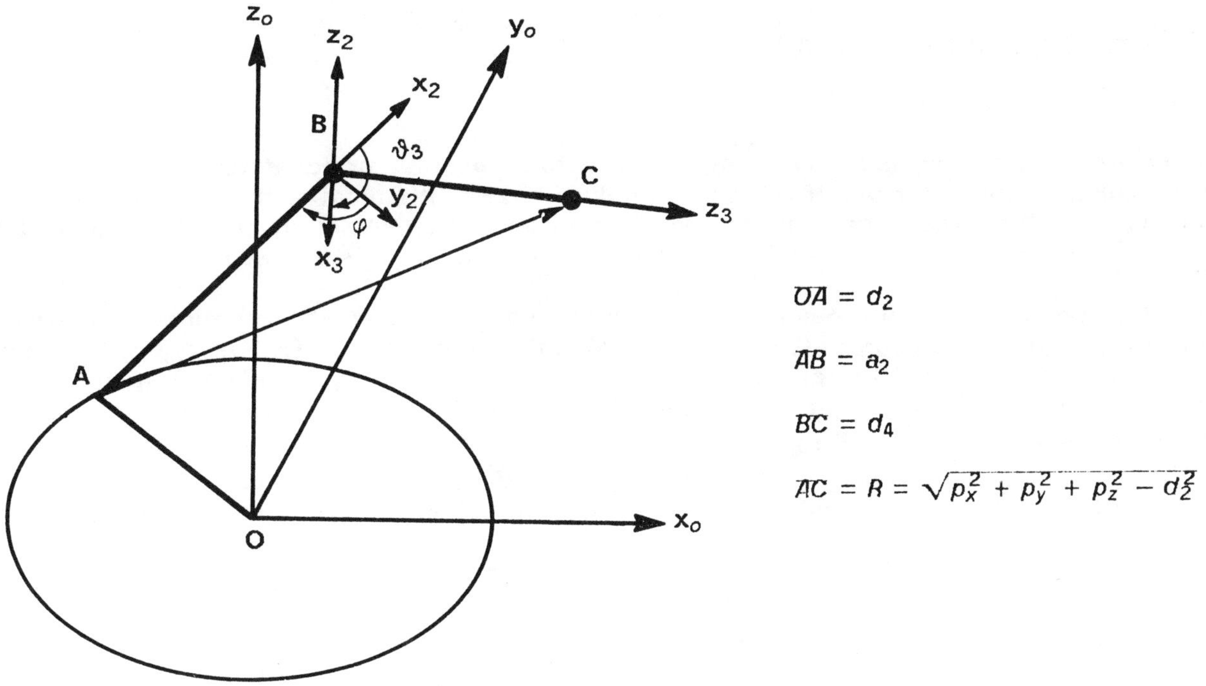

Figure 8. Joint Three Solution

$$-\cos\varphi = \frac{R^2 - a_2^2 - d_4^2}{2a_2d_4} = \sin(\frac{3\pi}{2} - \varphi) \qquad (31)$$

$$\sin\vartheta_3 = \frac{p_x^2 + p_y^2 + p_z^2 - a_2^2 - d_2^2 - d_4^2}{2a_2d_4} \quad ; \quad \cos\vartheta_3 = (ARM \cdot ELBOW)\sqrt{1 - \sin^2\vartheta_3} \qquad (32)$$

From Eq. 32, we find the solution for ϑ_3:

$$\vartheta_3 = atan2\left[\frac{\sin\vartheta_3}{\cos\vartheta_3}\right] \quad ; \quad -\pi \leq \vartheta_3 \leq \pi \qquad (33)$$

4.3. Arm Solution for the Last Three Joints of A PUMA Robot Arm

Knowing the first three joint angles, we can evaluate the T_o^3 matrix which is used extensively to find the solution of the last three joints. The solution of the last three joints of a PUMA robot arm can be found by setting these joints to meet the following criteria:

(1) Set joint 4 such that a rotation about joint 5 will align the axis of motion of joint 6 with the given approach vector (**a** *of* **T**)

(2) Set joint 5 to align the axis of motion of joint 6 with the approach vector.

(3) Set joint 6 to align the given orientation vector (or sliding vector or $\mathbf{y}_6$) and normal vector.

Mathematically the above criteria respectively mean:

$$\mathbf{z}_4 = \frac{\pm(\mathbf{z}_3 \times \mathbf{a})}{||\ \mathbf{z}_3 \times \mathbf{a}\ ||} \quad ; \quad given\ \mathbf{a} = (a_x, a_y, a_z)^t \qquad (34)$$

$$\mathbf{a} = \mathbf{z}_5 \quad ; \quad given\ \mathbf{a} = (a_x, a_y, a_z)^t \qquad (35)$$

$$\mathbf{s} = \mathbf{y}_6 \quad ; \quad given\ \mathbf{s} = (s_x, s_y, s_z)^t \text{ and } \mathbf{n} = (n_x, n_y, n_z)^t \qquad (36)$$

From Eq. 34, the cross product may be taken to be positive or negative. As a result, there are two possible solutions for ϑ_4. If the cross product is zero, it indicates the degenerate case. This happens when the axes of rotation for joint 4 and joint 6 are parallel. It indicates that at this particular arm configuration, a five-axis robot arm rather than a six-axis one would suffice.

Joint Four Solution. Both the noflip and flip orientations of the wrist are defined by looking at the orientation of the hand coordinate frame $(\mathbf{n}, \mathbf{s}, \mathbf{a})$ with respect to the $(\mathbf{x}_5, \mathbf{y}_5, \mathbf{z}_5)$ coordinate frame. The sign of the cross product cannot be determined without referring to the orientation of either **n** or **s** with respect to $\mathbf{x}_5$ or $\mathbf{y}_5$, respectively, which have a fixed relation with respect to the $\mathbf{z}_4$ unit vector. (From Fig. 2, we have $\mathbf{z}_4 = \mathbf{y}_5$)

We shall start with an assumption of the sign (positive) for the cross product in Eq. 34. This will be corrected later using the WRIST indicator and an orientation variable Ω which is defined to be:

$$\Omega = \begin{cases} 0 & ; \text{ if in the degenerate case} \\ \mathbf{s}\cdot\mathbf{y}_5 & ; \text{ if } \mathbf{s}\cdot\mathbf{y}_5 \neq 0 \\ \mathbf{n}\cdot\mathbf{y}_5 & ; \text{ if } \mathbf{n}\cdot\mathbf{y}_5 = 0 \end{cases} \qquad (37)$$

If both the orientation variable Ω and the WRIST indicator have the same sign, then the assumption of the sign in Eq. 34 is correct. From Fig. 2, $\mathbf{y}_5 = \mathbf{z}_4$ and from Eq. 34, the orientation variable Ω can be rewritten as:

$$\Omega = \begin{cases} 0 & ; \text{ if } \textit{in the degenerate case} \\ \mathbf{s}\cdot\dfrac{(\mathbf{z}_3 \times \mathbf{a})}{\|\mathbf{z}_3 \times \mathbf{a}\|} & ; \text{ if } \mathbf{s}\cdot(\mathbf{z}_3 \times \mathbf{a}) \neq 0 \\ \mathbf{n}\cdot\dfrac{(\mathbf{z}_3 \times \mathbf{a})}{\|\mathbf{z}_3 \times \mathbf{a}\|} & ; \text{ if } \mathbf{n}\cdot(\mathbf{z}_3 \times \mathbf{a}) = 0 \end{cases} \tag{38}$$

From Eqs. 12 and 37, we can construct various signed values of WRIST and Ω in the following table.

Wrist Orientation	$\Omega = \mathbf{s}\cdot\mathbf{y}_5$ or $\mathbf{n}\cdot\mathbf{y}_5$	WRIST
NOFLIP	≥ 0	+1
NOFLIP	< 0	+1
FLIP	≥ 0	-1
FLIP	< 0	-1

Table 3. Various Orientations for The Wrist

It can be shown that the followings are true:

$$\sin\vartheta_4 = -(\mathbf{x}_3 \cdot \mathbf{z}_4) \quad ; \quad \cos\vartheta_4 = (\mathbf{y}_3 \cdot \mathbf{z}_4) \tag{39}$$

where $\mathbf{x}_3$ and $\mathbf{y}_3$ are the *x* and *y* column vector of $\mathbf{T}_o^3$ respectively. Thus the solution for ϑ_4 with the orientation variable is

$$\vartheta_4 = atan2\left[\frac{K\cdot\sin\vartheta_4}{K\cdot\cos\vartheta_4}\right] = atan2\left[\frac{K\cdot(C_1a_y - S_1a_x)}{K\cdot(C_1C_{23}a_x + S_1C_{23}a_y - S_{23}a_z)}\right] ; \; -\pi \leq \vartheta_4 \leq \pi \tag{40}$$

where $K = WRIST\cdot sign(\Omega)$. Ω is defined in Eq. 38 and the sign function is defined as:

$$sign(x) = \begin{cases} +1 & ; \text{if } x \geq 0 \\ -1 & ; \text{if } x < 0 \end{cases} \tag{41}$$

If the degenerate case occurs, any convenient value may be chosen for ϑ_4 as long as the FLIP or NOFLIP option is satisfied. This can always be ensured by setting ϑ_4 = current value of ϑ_4 for the NOFLIP condition and ϑ_4 = current value of ϑ_4 + 180° for the FLIP condition.

Joint Five Solution. To find ϑ_5, we use the criterion that aligns the axis of rotation of joint six with the approach vector (or $\mathbf{a} = \mathbf{z}_5$). Using the fact that:

$$\sin\vartheta_5 = \mathbf{x}_4 \cdot \mathbf{a} \quad ; \quad \cos\vartheta_5 = -(\mathbf{y}_4 \cdot \mathbf{a}) \tag{42}$$

where $\mathbf{x}_4$ and $\mathbf{y}_4$ are the *x* and *y* column vector of $\mathbf{T}_o^4$ respectively and $\mathbf{a}$ is the approach vector. Thus the solution for ϑ_5 is

$$\begin{aligned} \vartheta_5 &= atan2\left[\frac{\sin\vartheta_5}{\cos\vartheta_5}\right] \\ &= atan2\left[\frac{(C_1C_{23}C_4 - S_1S_4)a_x + (S_1C_{23}C_4 + C_1S_4)a_y - C_4S_{23}a_z}{C_1S_{23}a_x + S_1S_{23}a_y + C_{23}a_z}\right] ; \; -\pi \leq \vartheta_5 \leq \pi \end{aligned} \tag{43}$$

If $\vartheta_5 \approx 0$, then the degenerate case occurs.

Joint Six Solution. Up to now, we have aligned the axis of joint 6 with the approach vector. Next we need to align the orientation of the gripper to ease picking up the object. The criterion for doing this is to set $\mathbf{s} = \mathbf{y}_6$. Using the fact that:

$$\sin\vartheta_6 = \mathbf{y}_5 \cdot \mathbf{n} \quad ; \quad \cos\vartheta_6 = \mathbf{y}_5 \cdot \mathbf{s} \tag{44}$$

where $\mathbf{y}_5$ is the y column vector of $\mathbf{T}_o^5$ and $\mathbf{n}$ and $\mathbf{s}$ are the normal and sliding vectors of $\mathbf{T}_o^6$ respectively. Thus the solution for ϑ_6 is

$$\begin{aligned} \vartheta_6 &= atan2\left[\frac{\sin\vartheta_6}{\cos\vartheta_6}\right] \\ &= atan2\left[\frac{(-S_1C_4-C_1C_{23}S_4)n_x+(C_1C_4-S_1C_{23}S_4)n_y+(S_4S_{23})n_z}{(-S_1C_4-C_1C_{23}S_4)s_x+(C_1C_4-S_1C_{23}S_4)s_y+(S_4S_{23})s_z}\right] \quad ; \quad -\pi \le \vartheta_6 \le \pi \end{aligned} \tag{45}$$

If the degenerate case occurs, then $(\vartheta_4 + \vartheta_6)$ = total angle required to align the sliding vector ($\mathbf{s}$) and the normal vector ($\mathbf{n}$). If FLIP option is on (i.e. WRIST=-1), then $\vartheta_4 = \vartheta_4 + 180^o, \vartheta_5 = -\vartheta_5$, and $\vartheta_6 = \vartheta_6 + 180^o$.

In summary, there are four solutions to the inverse kinematics problem - two for the right shoulder arm configuration and two for the left shoulder arm configuration. For each arm configuration, Eqs. 22, 29, 33, 40, 43 and 45 give one set of solution $(\vartheta_1, \vartheta_2, \vartheta_3, \vartheta_4, \vartheta_5, \vartheta_6)$ (WRIST = +1) and $(\vartheta_1, \vartheta_2, \vartheta_3, \vartheta_4+\pi, -\vartheta_5, \vartheta_6+\pi)$ (WRIST = -1) give another set of solution.

5. Decision Equations for the Arm Configuration Indicators

To evaluate the LEFT/RIGHT arm configuration, a decision equation is devised. The signed values of the decision equation (positive, zero, or negative) provide an indication of the arm configuration as defined in Eq. 10. From the RIGHT/LEFT ARM definition, the decision equation for the ARM indicator is:

$$\begin{aligned} g(\vartheta,\mathbf{p}) &= \mathbf{z}_o \cdot \frac{\mathbf{z}_1 \times \mathbf{p}'}{||\mathbf{z}_1 \times \mathbf{p}'||} = \mathbf{z}_o \cdot \begin{vmatrix} \mathbf{i} & \mathbf{j} & \mathbf{k} \\ -\sin\vartheta_1 & \cos\vartheta_1 & 0 \\ p_x & p_y & 0 \end{vmatrix} \cdot \frac{1}{||\mathbf{z}_1 \times \mathbf{p}'||} \\ &= \mathbf{z}_o{}^t \begin{bmatrix} 0 \\ 0 \\ -\sin\vartheta_1 p_y - \cos\vartheta_1 p_x \end{bmatrix} \cdot \frac{1}{\sqrt{\sin^2\vartheta_1 p_y^2 + \cos^2\vartheta_1 p_x^2}} = \frac{-\sin\vartheta_1 p_y - \cos\vartheta_1 p_x}{\sqrt{\sin^2\vartheta_1 p_y^2 + \cos^2\vartheta_1 p_x^2}} \end{aligned} \tag{46}$$

where $\mathbf{p}' = (p_x, p_y, 0)^t$ is the projection of $\mathbf{p}$ onto the $\mathbf{x}_o-\mathbf{y}_o$ plane, $\mathbf{z}_1 = (-\sin\vartheta_1, \cos\vartheta_1, 0)^t$ from the third column vector of $\mathbf{T}_o^1$, and $\mathbf{z}_o = (0,0,1)^t$.

- If $g(\vartheta,\mathbf{p}) > 0$, then the arm is in the RIGHT arm configuration.
- If $g(\vartheta,\mathbf{p}) < 0$, then the arm is in the LEFT arm configuration.
- If $g(\vartheta,\mathbf{p}) = 0$, then the criterion for finding the LEFT/RIGHT arm configuration cannot be uniquely determined. The arm is in a *singular* position. This happens when the wrist is within the inner cylinder of radius d_2 in the workspace (see Fig. 6). In this case, it is default to +1 (RIGHT arm).

Since the square root of the above decision equation is always positive, the determination of the LEFT/RIGHT arm configuration is reduced to checking the sign of the numerator of $g(\vartheta,\mathbf{p})$.

$$ARM = sign(g(\vartheta,\mathbf{p})) = sign(-p_x\cos\vartheta_1 - p_y\sin\vartheta_1) \tag{47}$$

where the sign function is defined in Eq. 41. Substituting the x and y components of $\mathbf{p}$ from Eq. 14, Eq. 47 becomes

$$ARM = sign(g(\vartheta, \mathbf{p})) = sign(g(\vartheta)) = sign\left[-C_1^2\left[a_2C_2+d_4S_{23}\right]+d_2S_1C_1 - S_1^2\left[a_2C_2+d_4S_{23}\right]-d_2C_1S_1\right] \tag{48}$$

$$= sign(-d_4S_{23} - a_2C_2)$$

Hence from the decision equation in Eq. 48, one can relate its signed value to the ARM indicator for the RIGHT/LEFT arm configuration as:

$$ARM = sign(-d_4S_{23} - a_2C_2) = \begin{cases} +1 & ; \text{ if } g(\vartheta) \geq 0 \quad ==> \quad RIGHT\ arm \\ -1 & ; \text{ if } g(\vartheta) < 0 \quad ==> \quad LEFT\ arm \end{cases} \tag{49}$$

We can evaluate the decision equation for the ELBOW arm indicator from Table 2. The decision equation for the ELBOW indicator is:

$$ELBOW = ARM \cdot sign(\cos\vartheta_3) = \begin{cases} +1 & ; \text{ if } ARM \cdot sign(\cos\vartheta_3) \geq 0 \quad ==> \quad ELBOW\ above\ wrist \\ -1 & ; \text{ if } ARM \cdot sign(\cos\vartheta_3) < 0 \quad ==> \quad ELBOW\ below\ wrist \end{cases} \tag{50}$$

If $\cos\vartheta_3 = 0$, then the elbow joint is within the inner cylinder of radius d_2 and the ABOVE/BELOW arm configuration is indistinguishable. In this case, the EBLOW is default to +1 (ABOVE arm).

For the WRIST indicator, we follow the definition of NOFLIP WRIST to obtain a positive dot product of **s** and $\mathbf{y}_5$ (or $\mathbf{z}_4$):

$$WRIST = \begin{cases} +1 & ; \text{ if } \mathbf{s} \cdot \mathbf{z}_4 > 0 \\ -1 & ; \text{ if } \mathbf{s} \cdot \mathbf{z}_4 < 0 \end{cases} \tag{51}$$

$$= sign(\mathbf{s} \cdot \mathbf{z}_4)$$

If $\mathbf{s} \cdot \mathbf{z}_4 = 0$, then the WRIST indicator can be found from:

$$WRIST = \begin{cases} +1 & ; \text{ if } \mathbf{n} \cdot \mathbf{z}_4 > 0 \\ -1 & ; \text{ if } \mathbf{n} \cdot \mathbf{z}_4 < 0 \end{cases} \tag{52}$$

$$= sign(\mathbf{n} \cdot \mathbf{z}_4)$$

Combining Eqs. 51 and 52, we have

$$WRIST = \begin{cases} sign(\mathbf{s} \cdot \mathbf{z}_4) & ; \text{ if } \mathbf{s} \cdot \mathbf{z}_4 \neq 0 \\ sign(\mathbf{n} \cdot \mathbf{z}_4) & ; \text{ if } \mathbf{s} \cdot \mathbf{z}_4 = 0 \end{cases} = \begin{cases} +1 & ; \ NOFLIP\ the\ wrist\ solution \\ -1 & ; \ FLIP\ the\ wrist\ solution \end{cases} \tag{53}$$

These decision equations provide a verification of the joint solution. We use them to preset the arm configuration in the direct kinematics and then use the arm configuration indicators to find the inverse kinematics solution.

6. Computer Simulation

A computer program was written to verify the validity of the inverse solution of the PUMA robot arm shown in Fig. 2. The software initially generates all the locations in the workspace of the robot within the joint angles limits. These joint angles are "fed" into the direct kinematics routine to obtain the arm matrix **T**. These joint angles are also used to compute the decision equations to obtain the arm configuration indicators. These indicators together with the arm matrix **T** are fed into the inverse solution routine to obtain the joint angle solution which should correspond to the joint angles fed into the direct kinematics routine. Fig. 9 shows the computer simulation block diagram.

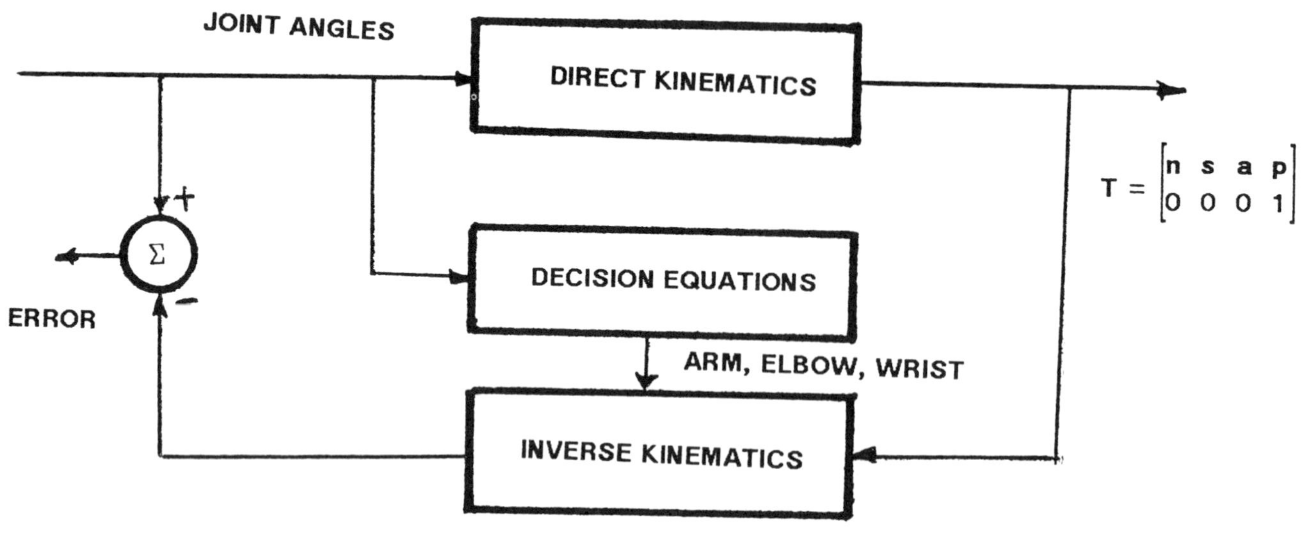

Figure 9. Computer Simulation of Joint Solution

7. Conclusion

The kinematics and inverse kinematics problems of PUMA robots have been discussed. The inverse solution is determined with the help of three arm configuration indicators (ARM, ELBOW, and WRIST). Computer simulation of the kinematics and inverse kinematics showed that the results are correct. This approach can be generalized to other robots with rotary joints. (T^3, ASEA etc)

8. Acknowledgements

The authors would like to thank Robert Horner who wrote a "C" program to verify the above direct and inverse kinematics equations. Readers may request the program from the first author.

This work was supported in part by the Robot Systems Division of the Center for Robotics and Integrated Manufacturing (CRIM) at The University of Michigan, Ann Arbor, MI. Any opinions, findings, and conclusions or recommendations expressed in this article are those of the authors and do not necessarily reflect the views of the funding agency.

9. References

1. J. Denavit, R. S. Hartenberg, "A Kinematic Notation for Lower-Pair Mechanisms Based on Matrices," *Journal of Applied Mechanics*, June 1955, pp. 215-221.

2. R. P. Paul, *Robot Manipulators: Mathematics, Programming and Control*, M.I.T. Press, 1981, pp 50-84.

3. D. L. Pieper, "The Kinematics of Manipulators Under Computer Control," Computer Science Department, Stanford University, Artificial Intelligence Project Memo No. 72, October 24, 1968.

4. R. P. Paul, B. E. Shimano, G. Mayer, "Kinematic Control Equations for Simple Manipulators," *IEEE Transactions of Systems, Man, and Cybernetics*, Vol. SMC-11, No. 6, June 1981, pp 449-455.

A New Approach to Coordinate Transformation: Parallel Computation

L. Mancini
Politecnico Di Milano
Milano, Italy

N. Nicolosi
Politecnico Di Milano

S. Scarselli
Politecnico Di Milano

G. Zoly
Politecnico Di Milano

Abstract

The solution of the inverse kinematic problem is certainly one of the most important in control systems for industrial and research robots. We discuss a new approach to this problem, based on parallel computation of each joint coordinate. An implementation of the proposed system on a five-degree-of-freedom robot, designed and constructed in our laboratory, with a multi-processor control system for parallel computation, is also shown. The advantages and drawbacks of such a system are considered, with special emphasis on the increase of computing speed and positioning accuracy.

1. INTRODUCTION

In robotics, and especially in the field of articulated robots, one of the most important problems is coordinate transformation, from world's cartesian coordinates to joints angular coordinates. The problem is important, because coordinate transformation requires a great amount of control programs computation time. The reduction of this computation time would yield some advantages, as a better dynamic precision and a greater speed, that are important for industrial applications.

To obtain this reduction, we propose a new algorithm for the solution of the inverse kinematic problem, based on parallel computation of each joint coordinate, that also allows the construction of a modular architecture for articulated robot control units.

Our work is not only a theoretical proposal: we present the Cyclope that is an implementation of the proposed method on a working robot system, consisting of mechanics, electronics and software.

2. PARALLEL COMPUTATION

In articulated robots, composed either only of rotational joints or rotational and translational joints, the transformation of cartesian coordinates into joint space coordinates plays a fundamental role, because the transformation algorithm and its efficiency strongly influence the architecture of the control system and its performance.

For a better definition of the problem we may say that we have a set of coordinates and angular informations that describe the final configuration of the robot arm in absolute coordinates (Xp, Yp, Zp of final point; A1, A2, A3 angles for hand orientation vector). Consequently we must drive each arm joint to obtain the defined final configuration (in other words we must obtain, for an n- degrees of freedom arm, n angles whose composition gives final configuration).

This transformation is usually obtained by a system of equations that describe the bindings between n frames (where n is the number of degrees of freedom and each frame is placed on one joint) as a set of vectors; the solution of this system (obtained by a matrix method) obviously depends both from the starting and the final configuration and is computed by a master computer [4,5].

Our method starts from a different standpoint: we describe the angular displacement of each joint in a way that aims at being as independent from the others as possible. This description is simple if referred to our mechanical structure, that has all joints on the same plane, but can also be applied to more complex structures.

We can say that the computation of the angles T2, T3 and T4 (fig. 1) may be done by means of Carnot's formulas, since we know the length of the sides L2 and L3, and Lws may be easily computed. The computation of the angles T1, T2 is simple because T1 depends only on the coordinates Xp, Yp of the final point and T5 is directly supplied as the hand orientation angle. We would like to point out that the method, that has been applied to a five degrees of freedom robot, can also be employed with a six degrees of freedom machine.

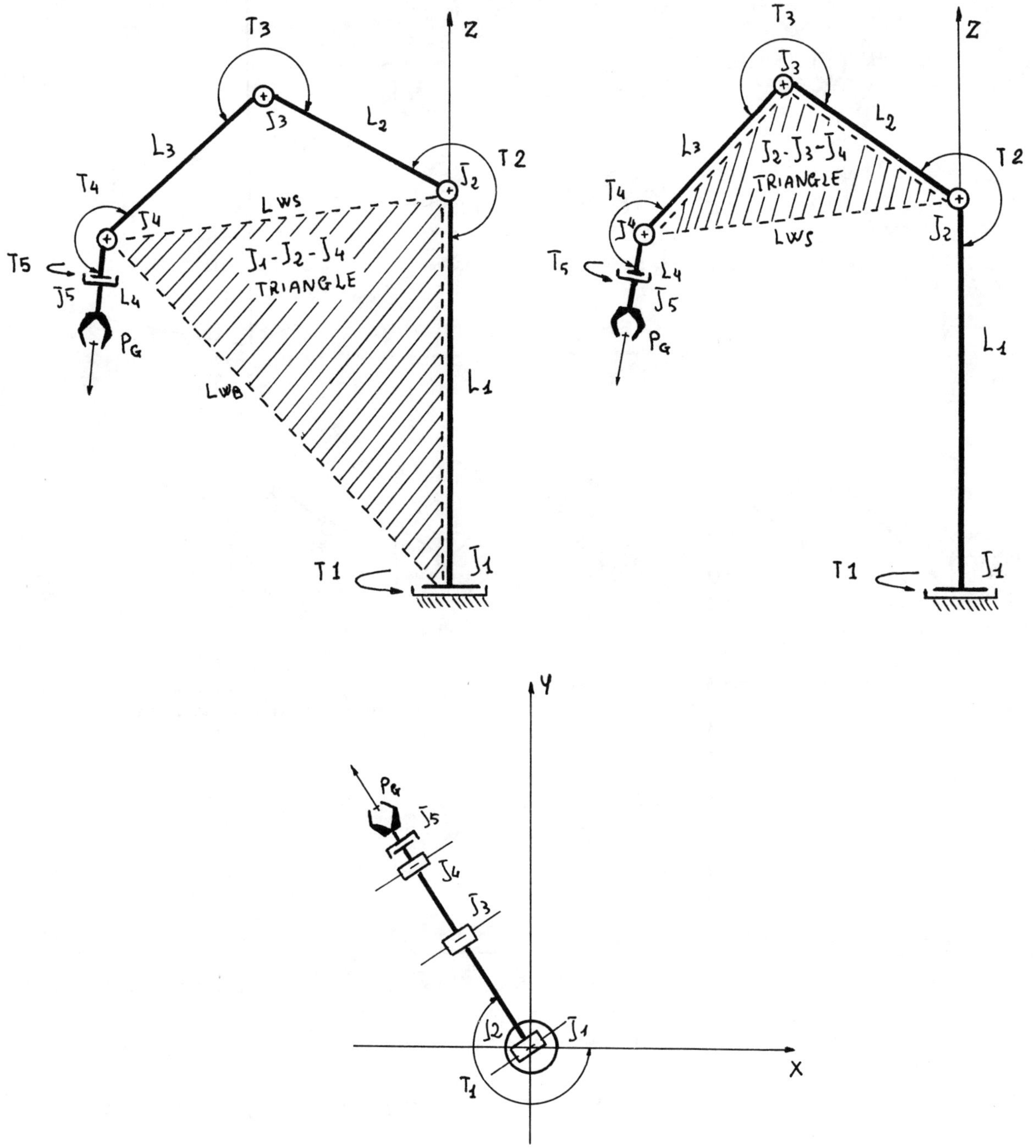

Fig. 1. Models for parallel computation.

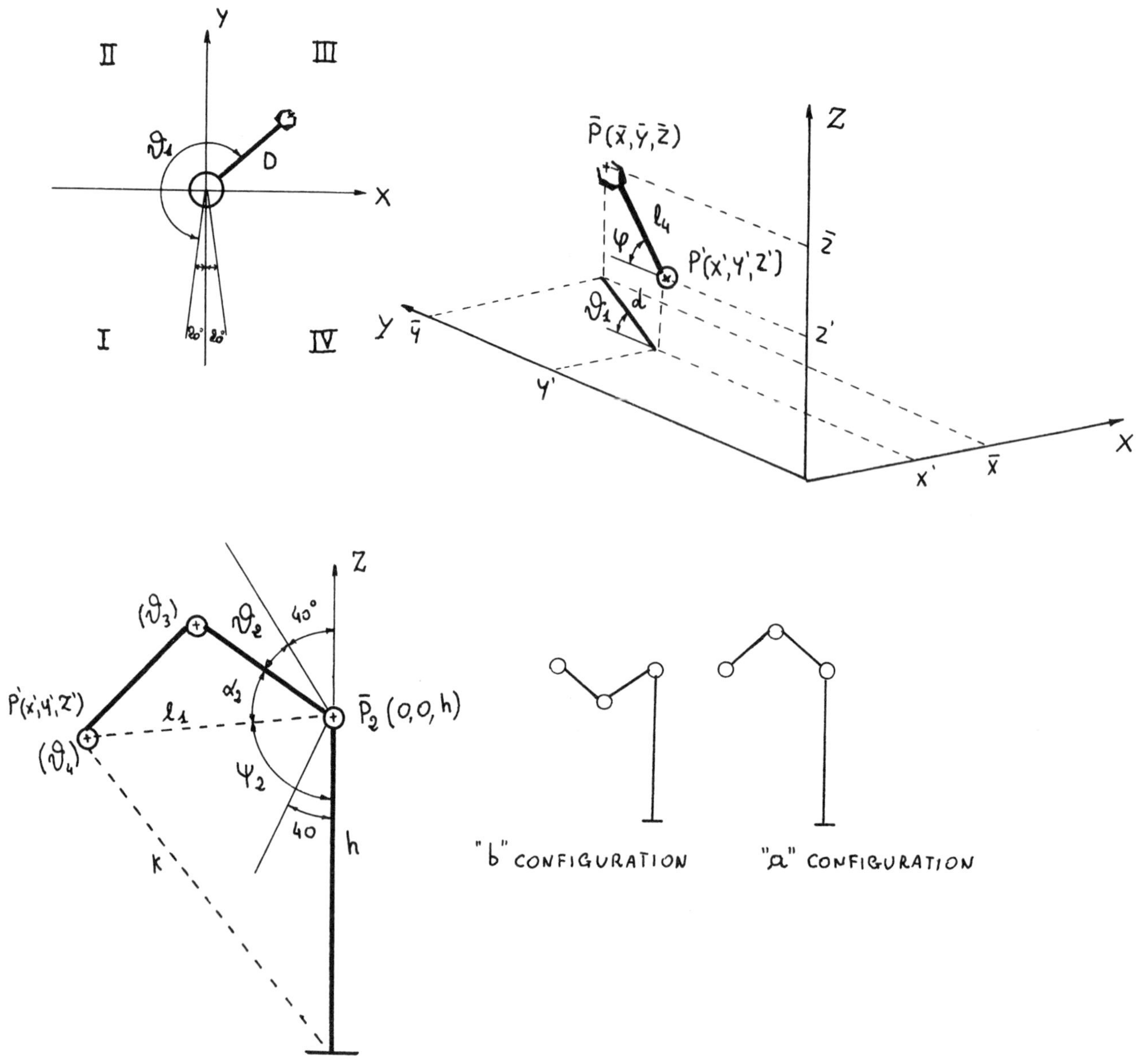

$$D = \sqrt{X^2 + Y^2}$$

$$\vartheta_1(I) = \sin^{-1}\left|\frac{X}{D}\right| - 20$$

$$\vartheta_1(II) = \cos^{-1}\left|\frac{X}{D}\right| + 70$$

$$\vartheta_1(III) = \sin^{-1}\left|\frac{X}{D}\right| + 160$$

$$\vartheta_1(IV) = \cos^{-1}\left|\frac{X}{D}\right| + 250$$

$$\vartheta_2(a) = 140 - \Psi_2 + \alpha_2$$

$$\vartheta_3(a) = 230 - \alpha_3$$

$$\vartheta_4(a) = 90 - \alpha_4 + \Psi_2 - \varphi - \gamma^*$$

$$l_1 = \sqrt{X'^2 + Y'^2 + (Z' - h)^2}$$

$$\alpha_2 = \cos^{-1}((l_1^2 + l_2^2 - l_3^2)/(2\, l_1 l_2))$$

$$\alpha_3 = \cos^{-1}((l_2^2 + l_3^2 - l_1^2)/(2\, l_2 l_3))$$

$$\alpha_4 = \cos^{-1}((l_1^2 + l_3^2 - l_2^2)/(2\, l_1 l_3))$$

$$\Psi_2 = \cos^{-1}((l_1^2 + h^2 - k^2)/(2\, l_1 h))$$

$$\vartheta_2(b) = 140 - \Psi_2 - \alpha_2$$

$$\vartheta_3(b) = \alpha_3 - 130$$

$$\vartheta_4(b) = 90 + \alpha_4 + \Psi_2 - \varphi - \gamma$$

$$K = \sqrt{X'^2 + Y'^2 + Z'^2}$$

$$d = l_4 \cos\varphi$$

$$Z' = \bar{Z} - l_4 \sin\varphi$$

$$X'(I,III) = \bar{X} \pm d \sin\vartheta_1$$

$$X'(II,IV) = \bar{X} \pm d \cos\vartheta_1$$

$$Y'(I,III) = \bar{Y} \pm d \cos\vartheta_1$$

$$Y'(II,IV) = \bar{Y} \pm d \sin\vartheta_1$$

Fig. 2. Models and computations

* γ IS AN OFFSET VALUE

3. CYCLOPE SYSTEM: A GENERAL VIEW

Cyclope is a complete robot system that has been developed by the authors at Milan Polytechnic Artificial Intelligence Project. While this paper is being written, the design of all system components is complete, all algorithms have been tested in a simulated environment, and hardware is being assembled.

It consists of the following main parts:

a. Mechanics;

b. Electronics, divided into analog part, for motor driving, and digital part, devoted to computation and control system;

c. Software, consisting of a set of control and computation programs (for the proper control system) and of an interpreter for a command language.

3.1. Mechanics

The mechanical part of Cyclope, named Polyfemus arm, is a five degrees of freedom general purpose manipulator, that is mainly based on two types of non reversible gearings, one consisting of a worm gear coupled with a helical geared wheel, and the other composed of a screw with its screw nut that forms a linear actuator. The most important reason for this choice is that a non reversible gearing does not need brakes, nor a complex control system for control position, and makes an easier dimensioning of motors possible (in other words we have no overheathing problems at zero speed). The arm is powered by six D.C. motors, five for the joints and one for the hand, that is also equipped with a force feedback system.

3.2. Electronics

The electronics are divided in two main parts:

a. The analog servo system and interface for motor driving and speed control, composed of a power system, that works in switch mode, and of a special circuit that implements the counter function for position loop and the analog speed feedback.

b. A digital part, consisting of a multiprocessor system with two levels. The lower level acts as a programmable interface: it is devoted to position control and is directly connected to the analog interface. The upper level is the parallel computation system: it provides coordinate transformation from cartesian to angular. A microprocessor for the management of the common bus, instructions interpretation and for other special tasks, is also included.

3.3. Software

The interpreter for the command language is a simple program that transforms, line by line, the high level instructions in low level commands; it has nor data structures, nor complex control structures (as loops) and receives the instructions from an external computer that plays an important role in the software system, as we will show in the sequel.

A general view of Cyclope system is presented in the next figure.

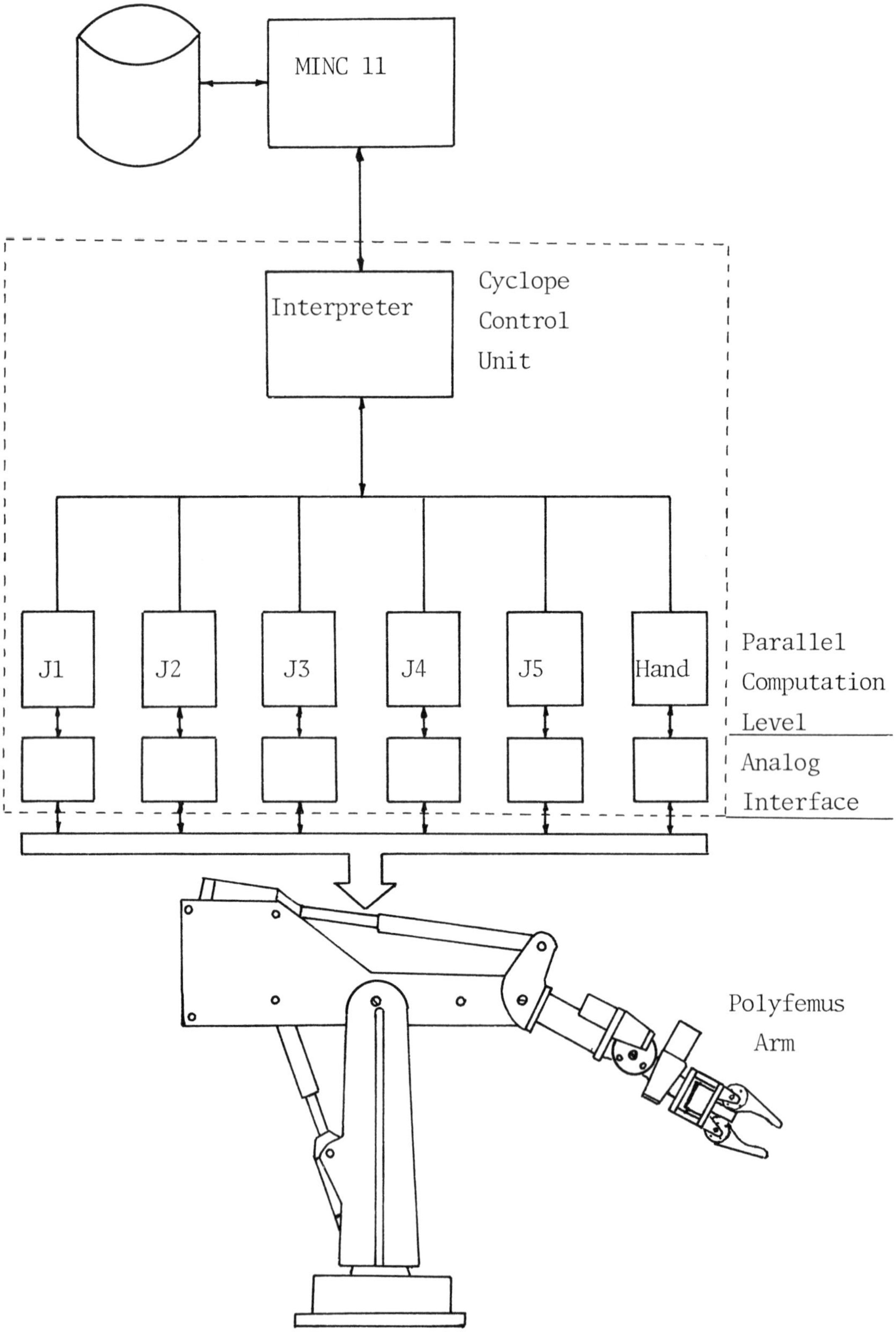

Fig. 3. General view

4. CYCLOPE CONTROL UNIT

On Cyclope system we developed a special hardware for the implementation of our parallel algorithm. The architecture of our control and computation system is strongly determined by the particular method we use for the solution of the inverse kinematic problem.

In our system the inverse kinematic transformation is distributed (as we showed before), and this allows the use of a new servo system [2].

This servo system is composed of two microprocessors on different levels: the first one performs computations that provide coordinate transformation for the correspondent joint; the second one is devoted to motor driving and is directly connected with the analog interface system.

In our architecture motion instructions are supplied to Cyclope control unit by means of a serial line, from an external computer.

Another microprocessor (as we will show later) translates these instructions and produces commands for the servo system. It is important to remark that the lower level commands are supplied to the servo system by means of the cartesian coordinates of the final point, of the hand orientation angles and of absolute velocity.

Another serial line, necessary for the robot manual controller, is connected with the microprocessor that translates the instructions.

A general view of hardware organization is shown in fig. 4.

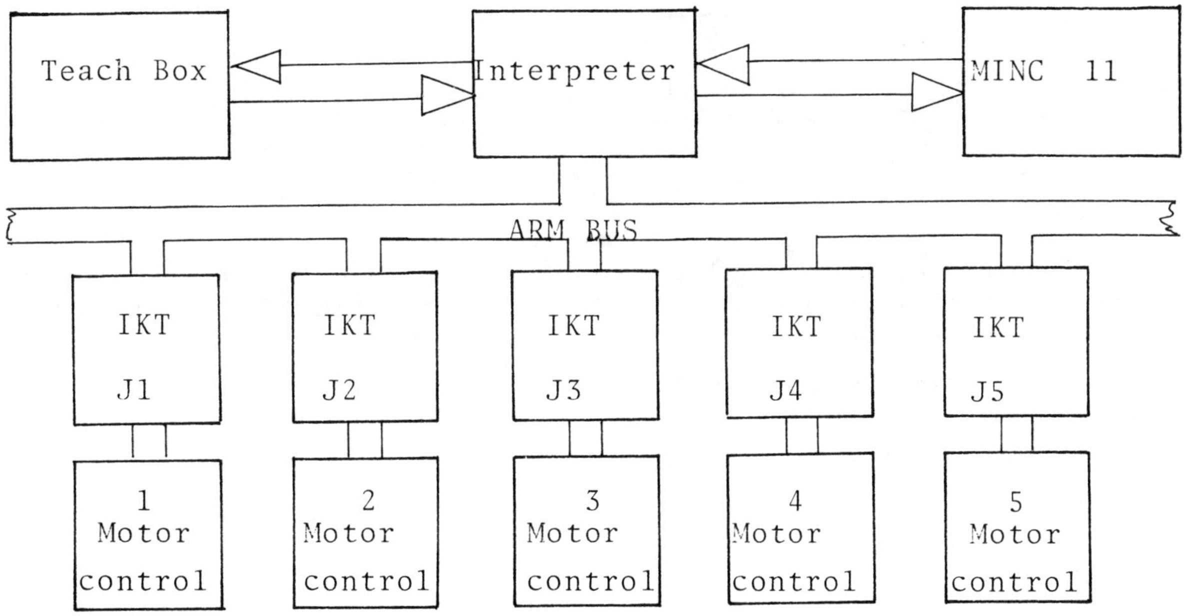

Fig. **4.** Hardware General View

The first microprocessor level, devoted to inverse kinematic transformation, is a custom board with:

- 8 bit microprocessor (8085);
- I/O controller;
- arithmetic processor unit (APU 9511);
- conventional RAM and EPROM memory.

The arithmetic processor is employed in order to increase the speed and the precision of the required computations.

It has a 3 Mhz clock frequency and 32-bit floating point arithmetics.

This computation level is connected to a common bus (named ARM BUS) with an eight bit bidirectional parallel port, and to the lower level, that is the motor control, with another eight bit bidirectional bus.

The motor controller is the lower level of our servo system: it is based on a 8 bit single chip microprocessor (8748).

Fig. 5 shows motor control organization.

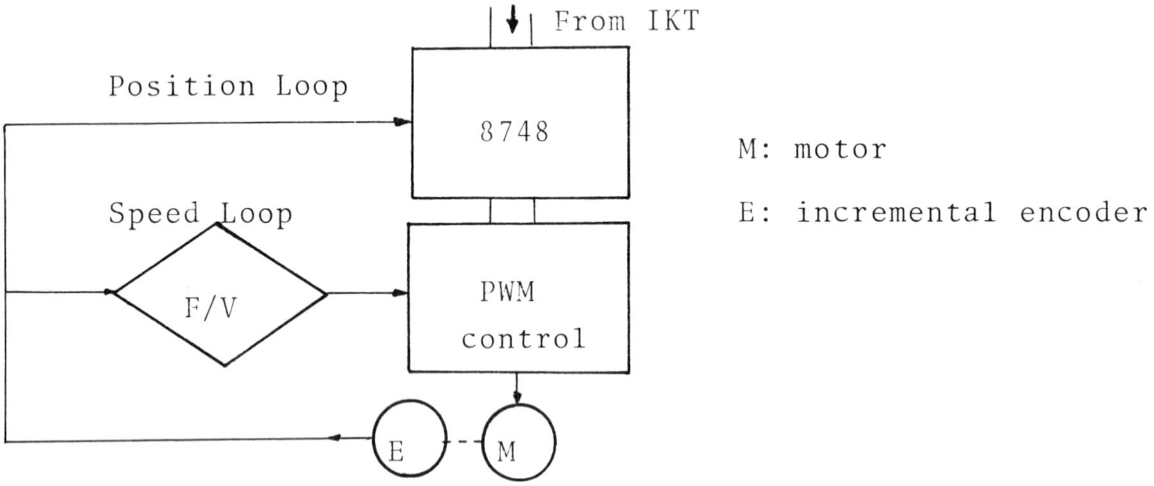

Fig. 5. Motor control

Speed control loop is implemented on a custom analog board, while the 8 bit single chip microcomputer provides position feedback and local acceleration - deceleration control.

The motors used in our system are conventional D. C. motors, and their maximum power is 90 watts.

We use, for position control, one 200 steps optical incremental encoder for each joint.

A F/V monostable converter and a speed error amplifier are used for speed control.

The microprocessor that translates the high level instructions and provides bus control is connected as shown in the following figure.

We can see that this microprocessor (that receives the instructions from an external unit) has a RAM memory for the local storage of point coordinates; it also computes the interpolation of straight line trajectories.

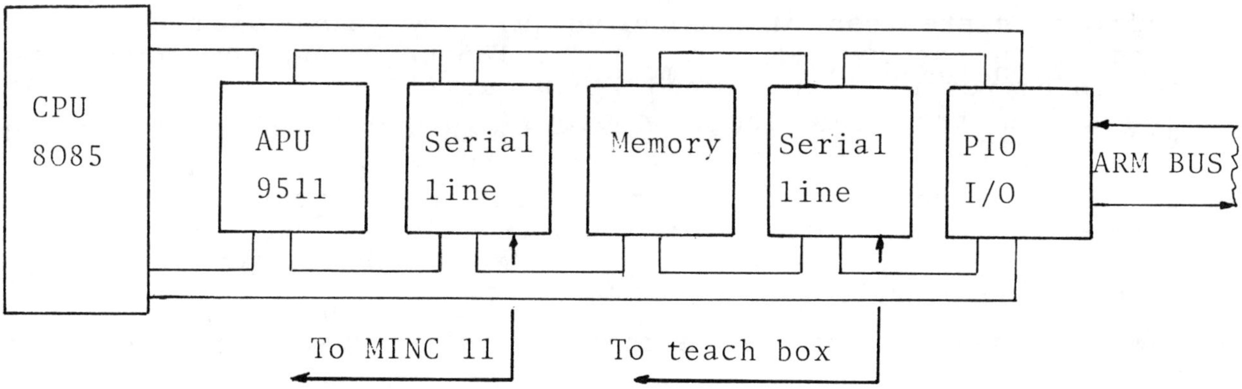

Fig. 6. Interpreter and bus controller.

A special instruction connects the control unit with a teach box that allows the manual control of the system and "teaching by showing" programming.

Our architecture allows the connection on its ARM BUS of other circuits (that can be used for special tasks, as sensor control or tools driving). In particular we have a microprocessor that controls a force proportional hand. This microprocessor receives the commands and realizes the hand motor driving and force sensor control.

The schematic diagram of the hand controller and its connections with ARM BUS are shown in the next figure.

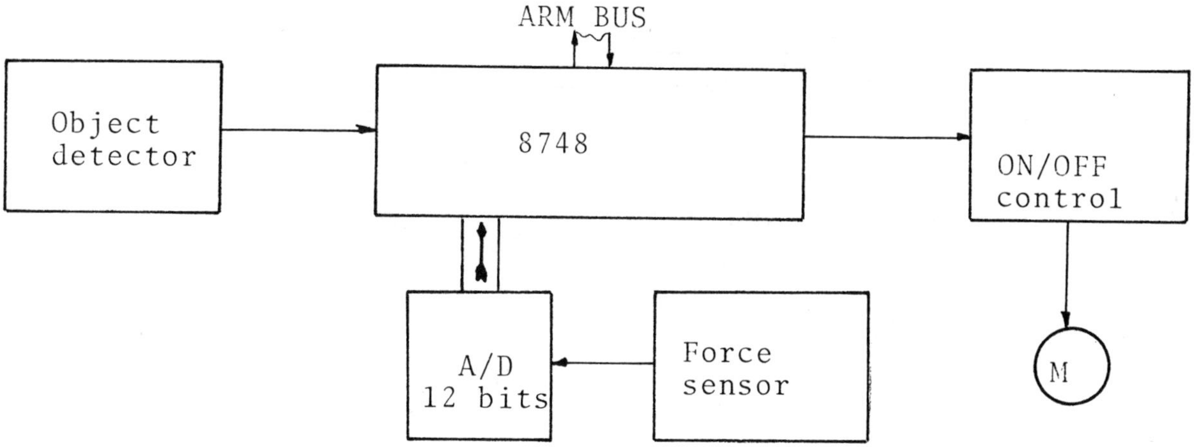

Fig. 7. Hand controller

5. SOFTWARE

Software system can be subdivided into two levels: the upper level consists of an interpreter for the VML language and the lower level is another interpreter for a command language. (1)

The VML interpreter (that is implemented on an external LSI 11 microcomputer) is devoted to control three robots and supplies to Cyclope control unit the instructions for its movement [3]. The lower interpreter receives these command instructions (that are written in a VAL-like language) and supplies parameters and low level commands to computation and control system.

The organization of the entire software system is shown in the following figure.

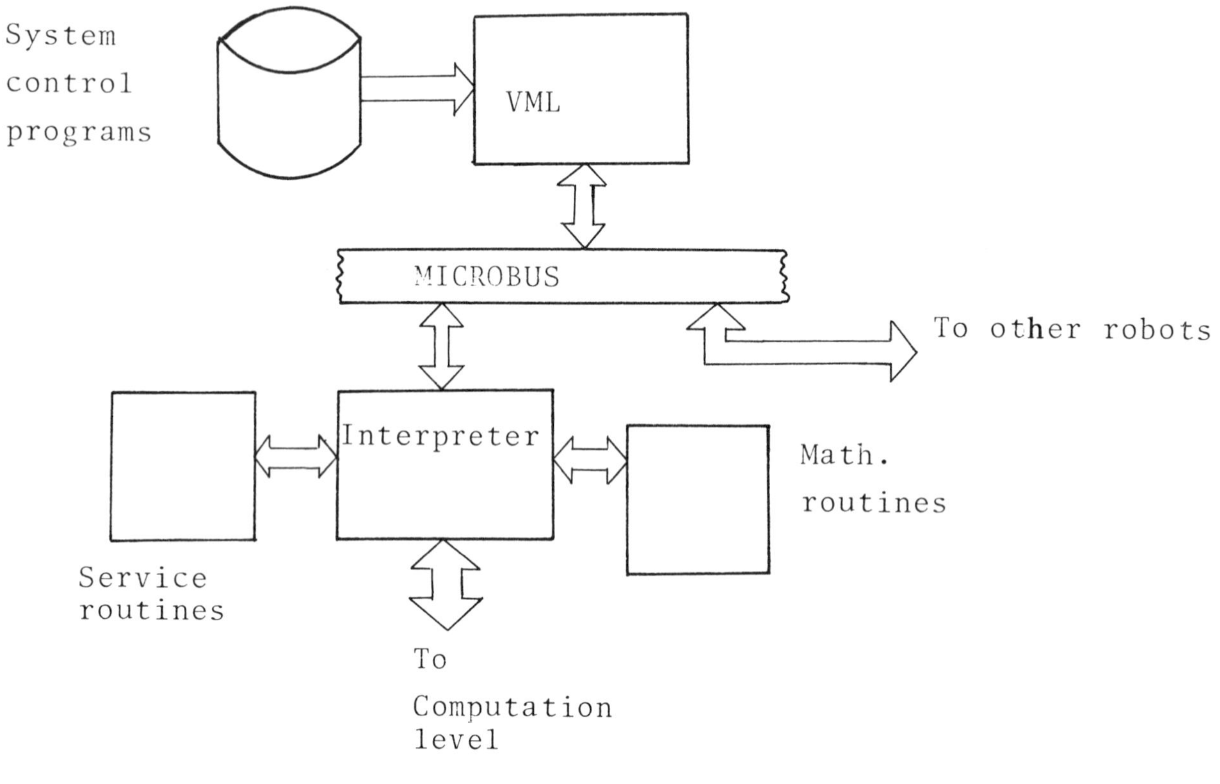

Fig. 8. Software organization

(1) VML (Virtual Machine language) is a language for robot system control and has been developed at the LADSEB laboratory of C.N.R. (Italian National Research Council) in Padova [1].

6. PERFORMANCE EVALUATION AND CONCLUSIONS

The entire Cyclope system design is based on the idea of parallel computation. In this way advantages and drawbacks of our system coincide with the advantages and drawbacks of parallel computation algorithm (this obviously happens only for control and computation unit).

The most important parameter for a robot control system is the amount of computation time spent for inverse kinematic transformation (I.K.T.). If this time is small, the robot can work quickly and it has a better accuracy in trajectory control. Our architecture has a computation time that is, in the worst case, 15 ms.

In the following table the main characteristics of Cyclope control unit are shown:

- Clock rate: 3 Mhz.
- A.P.U. instruction time: 3 us.
- C.P.U. instruction time: 3 us.
- Worst time for I.K.T.: 15 ms.
- Mean time for I.K.T.: 10 ms.
- Time for service routines: 5 ms.
- Mean point-to-point time: 15 ms.

An important problem in the present implementation is joint speed computation. We apply a non-parallel algorithm to solve this problem; this fact implies that the joint speed algorithm is not implemented at servo system level.

This problem could be solved with a control system we are thinking at, that will employ a common memory multiprocessor system.

7. ACKNOWLEDGEMENTS

Thanks are due to all persons helped us for the realization of this paper, and in particular prof. M. Somalvico, who made the realization of Cyclope project possible, dr. R. Cassinis, who suggested us the first idea of parallel computation, drs. G. Gini, M. Gini, M. Salmon and Prima Progetti spa for their fundamental contribution to our work.

REFERENCES

[1] Bison, P., Lorenzin, G., Pagello, E. - The formal definition of VML and a proposed portable implementation - Proceedings 11th I.S.I.R. - Tokyo, 1981

[2] Cassinis, R. - A distributed intelligence discrete process controller - Digital systems for industrial automation - Vol 1, N.1

[3] Cassinis, R. - Hierarchical control for integrated manufacturing system - Proceedings 13th I.S.I.R. - Chicago, 1983

[4] Kostantinov, M.S., Patarinski, S.P. - A contribution to the inverse kinematic problem for industrial manipulators - Proceedings 12th I.S.I.R. - Paris, 1982

[5] Paul, R.L. - Robot Manipulators: Mathematics, Programming and Control - MIT Press, 1981

Kinematics of Major Robot Linkage

V. Milenkovic
Ford Motor Company
Redford, Michigan

B. Huang
Ford Motor Company

Abstract

Controlling a robot to achieve numerically specified placement of the end-effector is becoming increasingly important in advanced robot applications. The method of solution is difficult to generalize to the variety of robot configurations in use. A step towards generalization has been achieved by a treatment which separates the kinematic description of the first three control axes from that of the wrist mechanism. In the first three axes, known robots use combinations of rotary and sliding joints, the consecutive ones being parallel or perpendicular. Only 12 potentially useful and mathematically distinct configurations have been found. A classification code has been assigned and transformation equations derived for each of these configurations. A computationally efficient methodology in obtaining complete transformation between absolute and joint coordinates for robots using any of the configurations is presented.

INTRODUCTION

Industrial robots used in manufacturing processes have advanced from point-to-point operation to controlled-path operation. In point-to-point operation, the robot moves from one programmed point to the next, stopping at those points. Programming consists of moving the robot through a set of points and registering joint angles or positions at each point. Positioning accuracy is maintained only at these located points. Elsewhere no synchronization is maintained between motions in various joints, resulting in a relatively unpredictable path.

The simplest path control added to point-to-point operation is one obtained by performing linear interpolation of joint coordinates between any two programmed points. This guarantees a relatively regular and predictable path between programmed points. The geometry of the path will still depend on the linkage configuration of the robot. However, an acceptable degree of approximation of a desired path is obtainable by programming of a sufficient number of nonstop points along the path.

A more advanced path control between programmed points, offered by a growing number of modern robots, is achieved by straight line interpolation in space coordinates rather than in joint coordinates. This guarantees straight line motion between programmed points, which is desirable in many applications. For certain robot applications it may be advantageous to employ curvilinear interpolation in space coordinates, as suggested in earlier work[1]. The execution of such a path control requires computation of joint coordinates, given the space coordinates at these interpolated points.

Furthermore, in some industrial processes, an object may be placed in a different location each cycle, or an object may move along a conveyer belt. Assuming that the new location of the object has been properly determined in terms of space coordinates, a robot cannot produce the desired motion without the corresponding new set of joint coordinates being computed.

Mathematical transformations from joint coordinates into space coordinates, customarily called the forward problem, are easily obtained by successive multiplication of homogeneous transformation matrices[2]. On the other hand, mathematical descriptions of joint coordinates in terms of space coordinates, customarily called the inverse problem, are generally cumbersome and can be derived only for certain robot configurations as outlined by Pieper[3] and Paul[4]. In use are many robots for which explicit inverse transformation formulas cannot be derived.

A rapidly converging iterative procedure to solve the inverse problem by computer is presented in this paper. The method conceptually divides the robot articulation into major and minor linkages[1]. The major linkage, consisting of the first three joints and the related links, serves to produce gross motion of the end-effector. The purpose of the minor linkage is to orient the end-effector. The motion of the major linkage often perturbs the orientation, and the articulation of the minor linkage often perturbs the position. The iterative procedure alternately computes and adjusts the joint coordinates of the major linkage to correct the displacement while holding the minor linkage fixed, and adjusts the coordinates of the minor linkage to correct the orientation while holding the major linkage fixed. The solution in which both the displacement and the orientation is correct is approached in the limit.

This method requires obtaining explicit formulas of joint coordinates in terms of space coordinates (inverse transformation) for the major linkage. Most significant variations among robots lie in the major linkage. Known robots use combinations of rotary and sliding joints, the consecutive ones being either parallel or perpendicular. The work presented here shows that there are only 12 useful and distinct categories of major robot linkages. A two-letter code is uniquely assigned to each category. Graphical representations are shown to aid visualization of the linkages. Transformation equations, forward and inverse, between joint coordinates and space coordinates have been derived for all 12 categories.

MAJOR ROBOT LINKAGE

The first three joints of a robot from the base are commonly designed to perform gross motion of the end-effector, and the remaining joints are used to accomplish orientation. The linkage consisting of the first three joints and the associated links is called "major linkage." The linkage of the remaining joints is called "minor linkage." This distinct feature of a robot manipulator which is referred to as separability increases its usability, modularity, and dexterity. Two robots are separable if the major linkage of one can be combined with the minor linkage (the "wrist") of the other, to yield another usable robot. Such combinations are actually offered by some of the robot manufacturers. A robot is non-separable if the motion produced by its first three joints is planar (two translations and one rotation in the same plane), rather than spatial. Such robots are theoretically possible, though none are commercially used to authors' knowledge.

The concepts of major and minor linkages apply only to separable robots. The last link of a major linkage is commonly called the arm, while the last link of a minor linkage is commonly called the hand.

Those linkages which use only rotary or sliding joints with the joint axes either perpendicular or parallel to each other are defined here as simple chains. A linkage which contains something other than rotary or sliding joints, or links joined to more than two other links, or multiple joints mutually constrained by gearing, is referred to as a compound chain. A linkage having the joint axes other than perpendicular or parallel to each other is referred to as an oblique chain. Compound and oblique chains are used only in minor linkages of some of the commercial robots. For optimum work space and simplicity, major robot linkages tend to use only simple chains.

Consider a major linkage which uses a simple chain having its three joints consecutively labeled first, second and third. There are two choices for each joint (rotary or slider), two choices for the relationship between the first and second joints, also between second and third joints (parallel or perpendicular). If the second joint is a slider and both other joints are perpendicular to it, then two choices in the relationship between the first and the third joint are considered: parallel or perpendicular. In all, there are 36 possible combinations of simple three-joint chains.

Nine among the 36 possible combinations degenerate into only one or two degrees of freedom, e.g., three sliders parallel to each other can produce translatory motion only in one direction. Seven of the remaining ones are planar mechanisms, e.g., three rotary joints parallel to each other can produce only planar motion of the arm. Thus, there are 20 possible spatial, simple-chain major linkages. Now let us consider a particular linkage in which the first joint permits rotation about the vertical axis, the second joint is a vertical slider (i.e., parallel to the first), and the third joint is a horizontal slider (i.e., perpendicular to the second). This major linkage defines a typical cylindrical robot. Changing the sequential order of the joints so that either (a) the vertical slider precedes the rotary joint, or (b) the vertical slider follows the horizontal one, will result in no change in the motion of the arm. In this case there are two linkages which are both "equivalent" to the standard cylindrical linkage. In all such cases where two or more equivalent linkages exist, the representative of the group will be the one in which the slider that is parallel to a rotary joint is in the middle (joint No. 2). Counting only one linkage to represent the group of equivalents will eliminate eight out of the 20 combinations.

The remaining 12 categories of major linkages are useful and distinct. "Useful" means that the linkage can produce gross motion in all three spatial dimensions. "Distinct" means that each linkage is kinematically unique among the twelve categories, in terms of the motion of the last link which it produces. Two or more equivalent linkages, being governed by identical coordinate transformation formulas, are not considered to be distinct.

Classification

To classify these major linkages, a two-letter code is assigned to each category: the first letter characterizes the first joint and its relationship to the second; the second letter characterizes the third joint and its relationship to the second.

Four letters are needed to designate all linkage combinations of interest:

- S - slider,
- C - rotary parallel to slider,
- N - rotary perpendicular to rotary,
- R - rotary perpendicular to slider, or rotary parallel to rotary

All 16 possible combinations of these four letters are tabulated in a matrix format as follows:

SS	SC	SN	(SR)
CS	CC	(CN)	CR
NS	(NC)	NN	NR
(RS)	RC	RN	RR

Among 16 combinations, NC and CN codes are inadmissible, because C requires the middle joint to be a slider but N requires the middle joint to be a rotary. Also, RS and SR codes give only planar mechanisms. The remaining 12 designations correspond to 12 categories of useful and distinct major robot linkages. The RR code is apparently ambiguous. It gives three possible combinations: three rotary joints parallel to each other; a slider between and perpendicular to two rotary joints which are mutually parallel; and a slider between and perpendicular to two rotary joints which are mutually perpendicular. Only the last combination is a spatial mechanism, and the designation RR will apply to it.

Graphical Representation

The following graphical symbols are used to aid visualization of simple-chain mechanisms. Each symbol represents a joint and its two associated links.

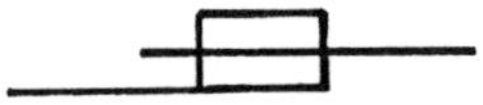

a slider (prismatic pair) with its axis in the plane of the paper.

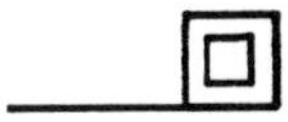

a slider with its axis perpendicular to the plane of the paper.

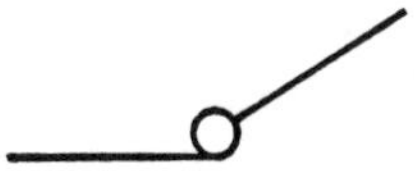

a rotary joint (revolute pair) with its axis perpendicular to the plane of the paper.

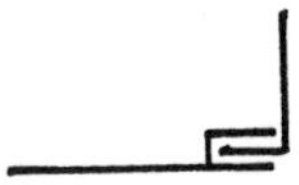

a rotary joint with its axis in the plane of the paper.

With the above four symbols, twelve major robot linkages are schematically shown in Figure 1. Known robots use major linkages which can be easily recognized as Rectangular (SS), Cylindrical (CS), Spherical (NS), or Jointed (NR). One commercial robot uses CC as its major linkage. Many robots use NN as wrist mechanism (minor linkage). Other linkages may have potential applications. As an example, RN can be used for maneuverability through an opening such as a door.

Kinematic Equations

The relationship between the position of the arm and the joint coordinates $\bar{e} = (e_1, e_2, e_3)$ in a major linkage is given by

$$\bar{x} = A\,\bar{x}' + \bar{a} \tag{1}$$

or expanded

$$\begin{pmatrix} x \\ y \\ z \end{pmatrix} = \begin{bmatrix} a_{11} & a_{12} & a_{13} \\ a_{21} & a_{22} & a_{23} \\ a_{31} & a_{32} & a_{33} \end{bmatrix} \begin{pmatrix} x' \\ y' \\ z' \end{pmatrix} + \begin{pmatrix} a_1 \\ a_2 \\ a_3 \end{pmatrix} \quad ; \tag{2}$$

$$a_{ij},\ a_i = \text{functions of } e_1,\ e_2,\ e_3, \tag{3}$$

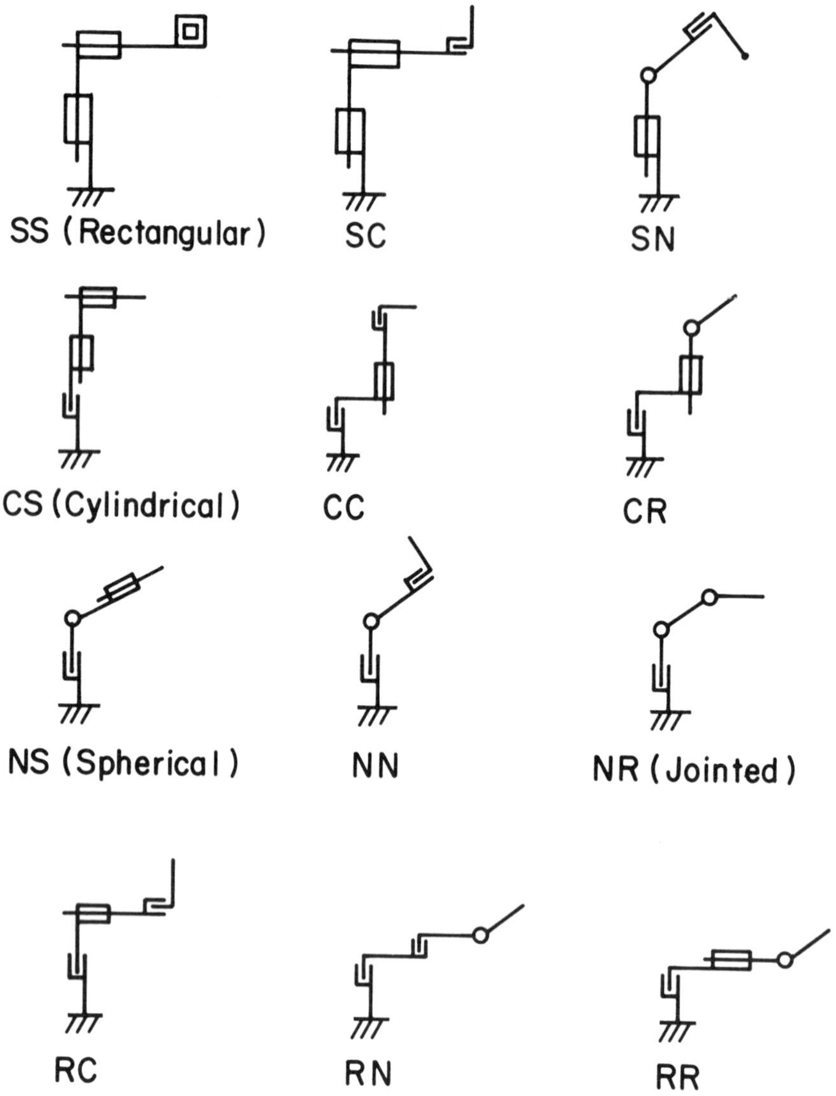

Figure 1. Schematics of Major Robot Linkages

where $\overline{x}' = (x',y',z')$ are coordinates of a point P fixed in the arm, expressed in a system attached to the arm, and $\overline{x} = (x,y,z)$ are coordinates of P relative to a system fixed in space.

Given a linkage, Eq. (3) can be obtained by successive multiplication of homogeneous transformation matrices[2]. Normally a_{ij} and a_i are expressed in combinations of sine and cosine of joint angles, and a_i is linearly related to sliding joint coordinates. To compute (x,y,z) from known values of (e_1,e_2,e_3) and (x',y',z'), or to compute (x',y',z') from known values of (e_1,e_2,e_3) and (x,y,z) constitutes a forward transformation. Eq. (1) and

$$\overline{x}' = A^T \overline{x} - A^T \overline{a} \qquad (A^T \text{ denotes the transpose of A}) \qquad (4)$$

are used for forward transformations. The rotational matrix A is othogonal ($AA^T = I$).

To compute (e_1,e_2,e_3) from known values of (x,y,z) and (x',y',z') constitutes an inverse transformation of the major linkage. One way to derive the inverse transformation formulas of a major robot linkage is by solving for e_1,e_2, and e_3 from Eq. (2). Since a_{ij} and a_i normally contain terms of products of slider coordinates and trigonometric functions of rotary joint angles, algebraic manipulations are quite cumbersome. Another way is by geometrical analysis. This method requires a good deal of intuition. The equations for both the forward and inverse transformations have been derived by the authors and tabulated in the Appendix for all 12 major robot linkages.

To find the transformation equations, one first identifies the type of major robot linkage with a two-letter code by employing the principles of classification. Then, one goes to the Table for Applicable Equations in the Appendix. One writes the general equations tabulated; replaces or defines the symbols in the equations by the proper parameters tabulated in the same column. This results in a system of forward and inverse transformation equations for the particular major robot linkage.

ROBOT KINEMATICS

Excluding power actuation and auxiliaries, a robot is a mechanism composed of an open chain of links connected consecutively by either a slider or a rotary joint. A robot having six independent joints are commonly referred to as a 6-axis robot. By the same token, there are 5-axis robots and 7-axis robots. An end-effector attached to the hand of a robot is a rigid body having six degrees of freedom, three translational and three rotational. A 6-axis robot gains complete control of a rigid body or an end effector in all six degrees of freedom. A 5-axis robot eliminates the control of the least important degree of freedom which is normally the roll of the end-effector in some applications. The redundant joint in a 7-axis robot is normally designed for the purpose of obstacle avoidance. Hereafter a 6-axis robot is used for the purpose of discussion.

The kinematic relationship between the position of the hand on which the end-effector is attached and the joint coordinates $\bar{e} = (e_1,e_2,\cdots e_6)$ can be written in a matrix form as

$$\bar{x} = R\,\bar{x}'' + \bar{r} \qquad (5)$$

or expanded

$$\begin{pmatrix} x \\ y \\ z \end{pmatrix} = \begin{bmatrix} r_{11} & r_{12} & r_{13} \\ r_{21} & r_{22} & r_{23} \\ r_{31} & r_{32} & r_{33} \end{bmatrix} \begin{pmatrix} x'' \\ y'' \\ z'' \end{pmatrix} + \begin{pmatrix} r_1 \\ r_2 \\ r_3 \end{pmatrix} \qquad (6)$$

$$r_{ij},\ r_i = \text{functions of } e_1,\ e_2,\cdots e_6 \qquad (7)$$

where $\bar{x}''$ = (x",y",z") are coordinates of a point H fixed in the hand, expressed in a coordinate system attached to the hand, and $\bar{x}$ = (x,y,z) are coordinates of H relative to a system fixed in space.

Eqs. (6) and (7), which can be derived by successive multiplication of homogeneous transformation matrices[2] between successive links, are used for forward problems to compute $\bar{x}$ from given values of $\bar{x}''$ and $\bar{e}$, or to compute $\bar{x}''$ from given values of $\bar{x}$ and $\bar{e}$.

Computing joint coordinates $\bar{e}$ from given coordinates $\bar{x}$ and $\bar{x}''$, and orientation of the hand constitutes an inverse problem. Solving the inverse problem facilitates kinematic control of a robot by prescribing the motion of the end-effector. Explicit expressions of e may be obtainable for robots with wrist mechanisms in which two or more joint axes intersect at one point, using a method trigonometric in nature outlined by Pieper[3] and Paul[4]. Generally an iterative procedure has become the only viable way of resolving the inverse problem for some robots.

Kinematicians have devised several iterative methods to resolve the inverse problem. These methods which involve some form of mathematical approximation can be used to solve inverse problems involving mechanisms having six degrees of freedom in general. Rate control method by D. E. Whitney[5], coordinate descent method by Malyshev[6], and discrete positions method by Konstantinov and Markov[7] belong to this group of methods. There is no guarantee that these methods would always converge or yield the correct solution.

In the following, we present an iterative method which can be used for robot manipulators. The method is based on the fact that known robots are separable. The advantage of this method, which is based on this distinct feature of robot kinematics, is that the solution is guaranteed and that the convergence is rapid, even for large changes of displacements and orientations from a known position. We give it the name "discrete linkage method."

Two kinematic equations in matrix form exist separately for major and minor linkages:

$$\bar{x} = A\,\bar{x}' + \bar{a}\;; \qquad A,\ \bar{a} = \text{function}\ (e_1,e_2,e_3) \tag{8}$$

where elements in A and $\bar{a}$ are functions of e_1, e_2, and e_3, only, and

$$\bar{x}' = B\,\bar{x}'' + \bar{b}; \qquad B,\ \bar{b} = \text{function}\ (e_4,e_5,e_6) \tag{9}$$

where elements in B and $\bar{b}$ are functions of e_4, e_5, e_6 only. Furthermore, the inverse transformations of all 12 categories of major linkages can be explicitly written

$$e_1,e_2,\ e_3 = \text{explicit function}\ (\bar{x},\ \bar{x}'), \tag{10}$$

By substituting Eq. (9) into (8) and comparing the result to Eq. (5) we obtain

$$R = AB. \qquad (11)$$

Inverse Problem Algorithm

The inverse problem consists of finding the values of six robot joint coordinates $(e_1,\cdots,e_6)$ which will bring the robot hand to a specified position and orientation. Position may be specified by prescribing both the local coordinates $\bar{x}''$ and the space coordinates $\bar{x}$ of an end-effector point of interest (e.g., the tip of a tool). At times it may be convenient to concentrate on the origin of the local system, $\bar{x}'' = (0,0,0)$, and to specify its space coordinates $\bar{x} = \bar{r}$, see Eq. (5). Orientation may be specified in one of several ways: yaw, pitch, and roll, or the conformal rotation vector which is convenient for interpolation[8], or other methods. Eventually, the numerical values of the nine elements of the rotation matrix R must be specified or computed. Mathematically, the inverse problem is then to find $\bar{e}$ when $\bar{r}$ and R, or $\bar{x}''$, $\bar{x}$ and R, are specified.

The discrete linkage method solves the inverse problem by means of the following iterative procedure:

1. Assume values of (e_4,e_5,e_6) by best available approximation, perhaps from a previously computed point.

2. Compute $\bar{x}'$ by Eq. (9).

3. Compute (e_1,e_2,e_3) by Eq. (10).

4. Stop iteration if new values of (e_1,e_2,e_3) are sufficiently close to those obtained in the previous iteration cycle. Alternatively, stop here after completing a specified number of iteration cycles.

5. Compute A by Eq. (8); then $B = A^{-1}R = A^{T}R$, see Eq. (11).

6. Solve the minor linkage relationship

   $$B = \text{function } (e_4,e_5,e_6), \text{ Eq. (9)}$$

   for new values of (e_4,e_5,e_6), using known matrix elements of B obtained in step 5 (three suitably chosen elements usually suffice), then compute $\bar{b}$.

7. Repeat the procedure, starting from step 2.

The physical counterpart of the above procedure is alternately to move the major and the minor linkage, one to satisfy the positional and the other to satisfy the orientational specification of the end effector, each time moving only one linkage while holding the other one fixed. This is commonly done during programming of some robots. In the algorithm, however, the robot will not be physically moved until the final result of the computation is available.

It should be pointed out that steps 3 and 5 depend on the specific kinematic description of the major linkage alone, whereas steps 2 and 6 depend on the description of the minor linkage alone. The robot as a whole is treated only in a general way, such as in the second part of step 5. There the numerical value of the minor rotation matrix B is obtained from the major rotation matrix A and the overall rotation matrix R by an operation that is identical in all robots.

Step 6 is an inverse operation involving the minor linkage, though a relatively simple one. It involves only the rotation matrix B, not the translation vector $\bar{b}$. Matrix B is practically identical in all wrist mechanisms which use the NN type simple chain (the sequential order of rows or columns of B may differ). Offsets between axes, which differ in different wrists, are reflected only in the vector $\bar{b}$. The method of solving B for the minor joint coordinates is the same for all gimbal-type wrists, and it is derivable from the information available in [4]. It has also been found that explicit solutions exist for compound and oblique-chain wrists constructed to date.

Convergence

In effect, the iterative procedure alternates between correcting the position of the robot hand but perturbing its orientation, and correcting the orientation but perturbing the position. Does this process converge towards simultaneous correctness of both position and orientation? It does so in all functionally well designed robots.

At the end of each iteration cycle (after step 6) the orientation is correct but the position is in error. The opposite is true at the end of step 3. Since step 3 perturbs the orientation and the subsequent step 6 corrects it, the hand rotations in these two steps are equal and opposite. Step 3 represents articulation of the major linkage alone, to produce a displacement of the hand which corrects the positional error at the beginning of the cycle. Step 6 represents articulation of the minor linkage to correct orientation at the expense of producing a residual positional error at the end of the cycle. Thus, the ratio of the new error to the old error, in one cycle, is the same as the ratio of displacements produced by the minor and the major linkages, respectively, when the respective rotations produced by the two linkages are equal.

In a properly designed robot, the major linkage is expected to produce substantial displacements with relatively little rotation, and the minor linkage to produce substantial rotation with relatively little displacement. Therefore, the ratio "ρ" between displacements due to the minor and due to the major linkage, at equal rotation, is expected to be substantially smaller than unity. This ratio is comparable to the ratio between respective distances from a point in the hand to the center of wrist rotation and to the center of arm rotation. The hand is supposed to be much closer to any of the wrist rotation axes than to any joint axes of the major linkage, thereby contributing to ρ being small. The rate of convergence of the discrete linkage method is of the nature of a geometric series, the error diminishing as ρ^n after n iterations. The rate of convergence may be estimated by inspection of geometry of any particular robot.

Any point in the robot hand may be chosen to represent the "position" of the hand. The local coordinates $\bar{x}''$ of the point are therefore subject to an arbitrary choice. The value of ρ, and hence the rate of convergence, will depend on this choice. In fact, if the three axes of wrist articulation intersect in a single point, and this point is chosen, then the ratio $\rho = 0$. In this case the algorithm terminates immediately and gives an exact, explicit solution. Conventional placement of the origin of the local system, $\bar{x}" = (0,0,0)$, tends to be in a place which minimizes the displacement ratio. Use of the origin in the procedure is recommended.

The discrete linkage method has been implemented in a jointed robot (type NR) with a compound-chain wrist. It has been found that four is a sufficient number of iterations to reach the required accuracy, and the number has been fixed in this implementation.

CONCLUSION

The variety of possible robot configurations is very large. Existing methods of kinematic description tend to treat each configuration as a special case, particularly when it comes to the inverse coordinate transformation problem. The inverse problem, one of finding how to control the robot in order to achieve a specified placement of the end-effector, has become increasingly important in the more advanced forms of robot control. It is also one of the most difficult problems, least amenable to generalization.

A step towards generalization has been made by emphasizing that robots of practical importance are separable into two primary subsystems: the major linkage which controls the position of the arm, and the minor or wrist linkage which controls the orientation of the working end (the end-effector) of the robot. Mathematical treatment of various robots may be modularized and thus greatly simplified by giving a separate description of various major linkages and of various minor linkages in common use. The emphasis in this paper has been in a comprehensive treatment of the major linkage, where most of the variety is found.

It has been discovered that there exists only 12 useful and distinct categories of major robot linkages which use simple chain mechanisms. The forward and inverse transformation of all major linkages have been derived and tabulated in the Appendix. A two-letter code has been assigned uniquely to each category. Only six categories have been found in existing robots. Others have potential applications.

A computationally efficient iterative method has been developed and described to solve the kinematic inverse problems of robots. The method takes advantage of the distinct features of robot manipulators that each robot is separable into major and minor linkages. So far, the method has only been used for offline editing of robot control parameters. Because of the requirement for real-time control of robot manipulators in some applications, some improvements in the iterative procedures are being considered. That will be the subject of future publications.

BIBLIOGRAPHY

1. Milenkovic, V., "Computer Synthesis of Continuous Path Robot Motion," Fifth World Cong. on Theory of Mach. and Mech. Proc., 1979 ASME, pp. 1332-1335.

2. Denavit, J. and Hartenberg, R. S., "A Kinematic Notation for Lower Pair Mechanisms Based on Matrices," J. of Applied Mechanics, Trans. ASME, V.77, June 1955, pp. 215-221.

3. Pieper, D. L., and Roth, B., "The Kinematics of Manipulators Under Computer Control," 2nd Int. Cong. on Theory of Mach. and Mech., Proc. 1969.

4. Paul, R. P., "Robot Manipulators: Mathematics, Programming, and Control," The MIT Press, Cambridge, MA, 1982.

5. Whitney, D. E., "The Mathematics of Coordinated Control of Prosthetic Arms and Manipulators," J. of Dynamic Systems, Measurement, and Control, T. ASME, Dec. 1972, pp. 303-309.

6. Malyshev, V. A., "A Method of Constructing Program Motions of a Manipulator," Trans. from Russian, 1981, Scripta Pub. Co.

7. Konstantinov, M. S. and Markov, M. D., "Discrete Positions Method in Kinematics and Control of Spatial Linkages," Mech. and Mach. Theory, 1980, V.15, pp. 47-60.

8. Milenkovic, "Coordinates Suitable for Angular Motion Synthesis in Robots," Robots VI, Conf. Proc., 1982, SME, MS82-217.

APPENDIX

In the following, we are tabulating and listing all of the equations needed for forward and inverse transformation of twelve major robot linkages. The Table immediately following the list of symbols serves as a guide to select a set of equations applicable to a particular linkage type, from the equations listed after the Table. Major linkage types CC, CS, CR, RR, NS, NR, and NN are found directly in the Table. All transformations for type SS are obtained from Eqs. (1abc). Solutions for types SC, RC, SN, and RN are obtained from equations for CS, CR, NS, and NR, respectively, by interchanging (x_o, y_o, z_o) and (x_3, y_3, z_3), and interchanging subscripts $()_1$ and $()_3$ in e_i (σ_i and/or s_i). The coordinates attached to the ground link (x_o, y_o, z_o) correspond to the space coordinates (x,y,z) in the text, and the arm coordinates (x_3, y_3, z_3) correspond to (x',y', z').

Coordinate System

e_i	the i-th joint coordinate, designated as s_i if slider, or σ_i if rotary.
s_i	the slider coordinate of i-th joint, equivalent to the translational distance from the space coordinate to the arm coordinate along the slider axis; being zero at null configuration of the linkage.
σ_i	the rotary coordinate of i-th joint, being zero at null configuration.
b	axis offset; usually between the 1st and 2nd joints.
h	axis offset; usually between the 2nd and 3rd joints.
(x_o,y_o,z_o)	space coordinates attached to the ground link.
(x_3,y_3,z_3)	arm coordinates coincident with space coordinates at null configuration.

The following table lists the direction of joint axes when the major linkage is in its null configuration.

	CC	CS	CR	RR	NS	NR	NN
e_1	y_o	z_o	z_o	z_o	z_o	z_o	z_o
e_2	y_o	z_o	z_o	x_o	y_o	y_o	$-y_o$
e_3	y_o	x_o	$-y_o$	$-y_o$	z_o	$-y_o$	z_o

TABLE

APPLICABLE EQUATIONS AND DEFINITION OF SYMBOLS

Forward Transform.	Major Linkage Type: CC	CS	CR	RR	NS	NR	NN
General Eqs.	←		2ab, 3ab, 4, 7				→
i(Equ. 2ab) =	1,3	1	1,3	1,3	1,2	1,2,3	1,2,3
e_1,e_2,e_3 =	σ_1,s_2,σ_3	σ_1,s_2,s_3	σ_1,s_2,σ_3	σ_1,s_2,σ_3	σ_1,σ_2,s_3	$\sigma_1,\sigma_2,\sigma_3$	$\sigma_1,\sigma_2,\sigma_3$
x,y,z Equs.	←		5abc, 8abc			→	2c,(6,9)abc
$x = x_3$; (#)y =	z_3	z_3	z_3	z_3	z_3	z_3	y_3
z =	$-(y_3+s_2)$	$-y_3$	$-y_3$	$-y_3$	$-y_3$	$-y_3$	z_3
h =	const.≠0	s_2	s_2	0	s_3	const.≠0	Const.≠0
b =	0	s_3	const.≠0	s_2	←const.(typ.0)→		const.≠0
S_o,C_o =	0,1	S_1,C_1	S_1,C_1	S_1,C_1	S_1,C_1	S_1,C_1	S_1,C_1
S,C =	$\sin(\sigma_3-\sigma_1)$ cos	0,1	S_3,C_3	S_3,C_3	S_2,C_2	$\sin(\sigma_3-\sigma_2)$ cos	$\sin(\sigma_1+\sigma_3)$ cos
S',C' =	S_1,C_1	0,1	0,1	0,1	C_2,S_2	S_2,C_2	$-S_1,C_1$*
S",C" =	S_3,C_3	0,1	S_3,C_3	S_3,C_3	1,0	S_3,C_3	S_3,C_3*

(*) Used in Equ. (3ab) to compute S,C

Inverse Transform.	Major Linkage Type: CC	CS	CR	RR	NS	NR	NN
P_1 (Equ.)	-	←		(11a)			→
P_3 (for y, see#)	(11b)	-	(11b)	(11b)	-	(11b)	(11b)
u_o (Equ.)	-	←		(13a)			→
v_o =	-	y_3	y_3	y_3	y_3	y_3	$y_3 - h$
P_2 (Equ.)	(14a)	-	-	-	(14a)	(14a)	(14b)
r_o =	x_o	-	-	-	$b-u_o$	u_o-b	$-u_o$
u_3 =	(15a)	-	u_o-b	(16a)	-	(15a)	(10)& (12abcd)
v_3 =	(15b)	-	(17a)	z_o	-	(15b)	
u_2 =	u_3	-	-	-	(18a)	u_3	$-[u_3+b]$
v_2 =	v_3+h	-	-	-	z_3	v_3+h	z_3
s_2 =	$y_0 - y_3$	z_0-z_3	z_0-v_3	u_0-u_3	-	-	-
s_3 =	-	u_0-x_3	-	-	$[-u\ ]-x_3$	-	-
j (Equ.20ab) =	1	-	-	-	2	2	2
S_1,C_1 (Equ.)	(20ab)	←		(19ab)			→
S_2,C_2 (Equ.)	-	-	-	-	←	(20ab)	→
S_3,C_3 (see #)	(21ab)	-	(21ab)	(21ab)	-	(21ab)	(21ab)

Major Linkage Equations (Forward Transformation)

(1) a. $s_1 = x_o - x_3$; b. $s_2 = y_o - y_3$; c. $s_3 = z_o - z_3$

(2) a. $S_i = \sin \sigma_i$; b. $C_i = \cos \sigma_i$ c. $V_2 = 1 - C_2$

(3) a. $S = S''C' - C''S'$ b. $C = C''C' + S''S'$

(4) $$q = (\ x\ ,\ y\ ,\ z\ ,\ h\ ,\ b\ )$$

(5)
$$\begin{array}{ll}
\text{a.} & x_o = (\ C_oC\ ,\ -C_oS\ ,\ S_o\ ,\ C_oS'\ ,\ C_o\ ) \cdot (q) \\
\text{b.} & y_o = (\ S_oC\ ,\ -S_oS\ ,\ C_o\ ,\ S_oS'\ ,\ S_o\ ) \cdot (q) \\
\text{c.} & z_o = (\ S\ ,\ C\ ,\ 0\ ,\ C'\ ,\ 0\ ) \cdot (q)
\end{array}$$

(6)
$$\left.\begin{array}{ll}
\text{a.} & x_o = (C-V_2C_1C_3\ ,\ -S+V_2C_1S_3, -C_1S_2\ ,\ S_1\ ,\ C_1C_2) \cdot (q) \\
\text{b.} & y_o = (S-V_2S_1C_3\ ,\ C+V_2S_1S_3, -S_1S_2\ ,\ -C_1\ ,\ S_1C_2) \cdot (q) \\
\text{c.} & z_o = (\ S_2C_3\ ,\ -S_2S_3\ ,\ C_2\ ,\ 0\ ,\ S_2\ ) \cdot (q)
\end{array}\right\} \text{NN Only}$$

(7) $$q_o = (\ x_o\ ,\ y_o\ ,\ z_o\ ,\ h\ ,\ b\ )$$

(8)
$$\begin{array}{ll}
\text{a.} & x = (\ C_oC\ ,\ S_oC\ ,\ S\ ,\ -S''\ ,\ -C\ ) \cdot (q_o) \\
\text{b.} & y = (\ -C_oS\ ,\ -S_oS\ ,\ C\ ,\ -C''\ ,\ S\ ) \cdot (q_o) \\
\text{c.} & z = (\ S_o\ ,\ -C_o\ ,\ 0\ ,\ 0\ ,\ 0\ ) \cdot (q_o)
\end{array}$$

(9)
$$\left.\begin{array}{ll}
\text{a.} & x = (C-V_2C_1C_3\ ,\ S-V_2S_1C_3\ ,\ S_2C_3\ ,\ S_3\ ,\ -C_3\ ) \cdot (q_o) \\
\text{b.} & y = (-S+V_2C_1S_3,\ C+V_2S_1S_3\ ,\ -S_2S_3\ ,\ C_3\ ,\ S_3\ ) \cdot (q_o) \\
\text{c.} & z = (\ -C_1S_2\ ,\ -S_1S_2\ ,\ C_2\ ,\ 0\ ,\ 0\ ) \cdot (q_o)
\end{array}\right\} \text{NN Only}$$

Major Linkage Equations (Inverse Transformation)

(10) $B = h^2 + b^2$

(11) a. $P_1 = x_o^2 + y_o^2$; b. $P_3 = x_3^2 + y^2$ ($y = z_3$ or y_3, see Table)

(12) a. $Q = (B+P_3+z_3^2-P_1-z_o^2)/2$; b. $R = \sqrt{BP_3 - Q^2}$
c. $u_3 = (hR - bQ)/B$; d. $v_3 = (bR + hQ)/B$ } NN Only

(13) a. $u_o = \sqrt{P_1 - v_o^2}$; b. v_o (see Table)

(14) a. $P_2 = z_o^2 + r_o^2$; b. $P_2 = P_1 + z_o^2 - v_o^2$; c. r_o (see Table)

(15) a. $u_3 = \sqrt{P_3 - v_3^2}$; b. $v_3 = (P_2 - P_3)/(2h) - h/2$

(16) a. $u_3 = \sqrt{P_3 - v_3^2}\ \text{sgn}\ (x_3)$; b. v_3 (see Table)

(17) a. $v_3 = \sqrt{P_3 - u_3^2}\ \text{sgn}\ (z_3)$; b. u_3 (see Table)

(18) a. $u_2 = -\sqrt{P_2 - z_3^2}$; b. u_2, v_2 (see Table)

(19) a. $S_1 = (y_o u_o - x_o v_o)/P_1$; b. $C_1 = (x_o u_o + y_o v_o)/P_1$

(20) a. $S_j = (v_2 r_o - u_2 z_o)/P_2$; b. $C_j = (u_2 r_o + v_2 z_o)/P_2$
($j = 1$ or 2, see Table)

(21) a. $S_3 = (v_3 x_3 - u_3 y)/P_3$; b. $C_3 = (u_3 x_3 + v_3 y)/P_3$
(y, see comment, Eq. 11b)

(22) $\sigma_1 = \tan^{-1} (S_i/C_i)$ (used when $C_i > 0$)

$\sigma_i = 2 \tan^{-1} \left[S_1/(1+C_i)\right]$ (general use)

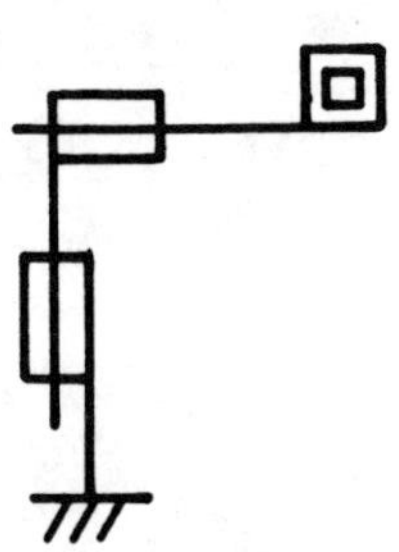

SS (Rectangular)

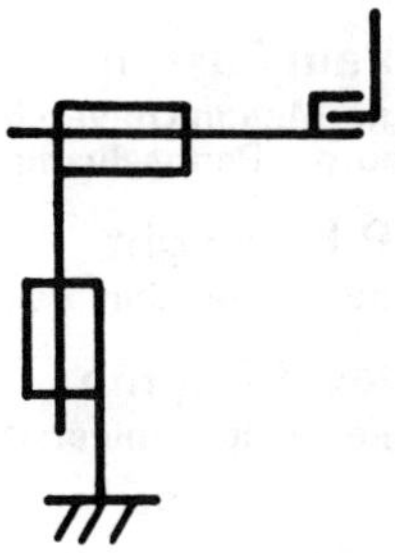

SC

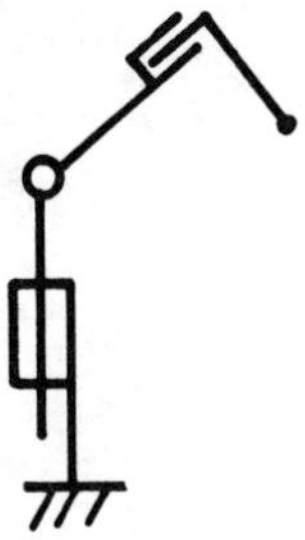

SN

CS (Cylindrical)

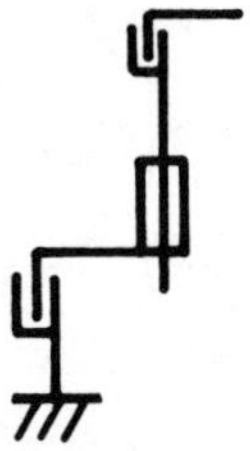

CC

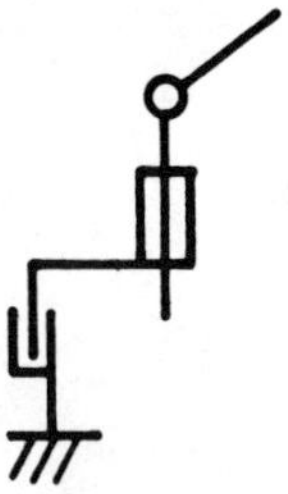

CR

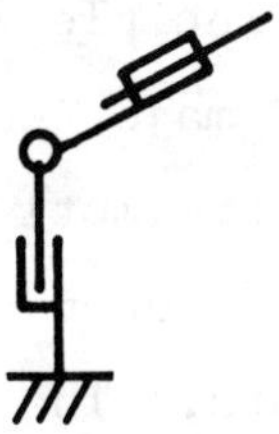

NS (Spherical)

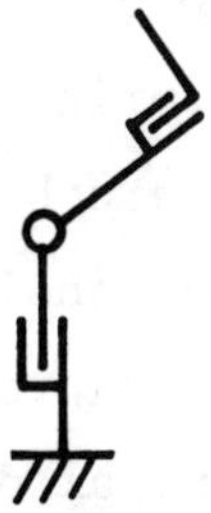

NN

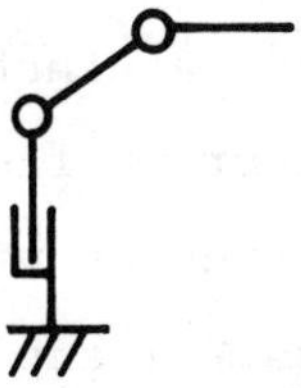

NR (Jointed)

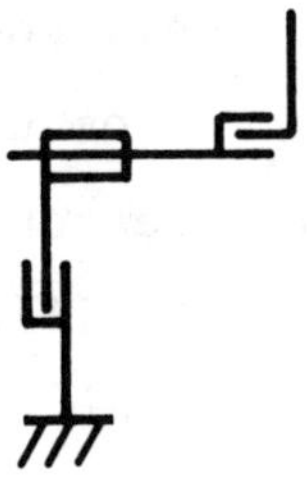

RC

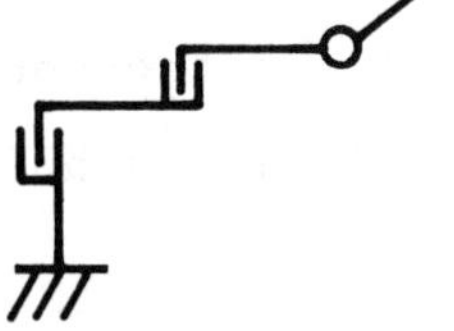

RN

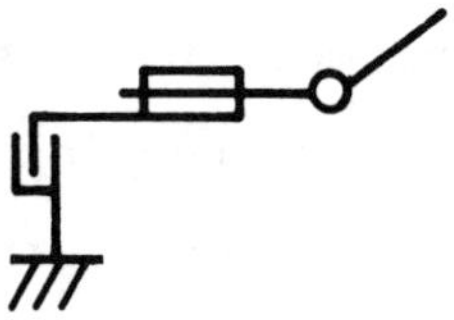

RR

A Design of a Controller as a Component of a Robotic Manufacturing System

Paul Fussell
Carnegie-Mellon University
Pittsburg, Pennsylvania

P.K. Wright
Carnegie-Mellon University

David Bourne
Carnegie-Mellon University

Abstract

This paper specifies the functional design of a robot or machine tool controller suitable for inclusion in robotic manufacturing systems. The principle interest is the unmanned manufacturing cells which are being developed as major components of new manufacturing facilities. The functional specifications are motivated by recent experiences with the creation of a prototype cell for an open die forging process. There are two key, technical design requirements. The first relates to the robot or machine tool being controlled. The controller must direct the machine actions, and, for unmanned operation, the controller must receive and process sensor information for process modification and for fault tolerance. The second relates to communication with the central cell computer (the host). The operation of an unmanned system requires a robust communication channel between the host and controller; the controller must also respond to a variety of instructions transmitted from the host. Additional controller design requirements are imposed by economics. The proposed controller can be used today, and can develop in an evolutionary manner to meet the needs of future manufacturing systems. The evolutionary development is made possible by modular design, organized in a hierarchical manner.

1 Introduction

The robot and machine tool controllers available today operate with frequent human operator intervention. This is inconsistent with the design goals of the robotic manufacturing systems of the future. The specification of a robot or machine tool controller which can be used in an unmanned manufacturing system is the topic of this discussion.

The authors and colleagues have recently been concerned with the implementation of a manufacturing cell [1] for open-die forging. This cell is a grouping of machines, including two robotic manipulators and two vision systems, managed and scheduled by a central computer (the cell host). The original intent was to build a system capable of unmanned operation for periods of time like a weekend. The original controllers for the robots, machine tools, and vision systems in the cell, though, did not support the functions necessary for this class of operation. For example, the movement of a billet from a loading rack to the outside of a furnace requires the cell host to ask a vision system to locate the part, instruct the robot to pick the part up, and then ask the robot to move the part to the furnace. Neither the controller for the robot nor the vision module was initially capable of receiving detailed instructions from the cell host, and then passing back intelligible data. To approach the design goal of unmanned operation for this cell, the controllers had to be modified either to the extent that the internal software was rewritten, or additional communication systems had to added to the controller so it could correspond with the host.

The difficulty with the earlier controller design rises from an incomplete view of the manufacturing environment that the controller should work in. The controllers are designed to be the *master* in any correspondence with a remote computer. This design makes the management of the robot and controller difficult from the remote computer. From our recent work, it has become clearer that the machine controllers must be able to correspond in two hierarchical directions, down the hierarchy to the machine and up the hierarchy to the manufacturing system computer. In another sense, the controller must control its immediate environment (the machine tool space, for example), but it must also be managed by the cell host.

The channels for communication to a remote host computer have traditionally been implemented in an ad-hoc manner. The most primitive communication is provided by tying existing controller input/output ports to parallel ports on the cell host. This is an inappropriate use of those ports, for the controller senses and controls the immediate physical environment with them. The immediate physical environment is below the controller in a hierarchical sense, whereas the cell host is above the controller. Using the parallel ports as a communication channel to the cell host is incorrect for this presents problems for extension of the communication channel and for the clarity of the supervision scheme. The next level of sophistication in communication channels is to add to the existing controller a bit-serial port (typically meeting the RS-232C specification). The controller, using the serial port, can receive and transmit some form of program from and to a remote computer. The correspondence between the controller and the remote computer generally requires human operator intervention. Unfortunately, the serial communication link is almost always restricted to program transfer.

The robotic system should also be able to adapt to changes in its environment. The robot should modify its behavior if the conditions under which its program was taught do not exactly correspond to the conditions at run time. The ability to adapt to changing conditions is extremely important if the robot is to be a component in an unmanned cell. The process of modifying the controller's behavior includes the reception of sensor data from the machine and process, the processing of that data, and then the formulation of decisions based on the data. Commercially available controller technology provides for input of binary switches, such as limit switches, but has difficulty in providing moderate processing power for more general inputs. The processing power to formulate decisions is beyond the current robotic controller.

The issues raised here of communication to cell hosts and reaction to sensor data can be solved in the research laboratories. The controllers developed for robotic research are generally full fledged computers in their own right [2, 3]. The tools (languages and operating systems) for setting up communication links and for processing sensor data exist on such computers. The problem becomes one of writing the programs to do these things. Controllers for commercial robots and machine tools, however, are computers with a different task. As a result, they have different capabilities. The approach to making each controller a fully programmable computer under a powerful language is attractive, but its cost may be prohibitive in the near and medium term. An alternative approach is to enhance the current controller with a good understanding of what is needed in the unmanned robotic manufacturing system. This approach holds promise for an economic controller in near and medium term.

The commercial development of controllers for robots and machine tools is an antecedent to the economically successful development of unmanned flexible manufacturing systems. In this paper, we review the requirements of the unmanned manufacturing system. The controller which fits into such a system can be developed in a modular way that is organized in a hierarchical structure. We illustrate this approach using examples from experimental work. A functional design for a general robot and machine tool controller that meets the requirements of an unmanned manufacturing system is then presented. The controller design can be implemented in an evolutionary way, so that it will be economical in today's market, yet meet the requirements of the robotic manufacturing systems as they are constructed.

2 An Unmanned Robotic Manufacturing System

A variety of robotic manufacturing cells have been developed in recent years and detailed descriptions can be found in Bjorke [4], Merchant [5], Weck [6] and Ennis [7]. The philosophy of the manufacturing cell suits many metal processing operations.

The way that the computers (host and controllers) are linked together in the cell constitutes the cell's *topology*. For example, Bjorke has used a three level control system: individual robot and machine tool controllers communicate with the cell host computer, and the cell host communicates with the factory level scheduling and/or numerical control program library computers. Other cells within the factory are constructed in a similar way, so the factory takes on an architecture schematically shown in figure 1. This scheme is one structure for unmanned manufacturing systems; it is, for example, the organization being used for the forging cell.

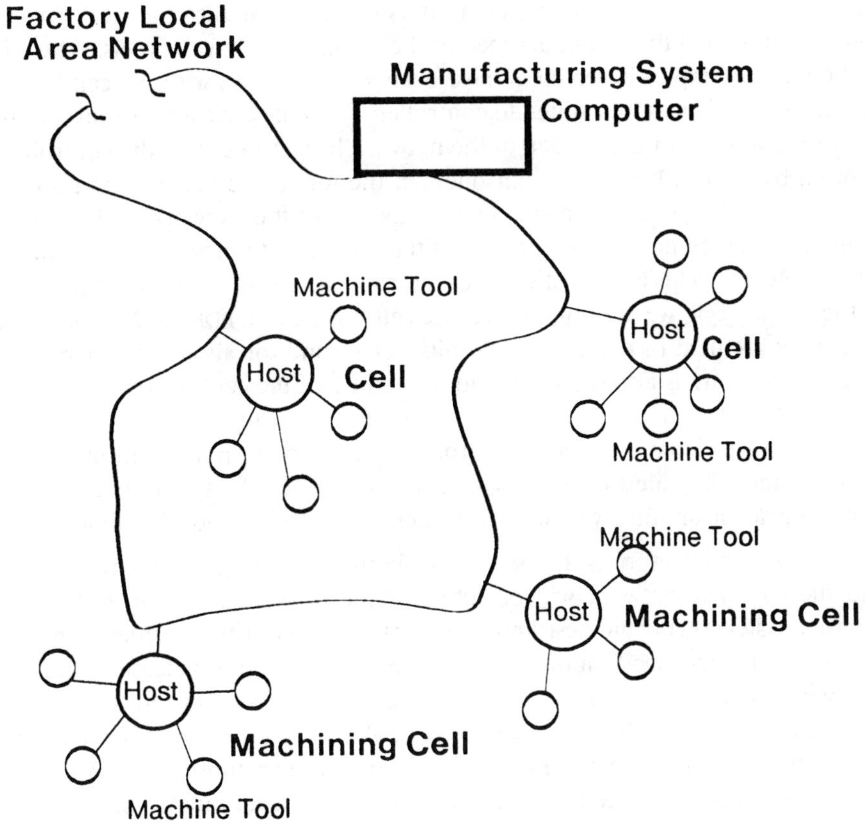

Figure 1: The Manufacturing System with Component Cells

A recent paper by Bourne and Fussell [8] outlines the approach used for the operating system of the forging cell. That approach has a number of salient features:

- The cell actions are programmed in a language [9] that allows independent activities in the cell to take place in parallel. If the two robots are ready to move and will not use the same space during their movements, then they are both told to move.
- The cell host maintains a database in the main memory of the host to represent the current status

of the cell. The data in the database is obtained on a timely basis from the controllers in the cell, so it dynamically changes with the state of the cell.

- The system is organized in such a way that new robots or machine tools can easily be added to the cell. The software on the host to support the new serial link, the new programs to command the new robot, and the database changes can all be developed and integrated in a simple and straightforward manner; the organization is modular, and the modules are placed into the system in a hierarchical way.

The computers which supervise the larger manufacturing system will communicate with the cell hosts (and other factory computers) over a factory wide communication network, the factory *local area network*. One manufacturing management system [10] is an expert system designed to schedule the factory and provide human managers with the information they need for their decisions about the overall operation of the factory. This Intelligent Management System requires the status of the individual cells and their components to supervise the complete factory.

As the cell is part of a larger system, it must be reasonably stable and it must be able to quickly respond to requests for data from the higher level factory computers. The cell architecture influences the tolerance to individual computer failures and the potential speed of the communication channels. The architectures of the forging cell and the other examples in section 3 have been set up as *stars* with the cell host at the center and the controllers at the points. This is well suited to controllers that only have a bit-serial communication link to the host for it requires that none of the burden of the network be supported by the controllers; the network is completely supported by the cell host. The controllers in the forging cell receive and send messages at will, and expect the cell host to always accept them. The bandwidth of this network, or how much information it can handle, is limited by the cell host for not only must the host react to messages, but it must also process the low level serial line protocols. This combination of duties can load the host, and limit the network bandwidth. For example, in the forging cell, we are finding that the cell host, a DEC PDP 11/23, can just keep up with the communications if they are held to a limit of 4800 bits per second (or about 500 characters per second) on each link. The star architecture is also susceptible to the failure of the central computer; when the host fails, none of the controllers in the cell can communicate. In the first manufacturing cells, this restriction is not particularly significant, for the controllers are not programmed to correspond with other controllers anyway, but an unmanned cell must be able to shut down with some grace. This can be achieved by allowing the controllers to communicate in a limited way when they detect that the cell host has failed.

An alternative network architecture is that shown in figure 2. The system is more tolerant to individual node failure, and the complete network will be robust if the network medium itself is not susceptible to failure. The Ethernet system [11] is the best known implementation of this architecture, termed a *broadcast cable-bus* system. The Ethernet specification uses a coaxial cable as the broadcast medium (the *ether*). This network structure will flexibly accommodate more processors on the system, but each processor must itself support the hardware interfaces and the network protocols. As a result the controllers and the host must share the burden of the network, so the controllers are more complex than those in the star network mentioned above. In the case of the Ethernet specification, much of the low level burden of communication over the network is offloaded to special hardware (Ethernet boards and chips) which process the raw data and provide complete and error free messages to the processor.

The controller in a manufacturing cell will greatly influence the architecture of the cell. The first implementations of the cell will use (as in our forging cell example) a star architecture. As the bandwidth requirements become more strenuous, and as the controllers become more advanced, we predict that the architecture of the manufacturing cells will change to the cable-bus system.

The principle requirements placed on a controller in an unmanned robotic manufacturing system are:

1. The controller should operate under the direction of the cell host with no intervention from a human operator.

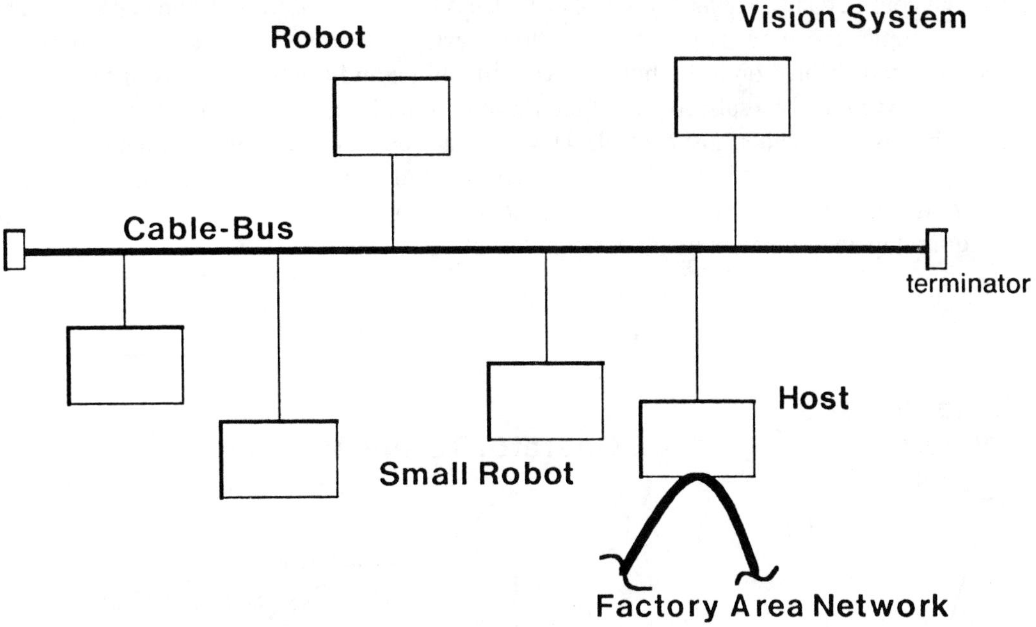

Figure 2: The Cable-Bus Network Architecture in a Manufacturing Cell

2. The controller should, at times, operate as an underling to the cell host. It must be designed such that it can receive instructions from the host and intelligently carry them out. The controller will act not as a slave, but rather as an intelligent assistant to the cell host.

A robot or machine tool controller's success at integrating into an unmanned cell will be judged by its ability to communicate and its ability to manage sensor information. The cell host will command the actions of each controller in the flexible manufacturing systems. The communication channel will be used for the passage of those commands. The exact nature of the needed commands is discussed below (*see* section 4).

The cell host must quickly receive from the controller the data that it requests. The speed of the response is a function of the communication channel speed and the controller's speed in responding to data requests. Even initial cell implementations, like the forging cell, need a large body of information from the controllers. The host maintains a representation of the cell that is being updated with data from the controllers, and the accuracy of the representation is dependent on the total time for controller response to data requests. Most of the forging cell links are operating at 4800 bits per second (*bps*), and this will probably not support the demands of the database when it is fully implemented. In addition to reasonably quick communication channels, the controller should also have the processing capacity to fulfill data requests quickly.

3 Examples of Experimental Controllers in Flexible Manufacturing Cells

The open die forging cell mentioned above is a collaborative research effort between the Westinghouse Corporation and Carnegie-Mellon University. The installation is at the Westinghouse Turbine Components Plant in Winston-Salem, North Carolina, USA. The cell produces stainless steel turbine blade preforms from raw billets. The basic hardware for the cell is a CNC hammer forge, a rotary furnace for heating the billets, two industrial robots, a cropper, a vision based station to locate parts, and a vision based gauging system. The host -- controller interfaces for the various cell components are shown in figure 3. Each robot and machine tool has its own commercially available controller that has been interfaced with the cell host. The robots in the cell, for example, are used as material handling devices. The cell host tells the robot to execute a particular subprogram (stored on the robot) knowing the initial and final conditions of the robot. (The host only requests the execution of a subprogram if the robot is in the initial condition.) At times, the cell host also requests the robot to move under host control. This message consists of a request for a move from the current position to x, y, and θ positions. Both of these instructions, and many others, were implemented by rewriting the controller software [12]. The original robot controller simply could not support these functions, all of which are required in an unmanned manufacturing system.

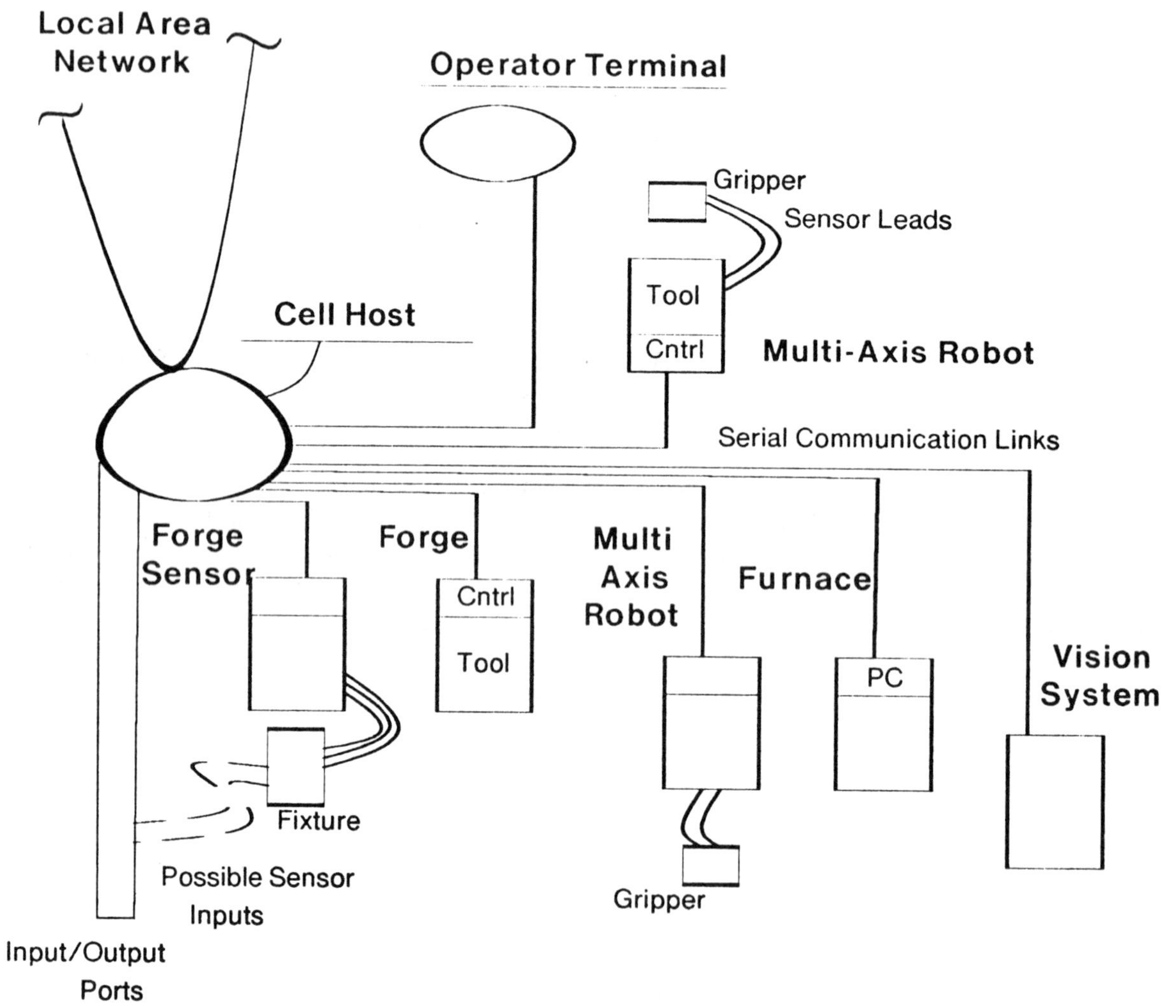

Figure 3: Cell Host -- Controller Communication Links

As a second example, in 1978, the Air Force contracted three aerospace companies to develop flexible automation concepts. The McDonnell-Douglas Corporation developed a cell to rivet portions of an airframe for the F-16 fighter. This cell includes a riveting machine and a Cincinnati-Millicron T^3 robot (among other components). The cell host is a DEC PDP 11/34. The cell design specified a uniform communication scheme between the cell hosts and the component controllers running under the RS-232C interface standard. The devices in the cell are guided by a series of common commands from the cell host. The communication scheme designed for the cell, however, was different from the communication link supported by the T^3 robot controller; the T^3 link is based on the data link layer of Digital's DECnet, DDCMP. To overcome this difference, the McDonnell-Douglas researchers used a μ-computer (an M6800) to interface the cell communication system and the T^3 link as well as provide access to the parallel inputs and outputs on the robot controller. Even with this interface, the functional control that the cell host had over the robot was constrained by restrictions in the controller [13]: an entire sequence of points (not a single point) has to be transferred, the number of points in the sequence can not be changed at the remote computer, sequence entry and exit points are constant, the transfer can not be invoked on a sequence that is being executed, and the sequence must be transferred within a preprogrammed time or the communication channel is aborted at the controller end.

As a third example, a prototype machining cell in the author's laboratories is being used for a variety of research in unmanned cell concepts. Cutkosky [14] has designed a position sensing wrist that attaches to the end-plate of a T^3 robot. The wrist supplies data about the relative difference between where the end-plate is and where the desired goal is. For example, the robot has an accuracy of about ±0.050 inches (±1.3 mm). In practice, the robot can be programmed to positions more precise than this, but one must expect that the robot may drift with time or that the external environment has shifted with respect to the robot, as a machine tool table might. The addition of a sensor unit, such as the position sensing wrist, allows the robot end-effector to repeatably approach a precise location (±0.010, say). The wrist system can also detect, and compensate for, minor shifts or time dependent shifts of the external world with respect to the robot. At the time that the work piece contacts the fixture (for an assembly operation), the wrist detects the relative movement of the end-effector and supplies that data to a μ-computer which is programmed to recognize whether the robot is correctly placed or not, and what action to take to correct the placement. The robot's position can be corrected in a number of ways. The offending point can be modified by moving the robot to the correct position, or the point can be remembered by the μ-computer and modified at some later time before the robot returns to the position. The former approach requires the ability to request an immediate change in the robot position, while the latter requires the time to upload, change, and download the robot program. The computational requirement for deciding how to change the path, or move the robot, is beyond the ken of the controller itself, so that burden will be placed on the separate μ-computer.

Cutkosky's work is oriented towards improving the accuracy of large industrial robots. The technique of obtaining sensor data, processing it, and then making a decision from that data is generic to robot and machine tool applications. The potential for improving the performance of the system by sensing and controlling new variables (such as robot drift) is integral to the application of robots in unmanned manufacturing cells.

4 Controller Functional Specification

4.1 Controller Programs

The different device controllers (robot, machine tool) all sequentially execute a set of program instructions. The notation for this set of instructions varies from implementor to implementor, so we choose to call the set of instructions a **program**. The purposes of the manufacturing cell are better served if the controllers can each store several programs and execute them in any order. For example, the total contents of the data tables in random access memory explicitly does not constitute ***a*** program in our scheme. The internal organization of the controller must be more complex than that.

4.2 Naming of data locations

Each of the controllers will gather data for its own operation and for the operation of the cell. The cell host should have a consistent technique for accessing the data throughout the cell.

The format for accessing the data varies from controller to controller. If the controllers were all programmable computers, then a uniform naming convention would be possible. As it is, the naming convention either depends on the internal representation for data locations in the controller (for example, addresses like $85C3_{16}$, or relation names like *furnace.temperature*) or it depends on byte position in data streams (for example, bytes 4 to 15 of a data point in a sequence stream are the robot coordinates for the T^3 controller).

In the forging cell, consistency is gained at the cell host by forming a relational database in the host that provides a mapping between variables such as *furnace.temperature* and the location of that data in the furnace controller. This scheme, as with the cell network, places the burden on the cell host.

4.3 Supervisory Functions

Supervisory functions are those capabilities that the controller supports so the cell host can manage the action of the cell. The management of the cell logically separates into areas about cell operation during a parts run, operation during a part change-over, and operation at quiescent times.

The commands that the controller should receive and understand so the cell host can manage the cell during a parts run are in table 1.

The controller needs to support an additional set of functions if the host is to change the cell from one parts style to another style without help from an operator. In particular, the host will need to manage the set of programs that are in the controller. Table 2 lists these functions.

During cell operation the controller should be able to accept the instructions in tables 1 and 2 at any time. The speed of the response may vary with the load on the robot controller (continuous path control, for example, might be expected to occupy much of the controllers capacity), but the controller should receive and process the command even if it is an incorrect command, such as delete a program that is currently being executed (in which case the controller would return an error message).

The logical link between the host and the controller should be bi-directional so that the controller should be able to send messages as well as the receive them. This makes the software at either end more complex, but the benefits are considerable. In particular, the integration of the controller with an intelligent management system requires this level of complexity. Two operations for illustration:

1. Frequently, a process on a device (especially a sensing device like a vision system) will pass data to the cell host. With the suggestion above, the controller can initiate a communication to the host, and send the message. The host must, of course, have a technique for receiving the message, and queuing it to the process that is expecting it. Alternatively, the host may poll the controllers, in which case the controller must have the capability to queue its messages to the host.

   The most important message would be an alert to the cell host (and potentially to the factory management system) that the process has in some way failed. The hydraulics have shut down owing to overheat, or a tool has broken, for example.

2. The size of main memory in a controller will be limited. Moderately complex parts can require parts programs (written in APT, say) greater in size than the available controller memory. If the controller can request a program download from the host, then arbitrary length programs could be developed for any of the controllers. Primitively, a special statement (*chain* to use BASIC's lexicon) could request the deletion of a program in the controller memory, and then the

Table 1: Run Time Functions

- ***begin program execution*** -- starting at the beginning of a stored robot program.
- ***emergency stop*** -- the cell host must be able to stop the motion quickly. In industry today, this generally has the connotation of turning the hydraulic pressure off in hydraulically driven robots. The quickness of this operation implies that the command should be given over a direct line to the controller instead of over the host -- controller communication channel.
- ***stop execution*** -- now. This command would stop the robot program at the robot's current position. The "state" of the robot is not lost, so the program can be re-started.
- ***gently halt execution*** -- soon. The controller will "hold" the robot program at the next "convenient" place. That might be at the next point, the end of the current sub-routine, or at the completion of a path. This is less harsh than **stop execution**, for it allows the controller to ramp the velocity of the arm.
- ***complete program and stop*** -- the normal way for the host to request the controller to stop work.
- ***restart execution*** -- used by the host to start the robot moving again after a **stop** or **gently halt** instruction. This should be undefined if the robot was last stopped with a **complete program and stop** command.
- ***status*** -- message. The operating parameters and controller status. For example, the location of the robot end-effector and the joint locations as well as controller flag status, execution status (program running, stopped, halted), and other relevant data.
- ***transmit data*** -- This is distinct from ***status***. The information contained in response to this message is more free-form, and will vary from controller to controller. The content will be defined by the software at either the controller or the host.
- ***receive data*** -- The same message as above, but in the other direction.

Table 2: Parts Style Change Functions

- ***download a program*** -- from the host to the controller.
- ***upload a program*** -- from the controller to the host. The utility of **upload** is important in the initial stages of cell development.
- ***delete a program*** -- from the controller memory.
- ***catalog of controller programs*** -- this is dependent on the notation that the controller uses for its programs.

downloading of a program from the host. Several segments could be *chained* together to form the complete parts program.

The same solution could be implemented by putting an on-line mass storage system with the controller, and given the decreasing costs in storage technology, this alternative is becoming more attractive.

The controller will need to initiate a sub-set of the messages that the cell host sends. Table 3 lists these functions.

Table 3: Controller Initiated Functions

- ***process exception*** -- at the machine level. The controller must be able to inform the host of an exception in the process. The content of the message will define the nature of the exception (for example, tool breakage, or gross part misalignment).
- ***upload a program*** -- from the controller to the host.
- ***download a program*** -- from the host to the controller.
- ***receive data*** -- from the controller to the host.
- ***send data*** -- a request for data by the controller from the host.

4.4 Sensor Input Functions

Existing controllers do not provide the raw capability to support sensor input, nor do they modify their behavior based on the sensor data (except for primitive sensor inputs, like limit switches). Part of the problem is based on the large variety of sensor information that is useful. Much of the sensor information is critical to particular operations. Varying weld torch parameters like current and consumable electrode feed rate, for example, are crucial for the success of a robotic welding robot's operation. Such parameters are meaningless in other contexts though. The answer to handling the variety of sensor information is in the intelligent design of the controller itself. The core to that answer will be in the programmability of the controllers.

A list of specific sensor driven commands for the variety of sensor inputs is well beyond the scope of this discussion, but there are some well motivated, general commands that would be desirable for the controller in the flexible cell. Table 4 shows these commands.

Table 4: Robot Sensor Functions

- ***move the robot.*** Either the move could be calculated in small changes from the current position, or the move could be prescribed in some more global coordinate system (robot or world). The power of this instruction is great for the majority of the sensors under development now are oriented toward adjusting the position of the robot.
- ***modify a point*** -- change the value of a program point in the robot controller. If the adjustments made in a robot's position are systematic, then the stored point should be modified so that future adjustments can be minimized. If the points are accessible by some naming convention, then this operation is simple, otherwise it may be limited to points like *next* or *last.*

The robot is not alone in getting benefit from sensor driven commands -- functions to modify the machine tool's action are listed in table 5.

The ***emergency stop*** function listed above (section 4.3) can be a sensor driven action. If the sensors on, say, a machine tool detect that the tool has just broken, the controller may elect to ***stop.*** In the unmanned cell, the

Table 5: Machine Tool Sensor Functions

- ***adjust feedrate*** -- Accompanying this command would be a new feedrate.
- ***adjust spindle speed*** -- or change the rate that the cutter is turning at.

controller should then be able to withdraw the broken tool, and change it out. This will require a controller complexity greater than just ***stopping***. One approach is to provide a *hook* for a process to grab when some condition is met. If the condition is *tool break*, and the hook allows the process to execute a special procedure that stops the spindle, draw the tool to a safe position, change the tool, and inform the host that the work piece has been damaged, then operation can begin to be unmanned. The complete action should be programmed in the controller itself.

The commands to change the movement of the robot have considerable utility in a robotic cell. The position sensing wrist [14] could provide data to a μ-computer which then could command the robot controller. Either final assembly operations can be controlled using the ***move*** function above, or some portion of the sequence of points that the robot passes through can be changed by the ***modify*** instruction.

4.5 Controller Error Handling

Errors will occur during host--controller communication. The most important errors are not the type that occur in the transmission medium (which can be caught and corrected in well understood ways [15]), but errors that occur when the controller is interpreting the command from the host. For example, the host may request something incorrect, like the deletion of a non-existent program. Typically, controllers today will react to these errors by kicking to operator level. This makes unmanned operation very difficult.

Error handling is a topic discussed in operating system design , but even simple solutions can be useful. For example, if an error occurs, then the controller might jump to an error recovery routine. That routine would be responsible for deciding what action would be necessary. If the error was caused by an incorrect cell host request, then the controller should return an error message to the host, wait for an acknowledgement of the error, and then go to the normal command reception state waiting for the next host message.

It is equally important for the controller to react in a similar fashion to errors that occur during the normal execution of a stored program. The vision system in the forging cell, for example, uses a strobe to light the part that it is inspecting. Occasionally, the strobe mis-fires, and the vision system fails to obtain a valid picture. The vision system internal software reacts to this condition by declaring a major fault, and exiting to the operator level, thus aborting any programs that were operating on the vision system. This reaction makes unmanned operation impossible.

5 Evolutionary Development of the Sophisticated Controller

The movement from the current state of the art to more sophisticated equipment will take several years and will result in a series of product developments. These developments are driven by the needs of the future robotic manufacturing systems.

5.1 Current Requirements -- Embryo Manufacturing Cells and Stand Alone Machines

The first requirements are the communication links with the cell host and, indirectly, the other controllers in the cell. A minimum set of functions are also needed by the cell host. These additions to existing controllers are providing components to a cell that has a greatly reduced operator workload [1]. The existing controllers are not provided with the power to receive complex sensor input and modify their own behavior; the implications for the changes to the internal structure of the controllers appear to be too great.

The communication channel to the cell host should be implemented as a communication link between computers. Experience shows that a bit-serial channel using physical and electrical interface standards such as RS-232C, RS-423, or RS-422 is appropriate for the near term. RS-422 has a greater noise immunity, but RS-232C and RS-423 are more widely available. The implementation of this channel, however, requires a change in motivation on the part of the controller builders. The context from which the controller is designed must be altered from the view that the controller is a stand-alone unit which controls a machine tool to a view where the controller can also be managed by a higher level computer. This, of course, must be done in such a way that the new controller can operate in a stand-alone "mode" for compatibility with existing equipment and installations. The most obvious way to do this is to provide a front panel switch to change from "manual" mode to "remote" manage mode. A much better implementation would allow the mode to be changed by commands from the remote computer (host). The functions that the controller should support are in table 1.

5.2 The Medium Term Requirements -- With the New Language

A medium term cell, circa 1985, will be unmanned for normal operation. Human operators will only be required to handle exceptions during a part run or make changes between part styles.

The controllers for this cell will need all the functions described in section 4.3. The cell host will be maintaining the programs in the controller, and it will need to control the execution of those programs during a part run. The cell host will also maintain a representation of the cell for its supervision algorithms. The representation is easily formed as a dynamic database which receives timely data from the controllers in the cell.

The sensor data for controller modification will be handled by a separate computer. The data acquisition and processing power is best left to a computer that acts in concert with the controller until the requirements for particular applications are better understood. The generic nature of the robot is an excellent illustration of the difficulty in developing a general sensor processor for a device. There are a great number of possibilities for a single, economic controller. The separate computer should be programmable to provide the structure for applying the power of the system to particular tasks.

The second computer will act with the controller to provide the full set of functions in section 4. In particular, the functions in table 3 can be programmed on the computer and the cell network can be implemented to the computer in a more systematic way than ad-hoc serial lines.

The requirements placed on the controller by this time imply that the controller must operate under a real time, multi-tasking operating system. The power provided by this type of system will also aid in the development of the advanced controller.

5.3 The Requirements of Unmanned Manufacturing Cells -- High Bandwidth Communication and Sensor Data Processing

The unmanned manufacturing cell of 1990 will require sophisticated components. They will be required to provide operational data to the cell hosts and the factory intelligent management systems.

The amount of data that will pass between host and factory computers and the controllers in the cells will be great . The network in the cell will need high bandwidth channels to support the load.

Sensor data will, by this time, come from a variety of sources and provide good understanding of the interaction between machine and environment. The controller will intelligently react to that sensor data in preprogrammed ways. The data from the robot and environment sensors should be processed on the controller and used locally for robot control. This data should also be provided to the cell host and factory management systems.

6 Conclusions

The unmanned robotic manufacturing system presents great potentials for productivity increases. The system depends on correctly designed and implemented components, one of which is the controller of individual robots and machine tools. These controllers currently do not support the functions that are needed for unmanned cells. Commercially available controllers just meet the requirements of flexible manufacturing cells that use human operators. Even these controllers must be modified to support a communication link to the cell host. The presentation above has been a description of the functional characteristics of a controller that can participate in a future manufacturing cell.

Implementation of these specifications can be difficult in the current economic market. We propose a controller built of modules that can develop with the flexible manufacturing systems. As new requirements are added to the controller, new modules can be added to increase the capability of the controller. The principle capabilities in the near term are associated with how the cell host manages the components in a manufacturing cell. The longer term capabilities are oriented toward increasing the controllers ability to sense and react to a changing environment. The result is a powerful component of a manufacturing system.

7 Acknowledgements

Thanks are due to Professor D.R. Reddy, Director of the Robotics Institute at Carnegie-Mellon University for encouraging this work. We thank the Westinghouse Electric Corporation for financial support and our colleagues J. P. Colyer, G. C. Schatz, and J. Isasi at the Turbine Components Plant for their contributions to this work. M. Cutkosky has also kindly provided technical support throughout this work.

References

1. P.K. Wright, D.A. Bourne, J.P. Colyer, G.C. Schatz and J.A.E. Isasi, "A Flexible Manufacturing Cell for Swaging," *The 14th CIRP Seminar on Flexible Manufacturing Systems*, Trondheim, Norway, June 1982.

2. R. H. Taylor, P. D. Summers, and J. M. Meyer, "AML: A Manufacturing Language," *International Journal of Robotics Research*, Vol. 1, No. 3, Fall 1982, pp. 19-41, discussion of AML

3. M. S. Mujtaba, R. Goldman, and T. Binford, "Stanford's AL Robot Programming Language," *Computers in Mechanical Engineering*, Vol. 1, No. 1, August 1982, pp. 50-57.

4. O. Bjorke, "Computer-Aided Part Manufacturing," *Computers in Industry*, Vol. 1, January 1979, pp. 3-9.

5. M. E. Merchant, "The Factory of the Future- Technological Aspects," *Towards the Factory of the Future*, L. Kops, ed., ASME, United Engineering Center, 345 East 47th Street, New York, NY 10017, 1980.

6. M. Weck, K. Zenner and Y. Tuchelmann, "New Developments of Data Processing in Computer Controlled Manufacturing Systems," *Society of Manufacturing Engineers Technical Paper*, No. MS79-161, 1979, .

7. Ennis, G. E., Eastwood, M. A., et al, "Robotic System for Aerospace Batch Manufacturing," Interim Technical Report IR-812-8, McDonnell Douglas Corporation, 1979.

8. D. A. Bourne and P. S. Fussell, "Designing Programming Languages for Manufacturing Cells," *Proceedings of Electro/82 -- 14*, IEEE, June 1982.

9. D. A. Bourne, "A Numberless, Tensed Language for Action Oriented Tasks," Tech. report CMU-RI-TR-82-12, The Robotics Institute, Carnegie-Mellon University, October 1982.

10. Mark S. Fox, "The Intelligent Management System -- An Overview," Tech. report CMU-RI-TR-81-4, Carnegie Mellon University, The Robotics Institute, August 1981.

11. John Shoch, Yogen Dalal, David Redell, Ronald Crane, "Evolution of the Ethernet Local Computer Network," *Computer*, Vol. 15, No. 8, August 1982, pp. 10-27.

12. David A. SyiekThe Robotics Institute, Carnegie-Mellon University, *User's Manual for Versat*, 18 August 1982 ed., Pittsburgh, Pennsylvania 15213, 1982.

13. G. E. Houston and L. B. Brown, "Robotic System for Aerospace Batch Manufacturing," Tenth Interim Technical Report IR-812-8(X), McDonnell Douglas Corporation, February 1981.

14. M. R. Cutkosky and P. K. Wright, "Position Sensing Wrists for Industrial Manipulators," *12th Internation Symposium on Industrial Robots*, SME, June 1982.

15. John. E. McNamara, *Technical Aspects of Data Communication*, Digital Equipment Corporation, 1982, ISBN 0-932376-18-5

Coordination and Dynamic Control of Robots by Microprocessors

M. Al Mouhamed
French Atomic Energy Commission
Gif-sur-Yvette Cedex, France

E. Csakvary
French Atomic Energy Commission

B. Dupeyrat
French Atomic Energy Commission

Abstract

We study here several fundamental features of the problem of the command conception of automatic assembly oriented robots. The problem of motion coordination for an articulated manipulator is studied. A force sensor (two grips) was built to measure the external forces exerted on the manipulated object. The information delivered by the sensor is used by functions which realize the forces regulation. We are also concerned with dynamic control. The structures we present here are implemented on performent microprocessors (INTEL 8086 with the floating point coprocessor 8087), which leads to a coordination computation time of 8 ms and 9 ms to compute the inertia and gravity forces matrix.

I - INTRODUCTION

We present in this paper several fundamental parts of the problem of the conception and the realisation of an automatic assembly oriented, programmed robot.

In this sense, the task of the robot is to pick the object to assembly in a certain way (problem of the coordination and of the exteroceptive sensors), to transport it at the presumed place of the assembly (avoiding eventual obstacles) to place it in a position convenient for the task (sentance interpretation and geometry description) and to realise appropriately the assembly (use of well adapted image informations and regulation of the exerted force).

The subproblems solved here are parts of the scene and are the following :

- General coordination of the movements
- Regulation in forces
- Dynamic and static control.

These different subjects are studied with a realistic point of view. The synthesized structures are meant to be implemented on performent microprocessors. We present, in our applications, results concerning the command cycle (in the case of coordination and dynamic mode), the structure of the global system and the real performances obtained. This study will be presented with some demonstration movics concerning assembly (insertion, extraction, place ...) and the dynamic command.

The interest of a coordination system is to allow to move the manipulated object in a coordinated or reflected way through the articular variables. The action referential is a referential that the user can by any mean (for example programming) translate or and rotate. The aim of the manipulation is to have an action on the objects, consequently, it is desirable to place this referential at the picking center. In the case of the assembly operations, the referential relatively to which corrections must be made, is situated in any location of the assembled object. Thus it is interesting to define a coordination system in which this referential is defined by program. Our system takes in account these considerations and disposes of several groups of motion functions :

- Transport of objects functions : The center of picking is the center of the action referential, the orientation of them is represented by the three Euler angles.
- Avoiding the obstacles functions : (Studies in progress) : Allows to determine a complex trajectory from the knowledge given by a small number of points. The generated trajectory has not stop points.
- Functions of declaration of translations or rotations of the action referential initially placed at the center of picking.
- Simple translations or rotations movement function acting on the action referential precedently defined.

A realisation of this system was made on a performing microprocessor unit (INTEL 8086 with floating point coprocessor 8087) to command a MA 23 manipulator (developped at the STEPD/EMH/ in the CEA). The elementary coordination time is 8 ms versus a duration of 30, 40 ms on a microcomputer without floating point device.

II - FORCES REGULATIONS

To deal with the imprecisions' dues to various problems in the case of articulated systems during the assembly operations [8] , the process must run in closed loop (using exteroceptive informations). The force regulation allows to realize fine assemblies when certain conditions are taken in account. The principal conditions are :

- Information about the forces must be of quality particularly the use of a general force sensor allowing to obtain decoupled informations. This one must be placed at the nearest location of the object to minimize coupling with the manipulator and consequently to avoid the effects dues to the mechanical defaults of this.

In general, the sensor behaves like an elastic component ; it is interesting to be able to program the tightness of this component (adaptation to the manipulated load). But this point seems particularly difficult to realize. The sensor we have developped is composed of two symmetrical grips admitting four independant degrees of freedom. These later allows to detect two forces components and two moments components.

- Programming the mechanism to afine the general position (regulation in forces) shall not be in charge of the user because of the complexity. A structuration of these regulations is possible according to the adapted definition of the simple movements (cf. 1). These regulations allow to correct the position and the orientation relatively with the definition of the action referential which is placed by programmation at the compliance center. The functions we have developed and experimented are divided into two groups.

1°- Global regulations in forces or in moments with an end of action condition. It is complete when the manipulated object will be translated or rotated from a given length or a given angle. Data for this function are consigned forces (componant of forces or moments). The insertion of a peg in a hole is a possible example. The forces and couples consigned are null except for the force component relatively to the axis of the peg wich allows to maintain a constant pressure to permit the insertion. The function which finishs when the peg will be entered in the hole on a certain length.

2°- Simple regulations relatively to one component of the force or the moment. The interest of this later is to allow a control in force along a particular direction and a control in position along the other components of the movements.

A mixed force-position control is obtained. In general these regulations concern tasks in which a trajectory is assigned along one or two axis, the remaining axis being under an external constraint that requires a regulation in force (for example to maintain a contact). We have experimented these regulations with the microprocessor system precedently described. The duration of the elementary regulation period is about 10 ms (including the elementary coordination period) when the force sensor is used.

III - DYNAMIC CONTROL

In an articulated system in motion, there exist appreciable dynamic [4] forces (inertia, centrifugal and Coriolis forces). These forces product at followed trajectory level, discards from the prescripted trajectories (in the case of an empiric regulation). Thus the aim requested is to build a command rule allowing to compensate these perturbations and to better control the poles of the system in a closed loop.

The generally followed step consists of in line computing the forces precedently mentioned from a model of parameters. This operation results from a very complex mathematical modelisation of the articulated system. It involves an important amount of computation wich sets a problem with the real time process.

Our step is oriented towards a more direct technic. We are interested in discrete modeles of parameters obtained from a direct identification phase on the articulated system. For this, we study the possibility to identify a parameter (for example the inertia coefficients) relatively to the articulated main variables, it is a function of. The methods concern articulated systems including reversible transmission.

The identification procedure is applicable when two conditions are both realized :

1) The fine knowledge [5] of a transfer model between the motor organ and the articulation wich takes in account the real effects of the system (elasticity of the cables, viscous frictions, ...). Such a model is necessary not only for the identification, but also for the control.

2) The knowledge of the respective evolutions of the input (the motor torque) and of the output (for example the motor angle). The excitation pattern applied on the input is determined so as to minimize the elasticity effects. Thus we search for an input which induces a quasi constant tension of the transmission cables.

The identification of the inertia terms A_{ii} and A_{jj} (diagonal coefficients of the inertia matrix) allows to deduce, by the same procedure, the A_{ij} terms. Applying this method using a quadratic criterion for the identification has allowed, for the above mentioned manipulator, to approach the coefficients with an error of about 5 % which is sufficient, we think, to elaborate the command. The coefficient tabulation function of the control articulary main variables and their interpolations (by a multi-variate technic) allows to obtain a computation time (inertia matrix and gravity forces) of 9 ms with the same microprocessor system (coordination and dynamic command in 17 ms). The control rule we have used is easily deduced [1, 4, 7] from the temporal equation of the movement in the following terms :

$$\Gamma_i = \sum_j A_{ij} \ddot{\theta}_{d_j} + A_{ii} G_i (H_i \dot{\varepsilon}_i + \varepsilon_i) - Q_i$$

where :

Γ_i	is the torque applied to the motor,
A_{ij}	is an inertial decoupling term of the articulations i and j
$\ddot{\theta}_{d_j}$	second derivative of the desired θ_d
$A_{ii}\ G_i$	equivalent gain
$A_{ii}\ G_i\ H_i$	equivalent damping parameters
ε_i	discard in position.

The programmation of the gain and the damping was obtained replacing the classical gain elements by analog multipliers (see the scheme of the command system).

Curves from figure 2 show the evolution of the torques due to the inertia relative to motor 1 to 6.

Curves from figures 4 and 5 show a comparison between desired evolution θ_{d_i} and the real evolution θ_{r_i} obtained with an empiric regulation an with our method for i varying from 1 to 6.

IV - CONCLUSION

In the three ideas developped above, our approach presents the following points of interest :

A - Coordination of the movements. Floating center of action, referential of the effector simply reconfigurable.

B - Coordination in forces. General forces sensor. Structural functions of regulation of the forces.

C - Dynamic control. Identification of the parameters of the comportment equation, tabulation and interpolation.

D - Implementation. Structures and applications an a performent microprocessor unit.

We think that the structures,we have developped, present a realistic and efficient approach for the problem of conceiving and realizing a programmable command in robotics.

ACKNOWLEDGMENTS

We thank M. J.VERTUT, chief of the Department of the Technics of the Protection Equipment and Dosimetry at the French Atomic Energy Commission (CEA) for his advices and for the robot he has lended at our disposal.

REFERENCES

1. M. AL MOUHAMED. "Contribution à la commande automatique en robotique - Application à la commande par microprocesseurs des manipulateurs articulés". Thèse d'Etat, Univ. Paris XI Orsay, Janv. 1982.
2. M. AL MOUHAMED. "Commande rapide pour le manipulateur M.A. 23". Le Nouvel Automatisme, Janvier 1982.
3. M. AL MOUHAMED, A. BASIRE, E. CSAKVARY, B. de COSNAC, B. DUPEYRAT. "Etudes d'heuristiques pour le contrôle-commande de robots". Conférence de l'AFCET sur l'Intelligence Artificielle et la Reconnaissance des Formes. Nancy, 16-19 Sept. 1981.
4. W. KHALIL. "Modélisation et commande par calculateur du manipulateur M.A. 23 - Extension à la conception par ordinateur des manipulateurs". Thèse de Doct. Ing., U.S.T.L., Montpellier, Sept. 1976.
5. P. COIFFET, A. LIEGEOIS, E. DOMBE, P. RIVES. "Manipulateurs flexibles et prise en compte de l'environnement". Comptes-Rendus des Journées d'Automatique du 30 au 31 Janvier 1979. Rapport IRISA n° 117.
6. D.E. WHITNEY. "The mathematics of coordination and purposeful motions". Journal of Cybernetics and Inf. Sciences, T1, n° 1, 1976.
7. R.P. PAUL. "The theory and practice of robot manipulator programming and control". 18th IEEE Conf. on Decision and Control, Fort Lauderdale, Déc. 1979.
8. J.L. NEVINS, D.E. WHITNEY. "The force vector assembler concept". Int. Conf. on Robots and Manip. Syst. Udine, Sept. 1973.

Figure 1

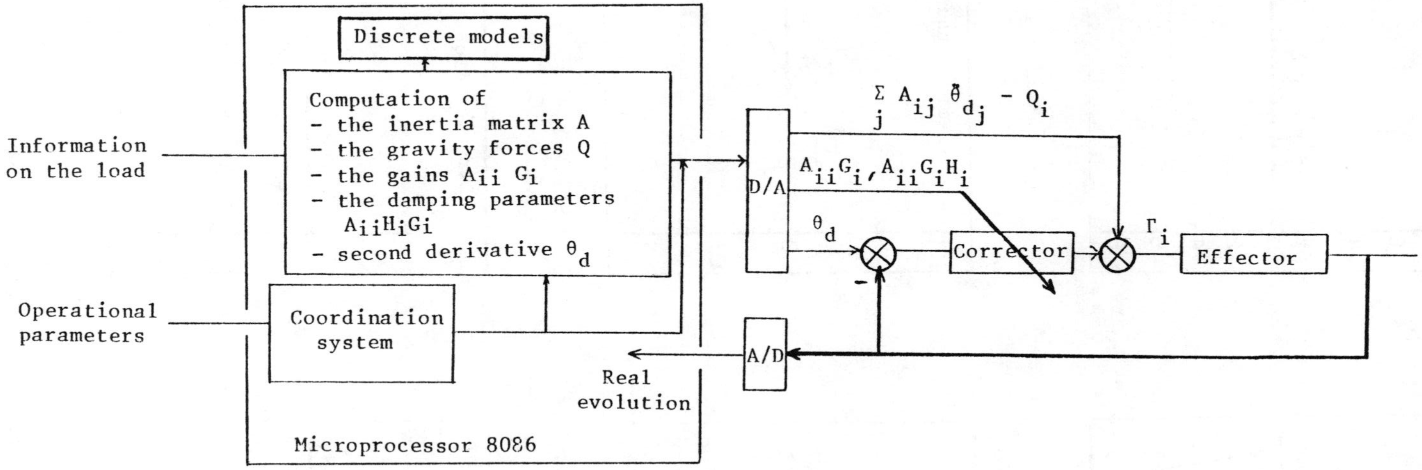

Figure 2 - System control scheme.

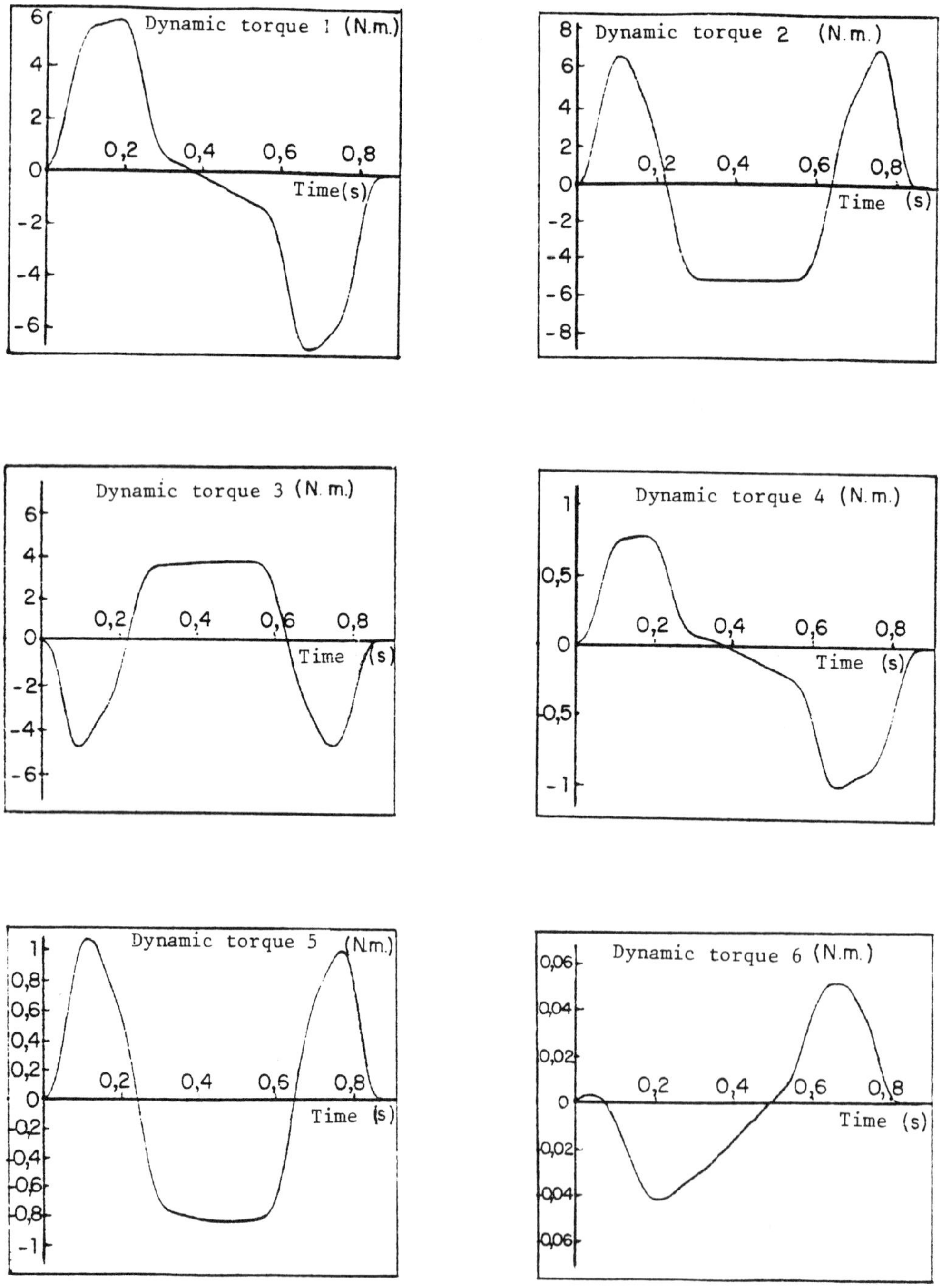

Figure 3 - Dynamic torques evolution during a movement.

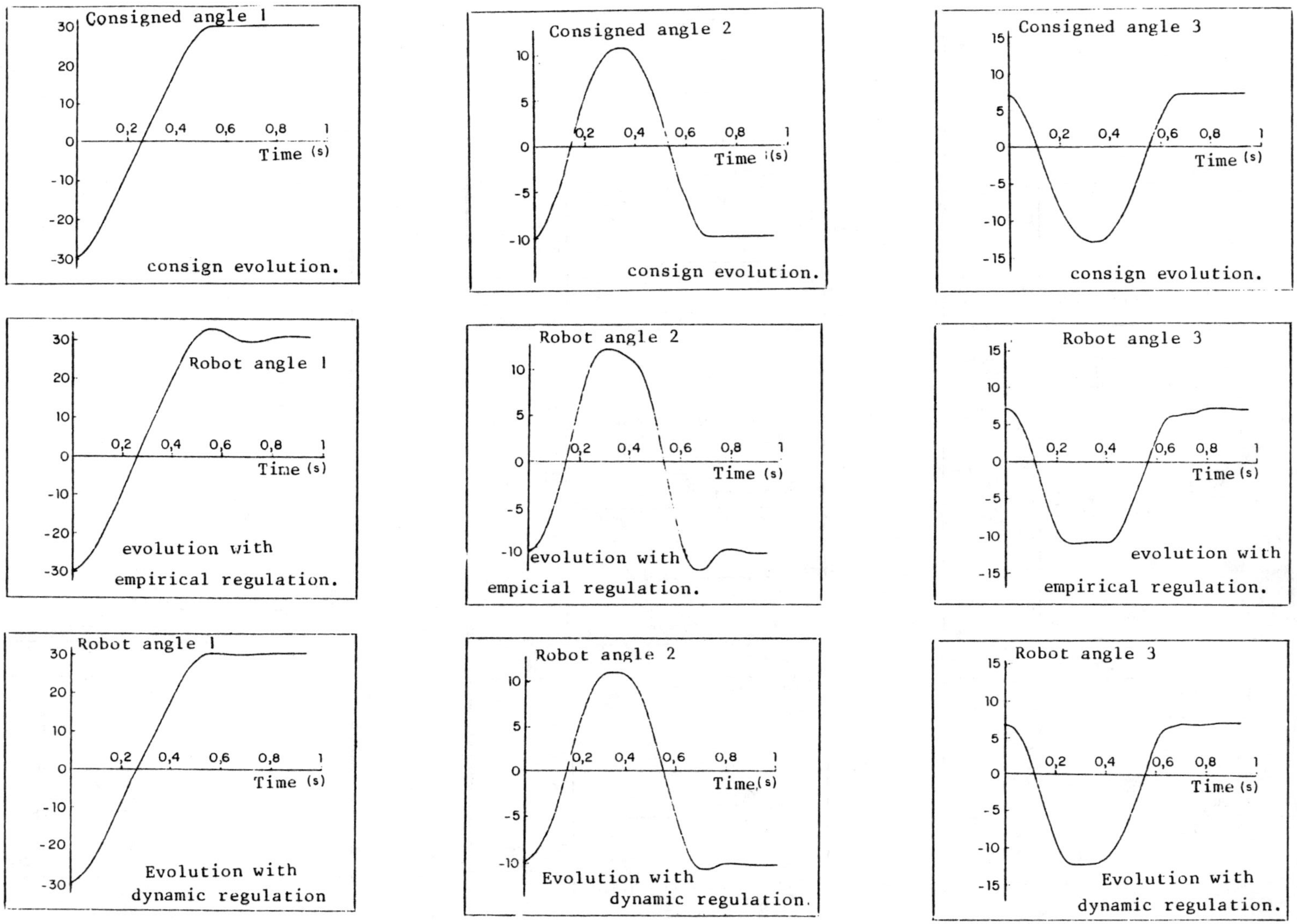

Figure 4 - Comparison between real evolution for the angles 1, 2 and 3.

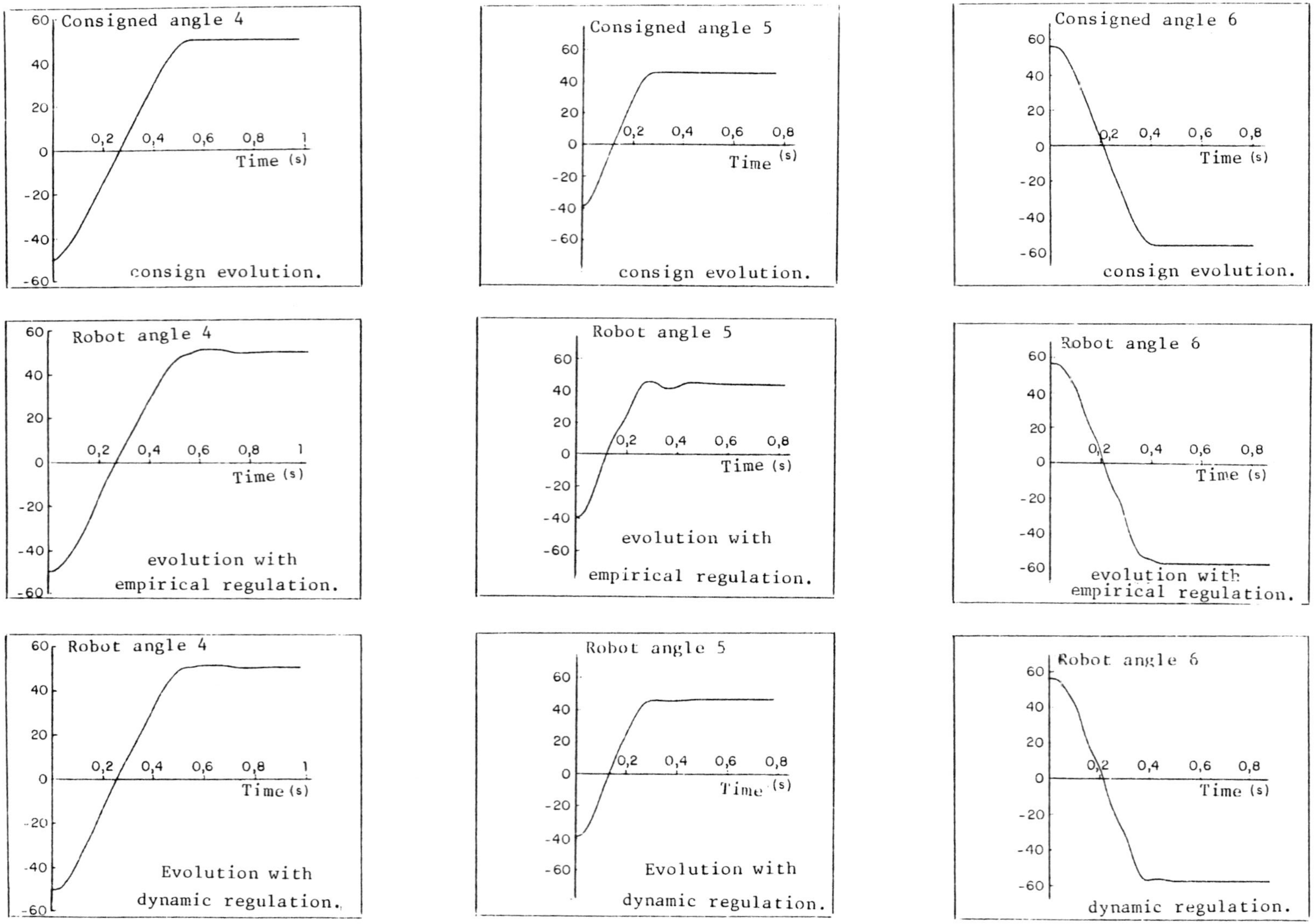

Figure 5 - Comparison between real evolution for the angles 4, 5 and 6.

The Experimental Method of Improving the Dynamics of the Grappling Process in Mechanical Hand of an Industrial Robot

Mariusz Olszewski
Technical University of Warsaw Poland

Marek Ziembicki
Technical University of Warsaw

Abstract

The application of a successive approximation algorithm in determining the shape of a binary signal which operates an electrop neumatic power member of the gripper kinematic system is discussed. Theoretical assumption and experiment results will aid in the evaluation and selection of a control process.

ABSTRACT

The dynamics of the grappling with power chucks process (as often as not with air-operated ones) affected both the quality and the effectivity of the automatization of handling functions in industrial process. An acquaintance of the grappling process dynamics, as well as method of shaping it (at least in part) makes it possible to protect an object grasped against dynamics overloads which may appear when grappling at simultaneous reduction of the working cycle of a robot. The effect is an improvement of the reliability of the handling machine as a whole.
The study deals with the application of succesive aproximation algorithm in determining the shape of a binary signal which operates an electropneumatic power member of the gripper kinematic system. The algorithm, accomplished by a microprocessor system, ensures automatically selection of the electropneumatic valve switching time of the gripper in such a way, that by properly setting of the time of accelerating the moving masses of the gripper (closing the fingers) to the time of braking it ratio, holding of an object in minimum time interval, at simultaneous reduction of dynamics forces, is obtained. A special attention is given to the discussion on theoretical assumptions and results of the experimental works, which are usable in selecting criteria for an evaluation of the control process quality of the gripper through evaluating effects of this process.

I. INTRODUCTION

In the most frequently found solutions of mechanical chucks employed has been an on-off controllable pneumatic motive element ("bang-bang"). The problem of simultaneous minimization of the dynamic overloads and the chuck response time is usually solved by design and consists in the use of a soft lining on fingers of the chuck. Such the approach reduces superficial loads and the time of oscillation of grappling force after the collision of chuck fingers with an object. Maximum values of the grappling force many times exceed its static values, only slight reduction being observed [1].
In the following study the problem of reducing dynamic forces of reaction of the chuck fingers in an object gripped, at simultaneously providing minimization of the grasping time, has been solved by formulating a suitable algothm of the on-off controlling the chuck drive, and not by design. The algorithm is based on the thesis [2,3] that an exponential generator of force which drives the chuck fingers may ensure (see also Appendix):

- oscillation-free character of grappling force function after collision of chuck fingers with an object gripped,
- reduction of dynamic forces appearing at the moment of collision of the chuck fingers and the object gripped,

- minimization of object grasping time, i.e. the high possible mean speed of bringing chuck fingers nearer to surface of an object as well as a rather short time of achieving nominal static value of grappling force after the contact of prehensile fingers with an object.

The above mentioned algorithm is accomplished by means of a digital control system based on a microprocessor. The system, on the basis of "learning" in the course programming the working cycle of the handling machine, generates a binary electric signal the shape of which is determined by duration of the acceleration and the braking phases of chuck fingers motion.

II. NOMENCLATURE

F_g, $F_{g\ stat}$, ΔF_g	N	- grappling force, grappling force rated value, grappling force tolerance,
F_m	N	- driving force,
F_r	N	- force of reaction,
b	kgs^{-1}	- coefficient of viscous friction,
k	Nm^{-1}	- coefficient of rigidity,
t_c	s	- time of collision,
t_{d1}, t_{d2}	s	- delay times of the control and driving system,
t_g	s	- grasping time,
t_o	s	- time of starting the process,
t_1, $t_{1\ me}$, Δt_1	s	- time of finishing of the phase of chuck fingers motion acceleration,
t_2, $t_{2\ me}$, Δt_2	s	- time of finishing of the phase of the motion braking and starting of the chuck fingers closing around an object, time t_2 defined in the procedure self-learning, tolerance of time t_2,
u	V	- binary, electric signal controlling the chuck operation,
v, v_c	ms^{-1}	- speed of motion, speed at the moment of collision,
x	m	- strain in the course of collision.

III. PHILOSOPHY OF CONTROL

We have assummed that a conventional design of the pneumatic drive system employed in the chuck kinematic strukture, as well as the binary character of pneumatic drive signals (Fig.1), remain invariable. These drive signals are provided by a bistable electro-pneumatic transducer, which is controlled by an electric binary signal u(t) generated by a control system. The one state of the control signal makes:

- accelleration of the chuck fingers motion towards object, if no contact between the fingers and the object occurred (closing of fingers),
- clipping of fingers under force $F_{g\ stat}$ on object gripped, if contact between the fingers and the object occurred.

The zero state of the control signal makes:

- braking of the chuck fingers motion, if they move towards the object,
- accelleration of the chuck fingers motion towards their maximum spacing (opening of fingers),
- maintaining the chuck in the state of maximum spacing of the fingers.

FIGURE 1: Scheme of the pneumatic drive system of the chuck

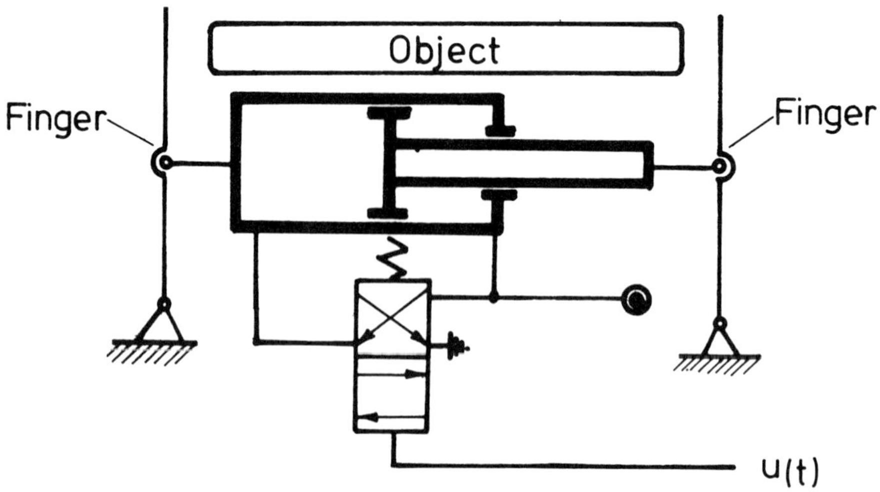

The grappling process has been divided into two phases as follows (Fig.2):

- phase 1: displacement of the fingers in direct neighbourhood of the object gripped (quasi-collision),
- phase 2: dynamic phenomena accomanying collision between chuck fingers chuck fingers and object, and then biting the object by generation of rated force of grappling - $F_{g\ stat}$.

In the first phase above, a control system should ensure displacement of the chuck fingers during minimum time interval. Contact between the object surface and the fingers should occur at precisely determined speed. This will provide a substantial reduction both in dynamic overload and the duration of grappling force oscillation F_g [1] in the second phase of the process, and by the same, the correctness and a short possible time of grasping - t_g will be assured.
To realize tasks in the first phase of the grappling process, motion of the prehensile fingers has been divided again into two phases (Fig.2, see also Appendix) as follows:

- accelleration of fingers (between t_o and t_1): u=1,
- braking of fingers (between t_1 and t_2): u=0.

Due to a rather large, as related to the prehensile fingers displacement time (f.e. for displacements of an order 10-20 mm, duration of the first phase of the grappling process t_2 may be between 40 and 50 ms), time delays in the measuring system, the control system and the pneumatic drive system, it was impossible to employ a feedback control system, i.e. a system in which switching of control signal u(t) can be initiated, for example, by means of a proximity sensor or a chuck fingers displacement sensor.
In view of the above we propose to use for controlling the chuck operation a control system which would be based on the principle of "selfteaching" and would incorporate a microprocessor system. The "self-teaching" procedures are executed once for every new object gripped and every new chuck of the handling machine, during experimental trials of gripping the object or model of it (f.e. in the case of perishable objects which can be damaged or destructed at dynamic overloads). In result, determined are times t_1 and t_2 related to control signal u(t) switching. These are stored and used then to generate the waveform of this signal u(t) after every initiation of the grappling process (that is in relation to time t_o).
During execution of the "self-teaching" procedures the control system

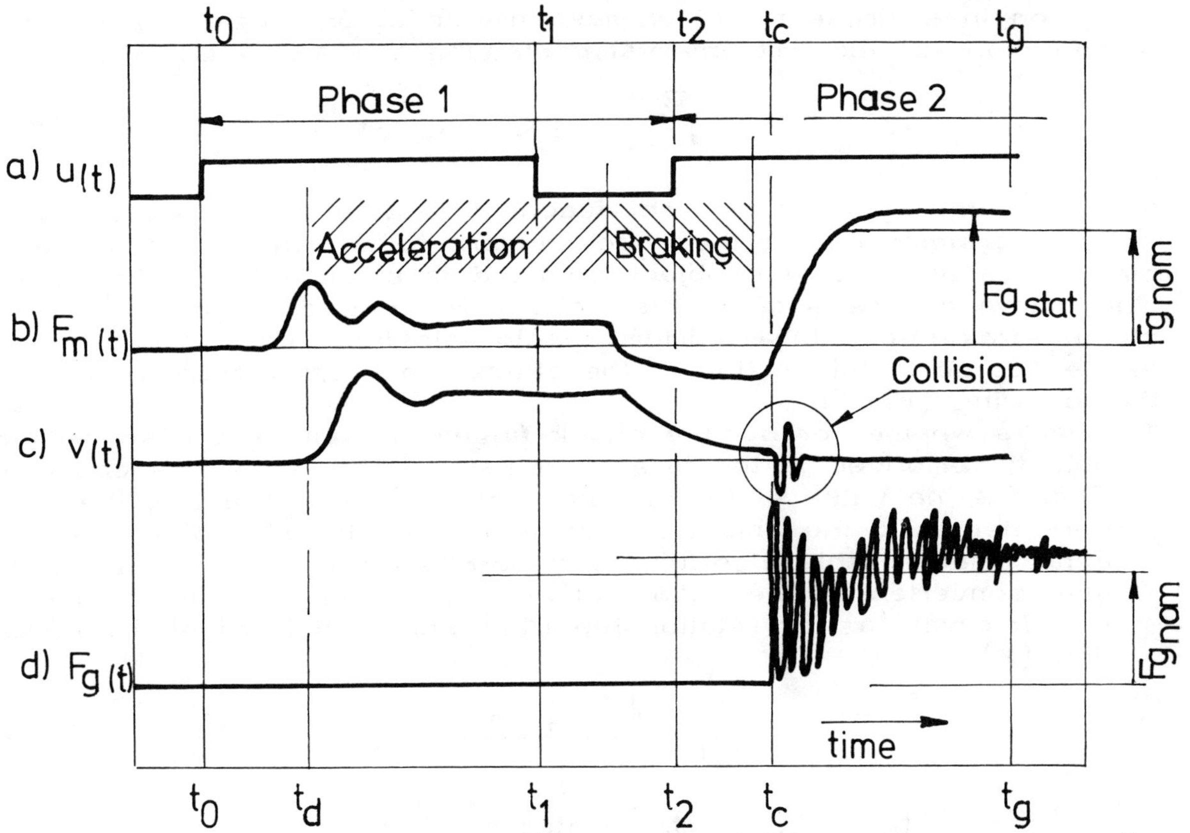

FIGURE 2: Control signal and grasping process variables versus time: a - control signal $u(t)$, b - drive force $F_m(t)$, c - chuck fingers motion velocity $v(t)$, d - grappling force $F_g(t)$

is connected to a sensor detecting collisions between the chuck fingers and the object gripped. Either an acoustic or accelerometer type sensing device with a settable function threshold means and binary output signal can be employes, having parameters suitable for the microprocessor system.

IV. SELF - TEACHING PROCESS

The "self-teaching" process is accomplished within three cycles:

- cycle 1 - assessment of time interval incorporating time t_1 (finishing of the accelleration phase and starting the braking phase of chuck fingers motion towards object),
- cycle 2 - accurate determination of time t_1,
- cycle 3 - accurate determination of time t_2 (finishing of the phase of braking chuck fingers motion towards object and clamping the chuck fingers with force $F_{g\ stat}$, on the object grappled).

Cycle 1

In the course of the first cycle performed are the following actions:

1. Accomplished is the trial of grasping an object without the phase of braking the chuck fingers motion ($t_1 \longrightarrow \infty$),
2. Memorized is the time of occurrence of the collision t_{c1},

3. Evaluated is (by means of a theoretical solution to the problem of motion in a simple model of mass transposition, neglecting any friction forces) the time interval containing time value t_1:

$$1/\sqrt{2} \cdot t_{c1} < t_1 < t_{c1} \qquad (1)$$

Cycle 2

In the second cycle of the "self-teaching" procedure accomplished are trials of bringing nearer the chuck fingers to an object, but no clamping of the finger on the object grappled is attempted (Fig.3, $t_2 \to \infty$). The following phases of fingers motion are consecutively realized: closing-braking-opening-maximum spacing-closing, and so on.

In the course of this action of the chuck the control system performs the following (Fig.4):

1. Checks whether or not the chuck fingers remain in contact with an object; memorized is the time of collision t_{c2} if such has occured,
2. Changes the value of time t_1 after every trial of bringing the fingers nearer to the object according to the algorithm of successive approximation (this algorithm is generally employed in analog-to-digital converters). The initial value of t_1 corresponds to the half time interval (as by relationship (1)) and is determined by equation (2)

$$t_1 = \frac{\sqrt{2}+1}{2\sqrt{2}} \, t_{c1} \qquad (2)$$

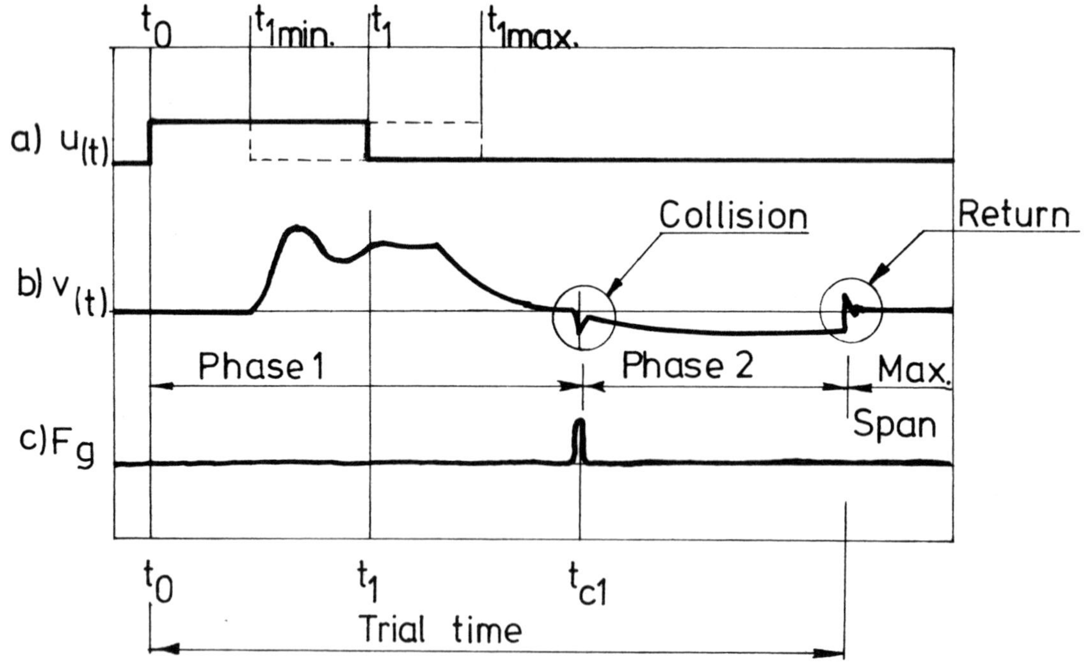

FIGURE 3: Time curves in the second cycle of the "self-teaching" procedure (single complete trial of bringing the chuck fingers nearer to the object):
a - control signal u(t), b - chuck fingers motion velocity v(t), c - grappling force $F_g(t)$

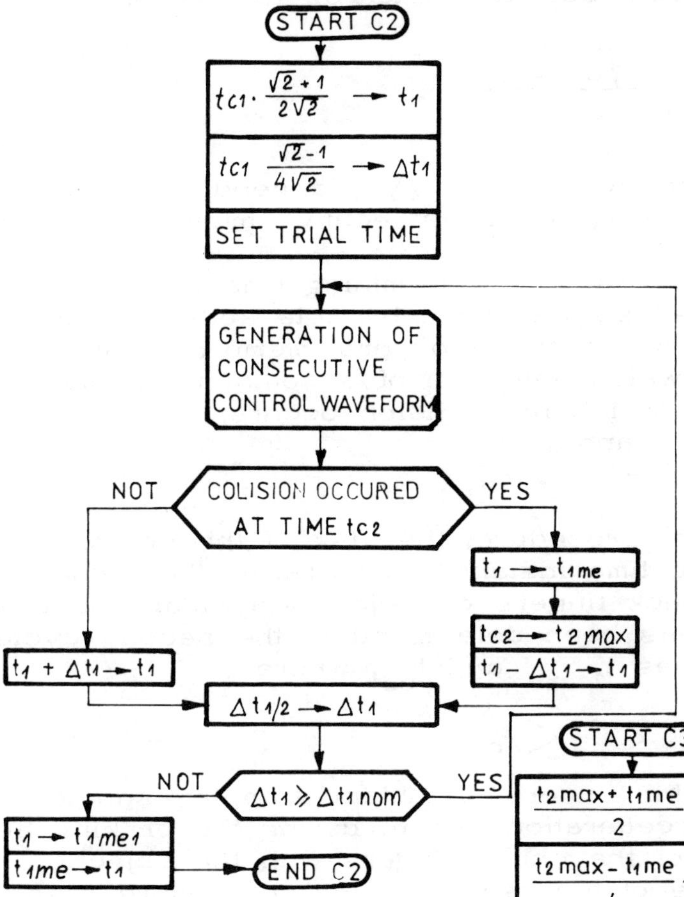

FIGURE 4: Flowchart of the second cycle of the "self-teaching" procedure.
Designations of storage registers:
t_1 - time t_1,
Δt_1 - approximation interval Δt_1,
$t_{1\ me}$ - final value of t_1, at which the detection of a collision occured,
$t_{2\ max}$ - time t_{c2}, maximum value of duration of the first phase of the grappling process.

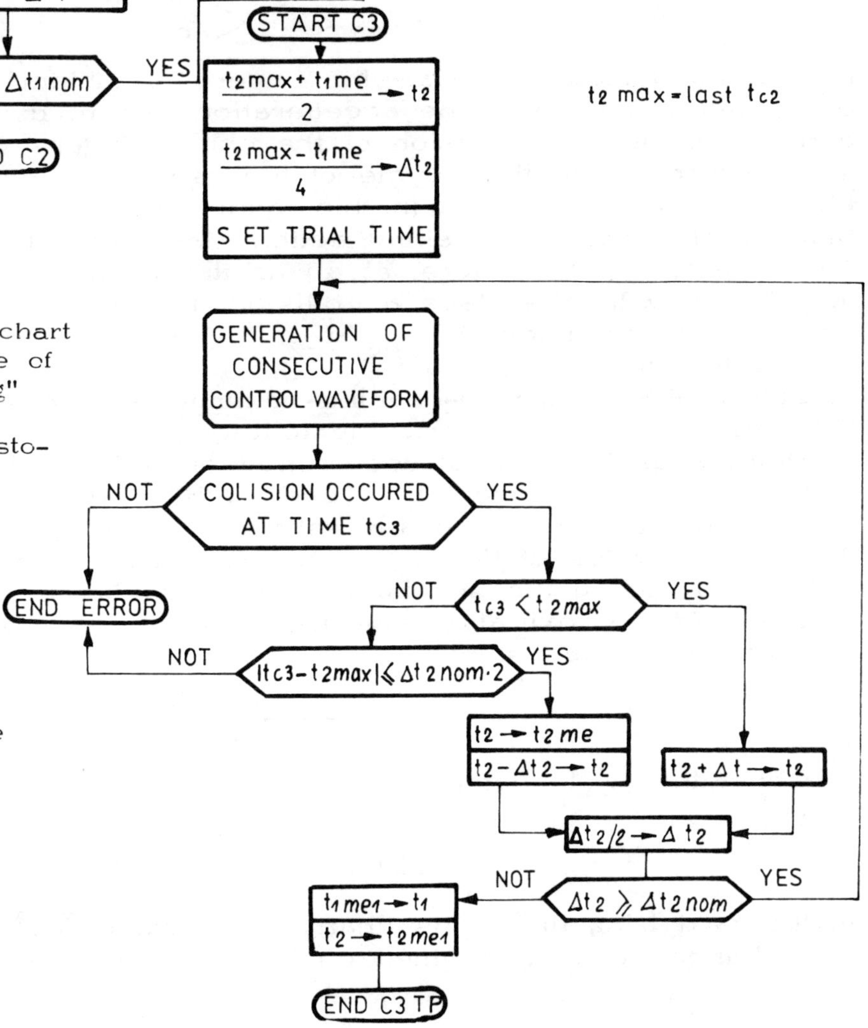

t2 max = last tc2

FIGURE 5: Flowchart of the third cycle of the "self-teaching" procedure.
Designations of storage registers:
t_2 - time t_2,
Δt_2 - approximation interval Δt_2,
$t_{2\ me}$ - final value of t_2,
$t_{1\ me}$ - value of t_1 determined in the second cycle of the procedure (Fig.4)

The initial value of Δt_1 equals 25% time interval, as by relationship (1),

$$t_1 = \frac{\sqrt{2} - 1}{4\sqrt{2}} \qquad (3)$$

In the next trial t_1 is changed by value $\pm \Delta t_1$ depending on whether collision occured (-) or not (+). In consecutive trials the values of time Δt_1 are bisected,

3. In the case when a set accuracy of determining time t_1 ($\Delta t_1 < \Delta t_{1nom}$) is reached so that no detectable collision between the chuck fingers and the object occurs, memorized are:

- last value of t_1 ($t_{1\,me}$) at which detection of a collision occurs,
- last value of t_1 ($t_{1\,me1}$) at which no collision occurs,
- time of this last collision t_{c2} and of concluding the cycle.

Cycle 3

In this cycle of "self-teaching" procedure, the task of the control system consist in determining the time instant of generation the signal which causes clamping of chuck fingers on object grappled. The time interval containing value t_2 has been determined in the second cycle; this being determined by values $t_{1\,me}$ and t_{c2}, where

$$t_{1\,me} < t_2 < t_{c2} \qquad (4)$$

Hence, the initiation of clamping fingers on object grappled should occur after concluding the accelleration phase, during the braking action and prior to a collision of the chuck fingers with the object. The course of the third cycle of the "self-teaching" procedure resembles the second one, but in this cycle trials of grasping an object being performed. The braking action starts at a time instant determined by value $t_{1\,me}$. Therefore, at an arbitrary value of time t_2 within interval (4), will take place a collision, the time of its occurrence being contained within interval $t_{c3} \leqslant t_{c2}$. In the course of consecutive phases of the chuck operation:
closing-braking-clamping-opening-maximum spacing-closing-and so on, the control system performs the following operations (Fig.5):

1. Determines the time of occurrence of collision between the chuck fingers and the object t_{c3} (eventually also any consecutive collision in the case when any rebounds occur),
2. Changes value of time t_2, after every trial of grasping an object, similarly as in the second cycle above, i.e. according to the successive approximation algorithm. The initial value of t_2 is determined by relationship:

$$t_2 = \frac{t_{1\,me} + t_{c2}}{2} \qquad (5)$$

whilst the Δt_2 determines analogically to (3), as by relationship (6):

$$\Delta t_2 = \frac{t_{c2} - t_{1\,me}}{4} \qquad (6)$$

In the next trial, time t_2 is changed by value $\pm \Delta t_2$, sense (sign) of this change being determined on the basis of the following criterion:

- when a collision occured at time $t_{c3} < t_{c2} - \Delta t_{2\ nom}$, it means that braking action was too short and in a consecutive trial t_2 will be increased (+),

- when the collision occured at time $t_{c2} - \Delta t_{2\ nom} < t_{c3} < t_{c2} + \Delta t_{2\ nom}$ and the control system detected the second collision after time t_{c2}, it means that the braking action was too long (fingers have been partially opened) and in a consecutive trial t_2 will be decreased (-).

In consecutive trials values of t_2 are bisected,

3. In the case when a set accuracy of determining t_2 ($\Delta t_2 < \Delta t_{2\ nom}$) is reached, memorized is time $t_{2\ me}$, changed is the value of t_1 from $t_{1\ me}$ to $t_{1\ me1}$ and the "self-teaching" procedure is concluded.

In any different case initiated will be an emergency procedure provided, which the procedure has not been discussed in this study.

FIGURE 6: Tested model of the chuck

V. GRIPPER, CONTROL, ACTUATORS

The control method presented was tested with a real chuck (Fig.6), designed for grappling objects of mass up to 1 kg and size up to 80 mm. In the course of these investigations no soft lining on the surface of prehensile finger has been employed.

The electric binary signal u(t) generated by the control system was converted to the pneumatic drive signal of value 0,63 MPa. The electropneumatic transducer employed in the experiment had the power output equal to ca 650 W (at signal pressure 0,63 MPa) and was connected to the chuck driving element by means of a pneumatic pressure conduct of length between 1 and 2 meters.

As the chuck driving element a pneumatic piston-type cylinder has been employed. The reverse motion of the piston has been effected by reaction of pressure on the actuator piston-rod.

The control function and the "self-teaching" procedures have been accomplshed by a microcomputer system provided with microprocessor MC 6800.

To detect collision between the chuck fingers and the object employed alternatively either acoustic or accellerometric type sensing devices, whereas recording the grappling force curves accomplished is with the use of strain-gauge sensors.

VI. CONCLUSIONS

Presented here method of controlling pneumatic drive mechanical chucks makes it possible to improve the dynamic of grasping process, i.e. to reduce detrimental (for example damage to the surface of object) or inadmissible (for example brittle objects, glass made objects) dynamic overloads, maintaining the same and even sometimes possible to be reduced, time of grasping (ellimination of oscillations of grappling force). The improvement is obtainable by means of providing suitable software for microprocessor control system, and not by improving design of the chuck. Since the chuck is a component part of a handling machine, nowadays controlled by means of microcomputer system, costs connected with the use of the method proposed equals costs of an extra storage and of a detector of collision between the chuck fingers and the object.

Procedures of "self-teaching" require to use as much as 1 kbyte of storage in the microprocessor system employed. The collision detector may be used as a standard sensing device for detection of grasping object, which the sensing device is necessary in overhelming majority of robot aided production processes.

During experimental tests of the chuck (Fig.6), controlled by a system programmed in the cycle of "self-teaching", nearly twentyfold reduction in the intensity of noise which arises at the time of gripping a steel object by means of chuck fingers free of the soft lining. This gives evidence of an abrupt improvement in the dynamic of gripping process. Times $t_{1\,me}$ and $t_{2\,me}$ were determined to the accuracy of $\pm$ 0,2 ms. This was possible due to the use of an electropneumatic transducer which the transducer can transduce electric signals, of minimum duration approximately equal to 20 ms, just with such the accuracy.

In this study we present the essentials of the method. Thus we do not discuss adaptation algorithms for variable environmentals, for example variation of supply pressure, friction in drive cylinder, and so on.

REFERENCES

1. Olszewski,M.,Ziembicki,M.,"A Contribution on the Dynamic Behaviour of Grippers", Proceedings of the 4th International Conference on Industrial Robots Technology / 8th International Symposium on Industrial Robots, Stuttgart, (May 30-June 6, 1978), p.426-443.
2. Shimada,K., "The Estimating Method of the Dynamic Behaviour of Pneumatic Cylinder for Adjusting the Active Condition to be at Optimum", Preprints of the IFAC Symposium Pneumatic and Hydraulic Components and Instruments in Automatic Control, Warsaw, (May 20-23, 1980), p.145-150.
3. Baily,E.M., McIlvanie, W.A., Trash,W.T., Winterrowd,D.B., "Optimizing the Performance of an Electromechanical Print Mechanism", Hewlett Packard Journal, November 1978, p.23.

APPENDIX

We discuss the behaviour of a simplified model of a mechanical chuck in which masses of all movable structural components has been concentrated in the mass m, which the mass is effected by a drive force $F_m(t)$ generated by an ideal force generator (value of the force F_m is independent of motion parameters of mass m).
The collision of chuck fingers with an object (Fig.7) is described by Kelvin-Voigt model in which the force of reaction F_r:

$$F_r = k \cdot x + b \cdot \dot{x} \tag{7}$$

The task consists in determining function $F_m(t)$ which ensures oscillationfree character of the collision process.
For the collision velocity $v_c \neq 0$ and the drive force $F_m(t) = 0$ the complex value of the reaction force may be defined as:

$$F_r(s) = m \cdot v_c \frac{bs + k}{ms^2 + bs + k} \quad \text{or} \tag{8}$$

$$F_r(s) = m \cdot v_c \cdot H(s) \tag{9}$$

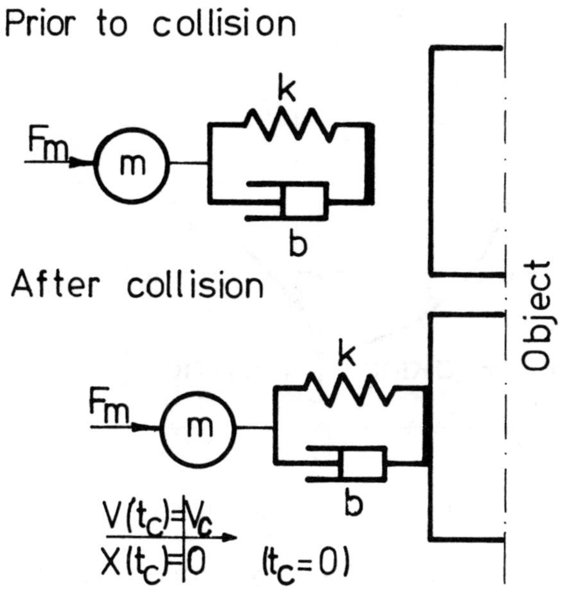

FIGURE 7: Simplified model of chuck and of the object grappling process

where: $H(s)$ - Laplace´s transform of response of the system to unit pulse function at initial conditions ($\dot{x}_{t=0} = v_c$, $x_{t=0} = 0$, Fig. 7)
For a given set function $F_r(s)$ the function $F_m(s)$ wanted may be determined making use of convolution:

$$F_r(t) = F_m(t) * H(t) \Longrightarrow F_m(s) = \frac{F_r(s)}{H(s)} \tag{10}$$

For a given constant valency force function $F_r(t) = F_{max} \cdot \mathbb{1}(t)$ a drive force $F_m(s)$ should be applied to mass m as follows:

$$F_m(s) = F_{max}\left(\frac{m}{b} \cdot 1 + \frac{1}{s} - \frac{km}{b^2} \frac{1}{s + \frac{k}{b}} \right)$$

The need for generating a pulse of infinitely large amplitude (complement $F_{max} \cdot \frac{m}{b} \cdot 1$) may be substituted by non-zero initial condition selected so, that for an assummed model of collision obtained is:
$F_{max} = v_c \cdot b$, i.e.

$F_r(t) = v_c \cdot b \cdot \mathbb{1}(t)$, hence:

$$F_m(s) = v_c \cdot b \left(\frac{1}{s} - \frac{km}{b^2} \frac{1}{s + \frac{k}{b}} \right) \quad \text{and} \tag{12}$$

$$F_m(t) = v_c \cdot b \left(\mathbb{1}(t) - \frac{km}{b^2} \cdot e^{-\frac{k}{b}t} \right) \tag{13}$$

From equation (13) it results that to ensure constant value of grappling force F_{max} (since the time instant at which fingers of the chuck contact the object) it is necessary to have on disposal an exponential generator of drive force having time constant $T = b/k$ determined by parameters of surface participating in a collision.

It is possible to show that for the drive force generator of present time constant $T = \alpha - 1$, the oscillation free collision process is obtained for the drive force:

$$F_r(t) = F_{max} \left[\mathbb{1}(t) - \left(1 + \frac{m}{k}\alpha^2 - \frac{b}{k}\alpha \right) e^{-\alpha t} \right] \tag{14}$$

the respective collision velocity being:

$$v_c = \frac{F_{max}}{k} \alpha \tag{15}$$

For $\alpha = k/b$ the equation (14) will take on the form (13).

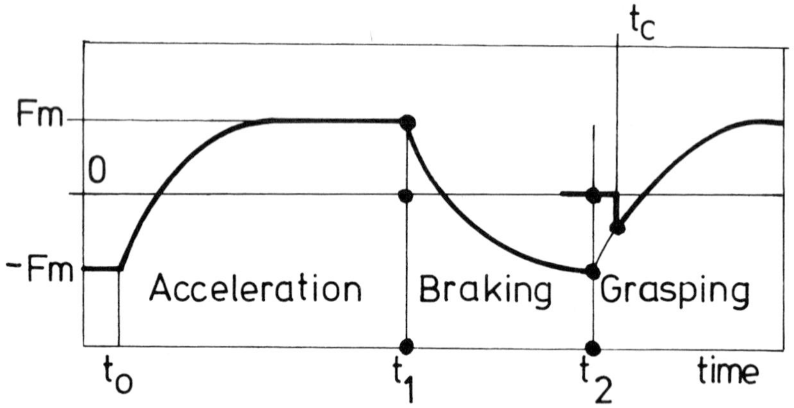

FIGURE 8: Chuck fingers driving force $F_m(t)$ versus time necessary to ensure oscillation free character of the grappling force F_g versus time curve (fragment of curve corresponding to equation (13) emphasized has been by means of the heavy line)

The shape of the charakteristic of the drive force $F_m(t)$ prior to the time collision does not affect the process of shaping the reaction force $F_r(t)$ after the collision provided that a system will attain necessary initial conditions.

Proposition: Having on disposal an exponential generator of the drive force $F_m(t)$ of an arbitrary time constant T (Fig.8) it is feasible to attain a collision free character of the collision process. The proposition was proved by means of the digital simulation method.

Interactive Procedure for Computer-Aided Design of Industrial Robots Mechanisms

V. Potkonjak
Electrical Engineering Faculty
Belgrade, Yugoslavia

M. Vukobratovic
"Mihailo Pupin" Institute
Belgrade, Yugoslavia

D. Hristic
"Mihailo Pupin" Institute

Abstract

Computer-aided design of industrial manipulators are discussed. A computer procedure is derived for the calculation of all relevant dynamical characteristics of manipulator motion based on the input manipulator configuration and input manipulator task. Further, we derive a practical interactive procedure for the choice of manipulator parameters such as: dimensions, actuators, etc. The significance of such a procedure is that it uses the complete dynamic model; it is thus the first time that the manipulator parameters can be chosen on the basis of exact dynamics.

INTRODUCTION

Recently, the computer oriented methods for forming and solving a dynamical model of active mechanisms have been shown to be a very powerful tool in many problems of manipulation robotics. These methods were the basis for the development of algorithms for the complete dynamical analysis of manipulator motion. Such algorithms have made possible a wider application of computers to the design of industrial manipulators. In this paper we present the methodology and some results of such application.

COMPUTER-AIDED ALGORITHM FOR THE DYNAMICAL ANALYSIS OF MANIPULATOR MOTION

Let us consider a manipulator as an open chain consisting of n rigid bodies interconnected by rotational and linear joints (Fig. 1). To define the manipulator position we introduce n internal coordinates $q = [q_1 \ldots q_n]^T$. If a joint S_i is rotational then the corresponding internal coordlinate is the rotation angle and if S_i is a linear joint then the linear displacement in the joint represents the corresponding internal coordinate. Driving forces and torques (drives in the subsequent text) are acting in manipulator joints: forces in the linear and torques in the rotational joints. Now, the dynamics of such a mechanism can be described in terms of a differential equations system in matrix form

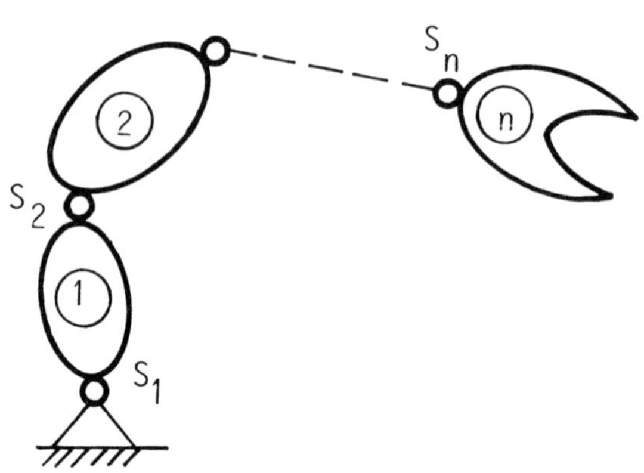

Fig. 1. Manipulator as an open chain

$$W(q) \cdot \ddot{q} - U(q, \dot{q}) = P \qquad (1)$$

where P is an n-vector of the drives, W is an nxn inertial matrix and the nx1 matrix U is due to Coriolis, centrifugal and gravitational effects.

Let us consider the calculation of the drives which are necessary to produce some prescribed motion of manipulator. If q, q, $\ddot{q}$ are known in each time instant (time is considered discretized), then this can be done by using the equation (1). It can be concluded that only $\ddot{q}$ has to be the input value since q, $\dot{q}$ in each time instant can be obtained by integration starting from the previous time instant. Only the initial state $q(t_o)$, $\dot{q}(t_o)$ must be given. A computer procedure has been developed [1, 2, 8, 9] for the computation of the matrices W, U for the known state q, q.

The problem with the procedure described is that manipulation tasks in practice are usually not given in terms of internal coordinates but in terms of some external coordinates X (e.g. the law of manipulator tip motion and gripper orientation). There is usually a relatively simple numerical procedure

$$X = \eta(q) \qquad (2)$$

where X is an n-vector of external coordinates. But the solution of q from such eqation system (2) is extremely complex. Since only $\ddot{q}$ is needed

(q, $\dot{q}$ are computed by integration) we use the Jacobian form

$$\ddot{X} = J(q)\ddot{q} + A(q, \dot{q}) \tag{3}$$

where J is an nxn Jacobian matrix and $A(q, \dot{q})$ is an nx1 adjoint matrix. A computer procedure has been developed [3-6, 8] for the computation of the matrices J, A for the known state q, $\dot{q}$, $\ddot{q}$ can now be obtained:

$$\ddot{q} = J^{-1}(\ddot{X}-A) \tag{4}$$

Now the time-iterative procedure for the dynamical analysis can be described by a block scheme given in Fig. 2.

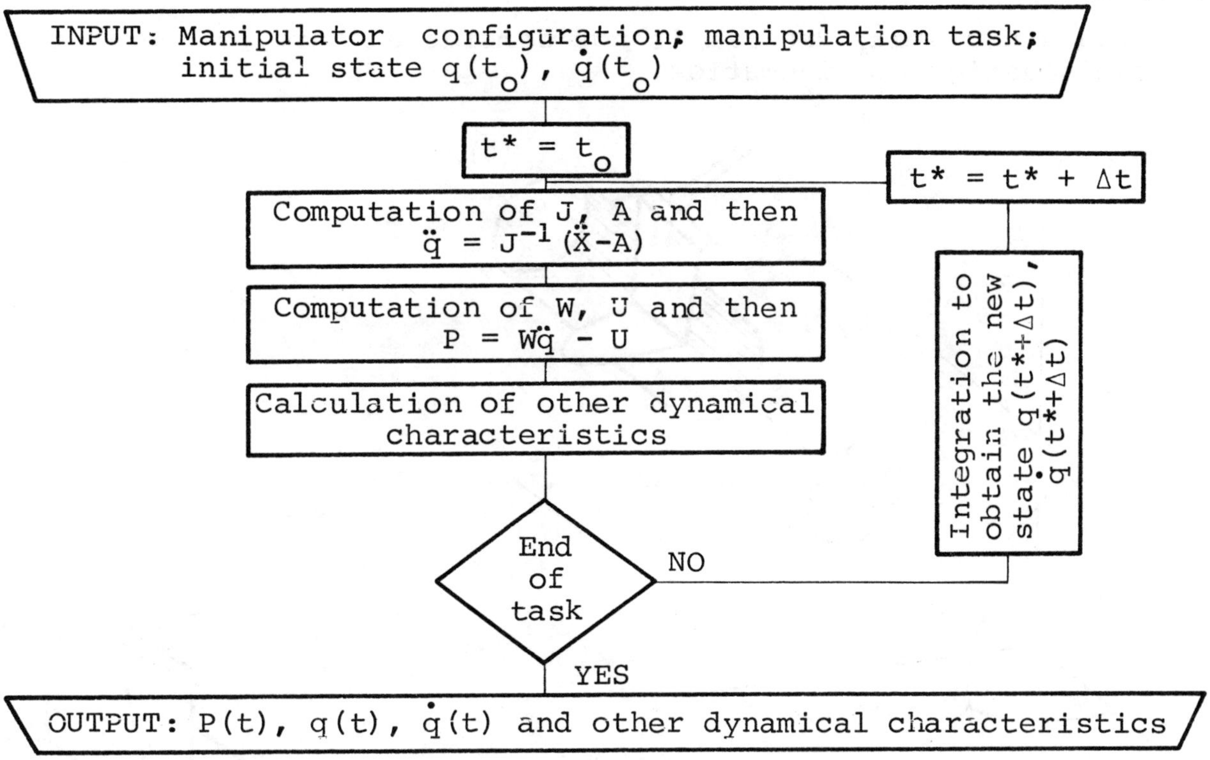

Fig. 2. Block scheme of the dynamical analysis algorithm

<u>External coordinates</u>. Let us descuss the external coordinates defining the manipulation task. We should first define some notions. By *positioning* we mean the motion of some given gripper point (e.g. center of gravity) along a prescribed trajectory according to a prescribed motion law. *Partial orientation* of the gripper means that the given gripper axis coincides with the prescribed direction in space ((b) in Fig. 3A). *Total orientation* means an exactly determined angular position of the body in space. For instance, the total orientation can be defined in terms of partial orientation plus the rotation ψ around the direction (axis) considered, or, in terms of two directions ((b) and (c) in Fig. 3B).

Positioning is usually defined via Cartesian coordinates (x, y, z). A direction can be defined in terms of two angles θ, ρ (Fig. 3) or in some other way.

If we consider a task of positioning along the task of partial orientation (Fig. 3A) then 5 degrees of freedom (d.o.f.) are needed. The external coordinates are $X = [x\ y\ z\ \theta\ \rho]^T$. To solve the positioning task along with the task of total orientation (Fig. 3B) six d.o.f. are needed. The external coordinates are then $X = [x\ y\ z\ \theta\ \rho\ \psi]$. This was not the complete, discusion on kinematics. We give here only some ideas and for the additional discussion we refer to [3], [8].

Such definition of manipulation task (position plus a set of angles defining the orientation) is convenient for users of the algorithm. A user prescribes the characteristic points on the desired trajectory and the orientation in these points. He choses some velocity profile (triangular, trapezoidal or some other) for the motion between these points, and also choses the execution time. The computer performs all other jobs. We consider this easy handling and convenience for users to be very important.

The singularity problem is not considered here, since it requires a large discussion on kinematics.

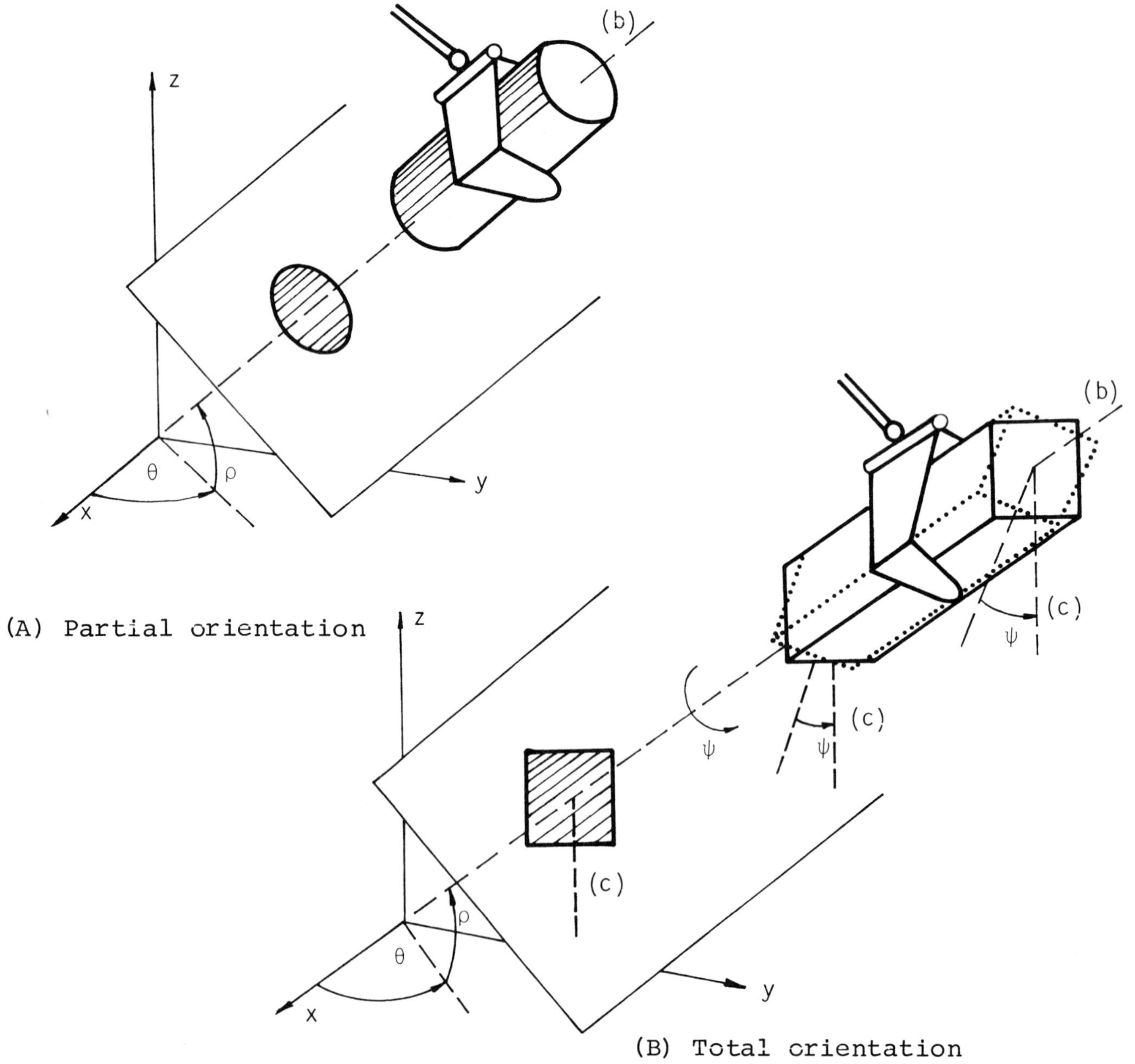

Fig. 3. Insertion of a peg into a hole

It was said that the algorithm computes the driving forces and torques P necessary to produce the prescribed motion (to perform the given task). With the P computed, many other dynamical characteristics can be obtained. Hence, we say that the output of the algorithm is: time history of the drives P(t), trajectory in the state space q(t), $\dot{q}(t)$, and other dynamical characteristics.

Other dynamical characteristics. The algorithm computes the reaction forces and moments in manipulator joints. On the basis of drives and reactions it also computes the stresses (bending and torsion) in manipulator segments.

Power requirements and energy consumption are also computed.

For each actuator (driving motor) the diagram of torque P vs. motor rotation speed (r.p.m. = rotations per minute) is obtained. Such characteristic is very useful when choosing the actuator. It can be compared with the maximal torque characteristics P^{max} vs. r.p.m. which can be taken from the catalog of motor manufacturer. This maximal characteristic represent the maximal motor torque depending on motor rotation speed (e.g. a permanent magnet D.C. motor characteristic has an almost rhombic form). By comparing the necessary characteristic obtained by the algorithm described with the one from the catalog one can decide whether the chosen motor suits its application. It should be pointed out that the reduction ratio has a strong influence on this characteristic.

A special block is developed [7, 8] which calculates the elastic deformations of manipulator segments. These deformations result in certain position and orientation errors.

Some other dynamical characteristics can also be computed, including some nominal control variables.

<u>Constraints and tests of dynamical characteristics</u>. The first test deals with actuators. The maximal characteristic P^{max} vs. r.p.m. defines the permissible domain. The diagram P vs. r.p.m. should be wholly within this domain. Fig. 4. illustrates this test (for a permanent magnet D.C.

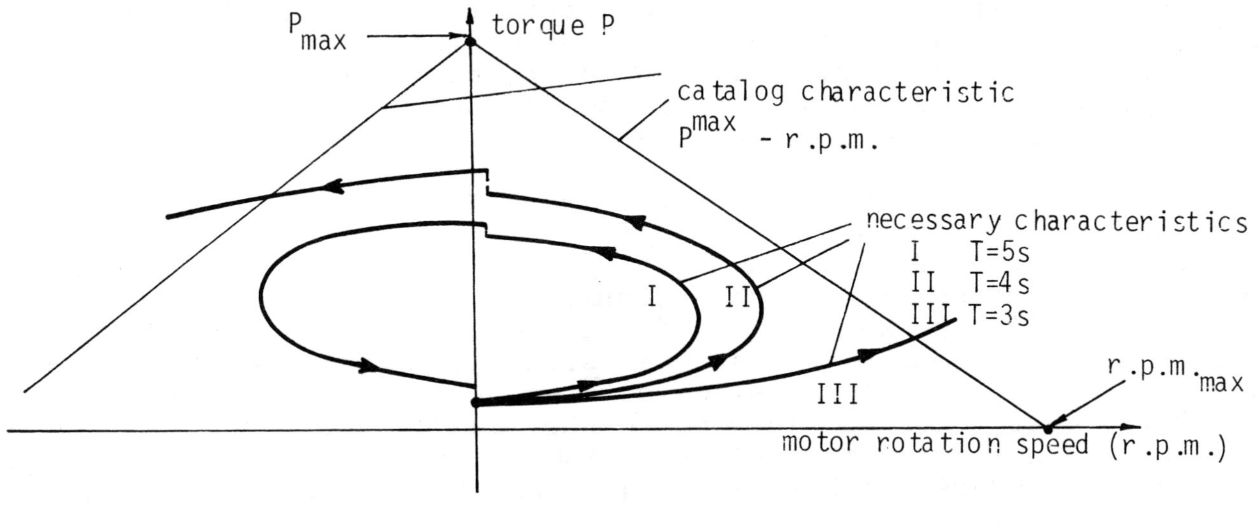

Fig. 4. Test of D.C. motor

motor). For the execution time T=5s the diagram is wholly within the permissible domain. It means that the chosen motor can produce manipulator work at that speed. For T=4s the diagram extends beyond the domain, i.e., the constraint is violated. For T=3s the constraint is violated considerably earlier. The violation can be corrected by choosing a new motor or, sometimes, by choosing a better value of reduction ratio. The motors can also be tested from the standpoint of power requirements. For practical reasons we did not consider the possibility of allowing intermittent motor overload.

The next constraint is the request that the stresses in manipulator segments be smaller than some given value.

There is also a test of elastic deformations. The position error and the orientation error due to elastic deformations have to be smaller than some given value. Some other constraints can also be introduced.

COMPUTER-AIDED DESIGN BY MEANS OF INTERACTIVE WORK WITH THE DYNAMICAL ANALYSIS ALGORITHM

The dynamical analysis algorithm offers the possibility of calculating all relevant dynamical characteristics in advance, i.e., before the manipulator is built. Then, corrections can be made in order to obtain the device with the desired characteristics. Thus, the algorithm described represents a very useful tool for a designer when designing a new manipulator device or correcting the characteristics of an existing one. The algorithm is convenient because it operates for arbitrary manipulator configuration and arbitrary manipulation task, i.e., these are input values. The procedure of systematic choice of manipulator parameters is described in [4, 5, 6, 8]. Here, we propose a practical design procedure based on using the dynamical analysis algorithm This procedure is called design game. For this procedure, a special comunication is programmed.

Let us discuss the tests included in the algorithm. There is a test for each constraint imposed. If a constraint is violated, there exist several possibilities for corrections. If the actuator constraint is violated, we can correct it by: 1) choosing a new actuator; 2) changing the reduction ratio; 3) reducing the mass of working object or some masses of the manipulator itself; 4) increasing the task execution time (slower manipulator work). If the constraint of elastic deformation is violated, it can be corrected by: a) increasing the cross-section dimensions of manipulator segments; 2) reducing the mass of working object or some concentrated masses of the device (motors, reducers etc.). Thus, for each test there are several options for corrections.

Let us describe the design procedure. The algorithm starts with some chosen manipulator parameters. If a test is negative (the corresponding constraint violated), then the computer prints the options for corrections. We choose one of the options, then we type the new values of the corresponding parameters and start the algorithm again. This is repeated until all constraints are satisfied (all tests positive). Then we check whether the characteristics obtained satisfy all our wishes. If it is not the case (e.g. we may wish further reducing of energy consumption), we introduce some new corrections of the relevant parameters and repeat the whole procedure. Such a design game may be represented by the scheme given in Fig. 5.

Such computer-aided design is illustrated by a detailed example

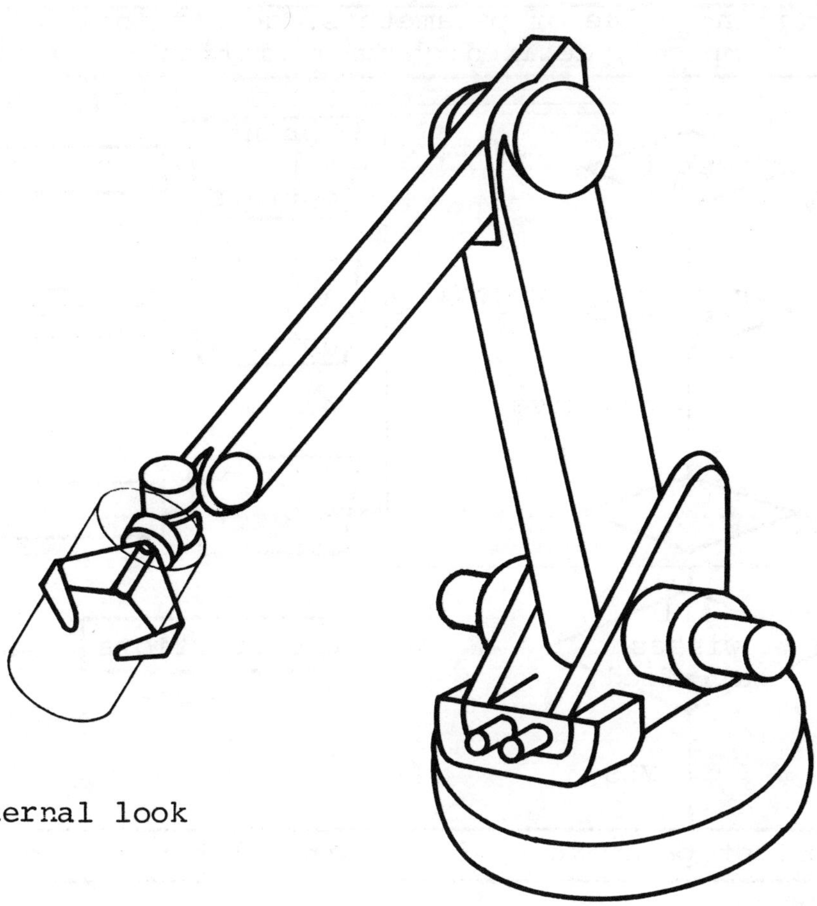

(a) External look

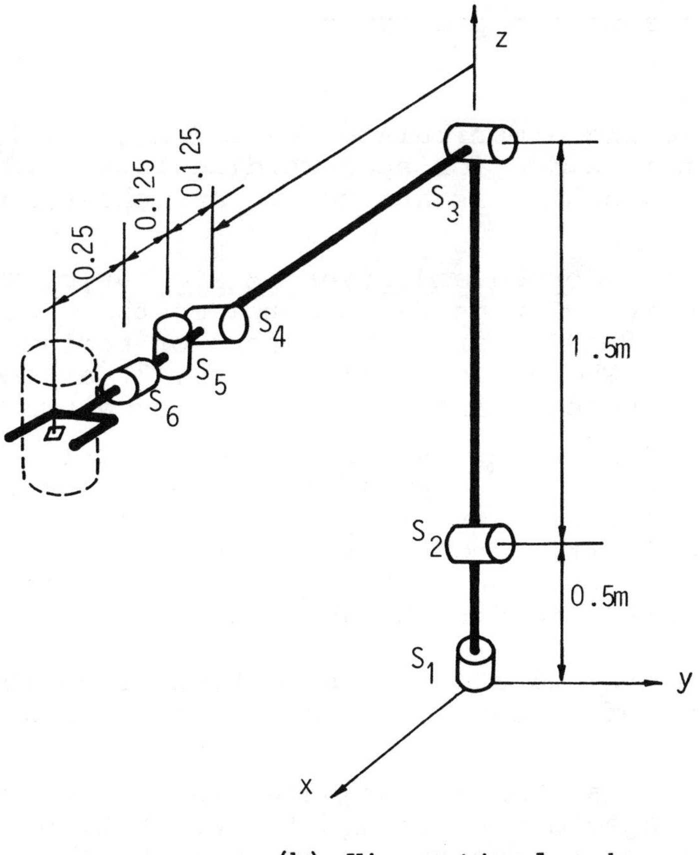

Manipulator data				
Segments	1	4	5	6*
m_i [kg]		4	4	35
I_{xi} [kgm^2]		0.02	0.02	0.17
I_{yi} [kgm^2]		0.02	0.02	0.17
I_{zi} [kgm^2]	3.2	0.02	0.02	0.12

* Working object included

(c)

(b) Kinematical scheme-initial position

Fig. 6. Arthropoid manipulator VE-2

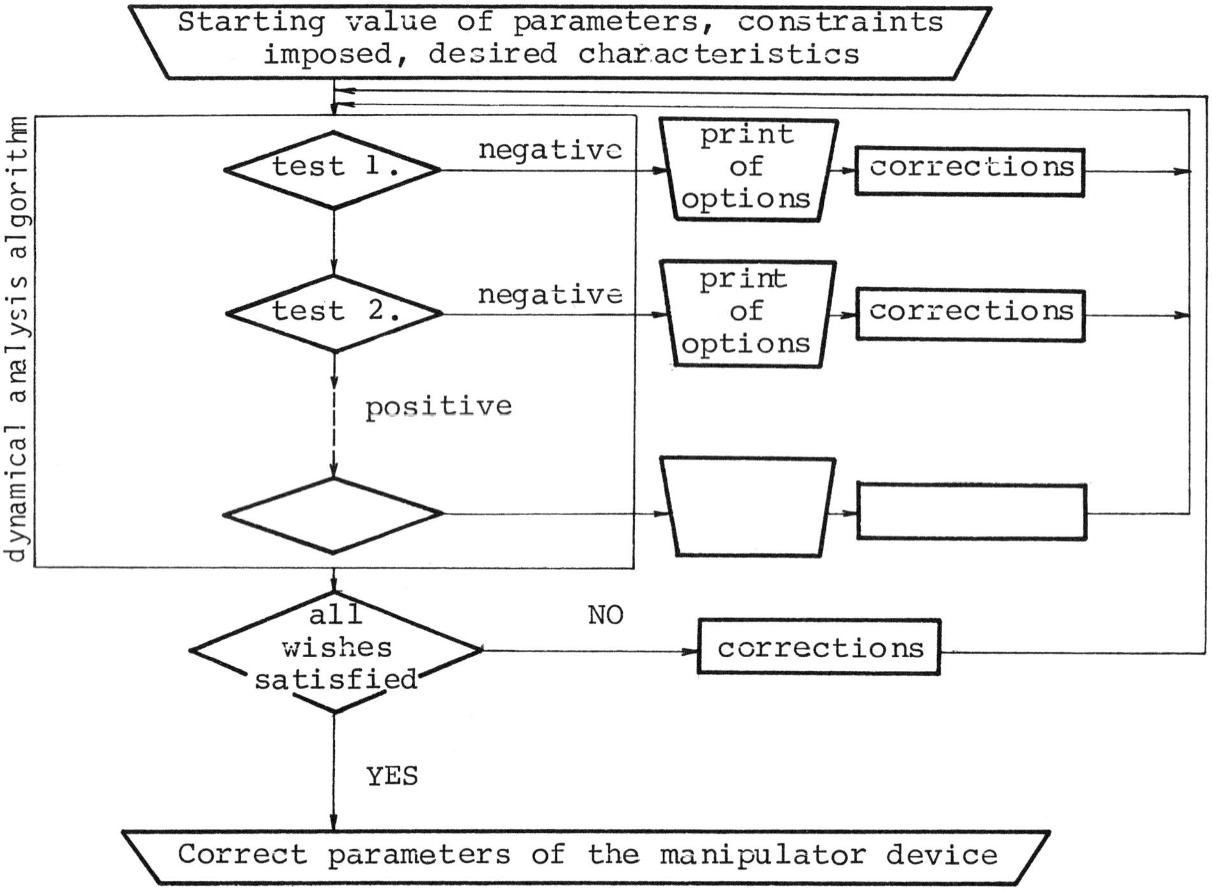

Fig. 5. Scheme of a design procedure

EXAMPLE

We consider a problem of designing a manipulator VE-2 (Fig. 6a) having six degrees of freedom. It is intended for spot-welding tasks and some assembly tasks. Its kinematical scheme is arthropoid and shown in Fig. 6b.

Some manipulator parameters are adopted and given in Fig. 6b,c. The other parameters have to be determined. Let us first consider the question of segments geometry. Segments 2 and 3 are in the form of rectangular tubes made of light alloy AℓMg3. The cross sections (Fig. 7) are defined by H_{x2}, H_{y2} and H_{x3}, H_{y3}, so these parameters have to be determined.

The actuators which drive the joints S_2, S_3 are placed in manipulator base. The torques are transported to the corresponding joints where there are Harmonic Drive reducers. The reducers in joints S_2, S_3 have the masses of 7kg and the mechanical efficiencies $\eta=0.8$. It is necessary to choose the actuators which will drive the joints S_2, S_3. The manipulator is equipped with spring compensators for the joint S_2.

Thus, the parameters to be determined are: cross section dimensions of segments 2 and 3, and parameters of actuators S_2 and S_3, and, finally, the reduction ratio.

The manipulator is tested on the assembly task shown in Fig. 8. The object should be moved along the trajectory $A_0A_1A_2A_3$. The velocity profile on each straight-line part of the trajectory is triangular. The changes in orientation are also shown in Fig. 8. The execution time is

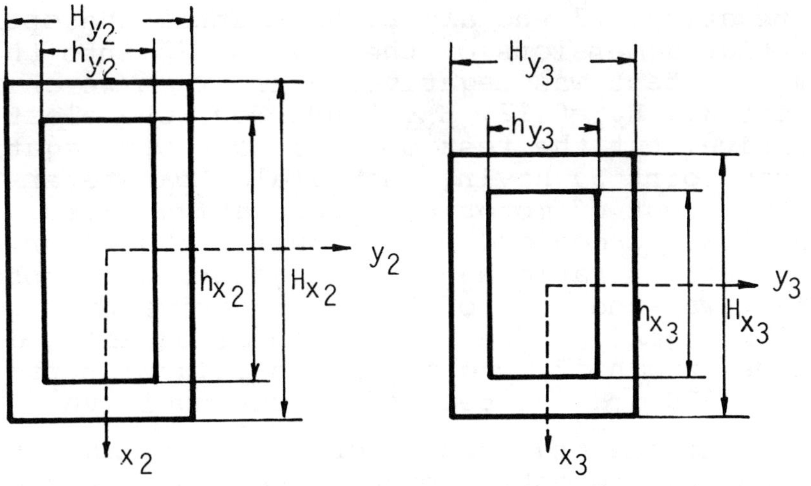

$$\frac{h_{y2}}{H_{y2}} = 0.9 \qquad \frac{h_{x2}}{H_{x2}} = 0.85 \qquad \frac{h_{y3}}{H_{y3}} = 0.9 \qquad \frac{h_{x3}}{H_{x3}} = 0.85$$

Fig. 7. The cross sections of segments

$T=T_{A_oA_1}+T_{A_1A_2}+T_{A_2A_3}$, $T_{A_oA_1}=2s$, $T_{A_1A_2}=1s$, $T_{A_2A_3}=1s$.

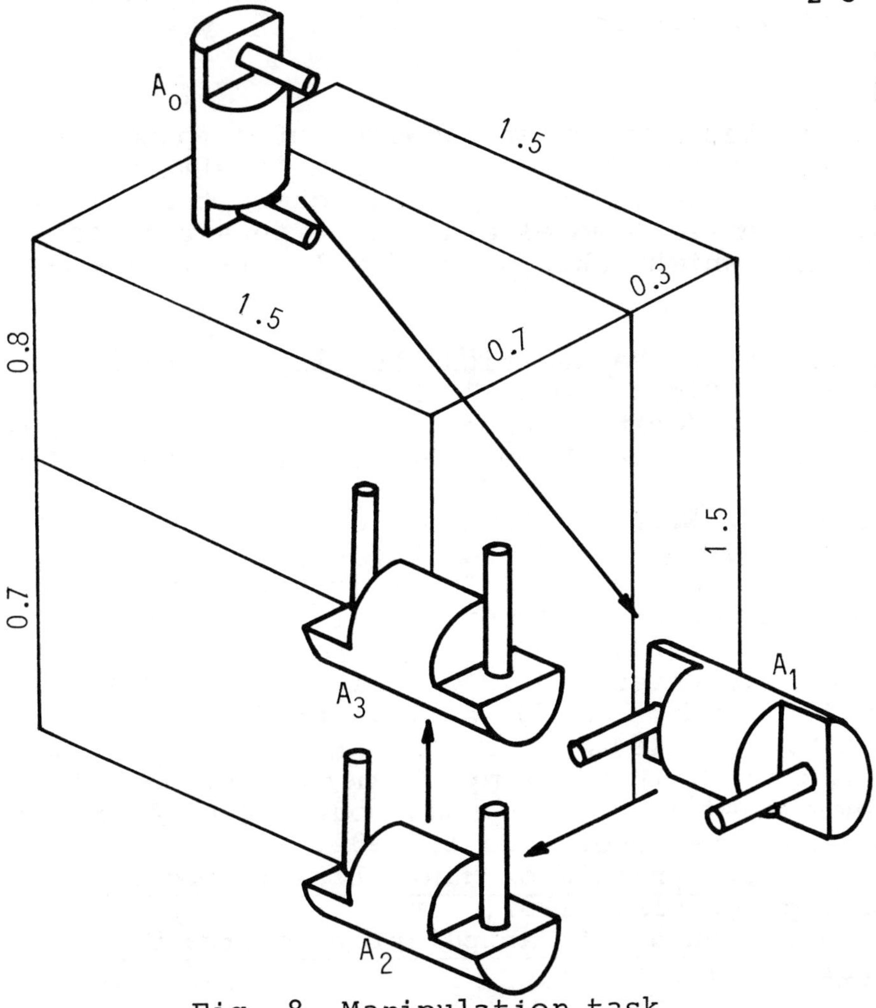

Fig. 8. Manipulation task

The manipulator starting position is given in Fig. 6b.

The constraint is that the position error due to elastic deformations be smaller than 0.002m.

The design (i.e. determination of parameters) is carried out by using the design game procedure. The procedure started with the initial cross section dimensions: $H_{x2}=0.25m$, $H_{y2}=0.13m$, $H_{x3}=0.16m$, $H_{y3}=0.08m$. We first chose, the 2kW D.C. motors as the actuators for the joints S_2, S_3. Catalog parameters (maximal torque P_{max} and maximal rotation speed $r.p.m_{max}$) were: $P_{2max}=10Nm$, $r.p.m._{2max}=2500$, $P_{3max}=10.Nm$, $r.p.m._{3max}=2500$. The reduction ratios were $N_2=150$ and $N_3=150$.

The procedure started with these initial values. The elastic deformation test was negative and the algorithm printed the options. We changed the cross section dimensions of the segment 2. into the values: H_{x2}= 0.26m, H_{y2}=0.14m. The test was negative again. Then we changed the dimensions of segment 3.: H_{x3}=0.17, H_{y3}=0.09. Now, the elastic deformations test was positive, but the test of actuators was negative. We chose a new motor for the joint S_2 having the catalog parameters: P_{2max}=12Nm, $r.p.m._{2max}$=3000. The test of motor S_2 was negative again, but since it was near to being O.K. we decided not to change the motor again but to change only the reduction ratio N_2 in that joint; we adopted the new value N_2=200. In that way the test of S_2 was positive but the test of actuator S_3 was still negative. The results obtained allowed us to conclude that it would be enough if we changed the reduction ratio N_3. We chose the new value N_3=200 and the test of S_3 was positive.

The game was finished and thus we obtained the manipulator parameters satisfying the requirements imposed. The cross section dimensions of segments are: H_{x2}=0.26m, H_{y2}=0.14, H_{x3}=0.17, H_{y3}=0.09. The actuator which drives the joint S_2 has the catalog parameters: maximal torque P_{2max}=12Nm and maximal rotation speed $r.p.m_{2max}$=3000. The reduction ratio in the joint S_2 is N_2=300. For the joint S_3 the motor parameters and the reduction ratio are: P_{3max}=10, $r.p.m._{3max}$=2500, N_3=200.

This simplified example illustrates the proposed procedure of computer-aided design. The results obtained by such game procedure are not considered as completely optimal, since no optimality criterion was used. To obtain the true optimal values, further corrections are performed according to the chosen optimality criterion [4, 5, 6, 8].

CONCLUSION

The procedure proposed enables a wider application of computers to the design of industrial manipulators. Such an algorithm is simple to use and is thus a very useful tool for designers. Its main characteristic is that it is based on the exact dynamical model and thus offers the possibility of a more advanced design which takes care of all dynamical effects.

REFERENCES

[1] Vukobratović M., Stepanenko Yu., "Mathematical Models of General Anthropomorphic Systems", Math. Biosc. Vol. 17, 1973.

[2] Vukobratović M., Potkonjak V., "Contribution to Computer Construction of Active Chain Models via Lagrangian Form", Trans. of the ASME, Journal of Applied Mech., Vol. 46, No 1, 1979.

[3] Vukobratović M., Potkonjak V., "Transformation Blocks in the Dynamic Simulation of Manipulators", Mech. and Mach. Th., No 3, 1982.

[4] Vukobratović M., Potkonjak V., Hristić D., "Dynamic Method for the Evaluation and Choice of Manipulators", Proc. 9th Intern. Symp. on Industrial Robots, Washington, 1979.

[5] Vukobratović M., Potkonjak V., Hristić D., "Contrib. to the C.A.D. of Manipulators", Proc. of 11th Intern. Symp. on Industr. Robots, Tokyo '81.

[6] Vukobratović M. Potkonjak V., Hristić D., "C.AD of Manipulators via their Dynam. Properties", Proc. of IV IFToMM Symp. "Romansy", Warsaw, 1981.

[7] Potkonjak V., Vukobratović M., "C.A.D. for Dynam. Modeling of Flexible Manipulators", Mech. and Machine Theory, No 1, 1982.

[8] Vukobratović M., Potkonjak V., Dynamics of Manipulation Robots: Theory and Application, Springer-Verlag, 1982.

[9] Vukobratović M., Stokić D., Control of Manipulation Robots: Theory and Application, Springer-Verlag, 1982.

CHAPTER 17

ADVANCEMENTS IN NON-BINARY VISION

A Hierarchical Vision System for Object Identification and Localization

A. Riad
Laboratoire d'Automatique
Toulouse, France

M. Briot
Laboratoire d'Automatique

Abstract

The ability of a robot to recognize and position objects is necessary for many industrial tasks. An experimental vision system which deals with this problem is presented. We especially concentrate on the case of partially observed objects. Local descriptions are needed, together with syntactic methods. The image data from a digitized picture is reduced to its outlines. A representation in terms of syntactic primitives and relationships among them is sought and a multilevel model is built. The identification phase consists in searching a pattern prototype so that it matches the primitive data extracted. The geometric information is used to guide the analysis and to simplify it.

I. *INTRODUCTION*

Automatic recognition of industrial parts is one of the important applications of pattern recognition. This paper is concerned with identification and sorting of small mechanical parts within the framework of an overall project of an automated assembly line.

Industrial vision systems that are currently available, like /7,10/, can recognize and locate isolated parts only against a constrasting background. The hypothesis of the separability of parts is generally accepted, yet there are many tasks which do not verify it. We consider here the problem of partially observed planar objects, as also the case of overlapping ones.

In pattern recognition, shape has been found to be the most discriminant factor of an object. In preprocessing, the binary image in study is reduced to its contour-lines. Since parts overlap, their global properties are not always available. The local properties of their contour-lines are then searched and introduced in an interpretation-decision procedure.

Object recognition involves matching an object model to a same description to determine the object's identity and its location in space.

The metric nature of the contours makes possible the introduction of a distance notion. Most previous studies concentrated on this approach /1-4,9/. The information gained from the contours is directly used ; the recognition amounts to a simple correlation-like comparison of the local descriptor used, such as the curvature function, with a reference one calculated during a previous learning phase. The advantages of this method, which is attractive because of the ease of its implementation, are very quickly limited when the number of parts considered increases and above all when internal details and structural relations between object components are important /5,6/. Indeed, when great defects appear, (false primitive insertion, primitive destruction or deformation,...), these techniques usually fall down.

Objects are viewed as complexes of structural elements (primitives) and relationships among them. These primitives and relationships may be both monodimensional or bidimensional. Hence, corresponding models have to reflect this fundamental nature of the objects.

Formally, such relational models can be represented as a graph whose nodes are primitives and whose arcs are relations. The recognition can then be formulated as a graph-matching problem. In principle, a straightforward search can be used to find a mapping of nodes representing object primitives onto nodes representing scene features, such that both the properties and relationships of the corresponding nodes are compatible. However, since scene descriptions may include partially occluded parts, it is necessary to match parts of the scene graph with parts of object graphs. Unfortunatly, such subgraph matching is known to be potentially combinatorially explosive.

When the degree of match of attributes and relations can be expressed in boolean terms, then much of the overhead of the tree search can be reduced by transforming the subgraph isomorphism problem into one of finding maximal cliques of a new graph, called the correspondance graph /12,16/.

Recently, a method was developed in /11/, called local-feature-focus (L.F.F.). The basic principle is to find one feature in an image, referred to as the focus feature, and use it to predict a few nearby features to look for. A maximal clique technique is then used. Since the list of possible features has been reduced, the correspondance graph becomes relatively small.

Efficient algorithms exist for finding maximal cliques /16/. They avoid much redundant effort, but the problem is still inherently exponential.

The probably most effective way to defeat the combinatorics of search is to decompose object models hierarchically into levels /6/. A level is used during the identification stage only when all preceding ones have failed or have given insufficient results (eg. for confirmation). During identification phase, an hypothesis-verify-like method uses the geometric information about the primitives already found to guide the procedure and so to simplify it.

The following sections will present the preprocessing phase, the learning and recognition process and some of methods used for primitive extraction and identification.

II. IMAGE PROCESSING

II.1. Signal processing

The sensor which is used in this study is a television camera (National New-vicon WV 1350) with normal scanning mode functionning.

The choice of the signal processing technique was made with the aim of having a reliable low-cost system. Thus a two-level quantization of the continuous video signal was implemented which yields a binary image of the processed objects.

II.2. Contour extraction

An object in a picture may be represented using either the surfaces forming it or using the borders of such surfaces.

The next methodological step consists of processing the binary image in order to reduce it to its contour points. This operation is in fact achieved simultaneously with the image acquisition step, the successive image lines being processed in a recursive way. This only takes a few complete lines to be stored /17/. The used method preserves the connexity property of contours. Moreover object segmentation, inclusion relations and a test for hole contouring are direct results of the method.

III. CONTOUR SEGMENTATION

If there is no noise, a simple discretization of the contour into north-south and east-west or better following Freeman's directions (Fig. 1) would allow the immediate recognition of a part. Unfortunately, any noise together with sensitivity to translations and rotations make this method inapplicable, except if these parameters are either filtered or very much more specific to an object. A processing is thus necessary so as to overcome these disadvantages and obtain as faithful a representation as possible of the object and its local properties.

In polyhedral objects, contours are always straight and hence are easily represented using their and points and slopes /14/. With curved objects representation of contours becomes a nontrivial problem. It is usually desirable to represent the contour lines after breaking them into meaningful pieces, say straight lines or circular arcs.

This has two clear advantages : 1) representation becomes easy and compact ; and 2) description is simplified, allowing more efficient matching at later stages.

There are many approaches to curve approximation /8,9,15/. We consider here two curve approximation schemes based both on the curvature function search :

(i) The first method is based on the observation that if a chain code representation of a curve is modified so that its graph against arc length is continuous, then circular arcs of the image result in straight graph lines of slope proportional to the curvature of the arc /2/.

The used code is the Freeman one. When plotting the graph of the code versus arc length, care is taken to rectify distortion due to the rectilinear nature of the digitization grid. The effect of the choice of the starting point on the line fitting process is eliminated. The resulting graph is smoothed by averaging each code value over a nine point window.

(ii) In this second method, the curvature function, R(S), is obtained by derivation of the tangent angles at the contour points (Figure 3).

Because of the double derivation by which it gets there, this function is noise sensitive ; it is true that the considerable advantage of the rapid acquisition of contours is accompanied by a quantification noise. This is drastically reduced by the application of an iterative curvilinear filtering (Fig. 2). In practice, the problem reverts to one of finding a good estimate of the curvilinear abscissa s /9/. The procedure can be described as follows :

if $v = \binom{x}{y}$: contour-point $\hat{v}^{(o)}$: observed variable

$\hat{v}^{(n)}$: same variable after n iterations

$$\hat{x}^{(n)}(\hat{S}^{(n-1)}) = R(\hat{S}^{(n-1)}) * \hat{x}^{(o)}(\hat{S}^{(n-1)})$$

$$\hat{y}^{(n)}(\hat{S}^{(n-1)}) = R(\hat{S}^{(n-1)}) * \hat{y}^{(o)}(\hat{S}^{(n-1)})$$

$$\text{with : } \hat{S}^{(n)} = \int_0^{\hat{S}^{(n-1)}} \frac{d\hat{S}^{(n)}}{d\hat{S}^{(n-1)}}\, d\hat{S}^{(n-1)} \,,\quad d\hat{S}^{(n)} = \sqrt{\left(\frac{d\hat{x}^{(n)}}{d\hat{S}^{(n-1)}}\right)^2 + \left(\frac{d\hat{y}^{(n)}}{d\hat{S}^{(n-1)}}\right)^2}\; d\hat{S}^{(n-1)}$$

A simple low-pass filter, often called "averager", is used ; h(s) is thus a rectangular window which can be made dynamic so as to overcome the time-information conflict arising when adjacent contour-points are either very near or very far.

IV. LEARNING SYSTEM

The learning system is divided into two major parts : model acquisition and decision rule generation. The model acquisition step produces models of objects to be recognized, while the rules generation step organizes the search for the recognition procedure. First, we present a scheme of the complete model.

IV.1. *Object representation*

As previously discussed, the object models are hierarchical. They are composed of the following four levels :

1. A contour shape description by a state-space representation using global features : $F = (F_1, F_2, \ldots, F_n)$.

F is an a-priori given set of characteristic functions computed on the complete object : external contour and holes. Examples are arcs, perimeter, number of holes, inertia axis, max and min radius, holes area,...

2. A contour shape description by a syntactic representation using local primitives

- $P = (p_1, p_2, \ldots, p_m)$, a set of characteristic components of a contour. We assume that each contour (or internal line) may be decomposed in terms of syntactic terminals of a set T (Fig. 4). The primitives p_i are then specific subsequences of the obtained syntactic descriptions.
- " α ", a link between two consecutive primitives with length attribute. This link is in fact a generalization of the concatenation relationship.

3. Local surface analysis by bidimensionnal template using either :

- pixel-based template matching techniques,
- texture analysis and image enlancement techniques.

4. A structural representation : we define a " β " vector link with length and orientation attributes, which define the relative distance and orientation between

two elementary components of a description. These components may be either syntactic primitives of the external contour or 2-D patterns (holes,...). These links are restricted by semantic and geometrical constraints (proximity,...).

For the rest of this paper, we will only consider the second and fourth levels. The first one is used for isolated parts identification (his global nature makes it without great interest for partially occulted parts identification), while the third one is not yet well studied. However, it is provided for in a near stage of our work.

IV.2. *Model acquisition*

The purpose of this step is to construct models of the objects to be recognized. A traditional training-by-showing technique is used. The object to learn is shown to the system several times under different experimentation conditions (rotation, translation,...), but is always isolated and completely in the camera field.

The model of each class of objects is built by first finding a complete syntactic representation and then adding structural representation. The object of figure 5 is represented by :

1) $p_1 \alpha_1 \, p_2 \alpha_2 \, p_3 \alpha_3 \, p_4 \ldots$
2) $p_1 \, (\beta_5 \, L_1) \, \alpha_1 \, p_2 \, ((\beta_4 \, L_2) \, ; \, (\beta_8 \, L_8)) \, \alpha_2 \, p_3 \ldots$

The 2-D patterns L_j are directly obtained from the contour extractor. The principal difficulty of this step is to find the primitives p_i. These are defined as specific factors of a description (subsequences which belongs to only one object description). We may also define a second kind of primitives which are common to only two descriptions or three descriptions. In /13/, a fast substring matching algorithm was developed which dealswith this problem.

Actually, in parallel with this automatic method, and because of the unstability of obtained descriptions, the operator interactively choose the primitives p_i. He first define those of the 1st object in study. These are searched onto the second object and the description eventually completed by news primitives. The process is iterated for all objects. Each primitive is then defined by a curvature sequence.

IV.3. *Generation of decision rules*

This step considers the identification phase mainly as a substring matching problem for the syntactic representation and a sub-tree isomorphism problem for the structural one.

It organizes this isomorphism and matching search by sets of decision rules and compiles each of these sets into an efficient decision tree.

Two types of rules are used :

- rules using only syntactic representation. They use highly discriminating primitives sub-sequences.

  Example : If $(p_{i1} \alpha_i \; p_{i2} \alpha_{i2} \; p_{i3})$ Then object K

- rules using syntactic and structural information :

  Example : If $(p_{i1} \, ((\beta_{i1} \, L_{i1}) v (\beta i_2 \, L_{i2})) \alpha_{i1} \ldots \beta_{i3} \, L_{i3}))$ Then object J.

V. RECOGNITION SYSTEM

V.1. *Partially observed object identification*

The aim is to find a prototype which best matches the shape considered. A heuristic research is used to overcome the problem of the enormous calculation time involved, and the inevitable combinatory explosion if all the different possibilities for the construction of a prototype are considered.

The identification program essentially consists in :

- extracting syntactic primitives : if a matching is found then confirmation by one or few structural relations,
- adding structural description and looking for sub-tree isomorphism.

In practice, an hypothesis-verify method is used. The most pertinent primitive is first sought. An hypothesis about the unknown object nature is generated and an evaluation function of the matching is calculated. Following the value obtained, and in accordance with the possible relations model (decision tree), another primitive is searched. The geometric information already disponible constrains and guides the search : the primitive is to be identified only in a limited interval of data. The object is recognized when the evaluation function (or similarity measure) passes a predefined confidence threshold (Fig. 6).

The similarity measure is defined as the sum of fitting measures which evaluate how well the model primitives match the input pattern data (corresponding curvature function).

A correlation-like comparison method is used for primitive recognition. Let the curvature sequence of the searched primitive be $R = r_1,\ldots,r_n$ and the sequence of the (unknown) object $S = S_1,\ldots,S_m$. Then the degree of matching can be measured by :

$$d(j) = \sum_{i=1}^{n} \left| r_i - S_{i+j} \right| \qquad j = 1, m-1$$

The best fit is obtained at the minimum of the function d(j) = AM. AM occurs at the shift of k places of the sequence S. It can be used for rotation estimation.

When it is run on a partly observed object, the described program consideres only the sub-contour in the field of view. For overlapping objects : with one internal line, it processes each sub-contour independently, adds the border line to the most likely sub-contour, and confirms the hypothesis.With more internal lines, it consideres in turn the different possible hypothesis.

This program is only a part of a general desional process which is briefly presented below.

V.2. *The general decision process*

It is a multilevel hierarchical structures of several decision processes, including :

- identification using a state-space representation level and global features,
- substring matching on syntactic primitives,
- bidimensional analysis with template matching,
- tree isomorphism on structural description.

Each decision level is defined by a set of decision rules generated by the learning procedure. These sets of decision rules are compiled into near-optimal decision trees (taking into account costs, processing tume, frequencies).

Transitions between levels of the decision process are defined by sets of decision rules, called transition rules (Figure 7).

Examples of transition rules are :

root node r {
- . If the object is occulted Then go to level 2
- . If the outline is complete Then go to level 1

Level 1 { . <u>If</u> (object is identified) ∧ [(confirmation is needed) ∨ (miror symetry must be determined)] <u>Then</u> go to level 4
. <u>If</u> (object is not recognized) <u>Then</u> go to level 2 }

VI. CONCLUSION

We have presented in this paper a multilevel hierarchical system for object identification using a television camera. This system provides much of perception system qualities (learning ability, speed, reliability,...) in a fairly large class of problems : isolated and partly occulted objects with near plane surfaces.

This last case is the one on which we particularly concentrated. The method described combines the advantages of two classical approaches : reliability of the parallel approach and much of the speed of the sequential one. This speed is obtained by the use of an (identified) pertinent primitive to define a co-ordination system inside which the others are localizated.

Further developments and extensions of this system are its integration to assembly experiments and the analysis of more complex scenes (3-D, objects in motion,...).

BILBIOGRAPHY

1. Y Shirai, "Edge finding, segmentation of edges and recognition of complex objects", in Proc. 4th Int. Conf. on Artificial Intelligence, pp.674-681, Tbilissi, Georgia, URSS, 1975.
2. J.W. McKee, J.K. Aggarwal, "Computer recognition of partial views of curved objects", IEEE Trans. Compt., vol.C-22, pp.790-800, Sept. 1977
3. M. Yashida, S. Tsuji, "A versatile machine vision system for complex industrial parts", IEEE Trans. Compt., vol.C-26, pp.882-894, Sept. 1977.
4. W.A. Perkins, "A model-based vision system for industrial parts", IEEE Trans. Compt., vol.C-27, pp.126-143, Feb. 1978.
5. K.S. Fu, "Syntactic methods in pattern recognition", New-York Academic, 1974.
6. W.H. Tsai, K.S. Fu, "A syntactic-statistical approach to recognition of industrial objects", Proc. 5th Int. Conf. on Pattern Recognition, pp.251-259, Miami Beach, Florida, USA, Dec. 1980.
7. G.J. Agin, "An experimental vision system for industrial application", 5th Int. Symp. on Ind. Robots, Chicago, USA, Sept. 1975.
8. A. Giraud, "Une méthode d'approximation de contours par droits et arcs de cercle", Note Interne LAAS-CNRS.
9. J.D. Dessinoz, "Visual identification and localization in a multi-object environment by contour tracking and curvature description", Proc. 8th Int. Symp. on Industrial Robots, Stuttgart, May-June 1978.
10. F. Stuck, "Réalisation d'un système autoadaptatif de traitement d'images pour l'identification et la localisation de pièces en Robotique", Thèse de 3ème Cycle, U.P.S., Toulouse, 1980.
11. R.C. Bolles, "Locating partially visible objects : the local-feature-focus method", Proc. 1st Annual National Conf. on Artificial Intelligence, pp.41-43, Stanford, Aug. 1980.
12. A.P. Auther and al, "A versatile computer-controlled assembly system", 3th IJCAI, Aug. 1973, Stanford, California, USA.

13. M. Ghallab, "Optimisation de processus décisionnels pour la robotique", Thèse d'Etat, U.P.S., Toulouse, 1982.

14. T. Pavlidis, "Structural pattern recognition", Springer-Verlag, 1977.

15. A. Riad, M. Briot, "Identification and localization of partially observed parts", 11th ISIR, Tokyo, Japan, 1981.

16. A.P. Ambler, H.G. Barrow and al, "A versatile computer-controlled assembly system", 3th IJCAI, Stanford, California, USA, Aug. 1973.

17. C. Bourdeau, M. Briot, J.M. Pons, J.C. Talou, "Etude et réalisation d'un extracteur rapide de contours fermés d'une image binarisée", Congrès AFCET-INRIA "Intelligence Artificielle et Reconnaissance des Formes", Nancy, Sept. 1981.

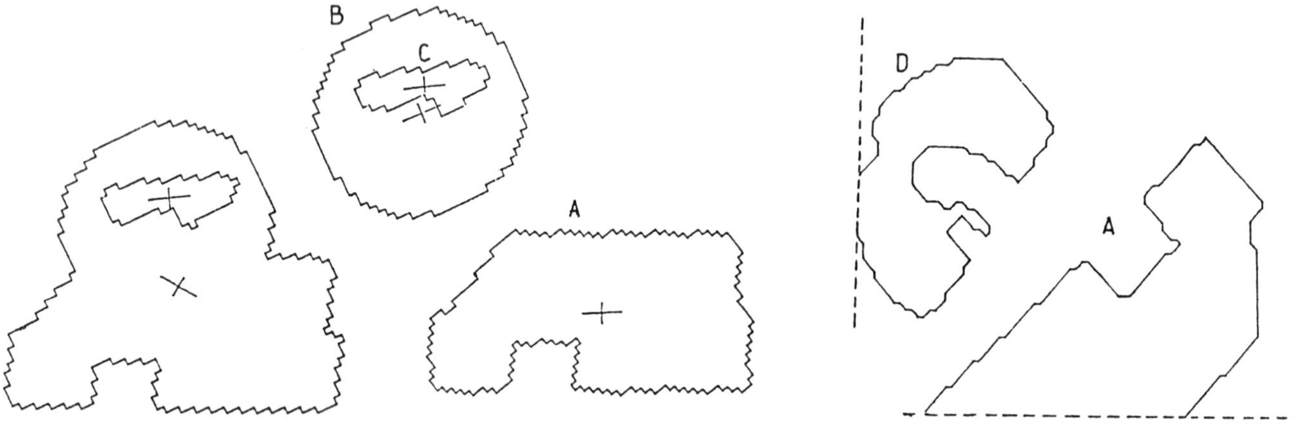

a) N.S.-E.W. directions b) Freeman's directions

FIGURE 1. Contour extraction

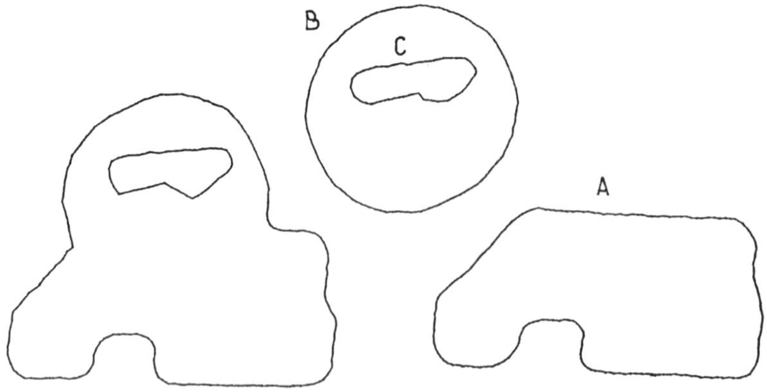

FIGURE 2. Filtering applied to Figure 1.a.

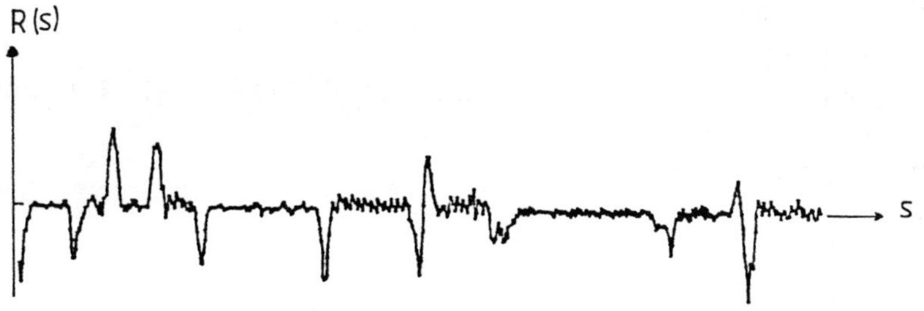

FIGURE 3. Curvature function of contour A (Figure 2)

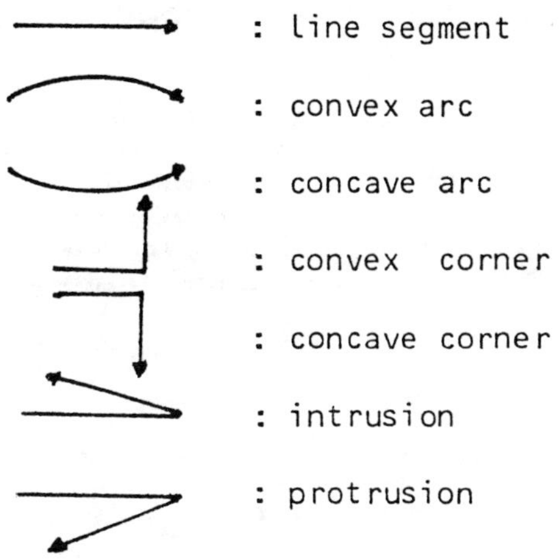

FIGURE 4. T : Set of syntactic terminals

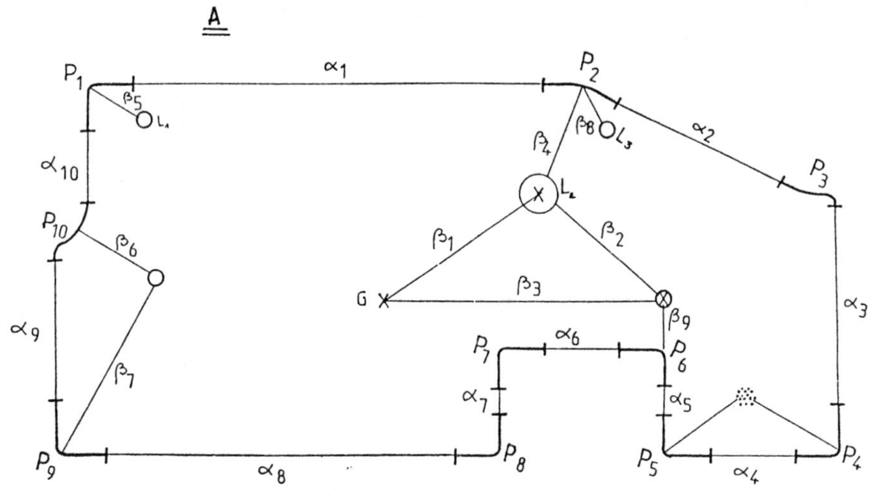

FIGURE 5. An example of object description

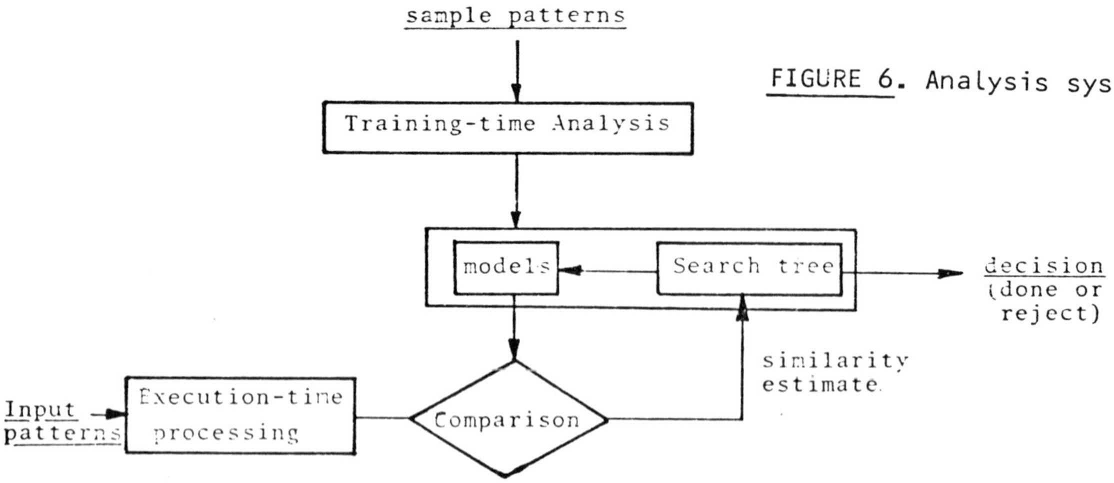

FIGURE 6. Analysis system scheme

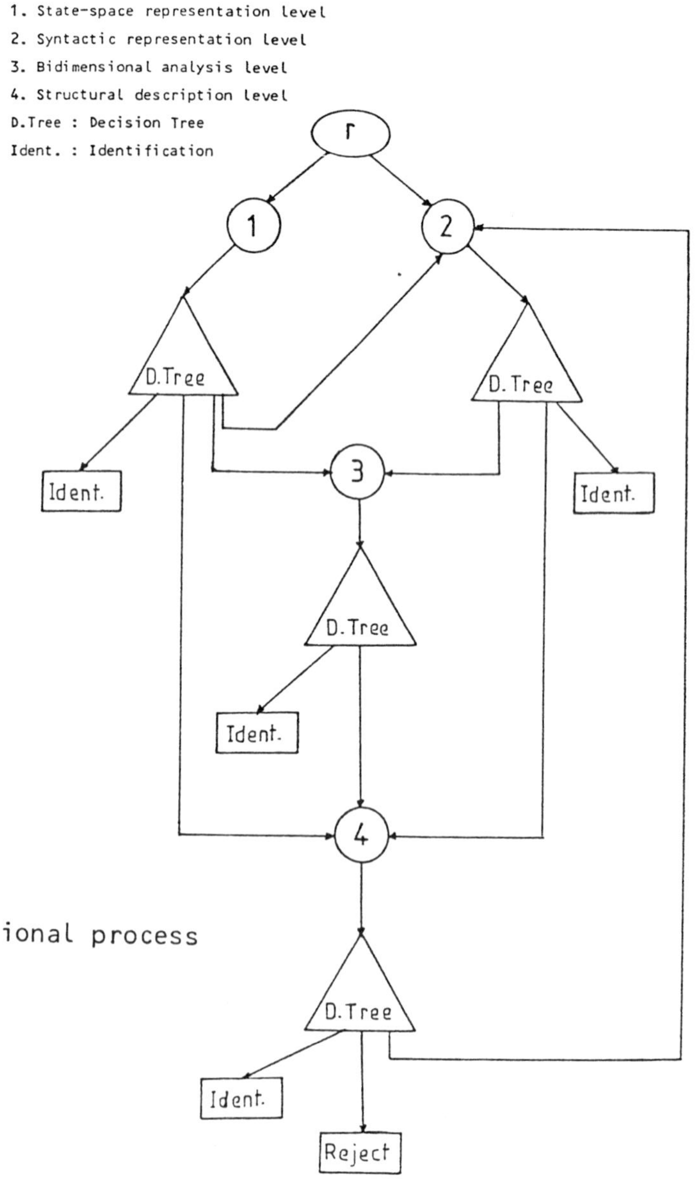

FIGURE 7. General decisional process

An Interactive Software System for Developing Robot Vision Algorithms

P.G. Selfridge
Bell Laboratories
Holmdel, New Jersey

Abstract

Software is becoming increasingly important in industrial robotics. The need for flexible and robust algorithms is especially evident in the area of visual processing. A flexible, interactive software system, running on a Vax, has greatly enhanced the ability to develop such algorithms. This system features: (1) a very user-friendly command interface; (2) a number of image variables for combined processing of images and retention of intermediate results; (3) a modular organization allowing new image processing algorithms to be added easily; (4) access to segmentation and region data; and (5) a macro capability that allows grouping of commands for storage and execution.

ABSTRACT

The need for algorithms to process visual information is shared by the fields of industrial robotics and computer vision. In industrial robotics, however, the need for fast algorithms amenable to implementation in a variety of environments is especially important. Developing such algorithms is hampered by a lack of interactive access to image processing algorithms and related primitives.

This paper describes an interactive image manipulation and processing system with several powerful features. It was developed to aid in the design of vision algorithms by giving one *interactive* access to vision primitives. This system allows digitization of images from camera stations, application of simple image processing algorithms, derivation of region data, graphical display of images, histograms, and regions, storage of processed images for later analysis, and a variety of other useful features.

This system is described in some detail, including design considerations, current capabilities, an example of its use, and future extensions. The system is running on a Vax 780, is written in 'c', and is inherently portable.

1. **Introduction**

Image processing and manipulation is an important area of computer science. Image understanding or computer vision is a related field which seeks techniques by which a computer program can identify and locate objects in image data. Many techniques for these fields have been developed over the years [1, 2, 3].

Robotics, on the other hand, is a relatively new field of exceptional industrial and economic importance. Robotics, as a research field, seeks materials and techniques which will enable robots to perform with more flexibility, robustness, and speed. One area of major concern is that of sensors: giving robots hardware and software to enable them to incorporate sensory data into their operation, such as output from visual [4] and tactile [5] sensors.

The impact of visual processing in robotics will be significant in many inspection and assembly operations. Flexible vision can be expected to improve inspection operations, allow parts handling in new and noisy environmental situations, allow existing systems to be changed without a "re-tooling" of the line, and allow entirely new manufacturing processes to be "roboticized". Current visual techniques are primitive, consisting primarily of binary vision [6], and present commercial robots are not adequately equipped with visual processing capabilities [7].

While further work in visual sensors and illumination systems is desirable, many of the problems in visual processing for robotics are algorithmic. The kinds of image processing algorithms, control structures and data structures applicable to a particular task or group of tasks is far from clear. Furthermore, developing such algorithms is difficult without access to a suitable system for *experimenting* with image algorithms.

With this problem in mind, I have developed an interactive image processing and manipulation system with considerable power and ease of use. This system enables one to digitize images, perform image processing operations on them, display images, apply segmentation algorithms to create region data structures, and a variety of other useful operations. The system features: 1) a very user-friendly command interface; 2) a number of image variables for combined processing of images and retention of intermediate results; 3) a modular organization allowing new image processing algorithms to be added easily; 4) access to segmentation and region data; and 5) a macro capability that allows grouping of commands for storage and execution.

The rest of this paper describes this system, called **SIMPS** (Simple Image Manipulation and Processing System), in some detail. Current capabilities of the system are first summarized. Design considerations are discussed, and the current capabilities of the system described with examples. Then, a simple example of the system in action is shown. Finally, implementation details and future extensions are discussed.

2. Summary of Capabilities

SIMPS allows the user to easily apply image operators to an image, creating new derived images. For example, the user can apply an edge operator, a gaussian filter, or a blurring function to an image, putting the resulting image into a new image variable. Images can be compared with a differencing operation which identifies parts of the images which are different (this can be used as a very crude but effective segmentation procedure.) Images can be thresholded producing a derived image containing regions. Small "noise" regions and "spurs" of a few pixels can be eliminated with a cleaning operation.

Once a thresholded image is created, it is useful to actually examine properties of the remaining regions. SIMPS allows the creating of region structures from a thresholded image. These region structures contain properties of the regions such as the size, major and minor axes of an approximating ellipse, boundary chain code, and the like.

It is very important that a user be able to view the results of operators. SIMPS allows the display of several kinds of information, including images and regions. Regions are displayed using their chain-coded boundary, which is overlaid on the original image. This allows visual checking of the accuracy of the region. In addition, intensity histograms can be displayed. This is an aid to the determination of a threshold to be applied.

As will be discussed, other capabilities of the system make the computational primitives very easy to use. These include simple and effective command interaction, the use of image variables, and a macro capability.

3. Design Considerations

3.1 Ease of Use

There are several important design considerations that went into the development of SIMPS. In many ways the most important, especially for a system for use by others, is ease of use and simplicity of user interaction. SIMPS was designed so that typing by the user is minimized. For example, to specify a command, the user has merely to type the minimum number of characters that will disambiguate that command from all the others. Typing less will cause the system to complain, and the user can try again. In addition, the system fills out the command so that legibility is maintained. Figure 1 shows several command interactions illustrating these features. As in later figures, what the user types is italicized, and what the system types is in bold. (Not shown is the space or carriage return that terminates the command input.)

Command: *;lkj*
No such command, try again.

Command: *re*
Ambiguous, try again.

Command: *re***adimage from file:** *test.img* **to image no:** *;lkj*
 No, an integer is needed.
  to image no: *0*

Reading file . . .

Figure 1: Command Interactions in SIMPS

Figure 1 also illustrates two other feature of command interaction in SIMPS. The first is that arguments needed by the command are prompted for individually; this increases legibility and helps the user know what is needed. This is in contrast to the naive command interaction where lines like "read test.img 0", or "repa 0 1 0 255" are used and the opportunity for incorrect commands is obvious. The second feature is that if the user mis-types, the system will complain and ask again. The system does type checking on each argument, guaranteeing that what the user types is correct and has the proper type.

Another method of minimizing typing is the use of default parameters. SIMPS maintains default parameters for **all** command arguments, and has methods for specifying defaults for each arguments, defaults for all arguments, and changing the defaults. This is done as follows. When the system prompts for an parameter, if the user types a space, the default for that parameter is automatically typed (for legibilty), and the next parameter is prompted for. If the user types a dollar sign ('$'), all defaults are typed and used for that command. If the user wishes to change a default argument, he types an asterick ('*') before the value, and that value is then used as the default argument in later calls. These interactions are illustrated in figure 2.

Command: *define* **image no:** *2*, **xsize: 256, ysize: 256**
owner: *pgs*, **title:** *test1*

(a space is typed after the "xsize" and "ysize" above, indicating the default is desired)

Command: *define* **image no:** *1*, **xsize: $256, ysize: 256**
owner: pgs, title: test1

(in the above commands, the system backspaces over the user input so the " " and the "$" are not seen)

Command: *define* **image no:** *2*, **xlsze:** **128*, **ysize:** **128*
owner: *pgs*, **title:** *test2*

Command: *define* **image no:** *3*, **xsize: $128, ysize 128**
owner: pgs, title: test1

Figure 2: Use of default arguments in SIMPS

3.2 Image Variables

Figure 2 illustrates another design consideration incorporated into SIMPS: the use of a number of image variables. In an experimental system it is important that a number of images be available and that those images not be fixed in size. For example, the user may wish to deal with images of different resolution. In addition, some image processing commands use two or more images, and it is important that these be supported. Finally, it is very convenient to keep intermediate images around. SIMPS provides a number of image variables (currently 10, but easily expanded) that are referred to by number. The actual images are dynamically allocated when defined as above, and de-allocated when replaced. This means the system does not carry the additional space burden unless the space is actually needed.

3.3 Macro capability

Another important capability, especially with a system that is used over time, is the ability to encapsulate a series of commands that are often used together. This is called **macroing**, and is currently incorporated into SIMPS. After a user has determined the desirability of macroing a series of commands, he executes the command "mdefine", for 'macro define', and gives it a name. He then enters the commands he wishes to macro, terminating with a blank line. That macro can be printed, stored on disk, and of course, executed. Macro execution is done with the "mexecute" command. Each line of the macro is passed to SIMPS as if it had been typed directly by the user, except a parenthetical line is typed indicating the system is in "macro mode." Figure 3 illustrates a series of macro command interactions in SIMPS.

Command: *mdefine* **macro called:** *initimages*
Type the macro:
def 0 256 256 pgs test0
def 1 256 256 pgs test1
def 2 256 256 pgs test2
def 3 256 256 pgs test3
def 4 256 256 pgs test4

Command: *me*xecute **macro called:** *ini*t**images**
(macro mode)
define image no: 0, xsize: 256, ysize: 256,
owner: pgs, title: test0

(macro mode)
define image no: 1, xsize: 256, ysize: 256,
owner: pgs, title: test1

(macro mode)
define image no: 2, xsize: 256, ysize: 256,
owner: pgs, title: test2

(macro mode)
define image no: 3, xsize: 256, ysize: 256,
owner: pgs, title: test3

(macro mode)
define image no: 4, xsize: 256, ysize: 256,
owner: pgs, title: test4

Command: *mw*rite **macro called:** *init*images **into file:** *init.mac*

Figure 3: Macro commands in SIMPS

Note that, as with normal commands, the user has only to type the minimum number of letters to specify a macro.

3.4 Extendability

Another design consideration that plays a part in any system such as SIMPS is extendability. Basically, in a system of this kind it must be easy to add new operations with their appropriate commands, without doing a lot of re-organizing. SIMPS is written with this in mind. The commands all have a similar structure, which is <command keyword> <arg1> <arg2> . . <argn>. In addition, the specification of arguments is done with an **argment template**, which specifies the argument prompt, the type, and the default value. This allows a single routine (called GetArg) to fetch an argument, check its type, and return the default value if called for. To add a new command necessitates only specifying the argument templates, adding another case to the case statement, and, of course, linking in the new routine.

3.5 Access to Region Data

For a system to have the capability for actual image understanding, as opposed to just image processing, it must have the ability to evaluate the results of its processing. Such evaluation is conveniently done by examining the properties of regions isolated or extracted by this processing. A region is a connected component, or blob, in an image, and represents the hypothesized important areas of the image. In order to use regions, a system must be able to compute them and examine various statistical and shape properties of the regions. This enables algorithms to be developed which take into account the regions created, and allows "closed loop" algorithms which can improve their results by adapting parameters to create "better" regions [8]. SIMPS has the capability to create and examine regions. Once an image has been processed, a simple command will create a list of **region data structures** describing the properties of the isolated regions. Current properties include location information, pixel statistics crude shape information. As discussed later, these features will be expanded, as will the ability to use them in later computations.

4. Current Capabilities

The current capabilities of the system fall under several categories: image manipulation, region creation, graphics, macro commands, and miscellaneous. These commands are listed with their arguments under their respective categories. The characters "<>" stand for user input. Short explanations are provided where appropriate. These commands represent the current capabilities at the time of writing only.

4.1 Image manipulation

read into image no: <> **from file:** <>
reads from an image file into an image variable, does nothing if the file cannot be found

write from image no: <> **into file:** <>
writes an image file; will overwrite a previous file

define image no: <>**, xsize:** <>**, ysize** <>**, owner:** <>**, title:** <>
define an image with a certain size; deallocates pre-existing
image, produces a pristine (all zero) image

examine image no: <>
allows examination of actual pixel values in 15 by 15 blocks, prompts user for the x and y of the upper left corner of the block

copy image no: <> **to image no:** <>
copies one image to another

clear image no: <>
clears an image, i.e. sets to all zero

negate image no: <>
sets all pixels to 255 minus the pixel value

thresh image no: <> **into image no:** <> **with lower threshold:** <> **and upper threshold** <>
create a new image where all pixels in the old image not between the thresholds are set to zero; as with all operations which result in new images, the images must be defined first

bthresh image no: <> **into image no:** <> **with lower threshold:** <> **and upper threshold** <>
the same as thresh, but pixels between the thresholds are set to 255

edge from image no: <> to image no: <>
creates an edge image using Roberts' cross operator; ignores edges of the image

repack from image no: <> to image no: <> using new min: <> and new max: <>
does a linear repacking of pixels so that they all lie within the new range; assumes that the min and max pixel values in the image header are correct

subtract image no: <> from image no: <> into image no: <>
does a pixel by pixel subtraction

difference image no: <> from image no: <> into image no: <> ignoring differences less than: <>
all pixels who differ by more than the minimum are set to 255, the rest are set to zero

clean image no: <> into image no: <> preserving those with number of neighbors: <>
preserves pixels with minimum or more number of neighbors, zeros others

average image no: <> into image no: <>
do a 3 by 3 window average from the first image into the second, ignoring edges

4.2 Graphics Commands

gopen with camera: <>
opens the Ikonas display, prompts for a camera number to use in digitization; fails if Ikonas is in use

gclose
closes the Ikonas display

erase
erases the Ikonas display

colormap normal? <>
if reply is yes, normal grey scale color map is loaded. Otherwise, special map is loaded where values from 0 to 249 are treated as normal grey scale, and values from 250 to 255 are mapped into a number of colors

getimage into image no: <> from xstart: <>, and ystart: <>, xminification <>, yminification: <>, xsize: <>, ysize: <>
digitize an image from the current camera starting at world coordinates xsize and ysize, with the given minification. It then store the image into the given image no., with the given x and y sizes.

showimage no: <> at x: <> and y: <>
displays the image on the Ikonas starting at the given x and location

4.3 Region Manipulation

makeregions from image no: <>, minimum threshold: <>, maximum threshold: <>
create a region list for all regions (connected components) in the image between the thresholds

list regions
list the region list on the terminal

drawregions starting at x: <> and y: <>, in color: <>, minimum size: <>
draw the boundaries of and centers of gravity of each region larger than the minimum size on the screen, starting at x, y, and with the given color

4.4 Macro Commands

mdefine macro called: <>
define a macro. The macro is entered, line by line, and terminated with a blank line

mprint macro called: <>
print the given macro

mexec macro called: <>
execute the given macro

mread macro from file: <>, call it: <>
read a macro file and name it

mwrite macro called: <> into file: <>
write a macro out to a file

4.5 Miscellaneous

?
print out a list of commands and current macros

systemcall with command line: <>
does a system call with the given command line

ramp image no: <>
puts a ramp into the image

minmax image no: <>
computes the minimum and maximum pixel values, prints them out, and stores them into the header

putbox into image no: <>, xstart:<>, ystart: <> xsize: <>, ysize: <>, of value: <>
writes a box into the image starting at xstart, ystart, of size xsize, ysize, and value value

gauss image no: <>, xstart: <>, ystart: <> xsize: <>, ysize: <>, with mean: <>, and sd: <>, add values? <>
writes a gaussian box into the image, adds the values to the existing pixels if the last argument is "yes"

noise image no: <>, xstart: <>, ystart: <>, minimum value: <> maximum value: <>, add values? <>
writes a box of uniform noise into the image, adds the values if the last argument is "yes"

histogram image no: <>
computes an intensity histogram of the image. If the graphics option is on, the user is prompted for an x and y position to display it, as well as a color; otherwise, the histogram is displayed on the terminal

5. An Example

This section presents one simple example of the kinds of results that can be obtained with SIMPS. It is important to emphasize the **interactive** nature of the system, which allowed the following example to be generated **at the terminal**, and intermediate results examined and processing changed as necessary. This example is shown as it was generated, with SIMPS commands and displays.

In this example, a printed circuit board inspection problem was examined and a very simple approach was taken. The problem was to detect misplacement of standard components after automatic insertion on the board. The approach was to backlight the board, image it, and compare it with an image of a good board. Any differences between the two would be indicative of a potential failure. This idea was conceived, and SIMPS used in a preliminary test. A board was taken and imaged; the commands and the image are shown in figure 4:

Command: *go***open - ikonas open**

Command: *get***image into image no: $0, xstart: 0, ystart: 0**
xminification: 1, yminification: 2, xsize: 256, ysize: 256

Command: *er***ase - ikonas cleared**

command: *sho***wimage no: $0 at x: 0 and y: 0**

Figure 4: Reading in the image

Then, a single component was removed to simulate an insertion failure, and the same board imaged again, using the same commands but with image number 1. This image is shown in figure 5:

Figure 5: Image of the "bad" board

Then, a pristine image was defined in image 2, and a differencing operation done between image 0 and image 1, and the result put into image 2. The commands and the results are shown in figure 6.

Command:*define* **image no: 2, xsize: $256, ysize: 256**
owner: pgs, title: test1

Command: *difference* **image no: $0 from image no: 1**
into image no: 2
ignore differences less than: 50

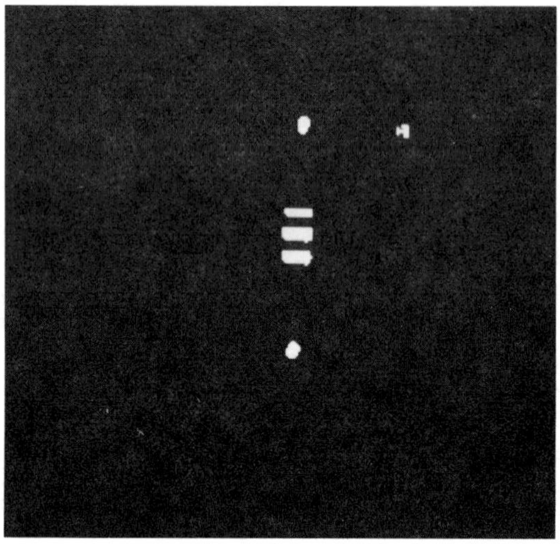

Figure 6: Computing a difference image

Finally, the difference image can be used to compute the actual regions of difference, and these can be examined and displayed. Figure 7 shows a composite display, with the original in the upper left, the difference regions displayed on top of it, the image of the bad board in the lower left, the difference image in the upper right, and an intensity histogram of the original in the lower right.

Figure 7: Composite display for the inspection problem

Other applications investigated with SIMPS include dual in-line package location and identification, quick texture analysis and adaptive segmentation experimentation.

6. Discussion

SIMPS has been implemented in an extremely reasonable manner, which was made possible by other well-designed packages. Our image format is quite straightforward, and images are accessed through a robust interface. Basic pixel access is through two routines, **GetPixel** and **PutPixel**, which do bounds checking on the passed pixel coordinates. Image processing routines, contained in a package called **imgpak** (image package), are modular and straightforward. They also do a variety of bounds and type checking, with the goal that nothing the user types will cause the system to crash. Display access is done through a variety of routines, contained in a package called **ikpak** (ikonas package), which allow straightforward calls to display images, digitize images, et cetera.

SIMPS presently represents a fairly powerful and easy-to-use image processing and manipulation system. However, it is really only the beginning of a very sophisticated tool which, when completed, will be a **language** for writing computer vision algorithms of a certain kind. The eventual goal of this system is the ability to write, interactively, computer vision algorithms that use feedback from the environment to adjust, or adapt, their processing. One claim of my previous work [8] is that this ability is crucial to the kind of robust performance desired. SIMPS will allow this ability by providing information about the results of processing in a form that will make it easy to use, interactively and algorithmically. The current SIMPS provides that information interactively in the form of **region lists** of regions formed by segmentation of a processed image. The planned extensions will include providing these as accessible variables, along with the needed control statements and other statements necessary to make the system a useful programming language.

7. Conclusions

SIMPS is an image processing and manipulation system with considerable interactive capabilities. It has a number of features that contribute to its usefulness, including a macro capability, image variables, a nice human interface, and others. It is currently running on VAX 11/780 and has already proved its usefulness in a number of instances. Future extensions will make it even more useful as an interactive tool.

8. References

1. Ballard, D.H. and C. M. Brown, **Computer Vision**, Addison - Wesley, 1982

2. Aggarwal, J.K., Duda, R.O., and A. Rosenfeld, **Computer Methods in Image Analysis**, IEEE Press, 1977

3. Artificial Intelligence Journal, Vol. 17 no. 1-3, **Special Issue on Computer Vision,** 1981

4. Birk, J. et al., **General Methods to Enable Robots with Vision to Acquire, Orient, and Transport Workpieces,** report prepared for the National Science Foundation, 1979

5. Harmon, L. D., **Touch Sensing Technology: A Review**, SME Tech. Report MSR80-03

6. Gleason, G.J. and G.J. Agin, **A Modular Vision System for Sensor-Controlled Manipulation and Display,** SRI Technical Note 178, SRI International, Menlo Park, Ca.

7. Robots 6 Conference, Robotics Institute of the Society for Manufacturing Engineers, March 2-3, Detroit, Michigan 1982

8. P. G. Selfridge, **Reasoning About Success and Failure in Aerial Image Understanding,** PhD Thesis, University of Rochester, Rochester New York

Industrial Computer Vision by Mathematical Morphology

Stanley R. Sternberg
Cyto Systems Corporation
Ann Arbor, Michigan

Abstract

A methodology for automated visual industrial inspection characterized by its simplicity, accuracy, speed, and flexibility is introduced. Mathematical morphology is an algebraic language of image processing, in which complex visual inspection problems can be efficiently solved, programmed and run. The morphological approach is introduced through an example, the inspection of gear teeth for defects. A two-line morphological program performs this inspection.

Mathematical morphology is a way of thinking about vision. Mathematical morphology is the algebraic treatment of shape. Computer programs based on morphological image analysis recognize the shapes of the objects they see in a precise manner which is programmed by the user. Special computer architectures incorporating parallel processors are capable of carrying out the morphological computations at rates compatible with the real-time requirements of manufacturing systems.

Image analysis by morphological methods is not new. The morphological treatment of images has been studied for more than 50 years, primarily in Europe. In the area of geological sample analysis, morphological image processing has been a standard technique for many years. The Leitz Texture Analyzer System (TAS) [1] whose operation is based on mathematical morphology, has been marketed worldwide for more than 10 years. Conceived by J. Serra [2] and G. Matheron [3] at the Ecole des Mines, Fontainebleau, France, the TAS design was based on a sparse literature of rigorous mathematical treatments of images by such stalwart mathematicians as H. Minkowski (Einstein's math teacher), G. Cantor and H. Poincare. Since mathematical morphology is first of all a mathematics of image transformation and analysis, it forms the basis for a computer language of image processing. Because the variables in the algebraic expressions which form a morphological image processing program are whole images, this very powerful language can be understood visually through the use of pictures.

Images in morphological image processing programs are of two types, either basic or structuring. A basic image is one that is taken from a camera or an image storage device. The basic image is the subject of the analysis. A structuring element (we say element rather than image for reasons which will become immediately apparent) is an imaginary object which serves as a probe to test the basic image. Structuring elements are created by programmers; they are not basic images, rather, they are stored as strings of instructions. The shape of a structuring element reflects what the programmer knows about the basic images he is processing. A structuring element interacts with the basic image and transforms it in useful ways. The result of this transformation is a basic image which displays the desired information or which can be further transformed by additional structuring elements until it displays the required result.

At this point an example would be most helpful. The basic image of Figure 1 shows a silhouette of a randomly positioned and oriented gear. The image is binary, as might be obtained from suitable back-lighting of the part. As we can see, some of the gear teeth are broken or missing. The function of the image processing is to locate the defects.

The treatment of this problem requires two morphological transformations of the basic image involving two structuring elements, both of them disks. The radius of the first disk, r_1, is selected by the programmer in relation to the gear teeth spacing. An appropriately dimensioned disk is shown in Figure 2 in relation to the gear. In Figure 3 we show many such disks, each one positioned in the spaces between gear teeth. In the limit we may imagine surrounding the gear, inside and out, with an infinite multitude of overlapping disks, each one touching the gear. We can clearly see that where there are defective teeth, the space between the teeth not covered by a disk is smaller than the space between two good teeth. Figure 4 shows a binary image of all the residual spaces. Figure 4 is the result of the first morphological transformation which is called a closing.

Notice how this morphological transformation, the closing, has transformed the problem. Instead of examining the teeth, it examines the space between the teeth. A shift in viewpoint like this is called a duality in mathematics. Mathematical morphology makes no distinction between the shape of the object and the shape of the space surrounding the object. Any transformation of one is the dual transformation of the other.

The second transformation in our example inspection program is called an opening. The opening examines the shape of each of the spaces between the teeth left by the closing transformation (the black regions of Figure 4). We see that the spaces are of two kinds, those which are large enough to contain a second disk whose radius is r_2 and those which are not. Again, r_2 is a programmed value which depends upon the particular geometry of the gear, the magnification of the camera, as well as how particular we are about what constitutes a good gear tooth.

Figure 5 shows the result of opening the binary image of Figure 4 by a disk of radius r_2. Where a disk will fit inside the black regions of Figure 4, we see the disk; where it doesn't fit, there is no disk. Here our processing example stops, for Figure 5 satisfies our immediate requirements.

Each morphological transformation can be expressed in a succinct way in the notation of mathematical morphology. For example:

<u>figure4</u> = (<u>figure1</u> close disk(r1)) - <u>figure1</u>

is a line of program code which instructs a computer to perform the illustrated transformation on whatever constitutes the basic image <u>figure1</u> and save the result in a variable called <u>figure4</u>. Notice how the morphological language deals in variables that are whole images and operations which affect the entire image rather than just a few pixels.

The second transformation of our gear example is expressed as:

<u>figure5</u> = <u>figure4</u> open disk(r2).

Here the expression instructs the computer to position a disk of radius r_2 in the black regions of <u>figure4</u> when it will fit. Thus these two expressions are sufficient to program a computer vision system to produce the illustrated result for any gear of the form shown in Figure 1.

What we would do next if this were an actual inspection application depends very much on exactly what action is to be taken as the result of the detection of gear defects. Somewhere in the course of our program, however, we will ultimately want to make a decision which is based on counting something, like the number of disks in Figure 5. Also, we might want the positions of certain features in the image, like the small disk in the keyway of Figure 6. In either case, the data we want is not an image but rather a quantity derived from an image - in other words, a measurement.

The gear example has allowed us to illustrate several basic properties of the approach that mathematical morphology takes to vision systems. First, mathematical morphology is a language for image transformation which can be implemented on any computer. Mathematical morphology is not an algorithm, such as the SRI algorithm which is the basis of the majority of commercial vision systems today. Rather, it is a language for expressing algorithms for a tremendous number of useful image transformations, like skeletons or convex hulls or internal holes or whatever it is the user wishes to detect, locate or measure.

Secondly, the transformations of mathematical morphology operate in parallel everywhere in an image. It does not matter how many teeth the gear has, just as it does not matter how many gears are in the basic image; the transformations apply uniformly across the entire image. This creates an opportunity to employ parallelism in computer architectures designed especially for implementing the morphological transformation.

Mathematical morphology opens up a far greater number of possibilities for applying computer vision systems than are permitted by the fixed-program SRI algorithm vision systems. Being able to program into a vision system a prior knowledge about part shapes the machine expects to see is a powerful advantage which is inherent in this approach.

Computer vision systems incorporating programmable morphological image analysis will shortly become commercially available. I have already mentioned the TAS which is a microscope based image analyzer used primarily in geology and materials sciences. A more recent development is the Cytocomputer, a parallel pipelined morphological processor developed at the Environmental Research Institute of Michigan [4]. As this method of image processing becomes better known, there will undoubtedly appear many more systems incorporating the features of mathematical morphology.

BIBLIOGRAPHY

[1] Klein, J.C., Serra, J., "The Texture Analyzer", J. Microscopy 95: 349, 1972.

[2] Serra, J.P., Image Analysis and Mathematical Morphology, Academic Press (London), 1982

[3] Matheron, G., Random Sets and Integral Geometry, Wiley (New York), 1976.

[4] Longheed, R.M., McCubbrey, D.L., "The Cytocomputer: A Practical Pipelined Image Processor", Proceedings of the 7th Annual International Symposium on Computer Architecture, 1980.

Figure 1: Binary Image of Defective Gear

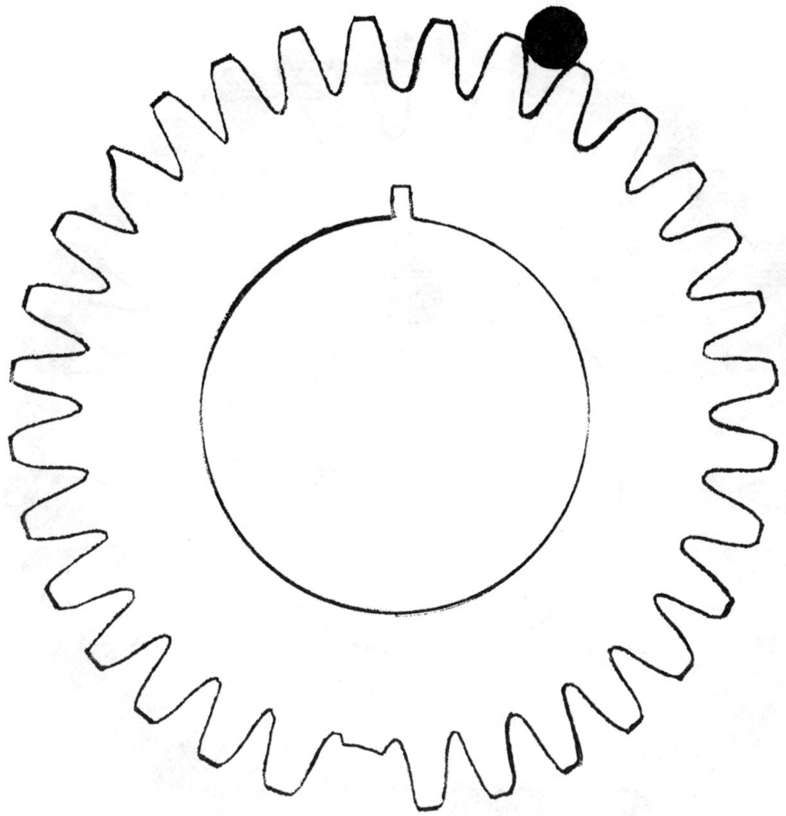

Figure 2: Disk Structuring Element Probes the Space Between Gear Teeth.

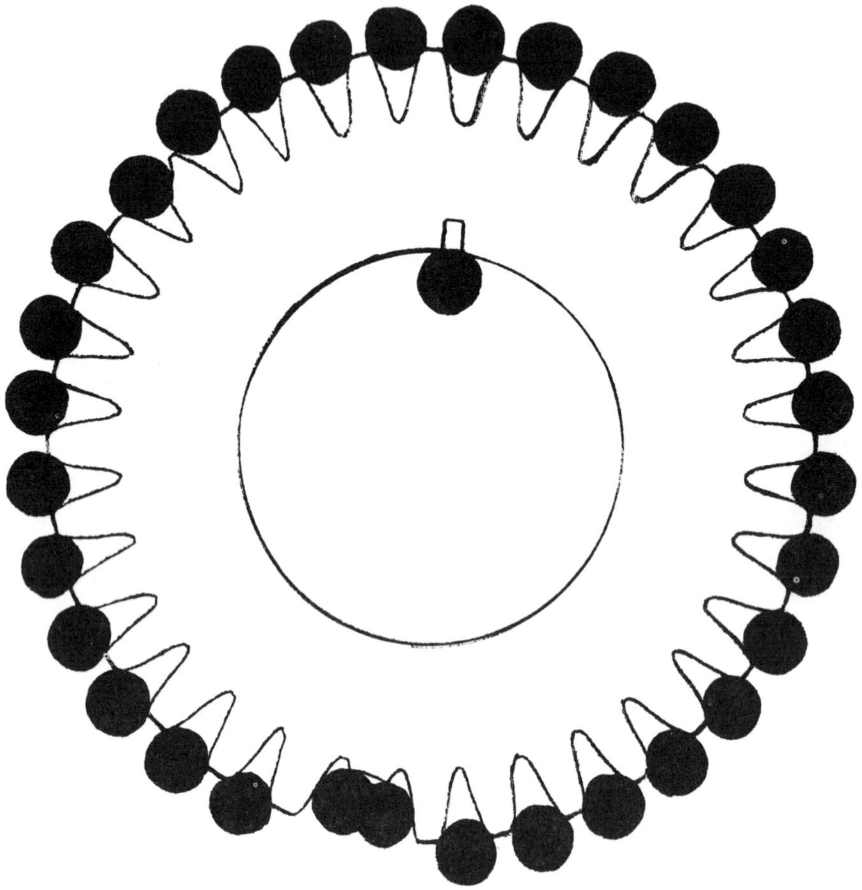

Figure 3: The Morphological Closing Transformation Probes All Spaces in Parallel.

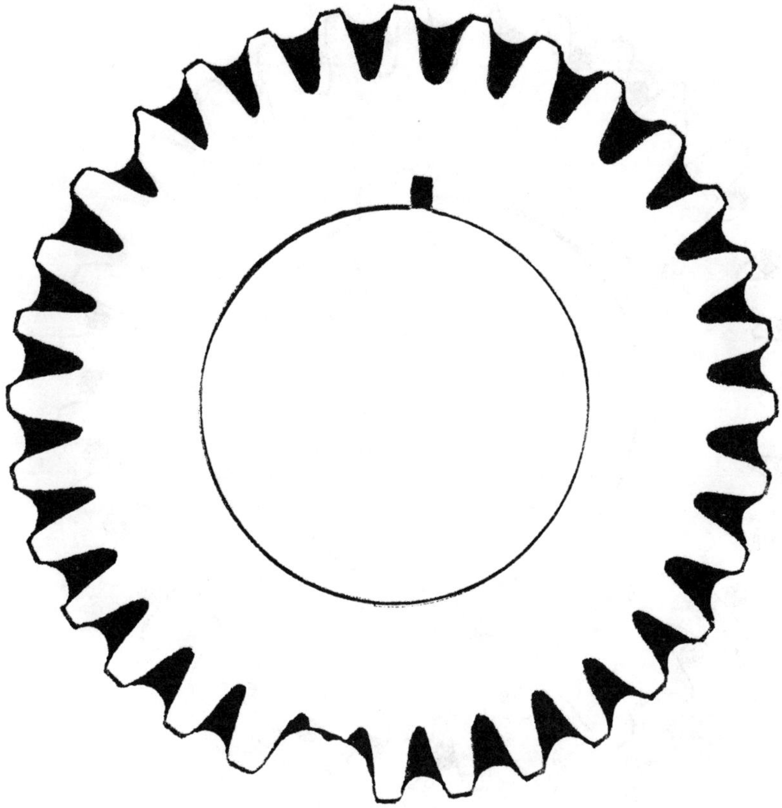

Figure 4: The Residue of the Closing is Uniform Only Between Good Gear Teeth.

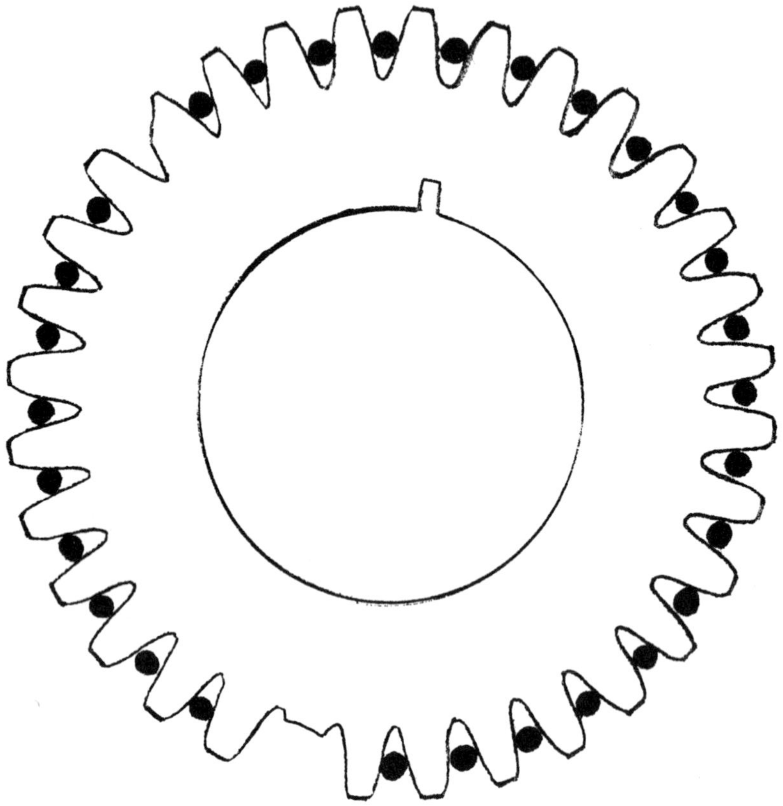

Figure 5: The Opening of the Residues Positions a Small Disk in the Space Between Good Gear Teeth.

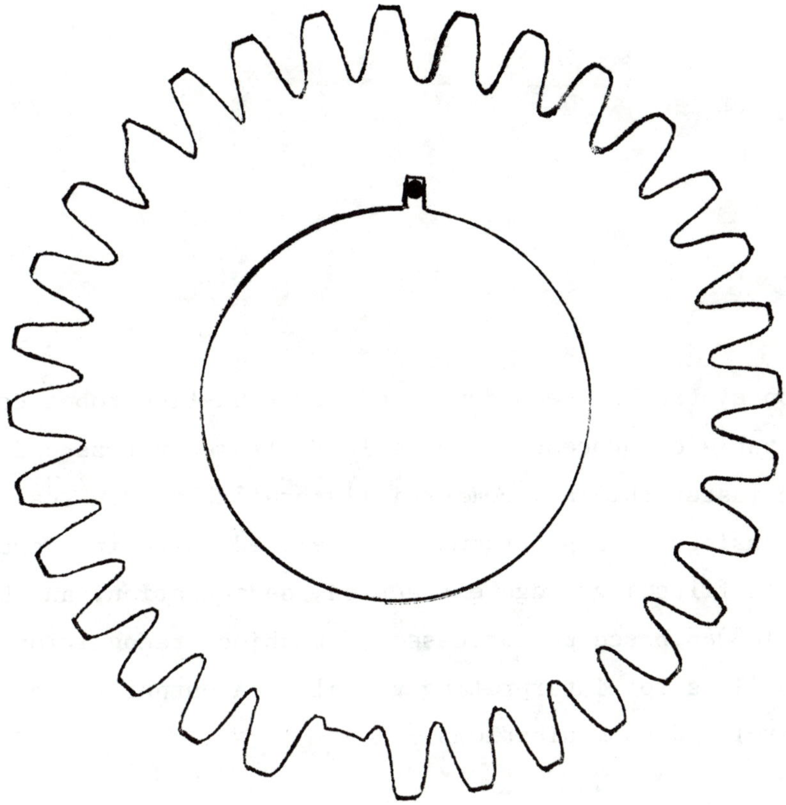

Figure 6: A Further Opening Locates the Gear Keyway.

Grey Scale Robot Vision for Real-Time Inspection and Assembly

Arkady G. Makhlin
Digital Equipment Corporation
Westboro, Massachusetts

Abstract

A 64 grey-level vision system has been developed for real-time robot control and inspection. It is capable of processing up to 12 TV frames per second which is an order of magnitude faster than any commercially available grey scale vision system. The system consists of a programmable pipelined, hardwired preprocessor for image conditioning, filtering, edge extraction, segmentation, and blob tracking, and an Intel 8086 based postprocessor for object recognition, positioning, and orientation, as well as for interpreting visual data output for a robot or external systems control and communications.

INTRODUCTION

Current robot vision systems rely primarily on the processing of binary images; that is, a threshold is applied to the output of the camera so as to provide a black and white image. In this image a correctly illuminated object will appear as a silhouette on a uniform background. For simple configurations, the binary image offers easy determination of parts measurements and locations, and a minimal processing load. In more complex configurations, however, the binary image is inadequate because:

a) Closely spaced parts are joined together;

b) Coloring of parts and background does not permit separation by a unique threshold;

c) Available illumination does not permit a clean separation of part and background.

For these reasons it is desirable to make use of the density information available from most cameras. With this information, as well as the ability to extract such image features as edges and "blobs", operations can be carried out on complex scenes under difficult and variable lighting conditions.

The Westinghouse Grey Scale Vision System offers gray scale processing at real time video rates.

HARDWARE AND OPERATION

The output of the TV camera, as provided for visual inspection and control, usually consists of a uniform two-dimensional array of density values (pixels). Useful information, such as the outline geometry of objects of interest, must be extracted by computations involving both pixel density and geometry. When these computations have been performed, there is the prospect that the desired information can be expressed at a greatly reduced bandwidth (one or more orders of magnitude) from the original image data.

As a result of the preprocessing operation, we would like to have a collection of image primitives; that is, basic image descriptors, along with their specifications and location in the image. The primitives should be easy to extract, should be consistent in value in spite of noise, and should be simply stated, so as to maintain a low data bandwidth. Furthermore, they should be useful for a variety of applications. A primitive is an information package extracted from the raw image data. It should serve as a simple building block for later image operations.

For the operations described in this paper, only two primitive geometric features are used. The first feature is a straight line contour of density gradient (an edge). Edges are defined by their endpoints and polarity (relationship of light and dark sides). The second feature represents a small region in the image which is lighter or darker than its surroundings. This is termed a blob. Blobs are defined by their location, their height and width, and their relative density with respect to the background.

Since the extraction of primitives is common to many sensors and applications, and since it results in the compression of high bandwidth image data, it is a logical candidate for incorporation into special-purpose digital hardware. For many applications, the data bandwidth reduction of the digital image preprocessor permits further operations to be performed in a programmable general-purpose processor, with the advantage of great flexibility in the range of problems handled. A Westinghouse grey scale vision system (Figure 1) includes the image preprocessor and the programmable image postprocessor. The image preprocessor consists of a pipeline series of computations operating throughout the image on 4 or 5 neighboring pixels. So that computations can be performed in both dimensions, lines of data across the image are stored in a series of shift registers. The computations include the extraction of density gradient, gradient thinning, and the tracking of similar gradient characteristics from point to point to extract edges and blobs.

Details of the grey scale vision are shown by Figure 2. Interfacing devices, such as the TV camera and display, as well as a robot, are shown at the left. Following synchronization and digitizing of the incoming video signal the data is entered into a full-frame buffer and then extracted from the buffer on a line-by-line basis for entry into the image preprocessor. The image preprocessor accepts a standard RS-170 TV signal.

The configuration and adjustment of the image preprocessor is under software control on a frame-to-frame basis. A variety of pixel conditioning options are offered, including averaging filters, a median filter, level slicing, and rate compression. Level slicing at two levels is provided so that simultaneous isolation of both light and dark regions may be accomplished for dynamic threshold adjustment. Rate compression reduces the flow of image data by selection of alternate pixels.

Pixel conditioning is followed by gradient extraction to obtain both amplitude and direction for a two-by-two pixel array. When this operation is preceded by averaging filters, the result is equivalent to larger operators with more smoothing but lower resolution. The gradient operation is a first derivative computation on the image data. All pixels with gradient values above an adaptively selected amplitude threshold are processed by the maximize function.

The blob tracker detects conditions which indicate the top of a potential blob. Left and right outlines are then tracked through the image on a line-by-line basis. The length of the vertical path is retained, as well as the extent of the left and right excursions. The blob is detected when a matching left and right outline connect.

Maximize is a second derivative operation which thins the gradient data into the fine lines of the target, including both the outlines and the internal detail. It compares neighboring gradient values in both the x and y directions. In the process, edge sharpness and amplitude are also derived.

The segment tracker connects pixels with compatible features defining the endpoints and gradient direction of image edges. This will occur for exterior edges and interior contrast lines associated with objects, as well as for background definition.

Group formation is dedicated hardware which reduces the post processor load by collecting all data associated with each object. This sorting function is assisted by identifying tags which are applied to blobs and edges as they are generated. Output from the group formation logic is provided to the image postprocessor. Outputs include groups which represent objects (work pieces) and edges associated with background.

Operations within the image postprocessor, which are fully programmable, may include:

--Frame-to-frame computation;
--Feature computation;
--Statistical decision making;
--Feature matching for final classification;
--Scene analysis;
--Object prioritization;
--Output synthesis;
--System control.

Detailed operations of the preprocessor are shown by Figure 3. The 50-by-50 array of density values is shown at the top. For ease of interpretation, only 16 grey scale levels are printed. Of the 16 levels, values above 9 are indicated by an overprinted slash (/). Gradient directions are also quantized to 16 directions, with an 1 indicating a change from light to threshold (the sensitivity threshold on the system), no output is shown, and the location is marked with a dot. The result of a line-thinning operation is indicated at the left, where some of the previous gradient outputs have been deleted if they are adjacent to a similar output of larger amplitude, and if their presence in edge formation is unnecessary. Edges are formed by tracking the thinned gradient outputs, and by recording both the beginning and endpoints of each track. They are shown on the graphical plot at the bottom of Figure 3. The window at right indicates the accumulated angles associated with tracking around a blob. The resulting blob symbol, a cross, is shown on the graphical plot, along with the edges. The height and width of the blob are indicated by the arms of the cross.

VISION SYSTEM SOFTWARE

The vision system software has been written generally in PL/M using structured-programming concepts. It has been written as system-independent as possible with the exception of disk and terminal I/O, where operating system service routines are used. The basic interactive routines between the image preprocessor and the postprocessor are programmed in the Intel assembly language. The structure of the software is shown in Figure 4. This software has been written on an Intel 8086 based general purpose computer running under RMX-86, utilizing floppy disk as the system's device and primary backup.

Command String Interpreter (VISCSI)

This module controls communications with the operator and controls the execution of all other modules. It calls on the image-processing System Nucleus (SYSNUC) which controls the interface to the image preprocessor.

System Nucleus (SYSNUC)

Commands the image preprocessor to either:

a) take a picture and read a preprocessed camera image to interface buffer;
b) take a picture and read the buffer to the postprocessor memory;
c) read the buffer to the postprocessor memory;

d) take a picture, read the buffer to the postprocessor memory, and calculate base feature values;
e) calculate base feature values from a given buffer;
f) as d) and add evaluation of the main object.

Feature Selection Module (VISFEA)

This module, based on operator input, selects feature value calculation routines from a library of such routines and controls the set-up of working storage for each feature selected.

Recognition Training Module (VISREC)

This module is called to establish a set of feature statistics for a part. The same object is presented in the vision field several times and the results of feature analysis are accumulated as statistics. Object recognition can be tested; now, the features of a presented object are verified against the statistics, and the program returns a message whether the object is, or is not, recognized.

Production Training Module (VISTRA)

This module is the responsibility of the user-programmer preparing a production program. It serves for training and refers both to system utility routines and to routines specific to the application under consideration. Essentially, it differs from the production module only in that it simulates robot commands through operator input, and outputs to the console the responses that will be sent to the robot. It can also serve to store in memory certain part-specific parameters or their retrieval addresses.

Production Module (VISPRO)

This module, like the preceding and the following ones, is the responsibility of the user-programmer. Those three modules constitute what is here named "specific application programming". The vision system is thus adaptable to the varied applications of the robot. The Production Module receives its commands from the robot through the robot-vision interface, ROBVIS.

Specific Application Utility Routines (APPUTL)

The routines of this module are those subdivisions of the specific programming that are common to the training and production modes.

Robot-Vision Interface (ROBVIS)

Entering this module causes the vision system to switch to a passive state, waiting for commands issued by the robot. These commands are decoded and transmitted to the production module for execution and response. The ROBVIS module has access to all the specific application routines of the production test module, and to all those of the system nucleus and system utility modules. This non-interactive robot-vision interface only returns control to the interactive command-string interpreter when a recognized error has been detected by the vision system but has not been processed successfully by the production module. Such an "error" can be a specific code transmitted by the robot to release the vision system to its own control. The ROBVIS module is obviously different for different robot-vision systems.

System Utility Routines (SYSUTL)

This module has many entries: all the vision system routines accessible both in training and production modes.

ASSEMBLY UNDER VISION CONTROL

The set-up for assembly of an electromechanical relay under vision control is shown in Figure 1. The PUMA 250 industrial robot was used to assemble a four-piece electromechanical relay. All work pieces were located in the field of view of a Westinghouse TV camera without overlapping with spaces between work pieces. Vision-robot communications were realized through the RS-232 Serial link. A PUMA assembly program and a vision-robot communications protocol are illustrated by Figure 5. The program ASSEMBLY has been written in VAL robot programming language, and the PUMA robot was communicating with the image postprocessor through its controller's CRT port as if an operator typed VAL instructions in the CRT keyboard.

The vision system commands the robot to start ASSEMBLY program (EXECUTE ASSEMBLY). The program stops at step 2, and, at this point, the robot is being asked to display a point A (an arbitrary location of a work piece within the field of view). The vision system, then modifies the location A by sending to the robot values of its position (X,Y,Z) and orientation (O,A,T), calculated on the basis of visual information. Consequently, the robot moves into the newly defined location A, picks up a work piece and brings it to the part predefined assembly point B. The program returns then to the step 2 and the robot receives the next location A. etc.

The use of the Westinghouse grey scale vision system allows assembly of combinations of similar parts under poor lighting conditions. It is also suitable for real time inspection, where complex parts configurations exist, and lighting is restricted.

BIBLIOGRAPHY

1. Pratt, W. K., Digital Image Processing, John Wiley, 1978.

2. Rosenfeld, A., Kak, A. C., Digital Picture Processing, Academic Press, 1976.

3. Makhlin, A. G., "Westinghouse Visual Inspection and Industrial Robot Control System", Proceedings of the 1st International Conference on Robot Vision and Sensory Controls, Stratford-upon-Avon, UK, April 1-3, 1981, pp. 35-46.

4. Tisdale, G. E., "A Digital Image Processor for Automatic Target Cueing, Navigation, and Change Detection," SPIE, Vol. 101, Airborne Reconnaissance - Tactical/Real Time, 1977.

5. Helland, A. R., Tisdale, G. E., "The AUTO-Q Digital Image Processor for Autoprocessing of Reconnaissance Images", NAECON '80, Proceedings, p. 102.

6. Tisdale, G. E., "Automatic Registration of Points in Two Separate Images," U. S. Patent No. 3,748,644, July 24, 1973.

7. Tisdale, G. E., "Preprocessing Method and Apparatus for Pattern Recognition," U. S. Patent No. 3,636,513, January 18, 1972.

8. Pincoffs, P. H., et. al., "Classification Method and Apparatus for Pattern Recognition," U. S. Patent No. 3,638,188, January 25, 1972.

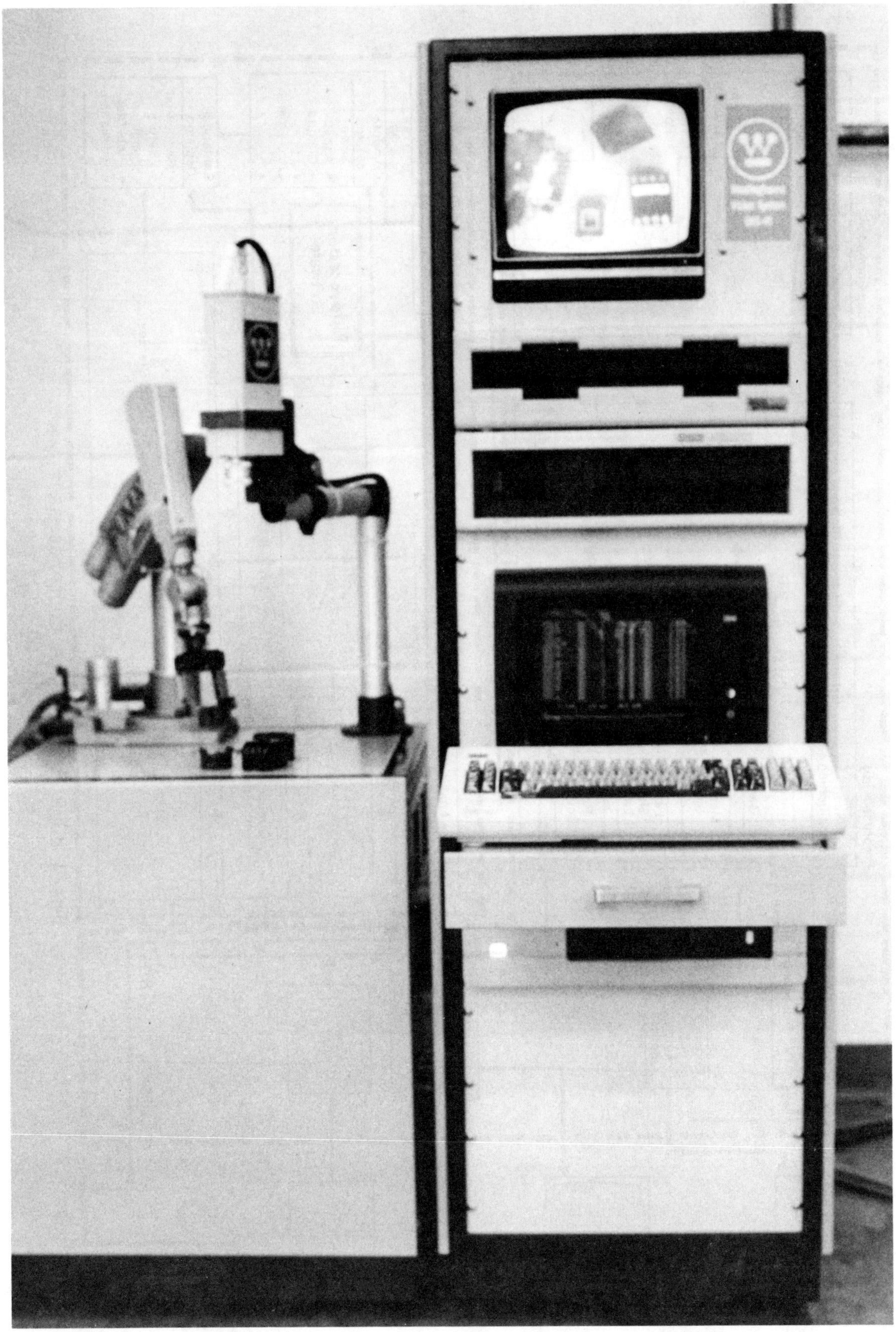

Fig. 1. Robot - vision set-up for assembly.

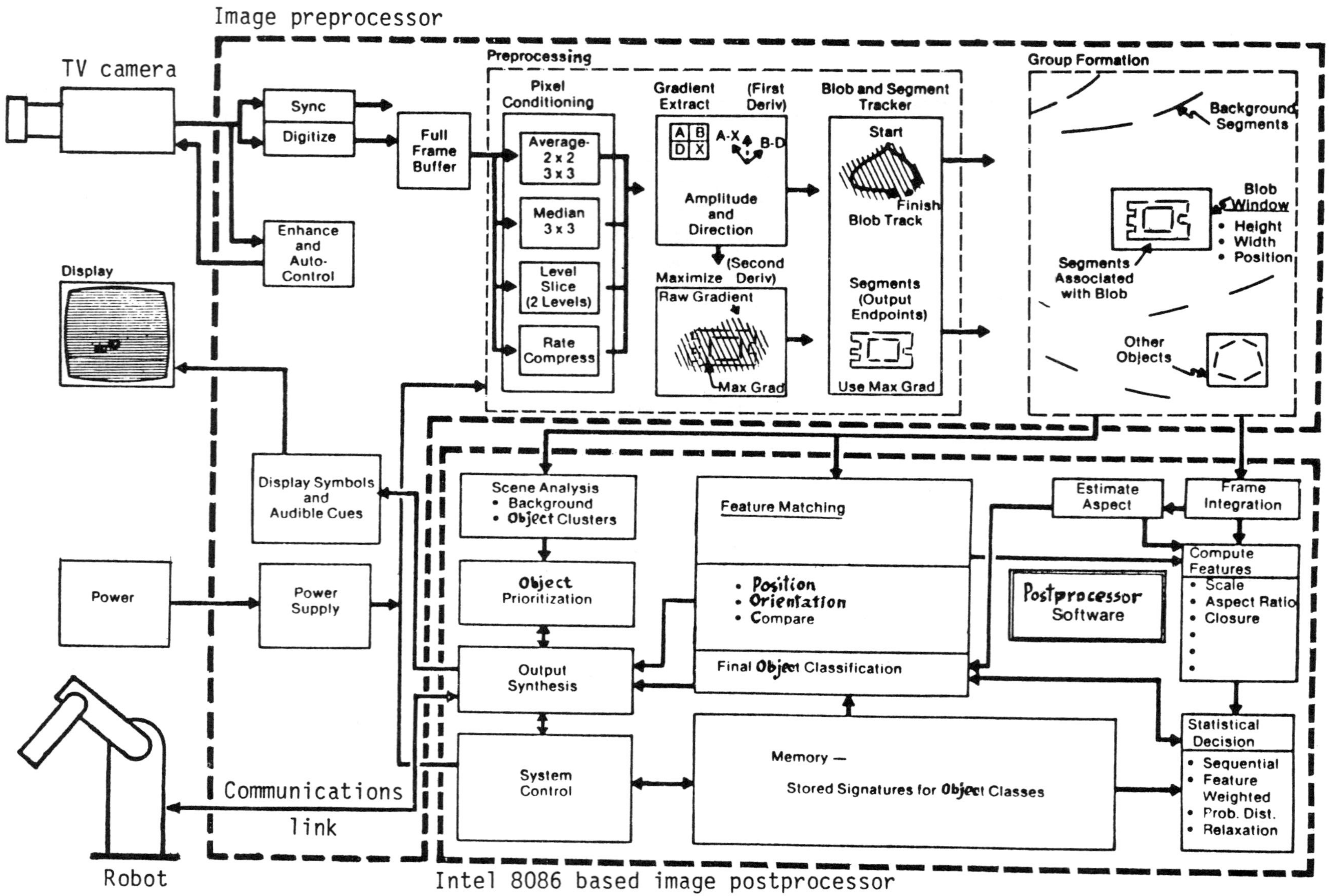

Fig. 2. Westinghouse grey scale vision system.

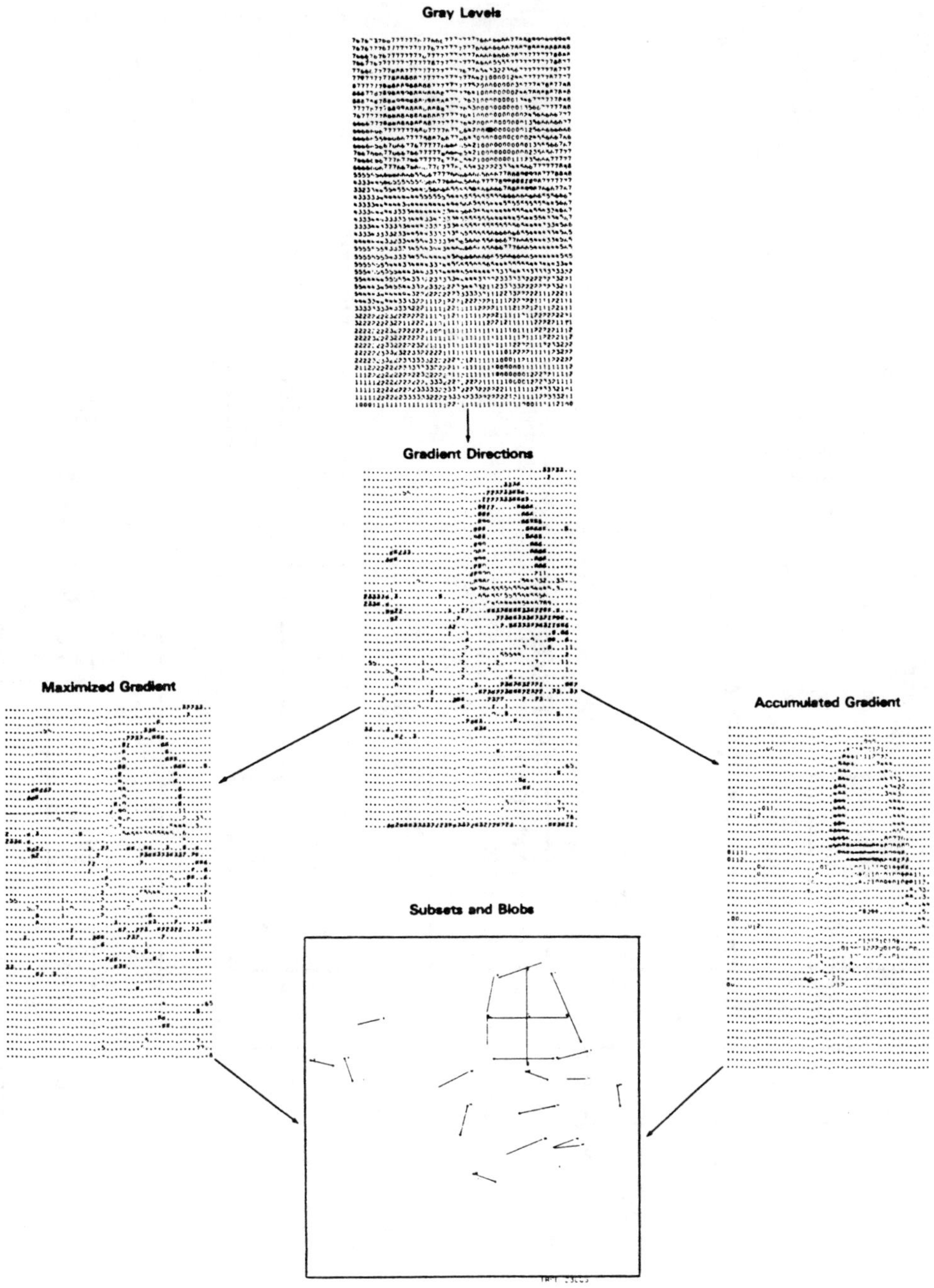

Fig. 3. Steps in the preprocessing of an image.

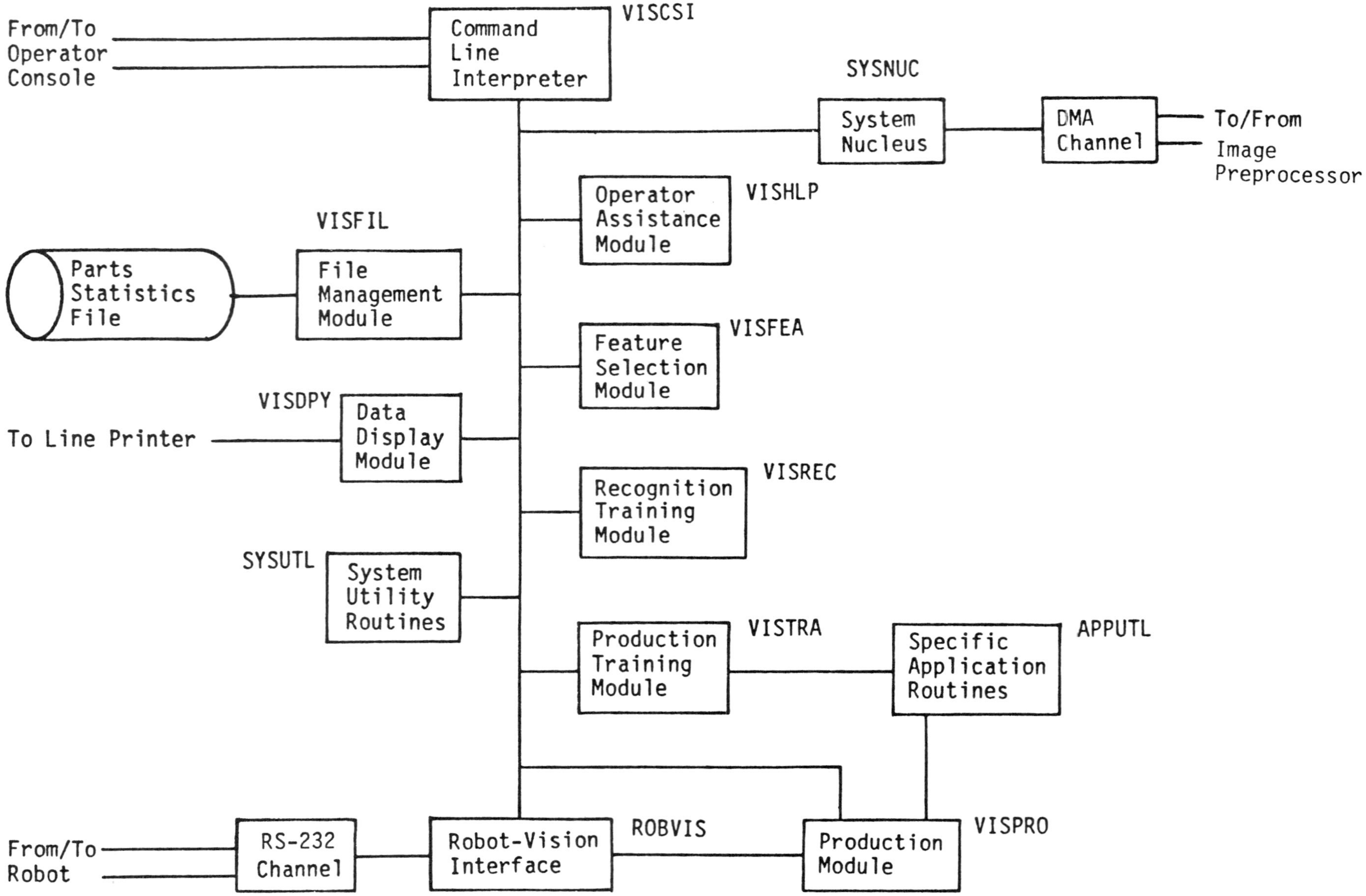

Fig. 4. Vision system software structure.

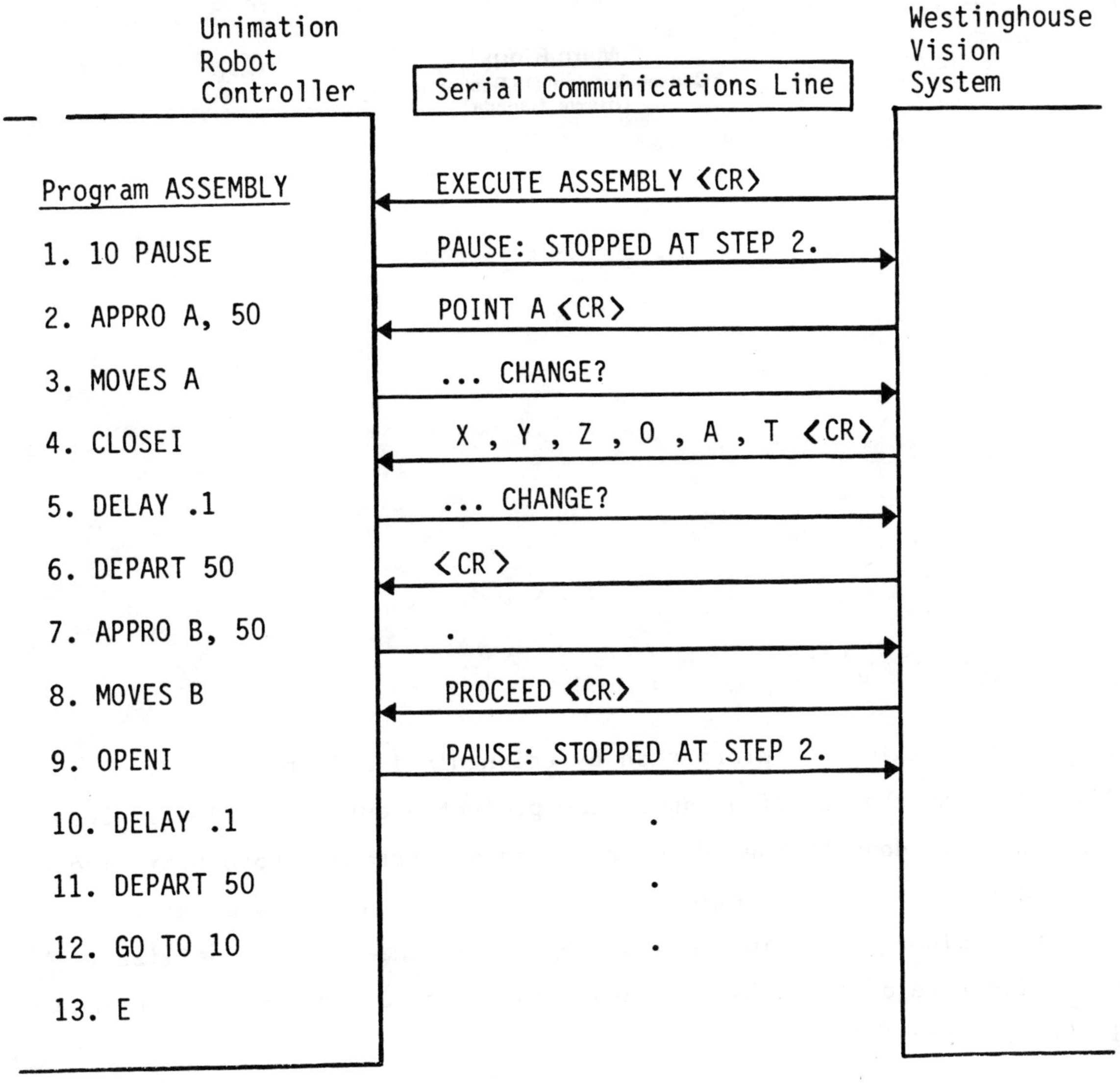

Fig. 6. Vision - Robot communications protocol and a PUMA assembly program.

3-D Camera Based on Synchronized Scanning

Marc Rioux
Conseil National De Recherches
Ottawa, Canada

Abstract

A new geometry to obtain 3-D information by triangulation is presented. The approach is based on the use of a laser beam projected onto the object. Scanning and measurements are done through scanners which are arranged to synchronize the projection and the observation angle. This arrangement allows wide-angle fields of vision, a large volume of sensing, and a compact 3-D camera to be utilized. A data acquisition rate of 1 MHz is obtained using a one dimensional position sensor (lateral effect photodiode).

INTRODUCTION

Extraction of tridimensional information from images have proven to be useful in medecine, geology, cartography and military activities. Sciences involved are photogrammetry and artificial intelligence. The former involved with the extration of 3-D coordinates from images, the latter interested in automatic interpretation of 3-D images. The importance of 3-D vision in robotics has been recognized and research activities in this field is increasingly growing. The industrial environment provides new constraints and limitations such as difficult environment, cost, compactness, etc. to the applicability of usual techniques, but, on the other hand, the proximity of objects allow active methods to be used in order to get 3-D data much more easily than passive techniques. Also, objects to be identified and manipulated are usually geometrically quite simple. One other property is that objects of the same class are quite similar to each other (in contrast to natural objects in natural scenes) so that discrimination between classes should be easier than in natural scenes specially with use of 3-D measurements of their surface profile or some sort of compact geometrical description.

Active methods where a beam of light is superimposed to the naturally lighten scene simplify a lot the signal processing to be done in order to recover distance information. Among the various techniques described in literature[1-15] we select active triangulation as an attractive approach that has the potential to evolve toward a low cost 3-D camera. Figure 1 shows the basic elements of such a system. It consist of a light source(S) that is usually a laser, a scanning mechanism(M) to project the light spot onto the object surface and a position sensor(D) with a collecting lens(L) looking off-axis for the light spot. Distance measurement is done by trigonometric algebra applied to the projection direction (scanner angular position) and to detection direction made by light spot position on the sensor with the principal point of the collecting lens. One property of the arrangement shown in figure 1 is that you can increase measurement resolution by increasing the off-axis angle (θ) but this is done at the expense of compactness and field of view. Another problem is a shadow effect that increases with the angle. This is related to the fact that some part of the object "seen" by the projection mechanism is not seen by the position sensor. That shadow effect prevents continuous profile recording and is more severe as the off axis angle is increasing. What we propose here is a new scanning mechanism approach that allow compact camera head to be designed without compromising resolution and field of view, and also limiting shadow effects to a minimum.

GENERAL APPROACH

The basic idea is to synchronize projection and detection in a way that the detected light spot on the position sensor keeps its spatial position stable when the projected beam is scanning a flat surface (scanning being parallel to that surface, see figure 2). Distance measurement is done the following way; first, we defined a reference plane which is the flat surface position that makes the camera output signal level to be "zero". That is the position in space of the intersection of the projection direction and the detection axis. Second, any object surface that departs from that reference plane will make the camera read its position (P_0) relative to that reference plane (P_1) expressed as $\pm z$ depending on which side of the reference plane the departure is. By geometrical analysis we know that the amount of spot position shift onto the position sensor is proportional to Δx_0. Now, providing θ is small, (that is highly desirable to make compact camera head), then Δx_0 is related to "z" by the following equation:

$$\Delta x_0 \simeq z\theta$$

where "z" is the departure of the object surface from the reference plane, and θ the angle of intersection at the reference plane of the projection and detection axis. Now, because we are scanning both the projection and the detection, we can keep wide field of view keeping resolution along "z" axis independent of "x" and "y" scans. Resolution of the camera along "z" axis can now be modified by changing the focal length of the collecting lens without reduction in field of view. We can summarize properties of a synchronized approach as following:

- first, under limited field of view ($\leq 5°$), no trigonometric computation is required to get profile amplitude (along "z" axis). Measurement of the scanners positions provide both "x" and "y" coordinates. In other words, suppose the position sensor used can resolve "m" positions, then the measuring system with synchronized scanners will provide a resolution of "m" positions along the "z" axis, the relationship between the two being linear.
- second, with increase in computational requirement for signal processing, the camera can be designed to cover a very wide field of view using rotating mirror scanners (90°x90°).
- third, a zoom effect can be realized by using a zoom lens as a collecting lens. Depth of the volume sensed is then reduced proportionally to the effective focal length of the zoom lens. On the other hand, we still get "m" positions resolved along depth, that means we can measure small surface details that are not resolved at full depth of view.
- fourth, although the scanning is done following two axis (x, y), the position sensor required to measure the light spot shift has not to be a bidimensional device (in contrast with the static triangulation). Indeed, for line profiling as well as surface profiling all we need is a unidimensional position sensor. The response time of a linear device is about two orders of magnitude faster than a bidimensional device consequently the output rate can be improved with synchronization accordingly. On the other hand, we have to take care at the design geometry to get that property. A few examples will be given in the following section.
- fifth, considerable reduction of the shadow effect problem normally encountered with all triangulation approaches is obtained. It is well known that shadow effect problems increase with the off-axis angle "θ". As we can keep "θ" small, shadow effect problems are kept to a minimum.

DISCUSSION

The implementation of a synchronized scanning mechanism can be done electrically or geometrically. Figure 3 shows an arrangement to electrically synchronize two galvanometer driven mirrors G_1, G_2. In this special case the source signal is a ramp, then the output signal from a position sensor seen on an oscilloscope provides a direct profile reading for which the time axis is proportional to displacement along "x" axis. Amplitude of the deflection is proportional to departure of the object surface from the reference plane. One interesting feature of that arrangement is that the position of the reference plane can be set electronically by modifying the phase relationship of the two excitation signals. In order to get a surface profile measurement a third galvanometer driven mirror (indicated as G_3 on figure 3) is used to deflect perpendicularly to the page both the projected and the received beams. As mentioned previously that geometry do not require a bidimensional position sensor. Figure 4 shows an arrangement that is geometrically synchronized. Both mirrors F_1 and F_2 are fixed when the device is in operation. Angular adjustment of those mirrors sets the position of the reference plane in space. Synchronization is realized using a pyramidal rotating mirror "P". By geometrical analysis, we see that synchronization can be obtained by using two opposite facets of the scanner. That approach is simpler to implement than electrical synchronization and is also obviously more precise and stable. In order to help the reader understand figure 4, we can mention that the light beam incident on the pyramidal mirror is perpendicular to the page as well as the reflected beam from the mirror G. An interesting feature here is the ability of rotating mirrors scanners to go at very high speed (~10,000 lines/s) compared to galvanometers (~100 lines/s). The arrangement shown in figure 4 has been realized experimentally in our laboratory and will be discussed in detail in a following section. At this point it is worthwhile to mention a few design considerations related to the choice of a light source, a scanner type and a position sensor.

LIGHT SOURCE

Although any light source can be used, a laser beam has many advantages over conventional ones:

- interferometric filter can be used to improve signal to noise ratio at detection
- brightness is orders of magnitude higher than an incoherent source
- compact devices (specially diode lasers)
- cheaper and going to be much cheaper
- very large depth of field due to spatial coherence

Until recently, He-Ne's were a good compromise for such an application. The picture may change in favour of diode lasers that are becoming more reliable, more powerful and cheaper. Another advantage of diode lasers is that the infrared emission match very well the peak sensitivity of silicon detectors (most commonly used material for position sensor devices). Furthermore, the laser output intensity can be electrically controlled. Some of the new devices have an integrated photodiode that allows servo loop control of the laser output intensity.

SCANNERS

Choice of scanners to be used is application dependent. Parameters to be taken into consideration are scanning speed and resolution. Also, random access operation, no moving part requirement and cost are parameters that will influence on the scanner type choosen.

Scanners can be grouped into three types:

- galvanometer driven
- rotating polygon
- acousto-optic devices

Galvanometer types and acousto-optic devices can be addressed at random with typical response time of respectively 1ms and 10μs. Acousto-optic devices have also the advantage of having no moving parts. On the other hand, they are usually more expensive and also limited in resolution. Another drawback of acousto-optic devices is the difficulty to get large scan angles.

Polygon mirrors are a good choice for very high speed requirements but we cannot address them randomly. Typical speeds are 2 to 10 thousand lines per second with 2 to 5 thousand resolved spots per line. Scan angles cover typically 20° to 120°.

We should also mention the emergence of holographic scanners made to replace polygon scanners. Success in that direction should lead to very low cost high speed scanners.

POSITION SENSORS

There are various types of position sensors, there are unidimensional (or linear devices) and bidimensional geometries. As discussed earlier, a synchronized approach can allow the use of a linear device for both, line profile and surface profile measurements. So, we will limit our discussion to linear types. There are three main types of such sensors. The first one is an analog device referred to as lateral effect photodiode (LEP). It is made of a photodiode on top of a resistive layer. Interesting features of the devices is the response time that can be as low as 500ns for small devices, it is also a low cost sensor. Furthermore, it is insensitive to large amounts of defocussing and the signal output amplitude is proportional to the centroïd of the spot. Another interesting property is that we can get a reflectance map of the scene under scan that is in perfet registration with the profile (or 3-D) image. Usefulness of such a measurement has been described by Nitzan et al.[16]

The second type is the linear array of discrete photoreceptors (photodiode arrays, CCD, CID...). Main characteristics of those devices are:

- high sensitivity

- very high spatial resolution (up to 4096 elements available commercially)

- very good geometrical precision

They are more expensive than LEP, we also need to scan a full line to get position information (so the maximum data rate is in the order of 10KHz) and they need much more signal processing than LEP to extract position signals. They also are limited in dynamic range due to sensitivity to defocusing and the cost of a position sensor based on linear arrays is much higher. On the other hand, such a device is required for mapping large volume scenes or distant objects. As an example, a typical arrangement using a low power(~mw) laser to map 3-D coordinates of a scene using a LEP sensor will be limited to objects no more distant than about one meter from the camera. In contrast a CCD linear array can map objects as distant as 10 meters. A key figure to remember here is that the bandwidth of the 3-D camera reduces as the distance increases.

A third type is the dual element photodiode. Useless with usual triangulation geometry that type of sensor can be used advantageously with synchronized approaches. It consists of two photodetectors located side by side. It has the advantage of having a bandwidth of more than 30 MHz. Signal processing circuitry required to cope with such a speed is the limiting factor for that device. It is also sensitive to geometric distortion of the spot and has a very small displacement range. Nevertheless, its low cost makes it attractive for the development of a video rate 3-D camera. Beside these well-known position sensors, there are simple geometrical arrangements that have been suggested in the literature that provide high speed, high resolution position sensing using a single photoreceptor (see ref. 8 and 9). Those arrangements can also be used with synchronized geometries with the advantage of increased dynamic range along "z" axis.

EXPERIMENTAL RESULTS

An experimental set-up as shown in figure 4 has been realized in our laboratory. The laser source is a He-Ne having 15 milliwatts of output power. A pyramidal mirror is used as a synchronized scanner. It consists of six facets that provide 60° of scanning angle. Individual mirror surface is about $5cm^2$. A lateral effect photodiode is used to make measurements. Camera to object distance is 75cm. Experimental results are shown in figures 5, 6, 7 and 8.

Figure 5 shows a photograph of the scene to be scanned. It is a stepped pyramid of 10cm on the side and 8cm in height. Each step represent a 1cm change along "z" axis of the 3-D camera. Figure 6 shows an intensity image display of the 3-D image. To obtain that display, we convert the analog position signal to digital data. Those data representing the amount of departure along "z" axis of the object surface from the reference plane are stored in an image memory whose line "i" and column "j" correspond to "x" and "y" coordinates. Then the image data contained in the memory is displayed on a T.V. monitor. So, areas appearing white on the photograph correspond to elevated areas where dark areas correspond to surface near the reference plane. A more interesting way to display the 3-D data is shown in figure 7 where an isometric view of the same data is shown. Furthermore, as mentioned-above, in addition to the 3-D image, we can get, using a LEP sensor, an intensity image of the actual object perfectly registered with the 3-D data. This is shown in figure 8. Each pixel represents the amount of light reflected by the object surface at that particular address. At the same address in the 3-D image, we have the (x,y,z) spatial position of that surface. Both information are perfectly registered and can be used to get an absolute reflectance map of the object surface.

3-D CAMERA AND ROBOTIC

One of the objectives of our project is to develop a low cost, compact, high speed 3-D camera to be used in automatic assembly. Having the possibility to grab 3-D data of a scene in a fraction of a second has many practical advantages, specially in an industrial environment. It reduces interference problems; mechanical vibration as well as acoustic and electromagnetic. Furthermore, moving object can be measured. Coupled to a robot arm a 3-D camera can be used to identify parts in the scene. On the other hand, we have to remember that a 3-D camera provide only the raw data (with noise) and there are still a lot of fundamental research to be done in order to use efficiently the 3-D data for object recognition. At least, one advantage of a 3-D camera is that measurement of the object geometry is not a projection (as 2-D imagery), indeed, we are actually measuring without ambiguity the surface profile of the object under examination.

In term of signal characteristics, the output from a 3-D camera as described here is very similar to a T.V. like signal. In a 3-D camera the only difference is that the amplitude of the signal is related to geometrical characteristics of the object. In contrast 2-D cameras (like ordinary T.V. camera) provide an output signal amplitude that is not geometrically related to the object, instead it represents the surface reflectance properties of the object surface combined with the ambient light conditions of orientation, intensity and spectral characteristics. In addition the net result is affected by the orientation of the object and by proximity of other objects. That is why 3-D features extraction from a 2-D image of a scene is so difficult to realize. A 3-D camera using an active approach also provides good immunity to background noise. This is done by the use of an interference filter that allows only laser light to reach the sensor.

Consequently, for robot assembly tasks, we can expect that a low resolution 3-D image would be needed to provide good discrimination between known objects, as opposed to inspection task where you need to make fine comparison between an ideal part and the actual manufactured part.

As an example, a 3-D camera designed for resolution of 100x100x100 voxels (volume element) and using a lateral effect photodiode as a position sensor could grab the entire field in less than one hundredth of a second. That should be fast enough for most assembly tasks.

CONCLUSION

A laser range finder based on triangulation has been described. New geometries are suggested using synchronized and autosynchronized scanners. Experimental results show megahertz rate measurements. With the increasing development of diode lasers and holographic scanners, the new geometries proposed should lead to compact, low cost, high speed 3-D cameras to be realized. We see besides robotic applications, inspection of manufactured parts, as well as medical applications such as prostheses fabrication and in diagnostic instrumentation.

ACKNOWLEDGMENTS

The author wishes to thank P. Chartier and J. Domey for many valuable discussions and help during the implementation of the 3-D camera system. I would also like to thank L. Cournoyer and R. Misner for their excellent technical support in realizing the programming and the constructing of the experimental set-up.

BIBLIOGRAPHY

1 - V. Bodlaj and E. Klement
"Remote measurement of distance and thickness using a deflected laser
APPL. OPT.; 15(6); June 1976, pp. 432-6.

2 - J.D. Boissonnat and F. Germain
"A new approach to the problem of acquiring randomly oriented workpieces out of a bin".
Proc. 7th Int. Joint Conf. on artificial intelligence, 1981, pp. 796-802.

3 - M.D. Altschuler et al
"Laser electro-optic system for rapid three-dimensional (3D) topographic mapping of surfaces".
OPTICAL ENG. 20(6), 1981, pp. 953-961.

4 - M. Bernasconi et al
"Accuracy of measurement through stereo images".
5th Int. Conf. on Automated inspection and product control. June 24-26, 1980, p. 309.

5 - F. Bien et al
"Absolute distance measurements by variable wavelength interferometry".
APPL. OPT.; 20(3), Feb. 1981, pp. 400-403.

6 - J.M. Burry
"Contouring in real-time with Moiré interference".
Opt. Comm.; 41(4), April 1982, pp. 243-8

7 - Z. Fuzessy and N. Abramson
"Measurements of 3-D displacement: sandwich holography and regulated path length interferometry".
Appl. Opt.; 21(2), Jan. 1982, pp. 260-4

8 - D.P. Himmel
"A laser measuring system for automatic industrial inspection"
Proc. 4th Int. Joint Conf. on pattern recognition; 1978, pp. 952-4

9 - G. Indebetouw
"A simple optical non-contact profilometer"
Opt. Ing. 18(1), 1979, pp. 63-6.

10 - K.E. Morander
"The optocator. A high precision, non-contacting system for dimension and surface measurement and control".
5th Intl. Conf. on automated inspection and product control, 1980, pp. 393-6.

11 - C.J. Page and H. Hassan
"Non-contact inspection of complex components using a rangefinder vision system".
Proc. of the 1st Int. Conf. on robot vision and sensory controls, 1981, pp. 245-54.

12 - F. Pipitone
"A ranging camera for 3-D object recognition".
Midwest symp. on circuits and systems, 1979, pp. 339-343.

13 - T. Sawatari and R.B. Zipin
"Optical profile transducer"
Opt. Ing. 18(2), 1979, pp. 222-5.

14 - J. Taboada
"Coherent optical methods for applications in robot visual sensing".
SPIE PROC. VOLUME 283, 1981, pp. 25-9.

15 - T. Kanade, H. Asada
"Non-contact visual three-dimensional ranging devices".
SPIE PROC. VOLUME 283, 1981, pp. 48-53.

16 - D. Nitzan et al
"The measurement and use of registered reflectance and range data in scene analysis".
Proc. of the IEEE 65(2) 1977, pp. 206-220.

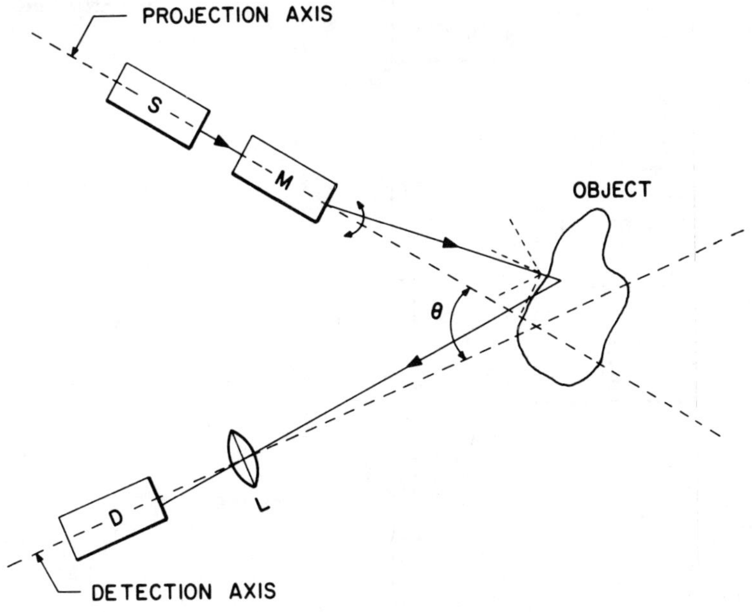

FIGURE I. LASER RANGE FINDER BY TRIANGULATION

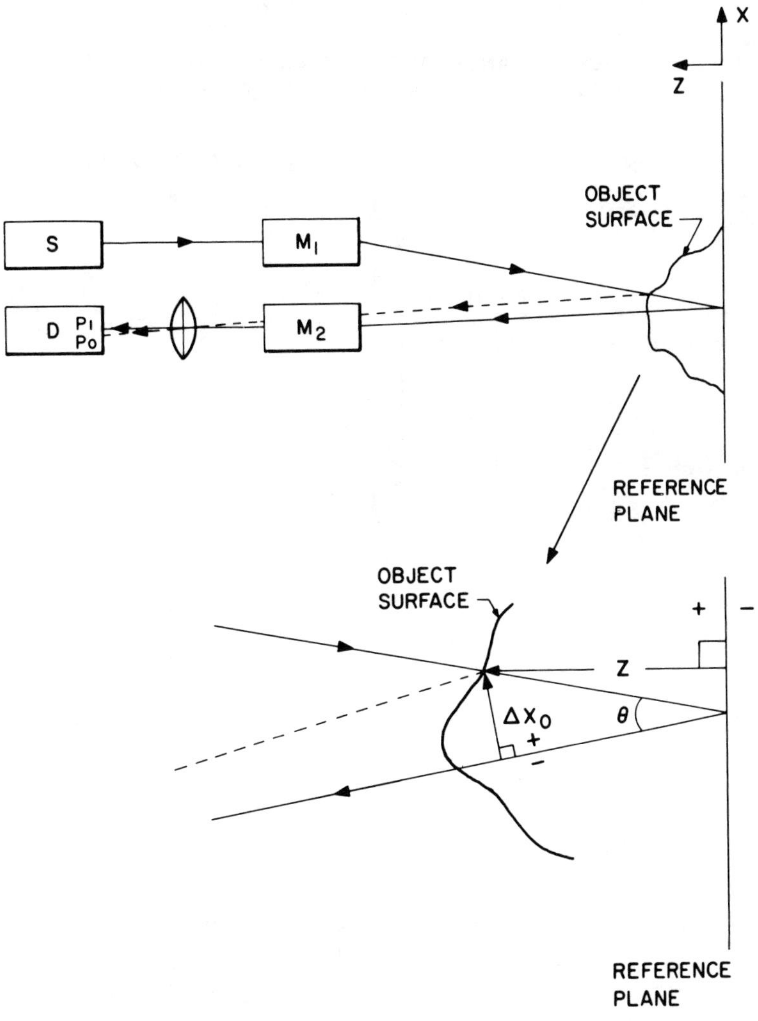

FIGURE 2. SYNCHRONIZED GEOMETRY

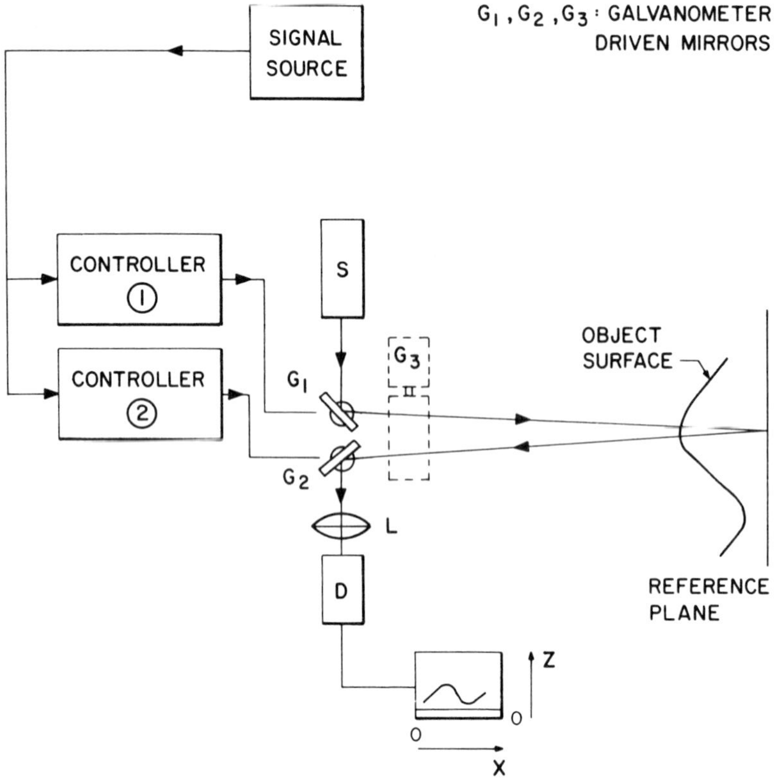

FIGURE 3. AN EXAMPLE OF AN ELECTRICALLY SYNCHRONIZED SCANNER

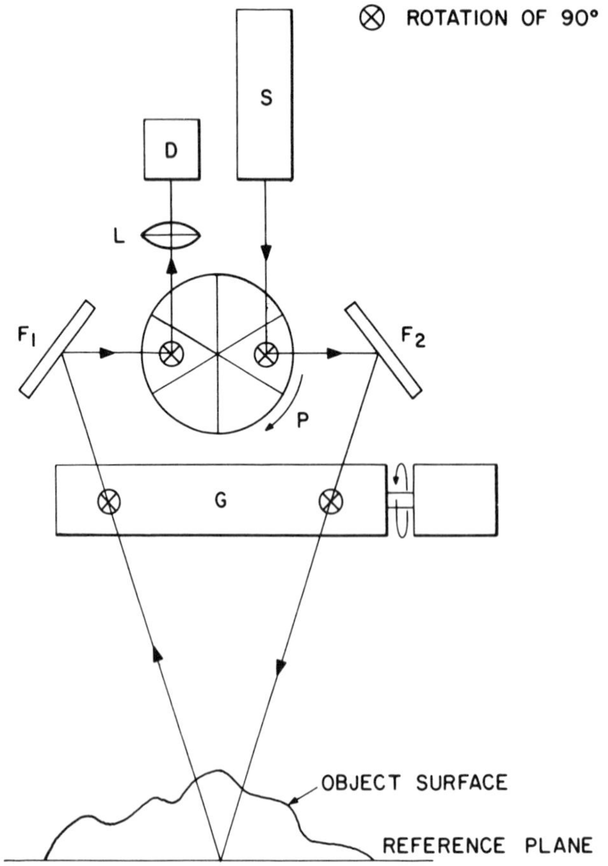

FIGURE 4. AN EXAMPLE OF A GEOMETRICALLY SYNCHRONIZED SCANNER

FIGURE 5. PHOTOGRAPH OF THE OBJECT

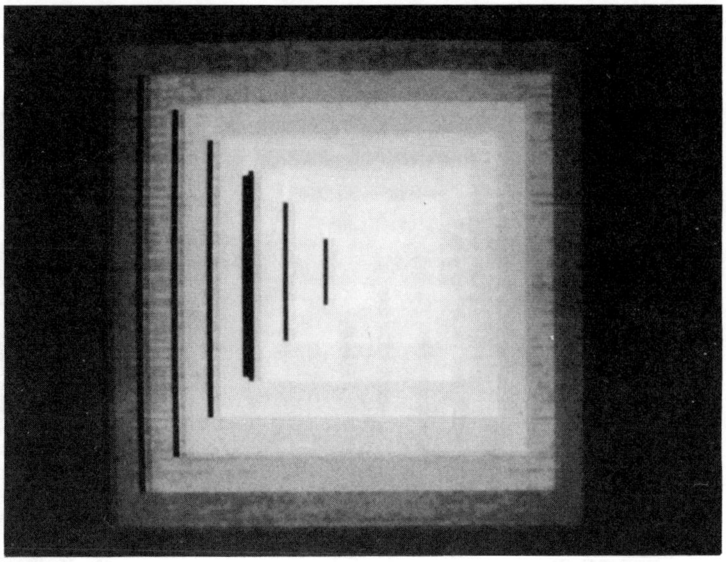

FIGURE 6. INTENSITY DISPLAY OF THE 3-D CAMERA OUTPUT

FIGURE 7. ISOMETRIC PROJECTION OF THE DATA SHOWN IN FIGURE 6

FIGURE 8. INTENSITY OF THE LIGHT AS DETECTED FROM THE SCENE

Differential Reflectance Functions and their Use for Surface Identification

P. Levi
University Karlsruhe
Karlsruhe, Federal Republic of Germany

E. Weirich
University Karlsruhe

Abstract

A rigourous vision model should incorporate a detailed understanding of the image-forming process. The radiometrical description of this process is formulated in a satisfactory way by differential (bidirectional) reflectance functions. The profitable engagement of these reflectance functions for the identification of surface properties (materials, defects) is a part of the first representational level of the early vision. We utilize a linear polarized laser beam (insensible to other natural light sources) in combination with a goniometer to measure the differential reflectance properties of typical industrial surface patches. The value of the reflected irradiance, the shape of the reflectance functions and the depolarization effects of material defects are employed to identify surfaces or surface defects.

1. INTRODUCTION

The two approaches widely used to solve computer vision problems are purely empirical (e.g. Hueckel-operator, /Hue 73/) or they are restricted to a toy world (e.g. block world, /Wal 75/. Representatives for these two approaches are edge-detectors. Until now, hundereds of edge-detectors have been developed, but only a few can be usefully empleyed in a systematic and rigorous vision approach, because these detectors are developed as an end in itself. A rigorous vision approach must include a model of all levels of image analysis (advanced vision). An excellent example for such a model is the computational vision representation of Marr /Mar 82/.

This model defines the vision process as a sequential evaluation of three representational stages: the primal sketch, the 2 1/2-D sketch and the 3D-representation. the main points of this approach are the sequential formation of these three representations and its analogy to the human visual perception. Our vision model integrates these three representational vision frameworks, but it abandons a strong sequential evaluation of the image descriptions and the analogy to the human visual sensation.

The image analysis should be performed in parallel with the set-up of the sequential representation. It starts at the first representational level of the early vision. These are the intensity changes of the image which are used for the description of the surface properties. The light source is a linear polarized laser beam.

Intensity changes are described by differential reflectance functions which are measured and parameterized. These experimental data are used to answer the following question: Can the numerical knowledge of the surface reflectance be applied satisfactoryly to the identification of surface properties (material, defect) within a reduced set of different materials (e.g. quality control)?

x) Supported by the Bundesministerium für Forschung und Technologie

2. EARLY VISION AND SURFACE IDENTIFICATION

As a part of a vision model the early vision is complex in itself. This complexity stems from the fact that the process of early vision must identify where the intensity (irradiance) values in an image are coming from. The following points are highly responsible for these values:

(1) surface geometry, (2) surface microstructure, (3) surface spatial organization (texture), (4) illumination (source position), (5) viewpoint (receiver position), (6) orientation, (7) discontinuities (depth, edges), (8) distance, (9) movement, (10) image-forming system (e.g. optical transfer function).

The purpose of the early vision is the sensation of the observed scene and the detection which irradiance changes are due to what functions. This goal must be achieved in different stages which represent different evaluating descriptions. The points (1)-(5) describe the surface properties and the points (6)-(9) define the surface dispositions under the condition that function (10) is known. So the process of early vision can be subdivided into two representational levels (properties, dispositions) which are further subdivided into sublevels (Fig. 1).

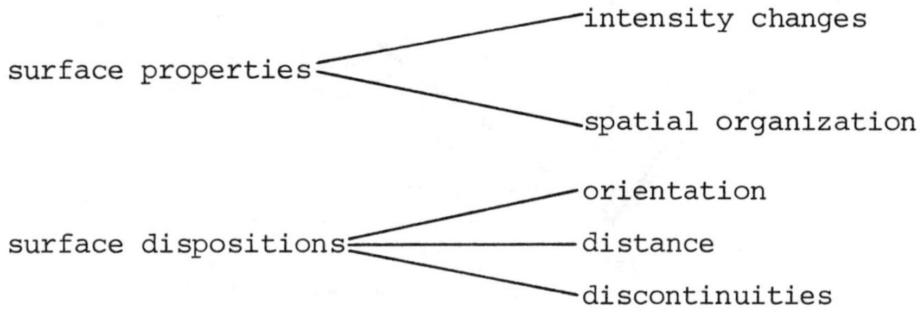

Figure 1 : The main representational frameworks of the early vision

The surface (object) identification is running in parallel with the sequence of representations done in the process of early vision. So an identification of surface properties (e.g. glass, sheet, rubber, etc.) may happen directly after the intensity changes are detected. But, this identification can only be done if the physical background of the image-forming process is integrated into the surface identification process. The relation of the geometry of image projection to the radiometric characterizations of the image formation is given by differential reflectance functions. These are bidirectional reflectance functions which relate the incident density flux to the reflected density flux.

Nowadays, the reflectance map techniques of Horn /Hor 79/ and Woodham /Woo 78/ renews the old photogrammetric knowledge of reflectance functions and relate the points (1), (3), (4) and (5) to the intensity changes in an image. But, the reflectance map technique is usually restricted to perfect diffuse surfaces and it results in the following two sinplifying assuptions:

(1) The incident light is coming in over a whole hemisphere.
(2) The image projection is an ortographic one.

The first restriction implies that the scene radiance can be expressed in a viewer-oriented system as a function of a two dimensional gradient space. The second restriction implies that the viewing direction is constant (all rays from the object are parallel). These two restrictions are too strong for most of the real robot applications in an industrial environment. Therefore, we quit these assumptions.

3. DIFFERENTIAL REFLECTANCE FUNCTIONS

3.1 Unpolarized Light

We assume that there is only one monochromatic source which produces a collimated unpolarized beam. If this beam strikes the surface patch dA from the incident direction θ_i, ϕ_i (Fig. 2) with a radiance L_i [watt $m^{-2} sr^{-1}$], then the incident radiant flux Φ_i [watt] is defined by

$$\Phi_i (\theta_i,\phi_i) = L_i (\theta_i,\phi_i) \cos\theta_i d\omega_i \, dA,$$

where $d\omega_i$ is the small incident solid angle. The factor $\cos \theta_i$ describes the foreshortening of the incident solid angle. In general, we will use the abbreviation $d\Omega = d\omega \cos\theta$ for the forshortened incident or reflected solid angle.

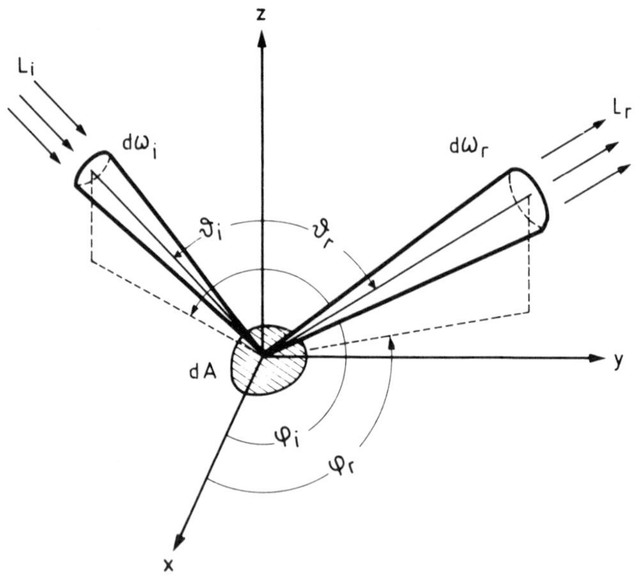

Figure 2: Geometry of differential reflectance

The incident irradiance E_i [watt m^{-2}] is defined by

$$E_i (\theta_i,\phi_i) = L_i (\theta_i,\phi_i) \, d\Omega_i$$

The reflected density flux is given by

$$\Phi_r (\theta_r,\phi_r) = L_r (\theta_r,\phi_r) \cos\theta_r \, d\omega_r \, dA = L_r (\theta_r,\phi_r) \, d\Omega_r \, dA$$

The differential reflectance function $d\rho$ is defined by the following formula

$$d\rho (\theta_i,\phi_i;\theta_r,\phi_r) = \Phi_r(\theta_r,\phi_r)/\Phi_i (\theta_i,\phi_i) = L_r (\theta_r,\phi_r) \cos\theta_r d\omega_r/E_i (\theta_i,\phi_i)$$

The quotient L_r/E_i specifies the reflectance in terms of the incident and the reflected beam. This expression is equivalent to a scattering amplitude (probability density) in a physical scattering experiment and was introduced by Nicodemus /Nic 77/ as the bidirectional reflectance-distribution function f_r [sr^{-1}]:

$$f_r (\theta_i,\phi_i ;\theta_r,\phi_r) = L_r (\theta_r,\phi_r)/E_i (\theta_i,\phi_i).$$

With the aid of this function the differential reflectance function per reflected solid angle $d\omega_r$ can be defined as follows

$$d\rho /d\omega_r = f_r (\theta_i,\phi_i;\theta_r,\phi_r) \cos\theta_r$$

It is obvious that

$$f_r = d\rho \,/\, d\omega_r$$

is the differential reflectance function per foreshortened reflected solid angle. There are two idealized approximations for f_r: a diffuse or "lambertian" surface element dA and a specular surface patch dA.

For a perfectly diffuse surface L_r is constant. So f_r is also constant if E_i is not changed and has the form:

$$f_{r,\,diff}\,(\theta_i,\phi_i\,;\theta_r,\phi_r) = a_{diff}\,(\theta_i,\phi_i\,;\,2\pi)/\pi$$

here a_{diff} is the albedo factor which is constant ($a_{diff}(\theta_i,\phi_i\,;\,2\pi) = a_{diff}(2\pi;2\pi)$) and which defines how much of the incident beam is reflected at all. To be more precise, a_{diff} is the directed hemispherical reflectance, which is generally defined by

$$a(\theta_i,\phi_i;2\pi) = \int_{2\pi} L_r\,(\theta_r,\phi_r)d\Omega_r/L_i(\theta_i,\phi_i)d\Omega_i$$

$$= \int_{2\pi} L_r\,(\theta_r,\phi_r)\cos\theta_r\,\sin\theta_r\,d\theta_r\,d\phi_r/E_i(\theta_i,\phi_i)$$

The integration over the reflected hemishere is symbolized by 2π.

For perfectly specular reflectances, f_r is defined by a Dirac delta-Function:

$$f_{r,sp}\,(\theta_i,\phi_i;\theta_r,\phi_r) = 2a_{sp}(\theta_i,\phi_i;\,2\pi)\delta\,(\sin^2\theta_r - \sin^2\theta_i)\delta\,(\phi_r - \phi_i \pm \pi)$$

Here, the specular albedo factor a_{sp} is again constant

$$a_{sp}\,(\theta_i,\phi_i;\,2\pi) = a_{sp}\,(2\pi,2\pi)$$

and is similiar defined as above

$$a_{sp}\,(\theta_i,\phi_i;\,2\pi) = L_r\,(\theta_r,\phi_r)/L_i\,(\theta_r,\phi_r \pm \pi)$$

Real reflectance functions are not just a combination of a delta-function (perfectly specular reflection) and a radianc which is constant in all direction (perfectly diffuse reflection). Real reflectance functions deviate substantially from perfectly specular and perfectly diffuse descriptions.

3.2 Polarized Light

If the incident, collimated beam is polarized and the polarization of the reflected beam is measured, then the differential reflectance functions d per reflected solid angle d_r can written

$$(d\rho/d\omega_r)^{p_i,p_r} = f_r^{p_i,p_r}\,(\theta_i,\phi_i;\theta_r,\phi_r)\cos\theta_r$$

Here, the indices p_i, p_r describe the polarization of the incident or of the reflected beam.

If we use the four Stokes parameter /Hech 74/ to define the polarizations p_i,p_r, then in the most general case we need two 4x4 matrices to define $d\rho/d\omega_r$. It is obvious, that especially for real time applications this is not possible. For this reason, the reflected polarization p_r was not measured (sections 3 and 4) or we measured only that part of the reflected beam which was transmitted by a crossed analyzer (section 5).

4. SURFACE IDENTIFICATION BY DIFFERENTIAL REFLECTANCE FUNCTIONS

The measurement of the differential reflectance functions was done by a goniometer which is shown in Fig. 3.

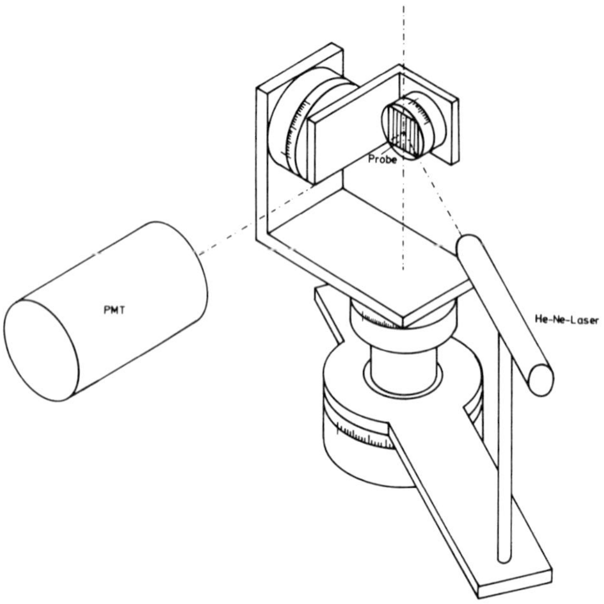

Figure 3: Constructed goniometer for measurements of reflectance functions

This goniometer can be used to define the four angles (rotary rings), which are needed for the measurement of the differential reflectance functions (see Fig. 1). The light source was a linearly polarized 5mW He-Ne laser. The receiver was an aperture photomultiplier.

The reflected radiance L_r is defined by the measured reflected irradiance E_r which strikes onto the receiver

$$L_r = M_r/d\Omega_r = (E_r/d\Omega_r)\ (dA_r/dA)$$

Here, M_r is the radiant exitance and the surface patches dA_r of dA define the surface elements of the receiver or of the reflecting surface. Then the bidirectional reflectance distribution f_r is given by

$$f_r(\theta_i,\phi_i;\theta_r,\phi_r) = L_r(\theta_r,\phi_r)/E_i(\theta_i,\phi_i) = E_r(\theta_r,\phi_r)/E_i(\theta_i,\phi_i)\ (dA_r/dA)$$

The incident irradiance E_i is defined by

$$E_i(\theta_i,\phi_i) = E_i(0)\cos\theta_i.$$

Figures 4-9 show the differential reflectance for some characteristical surface patterns, which are not textured. These materials are glas, sheet metal, rubber, plastic and white paper. The incident azimuth angle ϕ_i was equal 0° and the reflected azimuth angle ϕ_r was equal 180 0°. As mentioned above, the reflected polarization p_r was not measured in these series of experimental data. The error bound for the measured data is ± 2,5 %. The main sources are the photomultiplier and the angle adjustion at the goniometer.

Fig. 4 shows the well known delta-form of the differential reflectance function $(d\rho/d\omega_r)_{1lin}$ for a mirror. Here, as well as later on, the index lin means that p_i=horizontal linear polarized and p_r is not measured. The curve is normalized to 1 (division by the maximum value).

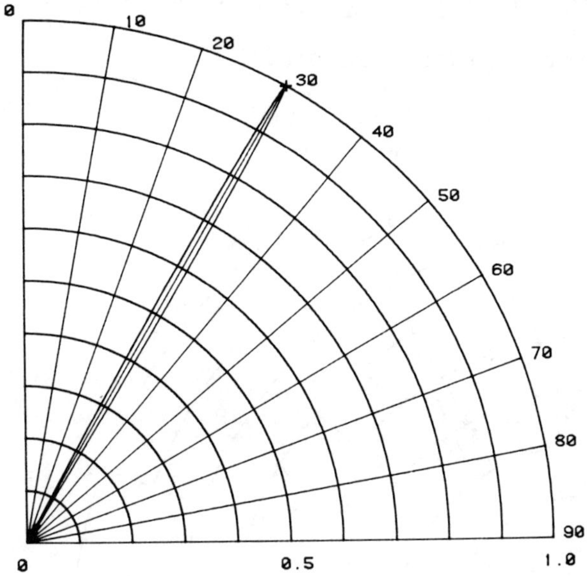

Figure 4: Polar diagram of $(d\rho/d\omega_r)^{lin}$ for specular reflectance ($\theta_i=30^0,\phi_i=0^0;\phi_r=180^0$; max. value = 12.8)

The reflectance function for specular reflectance may be described with an error of less than 1% as:

$$f_r^{lin}(\theta_r) = h * 8.65 * 10^{-6} * \cos(\theta_r-\theta_i)/(1+8.65 * 10^{-6} - \cos(\theta_r-\theta_i)),$$

where h = 12.8 is the maximum value.

Figure 5 summarizes the experimental results for a sample of sheet metal.

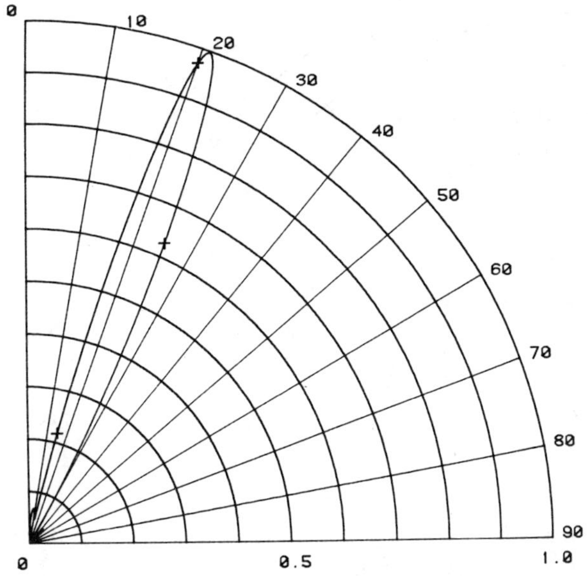

Figure 5: Polar diagram of $(d\rho/\omega_r)^{lin}$ for a sample of sheet metal ($\theta_i=21^0,\phi_i=0^0;\phi_r=180^0$; max. value = 8.44)

As for the mirror the reflectance function for the metal sheet can be described as

$$f_r^{lin}(\theta_r) = h * 0.0024 * \cos(\theta_r-\theta_i) / (1.0024 - \cos(\theta_r-\theta_i))$$

Fig. 6 shows $(d\rho/d\omega_r)^{lin}$ for a plastic sample.

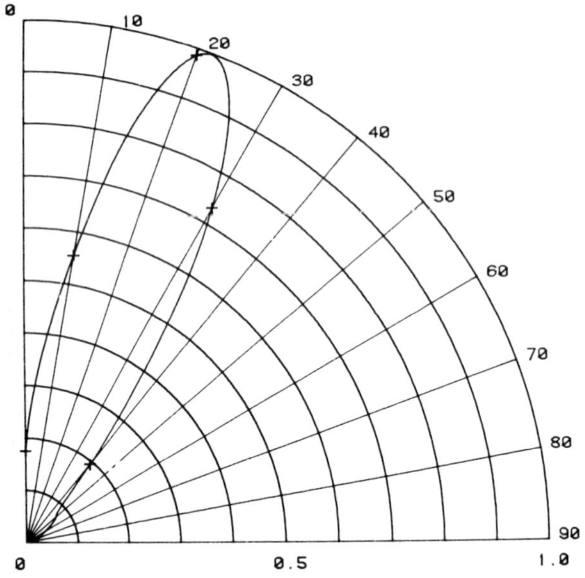

Figure 6: Polar diagram of $(d\rho/d\omega_r)^{lin}$ for a plastic sample ($\theta_i = 21^0, \phi_i=0^0; \phi_r=180$; max. value = 4.98)

Fig. 7 shows $(d\rho/d\omega_r)^{lin}$ for a manipulated (glossy) rubber sample.

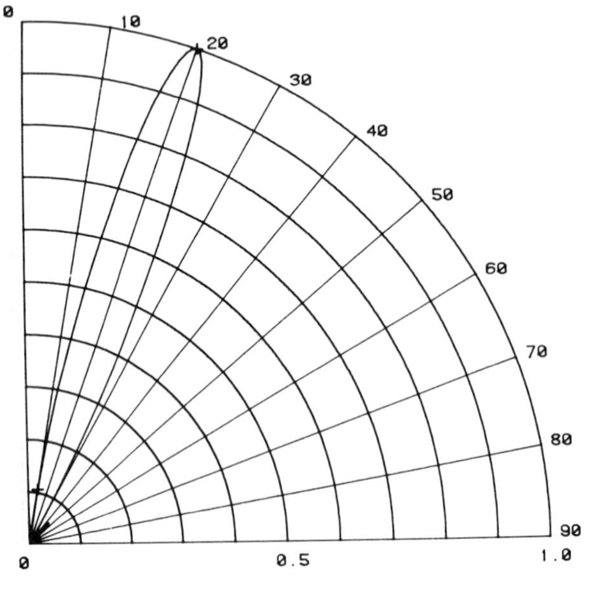

Figure 7: Polar diagram of $(d\rho/d\omega_r)^{lin}$ for a (glossy) rubber sample ($\theta_i=20^0, \phi_i=0^0; \phi_r=180$; max. value = 2.93)

Figure 8 represents the experimental result for white paper.

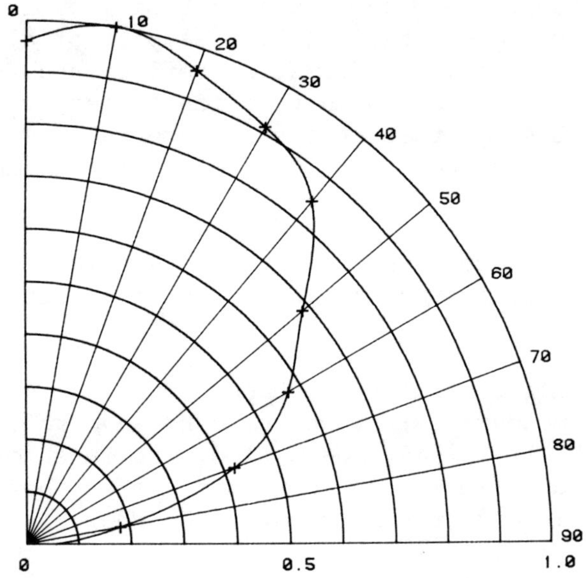

Figure 8: Polar diagram of $(d\rho/d\omega_r)^{lin}$ for white paper ($\theta_i=20^0, \phi_i=0^0; \phi_r=180^0$; max. value = 0.77)

Figure 9 shows the graph of this function. The deviations from a "lambertian" are obvious.

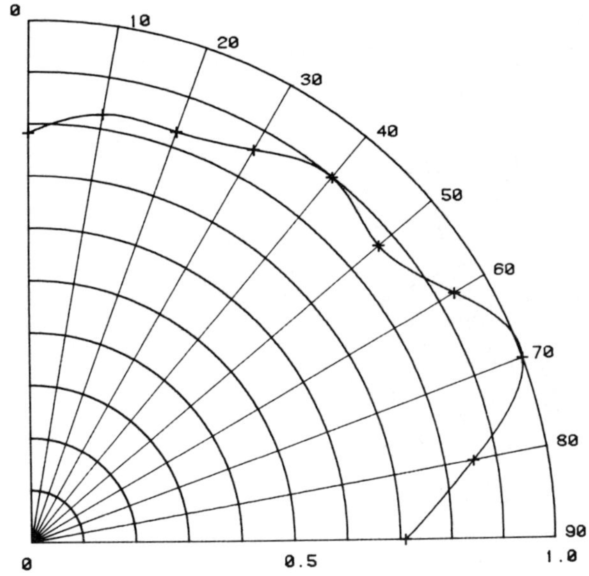

Figure 9: Polar diagram of f_r^{lin} for white paper ($\theta_i=20^0, \phi_i=0^0; \phi_r=180^0$; max. value = 0.77)

Summarizing our experimental results, we can state that usually the absolute value and the shape of the differential reflectance function are needed in order to identify non textured surfaces of different materials. Many of different surface patches show different, characteristical reflectancies. The identification process is easy, if the reflectance is not altered by rotations of a surface patch (no microstructure), and it is hard - but realizable - if there are prefered reflectance directions like for brushed aluminium or velvet (microstructure).

Fortunately, many surface layers have no directionality in their microstructure. Unfortunately, there are also some materials like samples of aluminium and polished sheet metal which may have identical differential reflectance functions (values and shapes). In these cases, the surface can only be identified by a multi-spectral scanning ("white" laser). But, besides these types of materials, we can say that differential reflectance functions can be usefully involved in assemblies, inspections etc. of non textured materials.

The identification of surface patches which are spatially organized must be executed in the second step (symbolic representation) of the description of the surface properties (see Fig. 1).

The knowledge of the differential reflectance functions cannot be only used for the identification surface properties; it can also be involved in producing synthetical images of objects which are already identified. The match of the real and the synthetical image of a surface sample can be used to detect defects. The differential reflectance functions of some defect surfaces will be shown in the next section.

5. DEFECT DETECTION BY DIFFERENTIAL REFLECTANCE FUNCTIONS

Surface defects change the reflectance properties of the surface. Defects absorb, deflect, scatter and alter the polarization of the incident beam. The depolarization will be described in the next section. Here, we present the modifications of the differential reflectance functions by defects for some of the surface samples represented before.

Figure 10 shows $(d\rho/d\omega_r)^{lin}$ for a scratched mirror.

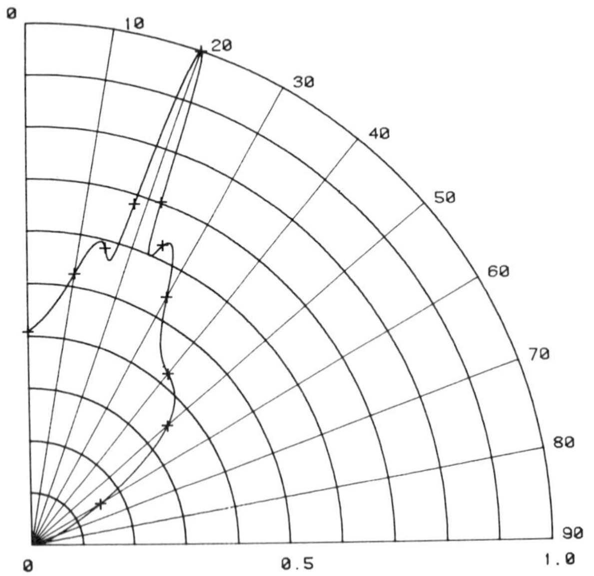

Figure 10: Polar diagram of $(d\rho/d\omega_r)^{lin}$ for a scratched mirror ($\theta_i=20^0,\phi_i=0^0;\phi_i=180^0$; max. value = 0.76)

Figure 11 demonstrates the differential reflectance functions for the metal sheet sample used for previous measurements, which is rusted.

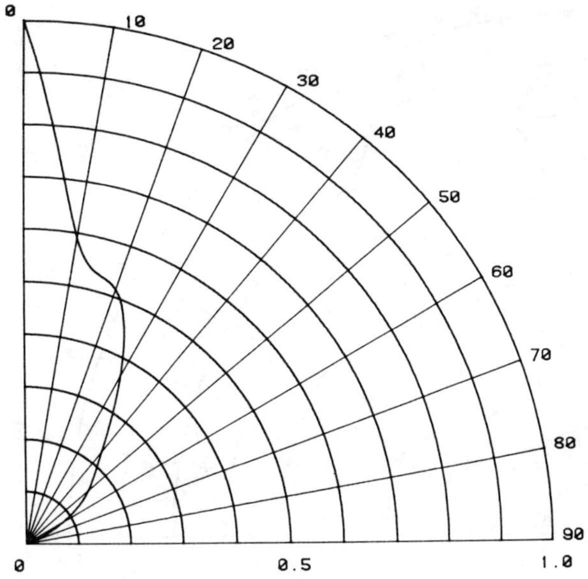

Figure 11: Polar diagram of $(d\rho/d\omega_r)^{lin}$ for a rusted sheet metal ($\theta_i=20^0,\phi_i=0^0;\phi_r=180^0$; max. value = 3.87)

The differences between figures 5 and 11 are clearly visible.

Figure 12 represents the differential reflectance functions for the plastic sample just used for measurements before, which is now scratched. The difference between figures 6 and 12 is obvious again.

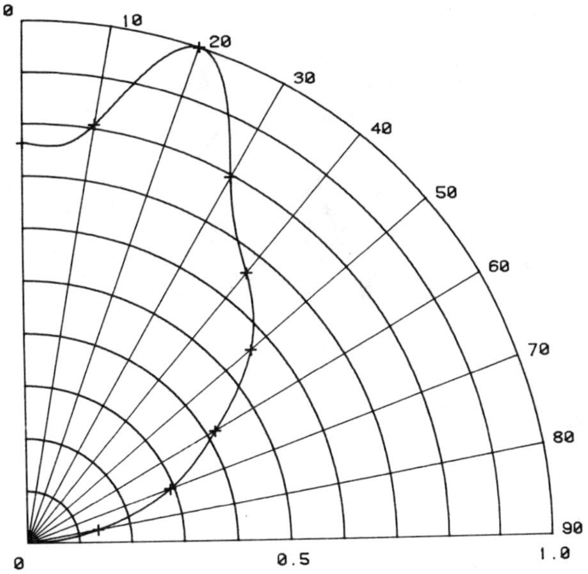

Figure 12: Polar diagram of $(d\rho/d\omega_r)^{lin}$ for a scratched plastic sample ($\theta_i=20^0,\phi_i=0^0;\phi_r=180^0$; max. value = 0.31)

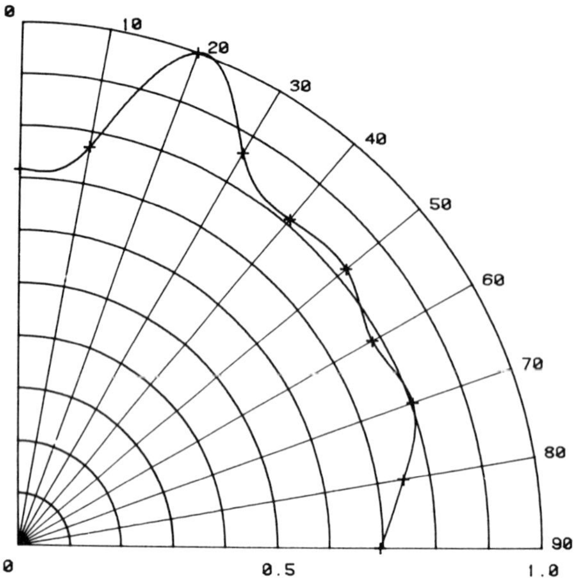

Figure 13: Polar diagram of f_r^{lin} for a scratched plastic sample (θ_i=20^0,ϕ_i=0^0;ϕ_r=180^0; max. value = 0.31)

Figure 14 shows the experimental result for $(d\rho/d\omega_r)^{lin}$ for a white paper sheet with a black ink point. Again, the differences between the figures 8 and 14 are obvious.

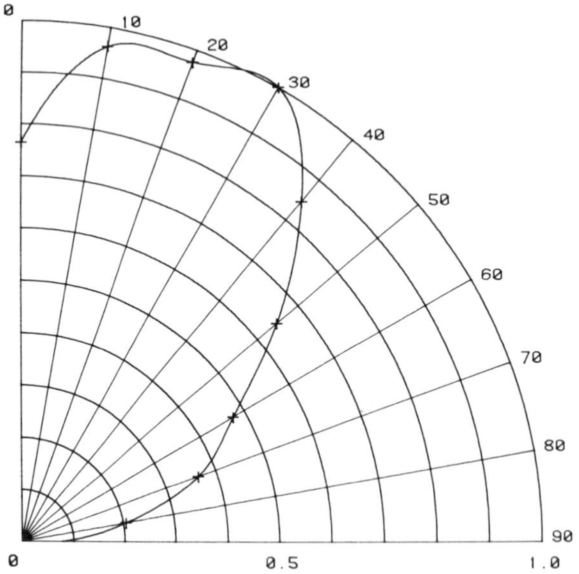

Figure 14: Polar diagram of $(d\rho/d\omega_r)^{lin}$ for a sheet of white paper with a black ink point

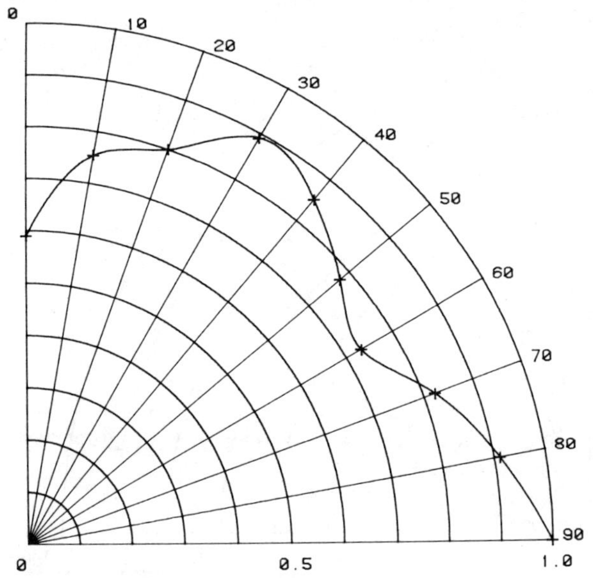

Figure 15: Polar diagram of f_r^{lin} for white paper with a black inc point ($\theta_i=20^0, \phi_i=0^0; \phi_r=180^0$; max. value = 0.51)

The experimental results of this section demonstrate that differential reflectance functions can be used for detecting possible defects on examined surfaces. Furthermore, we have measured, that some different defects for the same object also show characteristical values and shapes for the differential reflectance functions. But, it turns out that the identification of the main defect classes and sub-classes for one object is a real hard task and is far away from being completed /Nor 80/. In order to solve this identification problem the differential reflectance must be extended. The reflectance must be described by an additional microstructure expression, which represents for example the orientation and the size of a scratch on the mirror.

6. DEFECT DETECTION BY POLARIZATION MEASUREMENTS

Surface defects can also be detected if the alteration of the polarization state of the reflected beam is observed. Rusted sheet metal changes linearly polarized light into elliptically polarized light. The exact definition of the polarization of the reflected beam needs four measurements (Stokes vector). Therefore, we registered only that part of the reflected beam which passed a crossed analyzer. The experimental data of this section are all done with a crossed analyzer. Fig. 16 shows the implications of a scratch onto the plastic sample to the differential reflectance function.

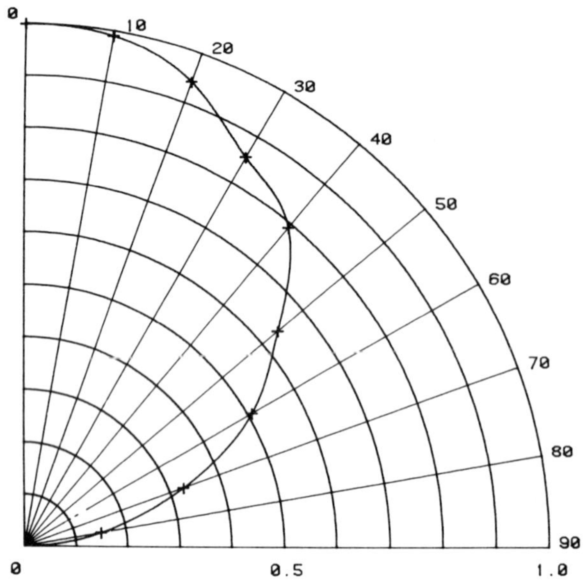

Figure 16: Polar diagram of $d\rho/d\omega_r$ (crossed analyzer) for scratched plastic ($\theta_i=20^0, \phi_i=0^0; \phi_r=180^0$; max. value = 0.23)

Figure 17 shows the experimental result for the rusted sample of sheet metal. As in the measurement before (scratched plastic) there are significant differences between the figures 5 and 11 or 17.

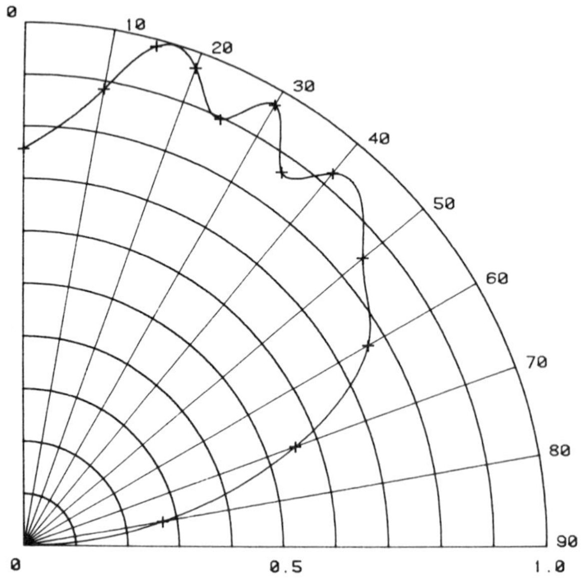

Figure 17: Polar diagram for $d\rho/d\omega_r$ (crossed analyzer) for rusted sheet metal ($\theta_i=20^0, \phi_i=0^0; \phi_r=180^0$; max. value = 0.05)

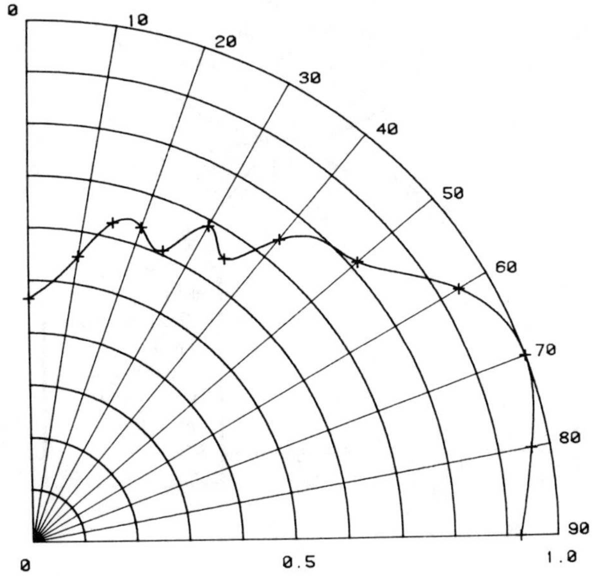

Figure 18: Polar diagram of f_r for rusted sheet metal
($\theta_i=20^0, \phi_i=0^0; \phi_r=180^0$; max. value = 0.05)

The three figures 16 to 18 sketch that surface defects can also be detected by polarization measurements. The reflected beam is depolarized, or elliptically (or circilar) polarized, or the incident polarization plane is rotated etc. by the defects.

7. SYNTHETICAL IMAGES AND LASER SCANNERS

With the aid of synthetical images the image-forming process is reversed. The synthetical image can be constructed if the following three points are known:

(1) differential reflectance functions.
(2) surface gradients (normales).
(3) source- and view point (incident and reflected angles).

The synthetical image will then be produced in two steps (perspective projection):

(a) estimation of the surface gradient and the incident or reflected angles.
(b) calculation of the image irradiance with the aid of point (a) and the differential reflectance functions.

These synthetic images can be used, as mentioned before, to detect defects on surfaces. If the two images are produced by cameras then the match of these two images needs a very time consuming coordinate transformation (translation, rotation, scaling). If the image-forming process is done by a laser scanner which measure range and intensity simoultaneously, then the comparison of the two images can be done directly. The synthetical image can be produced in parallel with the real image, and the pixels of both images coincide in their position. Figure 19 shows our laser scanner which measures range and intensity simultaneously.

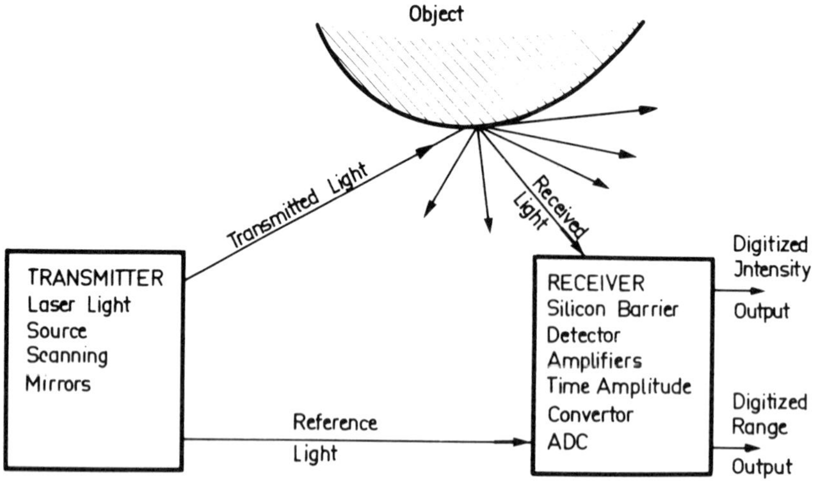

Figure 19: Block diagram of our range/intensity laser scanner

In order to meet the real time requirements the "heart" of this laser scanner is a very fast timing system which measures the flight time directly and which has a resolution of about 10 picoseconds /Lev 83/. The time jitter is decreased by a silicon barrier detector which replaces the conventional photomultiplier tube. With this apparatus we can store about every microsecond the digitized range (12 bits) and intensity (8 bits) data.

8. CONCLUSIONS

Reflectance functions are the first representation of the early vision. All materials which are not textured and which have different reflectance properties can be identified by their differential reflectance functions. All surfaces which are different, but have nearly identical reflectancies (for one frequency) cannot be identified by monochromatic laser light. In these cases, the surfaces must be identified with the aid of differential reflectance functions for a "white" laser. Surface defects can be detected by an alteration of the reflectance and by the change of the polarization state of the images, which are produced profitably by a simultaneous range/intensity laser scanner.

The problem of systematical classifications of surfaces and their defects for example by a syntactical approach are far away from being solved completely. This paper gives an input to solve the identification problems of surface properties by starting the image representation by differential reflectance functions.

BIBLIOGRAPHY

/Hech 74/
Hecht, E., Zajac, A.: "Optics", Addison - Wesley Publishing Company, 1974

/Hor 79/
Horn, B.K.P., Sjoberg, R.W.: "Calculating the Reflectance Map", Applied Optics, Vol. 18, No. 11, 1770-1779, 1979

/Hue 73/
Hueckel, M.H.: "An Operator which recognizes edges and lines", J. Assoc. Comp. Mach. 20, 634-647, 1973

/Lev 83/
Levi, P., Stiefvater, H.: ;A High Speed Timing System with Picoseconds Resolution for a Range/Intensity Laser Scanner", to be published in 1983 SPIE International Technical Conference/Europe

/Mar 82/
Marr, D.: "Vision", Freeman and Company, San Francisco, 1982

/Nor 80/
Norton-Wayne, L., Watts, C., Hill, W.J.: "The Automatic Visual Inspection of Steel Strip, in 5th international conference on Automated Inspection and Product Control, Stuttgart, 143-151, 1980

/Nic 77/
Nicodemus, F.E., Richard, J.C., Hsia, J.J.: "Geometrical Considerations and Nomenclature for Reflectance", NBS Monograph 160, 1977

/Spe 75/
Spencer, D.E. and Gaston, E.A.: "Current Definitions of Reflectance", Journal of the Opt. Soc. of America, Vol. 65, No. 10. 1129-1132, 1975

/Wal 75/
Waltz, D.: "Understanding Line Drawings of Scenes with Shadows", in: The Psychology of Computer Vision, P.H. Winston, ed. 19-91, McGraw Hill, New York, 1975

/Woo 78/
Woodham, R.J.: "Photometric Stereo: A Reflectance Map Technique for Determining Surface Orientation from Image Intensity", in SPIE, Vol. 155, 136-143, 1978

ACKNOWLEDGEMENT:

This work was carried out at the Institut für Informatik III (Research Group: Process Control Computer Technology, Prof. Dr.-Ing. U. Rembold) of the University of Karlsruhe, Federal Republic of Germany.

Automated Part Recognition Via Optical Character Recognition

Robert J. Shillman
Cognex Corporation
Boston, Massachusetts

Abstract

Commercially available vision systems can determine the position, orientation, and identity of parts moving down assembly lines. However, when a large variety of parts are involved, the probability of confusion increases. Furthemore, object recognition systems cannot distinguish among objects with identical silhouettes which have different internal components. A system is described which identifies each object, not by its physical attributes, but by reading its serial number. Because the system is based on studies of human perception, it can read serial numbers which are printed, etched, stamped, embossed or inscribed on a wide variety of materials under difficult industrial conditions.

1. INTRODUCTION

Many tasks currently performed by industrial workers depend heavily on the ability to see and interpret the wide variety of visual information present in factory environments. Automation of these complex tasks can only be accomplished with the integration of sophisticated vision systems and robots.

Vision systems are now commercially available which can accurately determine the position, orientation and identity of objects moving down assembly lines. This information can be communicated to robots which can then pick up selected items for further action.

Most vision systems identify objects by geometric features, such as circumference and area, which are extracted from a binarized two-dimensional view of each object. Although useful in a wide variety of applications, this technique is not appropriate when a large number of similar, but different, objects are involved, or when objects have identical silhouettes but different internal features. However, in these situations individual objects can be readily distinguished by automatically reading the serial number which is printed on it.

Recognizing an object by its serial number rather than by its shape avoids many of the difficulties introduced by projecting a three-dimensional object onto a two-dimensional image. A vision system capable of reading serial numbers has many additional advantages:

- The system need not be retrained for each new application or as different parts are added.
- The system could identify and discriminate between different shapes which happen to have identical two-dimensional projections.
- Identical items differing only in serial number could be accurately tracked (e.g., automotive engines).
- The system could identify objects having identical shapes but with different internal parts (e.g., 3 vs. 4 speed transmissions).

2. INDUSTRIAL APPLICATIONS OF OCR

The industrial applications which require the ability to read information from the surface of objects can be divided into four major groups:

1. Automated Assembly: In automated assembly applications the serial number on each part can be used to select the appropriate component for the assembly of products and to verify its placement. For example, in the automated manufacture of circuit boards, a vision system can instruct the handler to pick up the appropriate chip based on the serial number it reads from its surface and to then place the chip in the appropriate position on the PC board; see Figure 1 below.

Figure 1: Serial Numbers on Electronic Components

In another automated assembly application, the vision system will read serial numbers which are printed on the sides of aircraft wire prior to termination. The system will communicate this information to a robot which will select the appropriate connector for termination; see Figure 2 below.

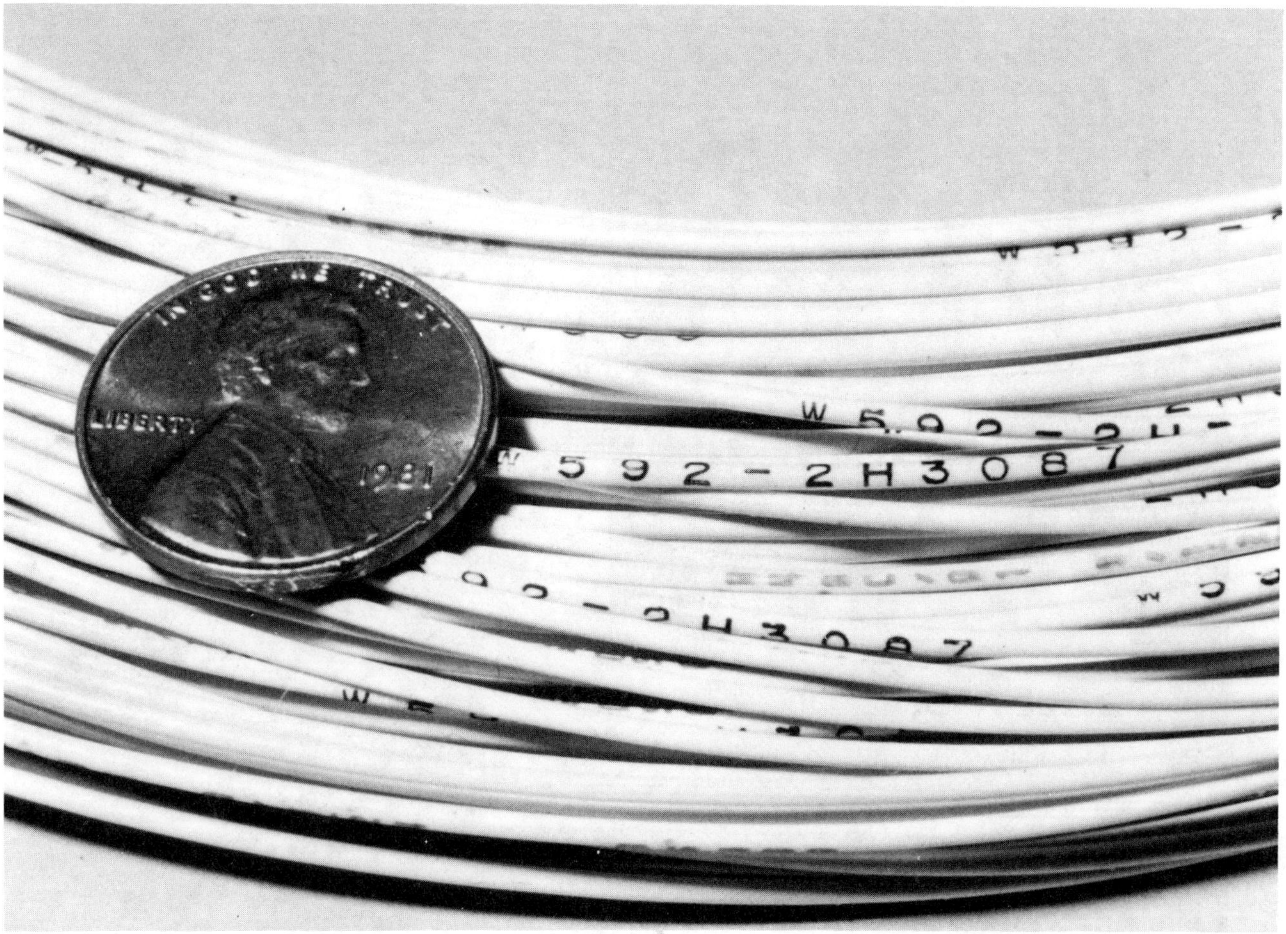

Figure 2: Serial Numbers Printed on Aircraft Wire

2. Materials Handling: In materials handling applications information can be automatically read from the surface of the item in order to control where the particular item is sent. For example, intermixed flavors of boxed frozen pastry can be automatically sorted and palletized by a robot after receiving dientifying information from the vision system which reads the preprinted labels on the corrugated cartons; see Figure 3 below.

Figure 3: Automated Sorting Via Character Recognition

3. Process Control: In process control applications information automatically read from the surface of the item can be used to direct subsequent processing of the item. For example, semiconductor wafers having certain lot numbers laser-etched on their surface may be converted into memories; whereas those with different lot numbers may be turned into CPUs. In each case the vision system reads the unique lot number and communicates this directly to process control equipment.

4. Tracking: In tracking applications the information read from the surface of each item can be sent to a host computer where a database is updated which contains information about the status of the item at any point in time. This information is useful, for example, in determining when a particular item, such as an automotive vehicle, will be available for delivery.

3. OPTICAL CHARACTER RECOGNITION TECHNOLOGY

Optical Character Recognition (OCR) has been an active area of research and product development for over two decades. The major thrust of OCR vendors has been in the office marketplace where machines for handling high quality material printed in a small number of fonts have been introduced. In contrast, factory applications require the ability to deal with difficulties in image capture, low print contrast, character misalignment and partially obliterated characters. The algorithmic techniques employed by office-grade OCR equipment are far too elementary to deal with these problems. For example, in a recent detailed experimental test of OCR products manufactured in the United States, Freudenreich and Shillman found that slight decreases in print quality which were easily ignored by humans often had noticeable effects on even the most advanced OCR machines. Unlike OCR machines which identify characters based on their overall shape, recent research in human perception indicates that humans can recognize degraded characters because they search for the distinctive features of each letter.

Cognex Corporation (Boston, MA) has incorporated the results of research on human vision into a vision product, DataMan. DataMan is a computer-based vision system which has the human-like ability to read letters and numbers under difficult factory conditions. The DataMan system consists of four components: a vidicon TV camera, a TV monitor, an industrial keyboard, and the DataMan processor. A complete DataMan system is illustrated in Figure 4 below.

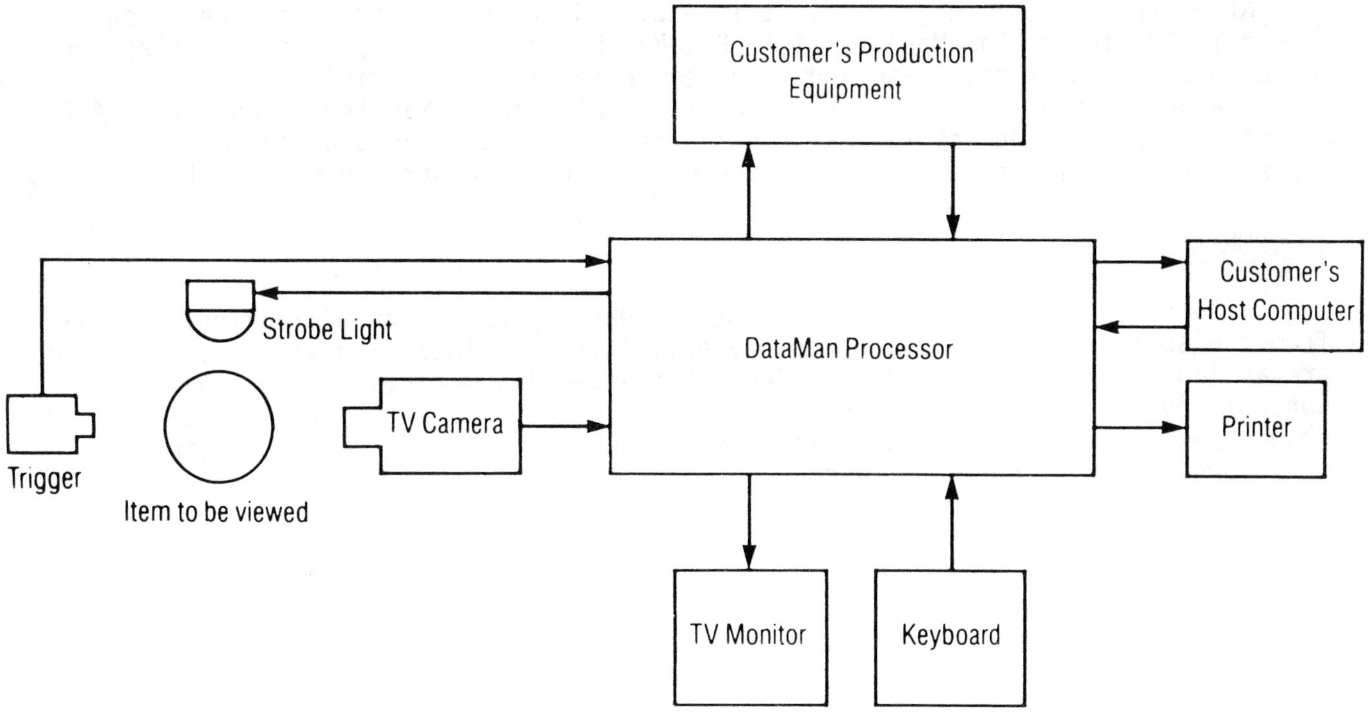

Figure 4: The DataMan Vision System

The system operates in an acquire-and-process cycle. Alphanumeric information marked on the surface of an item is presented to a vidicon camera. Vidicon technology provides high resolution images (480 x 640) to DataMan at a frame rate of 1/60th of a second. Although vidicons are known to introduce slight geometric distortion and photometric inaccuracy, DataMan has been designed to be insensitive to these. The system supports up to four cameras and can process images from objects as far as 1000 feet from the processor on a time-shared basis. Items can be either stationary or moving past the camera at high speed (for moving applications a strobe light is used to freeze the image).

Image acquisition and analysis can be initiated in three ways:

1. externally by a TTL signal from devices such as a photo-sensor

2. remotely under the control of a host computer

3. locally by an operator using a foot pedal

During image acquisition, analog video signals from the camera are sent to a "frame grabber" image capture subsystem located in the DataMan chassis. The frame grabber has a 6 bit A-D converter which digitizes the analog signal into a 64 grey level image. Because the system uses grey levels and not binary images, special lighting is usually unnecessary, and the intensity of the illumination need not be carefully controlled. The system can tolerate changes in illumination from one image to the next and even within one image, an important capability for continuous and reliable performance in industrial environments.

After the image is digitized, it is analyzed by proprietary software which is stored in ROM in the DataMan chassis. DataMan begins the analysis by locating the serial number within the image and then segmenting it into individual characters. Each character is analyzed to determine its identity; DataMan then reports the ASCII character code for each character. These results are displayed on the monitor and are simultaneously sent to other devices such as robot controllers or host computers.

4. SUMMARY

Vision systems now have the ability to read alphanumeric information directly from the surfaces of objects in factory environments. These industrial OCR systems are an important element in automating robot assembly, materials handling, process control and tracking. The systems are of particular interest where the identity of each object, rather than simply its class membership, must be determined in order for appropriate action to be taken.

CHAPTER 18

END EFFECTORS

End Effector Technology and Programmed Automatic Exchange

Allen J. Wright
Intelledex Incorporated
Corvallis, Oregon

Abstract

A major enhancement to any light assembly robot would be to allow it to change its end-effectors without assistance from an operator. A system developed at Intelledex allows the robotic arm to detach an end-effector into a tool rack, select another, and attach it without outside assistance. Two pneumatic and fourteen electrical connections are accomplished during a change operation. The overall operating system described has a large degree of flexibility and expandability.

INTRODUCTION

The use of robots for light industrial assembly is an emerging technology. One of the key goals in this area is to mimic capabilities of a human assembly person. This allows robots to be inserted into the assembly environment without redesigning the product or significantly altering assembly procedures. A major advantage of this approach is to allow a potential robotics user to bring a robot to his facility and easily gain experience and knowledge on this new type of soft automation. A second advantage is that the productivity improvement will allow for increased expenditure and effort in this area.

A key area of human versatility which is of primary consideration is the ability to use several tools at a workstation which might have vastly different operating characteristics (eg. a screwdriver and a retaining ring applicator). Two approaches to achieve this flexibility are to fabricate multifunctional end effectors or to allow the robot to change end effectors. While both approaches have their advantages, this paper will concentrate on the latter one because it is felt that a system such as this will be needed if robotics is to truly move into the factory of the future.

The end effector attach/detach mechanism described in this report was developed for a light (five pound) assembly robot. The configuration and intended use of this product was to allow it to be placed at an assembly station identical to those used by a human assembler. Much can be done to optimize the robot workspace but the cost advantages of these are not apparent without close working knowledge and experience with robotics. The system described here will bridge the gap for many users just entering the field and those who demand a very sophisticated assembly system for use in highly demanding work areas.

DESIGN OBJECTIVES

Several factors were considered necessary for the successful implementation of an end effector exchange mechanism. The following is a list of items, in order of priority, to be used as design objectives.

A. Light weight
B. Axial accuracy (Repeatibility)
C. Longitudinal accuracy (Repeatibility)
D. Rotational accuracy (Repeatability)
E. Coupling stiffness
F. Torque capacity (Stiffness)
G. Connection of two pneumatic lines
H. Full rotation capability of the pneumatic lines (Continuous rotation)
I. Connection of twelve electrical lines
J. Full rotational capability of the electrical lines
K. Easily adaptable to a variety of configurations

The placing of light weight in the number one design goal position was derived from the premise that any weight used in the coupling mechanism is subtracted from weight allowed for the end effector and the payload. The complexity of end effectors for assembly increases proportionately with either multifunctional or sensory adaptive tasks. It was therefore decided that the maximum amount of weight should be reserved for this area.

Axial, longitudinal, and rotational accuracy are all considered important for allowing the robot to function within its absolute world coordinate system. The repeatibility of these parameters allows the user to use a previously loaded program without having to reorient the robot each time a tool is changed. This becomes especially important in jobs where a lightweight item (such as component placement on a printed circuit board) is handled followed by a heavy or extended leverage type item (edge handling a printed circuit board).

Coupling stiffness and torque capacity must be adequate to prevent the need for excess deflection correction by the overall robotic system.

Connection of the pneumatic and electrical lines were determined to allow the most advantageous mix of power transmission, control flexibility, and future expansion capability. Figure 1 shows the schematic of the robot system to which the end effectors can be connected.

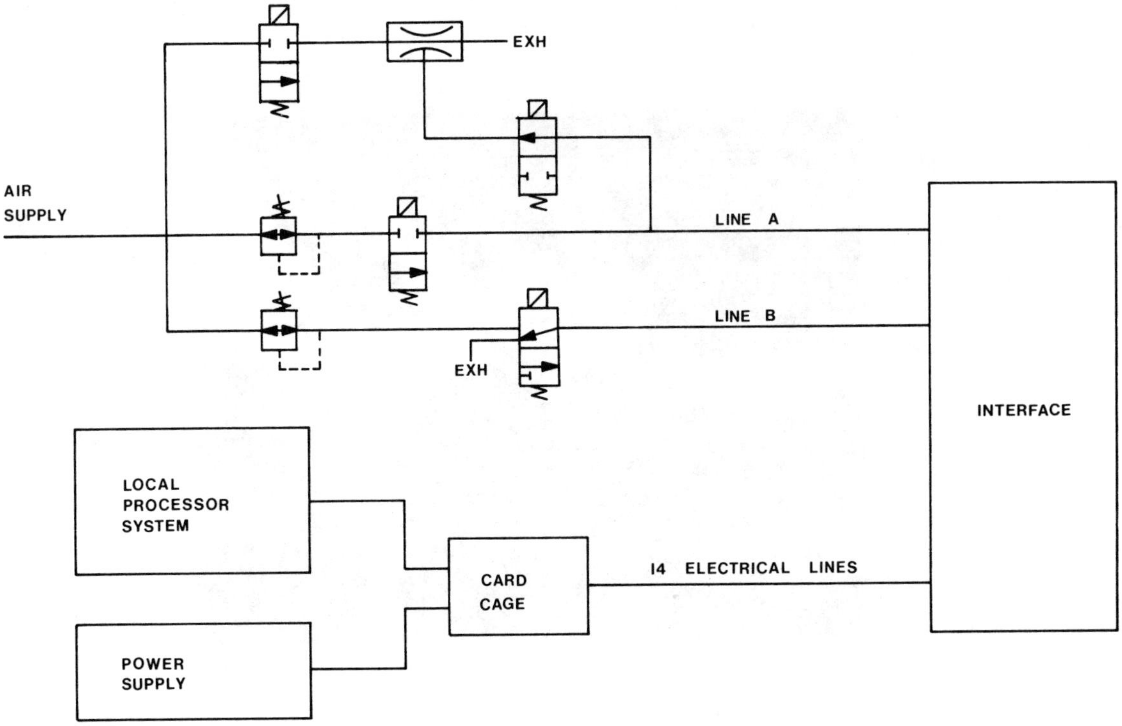

FIGURE 1

The pneumatic system can be program configured to the following operational modes.

A. Line A - Pressure/Vent
B. Line B - Pressure/Vent
C. Line A/B - Pressure/Vent (Four way operation)
D. Line B - Vacuum/Vent
E. Line B - Vacuum/Pressure Pulse Quick Release
F. Line B - Vacuum/Pressure
G. Line A/B - Pressure/Vent, Vacuum/Vent
H. External Vacuum System - High Flow Pressure Capability

Full rotational capacity and an easily adaptable configuration were areas which could be worked around in the short term but were considered highly desirable long term goals.

DESCRIPTION OF MECHANISM DEVELOPED

The mechanism which was developed is a cylindrical package which is roughly 2.3 inches diameter by 2.3 inches long. This interface mechanism, which is shown in Figure 2 attached to the robot, looks very similar to the tool attach surfaces used on most robots. A cutaway view in Figure 3 shows the interior construction of the interface.

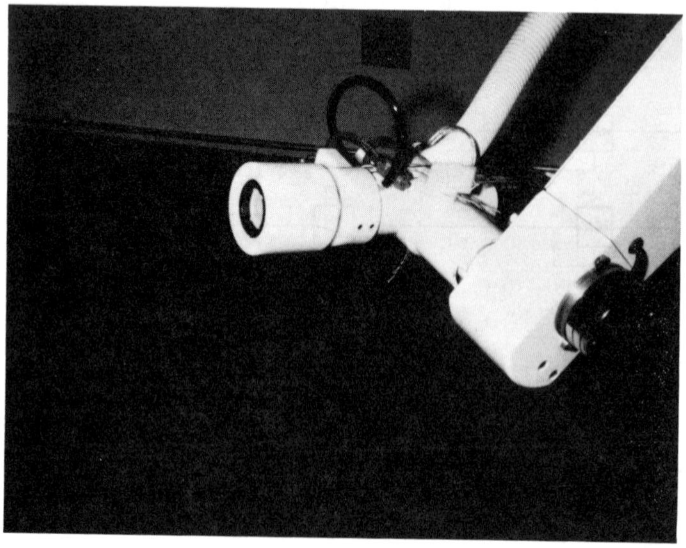

Figure 2

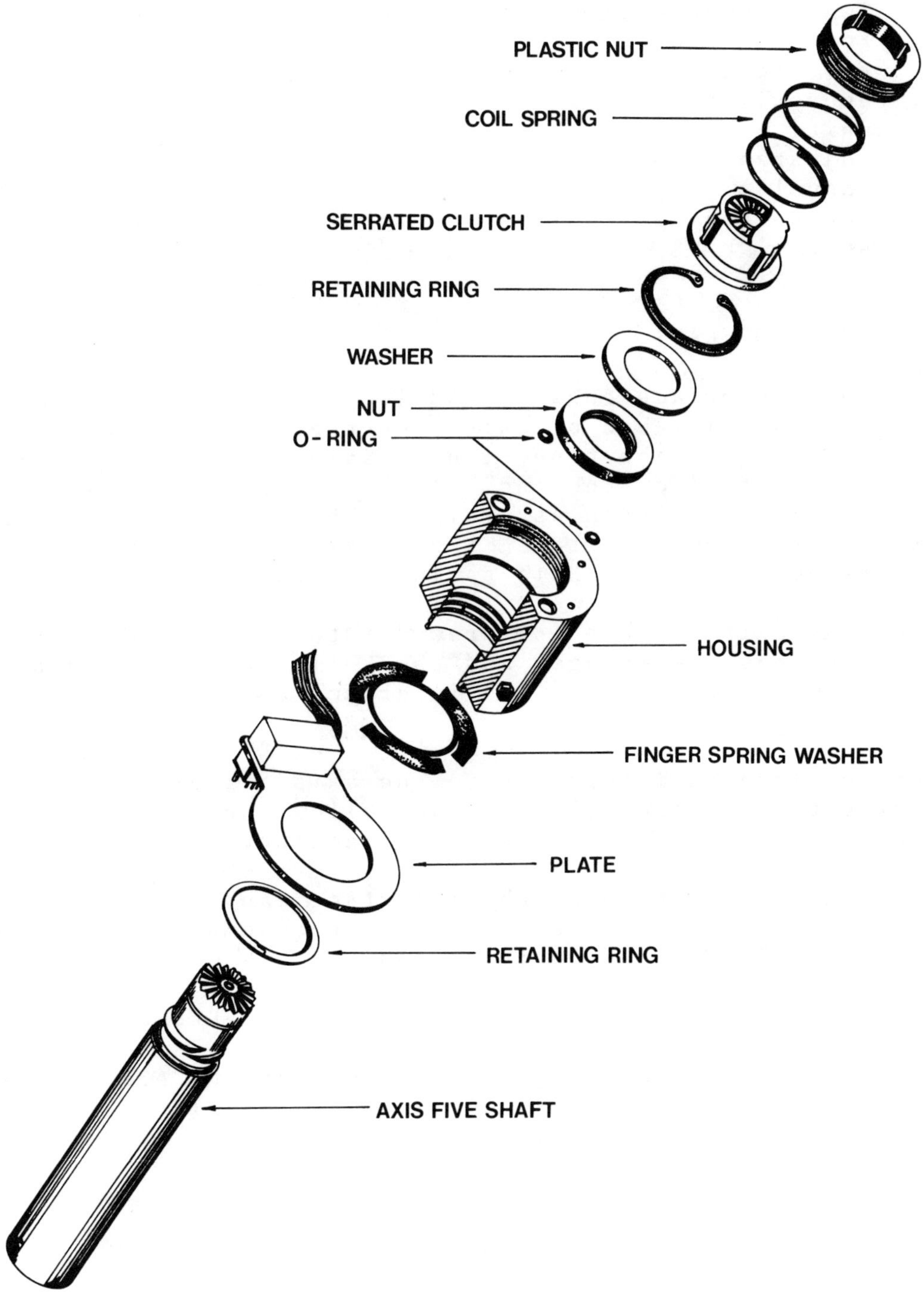
PLASTIC NUT
COIL SPRING
SERRATED CLUTCH
RETAINING RING
WASHER
NUT
O-RING
HOUSING
FINGER SPRING WASHER
PLATE
RETAINING RING
AXIS FIVE SHAFT

FIGURE 3

The design capabilities achieved in this design are as follows:

A. Weight - Electric 8 oz., Pneumatic 6 oz.
B. Axial and Longitudinal Accuracy - .001 inch
C. Rotational Accuracy - 0.096 Milliradians
D. Torque Capacity - 40 inch pounds nominal
E. Two pneumatic lines connected - 120 PSI maximum
F. Fourteen electrical lines (Electrical unit only)
G. Full rotation of pneumatic unit
H. Approximately +/-200 degrees of rotation on electrical unit
I. Adjustable rotational orientation
J. Manual or automatic attachment of a tool can be achieved but detaching requires the robot to be in an active state.

DESCRIPTION OF OPERATION

The tool interface engages and locks itself to the robot by the following means. The end of axis 5 on the robot has a shaft extending out with a short section of acme centralizing thread and a serrated clutch face on the end. The serrated clutch is connected to a one way roller clutch assembly which allows rotation in one direction only. The interface itself has a mating acme thread and a fixed serrated clutch which is spring loaded.

When the axis 5 shaft is screwed into the tool interface, the serrated clutches are engaged and the roller clutch prevents counterclockwise rotation so the interface cannot unscrew off the shaft.

A piston located internally to the axis 5 shaft can be extended to separate the serrated clutches and allow the unit to be removed by rotating in the counterclockwise direction.

The angular position of the interface can be adjusted by loosening the four setscrews which hold the acme nut and rotating the unit clockwise to the position desired.

The sequence for set up of the robot for automatic tool affix/detach operation is as follows:

A. Manually attach a tool to the end of the arm by pushing the tool on and rotating clockwise until the screw threads bottom in the tool (Figure 4).

Figure 4

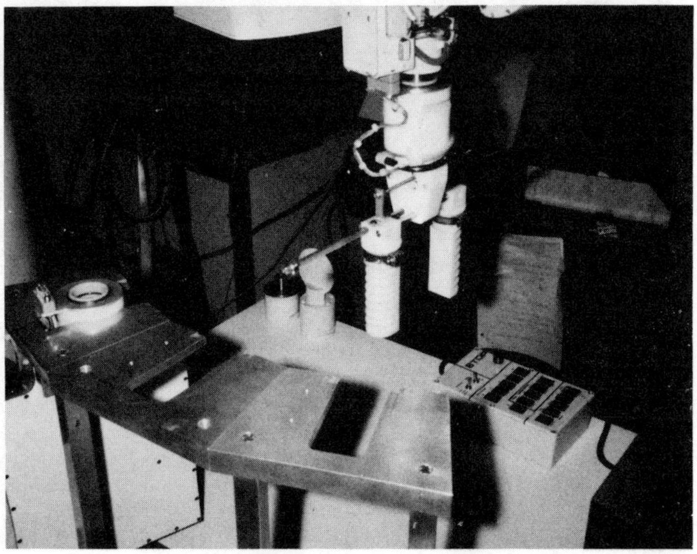

B. Position the tool vertically in front of the tool rack position desired (Figure 5), store this point as a world coordinate into memory.

C. Move the tool into the tool rack (Figure 6) while engaging either side of the tool interface body.

D. Activate the tool detach command.

E. Activate the thread command to run 2.50 threads/inch, CCW rotation, Z direction +.220 inches.

F. Store this point as a world coordinate into memory.

G. Move in the Z direction +2.00 inches (Figure 7).

H. Store this point as a world coordinate into memory.

At this point the operator can repeat this procedure for the rest of the tools desired to be loaded for this job.

Figure 5

Figure 6

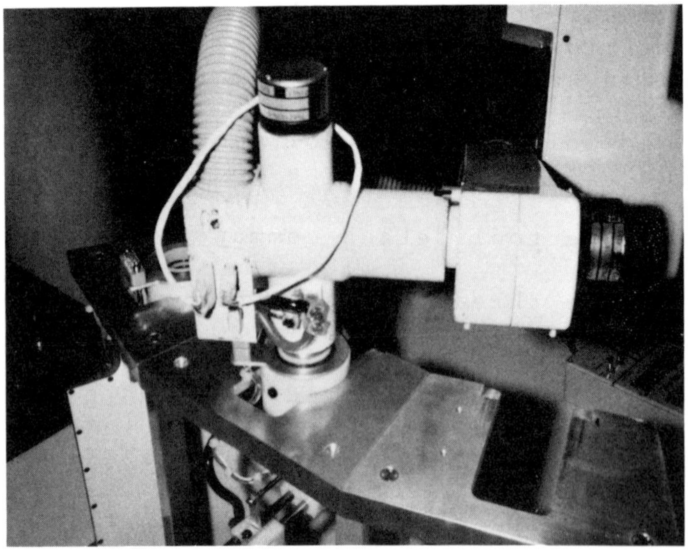

Figure 7

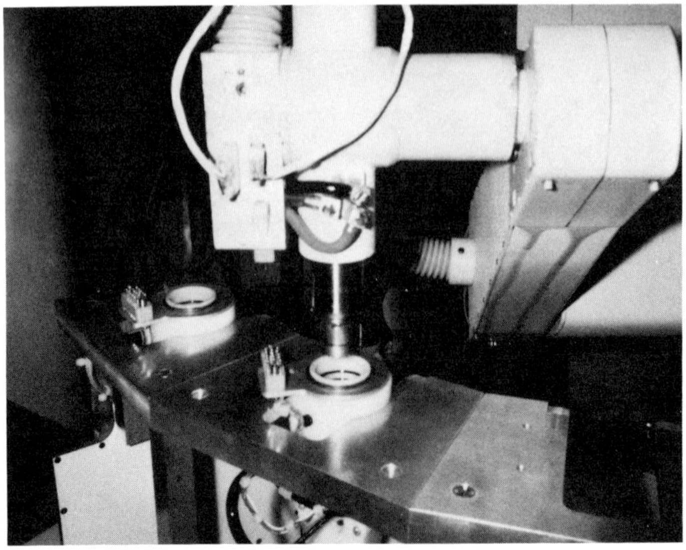

To reattach the tool we just placed into the rack, the following procedure would be used.

1. Move to position G
2. Move straight to position F
3. Activate the thread command to run 2.50 threads/inch, CW rotation, Z direction - .220 inches.
4. Rotate axis 5, 2 degrees CCW
5. Rotate axis 5, 3 degrees CW
6. Rotate axis 5, 1 degree CCW
7. Move to position B, the tool is now out of the rack and ready for operation.

Steps 4-6 in the above operation are used to set the serrated clutch and insure that all the backlash is taken out of the attach system.

The program sequence is the same whether the tool is electro-pneumatic or purely pneumatic. A key point to note is that the robot must place the tool into the rack itself before it can retrieve it. This is required because each tool can be oriented where desired with respect to the robots home position. Any change in the orientation of a tool after the program points have been memorized would require the operator to reposition the arm and record these points again.

This particular sequence of actions required to change a tool can also be performed by using a "TCHANGE" command in the robots programming language. To begin using this command the user will be prompted by the robot to enter the same points described previously to set up the subroutine. The use of this feature will greatly speed up some of the routine programming work.

PROGRAMMING AND LANGUAGE USED

The language used to operate the robotic arm system is "microsoft basic" which is used in the majority of the personal computers (Apple, IBM, etc.) on the market today. Approximately 140 verbs have been added to the language to help control the specific motions and operations of the arm and tools.

The position of the arm is defined by an array with seven elements. The definition of the elements is as follows:

(Array Name (0)) - X Coordinate
(Array Name (1)) - Y Coordinate
(Array Name (2)) - Z Coordinate
(Array Name (3)) - Theta Angle
(Array Name (4)) - Phi Angle
(Array Name (5)) - Rotational angle (tool from horizontal)
(Array Name (6)) - Ambiguity status (arm joint orientation, eg. underhand or overhand during a move).

A typical use of this array is to define a point for use in programming. The robot can be moved to a point with a joystick or teach pendant and a "where" command executed (where PT1). Array PT1 will now contain the position and status of the arm (PT1(0)=X coordinate, PT1(1)=Y coordinate, etc.). Another command available is "move". If we define another array named PT2 and assign parameters to it, we could then execute "move PT2" and the robot would move to this new position by the shortest time path possible. A few of the other commands available are listed below:

MOVE STRAIGHT	Straight line motion
IMOVE STRAIGHT	Incremental straight line motion
IMOVEX (or Y,Z,T,P,R)	Directional incremental move
HOME	Moves arm to home position
ROTATE	Rotate tool
THREAD	Rotates and precesses or retracts tool according to parameters specified.
INSERT	Straight line motion along tool rotation axis
ORIENT	Orients the robot to a new coordinate system.

The above commands, although not a complete listing, show some of the background language that will accompany the control of any tool. The above commands can be incorporated into any basic language program to yield an operational program for the robot.

The tool commands designed to operate in this system are as follows:

Tool Operations:

1. TON, Activates a tool function (operator defined)
2. TOFF, Deactivates a tool function (operator defined)
3. TMOVE, Alters position of a tool (eg. how wide jaws are open, etc.)
4. TOPEN, Activates an opening sequence
5. TCLOSE, Activates a closing sequence

Tool System Operations:

1. TDETACH, Activates detach piston
2. TAFFIX, Cancels TDETACH
3. TVACON, Activates system vacuum
4. TVACOFF, Deactivates system vaccum

Utility Routines:

1. TSTATUS, Checks status of switchs, optical pairs, electrical lines, time delays, PCB activations, etc. which are related to a tool.
2. TSELECT, Loads data tables of tool definitions.
3. TCHANGE, Calls a subroutine to change a tool and load associated data tables.

4. TDEFINE, Sets up internal definitions of the tool operations.

Each tool is assigned a number and each of these assigned numbers represents a new set of definitions for the tool operations. There may be eight tool numbers assigned to each tool in the system. A simple pneumatic end effector might have only two operations, TOPEN (1) or TCLOSE (1) to control it, whereas a complicated end effector might have several open and close sequences. There can be up to 180 possible operational definitions in the system at any one time. This allows considerable flexibility in the development of new tools in which the control modes needed are not clear at present.

The flexibility of this language system gives the user the ability to change a tool and redefine all the tool commands in the operational program. This allows the user to select tools of vastly differing operational characteristics in the same program without the loss of flexibiity or speed.

The card cage which is shown in Figure 1 accepts "personality" cards for complicated tools which can handle the signal decode and control sequences. This allows a user to build up an end effector which might interact with assembly fixtures or operate fixed tools. Several general purpose personality cards will be available to use with the system.

END EFFECTOR CONFIGURATIONS

In order to take maximum advantage of the interface sytem previously described, several end effectors have been developed following a modular or building block approach. This allows the user to configure a tool as simply as possible at the start and add complexity as needed to perform the work desired. Figure 8 show a basic pneumatic parallel jaw gripper (no jaws attached). This tool could be attached to an electrical interface if sensors were to be attached to the jaws. A second type of parallel jaw end effector is shown in Figure 9. This unit uses a double rack and pinion mechanism and is capable of very wide opening distances. Two load sizes are shown and the larger unit has special purpose pneumatic fingers attached.

A multi-tool head is shown in figure 10. This unit can carry up to five fingers for performing various operations. This unit is program controllable to move any finger tool in line with the robot arm's axis 5 shaft.

Several forms of vacuum tools are shown in Figure 11. Optical and switch sensors are used on these tools for part presence detection.

A three jaw planetary clamp is shown in Figure 12. This unit is capable of internal and external pickup of objects and will self center round objects with the robot's axis 5 shaft.

Figure 8

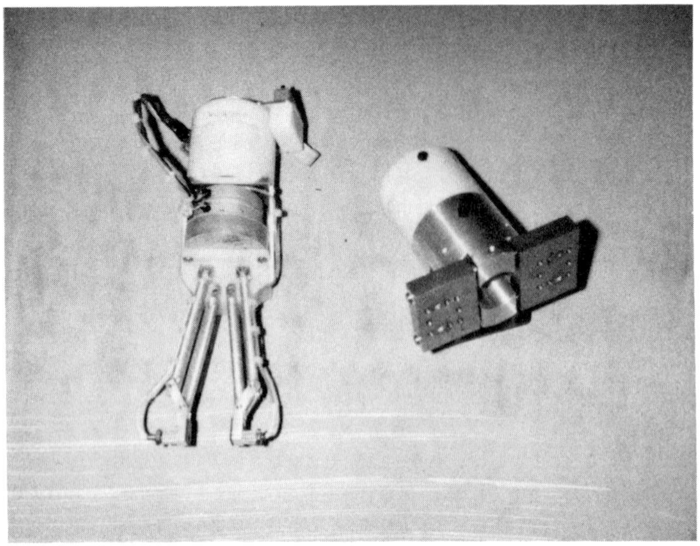

Figure 9

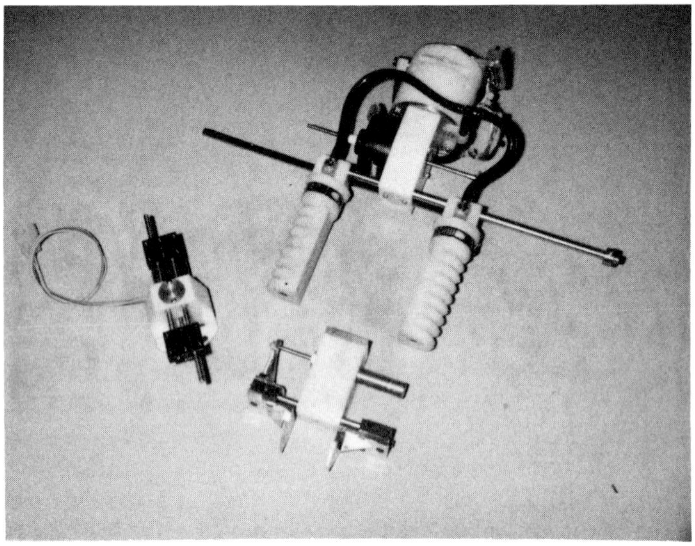

Figure 10

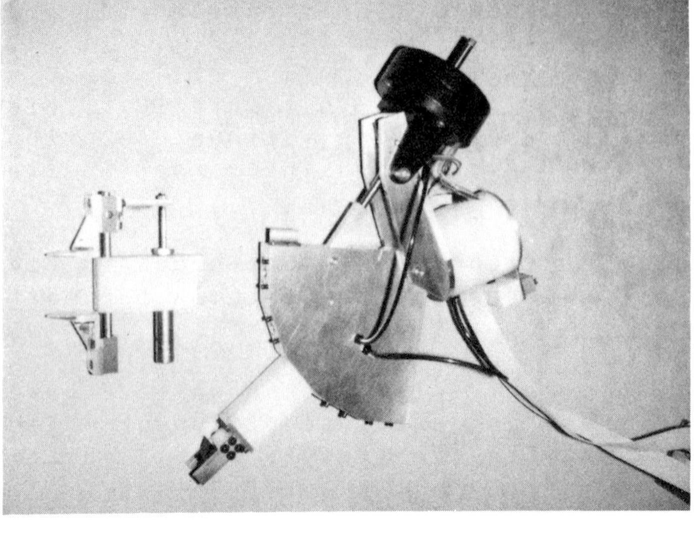

CLOSING REMARKS

This end effector interface which has been developed will hopefully allow for the high degree of flexibility desired by the experienced robotics user as well as letting a new user start simply. This integrated systems approach will now allow a variety of "standard" end effectors to be produced which are modifiable by the user to best fit his specific applications.

Figure 11

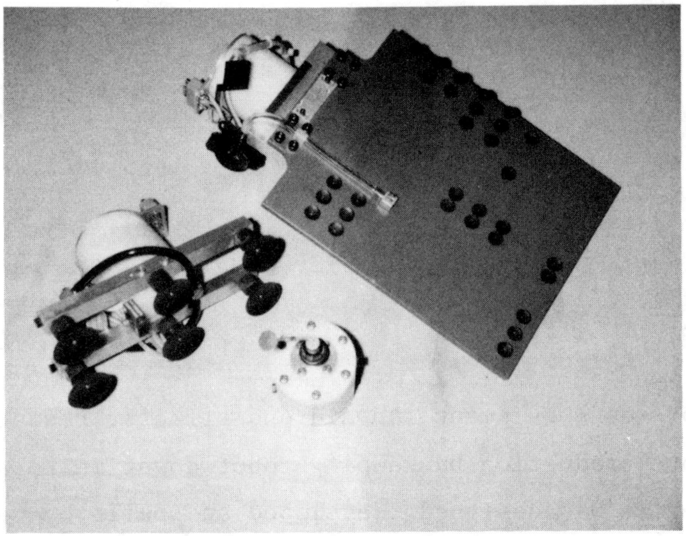

Figure 12

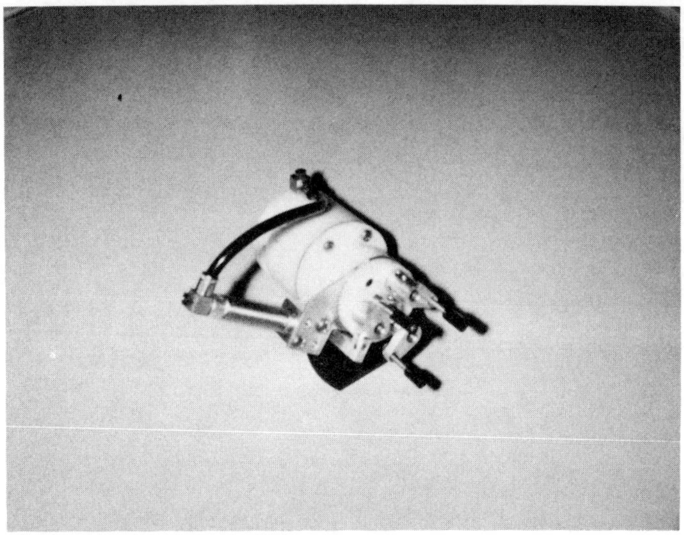

Robotic Drilling End Effector Test Equipment for Research and Development and Production Support

Jon M. Heywood
General Dynamic
Fort Worth, Texas

Abstract

Drilling end effector test equipment can aid end effector research and development activities and support production by keeping robot downtime to a minimum. The end effector test device was designed, developed and built by General Dynamics' Factory Automation and Robotics Laboratory personnel to allow drilling end effectors to be tested, debugged, fine-tuned and prepared for production in a laboratory environment.

INTRODUCTION

As the number of robots in a manufacturing facility increases, so does the need for specialized robotic support equipment. This case study describes how General Dynamics/Fort Worth Division uses end effector test equipment to support its end effector research and development activities and robotic production drilling programs. The equipment is used as a tool to facilitate preliminary testing, proof test new end effector designs and accessories, and to set-up and debug end effectors.

The robotics program at General Dynamics/Fort Worth Division is part of Manufacturing Technology's Computer-Aided Manufacturing Development Section's task. The objective of this extensive robotics program is to implement cost-effective robotic technology into the aerospace production environment; specifically, the manufacturing of the USAF F-16 fighter. For example, in drilling high tolerance fastener holes, the initial robotic application is yielding 3 to 4:1 productivity improvements along with an increase of hole quality due to the repeatability of the process. General Dynamics' robots are now drilling fastener holes in approximately 90% of the exterior surface skin of the F-16.

PROBLEM INVESTIGATION

In 1979, General Dynamics installed its first production robot to drill the F-16 vertical fin graphite-epoxy composite skin. During this time, extensive research and development was being conducted into drilling end effector design and improvement. Another robot in the Robotics Laboratory was being used for research and to test and debug the new prototype end effectors. When a new drilling application was found for the second robot, laboratory personnel could no longer use the robot to test the new end effector designs. After the second robot was put in production, the end effector redesign and improvement tests had to be carried out during the noon-hour, second shift or between production runs. By 1980, eight more drilling applications had been identified and more robots were ordered. As soon as each new robot arrived in the laboratory, work was begun to interface it to the work cell control computer. The build-up and interface work involved installing a valve box on the robot arm, providing a data link between the robot and work cell minicomputer, robot debug and diagnostics and interfacing the work cell safety system and part positioner (if applicable). Consequently, there wasn't adequate time available when these robots could be utilized for general research and development of end effectors. Time could only be taken to support the specific application if the project was to meet schedule. Thus, the need for off-line drilling end effector test equipment became apparent at this time.

THE SOLUTION

In the robotic production drilling operation, the robot positions the end effector in front of the desired hole location. At this point, the end effector then extends itself and begins the drilling process. Since the end effector test equipment need only "simulate" the drilling cycle, the test equipment will be referred to as the "simulator" from this point on.

A project was initiated in 1981 to design and build an end effector simulator that would be kept in the Robotics Laboratory to aid in the set-up, maintenance, and research and development of end effectors. The initial investigation into the design requirements revealed that the simulator must have the following characteristics:

1. Automatically or manually cycle the end effector and simulate the actual drilling process.

2. Adjustable enough so that all the various types of drilling end effectors would fit in it.

3. Flexible enough so that it could be modified for special tests.

The resulting design (Figure 1) consists of a floor-standing frame and fixture assembly in which the end effector is mounted. A "quick-change adaptor" is mounted on the tailstock to facilitate a fast and accurate means of locating the end effector. This "quick-change adaptor" is identical to the ones being used in actual production applications. The adaptor has features that allow it to be rotated 360°, pass pressurized air through it to the end effector, and transmit transducer signals from the end effector back to the work cell control computer. A magnetic switch also determines whether or not the end effector is firmly seated and located against the faceplate of the adaptor. A bushed hole in the faceplate of the adaptor mates up with a guide pin on the end effector so that the end effector rotates only when the adaptor rotates.

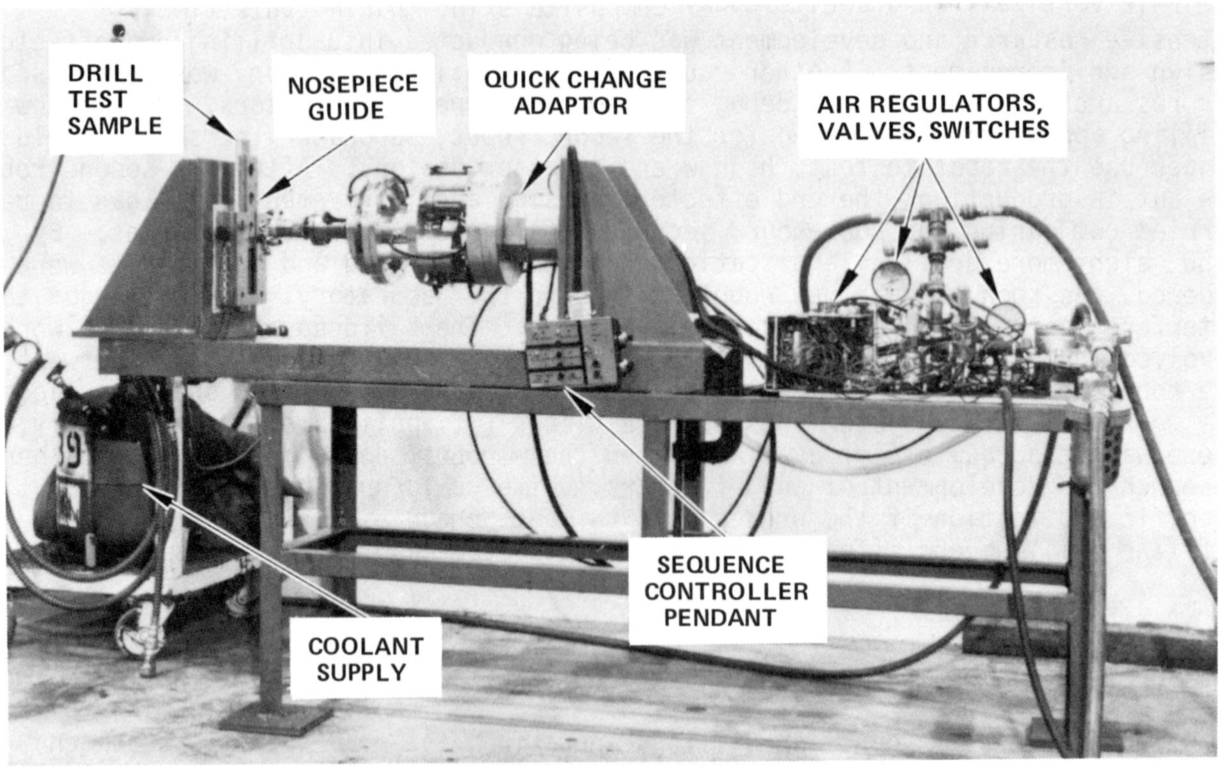

Figure 1 END EFFECTOR DRILL CYCLE SIMULATOR

The headstock can be moved back-and-forth approximately one foot to allow the mounting of different lengths of end effectors. The headstock can also be offset up to 3° from the tailstock-to-headstock centerline. On the face of the headstock is an adjustable nosepiece guide and test bar holder. The nosepiece guide plate slides up and down so that the required guide hole can be positioned immediately in front of the end effector. The nosepiece guide can be removed altogether if so desired. Test bars up to 8" long x 1/2" thick x 1 1/2" wide can be firmly located and held rigid inside the holder, behind the nosepiece guide plate. After a test hole is drilled, the test bar is unclamped and moved up or down so that another hole can be drilled.

The end effector can be sequenced either manually or automatically. The sequence controller pendant (Figure 2) actuates the solenoid valves, switches and transducer signals that make up the automatic drilling cycle. The flowchart for the drilling sequence controller is shown in Figure 3. If desired, the operator can override any part of the automatic drilling cycle to obtain a cycle that is part automatic and part manual. This flexibility is very useful when debugging and testing end effectors. Switches on the simulator fixture itself include a master electrical on/off, master air solenoid on/off, and another set of manual sequence control switches allows the simulator to automatically cycle two different styles of drilling end effectors whose electrical circuits are configured differently. Circuit test points allow the operator to easily monitor test parameters, such as the part contact signal and the analog drill feed transducer signal, with a voltmeter or oscilloscope.

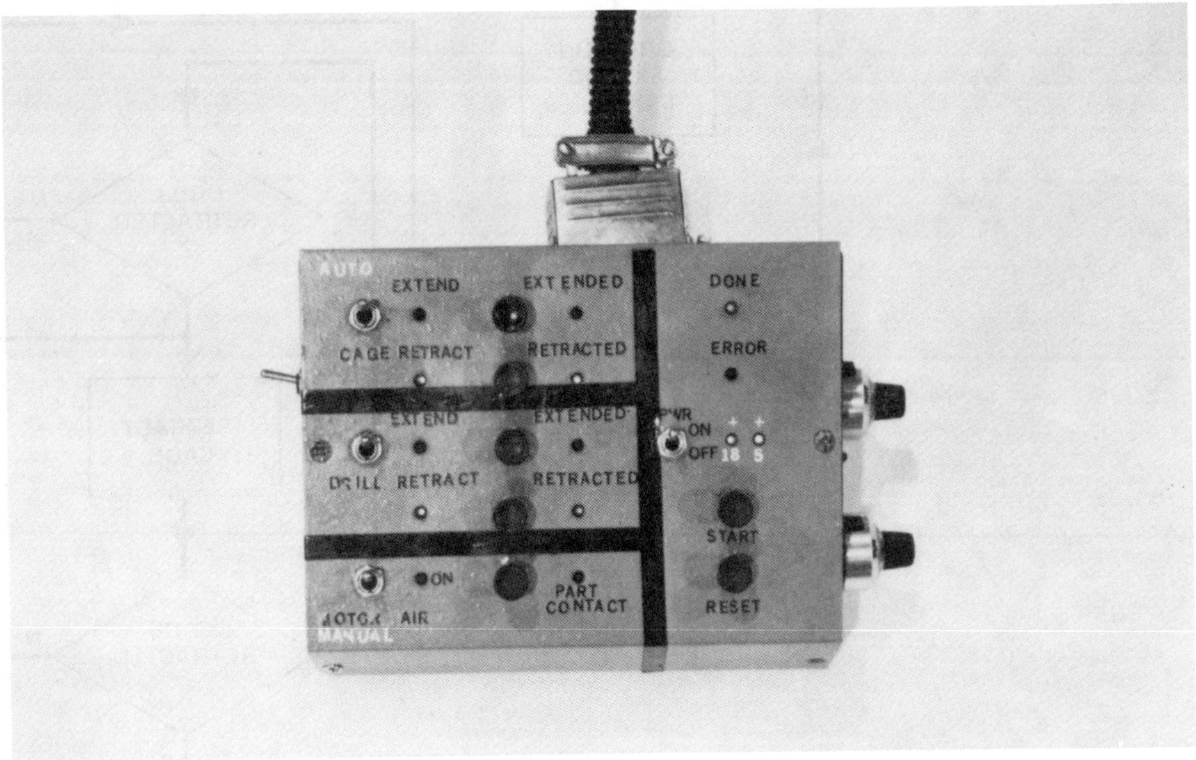

Figure 2 SEQUENCE CONTROLLER PENDANT

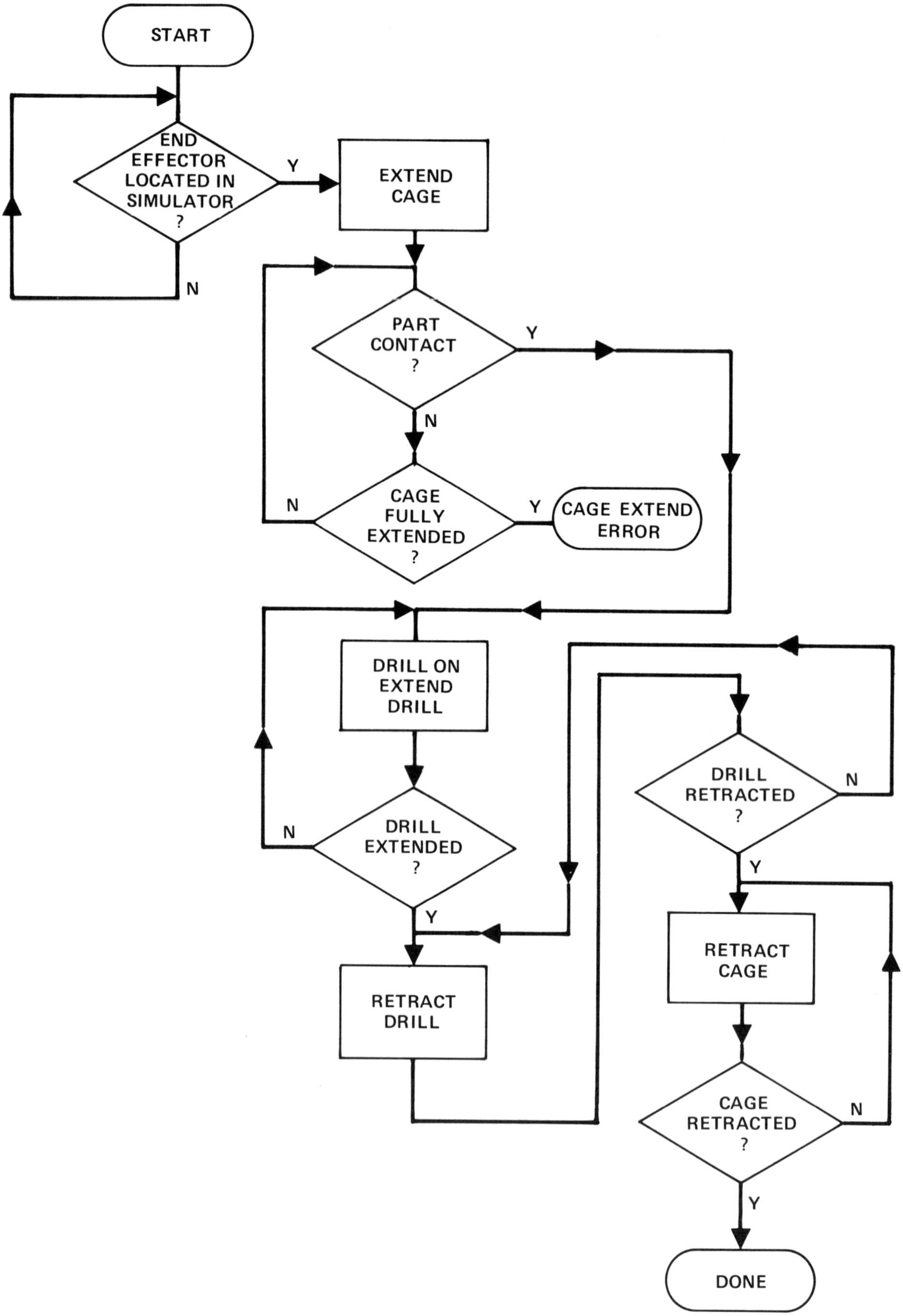

Figure 3 FLOW CHART FOR SEQUENCE CONTROLLER

END EFFECTOR R & D

At the same time that the simulator construction was being completed, a new prototype drilling end effector was being assembled. This new end effector incorporated a dampened drill feed stroke and an adjustable nosepiece to precisely control countersink depth to .001". The simulator allowed testing of the new design and the incorporation of some needed modifications before a robot was available in the laboratory. This method of testing and debugging end effectors off-line saved several weeks time.

The simulator has also been a useful tool in conducting tests and experiments. For example, in several robotic work cells, fiberglass drill templates use hardened steel drill guide bushings to insure that the location of the drilled hole is within the required accuracy. These templates were originally used in manual drilling. The steel guide bushings were needed because the operators tended to insert their drill nosepieces at an angle and abuse the hole diameter. The cost of precisely locating the guide bushings in the fiberglass is a cost driver of the total template fabrication cost. Observations in the robotic work cell indicated that there is considerably less misalignment between the nosepiece of the end effector and the drill guide bushings than between hand-held drills and the drill guide bushings. A test was carried out on the simulator to determine if bushingless fiberglass templates could be used with robotic drilling. The simulator's electrical circuit was modified so that the end effector's nosepiece was automatically cycled in and out of a fiberglass test specimen at various lateral and angular misalignments (Figure 4). An automatic cycle counter kept track of the number of insertion cycles. The compliance in the end effector took up the misalignment and no appreciable wear occurred after 6,000 cycles. Preliminary cost studies are now being performed to determine the cost savings likely to result from the use of this form of template tooling for future robotic drilling work cells.

The simulator has been used to gather data needed to develop a drill "breakthrough sensing system" (Figure 5). This sensing system would be useful in applications where the material being drilled varies in thickness.

Tests on new drill bit configurations, cutting/coolant fluids and minor end effector improvements have also been conducted. Currently, the simulator is being used to test, and further develop, a prototype blind riveting end effector.

PRODUCTION SUPPORT

The end effector used to drill and countersink holes through the aluminum alloys and/or composite material are custom-made high-precision tooling. Like any other tool of this nature, they require periodic maintenance and repair. As more and more end effectors were being utilized in production, the time spent on end effector set-up, maintenance and repair became a significant portion of the day-to-day activities of the laboratory personnel. There are now over 80 end effectors being used in production at the 10 robotic drilling work cells. A decision was made to transfer end effector production support responsibilities over the Tool Services Group, who currently have responsibility for all hand tools within General Dynamics. This would free the Factory Automation and Robotics Laboratory personnel to pursue more research and development work on other types of robotic applications.

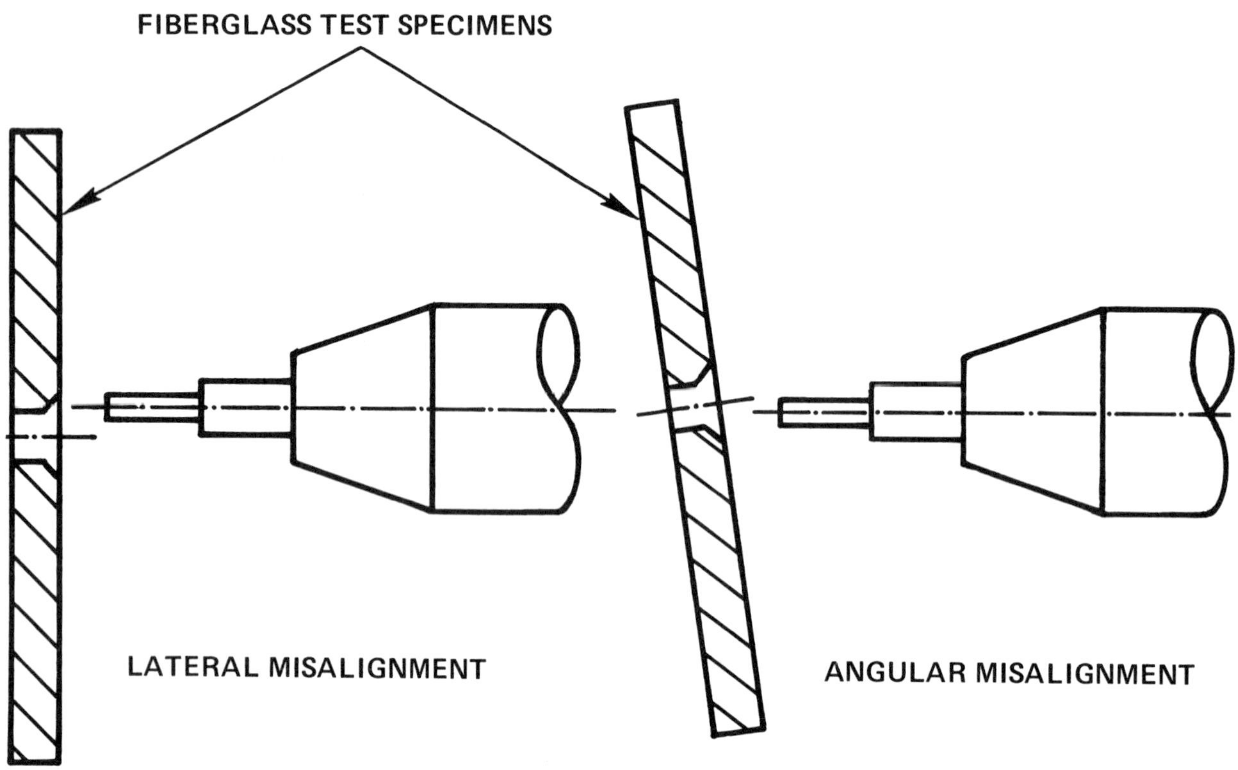

Figure 4 END EFFECTOR/DRILL TEMPLATE MISALIGNMENT SIMULATION TEST

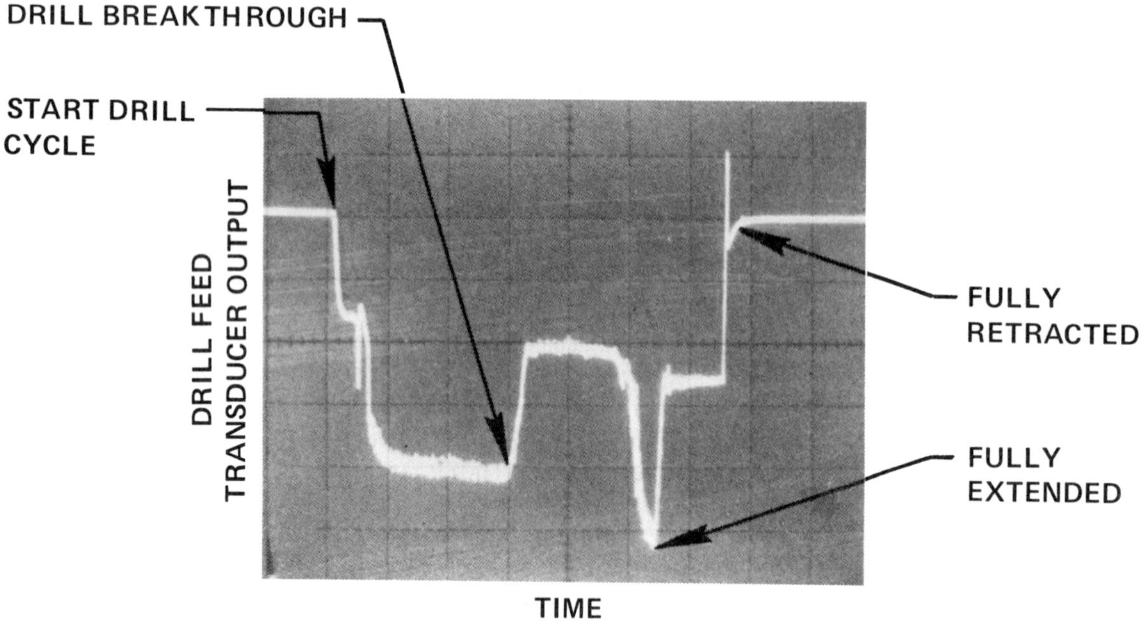

Figure 5 OSCILLOSCOPE PHOTO OF DRILL FEED CYCLE

The plan to transfer end effector production support to Tool Services consisted of seven requirements:

1. Provide Tool Services with a simulator.
2. Establish manpower requirements.
3. Write a maintenance manual for end effectors.
4. Procure a supply of commonly replaced parts.
5. Update end effector designs to the latest configuration.
6. Provide "hands-on" training.
7. Construct several spare end effectors.

Working from the original design of the laboratory's simulator, a new simulator was designed and fabricated for Tool Services. This new simulator is a more "user friendly" version of the first one. Some new features were added while others were deleted to make it more of a diagnostic aid rather than a research aid. This diagnostic simulator is now located in the Tool Services repair area.

Establishing manpower requirements involved recruiting those repairmen most qualified to do the work and developing a rotating schedule so that everyone has the opportunity to work with end effectors.

Documentation, in the form of a maintenance manual, was written and distributed to those who would be involved with the set-up, maintenance, and repair of end effectors. The manual covers five different styles of end effectors and contains step-by-step instructions with photographs of each maintenance or repair procedure. A troubleshooting guide, replaceable parts list, drawings and schematics are also included in the manual to make it an all-around reference guide.

Along with the maintenance manual documentation, Tool Services was provided a complete set of updated design drawings and bill of materials. From our end effector repair records, a year's supply of commonly replaced vendor items was estimated and ordered. These parts were then stocked in-house to minimize end effector down-time. Some parts on the end effectors have delivery dates several months after receipt of the order.

While the manual was being written, "hands-on" training classes were provided to familiarize the repairmen with end effectors. They were taught how to assemble, disassemble, diagnose, and repair end effectors and how to run the simulator. Many of the repairmen commented how much easier it is to work on end effectors compared to the hand-held drills because the parts are more accessible.

Spare end effectors were built so that should one production end effector need maintenance, the spare could take its place immediately. Before spares were built, production had to use another end effector that was set-up to drill some other size hole. This "switching around" required that the drill bits be changed and the feedrate reset each time. The spares are also used by the Factory Automation Group for use as aids in end effector design improvements and tests.

SUMMARY

End effector test equipment is very useful for testing both new and old end effectors off-line. As end effectors become more sophisticated and numerous in a manufacturing facility, the need for special support equipment increases. Although the cost savings resulting from the use of the end effector simulator are to some degree intangible, it is believed that this equipment has paid for itself several times over. End effector test equipment is best-suited for those applications where end effectors of the same type are being used extensively and/or where robots are required to have high production up-time.

A Sensory Gripper for Handling Textiles

D.R. Kemp
University of Hull
Humberside, United Kingdom

G.E. Taylor
University of Hull

P.M. Taylor
University of Hull

A. Pugh
University of Hull

Abstract

The major problem in introducing any kind of automation into the textile industry is that fabrics are inherently awkward to handle being non-homogeneous and behaving in a nonlinear and "live" manner. Many existing grippers and picking devices designed to separate a single ply from a fabric stack use "intrusive" methods that is they pinch, deform or even penetrate the fibres of the material which is clearly undesirable. A novel sensory gripper has been developed within the robotics group at the University of Hull. This gripper uses a non-intrusive technique which permits the top layer of the fabric stack to be lifted with minimal disturbances to the material fibres and to the remainder of the stack.

1: INTRODUCTION

THE CLOTHING INDUSTRY IN THE UK

The textile industry and in particular that part of it concerned with garment manufacture forms an important part of our national economy. The clothing textile and footwear industries currently employ a workforce of about 800,000 in the UK.

The manufacture of garments is highly labour intensive and subject to fierce competition from low wage economies. Hence, provided the technical and financial problems can be overcome, it should prove an eminently suitable area for increasing productivity by means of robot applications. Furthermore the production of garments is generally organised as a series of batch processes where fabrics and design often change rapidly. A programmable robot is well suited to work in such an environment Ref (1).

However fabrics are inherently awkward to handle being non homogeneous and behaving in a nonlinear and live manner. The industry therefore provides a difficult but challenging area for research into robot applications and so a collaborative project, funded by the United Kingdom Science and Engineering Research Council robots initiative, is being undertaken by the University of Hull Robotics Research Group, Marks and Spencer plc (a large British retail company) and Corah plc (one of their suppliers). The task is to pick up an individual piece of fabric from a stack and place on to this material with the aid of a vision system, Ref (2), a motif in the correct position and orientation. The motif is separately manufactured item with special backing material. The two items are then joined together under high temperature and pressure in a special purpose bonding machine. The material must then be removed and stacked.

A positional accuracy of about +/- 1mm and a rotational accuracy of +/- 1 degree is required together with a cycle time for the complete operation of the order of one minute.

2: SOME CHARACTERISTICS OF A STACK OF CLOTH

A major problem in handling fabrics with a robot presents itself when it is necessary to separate a single piece of fabric from the top of a stack. Some of the properties of the material on the stack are a direct result of the "laying up" and cutting stage in the manufacturing process.

At the laying up stage the cloth arrives in a large roll hundreds of metres long and approximately 3 metres wide. Sheets of material are then cut from these rolls and stacked on top of each other toan average depth of 50 layers.

The individual component pieces are then cut from the large stack producing the the smaller stacks that are used in the motif application stage and in the rest of the manufacture of the garment.

In some cases the large stack is held in place by vacuum sealed by a sheet of polythene over the top of the stack. The vacuum has the effect of compressing the layers together so that the fibres of each layer cling together. The cutting process can bind together the layers at the edges of the component stacks.

Rough handling (either intentional or unintentional) reduces the effects of edge binding and cling between layers, but it may mean that the component stack does not have vertical edges when it arrives at the motif application stage. This is an important factor since the unstacking device will have to accommodate discontinuities in the stack.

3: SOME METHODS OF SEPARATION OF A PIECE FROM A STACK

A few of the existing and proposed methods of separating a single piece from the top of a stack are detailed below with comments on the methods that have been investigated.

1. Mechanical pinching and "Intrusive" methods

Pinching devices (eg "Clupicker" manufactured by Cluett) pinch the fibres of the top layer attempting to avoid the second lay by good design and setting up.

Intrusive methods are those in which pins or fine wires are inserted into the uppermost piece on the stack. These methods rely on accurate engineering of the gripper and a consistent thickness of the material. Added to these disadvantages the gripper has to be accurately set up before any operations can commence, and may need regular readjustment.

These devices disturb the fibres of the material on the stack and in certain cases can cause irreparable damage to the fabric.

2. 'Velcro' (R)

'Velcro' (R) is the trade name for two strips of material consisting of fibrous hooks and loops. It was found that Velcro does not grip all materials. If the Velcro (R) can grip the material there remains the problem of removing the material from the Velcro (R) without damage to the material.

3. Adhesives

This method poses similar problems to the case above though it may be possible to find an adhesive that will pick up most fabrics.

4. Vacuum

Picking methods using various types of vacuum heads or nozzle have been rejected because the air flow penetrates two or more layers of material binding them together.

5. Air jets

An air jet can be arranged so as to produce vibrations in the uppermost layer of a stack of material. These vibrations are produced because of two factors:

1. The air flow from the air jet is turbulent.
2. The velocity of the air flow decreases the pressure above the material, this can have the effect of lifting the material.

The above two effects can combine to result in the material being lifted and vibrated. If two or more layers of material start to vibrate simultaneously because they are clinging together it is possible for the layers to separate simply because of the vibrations. This is the approach adopted at Hull to separate a single ply by applying an air jet and moving the extreme end of a gripper jaw under the vibrating edge of the material.

4: BEHAVIOUR OF MATERIALS UNDER AN AIR JET

There are both quantifiable parameters (eg air pressure to the air jet) and unquantifiable parameters (eg the amount of rough handling the stack of material receives) which affect the way in which the material vibrates. Consequently, a complete quantitative analysis has not been attempted but instead a qualitative analysis of some of the more useful relationships follows.

The positioning of the air jet to produce vibrations in the material with a minimum wastage of air is important. In many cases good results are produced if the air jet is directed at an angle of 30 degrees to the surface of the cloth with the point of contact of the air flow with the surface of the material approximately 30mm from the edge of the stack and the nozzle approximately 60mm from the point of contact.

For a given nozzle diameter in a fixed position a higher air pressure produces larger amplitude vibrations of the material.

The above effect is countered by two properties of the material in the stack. Firstly if the material clings together then higher pressure air is needed to start the material vibrating, this is analogous to stiction.

Secondly a heavy material will require higher air pressure to induce vibrations of the same amplitude.

The above variations also have the effect of altering the mean frequency of the vibrations. This frequency is difficult to quantify because of the random nature of the vibrations but from measurements carried out on jersey knit fabric using an air jet at 30 psi through a nozzle of 1.5mm diameter the vibrations were measured as approximately 100 Hz.

5: THE GRIPPER IN OPERATION

5.1 MECHANICAL CONSTRUCTION

The prototype gripper consists of two parallel jaws (Fig 1 450mm long and 20mm wide (a 40mm wide version is planned). The lower jaw of the gripper contains infra red light emitting diodes which supply the triggering for crossfire sensors mounted in the upper jaw. On the upper jaw is an adjustable mounting to direct an air jet at any desired angle on to the cloth and at the tip of the upper jaw is a mounting for a limit switch.

The gripper currently carries two sensors, this may in the future be increased to five, the two sensors are a crossfire sensor at the tip of the gripper jaws to sense when a piece of material is in between the gripper jaws and a stack top limit switch to sense when the lower jaw of the gripper is at the top of the stack. Other proposed sensors are two more crossfire sensors to assist the robot in final positioning of the gripper prior to grasping the material and possibly a sensor to detect the cloth vibrating and produce information on the amplitude of the vibration.

5.2 THE SET UP PROCEDURE

The gripper is positioned by the robot near the top of the stack, in setting the system up the operator positions the gripper at a convenient start point where the robot is to activate the air jet and move the gripper so as to insert the lower jaw of the gripper between the uppermost piece and the rest of the stack.

The start point in a simple case is where the lower jaw of the gripper is positioned near the centre of the edge of the stack with the tip of the lower jaw approximately 5mm above and 5mm over the edge of the stack (Figure 2). The air jet is then positioned as in figure 2 though the extract positioning, the type of nozzle and air pressure will be dependent on the type of cloth on the stack. The position of the gripper is then defined as the start position.

5.3 THE OPERATIONAL CYCLE

The cycle begins with the gripper at the start position with the air jet activated and a search initiated both horizontally and vertically in the pattern shown in figure 3. The air jet causes the top layer of material to vibrate and the tip of lower jaw of the gripper comes into contact with the vibrating cloth. The robot searches downward so that on continued incremental movement the fabric 'pops' above the lower jaw.

The sensor at the tip of the gripper jaw detects the cloth between the jaws and the air jet is shut off. The robot then checks the state of the sensor again to ensure that the cloth has not subsequently moved out of the jaws. If the sensor is falsely triggered, then the robot resumes the search pattern, otherwise the gripper is moved (depending on the type of cloth) approximately 3mm downwards (triggering the sensor occurs near the maximum height of the vibration of the cloth and thus the lower jaw is too high if it is immediately inserted between the layers of cloth). The gripper is then moved horizontally towards the stack thus inserting the lower jaw of the gripper between the layers. At approximately 40mm into the stack the sensor at the tip of the gripper is again checked to ensure the insertion between the layers has been successful. If the insertion has not been successful, the search routine is recommenced. Usually only minor disturbance to the stack has resulted which can be rectified by the air jet blowing the material back into place. If the insertion has been successful then the robot will continue to insert the gripper and position it at a convenient gripping point. At the gripping point the jaws are closed and the robot moves the gripper up and across the stack so as to peel the material away from the stack.

6: TEST RUNS

A prototype version of the gripper fitted with only a single crossfire sensor at the tip of the gripper has undergone a number of trial runs each consisting of unstacking a pile of approximately 50 pieces of cotton mixture,jersey knitted material. After the initial test runs the simple search pattern,used by the robot when separating the top piece from the stack, was modified to the pattern shown in figure 3. On the initial trial runs a number of developmental problems have been noted and solved. These included the following:

1. The robot failed to find the top piece on the stack in the search phase. This was because the stack did not have vertical edges and the robot was not capable of detecting the top of the stack so that it could modify

the search automatically.

2. The robot failed to separate a single piece from the stack. The robot had successfully inserted the lower jaw of the gripper between the layers of material but failed to overcome binding of the edge running parallel to the gripper at the gripping point. The result of this was that the top piece of material was held between the jaws and a second piece was clinging to the first around the lower jaw of the gripper.

3. The robot failed to pick up a piece of material from the stack because the lower jaw of the gripper was too high when it moved in between the layers to the gripping point. This resulted in the top piece falling out of the gripper.

7: EVALUATION

7.1 RELIABILITY

From the test results we find that at no point during the test runs have two or more pieces of material been detected between the jaws of the gripper. Including the modifications, previously mentioned and until more extensive test runs can be made the reliability of the prototype robot gripper system should be quoted as of the order of 98 percent (ie 2 failures in every 100). For an industrial system the reliability must be more than 99 percent and preferably the system should also be able to detect failures and reject the piece, rectify the fault, or halt and await rectification.

7.2 SPEED

At the time of writing the cycle time of the robot gripper in lifting a simple piece of cloth laying it down and then returning for another piece is 25 seconds. There are some modifications which can be made for example small air jets blowing out of the lower jaw to speed up the insertion of the lower jaw between the layers of material. Ideally the cycle time to be achieved is that of an experienced textile worker at the same task ie approximately 12 seconds. The two limiting factors in achieving this cycle time are the speed of the robot and the time taken for the robot gripper to find the next vibrating piece of material (during the search pattern).

7.3 MATERIALS

The gripper has been tested on stacks of the most commonly used materials. Further tests need to be undertaken on both lighter and heavier weight fabrics.

CONCLUDING REMARKS

A new sensory gripper has been designed for attachment to a robotic device in order to separate a single ply of fabric from a stack. Tests done up to the time of writing show that the device is applicable to several types of material. The gripper is non intrusive and therefore damage to the material is minimised.

The authors wish to acknowledge the advice given by Mr. T. Baker, Mr G. Bambury of Corah plc and Mr. C. Walter of Marks and Spencer plc and support by the Science and Engineering Research Countil under grant GR/B/80901. This work is the subject of a patent application.

REFERENCES

1. Kurt Salmon Associates, Altringham, Cheshire, England "The 1980's": The decade for technology? "Prepared for the commission of the European Economic Community, 1979.

2. G. E. Taylor, D. R. Kemp, P.M. Taylor and A. Pugh, "Vision applied to the orientation of embroidered motifs in the textile industry" 2nd Int Conf on Robot Vision and Sensory Controls, Stuttgart, Germany, 1982.

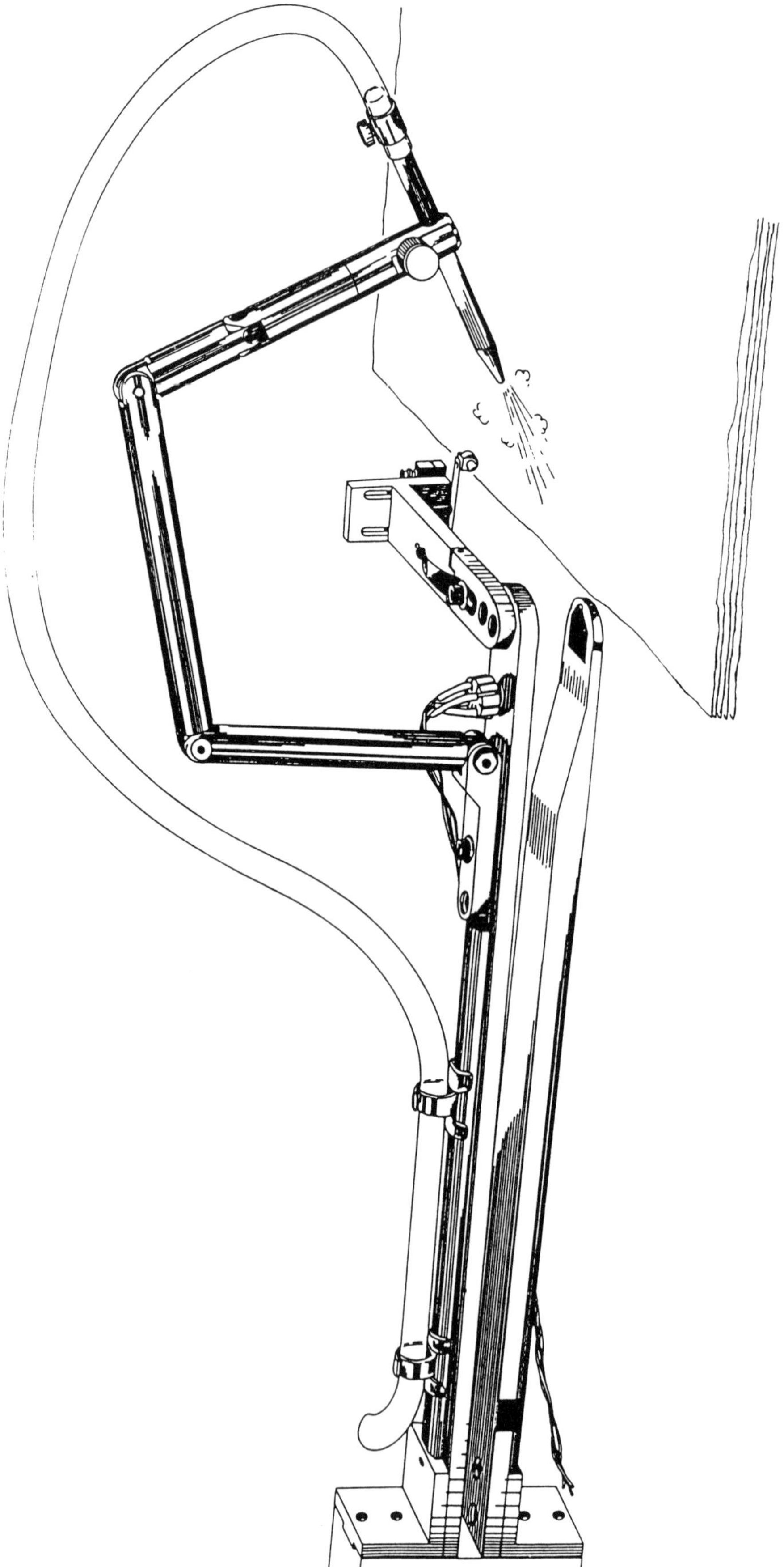

FIGURE 1a SCHEMATIC DRAWING OF THE GRIPPER

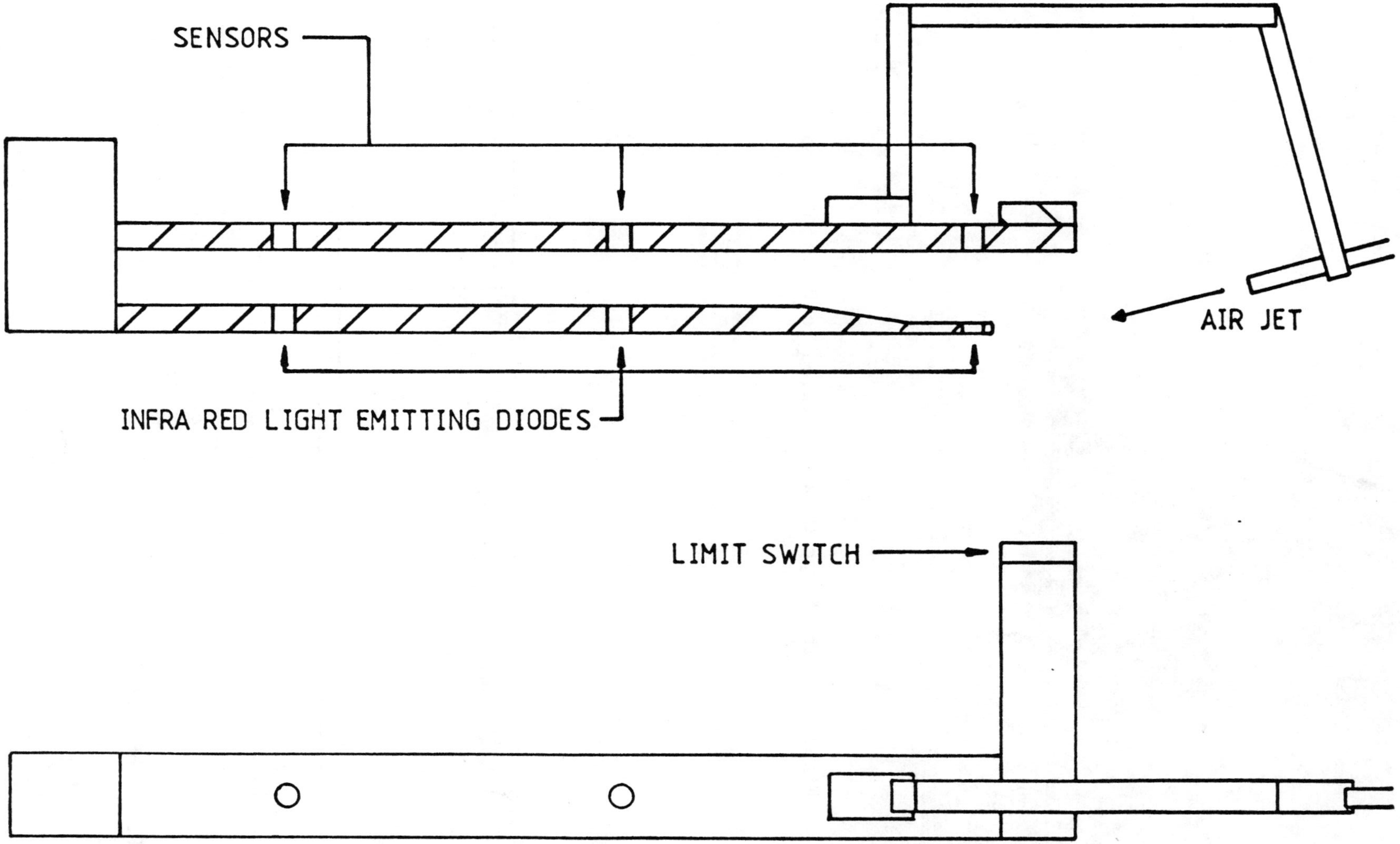

FIGURE 1b POSITION OF THE SENSORS

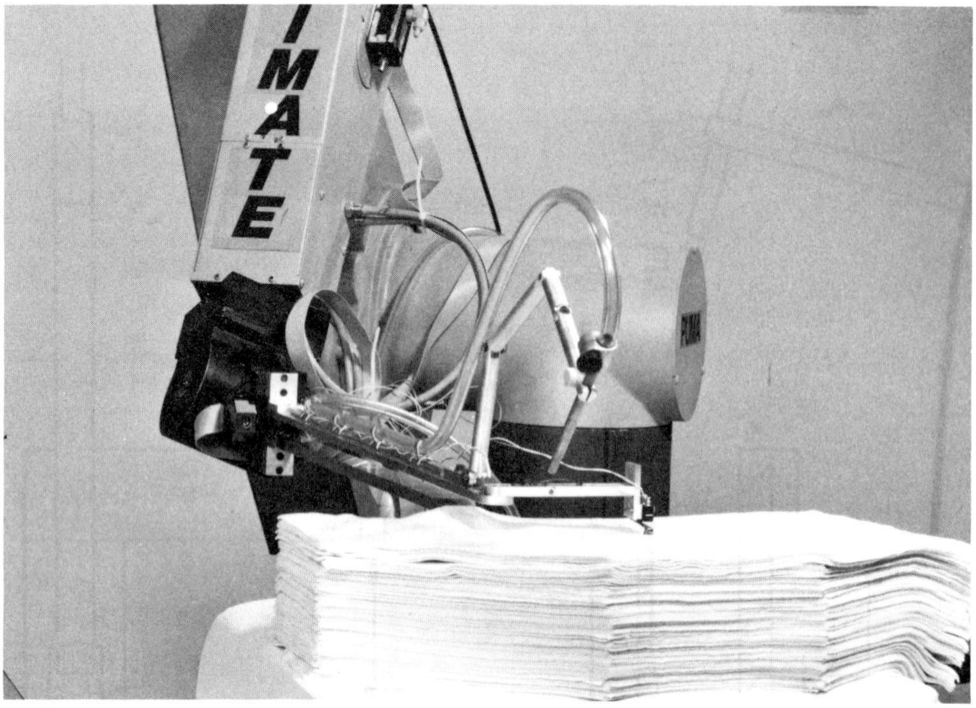

Figure 1c Prototype Gripper fitted to PUMA 560 Robot

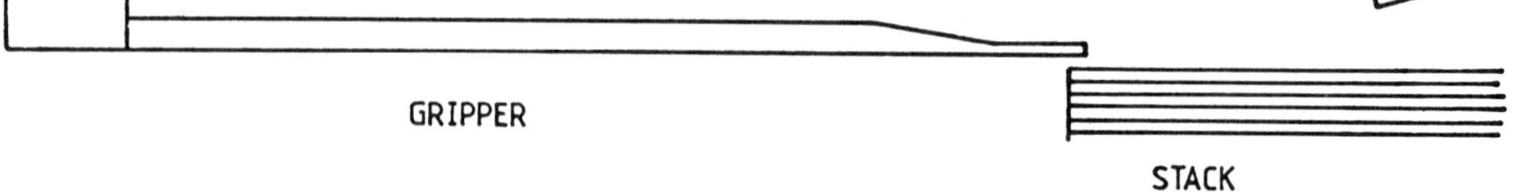

FIGURE 2. THE START POSITION.

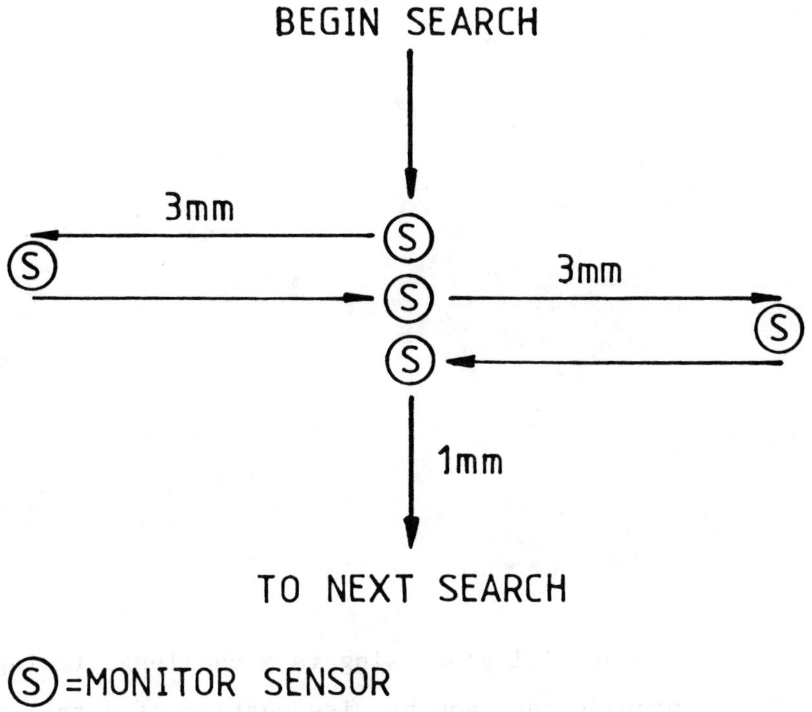

FIGURE 3. SEARCH PATTERN

Advanced Composite Material Handling Application

Kenneth B. Bubeck
Rensselaer Polytechnic Institute
Troy, New York

Abstract

The automation of composite material processing is a challenge facing the aerospace industry. Gerber Tables provide fast and precise cutting of detail parts with minimum waste when the nesting programs are used. However, unloading of composite materials is currently a slow and tedious task which can be facilitated by fitting a programmable robot with a variable span vacuum end effector that can transport pre-impregnated composite parts of varying sizes from the Gerber Table in a kitting operation. This paper describes the development of an adjustable span vacuum end effector that uses a microprocessor controller operating on an existing data base of detailed parts to determine the span.

INTRODUCTION

The cost saving potentials of automation are being recognized by the aerospace industry, specifically those applications related to composite material processes. Panel fabrication, one form of composite processing, is practiced extensively throughout the industry and because of its labor intensive nature it is a major contributor to total costs. However, by segmenting the production technique many aerospace companies have reduced labor and material costs considerably.

Wing skins, stabilizer skins, control surfaces, and fuselage components all belong to the "panel" family. They are formed by laminating pre-impregnated fiber reinforced cloth in a bonding fixture. The laminate is subsequently pressurized and cured in a specified curing cycle, debulked and trimmed, then assembled with other parts of the aircraft. The technique of putting the material in the mold form requires many hours of expensive handling. Oversize sheets of material are placed in the bonding fixture, and individually trimmed with a knife to match scribed marks on the mold form. The task of manipulating the large sheets into place and trimming off excess material make this a desirable application for automation.

Gerber Table Technology has developed equipment, shown in Figure 1, for the garment industry that accurately cuts patterns in large stacks of cloth. The patterns are digitized and stored in a data base which is operated on by nesting programs to minimize material scrap. Gerber Tables have now been adopted by aerospace companies and with some modification have successfully reduced material waste costs. However, the unloading of the composite material from the table is currently a slow and tedious manual task which could be facilitated by use of a robot. The robot would need a variable span vacuum end effector to transport parts in a sequence that would leave a kitted panel ready for the next assembly.

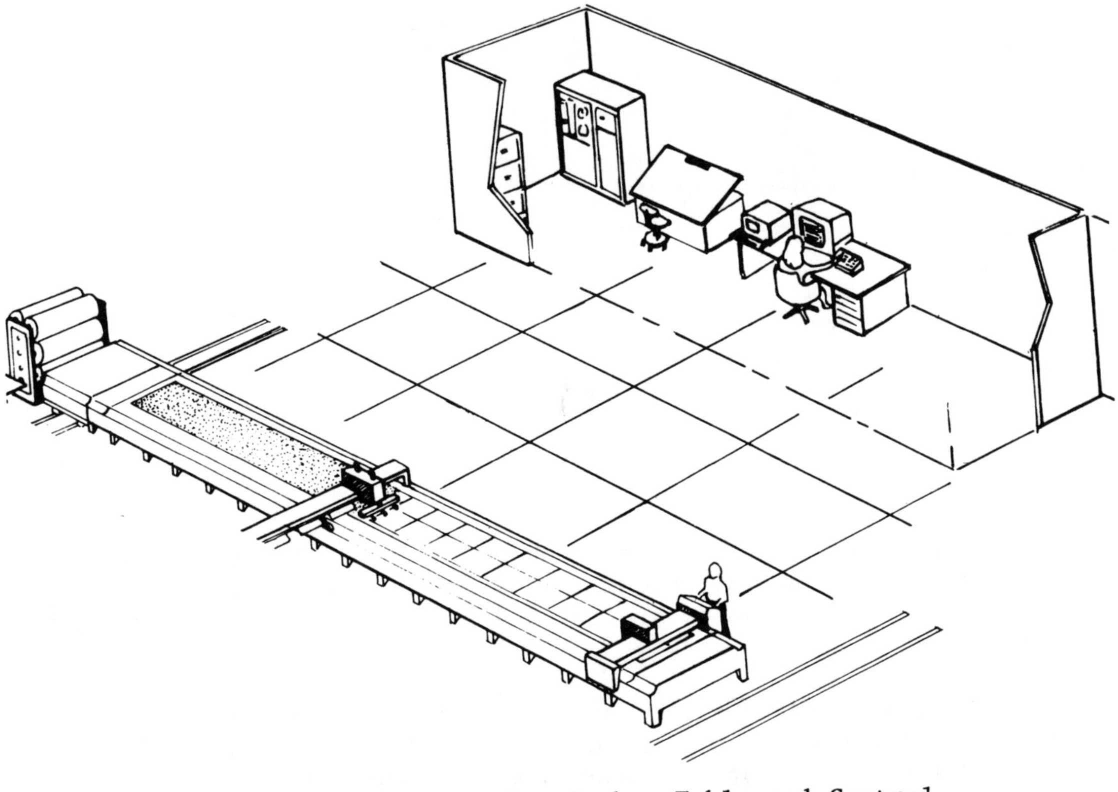

Figure 1: Gerber Table and Control

This paper describes the development of such an end effector. The device uses a microprocessor controller, operating on an existing data base of detail parts, to transport varying sizes of composite material in an ordered sequence. Future developments may lead to laying the material into the bonding fixture directly from the Gerber Table.

SYSTEM DESIGN

The computer used with Gerber Tables can,with some software modifications, be used as the controller for an unloading system. This could be accomplished by downloading to the robot from the N.C. program used for cutting the shapes. Figure 1 illustrates an overhead gantry arrangement but the actual mechanism which was developed was fastened to a Cincinnati Milacron T3 for the purpose of demonstration.

The system diagram, Figure 2, illustrates the closed loop feedback from the robot controller through an I/O module, to the microprocessor. The description of the system functions are listed below:

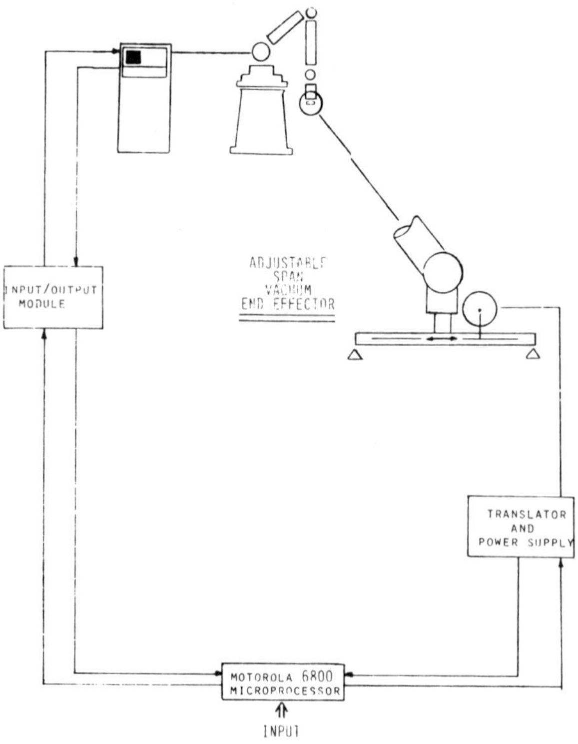

Figure 2: The System Diagram

- The data base of detail part geometries, location on table, points of pickup on the detail parts, assembly kit packages, order of stacking, and location of kit placement are entered as input to the microprocessor.

- The computer's software operates on this data and sends pulses in the forms of step functions and ramp functions to the input/output (I/O) module, and the translator and power supply respectively.

° The translator is a driver for the stepping motor mounted on the end effector. The power supply feeds the motor. The number of pulses sent through the translator to the stepping motor is a function of the size and shape of the detail part.

° The I/O module sends signals to the controller of the robot, which initiates the robot's motion. As the robot is "stepped-through", the program sends a signal back to the I/O module to acknowledge the motion (as indicated by the reversing arrow).

° The adjustable span vacuum end effector is a mechanical device that has four vacuum ports. The span is adjusted by a pair of movable racks and two vacuum ports on the end of the racks. These racks are set in motion by an arrangment of gears that are driven by the stepping motor. The vacuum ports are actuated by solenoid values that are controlled with the microprocessor.

Electrical Hardware:

The circuit diagram, shown in Figure 3, describes the integration of the electronics, the components were divided to fit into three (3) boxes (Figure 4 illustrates the front panels). The components are labelled on this circuit diagram and the following is a description of each (the numbering system corresponds to those in the circuit diagram):

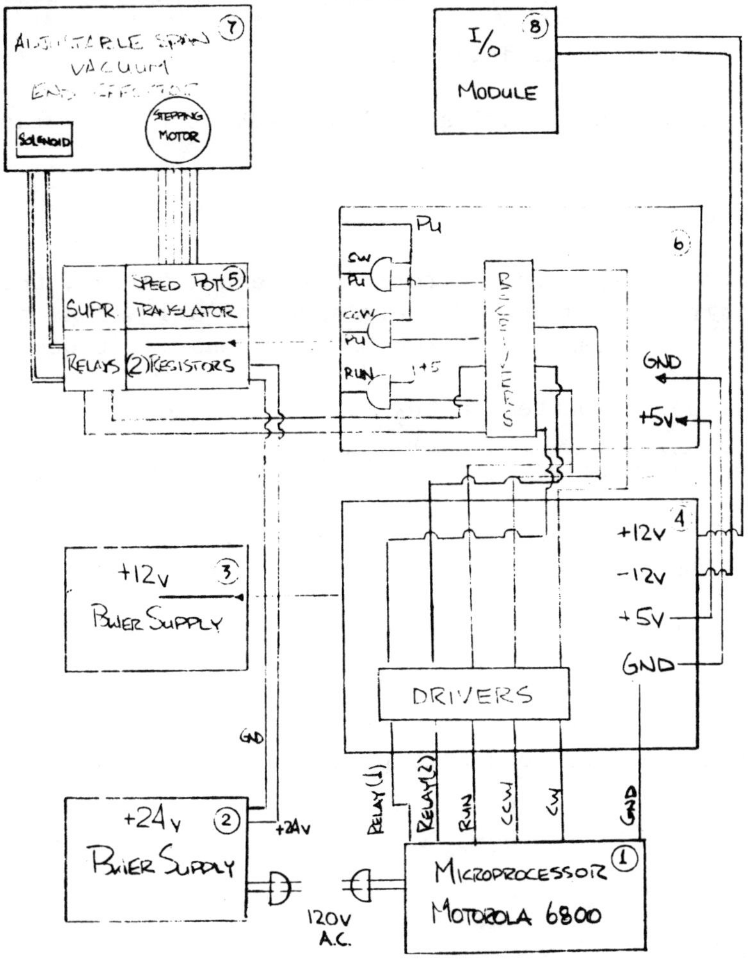

Figure 3: Electronic Circuit Diagram

1. Motorola 6800

Microprocessor, the devoted intelligence of the system. The data base is operated on by this component. The micro drives the stepping motor through the translator to adjust the span of the end effector, receives the robot position information and actuates the solenoid valves turning on the vacuum.

2. 24 Volt, 4.8 amp power supply (box #1)

This is used to supply power to the stepping motor.

3. 12 Volt, 1/8 amp power supply (box #2)

This is used to supply power to the solenoid valves and provide power for the ramping logic of the translator. The step functions delivered to the motor need to be accelerated gradually as if the full step is delivered, the motor would not react fast enough and probably miss an amount of steps. This situation is analogous to the tires of a car slipping when suddenly depressing the accelerator.

4. The circuit board mounted in box #2

This board contains the driver chips (1488) for the relays and the motor.

5. The interface control box (box #3)

It houses the relay switches for the solenoid valves, the translator card, the circuit resistors. This assembly allowed us to use a signal bypass and operate the mechanism manually if needed.

6. The circuit board mounted in box #3

It contains the receiver chips (1487) and the ramping logic.

7. The end effector which carries the material from the "cutting table" to the "stacking table".

8. I/O (input-output module of the robot controller)

This delivers and receives signals to and from the microprocessor as the sequence of operations take place.

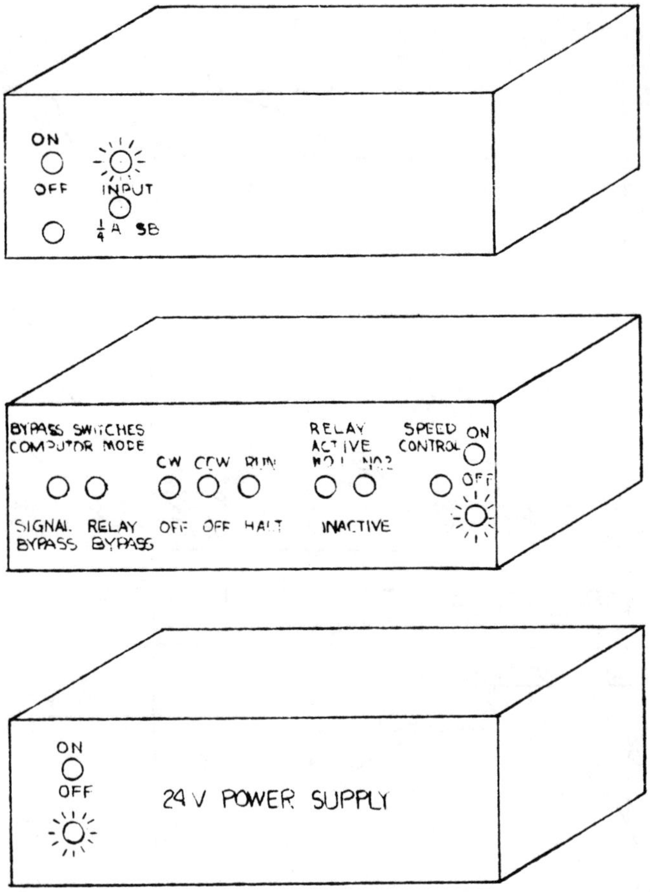

Figure 4: Control Boxes

Mechanical Hardware:

The adjustable span vacuum end effector was fabricated primarily with aluminum. The detailed drawings for a top view of the gear train and the front view can be seen in Figures 5 and 6 respectively.

The gear train was designed to provide opposing motion of the vacuum cups, having the racked shafts move in opposite directions. This is accomplished by reversing the velocity vectors of the gears labelled 3 in Figure 5. Number 20 indicates the solenoid valves used to control the application of vacuum. Number 6 is the stepping motor used to drive the train through the coupling, number 15. Number 8 is the cylindrical column used to attach the end effector to the robot (also see Figure 6).

Figure 6, the front view, illustrates the supports (number 50,51) for the movable shafts. Racks were fastened to the shafts and driven through a pinion arrangement. Numbers 36 through 44 show all the details used in the vacuum assembly.

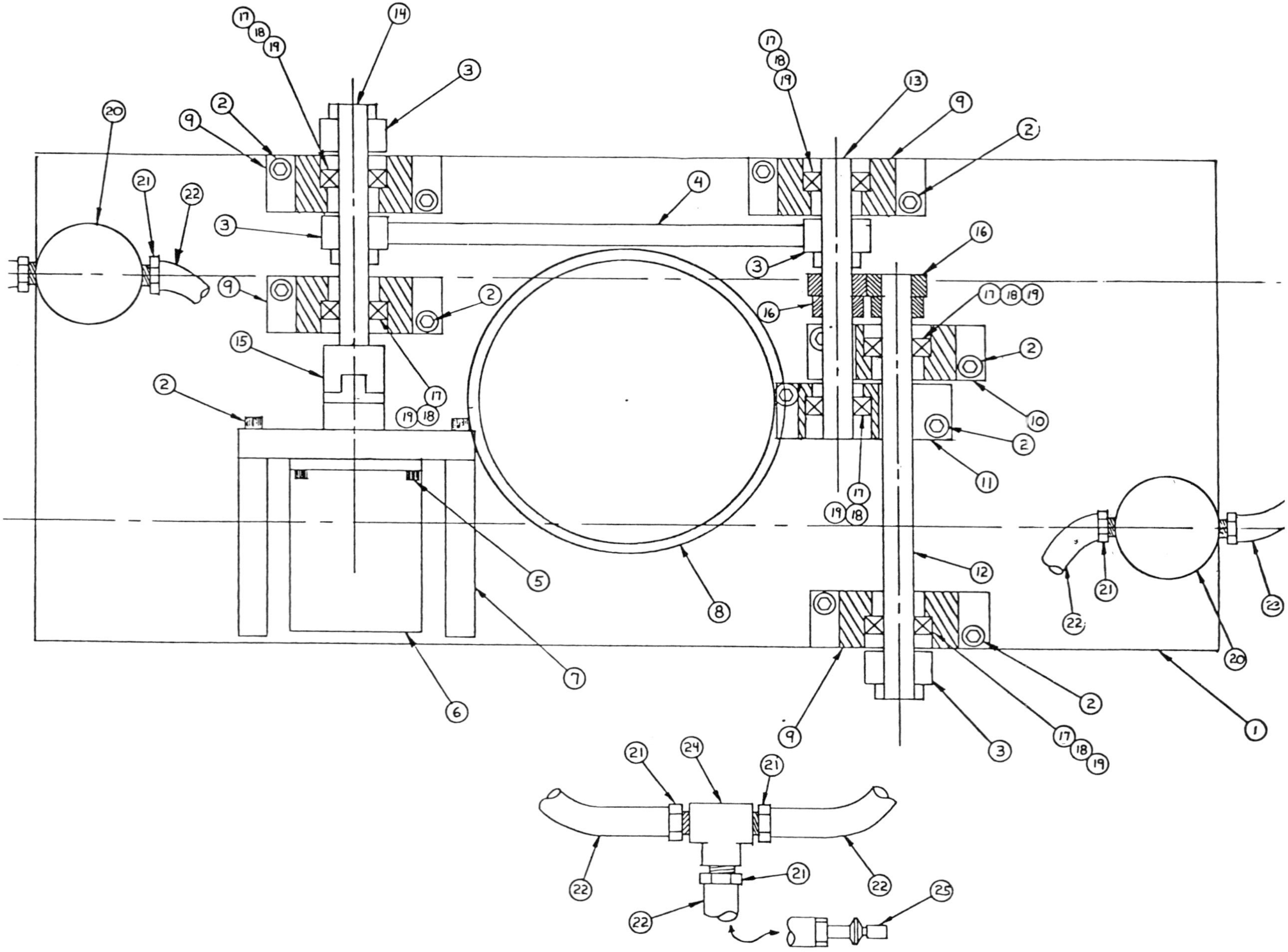

Figure 5: Top view of end effector showing the drive train

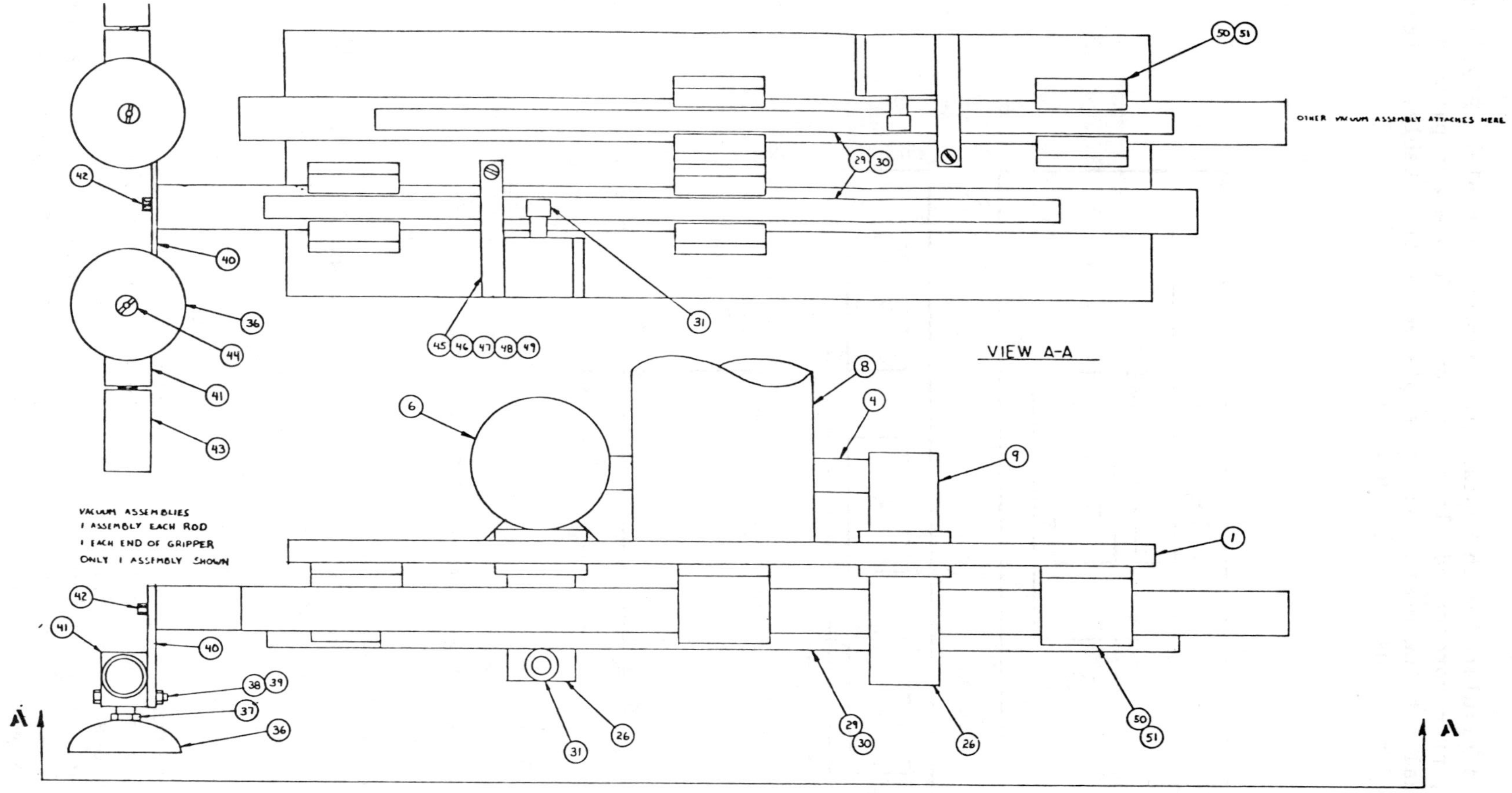

Figure 6: Side and bottom detailed views

Demonstration:

A Gerber Table was simulated with a standard laboratory table. Pieces of fiberglass reinforced, pre-impregnated cloth, cut out to represent patterns resulting from an actual cutting operation, were placed on this table. Figure 7 shows the dimensioned parts used in the demonstration.

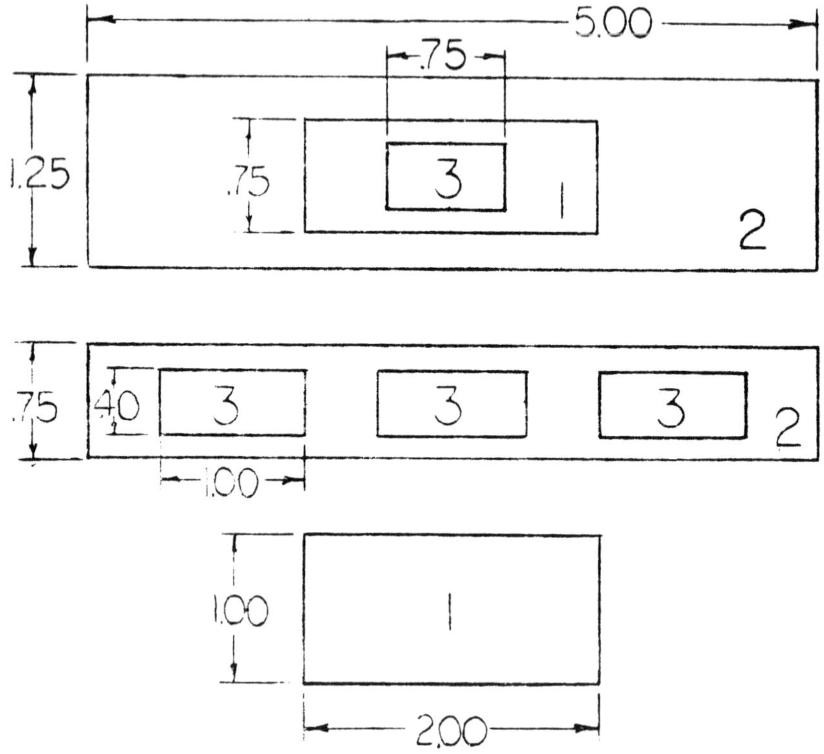

Figure 7: Arrangement of Fiberglass Parts

A data base was manually generated, read into the microprocessor, instructing the system how to sequence through all the steps. The data base consisted of part geometry, the order of material removal, location on the table, assembly kit packages, location of placement for the kitted material. The parts labelled with the number 1 represent the first kit, number 2 the second, and number 3 the third (see Figure 7).

The robot was programmed to locate the end effector over the respective part and then sent a signal back to the microprocessor to acknowledge position. The microprocessor, upon receiving that signal calculated the required span, adjusted the mechanism and then activated the vacuum. The part was lifted and transported to the removal table and then was released onto the unloading table. The operation continued until all the material was transfered in an ordered sequence.

CONCLUSIONS

The demonstration of this robotic application verified the feasibility of using a stored data base to drive a mechanical system. The stored data base can be found in the nesting algorithms produced in the Gerber Table computers. With some minor manipulation, a devoted microprocessor can be used to drive the mechanical system.

Applying this application to production would require many refinements. The end effector would have to be more complicated by adding features, thus increasing it's flexibility. As the first layer of material is removed, the height of the next part in that particular stack is now approximately .015 in. away. A servo drive to adjust the height of the vacuum cups would be required. Additional degrees of freedom of the vacuum cups, that is motion relative to each other, would enable the device to handle a larger range of parts. Although the nesting algorithms minimize material scrap, it is not eliminated, therefore provision should be made to remove the scrap before the next layer of material is removed. An overhead gentry robot would enhance the process. The software used to operate such a system would be very complicated and developing could take many months.

Capital expenditures needed to implement the refinements stated above can be easily justified because automating this phase of the material cutting process has significant cost savings, as shown by companies who have undertaken this task. Being able to facilitate a robot with an adjustable end effector, controlled with a devoted computer, will give the user the needed flexibility when automating material handling applications where the size and shape of the material are variables.

ACKNOWLEDGEMENT

I am grateful to all those engineers and students, working for the Center for Manufacturing, who made this application a success. Special thanks belong to Larry Ruff who fabricated the end effector, George Menz who engineered the electronics, and Chris LeMaistre the Associate Director of the Center for Manufacturing for his encouragement and support.

Figure 8

Piezo-and Pyroelectric Polymers Skin-Like Tactile Sensors for Robots and Prostheses

R. Bardelli
Centro "E. Piaggio"
Pisa, Italy

P. Dario
Centro "E. Piaggio"

D. De Rossi
Centro "E. Piaggio"

P.C. Pinotti
Centro "E. Piaggio"

Abstract

The use of piezo and pyroelectric polymers, such as poly vinylidene fluoride (PVF_2) as basic materials for the construction of an artificial skin is proposed. The main properties which render these polymers, suitable for reproducing the characteristics of the epidermis are outlined along with the theoretical and technological problems which could be encountered in the design of a multipurpose distributed skin-like sensor. A possible configuration for a PVF_2 based artificial skin is proposed. The results of some preliminary experiments, demonstrating the possibility of reproducing human sensitivity when touching materials having different thermal conductivity, are also presented.

INTRODUCTION

Though the antropomorphous approach is often deemed not necessary and sometimes even pregiudicial in robot design (1), duplicating human function is a matter of wide interest in many artificial intelligence research laboratories. We believe that, even if operation modes must be reconsidered when a human worker is substituted by a robot and a new task analysis is necessary, however the self adapting capabilities of biological systems still represent an ideal model in optimizing the subsystems which allow the interaction between machine and environment.

The design of sensing systems for robots has sometimes followed an antropomorphous approach. For example, the anatomical structure of the eye has been taken into account in optimizing the behaviour of artificial vision system (2) and others took inspiration from physiology in designing the electronic units for their tactile sensors (3).

The duplication of the human capability to recognize objects by manipulation alone, a process known in physiology as stereognosis (4), would provide substantial innovative features in the fields of robotics and prosthetics. We think that, to achieve such ambitious goal, research efforts should be oriented to a deeper comprehension of the mechanism of coordination between movements and perceptions and to the development of a suitable touch sensory system, capable of supplying complex information. The first issue has been studied by Hillis (5), whose artificial finger is purposely programmed to perform appropriate movements, against which it can interpret what it feels. The second issue concerns a problem which is attracting an increasing number of investigators, i.e. the construction of a really effective "artificial skin". Mechanical compliance and resistance (3) as well as sensitivity, linear response and good resolution in order to supply tactile information of complexity comparable to the human sense of touch (6) are considered the most important properties to be provided to this type of distributed sensor.

Conductive elastomers have been usually adopted for the construction of high resolution arrays; flexibility, low cost, potentiometer response to displacements seem to be the main qualities of these materials; while low sensitivity, non linearity, noise, drift, long time constant, hysteresis and poor fatigue resistance have been reported as their relevant defects (6).

In this paper we wish to point out the likeness between some mechano-and thermoelectric sensing properties of human epidermis and those of some synthetic piezo and pyroelectric polymers and to propose the use of these materials for the construction of an artificial skin.

HUMAN SKIN AND PIEZO-AND PYROELECTRIC POLYMERS

Classical physiology usually attributes the sensing properties of the human skin to the presence of sensory nerve endings, like Merkel discs, Meissner and Pacinian corpuscles, papillary nerve endings (4). However, the peculiar functions of each of these receptors are not well clear and the transducing mechanism, through which a mechanical or thermal stimulus results in a sequence of electric discharges, is not yet fully explained.

Piezo-and pyroelectric effects in some animal tissues, like tendon and bone, have been known since 1966 (7). These properties have subsequently been discovered in various materials of biological origin and in the human epidermis (8). Pyroelectric coefficients in the range between 2.1 and 27 $pC/cm^2 \cdot {}^\circ K$ and piezoelectric coefficients along the thickness direction (d_{33}) in the range between 0.02 and 0.19 pC/N have been measured in fresh skin samples. The same effects have been observed also in dead, dry skin preparations and this excludes the possibility that they could be due to other phenomena, such as ionic transport, associated with the living state (8). The close similarity between the electrical response of the epidermis to a rectangular heat pulse and the physiological reaction of a stimulated sensory organ suggests that the piezo and pyroelectric behaviour of the epidermis may be of fundamental physiological importance (7) and could play an important role in the transduction of external non electric stimuli into neuronic response (9).

Rheological properties and space-time resolution are other features to take into account, together with the transduction mechanism, when attempting to duplicate human sensory function.

Rheological properties of animal and human soft tissues and skin have been studied. In particular, skin has been shown to behave as a viscoelastic material (10).

The frequency response of skin receptors ranges from DC to ~ 400 Hz. It is usually accepted that these receptors can be classified into rapidly adpting, i.e. responding to skin indentation exclusively when the stimulus is changing, and slowly adapting, which exhibit a response also to steady stimuli (11).

The ability to individually resolve close object by tactile sensing, which is stronly dependent on the body district, has been evaluated by means of anatomical studies (4) and psycophysic tests (11). It has been demonstrated that a subject cannot discriminate on his finger tip two points less distant than 1.5 mm, even if the spatial distribution of the receptors is of the order of 30 per mm^2.

The discovery of a strong piezo-and pyroelectricity in some synthetic polymers, such as PVF_2, is quite recent (12). By means of suitable preparation procedures (13)(14), pyroelectric coefficients of the order of a few $nC/cm^2 \cdot °K$ and piezoelectric coefficients, d_{33}, of some tenths of pC/N can be obtained in PVF_2. The mechanical, electromechanical and structural properties of these materials, which are rugged, cheap and flexible, allow the construction of sensors of various shape, with a thickness ranging from about 6 μm to 2 mm. Membrane and thickness operation modes are allowed. High resolution complex patterns of electrodes on a single polymer sheet can be obtained either metallizing the film through suitable masks or by etching selectively a continuous metallization previously deposited on the film. The large bandwidth operation of PVF_2, ranging from about 0.01 Hz to hundreds of MHz, allows quasi static as well as ultrasonic response. A true DC response, inherently prevented to every piezoelectric material by its finite electric resistivity, can also be achieved with a particular technique which will be described in the next paragraph.

A POSSIBLE CONFIGURATION

The human skin is capable of supplying the nervous system with combined inputs such as contact, pressure changes, vibrations, slip, thermal exchanges. On the other hand, the extreme versatility of piezo and pyroelectric polymers may suggest a variety of sensor configurations in order to provide such a composite set of information. However, when using this class of polymers, it is not possible, in principle, to distinguish an electrical response originated by a thermal stimulus from one originated by a mechanical stress. This represents a severe obstacle to the construction of a polymeric artificial skin. Nevertheless, we propose here some solutions to overcome this problem.

Two methods are possible for limiting the pyroelectric sensitivity in order to obtain a sensor which is sensitive only (or mainly) to mechanical inputs: a) insulating the polymer sensor from the environment with a layer of material having low thermal conductivity; b) substituting the conventional polymer film with a microvoided one (15). The microvoided PVF_2 film can be obtained by special preparation procedures and it features a reduction of the pyroelectric coefficient compared to the piezoelectric one. A 4:1 reduction of the ratio λ/d_h (λ= pyroelectric coefficient, d_h = hydrostatic piezoelectric coefficient) has been reported (16).

The design of a PVF_2 sensor able to respond only to temperature changes is a much more difficult task; however its importance, in our case, is strictly connected with the performance which is demanded to the artificial skin and with the proposed configuration.

In a possible configuration, schematically depicted in fig. 1, a sandwich composed of two polymer sheets, carrying suitable electrode patterns, and of an interposed layer of rubber or other compliant material, is backed by a rigid support, for example the mechanical skeleton of an

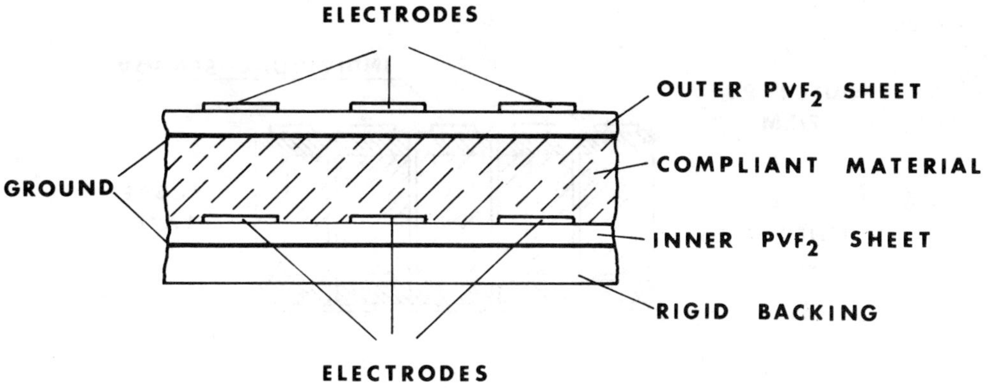

Figure 1 Scheme of the proposed configuration

artificial finger. The sensors on the inner polymer sheet will behave as mechanoreceptors only, while the outer sheet will be the thermoreceptor. The interposed rubber layer is designed to insulate the inner piezoelectric polymer sheet from thermal stimuli. Moreover, the rubber layer couples mechanically the outer polymer skin surface with the inner sensors, also providing the whole structure with the desired viscoelasticity. The rubber material improves the sensitivity of the inner piezoelectric film by inducing extension strain in it when pressure is applied. The amount of the induced strain in the PVF_2 film depends on the elastic modulus of the rubber material as well as on the manner the PVF_2 film is bound to the stiff backing.

The operation of the outer sheet is fundamental not only to detect environmental temperature changes, thus providing warning of potentially harmful situations, but also to recognize the type of material which is touched. In fact, the human perception of cold and warmth when touching materials having the same temperature, which is due to their different thermal conductivity, can be reproduced by skin-like pyroelectric sensors. If the artificial limb is kept at a costant temperature, higher (or smaller) than room temperature, the heat flowing through the composite transducer when the artificial skin touches an object depends on its thermal conductivity and it will be strongly different if the touched object is made, say, of metal or plastic. This difference can be faithfully detected by the outer pyroelectric sheet.

A solution to the problem of the noise represented by mechanical stimuli on the outer sheet could be based on a properly performed subtraction of the signal developed by the inner sensors from the one developed by the outer sensors. If identical and correspondent patterns of electrodes are deposited on both the inner and the outer polymer sheet and each pair of inner- outer electrode, with the interposed rubber layer, is rendered mechanically independent on the other pairs, as shown in fig. 2, each transducers pair could be separately calibrated for mechanical stresses only. In this way, subtracting the two signals and ta-

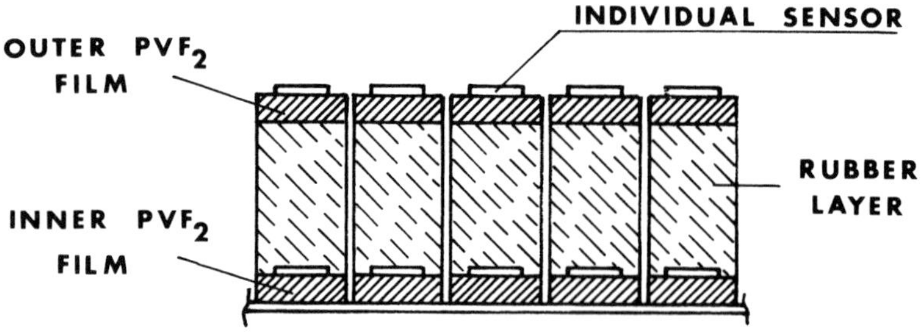

Figure 2 Scheme of the configuration for avoiding mechanical noise in the response of the outer polymer sheet

king into account the measured calibration factors, pyroelectric response only could be obtained.

This procedure would greatly complicate the construction of the artificial skin, also introducing constraints in the distribution of the outer and inner electrodes. However, we think that in active touch the problem of the mechanical artifacts would not be so relevant. In fact, robot movements will be programmed in such a way as to obtain a particular information. If we want, for example, to recognize material from its thermal conductivity, the robot will be programmed to touch the object with a predetermined pressure and it will be much easier to distinguish the piezoelectric from the pyroelectric response.

The proposed artificial skin will have no DC response, owing to the lower bandwidth limit of every piezoelectric transducer. The discharge time constant depends on the electronic circuit used to collect the signal but in no case it could be rendered lower than the intrinsic time constant $\varepsilon\rho$ of the material, which is about 100 s for PVF_2. A static response could in principle be obtained by using an ideal charge amplifier, but, in the actual case, drift problems are introduced.

However, DC response is useful in some robotic applications, for instance in grasping. In the proposed configuration, it could be achieved by using some of the inner sheet transducer as ultrasonic emitters and the corresponding outer transducer as receivers, following the method proposed by Shoemberg and co-workers (17). A steady indentation, i.e. a steady pressure applied to the skin, will be recognized by measuring the time of flight of the ultrasonic pulse transmitted through the rubber layer.

PRELIMINARY EXPERIMENTS

Some preliminary experiments have been carried out on a bilaminate PVF_2 transducer, shown in Fig. 3. The construction of this type of trans

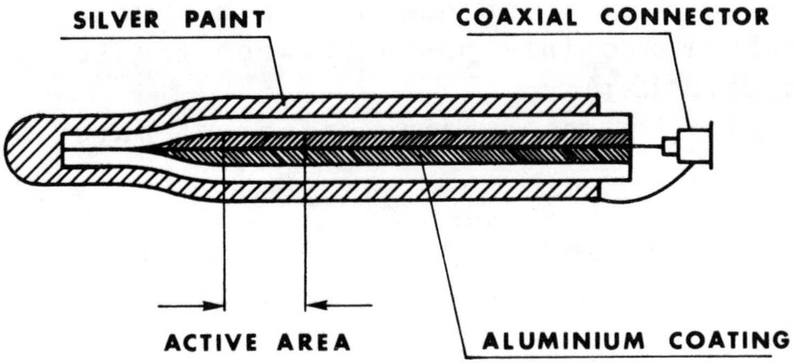

Figure 3 PVF_2 bilaminate transducer

ducer has been described in detail by De Reggi (18). Two identical halves, each having an active central area, corresponding to a circular electrode, 10 mm in diameter in our case, deposited on only one side of the film, have been bonded together and electrical leads have been provided in such a way that the halves are electrically in parallel and mechanically in series. The main properties of this type of transducer are increased capacitance, which is twice as high as the one of a simple layer sensor, and very effective electrical shielding.

Quantifiable mechanical stresses have been supplied to the PVF_2 transducer by means of a test device which comprises a bottom, metallic plate where the transducer is held, and an upper cylindrical rod with a screwable head. The rod is connected to a calibrated piezoelectric load cell (Kistler 9201; calibration factor 46.93 pC/N). This assembly is bolted to a support which is integral with the moving arm of a manually operated column, drilling machine. The output of the load cell is sent to a charge amplifier (Bruel & Kjaer, Type 2624), while the signal produced by the PVF_2 transducer is amplified by a high input impedance amplifier and then recorded on a storage oscilloscope. The assembly is shown in fig. 4.

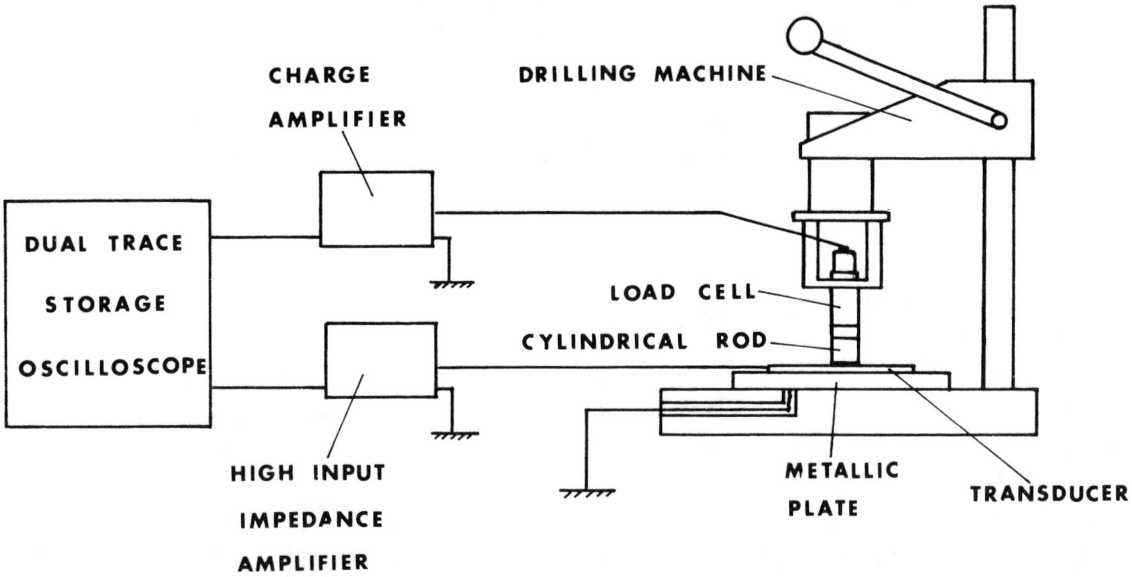

Figure 4 Scheme of the pressing apparatus

The experiments have been carried out to:
- discover how the electrical response of the transducer changes when layers of foam or rubber are interposed between the sensor and the rod of the pressing apparatus;
- assess the possibility of recognizing different materials from their thermal conductivity.

Fig. 5 shows the amplified responses of the transducer to fast pressure ramps, of $7.5 \cdot 10^5$ N/m^2 amplitude in four different conditions, i.e. applying the pressure:
a) directly;
b) through a 1.8 mm thick, natural rubber layer, having elastic modulus $E=6.8 \cdot 10^6$ N/m;
c) through a 4.5 mm thick, compact foam layer, with $E=4.2 \cdot 10^5$ N/m^2;
d) through a 3 mm thick, very soft foam layer, with $E=8.8 \cdot 10^3$ N/m^2.
The relevant difference between case a) and b) can be qualitatively explained considering that, applying the pressure directly with the metallic rod, the transducer operates mainly in thickness mode, while, interposing an elastic medium, extension modes largely contributes to the piezoelectric output. c) and d) represent intermediate situations. Extension modes were possibles because the transducer was not glued to the supporting metallic plate. Therefore, the choice of the intermediate layer and the way the inner polymer sheet is bound to the rigid backing will be very important in the artificial skin configuration proposed in the previous paragraph.

In a second set of experiments the transducer has been laid on the top section of a brass cyilinder, through which a circulation of hot water was maintained (fig. 6), while the pressing apparatus was kept at room temperature (18°C). A pressure ramp has been then applied when the

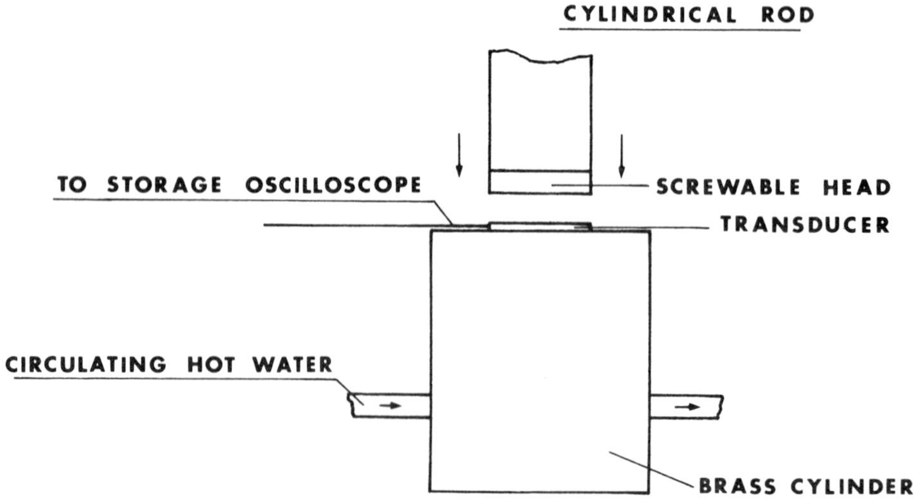

Figure 6 Scheme of the apparatus for assessing the possibility of recognizing different materials from their thermal conductivity

water temperature was 34°C. Afterwards, the screwable brass head of the pressing rod has been removed and substituted with a nylon head. The contact area was identical in the two cases. The test has been repeated supplying a pressure ramp of the same amplitude of the previous one. In fig. 7 the two responses are reproduced: it can be noticed that the response is considerably higher when the brass head is used. The experiment has been then repeated when the temperature of the water was 45°C and an even increased difference was observed (fig. 8). The water circulation has been stopped and the brass cylinder has been allowed to cool down to room temperature. Another test performed in these conditions did not show any difference when the nylon head or the brass head went in touch with the transducer.

CONCLUSION AND PERSPECTIVES

A model for the construction of a skin-like sensor, employing the attractive features of piezo and pyroelectric polymers is presented in this paper, along with some qualitative experiments which demonstrate the possibility of reproducing human cold and warmth perception when touching materials having different thermal conductivity.

The main problems that have been already encountered during these preliminary experiments are also discussed. We expect that further problems in the development of polymer skin-like sensors will be probably related to the collection of the electric signals, especially when large arrays must be scanned. In this case, rows by columns reading will be uneffective and a separate lead should be used for each sensor. As it has been already assessed by the authors during the development of a PVF_2 matrix sensor for robotic applications (19), the multiplexing of high impedance source arrays may introduce undesirable disturbances, due to stray capacitances. As the direct coupling polymer sensors with analog to digital converters is not allowed, a buffer will be necessary. However the excellent mechanical properties of piezo polymers, their multimode operation capabilities and their low cost suggest to take them into account in the future developments of the technology for the construction of an artificial skin, which is now taking its first steps.

ACKNOWLEDGEMENT

The valuable technical support of Mr. F. Vivaldi is gratefully ackowledged.

BIBLIOGRAPHY

(1) G. Liégeois, M. Bétemps, A. Jutard, P. Romand: "Task analysis for industrial robots use and design", Proc. of the 11th Int. Symp. on Industrial Robots, Tokyo, pp. 43-50, 1981.

(2) S. Gaglio, P. Morasso, G. Sandini, V. Tagliasco: "Antropomorphic fea-

che del polivinilidene fluoruro e di alcuni suoi derivati", Symp. on "Le immagini e le tecniche acustiche nella diagnostica medica, nelle prove non distruttive e nella sismologia", Rome, May 13-14, 1982.

(16) E.M.I.: Unpublished results, 1978.

(17) A.A. Shoemberg, C.D. Baker, D.M. Sullivan, H.E. Booth: "Ultrasonic sensory feedback transducer development", University of Utah Research Institute, Progress report TR4-536-008, 1982.

(18) A.S. De Reggi: "Piezoelectric Polymer Transducer for impact Pressure Measuremen", NBSIR 75-740, Final Report Covering Period December 15, 1973 to December 31, 1974: Section 2.2 "Transducer Construction", pp. 7-11, Available through NTIS.

(19) P. Dario, R. Bardelli, D. De Rossi, L.R. Wang, P.C. Pinotti: "Touch-sensitive polymer skin uses piezoelectric properties to recognize orientation of objects", Sensor Review, pp. 194-198, October 1982.

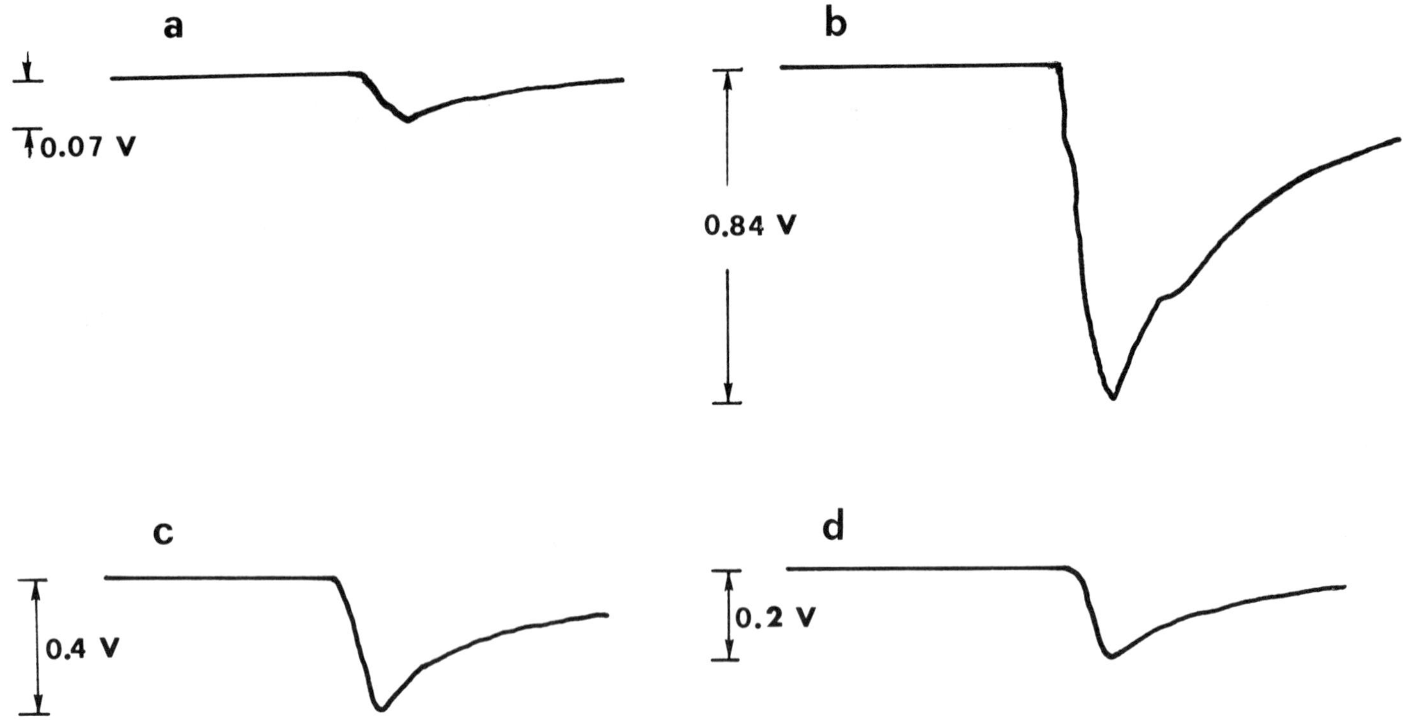

Figure 5 Responses of the bilaminate transducer when a pressure ramp was applied, a) directly, b) through a rubber layer, c) through a compact foam, d) through a soft foam

tures in artificial vision", Proc. of the 11th Int. Symp. on Industrial Robots, Tokyo, pp. 103-112, 1981.

(3) Z. Stojilykovic, J. Clot: "Integrated behavior of Artificial Skin", IEEE Trans. on Biomedical Engineering, vol. BME-24, No. 4, pp. 396-399, July 1977.

(4) T.A. Quilliam: "The structure of finger print skin", Active Touch, Pergamon Press, pp. 1-18, 1978.

(5) W.D. Hillis: "Active touch sensing", A-I Memo 629 of the M.I.T. Artificial Intelligence Laboratory, April 1981.

(6) L.D. Harmon: "Touch-sensing technology: a Review", Technical Report MSR80-03 for the Society of Manufacturing Engineering, 1980.

(7) S.B. Lang: "Pyroelectricity: occurrence in biological materials and possible physiological implications", Ferroelectrics, vol. 34, pp.3-9, 1981.

(8) M. Athenstaedt, M. Claussen, D. Shaper: "Epidermis of Human Skin: Pyroelectric and Piezoelectric Sensor Layer", Science, vol. 216, pp. 1018-1020, 1982.

(9) H. Fröhlich: "Non linear Dielectric properties of Biological materials", Rept. Conf. Electrical Insulation and Dielectric Phenomena (Natural Academy of Sciences, Washington D.C. 1978).

(10) Y.C.B. Fung: "Stress-strain-history relations of soft tissues in simple elongation." In: Biomechanics, its foundations and objectives, ed. Y.C.B. Fung et al., Prentice-Hall Inc., Englewood Cliffs, N.J., pp. 181-208, 1972.

(11) A.B. Vallbo, R.S. Johansson: "The tactile sensory innervation of the glabrous skin in the human hand", Active Touch, Pergamon Press, pp. 29-54, 1978.

(12) H. Kaway: "The piezoelectricity of poly(vinylidene fluoride)", Jpn. J. Appl. Phys. 8, pp. 975-976, 1969.

(13) G.M. Sessler: "Piezoelectricity in poly(vinylidene fluoride)", J. Acoust. Soc. Am. 70(6), pp. 1596-1608, 1981.

(14) E.L. Nix, L. Holt, J.C. McGrath: "Highly drawn poly(vinylidene fluoride) with enhanced mechanical and electrical properties", Ferroelectrics, 32, pp. 103-114, 1981.

(15) A. Nannini, C. Domenici, P. Dario, D. De Rossi: "Valutazione dell'influenza di alcuni procedure di preparazione sulle proprietà ultrasoni

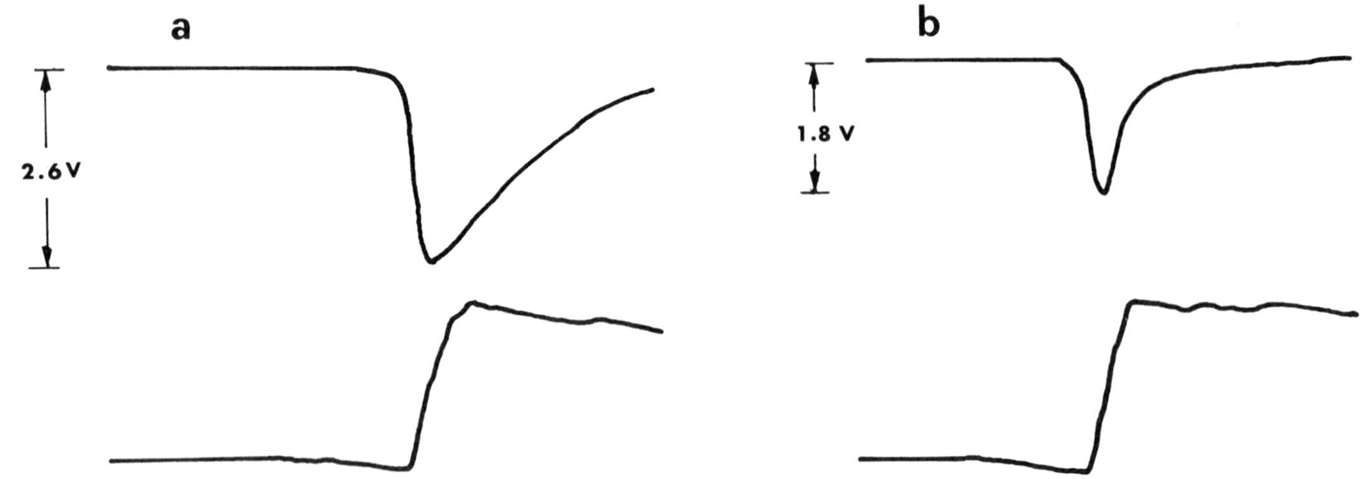

Figure 7 Responses of the transducer (upper traces) to a fast pressure ramp (lower traces) when a brass head (a) or a nylon head (b) were used. The temperature of the circulating water was 34°C

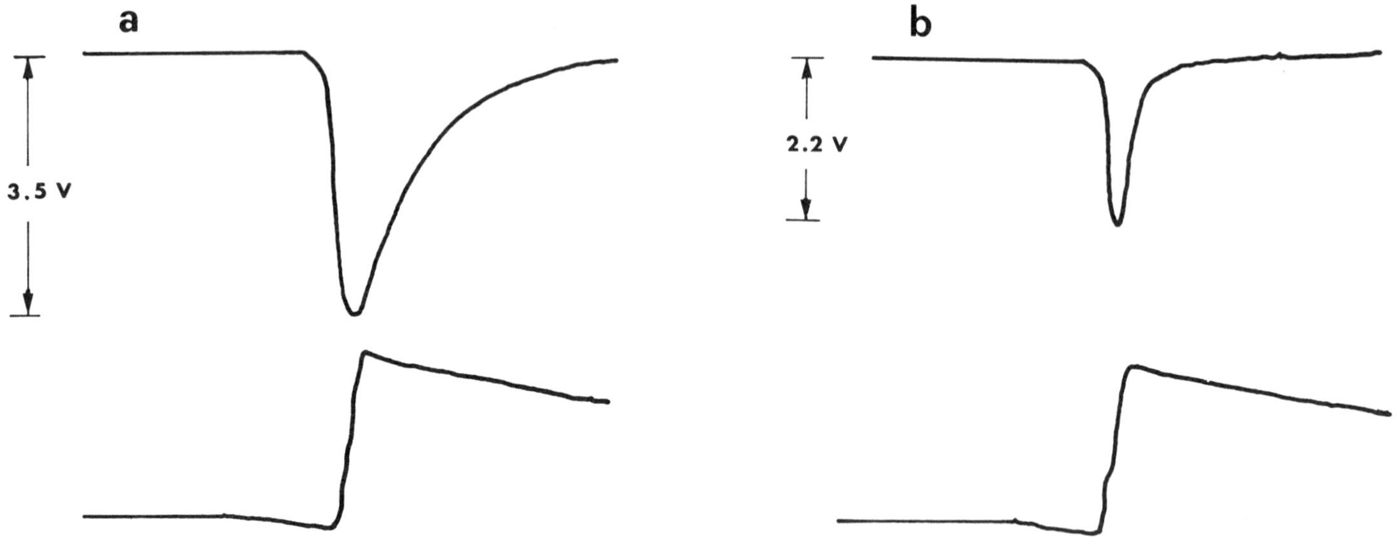

Figure 8 Responses of the transducer (upper traces) to a fast pressure ramp (lower traces) when a brass head (a) or a nylon head (b) were used. The temperature of the circulating water was 45°C

The Development of a Production Robot Tactile Position Sensor

Harry L. Benjamin
Centro Corporation
Dayton, Ohio

Abstract

This paper traces the development of a five-axis tactile position sensor. The early system is described as a five and one-half inch diameter device, 10 inches high, weighing 10 lbs., with a range of ± .050, ± .001 inches in the "Z", ± .065, ± .001 inches in the "X" and "Y" axes, and angularity of ± 1.0, ± .01 degrees. The production system is shown as a cylindrical device 2.3 inches in diameter, 5.5 inches high, weighing 1.25 lbs. with a range of ± .250, ± .001 inches in the "X", "Y" and "Z" axes and ± 3.0, ± .01 degrees. Operation and applications of the sensor system are outlined.

INTRODUCTION

Robots will become more clearly distinguishable from automated machines as they acquire sensory perception and the intelligence to respond to it.

With this capability the emphasis in robotic design could shift from high accuracy and repeatability to high resolution and intelligent response, eventually with dramatic cost savings impact on robotic systems and their utilization.

It was with these considerations in mind that Centro Corporation embarked on the development of its production RTS* - Series 200 5-axis tactile position sensor, and other tactile sensing systems.

The RTS* - Series 200 (See Figure 1) is a compact 1.25 lb device capable of measuring the position of a target in a plane in space relative to a robot's end effector. The measurements are provided in five axes. Three of these define the position of the target in the robot's "X", "Y" and "Z" axes (tool world) and the other two define the target plane's orientation with respect to the "Z" axis of the robot as measured in the "X" and "Y" planes.

The sensor's range is ±.250 of an inch in "X", "Y", and "Z", with accuracy of ±.001 of an inch and ±3 degrees in angularity, with an accuracy of ±.01 degrees.

Refer to Figure 2. This photograph shows the three major components of the sensor system. The visual display is on the left. The processing unit is seen in the center.

BACKGROUND

The sensor was not always this compact or lightweight. Figure 3 shows the RTS* - Series 100 prototype together with the production Series 200. The 100's capabilities are identical to those of the later unit. But as can be seen, it is almost twice as large and in fact weighs more than twice as much - 3 lbs.

Figure 4 shows the original earliest version of the 5-axis tactile sensor. It weighs 10 lbs, is 5 inches in diameter and 10 inches high. It's range in "X", "Y" and "Z" is only .065 of an inch and its angularity measurements are limited to 1 degree. Accuracy and repeatability of the system is the same as the production unit, ±.001 of an inch and ±.01 degrees in angularity.

* Patent Nos. 4,328,621; 4,338,723
Registration & Other Patents Foreign & Domestic Applied For

The visual display is shown on the console to the right, which also contains the processing unit. The display panel is 13 inches wide x 10 inches high and weighs approximately 15 lbs. The Series 200 visual display measures 4.5 x 4.5 inches and weighs only .7 of a pound and can easily be attached to a teaching pendant so that it and the display can be held in the operator's hand and the two viewed simultaneously.

EARLY DEVELOPMENT OF THE TACTILE SENSOR

The original sensor was designed several years ago as a robot teaching aide, permitting the robot operator to improve vastly his visual acuity and thereby reduce the labor associated with teaching a robot to do automatic drilling.

One such drilling operation is done by placing an automatic drill in a bushing with an associated lock fitting. The drill assembly is inserted into the bushing and given a quarter turn to lock it in place. Once placed in the drill bushing the drill's speed, feed and support are all provided by the bushing and drill combination; no other support or control is required.

A man or robot's task is simply to put the drill in the fixture, wait for the drill operation to be complete and then withdraw the assembly, ready for the next bushing. To teach a robot to do this task is labor intensive because of the positional accuracy involved.

Several years ago a manufacturer was teaching a T-3 Cincinnati Milacron robot to accomplish this task using two men, one a robot "teacher" or operator and the other a gauge operator.

The gauge consisted of two one-half inch thick steel ground discs of precisely equal diameter, about 5½ inches. (See Figure 5) Each disc had a shaft attached to its center, so that one disc could be mounted in a bushing on the fixture and the other on the end effector of the robot. Both shafts, with their attachments, were designed with proper offsets so that when the two were brought together by the robot, it had effectively been "taught" to insert the drill in the drill fixture.

To do this the teacher, using a teaching pendant, brought the robot's disc into contact with the fixture disc as close as he could by eye.

The gauge operator, using a .001 of an inch feeler gauge, would direct the robot "teacher" to move "up" or "down", "in" or "out", pitch "up" or "down" or roll "clockwise" or "counter-clockwise", all as appropriate to bring the faces of the discs together so that at no point around their peripheries could the "feeler" be inserted. At the same time the gauge operator would use his free hand to feel around the periphery. The position of the robot would be judged acceptable when the gauge operator was not only unable to insert his gauge but also when he could not tactilely sense any axial displacement between the two discs.

Because of the size of the discs these measurement criteria put the two discs into coincidence within .01 degrees of perpendicularity and .001 of an inch in position.

Referring again to Figure 4, it is clear that the tactile sensor replaces the top disc of the two-piece manual gauge shown in Figure 5. The fixture Plate for use with the tactile sensing system is the same as that of the manual gauge, except that it has at its center a ½ inch diameter 90° tapered hole.

Note that the precise diameter of the plate is no longer necessary. Flatness and perpendicularity of the shafts are of course still essential.

Refer to Figure 6. This is a schematic of the original sensor. There are five pins protruding from the bottom of the sensor. Four of these (only two are shown in the figure) are .125 of an inch in diameter and are equally spaced on a circle whose radius is 2.5 inches and whose center is in the axis of the sensor. Each moves independently perpendicular to the base of the sensor, and each is capable of a total travel of .100 of an inch. Note each pin is hemispherical at its ends.

The fifth pin or center probe is .250 of an inch in diameter and is also hemispherical at its end. It protrudes through a .380 inch hole in the base of the sensor for a distance of ½ inch. This probe not only can move vertically, but can also be deflected about its neutral axis a distance of .065 inches in any direction.

If the base of the sensor is brought down hard on the fixture plate, no harm is done to any of the pins since each can disappear into the base plate without damage. The force required to fully displace each of the four radial pins is 3 lbs. The force to fully displace the center probe vertically is 1½ lbs. Thus if the robot is brought down vertically on the fixture plate so that the center of the probe is not within the circumference of the ½ inch tapered hole, the center pin will simply move upward upon contact, and if the surface of the base of the sensor is brought down to the surface of the fixture plate, a total of 13.5 lbs will be imposed on the robot, as all pins disappear into the base of the sensor.

If further motion of the robot toward the fixture plate is continued, the entire bottom section of the sensor moves against springs into the top section of the sensor. A total motion of ½ inch is possible before the two sections are brought against mechanical stops. Just before this point a load of 20 lbs is imposed on the robot.

At the metal-to-metal mechanical stop the sensor is the equivalent of four solid ½ inch posts in compression. Obviously a "wreck" of this kind will not damage the sensor.

Referring to Figure 6 again it can be seen that the vertical pins each bear against a cantilevered spring. These springs are strain gauged with full bridges adjacent to their supported ends. Thus the springs not only insure motion of each of the pins toward its shouldered stop, but also each is a transducer whose output signal is directly proportional, within its range of deflection, to the vertical motion of the pin upon which it bears.

The center pin, as can be seen in the schematic, continues with a .250 inch diameter cross section for approximately 2.5 inches and then is inserted deeply into a 3/4 inch diameter hardened steel rod. (This rod rides in a linear ball bushing.) For the last inch before entering the 3/4" rod the 1/4" rod is ground square with .200 inch wide flats. The opposite flats are strain gauged with full bridges so that in effect, the center probe is two independent cantilevered rods capable only of orthogonal deflection. Motion of the single pin about its neutral axis thus is the vector sum of the two orthogonal deflections of the beam and the output of the gauges within their range is directly proportional to each of the orthogonal moves.

With the five transducers just described it is possible to define the position of the sensor relative to the Fixture Plate.

The differential summation of the signal output of each set of opposite vertical pin transducers will be directly proportional to the angle formed between the longitudinal axis of the sensor in each of the two orthogonal planes formed by the sets of vertical pins, and a line perpendicular to the plane of the Fixture Plate. If we call one of these planes "X" and the other "Y", we are defining the angularity of the Fixture Plate relative to the sensor in its two orthogonal planes, "X" and "Y".

The sign of the angle, plus or minus, is set arbitrarily by the sign given to each pin in the set. Then the sign of the angle will be that of the pin in each set with the greater deflection.

It is also clear that if there is no difference in length between the two pins of a set, that in the plane of that set the sensor is perpendicular to the Fixture Plate; and it is also clear that if this is the condition of both sets of pins, the sensor is then perpendicular to the plate.

Now it is appropriate to measure the distance of the base of the sensor above the Fixture Plate. Once the sensor is perpendicular to the fixture disc, the length of any pin defines the distance of the sensor to the plate. Since there will always be some slight difference between them, because the sensor will never in fact be absolutely perpendicular to the fixture disc, a more accurate measure of distance is the average length of the four pins. This is simply the summation of the buffered output of the four pins electrically divided by four.

If we arbitrarily set half of the pin deflection (.050 of an inch) as zero (gauge height), then when the pins are longer than .050, the sensor is above the desired gauge height, and when shorter, below it.

If, as is the case, we set the strain gauged flats of the center probe perpendicularly to the "X", "Y" planes formed by the pins, then any deflection of that probe can be defined by its vector components in the "X" and "Y" axes.

Therefore, if the sensor is brought in contact with the fixture disc and erected perpendicular to it and the robot moves so that the center of the center probe is within the perimeter of the ½ inch tapered hole, the probe will be spring driven down the sides of that hole, indicating the direction of offset from the hole's center at the bottom. When the pin is moved to within .065 of an inch of the hole's center, the pin will actually center itself in the bottom of the tapered hole.

Now the signal outputs of the two strain gauges in the "X" and "Y" planes are directly proportional to the distance the center of the sensor is from the center of the fixture plate and this information can be used to direct the two to coincidence.

It is now clear that analog electrical signals have been generated which completely define the relative positon of the sensor to the fixture plate.

EARLY DISPLAY DEVELOPMENT

To make use of the analog signals produced by the sensor in a teaching application it was decided that some form of non digital, semi pictorial presentation was required. Referring to Figure 4, it can be seen that this was done using three such presentations on a console representing "Perpendicularity", "Vertical Position" and "Horizontal Position". Figure 7 shows these presentations in detail.

Note they not only describe the robot's position relative to the fixture plate (the "target"), but instruct the robot operator in the proper corrective action to bring the robot and "target" into coincidence. Instructions are at the end of each axis of each display - "up", "down", "ccw", etc.

The format of the three displays is the same in any single axis of any one of them with five lights, one GREEN, two YELLOW, and two RED. The GREEN light on alone in an axis signifies the sensor is "on target" in that axis.

If the sensor is only slightly "off target", the appropriate YELLOW arrow and RED instruction light come on together. If the sensor is grossly "off target", the YELLOW arrow goes out but the RED instruction signal remains on alone.

Thus except when the sensor is "on target" in an axis or plane, an instruction light is always lit identifying the necessary corrective action.

Furthermore, the extent of the corrective action required is displayed at the same time in two ways. First by the "sensitivity switch." It sets the sensitivity for the three displays simultaneously. (See Figure 4; this switch is in the upper right hand corner of the display console.) It has three positions - "Fine", "Medium" and "Coarse." When the system is set with the sensitivity switch on "Fine" and the appropriate YELLOW and RED lights just come on together, the sensor is "off target" by .003 inches, and when the YELLOW light just goes out leaving the RED on alone, the sensor is "off target" by .006 inches. Similarly at this sensitivity setting the YELLOW-RED lights indicate at least .03 degrees, and the RED alone at least .06 degrees.

When the sensitivity switch is turned to "Medium", the range is doubled in distance and perpendicularity, and similarly when the switch is turned to "Coarse", it is increased by a factor of 10.

Thus when the GREEN lights are all "on" and the switch in on "Fine", the sensor is "on target" within .003 inches of true positon and perpendicularity within .03 degrees. Similarly, when the YELLOW-RED combination is on, the sensor is within .006 inches and .06 degrees of true position. When RED is on alone, the error is indicated in direction only until the sensitivity switch is turned to a finer level.

Clearly by use of the sensitivity switch in combination with the lights a great deal of information can be gained and utilized quickly.

EARLY SYSTEM OPERATION

The system was designed to be utilized in definite sequence:

First the sensitivity is to be set on "Coarse."

Then the sensor is to be brought by the robot operator to within the perimeter of the ½ inch target hole, and brought in "hard" contact with the "target" plate.

When this is done, the override signal will come on. This light can be seen in the center of the console, Figure 7.

Then the operator is to erect the robot to perpendicularity by eye and raise the robot until the RED override light goes off.

He then follows the instructions on the "Perpendicularity" display, using the other displays only to keep the sensor within its range to the "target."

He then directs his attention to the "Vertical Position" and brings the sensor to GREEN in the "Vertical Position" and then to GREEN in the "Horizontal Position."

Note he has been in the "Coarse" range during the previous operations so that when all lights are GREEN, he is in true positon within .030 inches and .3 degrees.

When he switches to the "Medium" range, he will probably find all signals go to RED only. He then simply follows the corrective action prescribed by the displays until all lights go GREEN again.

This will bring him to within .006 inches and .06 degrees of true position.

When he switches to the "Fine" range, he should find the displays reading YELLOW-RED. He then proceeds to the GREEN position in the three displays. He is "on target."

The ranges described above can be changed internally so that they are coarser or finer than those described. For manual operation these were thought to be as fine as practical.

<u>DEVELOPMENT OF THE FIRST PROTOTYPE - SERIES 100 TACTILE SENSOR</u>

The 5-axis tactile sensor just described was demonstrated to several potential users and placed on two experimental robot drilling systems - one at McDonnell Douglas Aircraft, St. Louis, MO; the other at General Dynamics, FT. Worth, TX.

Though the operation of the unit was as expected and gratifying, as a first effort, it was clear that its size, weight, force levels and range would seriously limit its application.

Furthermore the visual display was not practical for pendant teaching since it did not generally permit the operator to view his pendant and the display at the same time.

Therefore it was decided to develop another prototype system. This new system decreased the sensor's size even while greatly increasing its range without sacrificing accuracy. Also the system display was reduced in size so that it could be attached to a pendant and both held in the operator's hand, to be viewed simultaneously.

The Series 100 prototype is shown to the left of the production unit in Figure 3 and schematically in Figure 8.

It has all the features of the earlier model, but is one third the weight and only half its size. Position ranges in "X", "Y", and "Z" have been increased to $\pm$.250 of an inch - angularity to 3 degrees.

To increase range in "X" and "Y" the flexure section of the center probe was completely redesigned. To increase the range in "Z", the measurement of "Z" was decoupled from the measurement of perpendicularity.

To do this, a separate transducer was added to measure "Z" while at the same time two transducers were removed from the perpendicularity section and replaced by a single fixed pin.

Force levels in the system were reduced primarily by reducing the cross sections of the flexures. Thus "X" and "Y" forces were reduced from 1½ lbs to .06 lbs at full deflection. Perpendicularity was reduced from six lbs to 1½ lbs and the overload from 20 lbs to 5 lbs.

At the same time that the sensor was reduced in size, the display was also reduced to that shown in Figure 2.

Note the similarity in form between this small display and that shown in Figures 4 and 7. While the major change is in size, this configuration also provides a digital readout of "Z". This was done to facilitate the teaching process in two ways. First it allows an operator to come "hard down" on the "target" with information through ½ inch of travel and second, it allows him to accept an "offset" gauge height rather than a fixed gauge height, thus reducing teaching time.

Series 100, the prototype system was demonstrated to some potential customers but even as such presentations were made, they were aware of the fact that a much smaller unit, the Series 200, was the production configuration. Therefore, only one Series 100 unit was ever produced.

DETAILS OF THE SERIES 200 - 5-AXIS TACTILE POSITION SENSOR

The Series 200 - 5-Axis Tactile Position Sensor System is very much like the pre-production unit shown in Figure 2.

The differences are in small details. The visual display, for example, is modified slightly in format for ease of production. This format is shown in Figure 9.

The internal configuration of the Series 200 is basically the same as the Series 100 and so Figure 8 still represents it as well as the earlier unit schematically.

Since this is the configuration in production, we will now examine some of its features in more detail. Refer to Figure 8.

Center Probe ("X", "Y" Measurement)

First it should be noted that the Center Probe is mounted in the Perpendicularity Module in such a manner than only its motions in the "X" and "Y" planes are significant for measurement. Motion in the "Z" axis is provided to allow the probe to be used in "target" holes of varying depth within any tapered "target" no greater than .800 inches in diameter. It is also there to permit the probe to be undamaged by contact with a surface without any holes, or at a location in which a hole does not exist.

Furthermore, it is there to permit the probe to lift out of a tapered hole without damage when the sensor is swept out of the hole location.

And finally, it is there so that when the probe is brought to, or swept within, the hole location, the probe will automatically find its way, under spring load, precisely to the hole's bottom, thus defining the offset of the hole to the center of the sensor.

Referring again to Figure 8 it can be seen that the Center Probe has two distinct orthogonal flexures, as opposed to the square section originally utilized. This design greatly increases the range of the probe without introducing cross talk.

Perpendicularity Module (Angularity in "X" and "Y" Planes)

Looking at the Perpendicularity Module, it has been said that in the Series 100 and 200, perpendicularity is measured by only two movable and one fixed pin in the base plate. In both series a line drawn between the two movable pins is parallel to the "X" plane and is bisected by the "Y" plane of the sensor. The distance between them is 1.500 inches in the 200 Series. The fixed pin lies in the "Y" plane and is .800 of an inch from the "X" plane in the 200 Series.

It can be shown that the offset angle in the "X" plane (about the "Y" axis) is proportional to the differential sum of the signals from the two pins, with the sign of the angle being that of the sign of the larger signal. The offset angle in the "Y" plane (about the "X" axis) is similarly proportional to the sum of the two signals (with each signal having previously been set to zero when each pin is equal in length to the fixed pin) with the sign of the angle the sign of the resulting summation, with short pins represented by a negative signal and long pins the opposite. These measurements are easily dealt with electrically.

Referring to Figure 8 again, note that the three perpendicularity pins do not bear directly on the "target" surface but through a "wear" or "sole" plate. This configuration is desirable, so long as the target plates are flat. It not only protects the pins against damage and wear but eliminates errors due to a rough surface finish, local depressions or small holes.

Gauge Height ("Z" Measurement)

The gauge height of the system is the "Z" measurement at mid range, or in other words the fully extended length of the sensor less .250 of an inch. Note again that the movement of the Center Probe has nothing whatever to do with the "Z" measurement and conversely, "Z" measurement in no way influences the "X" and "Y" position. It is true, however, that neither "X", "Y" or "Z" measurements are completely accurate unless the sensor is erected perpendicular to the "target."

Details of Display

While the uses of the RTS - Series 200 have been extended far beyond that of a teaching aide, still manual pendant directed teaching, or guidance, is one of the important uses of the system and thus the visual display is still an important part of it.

As can be seen in Figure 9, the display is very little changed in basic format from the very earliest concept. However, it is different. For example, there are no longer lighted instructions or lighted arrows. The display is not intended to express pictorially where the robot is relative to the "target", although to an extent it does, but rather to simply instruct the operator in corrective action - to bring the sensor "on target."

Thus, if the sensor is pitched up in fact, a RED or YELLOW light (but not both) will come on indicating the written instruction adjacent - "PITCH DOWN" - is to be followed. The same reasoning is used in all the axes of all other displays.

Note there is still a "Fine", "Medium" and "Coarse" setting (F, M, C) in this display. It is set to the same limits as those already discussed, but these are arbitrary and can be factory set for other thresholds.

Finally, there are some new status features incorporated in this display which only are provided by the 200 Series sensor. First, if the sensor is not in contact with the "target" surface, all three displays will blink indicating the sensor is in space.

Second, if the sensor is in contact with the "target" surface, but the probe is not in the "target hole", the "Position" display alone will blink indicating this condition.

This is an important display feature because the "target hole" is only .800 of an inch while the sensor is over 2 inches in diameter. Thus the operator is not able to see whether or not the probe is in the hole once the sensor is in contact with the "target." Therefore, for example, if he is initially in the "target" and inadvertently sweeps out of it, he could not tell this immediately without this status signal. And since any signals from the sensor, when it is out of the hole, are meaningless, the operator must know when he is out of the hole and take steps to correct the situation.

Third, if the sensor approaches .020 inches of its mechanical stop in compression, the digital vertical display goes blank. This signal is provided as much to protect the robot as the sensor since the sensor in compression becomes essentially a solid tube, two inches in diameter with a quarter inch wall.

Discussion of the Other Readouts

Besides the visual display, data from the Series 200 system can be obtained in two other ways:

Analog Data

The processing unit provides five channels of buffered analog data. Each channel is updated every .01 seconds. The data is scaled 10 millivolts per .001 inches and .01 degrees. Full scale readings are ± 2.5 Volts for position and ± 3.0 Volts for orientation.

Digital Data

In addition to the analog data, the processing unit provides digital data in a 10 bit, 2's complement format together with the three status bits "out in space", "not in hole" and "overload". This data is provided at 19.2K BAUD through an RS-232 port.

The data is actually transmitted in two eight bit bytes with one start bit, two stop bits and no parity. The first byte contains two bits of the second byte, and three status bits; the balance of the bits in the byte are tied low.

The second byte connoted by bits 8 and 9 (MSB) of the first data byte make up the 10 bit data byte.

Data can be called up in a continuous stream of the five channels plus status or addressed for any channel of interest.

This RS-232 port of course permits direct computer call-up of data for intelligent closed loop response.

Modification to the Basic Configuration (See Figure 10)

Referring to Figures 1, 2, 3, 6 and 8, it is clear that the basic configuration of the Series 200 sensor has a straight Center Probe with a herispherical end. Obviously this probe must enter a tapered hole to automatically find center, lift out, and generally operate as previously described.

However, it is apparent that a reverse configuration is possible. That is one in which the taper is on the pin. This permits the sensor to operate in straight sided holes and such a configuration is not only possible but has been proven and sold. It is in fact probable that in the future this may be the more common configuration of the two.

DISCUSSION OF APPLICATIONS

The original use for the earliest 5-axis tactile position sensor was envisioned as a teaching aide.

It is now clear that the system as currently configured has many other applications.

Some of these are listed below:

- RESEARCH AND DEVELOPMENT OF ROBOTS

  To confirm robot performance - particularly accuracy and repeatability as a function of load, position, time, environment, etc.

- QUALITY CONTROL

  Robot Manufacturers - To confirm specified performance prior to shipment - or after repair and maintenance.

  Robot Buyers - To confirm specified performance upon receipt - after repair or maintenance.

- ON LINE OPERATIONS
  - Initial precise setup of fixtures to robot.
  - Continuous alignment of work to robot prior to programmed work routine of single or multiple operations. See Figures 11 and 12.
  - Periodic check of precise robot position to work station - daily, weekly, monthly (one, two or three point checks of position and orientation.)
  - Initial orientation for and correction to off line programs.
  - Continuous offset monitoring - using existing pattern of stamped holes to monitor operations at off locations.
  - Automatic and semi automatic teaching.
    Where automatic teaching is done by off line programming to rough positions automatically corrected on line, and semi automatic teaching is man directed to initial rough positions followed by automatic direction to precise positions.

CONCLUSIONS

The Centro RTS - Series 200 5-axis tactile position sensor in its present configuration presents many opportunities for improving the ability of robots to do precision machining and assembly tasks with reduced requirements for expensive tooling and orderliness in the work place.

Its most obvious drawback in on line application is the sensor's apparent conflict with the hand or the tool for the working position - but it would appear that a fixed offset, or active linear or rotary offset can very practically solve this problem without costly additions to the system. See Figures 11 and 12.

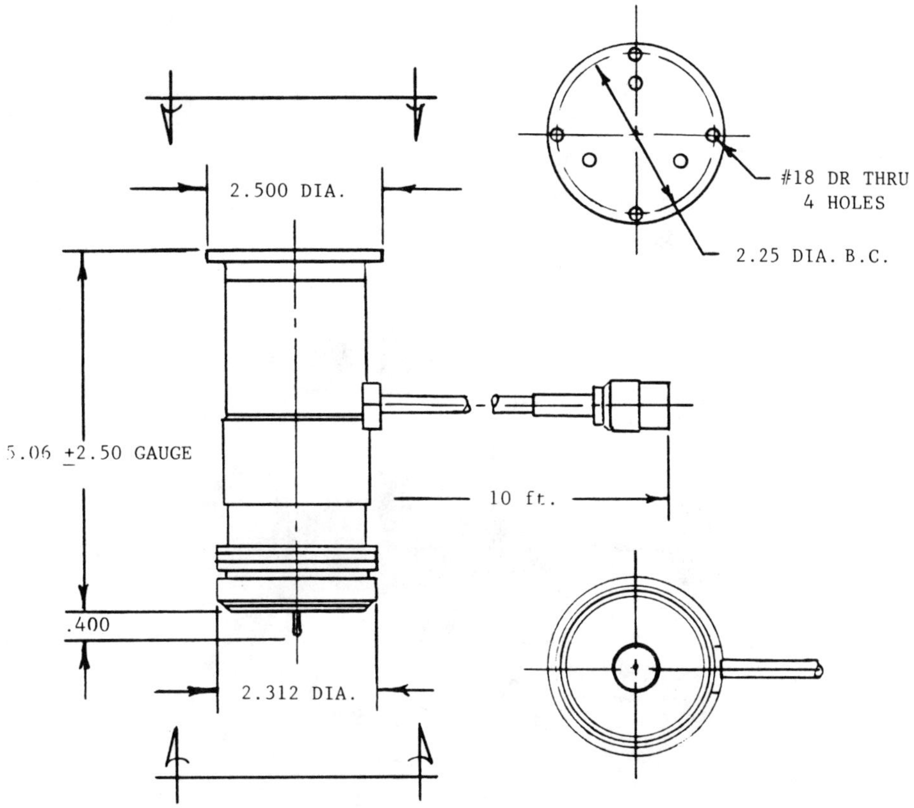

FIGURE 1. CENTRO RTS - SERIES 200 POSITION SENSOR

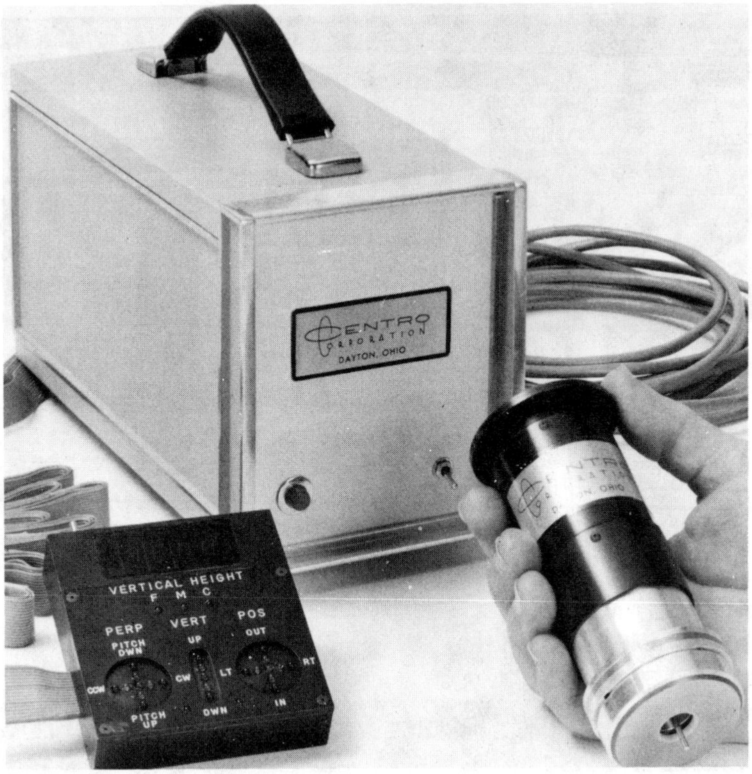

FIGURE 2. CENTRO RTS - SERIES 200
5-AXIS TACTILE POSITION SENSOR SYSTEM

FIGURE 3. RTS - SERIES 100 VERSUS RTS SERIES 200

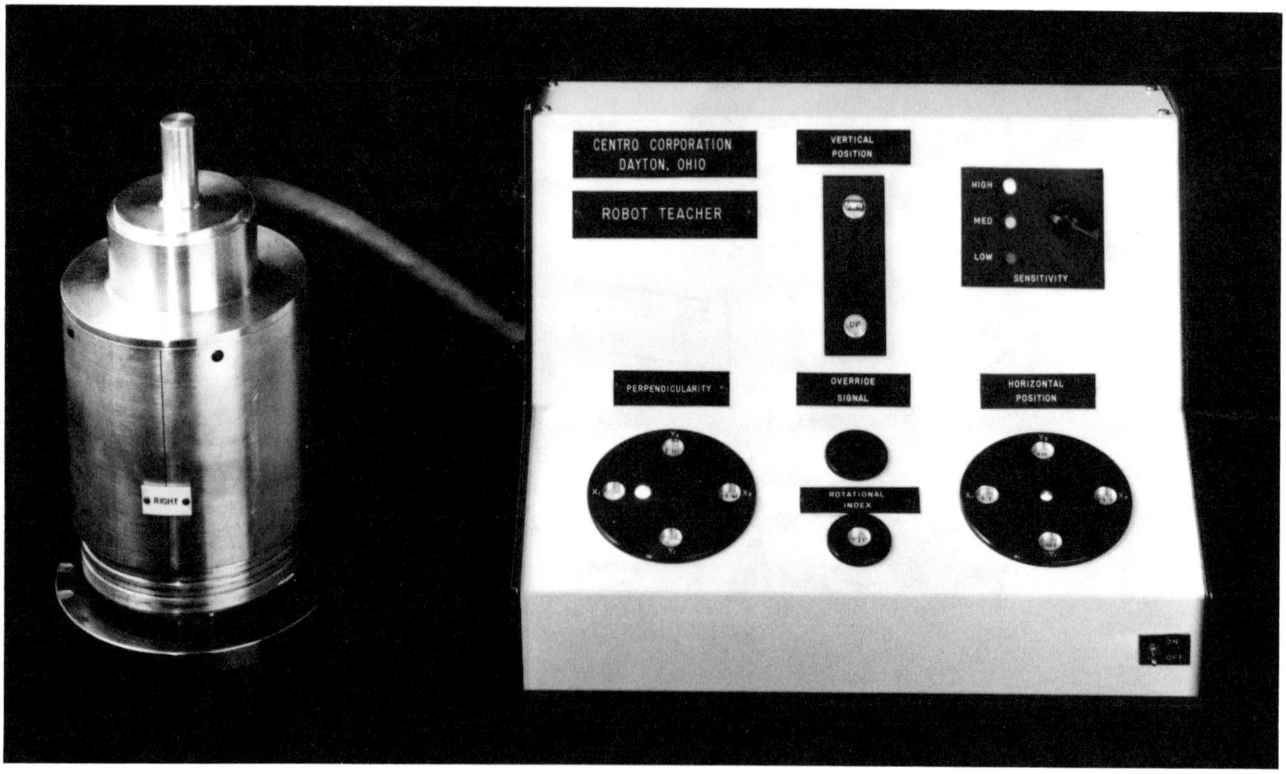

FIGURE 4. EARLY DEVELOPMENT CONFIGURATION
CENTRO 5-AXIS TACTILE POSITION SENSOR

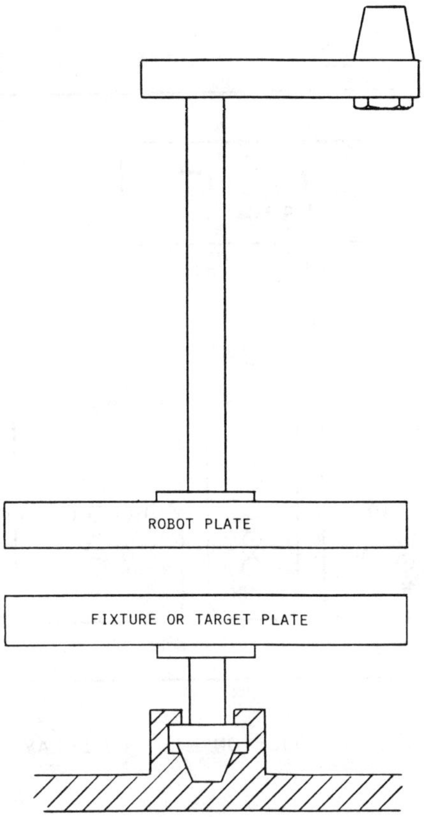

FIGURE 5. DRILL TEACHING GAUGE

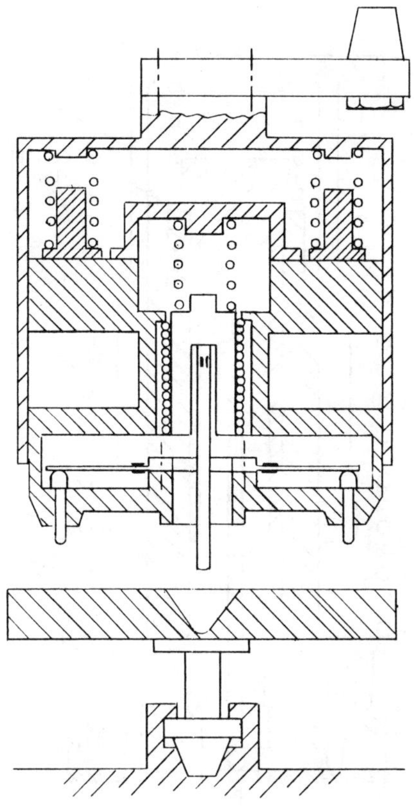

FIGURE 6. SCHEMATIC OF EARLY 5-AXIS TACTILE POSITION SENSOR

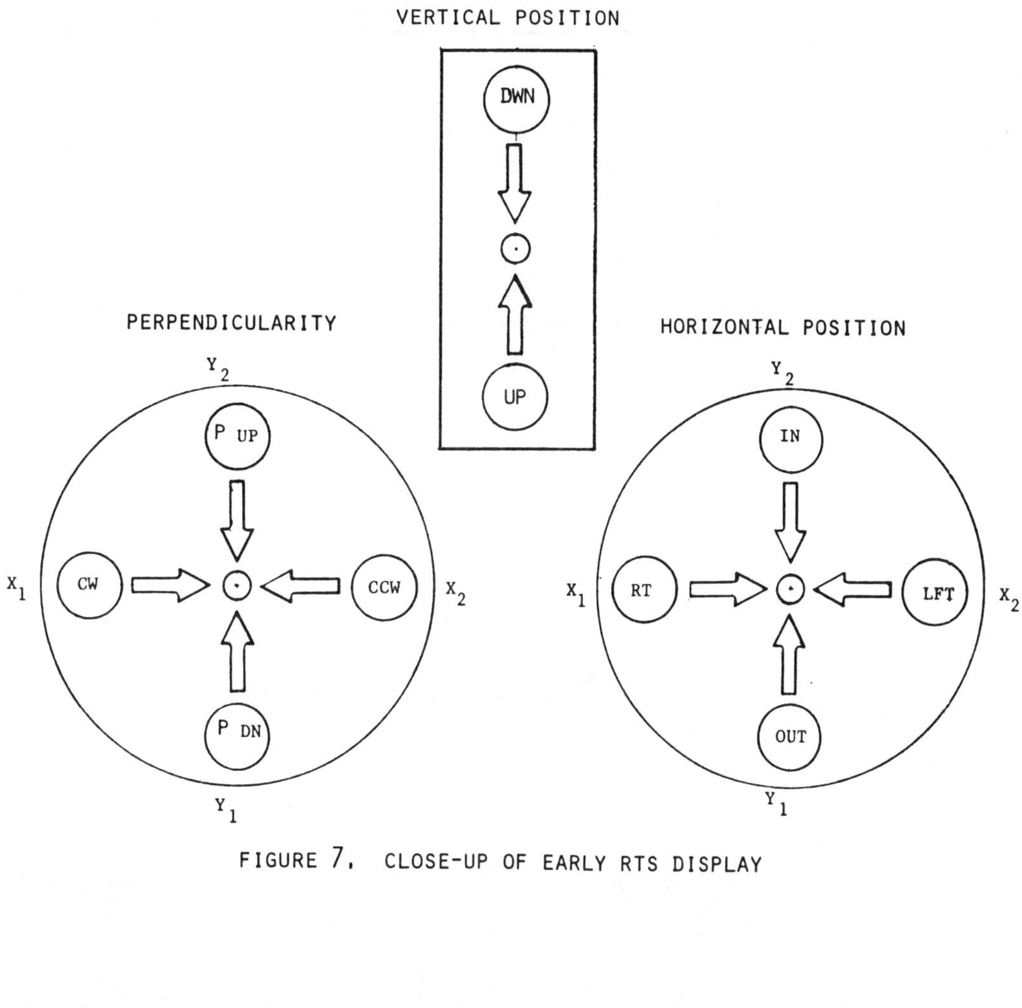

FIGURE 7. CLOSE-UP OF EARLY RTS DISPLAY

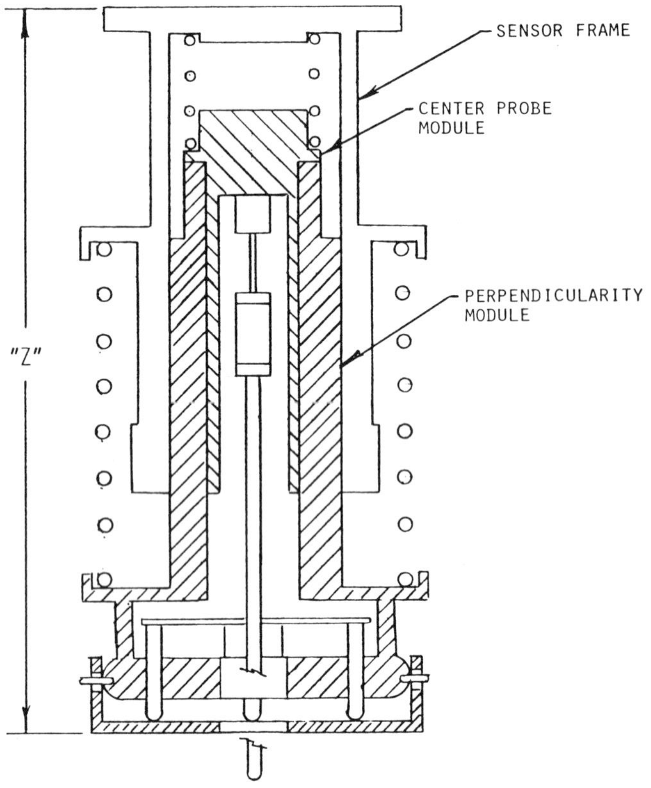

FIGURE 8. SCHEMATIC OF RTS - SENSOR SERIES 100 AND 200

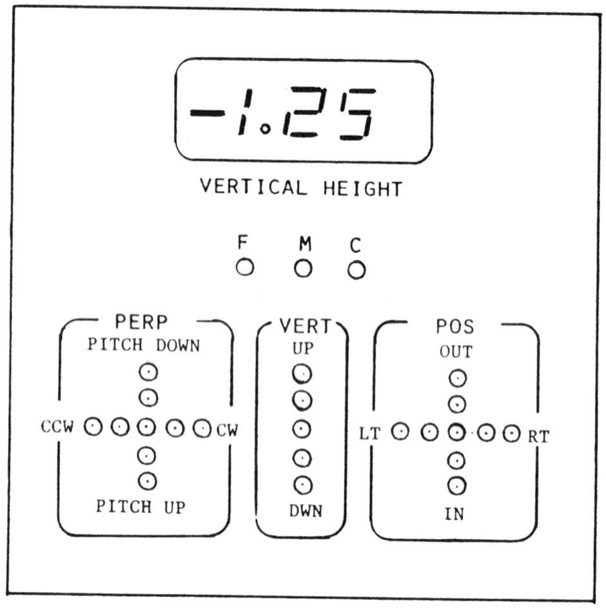

FIGURE 9. PRODUCTION RTS - DISPLAY

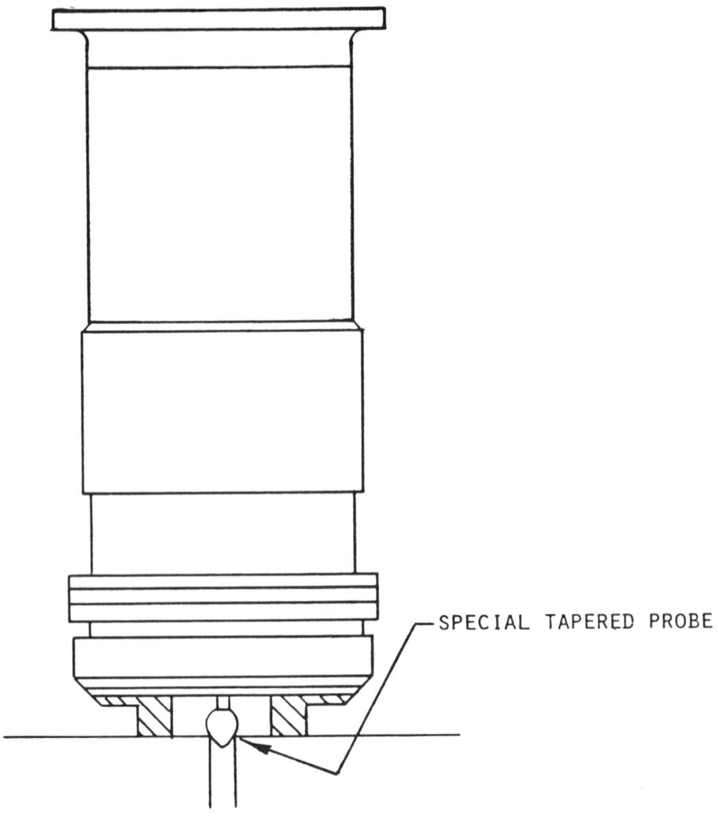

FIGURE 10. RTS WITH TAPERED PROBE

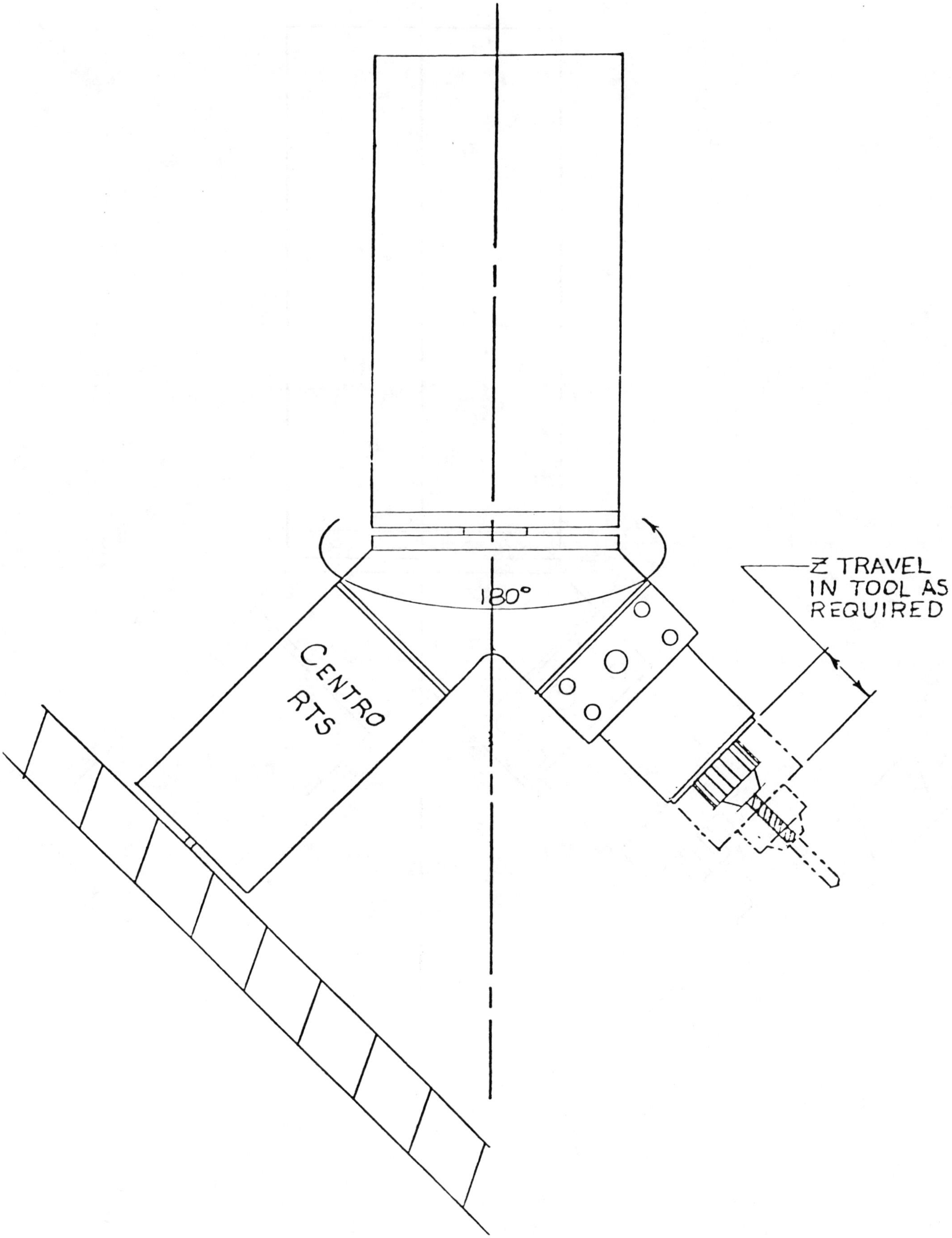

FIGURE 11

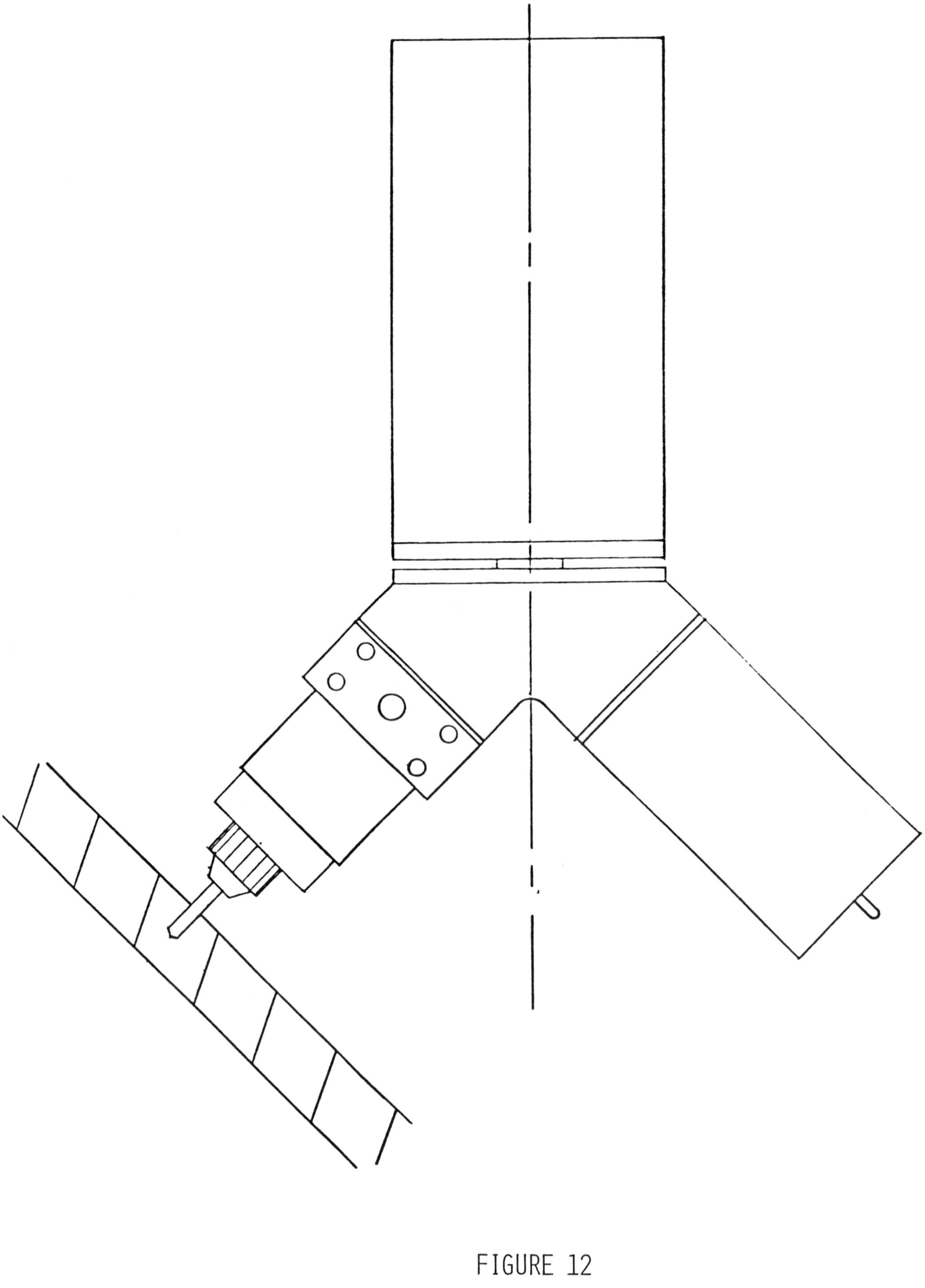

FIGURE 12

Mechanical Features of Gripper in Industrial Robots

I.B. Chelpanov
Leningrad Polytechnic
USSR

S.N. Kolpashnikov
Leningrad Polytechnic

Abstract

A theory for static interaction of an object with gripper working elements is described. The geometry of contact is essentially the principle factor defining the formalized statements and solution methods for the tasks of object interacting with working elements. In accordance with the stated classification of contact geometrical conditions, the principle results are discussed relating to the case of contact at discrete points, and new problems are posed for the mechanics in the cases of contact along lines and surfaces. Stable equilibrium positions are considered for a "gripper-object" mechanical system. Conditions are stated for rigidly fixing an object for grippers with one degree of freedom with floating working elements. Consideration of elastic properties permits solutions to be found when the "gripper-object" system cannot be statistically defined.

Abstract

The principles are described of the theory developed by the authors for static interaction of an object with gripper working elements. The geometry of contact is essentially the principal factor defining the formalized statements and solution methods for the tasks of object interacting with working elements. In accordance with the stated classification of contact geometrical conditions, the principal results are discussed relating to the case of contact at discrete points, and then new problems are posed for the mechanics in the cases of contact along lines and surfaces. Stable equilibrium positions are considered for a "gripper-object" mechanical system, with principal attention being paid to such equilibrium positions at which an object is rigidly fixed in position. Conditions are stated for the rigid fixing of an object for grippers with one degree of freedom, with floating working elements. Consideration of elastic properties permits solutions to be found when the "gripper-object" system cannot be statistically defined and new effects to be obtained. Elastic models are divided in the large-scale models (which take into account the bend and torsion of levers, the bend of thin walls in an object) and the local models (which consider the deformations in contact zones).

1. Mechanical arrangements of the grippers intended for grasping and holding objects to be manipulated are rather numerous. They are classified by many signs of indication 1 . Thus the causes can be different which maintain the hold-up of an object (weight forces, elastic forces of springs, and forces produced by off-line drives). Most simple are the grippers which do not have moving links and which are rigidly connected to the last links of manipulators. These are nothing more than movable supports whereon an object is free to stand in the vertical or horizontal position. For the purpose of industrial robots, more often is the use of simple mechanical grippers. They will take and hold an object on account of relative motion of links. The gripper mechanisms often have one degree of freedom. The grippers with one degree of freedom are designed to have one movable grasping link (Fig. 1), with two movable grasping links (Fig. 2), and with three grasping links (Fig. 3). In the second case, the links usually move symmetrically with respect to the middle plane. In the third case, symmetrically to the middle axis. Additional degrees of freedom are usually realized in the grippers with free or spring-loaded

floating elements (for instance, Fig. 4).

The static calculation of gripper mechanisms which is usually reduced to the determination of forces and moments in the assemblies and cross sections of links is rather simple and it can be conducted analytically by regular methods [2]. Principal and most interesting, in the authors' opinion, are the tasks of analysing the mechanics of interaction between an object to be manipulated and the gripper elements wherewith they are in direct contact. From here on, these will be termed as the working elements. Described in brief hereinunder are the basics of the mechanics, worked out by the authors, of interaction between an object and gripper working elements in various positions.

The principal factor defining both the formalization and the solution methods for the tasks of interaction between an object and working elements is the geometry of contact. For classification of the geometrical conditions of contact, reference should be made to Figure 5. Detected are the idealised models of contact at discrete points and along the lines. Under real conditions, common is contact on surfaces. However, if an object and working elements are considered to be absolutely rigid, while the shapes of their working surfaces are not coordinated, then it is necessary to consider contact at discrete points to be common. It is in such a statement the tasks of gripping statics have been posed and solved in the authors' papers [3], [4], [5].

Enumerated in brief hereunder in the first place are the basic results relating to the case of contact at discrete points and then new tasks are posed for the cases of contact at a greater number of points along the lines and over the surfaces.

2. Stable positions of an object in the gripper are considered, with maximum amount of attention being paid to such equilibrium positions at which an object is rigidly fixed in position. The conditions of rigid fixing on all coordinated, elastic compliance being neglected, are introduced as the conditions of the fact that application of restricted but finite forces and moments will not result in displacements from this position. Rigid fixing on all coordinates, depending on geometry, can be realised both regardless of friction forces and only due to friction forces.

Principal results have been obtained by the authors for the cases wherein rigid fixing is carried out, with friction forces being neglected. In these cases, the capabilities of rigid fixing are defined by a number m of contact points. The authors have shown that for a gripper device without movable links, the rigid fixing can be provided at m=6 also when some general conditions are imposed. In this case, normal reactions Ni at contact points are uniquely defined from six statics equations, as follows:

$$\sum_i N_i \vec{n}_i + \vec{F}_i = 0$$

$$\sum_i N_i \vec{r}_i \times \vec{n}_i + \vec{M} = 0$$

where $\vec{n}_i$ - basis vectors of the normals at contact points, $\vec{r}_i$ - radius-vectors of contact points in relation to a center being selected. Forces and moments applied are considered to be reduced to the center, so that F is the main center, while M - principal moment of forces applied. At m = 6, the system is statically definable, if all of the system equations (1) are linearily independent.

If the gripper mechanism has one degree of freedom, then rigid fixing is effected at m = 7. In this case, added to the system of equations (1) is one mechanism equilibrium equation, as follows

$$\sum_i \alpha_i N_i - S = 0$$

where S - force created by the gripper drive. In this case, the system also becomes statically definite. For grippers with extra l degrees of freedom (with floating elements) as shown in Fig. 4, adding one more of the new degrees of freedom will give an extra equilibrium equation, so that for rigid fixing and for static definability it is necessary to have (7 + l) contact points. Further on, regardless of a gripper type, the contact point number m_o, at which a system is statically definable and an object is rigidly fixed, will be called nominal. When the rigid fixing conditions are satisfied in the equilibrium position, the potential energy of an object-gripper system is at its minimum. This minimum, however, is not analytical and it can be obtained at a conventional point.

The forces of dry friction at contact points permit rigid fixing to be obtained in those cases when the number of contact points is fewer than the nominal number and when the coordinates are available which cannot be fixed without friction forces. At $m \geqslant 3$, the system is statically undefinable even in the presence of friction forces. However, for ultimate equilibrium, when at contact points the modules of friction force vectors assume maximum values Q_{i_o} = fN, where f - friction coefficient, the solution becomes unique. To find friction forces, use is made of the kinematic method. Given for this purpose are small displacements of an object which are defined by the vectors of motion of the center and of the smaller turn angle, and relative displacements are defined at contact points. Friction forces are derected in opposition to the vectors of these small displacements. If said small displacements are given to be compatible with the conditions of contact at all of the points or at some selected points, then at m m_o the number of unknown is equal to the number of equations, and the system becomes statically definable.

The capability of a gripper device to rigidly fix an object in the presence of forces and moments is clearly seen in the shape of a rigid fixing region in the six-dimension space of components F_x F_y F_z of the main vector $\vec{F}$ and of components M_x M_y M_z of the moment $\vec{M}$. If a depicting point, defining the loading conditions, is inside the region, then an object is rigidly fixed, and coming out to the border will correspond to the ultimate condition. At $m = m_o$, with friction forces being neglected, displacement of an object in relation to the gripper is only possible with contact being lost at one of the points, i.e. when the relevant normal reaction becomes nil. When contact is lost at one point, i.e. at $m = m_o - 1$, an object in the gripper will be able to move (one degree of freedom is obtained). Assuming $R_l = 0$ in the statics equations system, we obtain one equation which interconnects the components F_x, F_y, F_z, M_x, M_y, M_z, i.e. the equation for a fraction of the rigid fixing region boundary. Assuming successively that $l = 1, 2, \ldots m_o$, we can obtain the equations for all of the boundary portions. Since all of the statics equations are linear with respect to the forces, the rigid fixing region boundaries are piecewise-linear. The rigid fixing region can be closed and open. Capability of the appearance of regions going to infinity is associated with the fact that the gripper device mechanical model adopted does not consider the restrictions for forces acting upon the stationary parts (downward in Fig. 1), as well as with the fact that wedging is possible (as in Fig. 6 when moment is acting).

Practically speaking, more convenient for use and more comprehensive are two-dimensional and one-dimensional cross-sections of the rigid fixing region. The central one-dimensional cross-sections of the rigid fixing region or, for short, the ultimate forces and moments are obtained, if all of the components of forces and moments but one are assumed to be equal zero. For the grippers with movable elements, all the ultimate forces and moments are directly proportional to the effort S while the ultimate values, corresponding to the coordinates, which are only fixed due to friction forces, are also directly proportional to the friction coefficient f.

Two individual cases present the main practical interest.

(a) Surfaces of an object and of the working elements are prismatic with parallel generating lines directed along the axis Z (Fig 7). In so doing, it is a good idea to consider separately the state of plane problem of rigid fixing of an object cross-section at right angles to the axis Z (i.e. on coordinates X and Y and angle θ_z) and separately, on coordinate Z and angles θ_x and θ_y. If we consider only the state of plane problem on the plane XY, then for a gripper without movable elements the nominal number of contact points which permit rigid fixing of a cross section $m_o = 8$, (neglecting the friction forces, three normal reactions can be found from the three statics equations), while for a gripper with one degree of freedom, $m_o = 4$. The state of plane problem has its own significance and it is convenient for illustrating general statements. Pictures in Figure 8 is an example of rigid fixing with a cross sectional area in the shape of a circular segment. Figure 9 shows a rigid fixing region. Fixing on Z axis is effected only on account of friction forces.

(b) The gripper is plane-parallel so that the surfaces of working elements are plane and parallel to one another (Fig. 10). For the purpose of rigid fixing on the coordinate Y and the turn angles θ_x and θ_z four contact points are required. Fixing on X- and Z-coordinates and on the angle θ_y is effected only on account of friction forces.

If the number of contact points exceeds the nominal number ($m > m_o$) or if contact is effected along lines or over surfaces, the object-gripper system becomes statically indeterminable. Two quality-wise different situations should be distinguished which are dealt with in Clauses 3 and 4 below.

3. Let us assume that though some "extra" contact points exist in the initial position of rigid fixing, the number of contact points becomes equal (m_o - 1) during all minor displacements or during displacements of the class in question, as well as for the case of rigid fixing without "extra" points. Illustrated in Figure 11 is an example of the state of plane problem of rigidly fixing a cross-sectional area in the shape of a trapezoid by means of a gripper without movable elements. It can be seen that after displacement there will remain m = 2, i.e. two contact points get lost at the same time. For leftward displacement for $l > a$ the contact gets lost at the points A and D, for $l < a$ - at the points A and C. Since in the case of minor displacement there will remain (m_o - 1) contact points, then the population m_o of the statics equations will define one linear equation of association among the components F_x, F_y, F_z, M_x, M_y, M_z of the forces and moments applied. This equation assigns one of the portions of the piecewise-linear boundary of the rigid fixing region. Qualitative difference from the case of rigid fixing at $m = m_o$ lies only in the fact that at $m > m_o$ the region boundary has a great number of portions. For the models of rigid contacting bodies, transition to a contact at (m_o - 1) discrete points during displacements from the contact initial position at $m > m_o$ points, along the lines or over areas, should be considered to be a common case. This takes place when the lines or areas have variable curvature. Definition of those points at which contact takes place during movements of an object is in itself an independent problem. The practically convenient numerical method is based upon the piecewise-linear approximations of the surface areas if an object and of the working elements. In this case, fundamental is the sort-out procedure among the apexes of the convex surface of an object. Only at these apexes can a contact occur.

4. Let us assume that at minor object displacements (at least for some of the classes of displacement) the contact conditions do not vary. The "extra" contact points will re-

main, the contact on lines will remain the contact on lines, and the contact on surfaces will remain the contact on surfaces. In these cases, the system remains statically indeterminable, and for its calculation it is necessary to attract more complex models. Initial notions about the appearance of contacts of such a kind can be drawn proceeding from two different groups of notions.

(a) Accurate agreement of the geometry of surfaces of an object and of the gripper device working elements. Similarly, it is assumed that elastic deformations can be neglected. It is supposed that the contact on lines and over surfaces is obtained due to precision machining of surfaces. It is possible to detect the following typical kinds of surfaces for which the displacements are probable, with the contact type remaining the same:

- the prismatic surfaces with parallel generating lines (the contact occurs on rectilinear generating lines and this contact is not disturbed when moving along the guide-ways;
- the parallel planes of a flat-parallel gripper and an object (surface contact is not disturbed at any two-dimensional displacement of an object;
- the coaxial gyration surfaces fully coinciding or not coinciding as shown in Figure 8 (the contact on circumferences is not disturbed during rotation about Z-axis). The static indefinability for all of these problems is expressed in the fact that there are not enough statics equations in order to define the distribution of normal and tangent forces (tensions) on contact lines and contact surfaces. Nevertheless, in spite of the fact that the rigid fixing regions for these cases cannot be plotted by means of the methods described above, the maximum values can be accurately defined for some of the components of forces and moments, while for the other components it becomes possible to provide upper bounds. So, for the plane-parallel gripper and for an object in the shape of a parallelepiped (Fig. 10), the expressions for the maximum values of five of the six components of forces and moments can be computes in the elementary manner and these are illustrated in Table 1. Here S - gripping force, compared wherewith the weight force of an object can be neglected. To define the ultimate value for M_y determined in accordance with the expression:

$$M_y = 2\iint_{(\Omega)} f\,\sigma(x,y)\sqrt{x^2+y^2}\,dx\,dy \qquad (3)$$

it is necessary to know the distribution $\sigma(x,y)$ of normal tensions over the contact area Ω. However, it can be evaluated by the maximum:

$$max\{M_y\} = 2Sfa \qquad (4)$$

(it is assumed that a > b). The lower bound for this moment ($\min |M_y| = 0$) is of no practical interest. For an object and for the working elements with prismatic surfaces, the expressions for the ultimate values of all of the component forces and moments are obtained regardless of the distribution of forces on contact generating lines.

(b) Consideration of elastic and plastic properties of elements. Investigation into the elastic properties will not only permit the solutions to be found when the system remains practically indeterminable, but also new effects listed below to be obtained. Common for all of the cases is the fact that in considering the elasticity, the initial notions about rigid fixing become approximate. It has to be assumed a priori that either object position variations due to elastic deformations can be trivially neglected, or the limita of permissible values should be prescribed with subsequent checking if the movements are within the allocated ranges. For convenience purposes, the elastic models can be divided in large-scale and local. The large-scale models take into account factors as bend and torsion of the working elements and bend of the thin walls of an object. Calculation models in these cases are reduced to elastic beams, plates, and shells. The principal effects can be as follows:

1. Variation in the positions of contact discrete points. In case of small curvatures of object surfaces in the contact zone, displacements can be considerable (Fig.12).

2. Transition for surface contact to line contact and from line contact to point contacts. It is kept in mind that for the initially agreed shape of an object and of the surfaces of working elements, bending action will disturb this agreement. This is illustrated in Figure 13 using an example of the plane-parallel gripper. Elastic deformations in these cases can result in the same thing as the initial deformations of an object and elements.

The local models take into account the deformations in contact zones. If the mating surfaces are hard, then it is considered to be possible to utilize the statements and solutions of contact tasks of the elasticity theory for elastic half-space, sphere, and cylinder. If, on the other hand, it is required to consider the compatibility of specially introduced elastic spacers, then the use can be made either of more simple models (for instance, a Winkler base model) or of more complex non-linear models. The principal effects can be as follows:

1. Conversion of point contact into surface contact with the known (computed) distribution of normal tensions. In case of a small number of contact points, the finiteness of even a small contact zone can be substantial. For instance, in the common case of two-point contact (Fig. 14), the whirling friction moment is defined by the expression:

$$M_y = 0{,}22 f S^{4/3} \eta^{1/3} \left(\sqrt[3]{\frac{\rho_g' \rho_g''}{\rho_g' + \rho_g''}} + \sqrt[3]{\frac{\rho_o' \rho_o''}{\rho_o' + \rho_o''}} \right) \qquad (5)$$

where ρ_g' and ρ_g'' - radii of curvature for the surfaces of the upper and lower gripper jaws at a contact point; ρ_o' and ρ_o'' - radii of curvature for an object at the points of contact with the upper and lower jaws, respectively;

$$\eta = \frac{1-\mu_g^2}{E_g} + \frac{1-\mu_o^2}{E_o}$$

wher μ_g, μ_o - Poisson coefficients, while E_g, E_o - moduli of elasticity for the gripper material and the object material, respectively.

2. Partial separation or slippage on a portion of contact surface. In the event of non-uniform distribution of normal tensions on contact surface, slippage usualyy starts in the zones of smaller values of normal tensions. Noteworthy is to mention, however, that the ultimate values of forces and moments can often times be obtained approximately, with this effect being neglected, in the ultimate supposition that slippage occurs all over the contact area.

3. Contact compliance under certain conditiond can ensure the contact on lines and surfaces in those cases when the surfaces of an object and working elements are close to the compatible shapes. So, for the plane-parallel gripper in case of deviations of object surfaces from the planes, the surface contact can be restored in case of sufficient compliance of the elastic spacer. In so doing, the computation of contact tension distribution presents an independent problem. It can be noted, however, that a possibility of restoring the contact on lines and surfaces due to elastic properties is depending upon the gripping force and upon the forces applied.

The authors have posed and solved a number of problems wherin the influence is evaluated of the above-listed defects upon the ultimate values of forces and moments and upon the rigid fixing area configuration.

REFERENCES

1. Kolpashnikov S.N., Chelpanov T.B. Defining of the scheme and the parameters of an industrial robot gripper by a group of specifications. "Proc. 11th ISIR, Tokyo, 1981", IFS, 1981.

2. Konstantinov M.S. Law-Type Gripper Mechanism. Proc. 5th ISIR. Chicago, 1975, IFS, 1975, p. 329-334.

3. Колпашников С.Н., Челпанов И.Б. Задачи механики схвата манипуляционного робота. Депонировано в ВИНИТИ № 3094-80.

4. Колпашников С.Н., Челпанов И.Б. Обеспечение надежности захватнвания в динамических режимах. В кн.: "Робототехника", Л., 1980, с.80-85.

5. Kolpashnikov S.N., Chelpanov T.B. The problem of mechanics of industrial robot grippers. "Proc. 4th CISM-IFToMM Symposium "Romansy-81"? Poland, Warsaw, 1981.

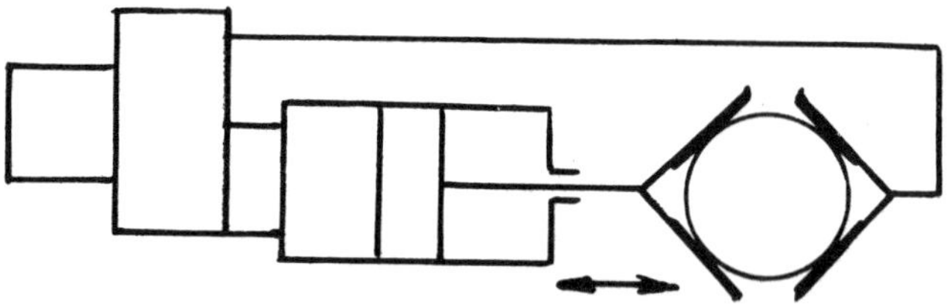

Рис. 1

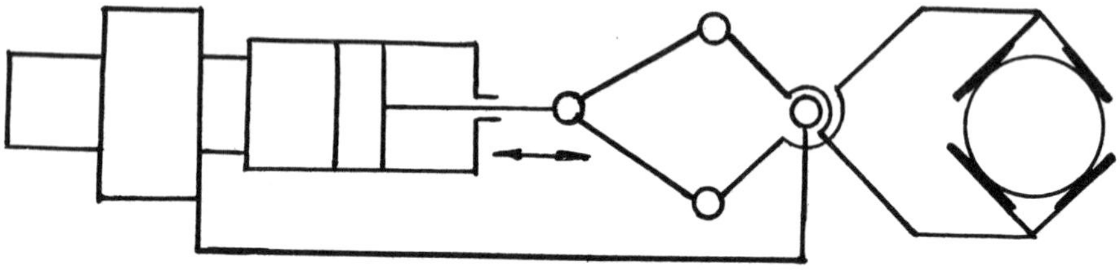

Рис.2

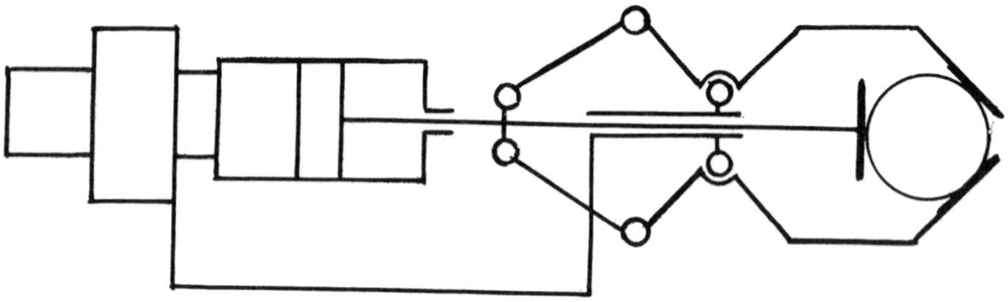

Рис. 3

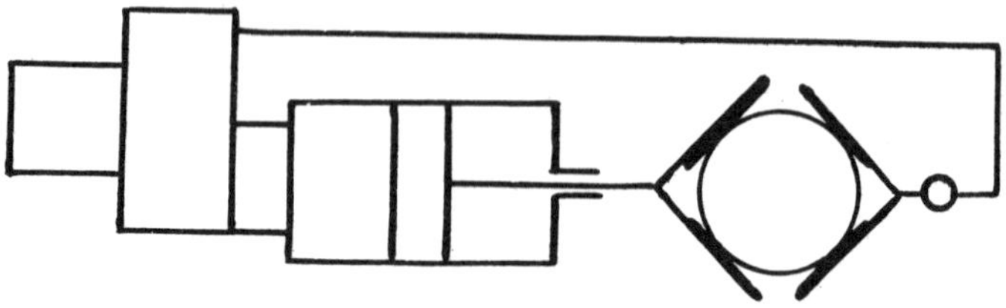

Рис.4

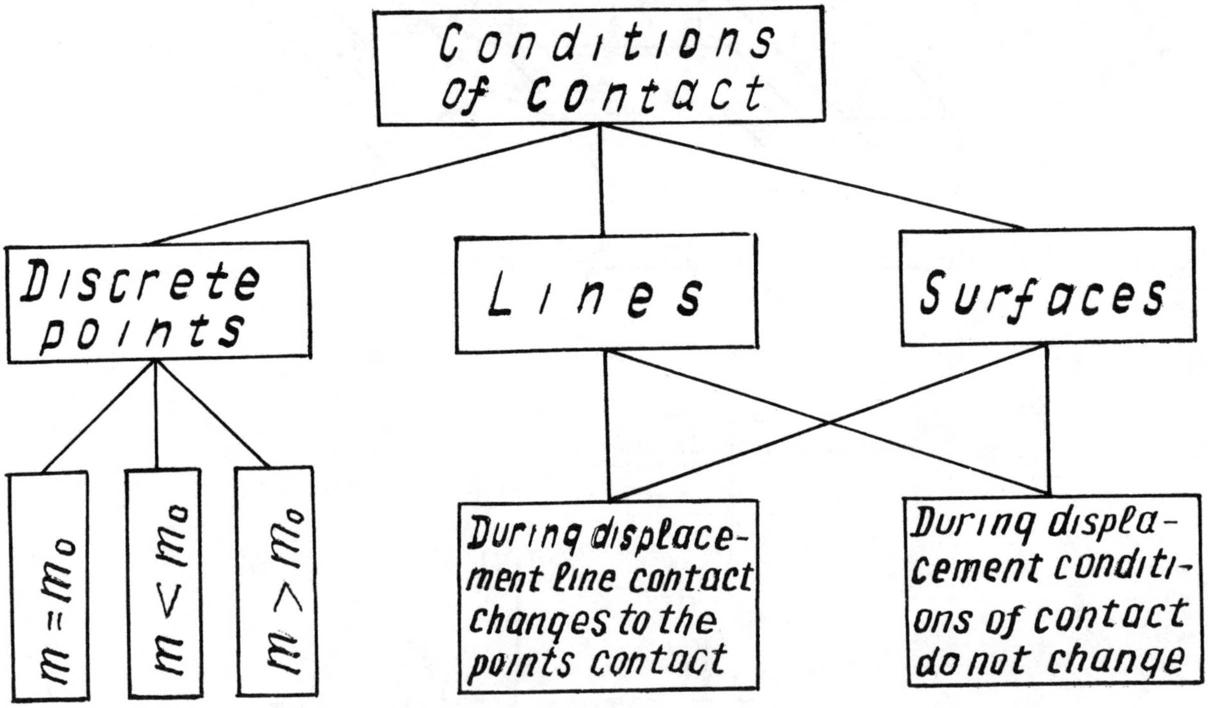
Conditions of Contact
Discrete points
Lines
Surfaces
$m = m_0$
$m < m_0$
$m > m_0$
During displacement line contact changes to the points contact
During displacement conditions of contact do not change

Рис. 5

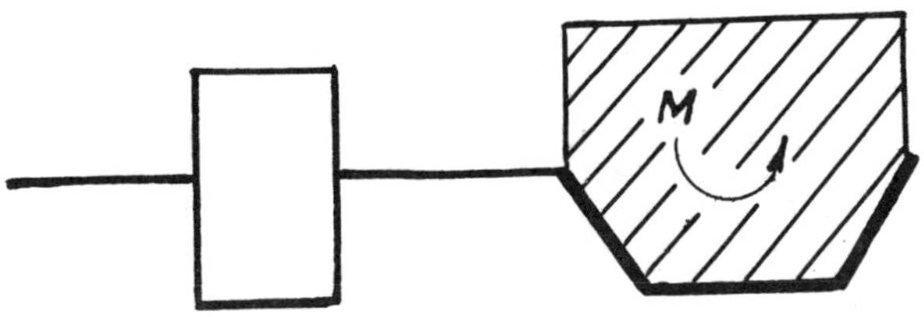
M

Рис. 6

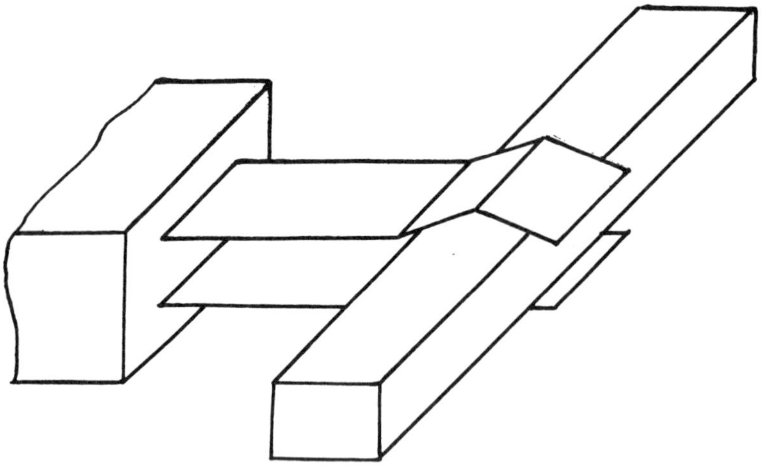

Рис.7

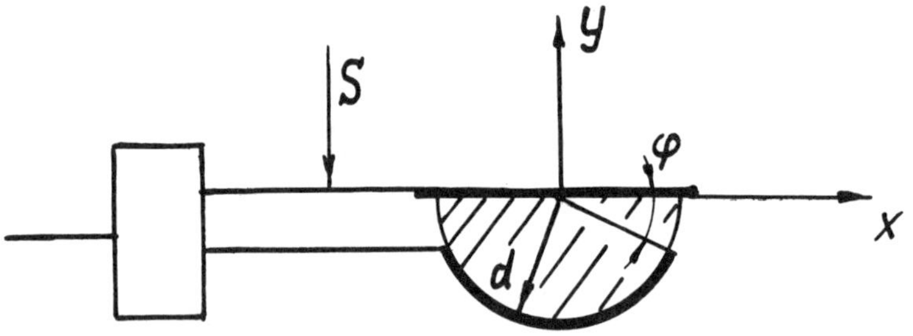

Рис.8

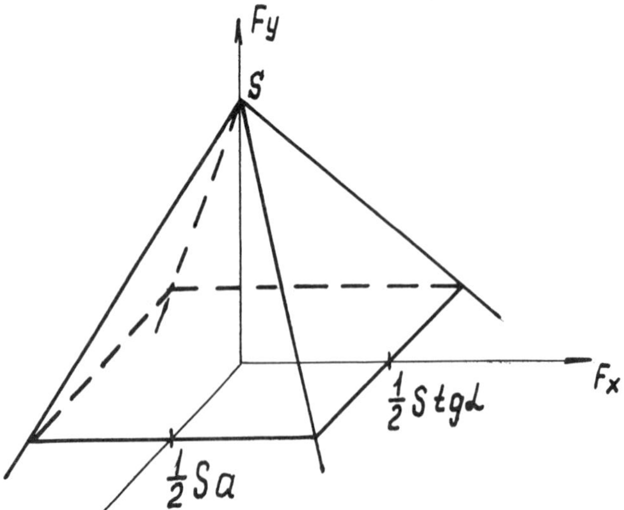

Рис. 9

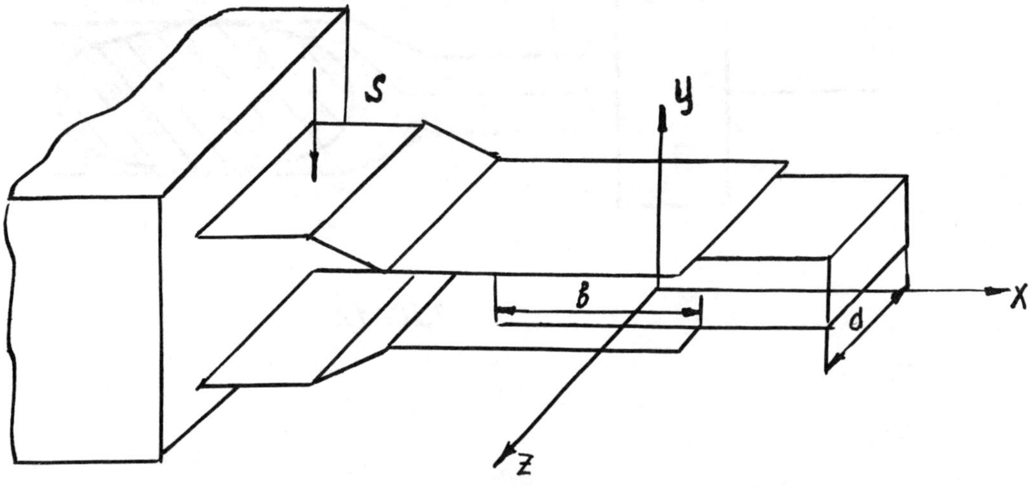

рис 10

Table 1

F_x	F_y	F_z	M_x	M_z
fS	S	fS	Sa	Sb

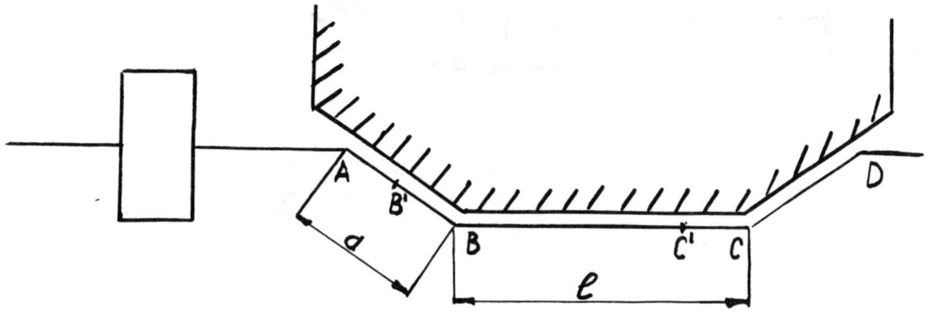

Рис 11

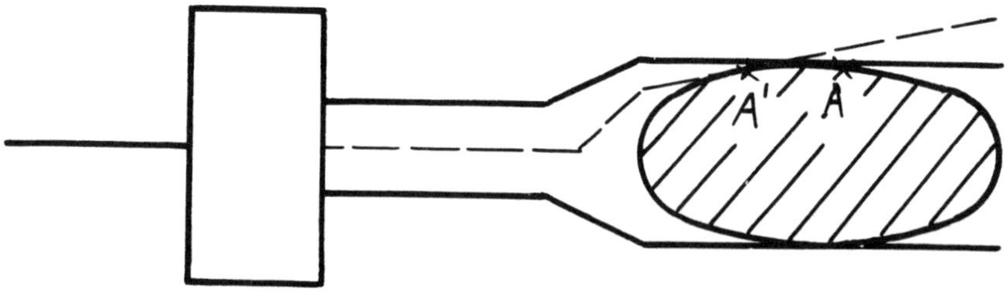

рис. 12

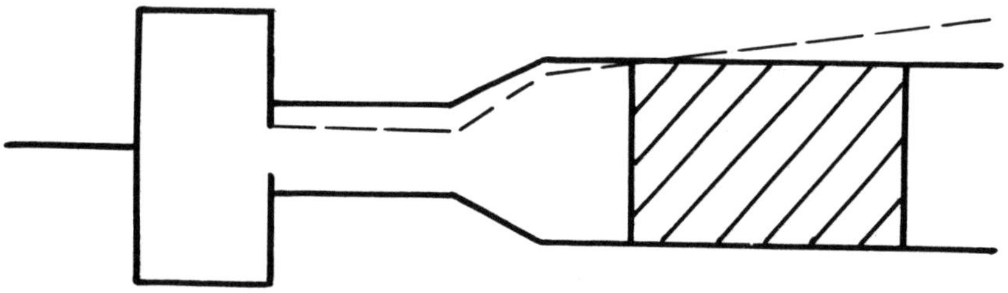

рис. 13

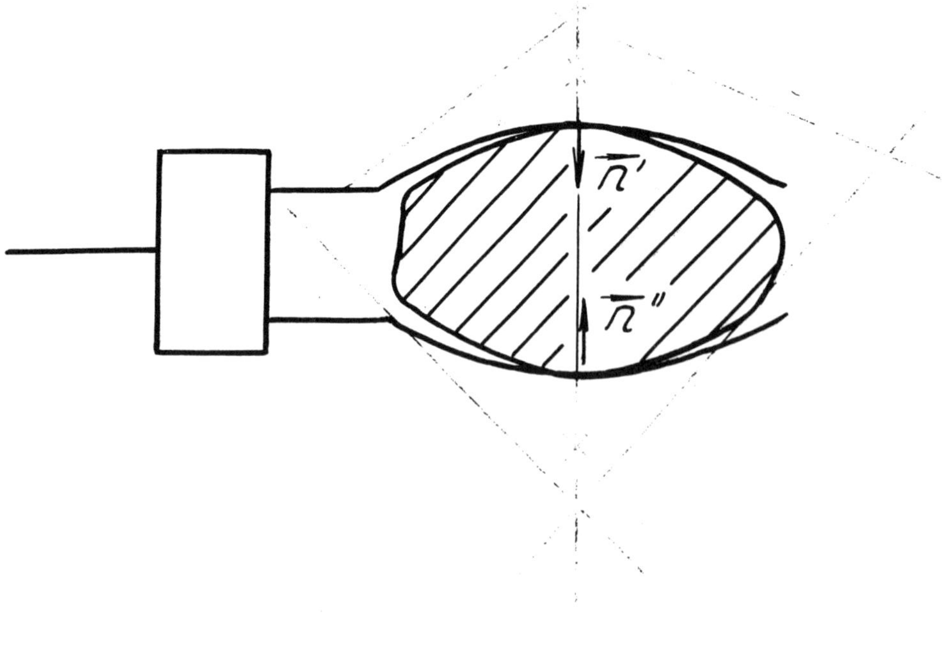

рис 14

A Three-Fingered, Articulated, Robotic Hand

Du Lian
University of Minnesota
Minneapolis, Minnesota

Steven Peterson
University of Minnesota

Max Donath
University of Minnesota

Abstract

Claims that robot manipulators are the key to flexible automation are ill-founded as long as there exists a need to custom-design a gripper for different applications. The human hand, in particular, serves as a model for the type of flexibility desirable for performing manipulative tasks. The question, however, is how to trade flexibility versus complexity. The design and implementation of a three-fingered, nine degree-of-freedom hand is described. The fingers are driven from motors on the forearm of the robot via cables. Position and force control strategies for the joint are examined.

INTRODUCTION

End effectors typically found on industrial manipulators today consist of two jaws which are opened or closed using electro-mechanical, pneumatic, or hydraulic actuators. Not only can't these grippers handle cylindrical, spherical, or other arbitrarily complex geometry parts, but they also don't do very well at grasping objects with flat parallel faces. As a result new grippers or jaws must be designed for each part geometry, or alternatively the part must be modified to accommodate the end effector. To assist in designing grippers, a large number of mechanisms generic to end effectors have been reviewed by both Lindstrom(1) and Chen(14).

An alternative approach to this problem is to design and build a flexible gripping device capable of a large variety of tasks. Several investigators have developed designs for such devices. Most of these designs are composed of three fingers which are controlled in a variety of ways. The simplest design uses four motors and a gear box to move the fingers into the different gripping positions(2). In this design one motor controls the grasping function of the gripper, and the other motors are used to flex and extend the individual fingers. More complex designs which try to emulate the human hand have up to eleven degrees of freedom(3), and each finger behaves similar to a small independent manipulator. The independent motions of the fingers allow these grippers to manipulate objects as they're being held.

This article presents the design of an alternative three fingered electro-mechanical hand. The purpose of the design was to build a hand that was as flexible as possible - allowing controlled movements of the object being handled - while striving for simplicity. The motivation was to reduce the total weight of the gripper, and also minimize the cost of building the device.

HAND CONFIGURATION

The configuration of the mechanical hand was based on the motions of the human hand (hence the name). Like other complex grippers(see references 2-12), it has only three fingers. A three fingered approach simplifies the gripper design (as compared to the human based five fingers) while maintaining a high degree of flexibility in the manner in which it can grasp an object. The hand design developed here is shown in figure 1. The fingers of the robot hand were designed to mimic the motions of the human hand. Fingers A and B correspond to the index and middle fingers, and the thumb was designed to correspond to the motion of the human thumb. The mechanical hand is symmetric about a vertical plane placed between fingers A and B which makes the design ambidextrous - either right or left handed. As a result fingers A and B switch roles depending on which side of the gripper the thumb is positioned. Each of the fingers and the thumb of the mechanical hand have three joints which allow the tip of the respective finger to be positioned in three dimensional space within its work volume. By controlling the torque at each joint, the force exerted by the finger's tip can also be controlled in up to three directions. This arrangement is different from the human hand in which the thumb and fingers have many more degrees of freedom.

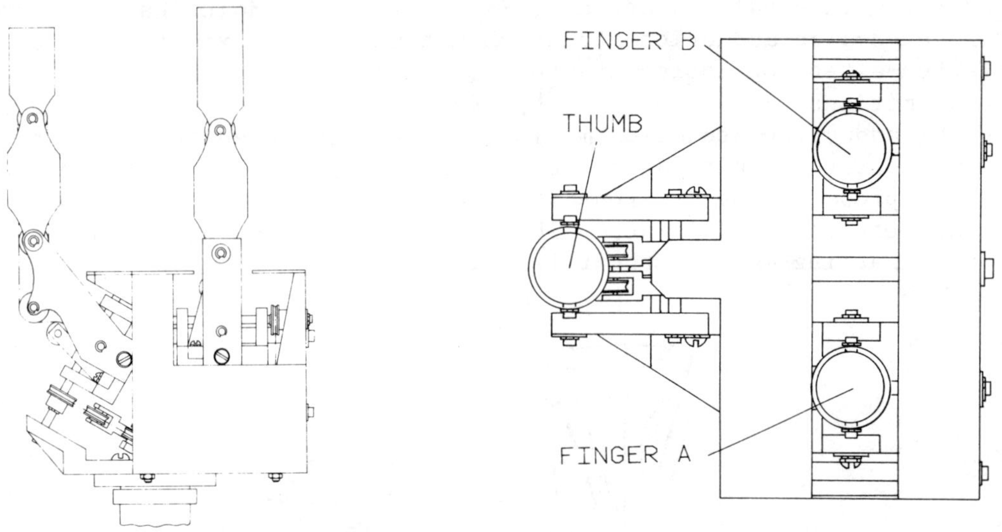

Figure 1. A) Side and B) Top Views of the Mechanical Hand.

If we assume that the tip of the finger maintains point friction contact (i.e. no slipping) with the object being handled, the finger can control the three dimensional position of that point. Specifying the positions of three non-colinear points on a rigid body specifies its position and orientation in space. Therefore if the hand can control the position of three points on the object, it can control the orientation of the object. This enables the hand to make adjustments in the orientation of the object, and if the placement of the joints is done properly, enables this simplified model of the human hand to reproduce many of the natural hand's grasping functions.

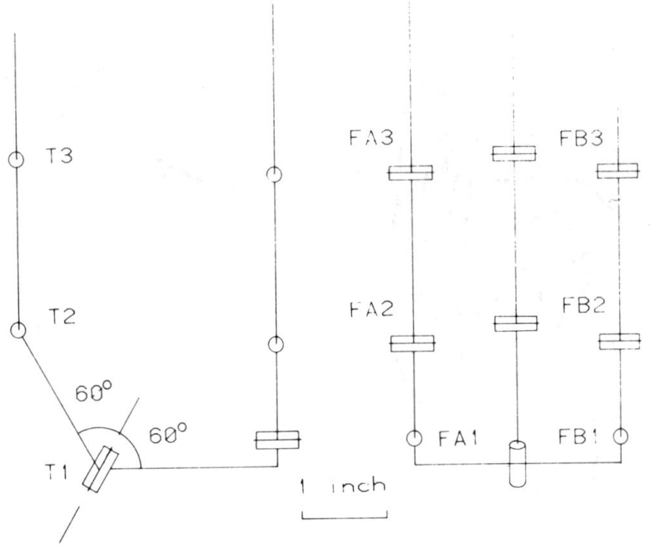

Figure 2. A Kinematic Diagram of the Mechanical Hand.

The motions of the robot hand's joints were defined using the human hand analogy. A kinematic diagram of the mechanical hand is shown in figure 2. Fingers A and B are identical, and the motions of their joints are defined using the index finger as a model. The index finger of the human hand has three joints. The proximal joint - the joint closest to the palm - is called the metacarpal-phalangeal (MP) joint. This joint has two motions (see figure 3): an adduction-abduction motion of approximately 30 degrees, and a flexion-extension motion of about 120 degrees. In the mechanical hand these two motions are provided by FA1 and FA2 (also FB1 and FB2). FA1 produces the adduction-abduction motion, and was designed to have the same range of motion as the natural MP joint. The second joint in the mechanical hand, FA2, reproduces the flexion-extension motion. Its range of motion is twice that of the MP joint. This allows the mechanical hand to grasp objects by expanding against their interior walls.

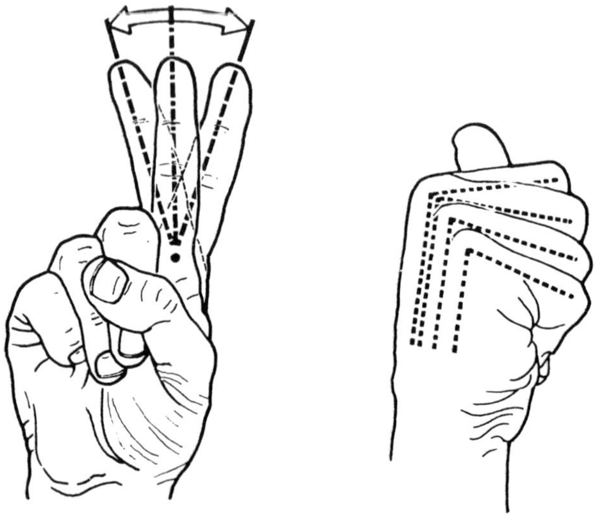

Figure 3. The Motions of the Index Finger's Metacarpal-Phalangeal Joint.[1]

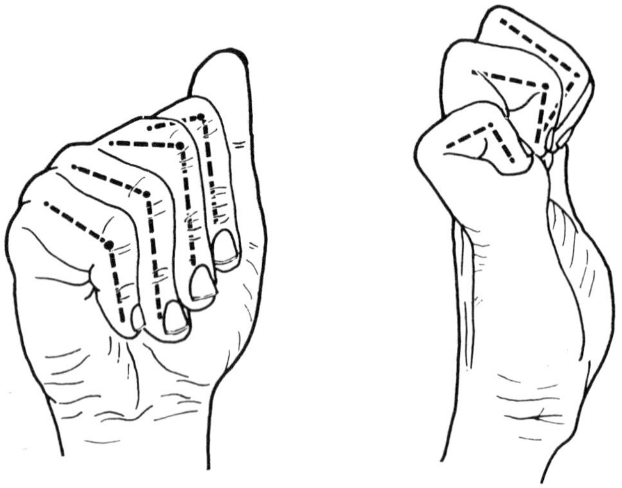

Figure 4. The Motions of the Interphalangeal Joints in the Index Finger.

---

1. Figures 3 through 8 were taken from The Physiology of the Joints Vol. I(13).

The next two joints of the human finger are the interphalangeal (IP) joints. These act like hinge joints with only one degree of freedom (see figure 4), and can move about 90 degrees about this axis. The mechanical hand models these two joints as one, joint FA3. This simplification cuts down the total degrees of freedom of the mechanical finger, but it should not significantly hamper the performance of the gripper. Joints FA3 and FB3 are allowed to move 180 degrees about their respective axes.

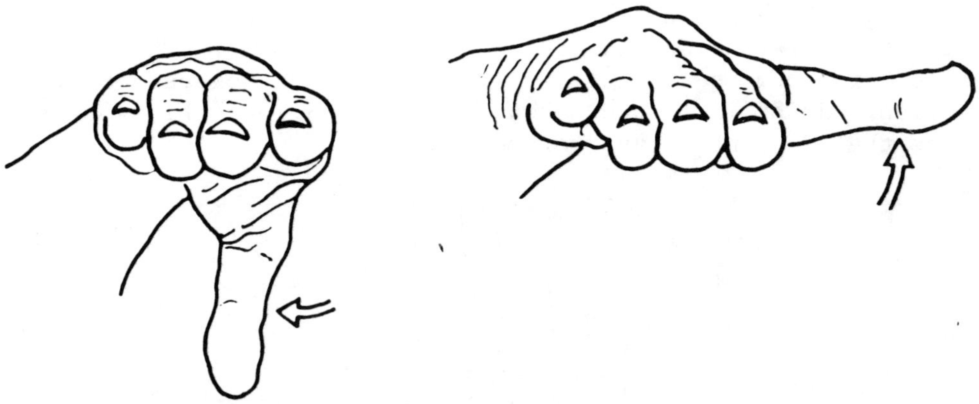

Figure 5. Adduction-Abduction Motion of the Carpometacarpal Joint in the Thumb.

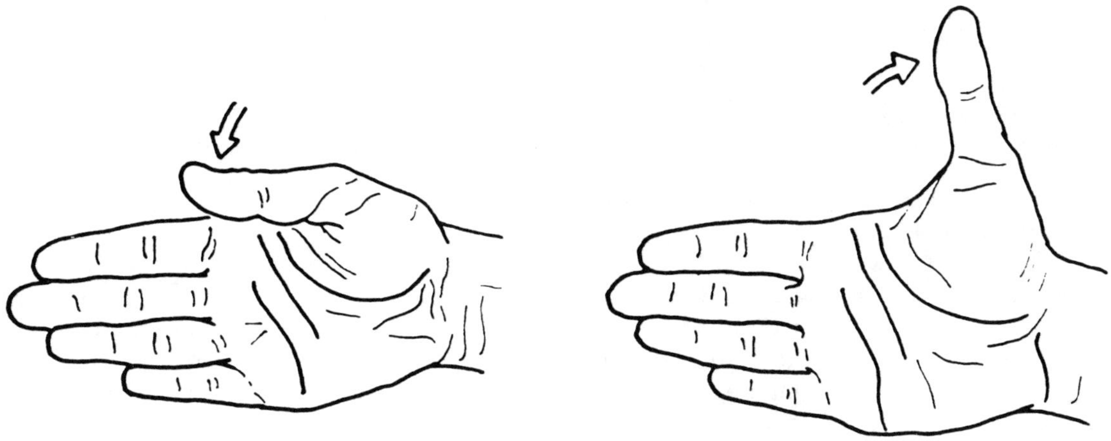

Figure 6. Flexion-Extension Motion of the Carpometacarpal Joint.

The human thumb is a very complex mechanism; therefore its structure was greatly simplified for the mechanical hand. The proximal joint in the human thumb is refered to as the carpometacarpal joint. This joint allows two degrees of freedom: an adduction-abduction motion shown in figure 5, and a flexion-extension motion shown in figure 6. The adduction-abduction axis of motion makes an angle with the palm of the hand. The orientation of this axis is not well defined. The mechanical thumb models these motions with one axis. The axis of T1 is inclined 60 degrees from a datum associated with the base of the gripper. This axis approximates the adduction-abduction axis of the natural thumb joint. The range of motion about T1 is 180 degrees. This allows the thumb to be positioned on the right or left side of the hand. The next joint in the thumb is the metacarpal-phalangeal joints which possesses three degrees of freedom. The joint's major motion is flexion-extension (figure 7) which is duplicated in joint T2 of the robot hand. Notice that T2 is placed at a 60 degree angle with respect to the axis of T1. Positioning T2 in this manner allows the mechanical thumb to oppose the fingers in a tip to tip and lateral fashion. These grasping functions are discussed later. Like the other flexion-extension joints, T2 has a range of motion equal to 180 degrees. The distal joint in the thumb is called the interphalangeal (IP) joint of the thumb. This joint has one degree of freedom (see figure 8) which allows 90 degrees of flexion-extension motion. Joint T3 in the mechanical thumb allows this same motion, but has twice the IP joint's range of motion.

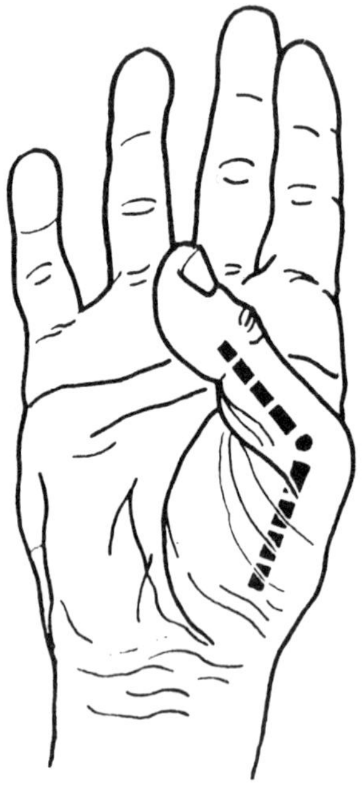

Figure 7. Flexion-Extension Motion of the Thumb's Metacarpal-Phalangeal Joint.

The exact dimensions of the finger links were found by placing the mechanical hand in several grasping positions. The normal grasping (prehension) functions of the human hand are well defined in biomechanics texts(13). Some of these functions are tip opposition - the tip of the thumb meets the tip of the index finger, lateral opposition - the tip of the thumb meets the side of the index finger, palmar prehension - the fingers and thumb wrap around a cylindrical shape, spherical prehension - the thumb and fingers wrap around a spherical shape, and digito-palmar opposition - only the fingers wrap around the object. The dimensions of the fingers were designed so the hand could grasp objects in each of these prehension modes. As previously mentioned, the fingers were also allowed to hyper-extend, which provided the mechanical hand an additional unhuman-like grasping capability. The mechanical hand can be inserted into hollow objects and expanded; grasping the object by pressing against the object's interior walls.

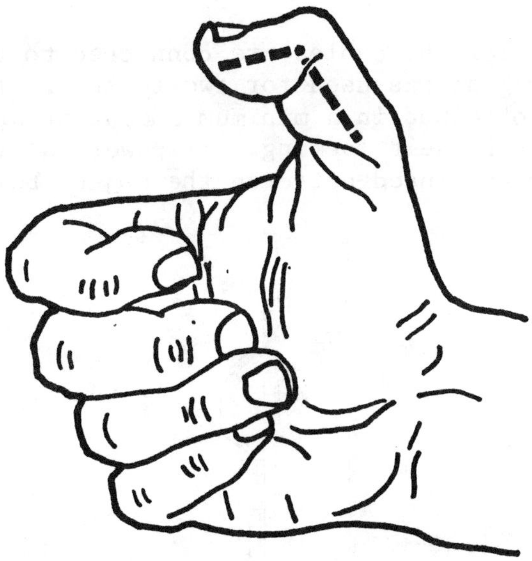

Figure 8. The Motion of the Interphalangeal Joint in the Thumb.

The final design was verified using a computer graphics program. The program was written for an Evans and Sutherland PS300 graphics system. This graphics system allows the programmer to enter the three dimensional configuration of an object, and then construct the connections that let the user interactively manipulate the picture on the screen. The configuration of the hand as well as the geometry of the finger links were entered into the program. Each of the finger joints were varied from the control panel of the PS300. The fingers were displaced about their joints until each of the prehension modes were verified.

HAND HARDWARE

The hand was designed to be as light as possible. The fingers are constructed out of hollow aluminum tubes. The fingers were designed with a cylindrical shape so they can approach the object from any direction. Since the fingers are hollow, the wires and cables needed for driving and sensing the motion of the fingers can be threaded through them; protecting the wires and cables from damage. The end of the finger can be fitted with a replaceable hemi-spherical tip which has a high coefficient of friction. The finger tip provides the friction contact the finger needs to control the orientation of the grasped object. The tip can house a tactile array sensor and can be suitably compliant to accommodate a wide variety of tasks. To lessen the weight at the end of the robot arm, the hand is driven by D.C. motors mounted on the robot's forearm (see figure 9). Power is transmitted to the finger joints by stainless steel cables that run through wire wound sheaths. The sheath eliminates the need for cable transfer points at each of the robot's joints, and keeps the cables from tangling with each other. By placing the motors on the forearm of the robotic manipulator, the inertial effects of the gripper on the motion of the manipulator are minimized, and as a result the actual payload size is maximized. In this design, the weight of the hand is less than two pounds (1 kilogram) while the weight of the motor housing on the forearm is approximately three pounds (1.5 kilograms). Thus this configuration significantly lowers the weight placed at the distal end of the robot arm. The hand was designed to carry a five pound (approximately 2.5 kilograms) load.

The drive pulleys for the cables are connected to the motors through a worm gear pair. A worm gear was used for two reasons. The first is that a high gear ratio can be obtained in a minimum amount of space. Second, worms gears can be designed to be self locking. If power is lost to the drive motors of the hand, the hand need not drop the object being held. Therefore

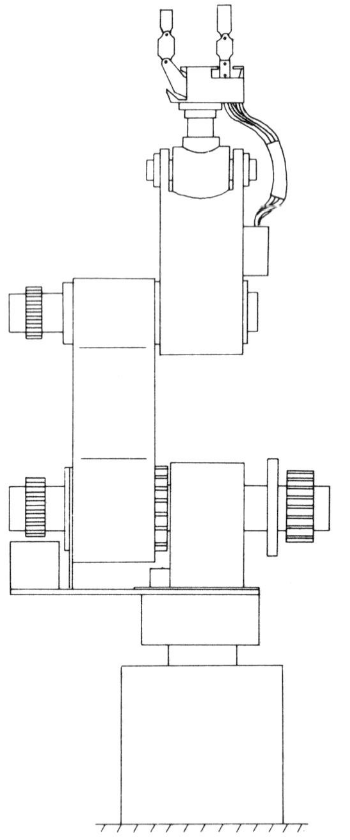

Figure 9. The Hand Mounted on an Industrial Robot Arm.

the transmission must be unidirectional (i.e. not back drivable). The self locking worm gear accomplishes this without needing extra brakes or clutches. The criteria for a self locking worm gear can be determined from a static force balance, and is given by the following inequality;

$$\tan \phi < \mu$$

where ϕ is the lead angle of the worm and μ is the coefficient of friction between the worm and its mating gear. This design has one major drawback; the efficiency of the transmission system will probably be less than one half. The low efficiency should not significantly limit the performance of this design.

JOINT CONTROL

The results of several other studies (6,8,10) have shown that the joints must be controlled in two modes: force and position control. A block diagram of the control system is shown in figure 10. In the finger drive configuration, the cable - which acts as a spring - lies between the drive pulley and the finger joint. To better describe the coupled dynamics of the system, state space techniques are used. The state vector of the system, $\underline{x}$, has four variables: the motor's velocity and position, the tension in the cable, and the velocity of the finger joint. The control system is defined by four matrices. Matrix A defines the dynamics of the system, B defines the input relationships, and K contains the sensor and controller gains. The system input variables are defined by the vector $\underline{u}$ which includes terms for the motor torque and static friction (Coulomb friction is introduced by the cable sheathing.) In the state determined linear regulator problem(15) state variables must be measured and fed back. Since space at the finger joint is severly limited, most of the sensing is done in the motor housing. Motor velocity is determined using a tachometer mounted to the motor's shaft, and the position of the drive pulley is monitored by an encoder mounted on its shaft. The actual variables that need to be controlled are the position and torque at the joint. The torque at the joint is measured using a force sensor that is placed as close to the finger joint as possible. Since the position of the joint cannot be measured, a state estimator or observer is introduced into the control scheme. The observer estimates the position or velocity of the joint using the tension in the cable and the position of the drive pulley. The cable tension can be related to the difference between the drive pulley and joint positions using Hooke's law. The observer incorporates this difference to the position of the drive pulley to produce an estimate of the finger joint's position. Static friction, caused by contact between the cable and the cable sheath, is a major source of non-linearity in the system. A feedforward signal may have to be incorporated into the control scheme to eliminate its effects on the system dynamics.

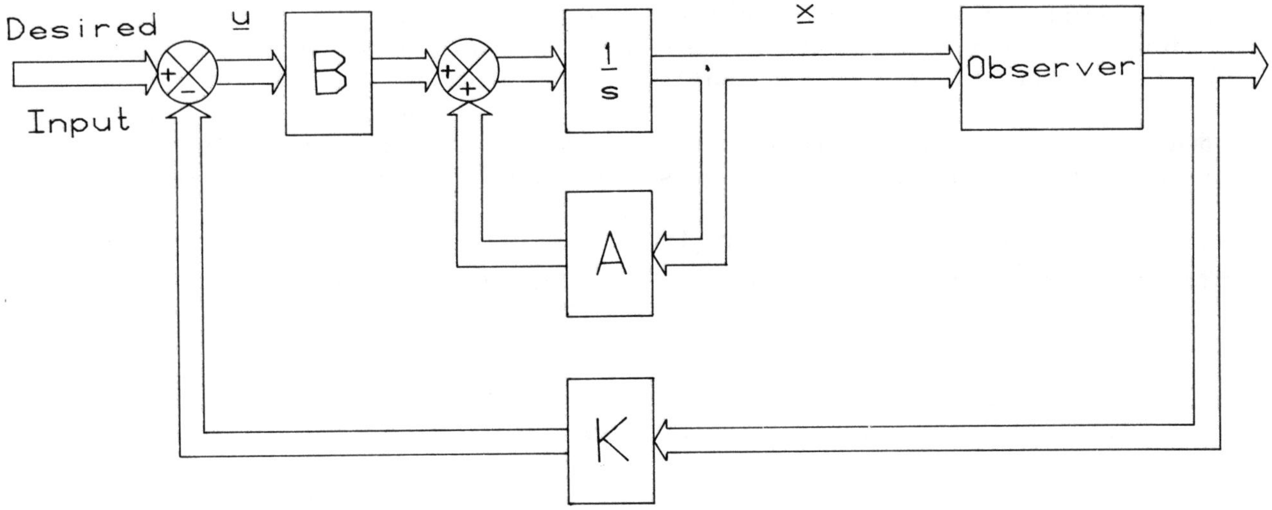

Figure 10. A Diagram of the Control System.

The force sensor measures the force in the cable by monitoring the strain in a cantilevered beam. The force sensor design is shown in figure 11. The beam is constructed out of stainless steel in order to maintain stiffness in the small configuration. The stiffness of the beam minimizes sensor non-linearities caused by the deflection of the beam's tip. At maximum load the tip of the beam deflects less than one thousandth of an inch (0.254 millimeters). The cable is preloaded so the sensor can measure both positive and negative cable forces. The strain in the beam is measured using four strain gages; two of the gages are placed of the top surface of the beam, and two are placed on the bottom surface. The axes of the strain gages are aligned with the length of the beam. The strain gage bridge circuit is designed so the sensor self compensates for force misalignments and temperature effects.

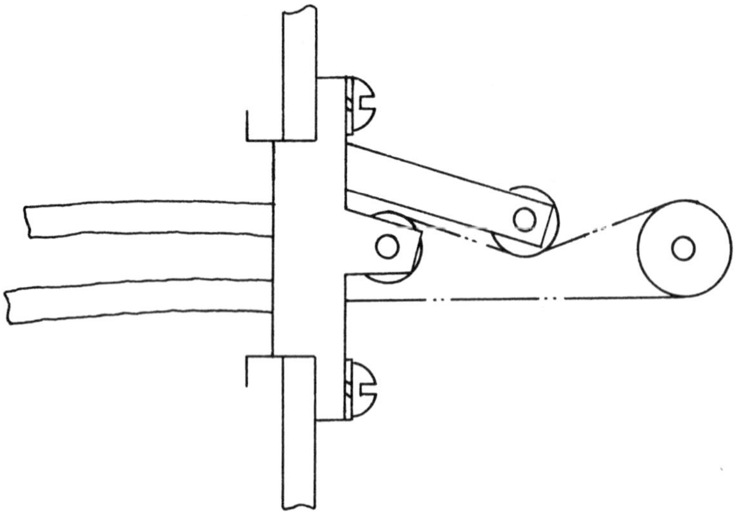

Figure 11. The Cable Tension Sensor.

CONCLUDING REMARKS

The design of a three fingered hand has been presented. The joints of the hand will be controlled in either force or position control modes. The motions of the fingers must be coordinated. Research along these lines are being conducted by Okada(6) and Salisbury(9), who have developed algorithms that coordinate the motions of three fingered hands. Future work will be focused on this problem. If an efficient means of coordinating the control of multi-fingered mechanical hands can be achieved, (it must balance the loading and avoid collisions) then the need for custom grippers to accomplish different tasks may be obviated, and a more flexible form of automation will be achieved.

BIBLIOGRAPHY

1 Lindstrom, G., Industrial Robots - Gripper Review, International Fluidics Services, LTD., Bedford, England, 1977.

2 Skinner, F., "Designing a Multiple Prehension Manipulator," Mechanical Engineering, Sept. 1975.

3 Okada, T. and Tsuchiya, S., "On a Versatile Finger System," 7th International Symposium on Industrial Robots, Oct. 1977.

4 Okada, T., "A Versatile End-Effector with Flexible Fingers," Robotics Age, Winter 1979.

5 Okada, T., "Object-Handling System for Manual Industry," IEEE Transactions on Systems, Man, and Cybernetics, Vol. SMC-9, No. 2, Feb. 1979.

6 Okada, T., "Computer Control of Multijointed Finger Systems," 6th International Conference on Artificial Intelligence, Tokyo, 1979.

7 Salisbury, J. and Craig, J., "Articulated Hands: Force Control and Kinematic Issues," The International Journal of Robotics Research, Vol. 1, No. 1, Spring 1982.

8 Salisbury, J. and Craig, J.,"Articulated Hands: Force Control and Kinematic Issues," 1981 Joint Automatic Control Conference.

9 Salisbury, J. and Roth, B., "Kinematic and Force Analysis of Articulated Mechanical Hands," ASME paper No. 82-DET-13, May 1982.

10 Takase, K., "Force Control of a Multi-Jointed Robot Arm," Robotics Age, Winter 1979.

11 Crossley, F. and Umholtz, F., "Design for a Three-Fingered Hand," Mechanism and Machine Theory, Vol. 12, 1977.

12 Edel, M. and Jolly, D., "Conception and Realization of an Intelligent and Autonomous End Effector," Robotics VI, Detroit, Mich., SME, 1981.

13 Kapandji, I., The Physiology of the Joints,Vol. I, Churchill Livingstone, New York, New York, 1970.

14 Chen, F. Y., "Gripping Mechanisms for Industrial Robots: An Overview," Mechanisms and Machine Theory, to be published.

15 Sage, S. and White, C., Optimum Systems Control, 2nd ed., Prentice Hall, 1977.

A Isostatic Six Component Force and Torque Sensor

Alain Gaillet
CERT
France

Claude Reboulet
CERT

Abstract

This paper presents an overview of a project dealing with the design and realization of a six-component force and torque sensor. The chosen mechanical structure includes a fixed and determined coupling between the different components of the sensor. The matrix, which transforms the measured forces into the applied sensor, depends only on the geometric characteristics of the sensor. The isostatic character of the structure results in numerous advantages for the sensor: an important reduction of the gaging phase for a sensor of this type. The sensitivity of the measurement to temperature variations is reduced to those of the sensing element. The internal stresses of the sensor material don't appear in the measurement. A modification of the measurement range may be obtained by changing the measurement range of sensing elements. The tooling of the sensor is very simple. The sensor's performance: 15gps sensitivity for a measurement range of 10kg, gives it an important role in the control loop for realizing an automatic insertion by a manipulator.

INTRODUCTION

The introduction of manipulators in some factories requires an extension of their possibilities beyong the tasks of pick and place, welding and painting. The automatic assembly of small clearance parts is one of these tasks and the manipulator which must achieve this task will process a partial intelligence which relates it to a real robot.

Numerous experimentations took place and it appears that the problems related to automatic assembly were of two types :

- the first one is technical and concerns the realization of a six component force and torque sensor for measuring the interaction forces between the two parts to assemble
- the second one, more theoretical, concerns the control of displacement of the wrist of the manipulator in the space of the task in response to forces and torques measured by the sensor.

Among the experiments, we can mention those of Goto (1), Watson and Drake (2), S.S.M.Wang (3), Groom (4), and those of Bejczy (5) and Nevins (6).

Most sensors are equiped with strain gauges stuck on a more or less complex mechanical structure. The design and the positionning of the strain gauges are calculated so as to uncouple as much as possible the forces and the torques applied on the gripper by assigning to each sensing element or to each pair of them the measurement of element of the tensor applied to the sensor.

This mechanical uncoupling simplifies theoretically the transfer matrix between the measured forces in the sensing gauges and the tensor. But in fact, the gauging of this type of sensor leads to add numerous correction terms available only by experimentation.

The design methodology of our sensor is based on a strong but known coupling between the different elements of the tensor. The choosen lattice work is inspired by the insertion module of McGallion (7).

BASIC STRUCTURE OF THE SENSOR

Figure 1 presents a rough drawing of the structure and the associated reference axes. The two disks A and B are joined by 6 articulated rods each carrying a measurement device. The development of the mechanical structure according to the directrixis presented on figure 2. It appears under the classical form of an isostatic lattice work.

The reaction forces on the articulations resulting from application of a force or a torque on the plate A are oriented in the direction of rods and measured by their sensing elements. The measurement of the stresses transmitted by the rods conversely provides the tensor applied on the A plate :

$$\vec{T}_{xyz} = M * \vec{F}_i$$

Where the Fi's are the vectors of the measured forces in the rods, the transfer matrix M between the forces Fi and the tensor T_{xyz} includes -for the forces : the projection coefficients of the forces Fi on the reference axes - for the torques : their lever arm with respect to their axes.

The values on the elements of matrix M depend only on the geometric characteristics of the sensor itself. In the prototype we realized, the triangles formed by two adjacents rods and the segments joining their articulations are equilateral and identical (see figure 2).

- the absence of gauging due to the nature of the coefficients of the matrix M,
- temperature sensitivity reflects only that of the transducers compensated in the signal processing device,
- a stability of the measurement because the structural deformations, thermic or others don't influence the outputs of the sensor when the geometrical model itself is not modified,
- an easy zero tuning of the sensor, reduced to those of the sensors on the rods.

The successive prototypes were tried on this testing bench which allowed us to evaluate the performances of the different versions, to detect their flaws and to correct them by changing the joints between disks and rods.

The swivel joints of the first prototype possess non linearities uncompatible with the required performances.

In order to find out these non linearities, we plotted one against the other on a x-y table, the analog outputs corresponding to the components Fx and Fy of the tensor.

When the sensor turns around its z axis, the end point of the force vector in the xoy plane ranges the locus represented on figure 5. The irregularities which appear on the locus correspond to changes of state of the swivel joints under applied forces.

Indeed with this first type of joint, the dry friction forces are important due to the nature of contacts between the swivel joint and the rotating part which surrounds it and the clearance between the two rotating parts.

These two types of mechanical non linearities imply also an hysteresis on the measurement. Following these experiments, we exchanged the swivel joints against ball bearing swivel joints. The clearance has been reduced but an important hysteresis always remains due to friction forces.

These experiments lead us to think again of the role of the joint between the rods carrying the force transducers and the two disks.

The theoretical swivel joints define a kind of joint such that any force whose direction don't coincide with the rotation center results in a rotation of the joint for canceling the created torque.

Now the deformations of the joint under applied forces is very weak (0.1 mm under 20 Dan compression force on each rod) and the swivel joints can be replaced by thin metallic threads oriented in the direction of the rods and housed in the disks (fig.6).

This device presents the following characteristics :

- almost infinite rigidity in the direction of the rod
- very weak rigidity in all other directions.

The setting up avoids the problems of clearance of swivel joint, the flexibility ensures a back of hysteresis. We must note that in fact very weak torques exist in this sort of joint. They can be minimized by being careful in the setting of flexible joints.

The remaining torques don't act as a drawback because they are translated into a shift of the sensor zeros which can be easily compensated.

$$
\begin{vmatrix} F_x \\ F_y \\ F_z \\ M_x \\ M_y \\ M_z \end{vmatrix}
=
\begin{vmatrix}
-\frac{R}{2\rho} & \frac{R}{\rho} & -\frac{R}{2\rho} & -\frac{R}{2\rho} & \frac{R}{\rho} & -\frac{R}{2\rho} \\
\frac{R\sqrt{3}}{\rho} & 0 & -\frac{R\sqrt{3}}{2\rho} & \frac{R\sqrt{3}}{2\rho} & 0 & -\frac{R\sqrt{3}}{2\rho} \\
\underline{h} & \text{"} & \text{"} & \text{"} & \text{"} & \text{"} \\
0 & -\frac{Rh\sqrt{3}}{2} & -\frac{Rh\sqrt{3}}{2} & \frac{R\sqrt{3}h}{2} & \frac{Rh\sqrt{3}}{2} & 0 \\
\frac{Rh}{\rho} & -\frac{Rh}{2\rho} & -\frac{Rh}{2\rho} & -\frac{Rh}{2\rho} & -\frac{Rh}{2\rho} & \frac{Rh}{\rho} \\
-\frac{R^2\sqrt{3}}{2\rho} & \frac{R^2\sqrt{3}}{2\rho} & -\frac{R^2\sqrt{3}}{2\rho} & +\frac{R^2\sqrt{3}}{2\rho} & -\frac{R^2\sqrt{3}}{2\rho} & \frac{R^2\sqrt{3}}{2\rho}
\end{vmatrix}
\cdot
\begin{vmatrix} F_1 \\ F_2 \\ F_3 \\ F_4 \\ F_5 \\ F_6 \end{vmatrix}
$$

Several symetrical elements appear in the matrix thus simplifying its computation and showing the absence of a more sensitive measurement axis.

The remark we can do is that the sensitivity of the components of the tensor is proportionnal to the radius of the sensor. This permits to size the sensor when knowing on the one hand the desired sensitivity of the sensor and its measurement range, on the other hand the sensitivity of the sensing elements on the rods and their measurement range.

And for a given structure, a change in the measurement range of the sensors on the rods involves a correlated change in the measurement range of the sensor.

REALIZATION AND TESTING

In a first stage, a set of plates was machined and on each rod joining them, a ready made one dimensionnal force sensor was inserted. On the first prototype, the theoretical swivel joints of figure 1 were realized by ready made swivel joints which are technical devices which best approximate the theoretical concept (figure 3). One can see on the picture of figure 3 that we inserted in the middle of the sensor, a compliant part like that of Nevins (6) which, when the sensor is mounted on the manipulator of the lab, will prevent possible overstresses on the six force sensors on the rods.

The first tests we have done have been made on this prototype. The six analog outputs of the signal processing device of the six sensors were connected to the analog inputs of a microprocessor Motorola 68000 on which the M matrix was previously programmed. Two kinds of outputs were available :

- digital outputs of the six components of the tensor
- analog outputs of two of them.

In a first serie of tests, the second ones were used. The testing bench is represented on figure 4. A fixed weight can be normally applied to the revolution axis of the sensor at different distances from the rotation center O of figure 1.

The acquisition and processing of the transducer's information is continuously made by the microprocessor.

The choosen isostatic structure involves several simplifications in the implementation of the sensor :

PERFORMANCES

The last version of our sensor has been tried on the testing bench ; the table of figure 7 shows the results we obtained for different orientations of the sensor and differents forces applied on it.

In the first trial for an applied force of 2 daN, the maximum deviation in a complete rotation of the sensor around its z axis is ±40gps, ±52Nm for the value of the torque.

Our trials also resulted, for any orientation, in a sensitivity of 15 gps in all directions and on the whole measurement range (10 kg).

This last characteristic is essential for resolving problems in the domain of automatic assembly. These performances added to a negligible hysteresis were obtained with force sensors whose sensitivity is 0.2% on a measurement range of 20 Dan.

CONCLUSION

Otherstrials are currently carried out on the sensor mounted on the manipulator of the lab. The two experimentations we plan now with the sensor are :

- the force and torque control : we will try to generate displacement and rotation of the wrist, the compliant part being active so as to apply a static and fixed force in a fixed direction on some object

- the contour's following : we will try to generate in paralell with the force and torque control, displacement and rotation of the wrist to follow the contour of a fixed object.

BIBLIOGRAPHY

(1) Goto T, Inoyama T. and Takeyasu K. (1974) - Precise insert operation by tactile controlled robot - Proceedings of the 2nd Conf. industrial Robot Technology Birmingham (UK), March 27-29.

(2) Watson P.C., Drake S.H. (1975) - Pedestral and wrist force sensor for automatic assembly - Proceedings of the 5th Int. Conf. Industrial Robots. ITT Research Institute, Chicago, pp 501-511

(3) Wang S.S., Will P.M. (1978) - Sensors of computer controlled mechanical assembly The Industrial Robot - March.

(4) Groome R.C. (1972) - Force feedback steering of a teleoperator system - Report n° T 575, Draper Laboratory, Cambridge.

(5) Bejczy K. (1980) - Smart sensor for smart hands - R.A. Beckenridge Editor, vol. 67 of Progress in Astronautics and Aeronautics.

(6) Whitney D.E., Nevins J.L. (1979) - The remote center compliance. What can it do ? Robotics today summer.

(7) McGallion H., Johnson G.R., Pham D.T. (1979) - Compliant device for inserting a peg in a hole - The Industrial Robot, June, pp 81-87.

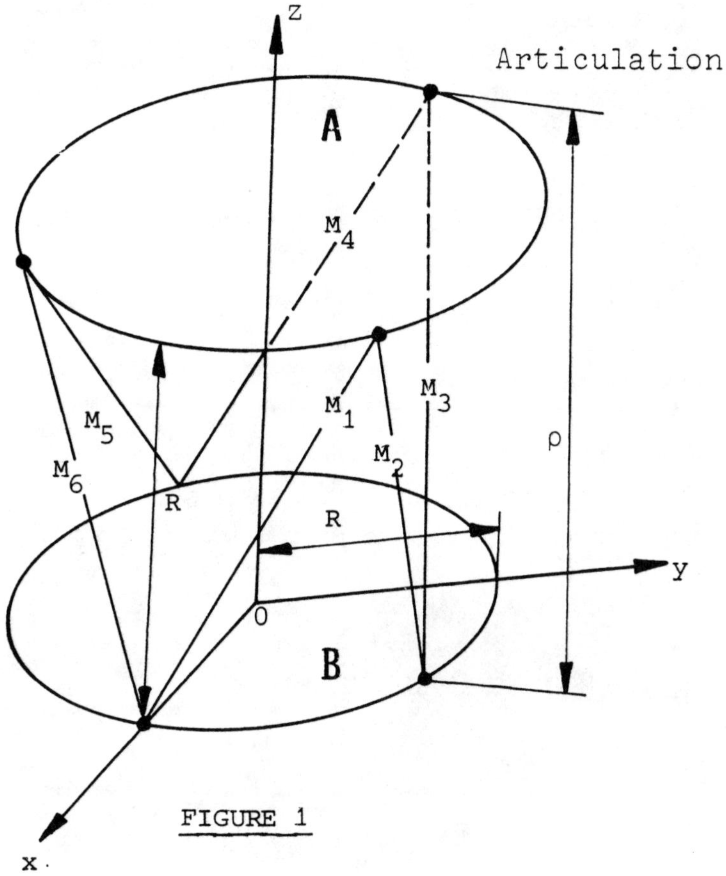

FIGURE 1

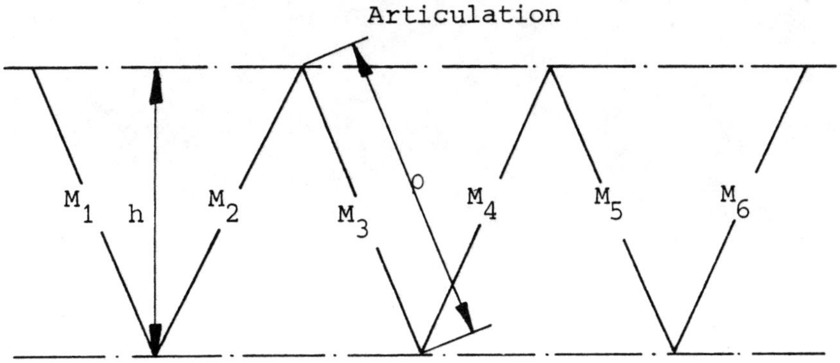

FIGURE 2

FIGURE 3

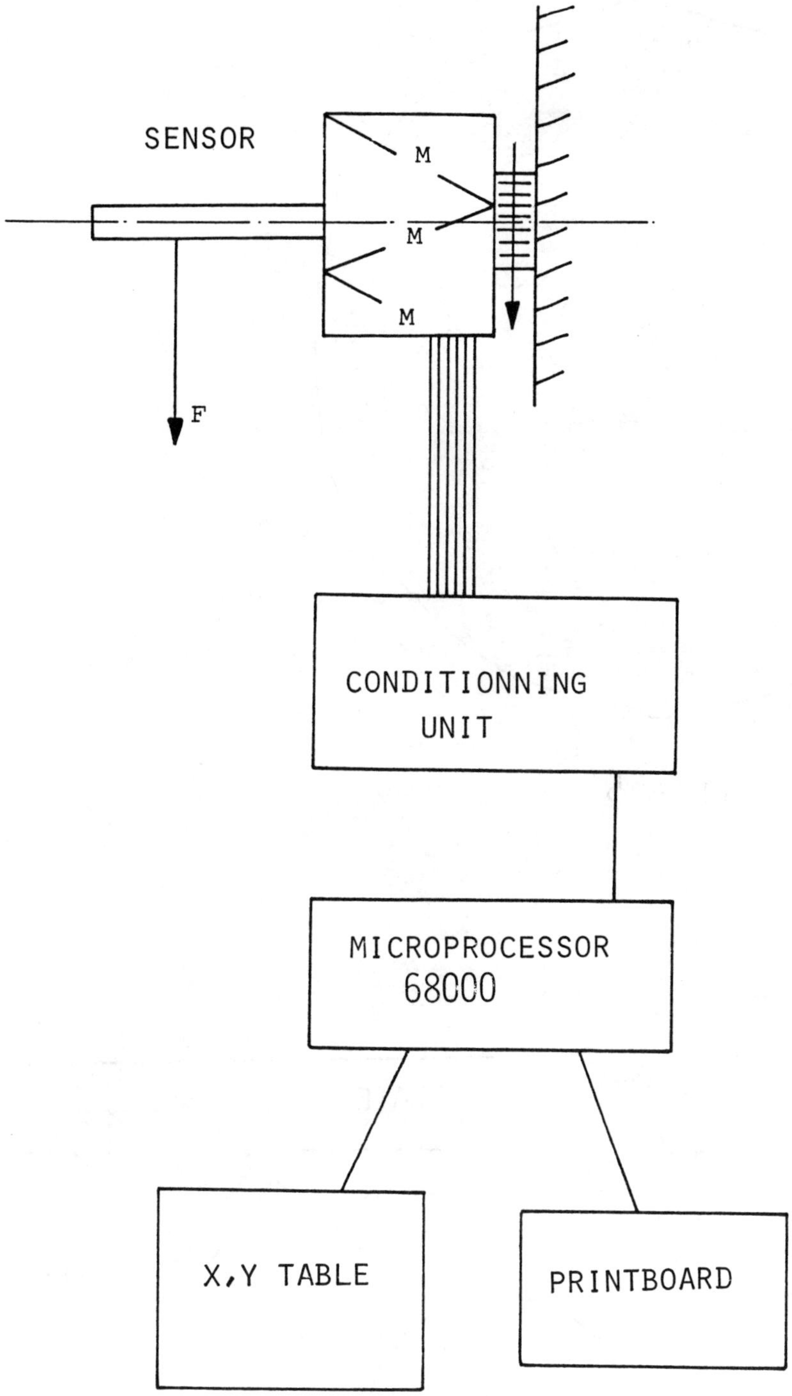
SENSOR
M
M
M
F
CONDITIONNING UNIT
MICROPROCESSOR 68000
X,Y TABLE
PRINTBOARD

FIGURE 4

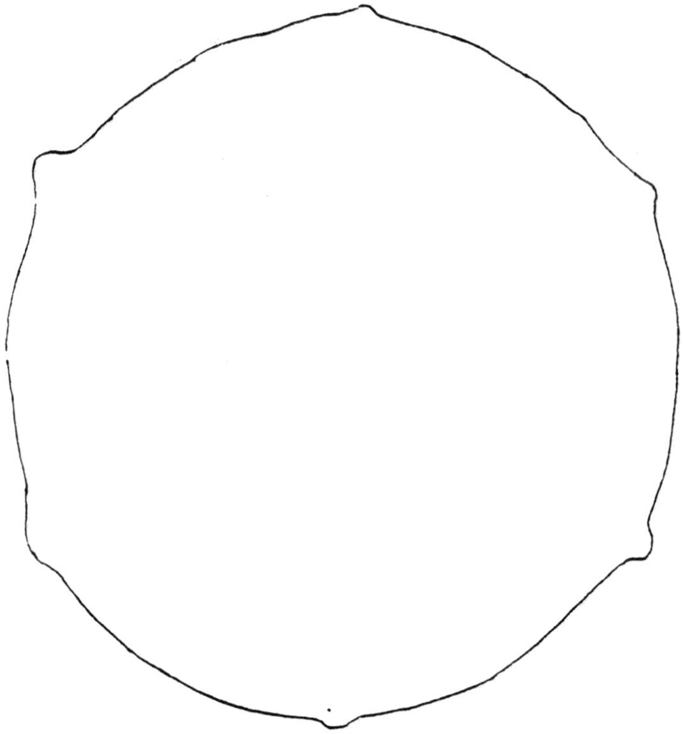

FIGURE 5

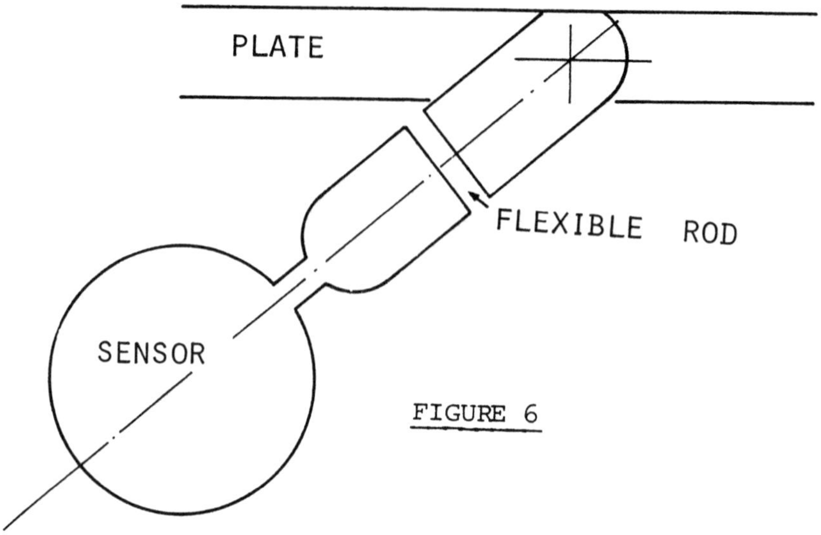

FIGURE 6

ESSAI 1 : APPLIED FORCE 1962 10^{-3} DAN - TORQUE APPLIED 7455 N.MM

Orientation	15	45	75	105	135	165	195	225	255	285	315	345
10^3 x measured force	1952	1974	1953	1929	1913	1918	1962	1985	1968	1933	1925	1939
measured torque	7474	7477	7422	7403	7363	7449	7473	7456	7427	7399	7428	7449

ESSAI 2 : APPLIED FORCE 1962 10^{-3} DAN - TORQUE APPLIED 2550 N.MM

Orientation	15	45	75	105	135	165	195	225	255	285	315	345
10^3 x measured force	1918	1936	1956	1940	1931	1916	1920	1940	1944	1947	1937	1931
measured torque	2412	2404	2419	2401	2409	2449	2435	2423	2435	2437	2454	2430

FIGURE 7

Evaluating the Effectiveness of Two-Finger Parallel Jaw Robotic Grippers

H. Steven Schafer
Iowa State University
Ames, Iowa

Eric M. Malstrom
Iowa State University

Abstract

This paper describes an evaluation of gripping effectiveness for a variety of standard geometric shapes using a Microbot Mini-Mover 5 robotic arm. The evaluation was performed on a two-finger, parallel jaw, robotic gripper. The effectiveness of this gripper was evaluated using a program designed to effectively lift and shake a variety of geometric shapes including a sphere, a cylinder, a rectangle, a variety of cones, and a variety of four-sided pyramids. The weight of each shape was systematically varied in small increments to determine the threshold at which part slippage occurred.

I. INTRODUCTION

The most common job the robot faces in industry today is the handling of pieces, parts, equipment, and materials. In all applications the use of the correct type of gripping mechanism is extremely important. No matter how adaptable its control system, how large its memory, or how many articulations it possesses, the robot's capability is ultimately determined by how well it handles objects. Generally speaking, grippers are designed to accommodate specific jobs or tasks. The variety in hand tooling attached to the wrist of the robot permits it to undertake a wide variety of tasks.

There are three main classifications of grippers: mechanical grippers, surface-lift devices, and tools (welders, spray guns, etc.). Finished products usually consist of parts which are geometric figures oriented in different ways. A robotic device must be capable of grasping these parts and positioning them properly.

In past robotic applications, it has been common procedure to design special gripping devices based on the specific, individual production applications. At present, the trend is to design grippers which are capable of performing a variety of tasks. The complete versatility of commercially available all-purpose grippers is somewhat questionable. Unfortunately, not much has been documented on the effectiveness and limitations of these grippers.

The principal thrust of the research documented in this paper has been to:

- Determine the types of standard geometric shapes that can be effectively grasped by a two-finger, parallel-type robotic gripper.
- Evaluate gripping effectiveness as a function of finger contact points on the shape being grasped.

The most common type of gripper available on today's market is the jaw-type gripper. This paper analyzes the limitations of a parallel-action jaw-type gripper with two opposing finger-like levers. A table-top robotic arm has been used to test various geometric figures in order to analyze the capabilities of this particular end-effector. This gripper is illustrated in Figure 1.*

II. EQUIPMENT DESCRIPTION

The primary equipment used in this analysis was a Microbot Mini-Mover 5 miniature robot and a variety of fabricated geometric shapes. Both equipment groups are described separately in the following sections.

*Used with permission of Microbot, Inc., Mountain View, California.

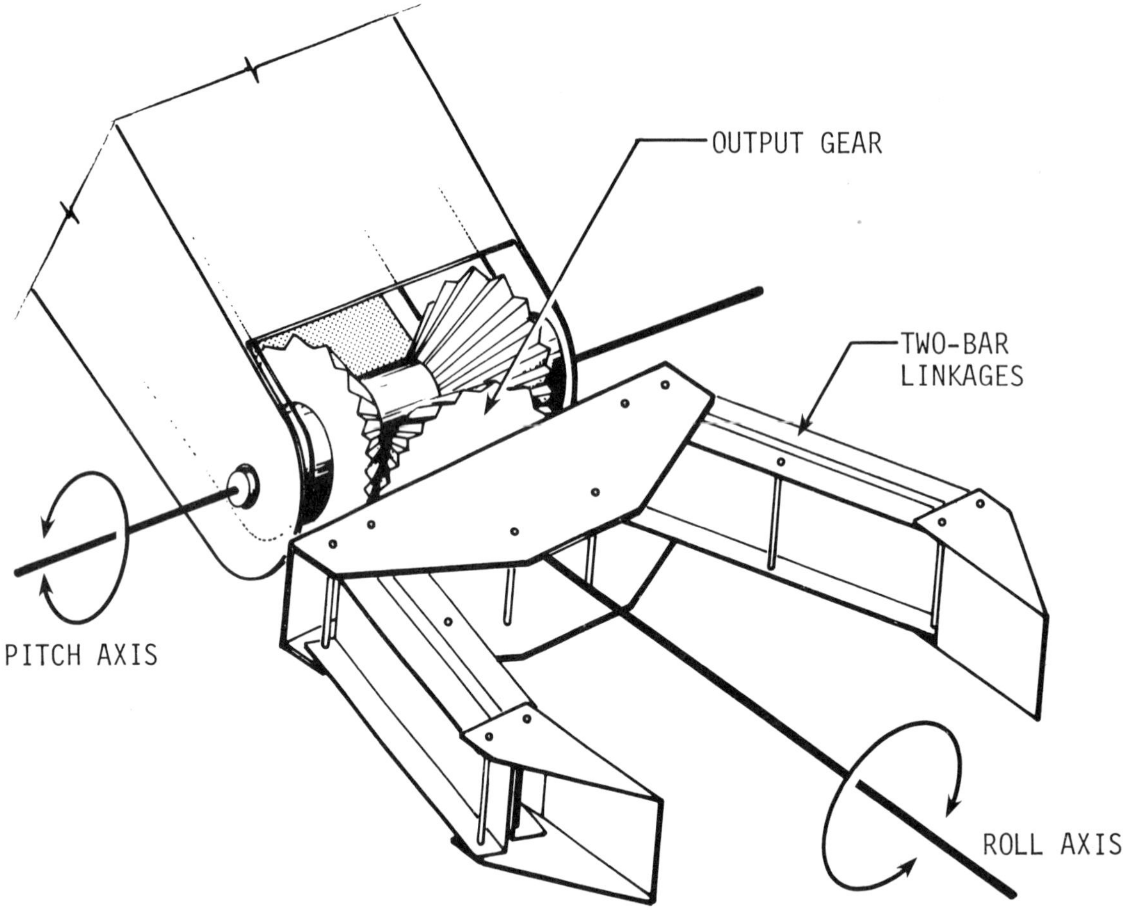

Figure 1. Two finger robotic gripper.

A. The Robotic System

The Microbot Mini-Mover 5 manipulator is a computer-controlled, five-jointed mechanical arm which has been designed to provide an unusual combination of dexterity at low cost [17]. Radio Shack's TRS-80 microcomputer (Model I with level II BASIC) was used to control the Mini-Mover 5 [27]. The arm is electrically connected to the computer, and ARM-BASIC, a special command language developed exclusively for the Mini-Mover 5, is loaded by cassette.

The Mini-Mover 5 has a lifting capacity of 8 ounces when fully extended. Each joint is controlled by a stepper motor mounted on the robot's body. The Mini-Mover 5's hand (gripper) has a maximum opening of 3 inches. A unique control system permits one stepper motor to control both the opening of the hand and its gripping force. A built-in sensor permits the gripping force to be controlled from 0 to 3 pounds. The sensor permits adaptive control so that the presence of an object as well as its size may be determined as the hand closes.

The motors and drive system are contained almost entirely within the robot's body. The stepper motors drive separate cable drums by means of individual gear reductions. Each articulation (base, shoulder, elbow, wrist and hand joints) is moved by cables attached to the cable drums. These cables are contained within the body.

To prevent slippage between the motor hubs (cable drums) and the cables, it is important that the rated load (8 ounces) not be exceeded. Excessive loads cause slippage and result in positioning and control errors.

B. Geometric Shapes

The geometric shapes that were analyzed included a sphere, a cylinder, a rectangular block, and a variety of pyramids and cones. All of the objects were fabricated from Balsa wood. Their overall size did not exceed 3 inches (the maximum gripper capacity of the Mini-Mover 5). Each was fabricated with an extremely smooth finish. This resulted in a minimal amount of friction between the gripping jaws and the object's surface.

The weight of each object was varied. This permitted a determination of gripping effectiveness as a function of part weight for a specific shape. Both of the following methods were used to accomplish this variation in weight. The first entailed filling a hollow cavity in each shape by inserting small lead weights, and then plugging the hole. In the second method, washers were fastened to a machine bolt which was attached to a plug that sealed the cavity. When each method was used, care was taken to place the weight at the centroid of each object.

The total weight of each object was varied up to 8 ounces. This is the rated maximum load for the Mini-Mover 5. The specific shapes that were tested included the following:

- A sphere (2-7/8 inch diameter)
- A cylinder (3 inches long by 3 inches in diameter)
- A rectangle (1 × 2 × 3 inches)
- Nine cones (each having one of the following angles: 10°, 20°, 30°, 40°, 50°, 60°, 70°, 80° and 90°)
- Five four-sided pyramids (each having one of the following angle combinations: 10°-20°, 30°-40°, 50°-60°, 70°-80° and 90°-90°). Each angle corresponds to each set of opposite sides on the pyramid.

III. PARAMETER SELECTION

Completion of the gripping analysis required that movement sequences and related speeds be defined. In addition, it was necessary to determine which additional variables were to be held constant from experiment to experiment. These topics are addressed in the following sections.

A. Motion Sequence

Both the weight of the geometric shape and the angle between the contact points of the gripper are critical in determining whether or not the gripper can handle a specific shape without dropping it (see Figure 2). Gripping effectiveness was evaluated in the following manner. Each shape was moved through a predetermined path using a computer routine called the "Shake Program." The program was designed to produce the most violent movements possible so that a critical situation was present in the transportation of the object. The robot moved each shape through this path for a 30 cycle run.

A complete shake pattern cycle consisted of five individual shake sequences. The first sequence moved the geometric shape from side to side, the second sequence moved the shape away from then toward the robot's base, and the third sequence moved the shape up and down.

In the above three sequences, the hand or gripper configuration was kept the same. The hand was pointed down, parallel to the front panel of the robot's body. In the last two shake sequences, the hand's configuration was changed, and the up and down motions of the third sequence were repeated. In the fourth sequence, the hand was parallel to the surface upon which the robot was placed. In the fifth and final sequence, the hand was in a vertical position perpendicular to the work surface.

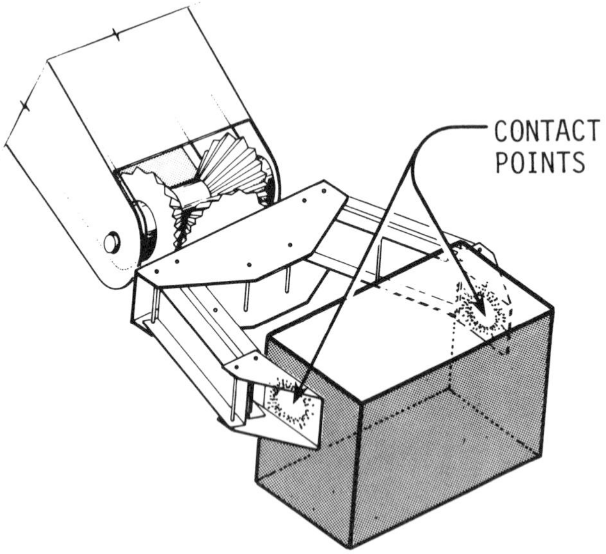

Figure 2a. Contact points for the rectangle.

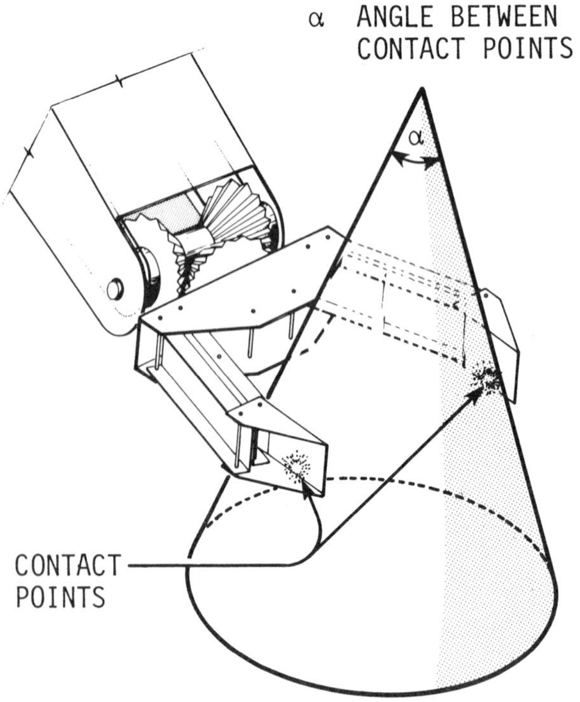

Figure 2b. Contact points for the cone.

Each of the five movement sequences was repeated five times within a shake cycle. The complete shake pattern consisted, therefore, of performing each of the five sequences five times. This complete shake pattern constituted one run or cycle.

The shake sequences were selected to place the most vigorous demands possible on the holding capability of the gripper. Repeating the first and second shake sequences for each of the hand configurations in the fourth and fifth sequences was felt to be redundant and was not performed in the gripping studies.

B. Speed Selection

The most difficult parameter to select in the data statements for the Shake Program was the speed setting. This parameter controlled the speed of all six stepper motors simultaneously for the specific movement of each data statement.

Selection of motor speeds was complicated by two factors. It was desirable to have the Shake Program produce the most violent movement path possible. Specifying arbitrarily high arm speeds was found to cause motor slippage at high load values. Consequently, the problem was to determine the maximum speeds possible at which slippage would not occur under conditions of maximum load.

The rated load used was an 8 ounce rectangle. This shape was used throughout the entire speed selection procedure. The procedure used to select the ultimate speed settings consisted of a two step process involving cursory and vernier speed selection. The cursory selection of the speeds entailed the use of a 5 cycle run of the Shake Program after each speed selection change. After each speed was evaluated, a check was made to determine whether slippage had occurred. A 30 cycle run was used to perform vernier speed adjustments. Checks for slippage were repeated. Successive trials were made until maximum no-slip speeds were found for each of the movement sequences. The selection procedure was largely one of trial and error and was one of the most time consuming operations in this research.

C. Analysis Variables

The variables that were held constant throughout the analysis include the following:

- The number of shake cycles executed for each shape
- The gripping pressure
- The type of gripper
- The type of gripping pads on the gripper

Both the surface of the shape and the surface of the gripping pads were cleaned prior to loading any shape. The gripping pressure was held constant at a maximum of 3 pounds. The standard parallel-lever-acting gripper that comes with the Mini-Mover 5 was used throughout the entire analysis.

Finally, flat rubber gripping pads were used instead of the "V" shaped pads provided with the Mini-Mover 5. The flat pads provided a larger gripping area for a greater variety of surface configurations.

The parameters which were varied included the shape's weight and the angle of contact for the specific geometric shape being analyzed.

IV. ANALYSIS PROCEDURE AND RESULTS

A. Gripper Approach

Each of the geometric shapes was grasped using the most natural approach possible. In other words, each shape was grasped in an orientation similar to that of a human hand in terms of initial placement position. The robotic hand was positioned downward while grasping the shapes.

B. Shape Variation

Each of the tests performed on the geometric shapes is described in the following sections.

1. Rectangle--For this shape a unique set of tests was performed. These tests consisted of grasping the rectangle in three different ways and varying the weight in each case. Grasping the rectangle by each of three sets of opposite sides constituted the gripping method. Each set of opposite sides was separated by a different distance (1, 2 and 3 inches). It was desirable to determine whether there was any significant difference in the gripping capability of the hand when the "closed position" distance between its fingers was varied.

For each test, the hand always grasped the shape in the center of the sides' faces. The gripping force was held at a maximum, and the weight was varied for each set of opposite sides. The tests performed on the other geometric shapes were similar. In each test, the angle of contact was varied. (The angle of contact is the the resultant angle produced by the two contact points which the gripper pads make on the shape's surface.) The weight was also varied.

2. Pyramid--To achieve a change in the angle of contact, various pyramids were needed. Five pyramids were fabricated, each with a combination of two angles. The angle combinations of the pyramids were 10°-20°, 30°-40°, 50°-60°, 70°-80° and 90°-90°. All together they produced nine different angles of contact: 10°, 20°, 30°, 40°, 50°, 60°, 70°, 80° and 90°. Each angle was formed by two opposite sides of the pyramid. The gripper contact points were positioned as close to each shape's centroid as possible for all cases.

3. Cone--The method employed in changing the angle of contact on the cone was similar to that used for the pyramids. Instead of five figures, nine were used, each with one of the following angles: 10°, 20°, 30°, 40°, 50°, 60°, 70°, 80° and 90°. The cylinder doubled as a 0° cone when set on its base. Again, gripper contact points were positioned as close as possible to the shapes' centroids for all cases.

4. Cylinder--All tests for this shape were performed while the cylinder was positioned on its side. The angles of contact tested were 10°, 20°, 30°, 40°, 50°, 60°, 70°, 80° and 90°. The gripper was positioned parallel to both bases, and it grasped the cylinder's sides at a point which bisected its length. Because of surface contour, the gripping pads had to grasp the cylinder by their outermost edges in order to achieve the proper angle of contact.

5. Sphere--The same procedures used for the cylinder were used on the sphere.

C. Test Procedure

Each of the shapes was analyzed separately. The Shake Program previously described was run for 30 cycles for each of the shapes. The weight of each shape was varied in increments of 0.25 ounces.

Small amounts of slippage on the surface of each shape could not be detected without some means of measuring the position of initial placement for the gripping pads. To achieve this measurement, a black line was drawn around the perimeter of the gripping pads with a non-permanent marking pen. The slightest slippage of the shape being held by the gripper could then be detected.

If for any reason a shift in shape position was detected after a 30 cycle run, it was considered a slip. When no slippage was observed, the weight increment of each shape was increased, and the Shake Program was re-executed. For all shapes except the rectangle, each analysis was repeated for ten different contact point angles.

D. Analysis Results

The slippage results for the rectangle are presented in Table 1. Inspection of the data shows that no slippage occurred, regardless of the weight transported or the distance between the gripper pads.

Table 1. Test results for the rectangle.

Size of Object Between Jaws (inches)	Maximum Weight Before Slippage Occurs (ounces)
1	8
2	8
3	8

Slippage results for all of the other shapes are presented in Table 2. Slippage is observed to increase with the angle between the contact points. As the contact angle was increased, the cylinder and the sphere became more difficult to grasp than the pyramids or the cones.

Table 2. Test results for the pyramid, cone, cylinder and sphere.

Angle Between Contact Points (degrees)	Maximum Weight Before Slippage Occurs (ounces)			
	Pyramid	Cone	Cylinder	Sphere
0	8.0	8.0	8.0	8.0
10	8.0	8.0	8.0	7.25
20	8.0	8.0	8.0	6.0
30	8.0	8.0	8.0	4.50
40	8.0	8.0	6.25	2.50
50	8.0	6.0	3.0	---
60	1.75	4.0	1.25	---
70	---	1.50	---	---
80	---	---	---	---
90	---	---	---	---

V. CONCLUSIONS AND EVALUATION OF RESULTS

Only one conclusion can be drawn from the test results for the rectangle. With a rated load of 8 ounces, there is no effect on the gripping capability of the robot's gripper when the distance between the gripper's fingers is varied.

The results of the tests for the other shapes require more interpretation. Inspection of the data in Table 2 reveals the following. First, the robot's gripper was able to lift the pyramid with a rated weight of 8 ounces at a greater contact angle (50°) than any of the other shapes. The gripper was able to lift an 8 ounce cone at a lesser contact angle (40°) than the pyramid, but still at a greater contact angle than for the other two shapes. At 8 ounce loads, the maximum contact angles successfully lifted were 30° and 0° for the cylinder and the sphere, respectively.

Lesser weights were lifted at contact angles greater than 50°. The largest obtained contact angle was 70° at a 1.5 ounce load for the cone. Maximum angles of 60° were obtained at 1.75 ounces for the pyramid and at 1.2 ounces for the cylinder.

Once slippage occurs, the gripper loses its gripping capability more rapidly on the pyramid than on any of the other shapes. If weight and the range of contact angle are considered, it may be concluded that initially pyramids and cones are the shapes most easily grasped. Changing the type of gripper or the type of gripping pads probably would have yielded quite different tests results.

There are several other commercially available, general purpose robotic grippers. Virtually no results have been previously published documenting similar analyses on the effectiveness of these devices. The need to complete such analyses for a variety of different grippers and robotic arms suggests the direction for extensions of the work completed in this research. These topics are currently under investigation by the authors.

VI. REFERENCES

1. Abraham, R. J. and J. F. Beres. "Cost Effective Programmable Assembly Systems." Proceedings of the First North American Industrial Robot Conference, Society of Manufacturing Engineers, Rosemont, Illinois, October 26-28, 1976.

2. Belcher, J. V., D. W. Hanify and K. G. Johnson. "Criteria for Establishing Industrial Robot Feasibility." Proceedings of the 4th International Symposium on Industrial Robots, JIRA, Tokyo, Japan, November 19-21, 1974.

3. Bublick T. "The Justification of an Industrial Robot." Proceedings of the Finishing '77 Conference and Exposition, Society of Manufacturing Engineers, Detroit, Michigan, October 18-20, 1977.

4. "Details of Construction, Performance, Prices and Applications." Industrial Robots--A Survey. International Fluids Services Ltd., Bedford, England, 1972.

5. Engelberger, J. F. "Robots Make Economical and Social Sense." Atlanta Economic Review, Vol. 27, No. 4 (July-August, 1977), pp. 4-7.

6. Friedrich, O. "The Robot Revolution." Time, Vol. 116, No. 23 (December 8, 1980), pp. 72-83.

7. Goldhamer, W. M. "A Systems Approach to Robot Use in Die Casting." Proceedings of the 8th International Symposium on Industrial Robots, International Fluids Services Ltd., Stuttgart, West Germany, May 5 - June 1, 1978.

8. Heroux, N. M. and G. Munson, Jr. "Robots Reduce Exposure to Some Industrial Hazards." National Safety News, Vol. 10, No. 7 (July 1974), pp. 82-88.

9. Hirose, S. and Y. Umetani. "The Development of a Soft Gripper for the Versatile Hand." Proceedings of the 7th International Symposium on Industrial Robots, JIRA, Tokyo, Japan, October 19-21, 1977.

10. Hohn, R. E. "Application Flexibility of a Computer-Controlled Industrial Robot." Proceedings of the First North American Industrial Robot Conference, Society of Manufacturing Engineers, Rosemont, Illinos, October 26-28, 1976.

11. "Industrial Robots: Better Than Man--Sometimes." Modern Material Handling, Vol. 35, No. 4 (April 1980), pp. 90-97.

12. "Investment Casting Takes a Step Toward Automation." Precision Metal, Vol. 32, No. 5 (May 1974), pp. 29-31.

13. Kjellberg, T. "A Method to Analyze Handling Problems in Robotic Applications." Proceedings of the 2nd Conference on Industrial Robot Technology, International Fluids Services Ltd., University of Birmingham, U.K., March 27-29, 1974.

14. Konstantinov, M. S. "Jaw-Type Gripper Mechanisms." Proceedings of the 5th International Symposium on Industrial Robots, Society of Manufacturing Engineers, IIT Research Institute, Chicago, Illinois, September 22-24, 1975.

15. Kusmierski, T. "Robot Application in Aerospace Batch Manufacturing." Proceedings of the Robot II Conference, Society of Manufacturing Engineers, Detroit, Michigan, October 31 - November 3, 1977.

16. Mattox, J. "Robots Don't Just Handle Things--They Do Things Too." Automation, Vol. 23, No. 11 (November 1976), pp. 69-72.

17. Mini-Mover 5: User Reference and Application Manual. Microbot, Menlo Park, California, 1980.

18. Mosher, R. S. "Handyman to Hardiman." Society of Automotive Engineers, Report Number 670088, January 1967.

19. "New Outlook on Industrial Robots." Material Handling Engineer, Vol. 29, No. 9 (September 1974), pp. 47-53.

20. Okada, T. "Analysis of the Motion and Handwork of Human Being's Fingers." Proceedings of the 4th Biomechanism Symposium, SOBIO, Tokyo, Japan, 1975.

21. Okada, T. and S. Tsuchiya. "On a Versatile Finger System." Proceedings of the 7th International Symposium on Industrial Robots, JIRA, Tokyo, Japan, October 19-21, 1977.

22. "Robot Welding of Car Bodies." Automotive Industries, Vol. 156, No. 4 (February 1, 1977), pp. 25-32.

23. Rovetta, A. "On Specific Problems of Design of Multipurpose Mechanical Hands in Industrial Robots." Proceedings of the 7th International Symposium on Industrial Robots, JIRA, Tokyo, Japan, October 19-21, 1977.

24. Skinner, F. "Multiple Prehension Hands for Assembly Robots." Proceedings of the 5th International Symposium on Industrial Robots, Society of Manufacturing Engineers, IIT Research Institute, Chicago, Illinois, September 22-24, 1975.

25. Tanner, W. R. "Basics of Robotics." Proceedings of the Robots II Conference, Society of Manufacturing Engineers, Detroit, Michigan, October 31 - November 3, 1977.

26. Tanner, W. R. "A User's Guide to Robot Applications." Proceedings of the First North American Robot Conference, Society of Manufacturing Engineers, Rosemont, Illinois, October 26-28, 1976.

27. TRS-80 Model III Operation and BASIC Language Reference Manual. Catalog No. 26-2112, Radio Shack/Tandy Corp., Ft. Worth, Texas, 1980.

28. Uhde, W. E. "Robot Automation--Today's Status." Proceedings of the 1978 International Engineering Conference, Society of Manufacturing Engineers, Philadelphia, Pennsylvania, May 8-12, 1978.

29. Von Muldau, H. "Greifer, die Programmierte Hand." Tagungsbericht. München: Verlag Moderne Industrie, November 1972.

CHAPTER 19

ROBOTIC EDUCATION AND TRAINING FORUM

Career Opportunities in Robotics and Education Requirements

William E. Weisel
Prab Robots, Inc.
Kalamazoo, Michigan

Abstract

This paper provides an overview of potential career opportunities in the field of robotics. Included are a brief description of general job categories that may be available, the types of potential employers, and the anticipated educational and experiential requirements for success in robotics careers.

Introduction

As the number of robots used in industry steadily increases, the potential need for people in jobs related to robots also increases. Getting people properly assigned to jobs they are prepared to effectively fulfill is an important element for successful robot application and utilization.

Let's look at where these jobs may be available, what they are, and what type educational and experiential background is needed.

Several types of employers may have potential jobs related to robots. The major areas are robot users and robot manufacturers. Other employers are in services related to robots; such as end-of-arm tooling, controls, components, marketing, education, and a variety of consultants.

The number of companies utilizing robots will continue to grow rapidly. Robots are being integrated into the manufacturing systems of almost every kind of industry. This growth will result in demand for retraining or adding people for involvement with robots.

Manufacturers of robots are another source of robot related employment. While the number of robot manufacturers is considerably less than that of the users, the variety of job assignments may be somewhat greater.

Another source of employment related to robotics is in the field of education. Many educational institutions are expanding their program content to include instruction and courses in robot related subject matter. Teaching personnel with education/experience in the field of robotics may be required.

Employment assignments in related services will be mentioned in subsequent descriptions of each job area.

The following general categories of job assignments can be foreseen in the field of robotics:

- Research/Development
- Design
- Manufacture
- Application
- Marketing
- Service/Maintenance
- Training/Teaching

Each of these assignments will be briefly reviewed and an opinion given of the educational background required and which type of employer may have needs.

Before making suggestions concerning educational background requirements for each job assignment, it seems imperative that consideration be given to the difference in individuals. The ability of each person is a result of their past education and experience. How the individual is able (or willing) to apply that knowledge may result in equivalent capability in individuals with dissimilar background. Conversely, those with similar backgrounds may perform vastly different in the same type work environment. Therefore, the general criteria outlined in each category must be considered just that --- general criteria.

It also should be understood that a listed degree level and program is meant to cover the basic knowledge that is gained from such a degree. Equivalent background knowledge acquired through formalized study and work experience may prepare an individual to successfully accomplish required results, without a formal degree.

Each of the potential job categories will now be reviewed and brief comments made about educational requirements.

Research/Development

Activities in research/development may take place at robot manufacturers, robot users, educational institutions, and consulting firms devoted to R & D in robotics.

Since research may range from physical research through applied research, a wide variety of educational background may be required for experimental analysis in all aspects of robotics. Advanced degrees will probably be preferred, particularly in educational institutions. Bachelor's degree level may be acceptable if the individual has knowledge unique to the specific area of research. The subject field will primarily be in the engineering or computer sciences related discplines.

Design

Included in the design category are job assignments in the design of the basic robot, the control, and related equipment.

Design of the basic robot will be done by the robot manufacturer. Educational background should include the broad base normally found in the BS level of mechanical engineering; such as the basic sciences, mechanics, materials, electrical/electronics, thermodynamics, fluid mechanics, design courses, and the general development of problem solving abilities.

Control design may be done by the robot manufacturer (who also produces controls), or by a manufacturer of controls. Educational background should be at the BS or MS level in electrical/electronic/control engineering. Application or modification of controls may be the responsibility of either the robot or control manufacturer; or in some instances, the robot user. Based on the complexity of the application or modification, a Bachelor's degree in electrical/electronics engineering/technology, or Associate degree level in that field with related work experience, may be sufficient for those aspects of design.

Design of the related equipment required to integrate a robot into a manufacturing system may be made by the robot manufacturer, the user, or by consulting firms and equipment suppliers. Design personnel should have educational background ranging from Associate degree level on up to Bachelor degree competency in engineering or technology fields of mechanical, electrical, industrial, or manufacturing. Prior experience in material handling, interface mechanisms, safety equipment, and other related manufacturing elements will be useful.

End-of-arm tooling will be designed by the robot manufacturer, the user, or by firms devoted to that business. Education requirements parallel that listed for related equipment design, except that design personnel will need greater depth of knowledge about the broad variety of applications of robots in many industries.

Manufacture

When considering educational background requirements for successfully fulfilling potential jobs in the field of manufacturing, two different points of view are necessary. One is in the process of manufacturing the robot and related equipment. The other is the utilization of the robot as an element of the manufacturing system within the plant.

Producing robots requires similar knowledge and skills as that required for production of machine tools, material handling equipment, and other types of capital goods. The knowledge and skills must be applied to the efficient and effective utilization of human resources, capital, time, materials, facilities, and energy.

A wide spectrum of educational background is necessary to accomplish the overall manufacturing task. That spectrum ranges from limited formal education beyond high school, through advanced degrees.

For example, in part manufacture, a simple lathe producing standardized parts may be operated by an individual with on the job training only. More complex machining processes, producing specialized parts, may be more effectively operated by an individual who has completed a certificate or Associate degree program in machine tool technology.

Individuals responsible for creating, setting up, and managing the overall part manufacturing system will be better prepared to meet that challenge if they have completed a BS/MS level degree program in manufacturing or industrial engineering.

Once again, we need to keep in mind that each individual is unique and applies their knowledge and skills in differing degrees of effectiveness.

Assembly of the robot will be done by the robot manufacturer. Educational requirements for assembly depends upon the complexity and special content of the robot. An individual assembling a robot with special content and complex modifications needs knowledge about interpreting engineering drawings, electrical diagrams, and other engineering instructions. That knowledge is available through appropriate associate degree programs; in fields such as mechanical, industrial, or electronics engineering technology.

When the robot is completely assembled, it will go through a test period where all elements will be run under power to determine if the system or robot and related equipment is operating properly. Individuals involved with that activity generally need the testing and instrumentation knowledge obtained by completion of an engineering technology associate degree level program.

Installation of the robot into an operational manufacturing system is primarily done by the user. When required by the user, robot manufacturer personnel will assist and direct installation procedures. In either case, the individual involved will usually require education at the associate degree level to analyze and solve problems that may arise in installation and start up. If the problem is highly complex or unusual, a higher level of educational background may be required.

While service and maintenance are part of the on-going manufacturing process, those subjects will be covered later in a separate category.

Application

Work related to planning and incorporating robots to the manufacturing process may be accomplished by the robot manufacturer, the user, or consulting firms in that field. Close coordination between the user, the robot manufacturer, and others supplying equipment or direction is vital for a successful application.

As previously stated in the manufacturing category, individuals planning and implementing an overall system should have the educational equivalent of a BS/MS level in industrial or manufacturing engineering. In addition, in-plant work experience will be vital.

Application design and/or specification of related equipment such as end-of-arm tooling, conveyors, pallets, positioning equipment, interface devices, etc. will require knowledge obtained from associate degree through BS level programs in engineering and technology. As the complexity and specialized nature of the application increases the need for greater educational and experiential background increases.

For example, applying a robot to a system for welding car bodies will require specific knowledge about the welding process, in addition to that education necessary to apply the robot. On the other hand, applying the robot to an automated system for processing investment castings calls for some background knowledge about that field.

Individuals involved in the application of robots may supplement their formal education and work experience by participating in appropriate robot related seminars, workshops, and classes that are now available. The Society of Manufacturing Engineers and Robotics International of SME, universities, and others conduct these sessions. There is a wide variety of subjects, such as Introduction to Robotics, End-Of-Arm Tooling, Applying Sensors, Machine Loading/Material Handling, and others. An increasing number of universities and community colleges are adding robotic subject matter to their programs. Such classes will be helpful for those responsible for applying robots.

Justification of the robot is another area of consideration. This activity requires an ability to analyze the basic cost factors involved and to include the impact of the robot application on productivity, quality, human factors, and safety. Educational background should be parallel that of a BS degree program in industrial engineering.

In any application of a robot to a manufacturing system, controls and programming are major areas of attention. The complexity may range from a simple application to a very sophisticated computer- integrated manufacturing system. Obviously, the educational level required to cope with control and programming problems ranges from the associate degree level through the BS/MS level in electrical/electronics engineering/technology.

In Summary, individuals engaged in the process of applying robots may require educational background and experience in a wide variety of degree levels and technology disciplines.

Marketing

Marketing of robots is a function of the robot manufacturers or their assigned sales representatives and distributors. Personnel engaged in the selling of robots and robot systems should have several fundamental strengths.

In-depth knowledge about robotics and a broad understanding of manufacturing processes are essential. That background can only be obtained by on-the-job involvement. Ideally, the individual should spend adequate time in the shop, in the design department, and in field service activites. The more hands-on experience the person has, the better prepared the person will be for marketing.

An associate degree in the engineering technolgy fields of mechanical, industrial, or electronics is preferred as a minimum. Additional courses related to human relations and communications will be most helpful. Obviously, an individual in marketing must be a self-starter and have above average skills in dealing with others. Ability to effectively communicate orally and in writing is of extreme importance.

Successful marketing requires an individual with a combination of strong technical competence and positive personality traits. Experience and education must be properly blended so that the person is competent to approach prospective buyers in a wide variety of industries with a complex diversity of robot applications.

Another phase of marketing is the field of advertising. Robot manufacturers, consulting firms, or ad agencies are potential employers for this function. Educationally, an associate degree or more is preferred in an adevertising related program. On-the-job involvement with robots will help give the individual necessary background knowledge.

Service/Maintenance

The terms service and maintenance used here are meant to indicate the work that is necessary to assure that the robot will continue to function properly, to solve and correct operational problems that may arise, or make adjustments and minor modifications.

Robot manufacturers have service personnel that are trained to analyze and rectify robot problems that may occur in the users' plants. These individuals must understand the entire robot system including the basic structure, power source, control programming, end-of-arm tooling, and other associated interfaces.

Educational background required is genrally a minimum of an associate degree in mechanical or electronics technology. Since the service person will need to be knowledgeable about the overall system, formal education must be supplemented by in-depth on the job training and experience. Those with education in the mechanical field will need to gain electronics and control background. Conversely, electronics graduates need to become well acquainted with the basic mechanical features of the robot.

Human relations are also very important in field service work, since the individual will be working with a wide variety of people at all levels and usually has to make decisions independently. Having completed courses in social studies and humanities will be helpful to the service person.

In addition to the service work previously outlined, the service personnel of a robot manufacturer may also be involved in the original installation of the robot or robotic system in the plant of the user.

Thos companies with a limited number of robots will probably take advantage of the broad experiential background of their existing maintenance personnel to take care of robot problems. Companies having large robot installations may assign maintenance personnel to robots only. In either case, it is necessary to broaden the individual's capabilities to include robots by giving them additional training associated with robotics. Educational background and experience should be at least equivalent to associate degree level, particularly as related to controls. Graduates of technical associate degree programs with robotic content may be good candidates for this type activity.

Training/Teaching

Educational activities concerning robotics are conducted by robot manufacturers, users, educational institutions, technical societies, and films specializing in robotic education. A brief description follows of the function of each area and the personnel involved.

Robot manufacturers are responsible for two types of training on their robots. They provide training for user personnel and for their own employees. Training for users is generally aimed at overall understanding of operation, maintenance, programming, and controls. Classes may be aimed specifically at one of these functions if appropriate. This instruction consists of classroom discussion, visual aids, documentation, and hands-on involvement with the robot. Training for employees of robot manufacturers will be on-the-job training supplemented with the same instructions given to user personnel.

The instructors for this training are usually employees who have gained broad knowledge about the product through active work experience with robot applications. Their formal education is likely to be a technical associate degree or more, plus in-depth involvement with the product.

Based on how many robot installations a user has, or anticipates, the training may be done in the users plant by one of the users employees. In most plants with major robot installations, a training function already exists in the company. Existing instructors may broaden their knowledge about robotics by attending classes and seminars. In many instances, these instructors have similar educational and experiential backgrounds to that just described for manufacturers instructors.

Multi-plant operations may send employees to a central training center for instruction or the instructor may go to the plant location. Due to turnover, the instructors probably have to repeat classes periodically.

Teaching robotics in educational institutions may range from an introductory overview at the beginning level to programs leading to the doctorate at the university level. At the various stages, the focus on robots can be related to the subject being taught. For example, in classes on Fluid Mechanics, Design, Controls, etc., the basic principles of the subject could be applied to the robot as a representative mechanism.

It seems unnecessary to outline requirements for teaching at various levels in educational institutions, since that is fairly well standardized. Hopefully, those instructing students on robotics will have had some real-time exposure to robots and robotic systems. Ideally, persons teaching robotics will have first hand exposure to robots, or will have done an effective job of correlating their specialty to the field of robotic applications.

Many schools are successfully using knowledgeable individuals from industry as adjunct professors. Their background dealing with operating robotic systems is proving very valuable in the teaching of robotics.

One of the needs being fulfilled by educational institutions is the need for retraining of those workers seeking to get involved in robotics. Included may be either presently employed workers or those who are unemployed. In some instances, a professor has gone to an employers plant and conducted a class on robotics for employees.

Robotics International of SME sponsors seminars and workshops on robotic subject matter. Those individuals conducting the sessions are either industrial representatives who have background and experience in robotics or are from firms that include education on robotics as one of their activities. Similar workshops and seminars are also conducted by educational institutions using their staff people and often include industry representatives as instructors.

Summary

The number of robot installations and robotic systems is steadily increasing. That growth will create more opportunities for people to work in the field of robotics. A wide variety of job assignments will exist throughout business, industry, and education. Personnel filling these varied career opportunities will need education, experience, and in-depth knowledge about robotics. While traditional programs in technology and engineering will provide a base for the various job categories, subject matter and lab work related to robotics should be incorporated. An increasing number of educational institutions are providing robotic course content.

These courses are available for not only full time students but for those workers interested in continuing their skills and employment potential or for career advancement.

Potential employers for those interested in robotics are robot users, robot manufacturers, educational institutions, and those companies providing goods and services associated with robotics.

Getting the right person in the right job, with sufficient background for success, remains a challenge. That challenge will be met through the coordinated efforts of educational institutions and business and industry in the field of robotics.

Automated Manufacturing—A Systems Training Concept

James A. Rehg
Piedmont Technology Center
Greenwood, South Carolina

Richard Upchurch
Piedmont Technology Center

Abstract

The State Board for Technical and Comprehensive Education in South Carolina initiated a plan to develop high technology training in five of the two-year technical colleges. As a part of this plan, called "Design for the 80's," Piedmont Technical College was selected as the site for the Center for Robotics. This paper describes the development of a two-year program in Automated Manufacturing Technology and the development of a training laboratory with 11 robots integrated into 10 instructional work cells. The identification of instructional competencies for a robot maintenance technician and robot application technician are included.

INTRODUCTION

In the spring of 1981 the State Board for Technical and Comprehensive Education in the state of South Carolina initiated a plan called Design for the 80's. Under this plan Piedmont Technical College in Greenwood, South Carolina, was designated as the Innovative Center for Robotics. In addition to Piedmont Tec five other schools in the South Carolina technical education system were identified as high technology centers to form a network of training programs which could provide high technology training. The primary training goal for all of the centers is the upgrading of two-year college instructors in the areas of robotics, microelectronics, computers, advanced machine tools and advanced office practices. In addition, specialized training programs for industry are offered, and curriculum programs in each high tech area are being developed for students with career interests.

This paper describes the development of the curriculum program, titled Automated Manufacturing Technology, to be offered by the Center for Robotics and will include a description of the laboratory used in the training of faculty, industrial representatives and curriculum students. The program is structured as an Associate of Applied Science degree with a two-year or six quarter time frame.

COURSE DEVELOPMENT

COMPETENCY IDENTIFICATION

One of the most difficult tasks in developing a new program is the identification of the necessary skills which participants completing the program must possess. After these skills, often called entry level competencies, are identified, a series of learning activities must be written which train for each skill required. These learning activities are then structured into a set of courses which constitute a curriculum. It is often possible to shorten this three-step process by using programs already in operation or data already gathered by other institutions.

In developing the program for the Robotics Center the first activity included a search for programs currently in operation at other two-year colleges, which could serve as a model for the curriculum planned at Piedmont. The search revealed only one program currently in operation and very little data with which to build program development. As a result a competency identification process was initiated called DACUM which stands for Develop A CurriculUM. The process includes the development of a competency matrix using experts from industry and a DACUM facilitator in a brain-storming session. The matrices developed for two types of robotics technicians, using this process, are illustrated in Table 1 and Table 2. The two jobs described are Robotics Maintenance Technician and Robotics Applications Technician. The matrix is structured with the major competencies listed down the left hand column and subcompentencies for each major competency listed from left to right in rows next to the major competency. The industry representatives participating in the DACUM process are listed at the bottom of Table 1 and 2.

The primary advantage of the DACUM process is its ability to develop competency lists for industrial job situations in a very short period of time. In addition, the validity of the competencies is established by active discussion amoung the participants. However, one disadvantage is that the competency statements are very brief and general in nature, which can be misleading when the statements are used to guide the development of training material. To overcome this problem the industrial participants were questioned after the process was completed by members of the Robotics Center about the meaning of each statement. The results of that interview are in Appendix A and include a staff interpretation of the competency statements which were judged to be too general or needing clarification.

COMPETENCY ANALYSIS

The results of the DACUM process were shared with five robot manufacturers and ten manufacturers using robots in production applications. The feedback from the reviewers can be summarized as follows:

- The entries were valid and would be supported by a majority of the persons reviewing the matrices.

- The Robot Maintenance Technician training program would produce a technician needed by the robot manufacturers. However, because the training does not include competencies directed at system level installation, maintenance and repair, the program would be too narrow for training robotic technicians needed for high technology work cells in manufacturing installations.

- The training of Robot Application Technicians in a two-year time frame was questioned. This was due to the broad training background required for an application expert in robotics and the responsibility normally placed on such an individual in industry.

- The competencies listed in the applications matrix were representitive of the type of skills needed in manufacturing situations where complete production systems are integrated to produce a product.

- The competencies from the maintenance technician

A DACUM MATRIX OF OCCUPATIONAL COMPETENCIES

ROBOTICS SERVICE TECHNICIAN: A Robotics Service Technician will: Identify problems and take appropriate action to remedy the problem quickly.

A. Communicate	A-1 Communicate Orally	A-2 Listen	A-3 Read Effectively	A-4 Write Technical Reports	A-5 Train	A-6 Follow Up		
B. Implement & Maintain Robot	B-1 Develop Thorough Operating Ability Of Robot	B-2 Practice Safety	B-3 Evaluate Robot Performance	B-4 Apply Programming Techniques	B-5 Implement and Maintain End Effectors	B-6 Implement and Maintain Sensors	B-7 Implement a Preventive Maintenance Program	
C. Troubleshoot Electronics Systems	C-1 Demonstrate Electrical Fundamentals	C-2 Demonstrate Electronic Fundamentals	C-3 Demonstrate Computer Fundamentals	C-4 Demonstrate Programmable Controller Fundamentals	C-5 Demonstrate Servo System Fundamentals	C-6 Read Electrical Schematics	C-7 Demonstrate Electronic Test Equipment	C-8 Utilize Diagnostic Aids
D. Troubleshoot Hydraulic and Pnuematic Systems	D-1 Demonstrate Hydraulic Fundamentals	D-2 Demonstrate Pnuematic Fundamentals	D-3 Demonstrate Servo System Fundamentals	D-4 Read Schematics	D-5 Operate Test Equipment			
E. Implement and Maintain Interface Elements	E-1 Practice Safety	E-2 Evaluate System Performance	E-3 Implement and Maintain Peripheral Equipment	E-4 Implement and Maintain Programmable Controller	E-5 Implement and Maintain Communications Links	E-6 Implement and Maintain Interlocks	E-7 Implement a Preventive Maintenance Program	
F. Utilize Problem Solving Skills	F-1 Apply Problem Solving Logic	F-2 Apply Statistical Methods	F-3 Maintain Records	F-4 Determine Cause of Failure				

DACUM CONSULTANT

Jack Harris, Stark Technical College

PANEL MEMBERS

Wayne Mechlin, Cincinnati Milacron
L. E. Ording, Ford Motor Company
Dick Wetherill, Caterpillar Tractor Co.

Sponsored by the S. C. State Board for Technical and Comprehensive Education and Piedmont Technical College.

A DACUM MATRIX OF OCCUPATIONAL COMPETENCIES

ROBOTICS APPLICATION TECHNICIAN: A Robotics Application Technician will: Identify, analyze, justify, acquire, and implement the robotic needs of the company.

A. Communicate	A-1 Communicate Orally	A-2 Listen	A-3 Sell self	A-4 Sell concept	A-5 Read effectively	A-6 Maintain records	A-7 Supervise
	A-8 Train	A-9 Write Technical Reports	A-10 Write business correspondence	A-11 Present effective presentations	A-12 Utilize A-V Material	A-13 Follow up	A-14 Apply Decision - making processes
B. Explain Manufacturing Processes	B-1 Explain Arc Welding Processes	B-2 Explain Resistance Welding	B-3 Explain Printing & Coating Processes	B-4 Explain material handling	B-5 Explain Assembly Processes	B-6 Explain Grinding & Finishing Processes	B-7 Explain Basic Machine Tool Operations
C. Explain Automation	C-1 Explain Batch and Dedi-cated processes	C-2 Explain In-line and Off-line processes	C-3 Explain Synchronous & Non-Synchronous Automation Systems	C-4 Explain Material Handling Automation	C-5 Explain Adaptive Processes	C-6 Explain CAD/CAM	C-7 Explain Flexible Automation
D. Utilize Robots	D-1 Practice safety	D-2 Analyze job requirements	D-3 Analyze robot capabilities	D-4 Analyze Robot Control Capabilities	D-5 Analyze Teaching/Program-ming Methods	D-6 Analyze Robot Interfacing Capabilities	D-7 Select appropriate Robots
	D-8 Select appropriate End Effecters	D-9 Select Peripheral Equipment	D-10 Apply Optimum Programming Techniques	D-11 Develop Preventive Main-tenance Program			
E. Interface Systems	E-1 Converse about Electronic Fundamentals	E-2 Converse about Pneumatics	E-3 Converse about Hydraulics	E-4 Converse about Computer Control Systems	E-5 Converse about Programmable Controllers	E-6 Utilize Computer Languages	E-7 Explain Communications Links
	E-8 Utilize contact & non-contact sensors	E-9 Apply ergomic interface	E-10 Utilize fail-safe practices				

DACUM CONSULTANT
Jack Harris, Stark Technical College

PANEL MEMBERS
Wayne Michlin, Cincinnati Milacron
L. E. Ording, Ford Motor Company
Dick Wetherill, Caterpillar Co.

Sponsored by the S. C. State Board for Technical and Comprehensive Education and Piedmont Technical College.

F. Conceptualize Systems	F-1 Analyze the work Environment	F-2 Establish a goal	F-3 Invent a concept	F-4 Plan the process	F-5 Design and Lay Out Process	F-6 Perform a time and motion study
	F-7 Estimate cost	F-8 Evaluate financial return	F-9 Evaluate practicality of project	F-10 Promote the project		
G. Interface Systems	G-1 Develop specifications	G-2 Assist in evaluation of quotations	G-3 Create project implementation plan	G-4 Coordinate plan elements	G-5 Try out and debug the system	G-6 Acquire System acceptance

and application technician matrices could be combined and the emphasis shifted to operation, installation, maintenance, repair and trouble-shooting of automated manufacturing systems. The result would be a technician in demand by both the user and the manufacturer of robots.

Using the DACUM competencies and the input from both industrial users and manufacturers of robots, a curriculum was designed to incorporate as many of the elements of the totally automated factory as possible.

CURRICULUM DESIGN

The proposed curriculum will train technicians to install, maintain and service the high technology systems which are currently being used in new automated manufacturing facilities and to prepare technicians for the systems which are planned for the factory of the future. The graduates of the program will be prepared to install, operate, analyze, troubleshoot and repair the following automated systems:

- Advanced manufacturing systems used in the shaping, forming and processing of raw materials into finished parts and products.

- Advanced assembly systems using automation, vision and robots to assemble parts into finished products.

- Automatic material handling systems which include conveyors, manless parts vehicles, robots and transporters.

- Automated warehousing and parts storage systems.

- Data communication networks of computers which include machine controllers, robot controllers, work cell computers, inventory control computers and manufacturing computers.

In addition to training to meet the needs defined above, the curriculum was also required to meet the following guidelines:

- The sequence of courses used in the program must cover the conpetencies identified by the DACUM process.

- The program must provide hands-on experience by emphasizing the use of the robotics laboratory and on-site training.

- The program must have a co-operative education segment.

- The program must meet the requirements necessary for ABET (Accreditation Board for Engineering Technology), accreditation or Certification by SME (Society of Manufacturing Engineers) and RI (Robotics International).

- The program must satisfy the requirements of the State Board for Technical and Comprehensive Education and the Commission on Higher Education in the State of South Carolina.

The two-year curriculum program designed to satisfy these guidelines is composed of the following courses offered on a quarter credit hour basis.

		class - lab - credit
CORE COURSES		
EET 160	D.C. Circuits I	3 - 0 - 3
EET 180	A.C. Circuits I	3 - 0 - 3
EET 205	Introduction to Microprocessors	4 - 0 - 4
EET 208	Microprocessor Applications	3 - 3 - 4
EET 210	Basic Electronics	3 - 3 - 4
EET 214	Pulse and Switching Circuits	3 - 3 - 4
EET 227	Electrical Machinery	4 - 0 - 4
ROB 100	Introduction to Robotics	3 - 0 - 3
ROB 110	Robot System Operation	3 - 3 - 4
ROB 120	Sensor and System Interfacing	3 - 3 - 4
ROB 130	Work Cell Design	4 - 0 - 4
ROB 140	Automated Manufacturing Systems	4 - 0 - 4
ROB 150	Advanced Robot Systems	3 - 3 - 4
ROB 160	Failure Analysis Techniques	3 - 3 - 4
ROB 170	System Project	0 - 3 - 1
ROB 180	CWU in Robotics I	4 - 0 - 4
ROB 190	CWU in Robotics II	4 - 0 - 4
ROB 200	CWU in Robotics III	4 - 0 - 4
MET 133	Manufacturing Technology	3 - 0 - 3
MET 215	Applied Mechanisms	3 - 3 - 4
MET 217	Fluid Power Systems	3 - 3 - 4
MET 224	Hydraulics and Pneumatics	4 - 0 - 4
EGR 200	Computer Programming	3 - 0 - 3
		Total 84
GENERAL EDUCATION		
MAT 111	Algebra and Trig. I	5 - 0 - 5
MAT 121	Algebra and Trig. II	5 - 0 - 5
ENG 141	English Comp. I	4.5 - 0 - 4.5
ENG 142	English Comp. II	4.5 - 0 - 4.5
ECO 111	Industrial Economics	3 - 0 - 3
PSY 230	Industrial Psychology	4.5 - 0 - 4.5
		Total 110.5

LABORATORY DEVELOPMENT

LABORATORY DESIGN GUIDELINES

The Robotics Center will offer training for the industrial community, two-year college faculty and students enrolled in the Automated Manufacturing Technology program. The Robotics Laboratory must provide hands-on training for participants from these three target groups; therefore, the hardware must possess the following capabilities:

- Permit training on large and small robot systems, on electric, hydraulic and pneumatic drive systems, on servo and non-servo models, on stop-to-stop, point-to-point and continuous path machines, on the three major arm geometries and on low and high technology systems.

- Permit training in basic hydraulics and pneumatics.

- Include some work cells which are designed to provide robotics training at as low a cost per station as possible.

- Permit training which imitiates actual industrial work cells.

- Permit training in the application of programmable controllers in automated work cells.

- Permit training in the application of work cell sensors using limit switches, photoelectric sensors, proximity sensors, pressure pads and vision.

- Permit training in interfacing the robot with sensors, vision systems and other computer-controlled devices.

- Include equipment for robot applications in welding, machine tending, palletizing, inspection, assembly and coating.

- Permit training in system and robot trouble-shooting.

- Permit laboratory class sizes of twelve students.

- Permit training in the programming of industral robots and programmable controllers.

- Demonstrate the use of safety devices to protect both the robot and the operator.

WORK CELLS IMPLEMENTED

The robotics laboratory has ten work cells for training in automated manufacturing concepts. The cells offer training in low and high technology robot systems with a variety of power sources and types of path control. The work cells currently implemented in the laboratory include three classified as high technology, three classified as low technology and four designed for basic concept training. The basic work cells and the simulated production processes in each are described below:

HIGH TECHNOLOGY GROUP

Large work cell with a Cincinnati Milacron T3 hydraulic robot with a Modicon programmable controller, conveyor system, welding system, machine mock-up, parts feeder and contact and non-contact sensors.

Medium work cell with a Cincinnati Milacron T3 - 726 small electric robot with a TI programmable controller, parts assembly mock-up and system sensors for interfacing.

Small assembly work cell with a Unimation Puma 260 electric robot, parts feeder, assembly fixture and system sensors.

LOW TECHNOLOGY GROUP

Pneumatic work cell with a Seiko 700 pick-and-place robot, Allenair rotary index table, parts feeder, conveyor, TI programmable controller, system sensors and system control panel. Machine loading and unloading is simulated.

Two work cells, each equipped with a pneumatic modular robot from Mack Inc., TI programmable controller, parts feeder, system sensors and system control panel. Material handling is simulated at these work cells.

TRAINING GROUP

Two work cells each equipped with a Sandhu Machine Rhino electric robot, conveyor system, parts feeder, machine mock-up and Apple computer system. Training in basic system concepts is emphasized at these work cells.

Two more work cells, similar but not identical to the two preceding stations, contain an additional linear base. Also, a rotary table replaces the conveyor.

SUPPORT HARDWARE

A Machine Intelligence Corporation VS-110 vision system

for use in any of the work cells where vision would enhance the manufacturing operation being simulated.

A Digital Equipment Corporation PDP 11/23 minicomputer networked to a VAX 750 for use with any of the work cells where host computer control is necessary.

LABORATORY TRAINING LEVELS

Each of the work cells will support many levels of training from basic machine operation to troubleshooting a complete automated work cell. Projects developed for each cell fall into one of six progressive training levels. Completion of each lower level is a prerequisite for the next higher level. A general description of each level follows:

LEVEL ONE - In the first level projects and activities are designed to train participants in the operation of the basic equipment which is included in the work cell. For example, learning the operation of the robot would be an objective for this level, along with programming both the robot and any programmable controller which is present in the work cell. In addition, the operation of all support and peripheral equipment in the cell would be covered in a level one project.

LEVEL TWO - Level two projects are designed to train participants in the integration of the robot and the basic work cell support equipment necessary to perform a manufacturing function. A typical project would include the integration of the robot with a gripper and parts feeder to permit the development of a palletizing program. In a low technology cell the project would include assembling a modular system, connecting the pneumatic lines, wiring the valves to the controlling device and writing the necessary program.

LEVEL THREE - Level three projects are built directly on the level two projects which preceded them. The work cell must be upgraded to include the sensors necessary for continuous untended operation. These devices may be contact sensors for some projects, proximity and photoelectric sensors for others and vision for still others. Also included at this level is the interfacing of the sensors to the work cell control system and the integration of the sensor information into the program driving the cell.

LEVEL FOUR - This level builds on previous project work in level three. The operating work cell with sensors interfaced to the controller does not have any control panel to display work cell conditions for an operator to use. Rather, this level will require the interfacing of controller conditions to a control display panel to indicate work cell conditions. For example, a robot may be halted because the part supply is low in the parts feeder.

LEVEL FIVE - Projects are included in level five which require interfacing of robot controllers to other control devices, such as programmable controllers and host computers.

LEVEL SIX - The last level includes projects which require troubleshooting of the work cell to identify the location of malfunctions in the total system.

SUMMARY

The application of complex automated manufacturing systems for production in American factories requires a trained force of technicians capable of working at technical levels higher than has ever been experienced. The Automated Manufacturing Technology program at Piedmont Technical College is designed to train technicians for this challenging new career area.

APPENDIX A

ROBOTICS SERVICE TECHNICIAN:

A-5 Train .. Good teaching techniques are required to teach operators and other manufacturing personnel how to operate and correct problems in the work cell.

A-6 Follow Up .. What is intended here is responsibility. What the group wanted was someone who would stay on top of problems and see a project through to completion.

B-3 Evaluate Robot Performance .. A complete knowledge of the robot operation and specifications is important so that any operation out of spec can be identified.

B-4 Apply Programming Techniques .. Programming was emphasized for two reasons. First, the skills learned when programming is taught in a structured manner provides good general training for troubleshooting and problem solving. Second, programming was a needed skill for robot implementation.

B-6 Implement and Maintain Sensors .. Sensors here refers to contact and non-contact type which include switches, photoelectric, proximity, ultrasonic, lasar, touch and vision systems.

C-1 to C-7 .. The term "Demonstrate" means a through knowledge of the area which is named in the competency.

C-8 Utilize Diagnostic Aids .. The technician should be able to utilize diagnostic information from error codes and test equipment when troubleshooting robot and computer hardware.

E-6 Implement and Maintain Interlocks .. Interlocks here means hardware and software which when activated prevents the occurance of situation where either a person or the robot may be injuried.

F-1 Apply Problem Solving Logic .. Produce good logical problem solvers which get to the heart of the problem by the shortest path, and see the problem through to a complete solution. When you find out how to do this, please, let us know.

F-2 Apply Statistical Methods .. Use basic mathematics to calculate Mean Time Between Failure, percentage down time, etc.

F-4 Determine Cause of Failure .. Emphasize in the training that the cause of the failure should be identified and corrected if possible along with the failure itself.

ROBOTICS APPLICATION TECHNICIAN

A-3 Sell Self .. It is important for those affected by a robot project to have confidence in the person directing the program. The ability to instill this type of confidence is what was intended by this conpetency term.

A-4 Sell Concept .. This competency stress the need to be able to sell to all personnel affected the benifits of the project in the face of all the problems that will arise with its installation.

A-14 Apply Decision Making Process .. This recognizes the fact that many decisions must be made in a robotic project and some technique must be employed to bring about a concensus.

E-1 to E-5 .. The action word used here is "converse" as conpared to the word "explain" used in previous competency statements. The words are intended to convey a different level of understanding on the part of the applications technician. While he must throughly understand the items under manufacturing processes he does not have to have the same depth of understanding about interface systems.

E-9 Apply Ergomic Interface .. This stress the need for training in the design of good man-machine interfaces.

E-10 Utilize Fail-Safe Practices .. Under some conditions a robot failure would pose on potential danger for the operator or workers, but in other applications a system failure could cause loss of life. The application technician should be trained to recognize the potential for danger and provide for the appropiate action with the design.

F-3 Invent a Concept .. Be creative! Quite often there is a need to find a solution in an unorthodox fashion. In robot projects an original solution is often required. Training people to do this is entirely another problem. Any ideas?

CHAPTER 20

ROBOT DESIGN FORUM

Kinematic Arrangements Used in Industrial Robots

James C. Colson
University of Texas at Austin
Austin, Texas

N. Duke Perreira
University of Texas at Austin

Abstract

A survey of currently used robots has shown that a limited set of kinematic arrangements are currently being manufactured and sold. The reasons for presenting a summary of these kinematic arrangements are threefold: to allow users to easily compare arrangement types for their application; to show that of the many kinematic arrangement possibilities, few are being manufactured; and to show that general purpose analysis, simulation, and control routines need only be developed to handle certain kinematic arrangements. The tables and figures are presented using standard notations in order to accomodate users, manufacturers, and researchers in this area.

INTRODUCTION

The purpose of this paper is to identify existing types of kinematic chains used in industrial robots. This information should prove useful to users, manufacturers and researchers in the field. It will provide users a way to easily identify and compare the various possible types of kinematic arrangements that could be purchased and used in a particular application. The user could use for example working volume, and dexterity to decide which configuration is most appropriate. A preliminary determination of the performance of the robot in a work environment based solely on geometric arguments is possible by simple examination of the linkage type. Once a particular kinematic arrangement is chosen, the user can then use a stand alone manipulator simulation program, a cost analysis outline and/or some other means to select between the alternatives. A complete evaluation of the robot performance in the work situation without actual implementation would of course require a work cell simulation.

The benefits of the results presented in the tables of this paper to manufacturers are twofold. It displays the names of the manufacturers that are currently marketing similiar kinematic arrangements. It also shows, by the process of elimination, what arrangements are not being manufactured. Many kinematic arrangements and link parameter selections are possible but manufacturers are currently marketing a very limited selection. This may leave the user in a compromised situation for although implementation of a robot may increase the work cell throughout the optimal kinematic arrangement required for his particular class of applications may not be manufactured.

The benefits of these results to research and development are many. For example, it is only necessary to model a few different types of kinematic arrangements to simulate and analyze most of the industrial robots currently manufactured. For the most general six degree of freedom arm there are 21 distinct classes of devices containing a total of 182 completely general kinematic arrangements. Each of these kinematic arrangements contains up to 18 infinitely adjustable geometric dimensions [1]. It has been shown that closed form polynomials of 32nd order can be used to describe the configuration of the general 6 degree of freedom manipulator for any particular end effector location [2,3]. Many, if not most, of the manipulators manufactured can be described by quartic or eighth order polynomials. There are many other kinematic arrangements not being manufactured that would lead to these lower ordered polynomials. Use of robots whose link positions and orientations which can be determined using these lower order polynomials can minimize the computational burden. These polynomials can of course also be used to design manipulators based on working volume, dexterity as well as sensitivity to manufacturing and other design constraints.

NOMENCLATURE

The physical elements that make up a single robotic manipulator and its work environment can be separated into six groups or types of linkage chains. The manipulator is located relative to ground by either a fixed "base" or a moving base such as a conveyor or other transport system. The first three links of the manipulator, the "body", are usually used to place the end of the manipulator at some location within the working volume. The next few links, or "wrist" of the manipulator are primarily used to define the orientation of the manipulator end point and to make small modifications of the location of the manipulator's end point. The "end effector" is usually a gripper, welder or powered device (for example a drill) that is connected to the wrist. In many cases the end effector holds a replacable "tool" such as a drill bit, or welding tips. The end effector and/or tool is then used to apply some work onto a "workpiece" which is either held to the base or forced to move relative to the base, for example by a conveyor system, or by using some type of motion control algorithm. The work can be manifested by some type of manufacturing or assembly process. This paper addresses the linkage types currently used to make up the body and the wrist of the base-body-wrist-end effector-tool-workpiece-base model of the manipulator-work environment model. The most efficient manipulator-work environment use and design criteria would of course necessitate a study of all of these components.

The location and orientation of any body or link in space requires the knowledge of six variables, three that determine the location of a point in the body relative to a second point in a second body and three that define the orientation of a line fixed in the first body relative to a line fixed in the second body. When two of these links are attached to each other in a way that allows relative motion between the two links the arrangement is known as a linkage of two elements. Typically the base, tool and workpiece are one element linkages, while the end effector is either a one or two element linkage. The body of all the manipulators examined in this paper have either two or three element linkages while the wrists vary from one to three element linkages.

The connection between two elements is known as the "pair" of connecting elements. There are three types of lower order surface pairs that allow only one degree of freedom between the connection elements (and thus the links). These three pairs are the "prismatic", the "revolute" and the "screw" pairs. Other pair elements such as the "global", "plane or flat", and "cylindrical" elements are made of combinations of the one degree of freedom elements. The revolute pair allows pure rotation of one link around another. A pure translation of one link with respect to another is allowed by the prismatic pair. When one link moves relative to a second with both translation and rotation in much the same way as a nut moves relative to a bolt a screw element is used. Since the angle of rotation and the magnitude of the translation are related by the lead of the "screw" only one degree of freedom exists. This is called a "screw pair". When the lead is infinite the screw pair reduces to the prismatic pair and when the lead is zero the screw pair reduces to the revolute. We are only concerned with revolute pairs and prismatic pairs since they are the only pairs used in robotic linkages.

When addressing a general pair type the symbol S_j is used, where j is used to represent the number of the joint. Joint numbering begins at the joint between the base and the body; it is called joint 1. A prismatic joint is labelled P_j while a revolute joint is labelled R_j.

There are various notations used to describe the kinematic chains used in robotic manipulators, the most common of which are described in references [4-6]. The notation used in this paper is most similar to that of reference [6] although there are some differences. The notation is defined in Figure 1. The axes of rotation for the revolute pair or for translation in the case of the prismatic pair are identified by Z_j. The common perpendicular between Z_j and Z_{j+1} defines the X_j axis. The distance between the Z_j and Z_{j+1} axes along the common perpendicular is $a_{j,j+1}$. The positive direction of the unit vector $\hat{z}_j$ along the Z_j axes is defined such that a positive rotation in the right hand sense by the angle, θ_j, about $\hat{z}_j$ locates the unit vector x_j relative to x_{j-1}. When $\hat{z}_{j+1}$ is not equal to $\hat{z}_j$ the positive direction of $\hat{x}_j$ is found by simply taking the vector cross product between $\hat{z}_{j+1}$ and $\hat{x}_j$. The unit vector y_j, not shown in the figure, is determined by the cross product of $\hat{z}_j$ and $\hat{x}_j$. The angle $\alpha_{j-i,j}$ is defined by a positive rotation about $\hat{x}_{j-1}$ from $\hat{z}_{j-1}$ to $\hat{z}_j$. The distance along Z_j between X_{j-i} and X_j is defined by d_j.

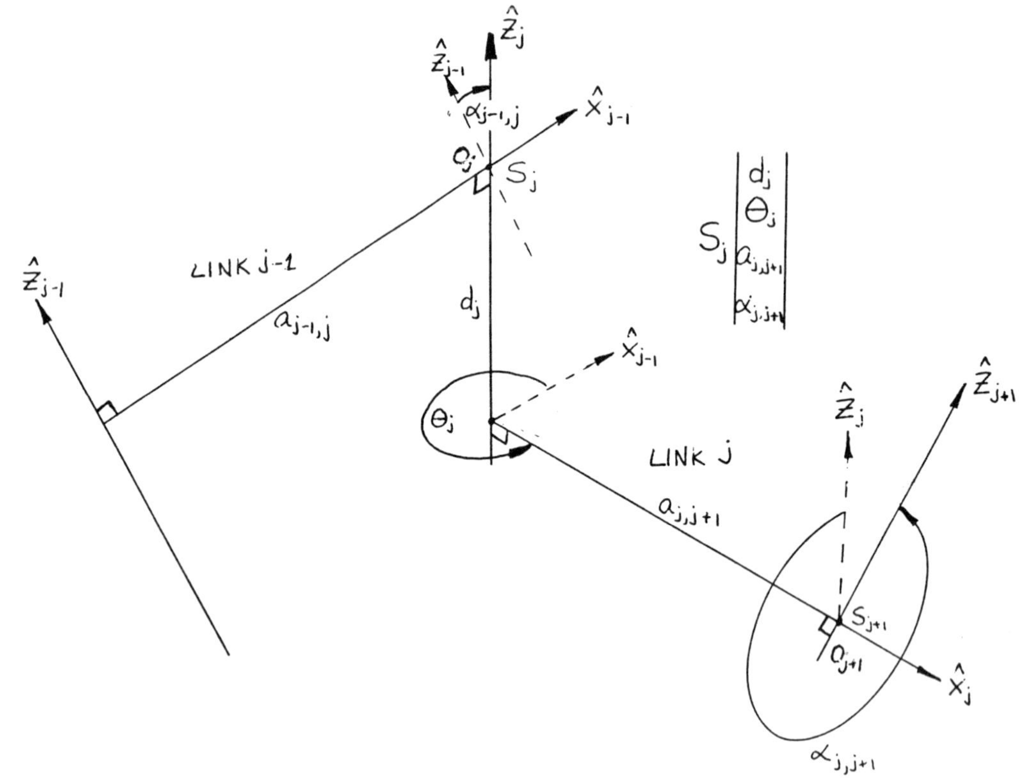

Figure 1 Kinematic Notation

The jth connection between the two links j-1 and j of a kinematic chain are represented by the linkage pair symbol S_j at the intersection of X_{j-1} with Z_j. The jth link is assumed to begin at the S_j joint and end at the S_{j+1} joint. The origin of the jth coordinate system fixed in the jth link, O_j, is located at the intersection of Z_j with X_{j-1}.

The location and orientation of Z_{j+1} at S_{j+1} relative to Z_j at S_j is found by prescribing d_j, $\theta_{j,j+1}$ and $\alpha_{j,j+1}$ [4]. For any link three of these four quantities are fixed and the fourth, being either θ_j or d_j, is allowed to vary and is called the pair variable. In the case of the prismatic pair, P_j, d_j is the pair variable and for the revolute joint, R_j, θ_j is the pair variable. The orientation of Z_{j+1} relative to Z_j and the pair type S_j is conveniently portrayed using a 1x4 link description matrix as shown in Figure 1a. A linkage chain with n links starting at S_j is defined by simply premultiplying the link description matrices for link j by link j+n-1.

Although the location of O_0 and the orientation of Z_0 is arbitrary, its choice influences the value of θ_1, d_1, $a_{0,1}$ and $\alpha_{0,1}$. In this paper it is assumed that O_0 is not located at O_1 and that neither Z_0 nor X_0 are parallel to Z_1 and X_1. In many applications this choice may not be the most expedient.

Figure 2 is an example of the notation described above. The first joint is a revolute, R_1, while the second is prismatic, P_2. Because the location of the zeroth or base reference frame is arbitrary the values for both d_1 and θ_1 shown in the 1x4 link description matrix can be specified as needed by the user. The linkage shown in the figure is assumed to be that of the manipulator body and thus, connects the base to the wrist. To show this, the linkage description matrices are appended by the words "Base" and "Wrist". The location of the revolute or prismatic joint S_j in a real manipulator does not have to occur at the intersection of the X_{j-1} and Z_j axes as shown in the kinematic representation. It is shown at this location only for convenience.

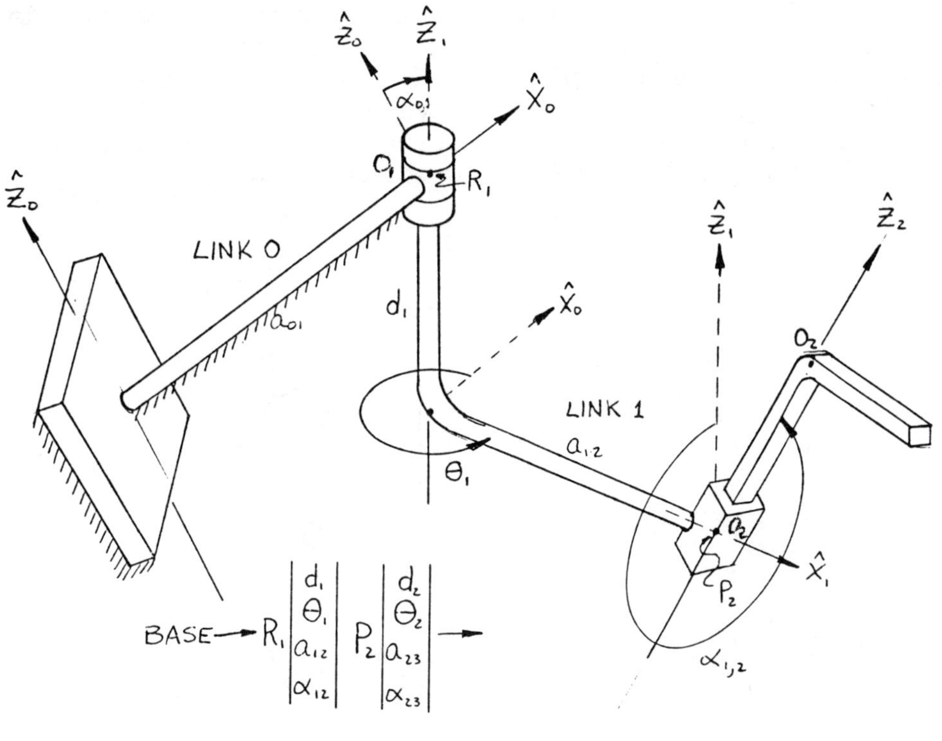

Figure 2 R_1P_2 Example Using Kinematic Notation

These representations are intended for use as design, analysis and prototyping aides they are not intended to be used as mockups or visualizations of the actual manipulator.

MANIPULATOR BODIES AND WRISTS

A survey of currently manufactured robotic manipulators, based on the data found in reference [7], indicates that very few of the infinite varieties possible (as defined by the 21 distinct classes of 6-degree of freedom manipulative devices, 32 possible kinematic arrangements and the infinite choices of each of the 18 geometric dimensions for each kinematic arrangement) are currently available. This rather large assortment of possibilities yyt rather small assortment of actual devices indicates a need for research to determine which of the infinite possibilities should be used given any particular manufacturing situation.

The first three degree of freedom of a manipulator are usually associated with the gross motion of the device. The associated three links are called in this paper the body of the manipulator. Of the eight possible kinematic arrangements only five are currently constructed. For each of the five that are currently constructed many of link pair variables have been chosen to be either zero for the case of d_j and $a_{j,j+1}$ or 0, 90 or 180 degrees for θ_j and $\alpha_{j,j+1}$. Usually attached to each body is a wrist. Currently wrists are manufactured with from one to three degrees of freedom. There are only six of 14 possible wrist configurations currently manufactured.

BODIES Listed in Table 1 are the various types of kinematic arrangements used in the bodies of a-l the manipulators surveyed. Linkage configurations not found include $R_1P_2R_3$, $P_1R_2P_3$ and $P_1P_2R_3$. Examples of the other body types listed in Table 1 are shown in Figures 3 thru 8. The body in Figure 3 is the simplest design having only two degrees of freedom being of a R_1P_2 configuration. The motion of the end of the body is restricted to a single plane. Although a three degree of freedom wrist would allow the user to obtain a variety of orientations of an end effector using this body type, finite length links in the wrist will not allow the end effector to be completely dexterious.

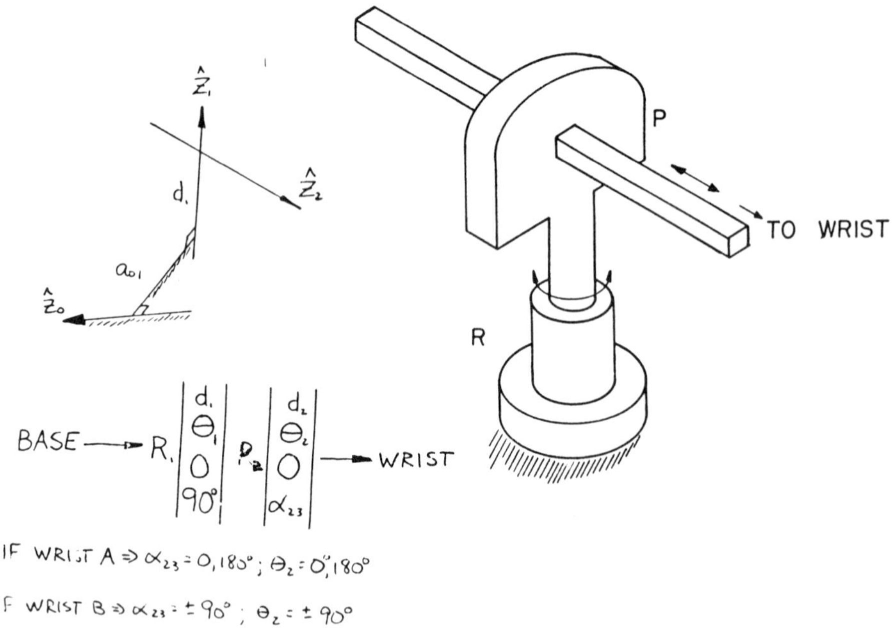

Figure 3 R_1P_2 Manipulator Body

The working volume of the $R_1R_2P_3$ manipulator body shown in Figure 4 is defined by two concentric spheres. This manipulator body is usually described as being a spherical type because the location of the end point of link 3 can be described using spherical coordinates centered about the intersection of the two revolute axes.

The manipulator shown in Fgiure 5 is commonly called an articulated or an anthropomorphic kinematic design. It could be called an $R_1R_2R_3$ Body with the last two joint axes being parallel. Its working volume is

TABLE 1

Kinematic Arrangements of Manipulator Bodies Used in Industrial Robots

1X4 Matrix Equivalence	$a_{3,4}$ θ_3 and $\alpha_{3,4}$	Vector Equiv.
$\rightarrow R_1 \begin{vmatrix} d_1 \\ \theta_1 \\ 0 \\ 90^\circ \end{vmatrix} R_2 \begin{vmatrix} d_2 \\ \theta_2 \\ a_{2,3} \\ 0 \end{vmatrix} R_3 \begin{vmatrix} d_3 = d_1 \\ \theta_3 \\ a_{3,4} \\ \alpha_{3,4} \end{vmatrix} \rightarrow$	IF WRIST A $\Rightarrow a_{3,4} = 0$; $\alpha_{3,4} = \pm 90^\circ$ IF WRIST B $\Rightarrow a_{3,4}$ = CONSTANT ; $\alpha_{3,4} = 0, 180^\circ$	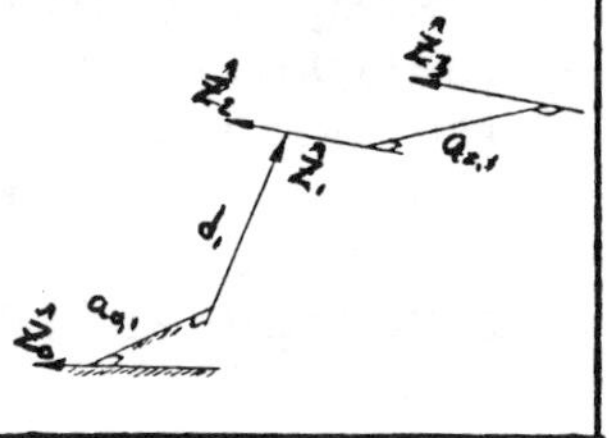
$\rightarrow R_1 \begin{vmatrix} d_1 \\ \theta_1 \\ 0 \\ 90^\circ \end{vmatrix} R_2 \begin{vmatrix} 0 \\ \theta_2 \\ a_{23} \\ 90^\circ \end{vmatrix} P_3 \begin{vmatrix} d_3 \\ \theta_3 \\ a_{34} \\ \alpha_{34} \end{vmatrix} \rightarrow$	IF WRIST A $\Rightarrow \alpha_{34} = 0^\circ, 180^\circ$; $\theta_3 = 0^\circ, 180^\circ$ IF WRIST B $\Rightarrow \alpha_{34} = \pm 90^\circ$; $\theta_3 = 0^\circ, 180^\circ$	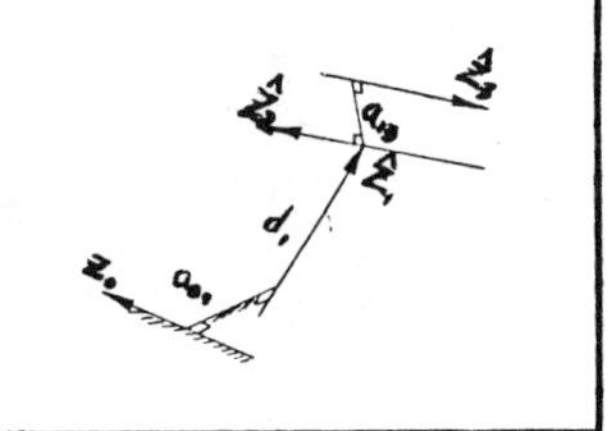
$\rightarrow R_1 \begin{vmatrix} d_1 \\ \theta_1 \\ 0 \\ 0^\circ \end{vmatrix} P_2 \begin{vmatrix} d_2 \\ 0 \\ a_{23} \\ 90^\circ \end{vmatrix} P_3 \begin{vmatrix} d_3 \\ \theta_3 \\ 0 \\ \alpha_{34} \end{vmatrix} \rightarrow$	IF WRIST A $\Rightarrow \alpha_{34} = 0, 180^\circ$; $\theta_3 = 0, 180^\circ$ IF WRIST B $\Rightarrow \alpha_{34} = \pm 90^\circ$; $\theta_3 = 0, 180^\circ$	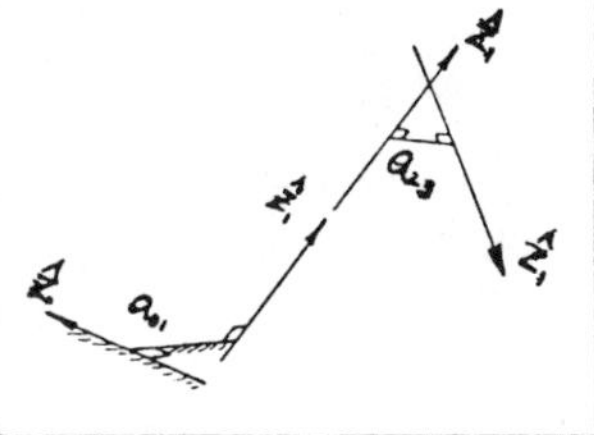
$\rightarrow R_1 \begin{vmatrix} d_1 \\ \theta_1 \\ 0 \\ 90^\circ \end{vmatrix} P_2 \begin{vmatrix} d_2 \\ \theta_2 \\ 0 \\ \alpha_{23} \end{vmatrix} \rightarrow$	IF WRIST A $\Rightarrow \alpha_{23} = 0, 180^\circ$; $\theta_2 = 0^\circ, 180^\circ$ IF WRIST B $\Rightarrow \alpha_{23} = \pm 90^\circ$; $\theta_2 = \pm 90^\circ$	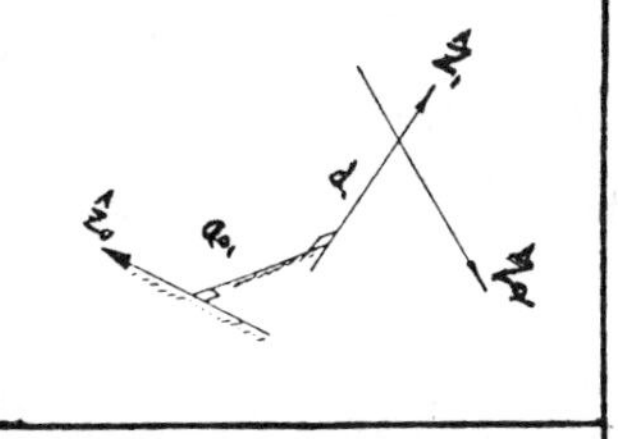
$\rightarrow P_1 \begin{vmatrix} d_1 \\ \theta_1 \\ a_{12} \\ 0 \end{vmatrix} R_2 \begin{vmatrix} 0 \\ \theta_2 \\ a_{23} \\ 0 \end{vmatrix} R_3 \begin{vmatrix} 0 \\ \theta_3 \\ a_{34} \\ 0 \end{vmatrix} \rightarrow$		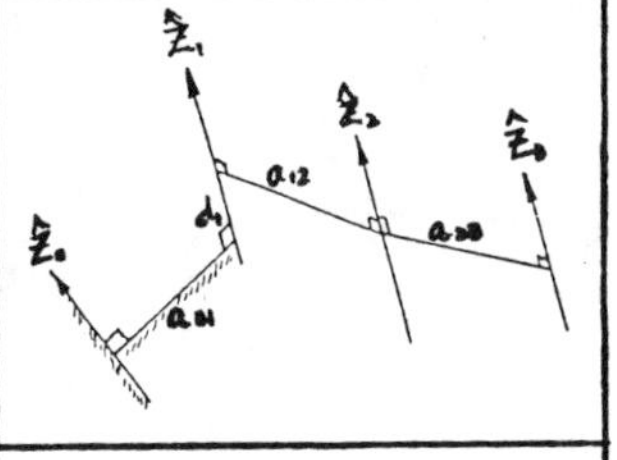
$\rightarrow P_1 \begin{vmatrix} d_1 \\ \theta_1 \\ 0 \\ 90^\circ \end{vmatrix} P_2 \begin{vmatrix} d_2 \\ \pm 90 \\ 0 \\ 90^\circ \end{vmatrix} P_3 \begin{vmatrix} d_3 \\ \theta_3 \\ 0 \\ \alpha_{34} \end{vmatrix} \rightarrow$	IF WRIST A $\Rightarrow \theta_3 = 0, 180^\circ$; $\alpha_{34} = 0, 180^\circ$ IF WRIST B $\Rightarrow \theta_3 = \pm 90^\circ$; $\alpha_{34} = \pm 90^\circ$	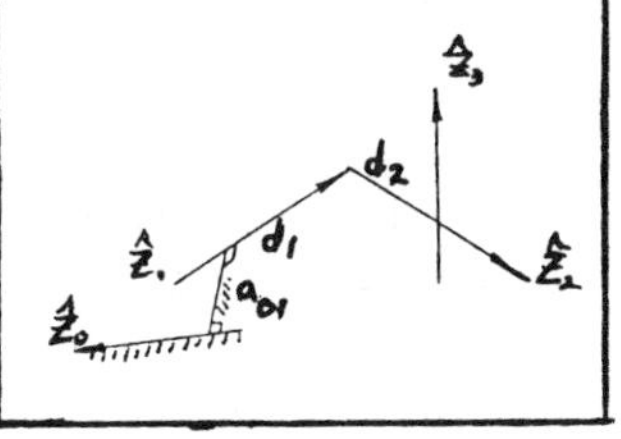

d_1 and θ_1 depend on 0th bases.

0th bases can be defined to make $\theta_1 = 0$ if P_1 or $d_1 = 0$ if R_1

$a_{3,4}$ is either zero or a constant depending on the wrist used.

$\alpha_{3,4}$ is 0° or 90° depending on the wrist used.

defined on the outside by a sphere and on the inside by scallops which are a function of the link lengths and the mobility (allowable angular positions of each of the joints). A common construction of this configuration is as parallel linkage. When constructed as a parallel linkage the working volume is decreased due to the decrease of the mobility of the linkage. The principle advantage of the parallel configuration over the simple configuration is that it lends itself very well to remote actuators. In the parallel configuration the actuators for links 2 and 3 are both mounted to link 1 whereas in the simple configuration the actuators for links 2 and 3 are located in links 1 and 2, respectively. This results in a larger inertia variance for link 2 in the simple fonfiguration and, therefore, makes it more difficult to control.

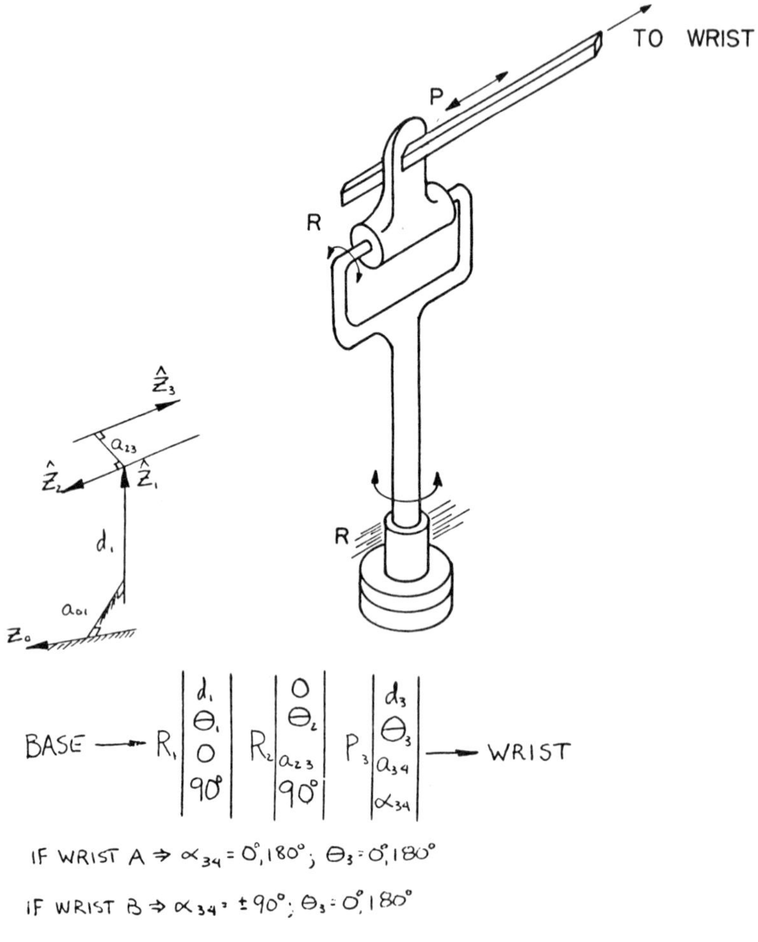

Figure 4 $R_1R_2P_2$ Manipulator Body

In Figure 6 is shown an $R_1P_2P_3$ body. Its' working volume is defined as the locus of points between two concentric cylinders. The location of the end of link 3 relative to the base can therefore be defined using cylindrical coordinates centered anywhere on the axis of the revolute pair.

Figure 7 shows a $P_1R_2R_3$ manipulator body configuration which is very similar to the configuration shown in Figure 5. Its working volume is similar to the $R_1P_2P_3$ body configuration discussed earlier. This design is typically used for pick and place operations where stiffness in the vertical axis is necessary. When it is used in this manner, the first joint is usually locked.

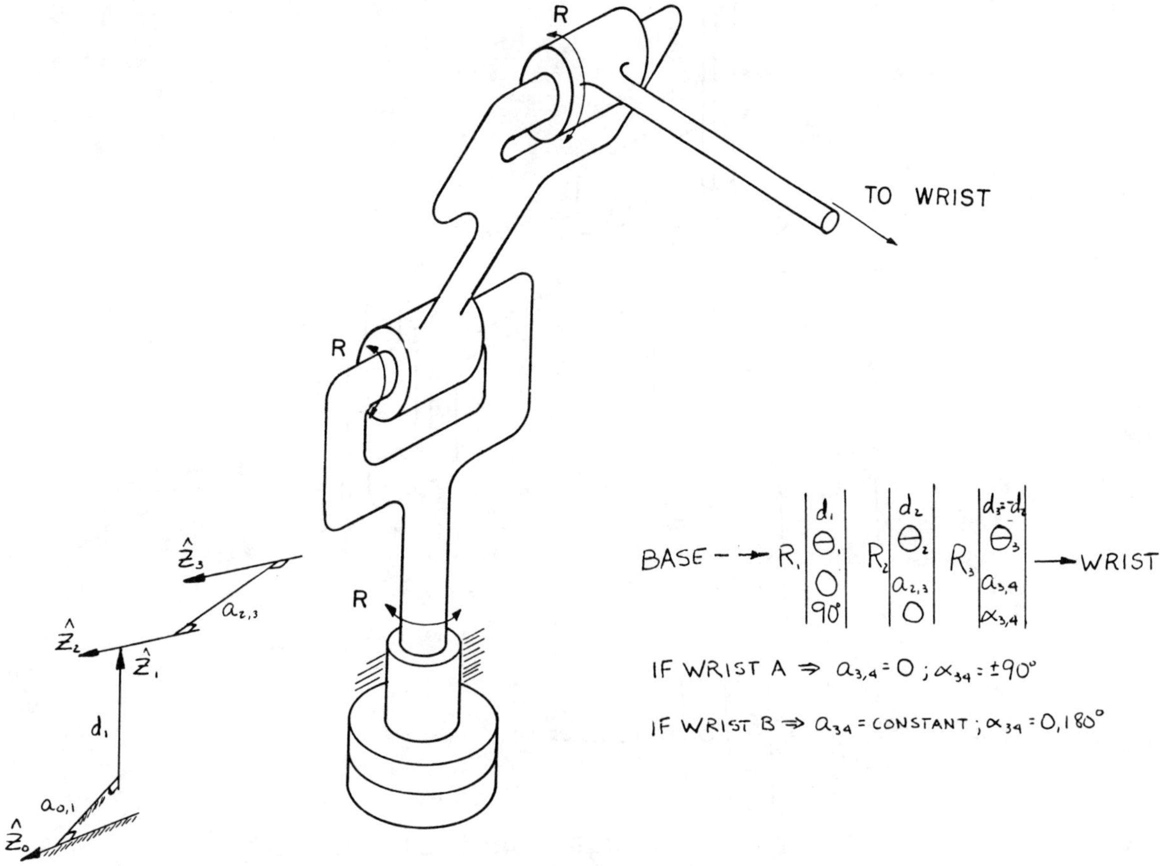

Figure 5 $R_1R_2R_3$ Manipulator Body

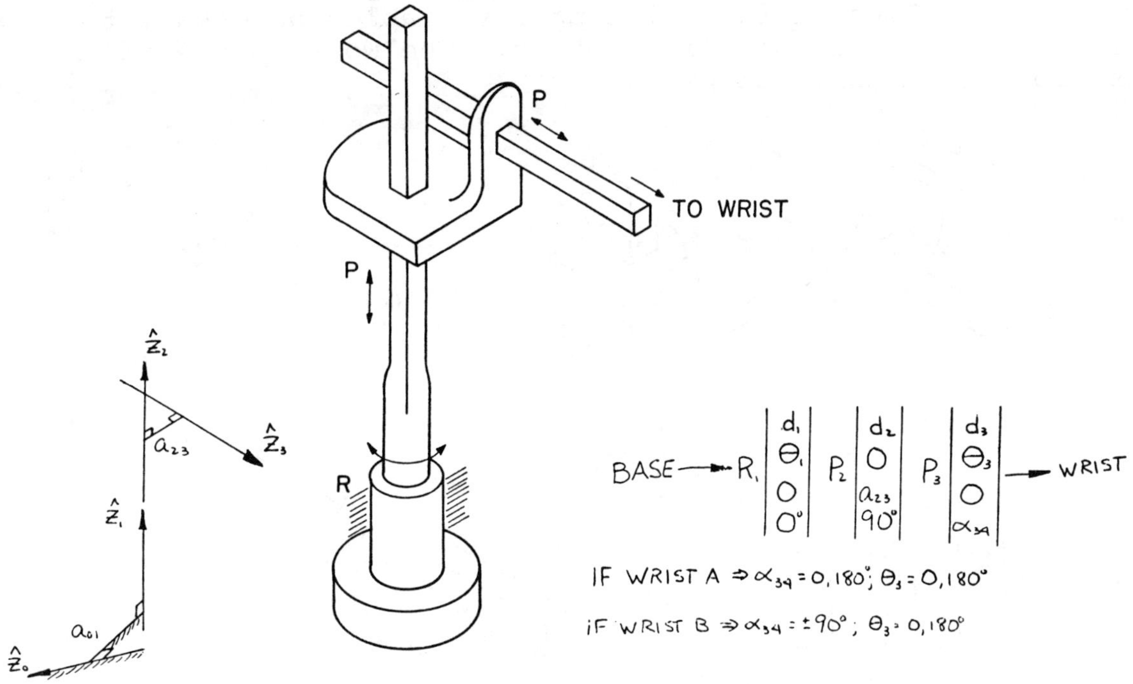

Figure 6 $R_1P_2P_3$ Manipulator Body

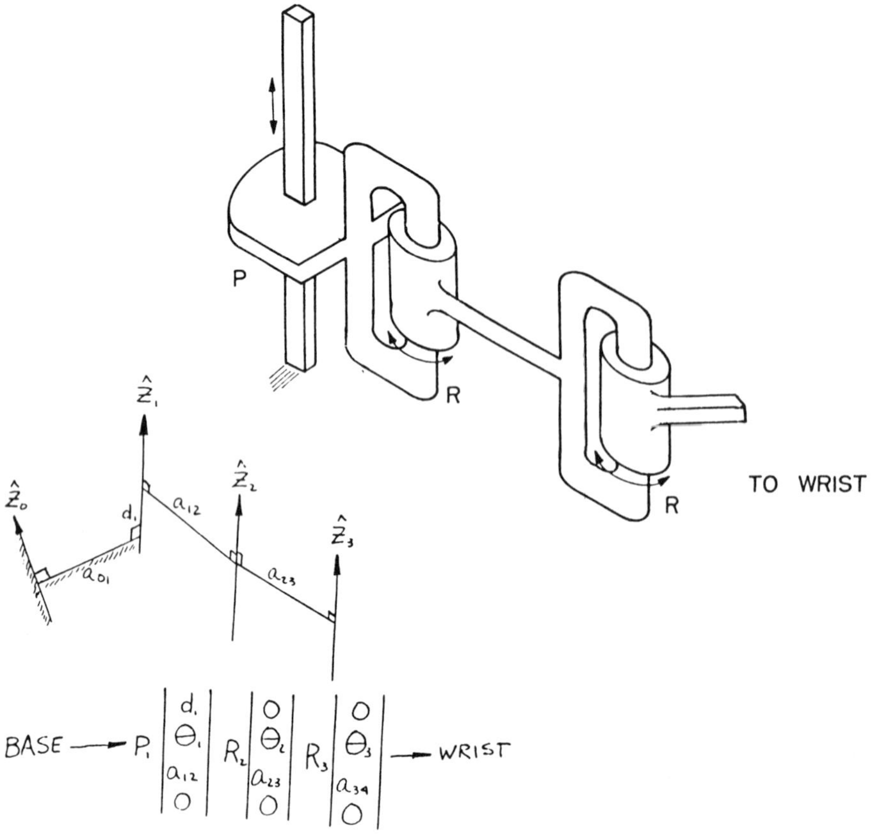

Figure 7 $P_1R_2R_3$ Manipulator Body

The final body configuration currently manufactured is a $P_1P_2P_3$ linkage. It is shown in Figure 8. It is very easy, in this case, to determine the values of the joint pair variables necessary to obtain the commonly used straight line motion. The work volume of this body is described by a parallelepiped.

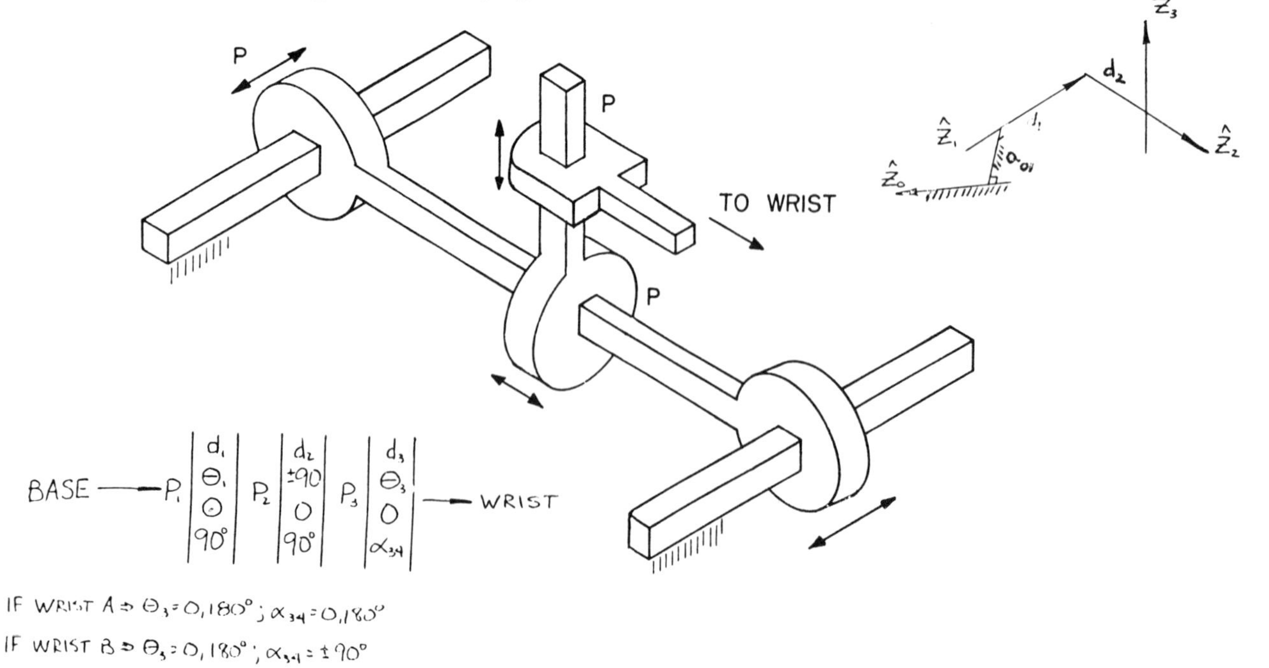

Figure 8 $P_1P_2P_3$ Manipulator Body

WRISTS Listed in Table 2 is a summary of the wrists that are currently used on robotic manipulators. It is assumed that the body has three links and thus, the first joint in the wrist is joint number 4. The wrists vary from one to three links and correspondingly one to three degrees of freedom. Some of the simpler robots do not have wrists. In many applications, object orientation is fixed with tooling and can be set such that a fixed "orientation" of the end effector is sufficient and no wrist is required. In the more general case, a wrist is necessary is required. In the more general case, a wrist is necessary to enable the robot to work in a variety of orientations.

The simplest wrists have only one degree of freedom and are always revolutes as shown in Fgiures 9a and 9b. Wrist 1A incorporates pitch or yaw about the last link of the body chain depending upon how it is connected to the body. Most commonly it is installed to incorporate yaw about the last link in the body.

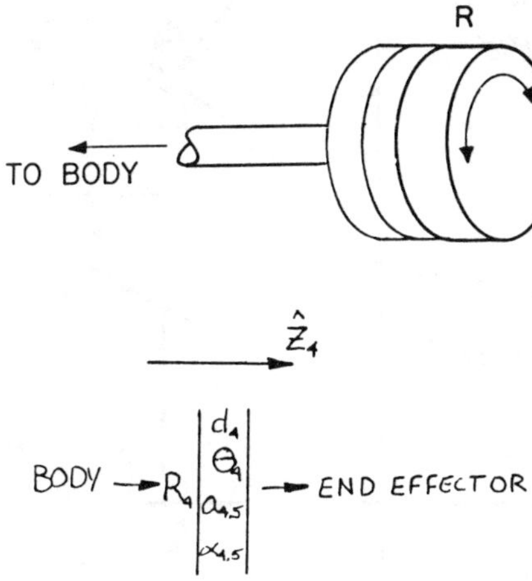

Figure 9a Wrist 1A

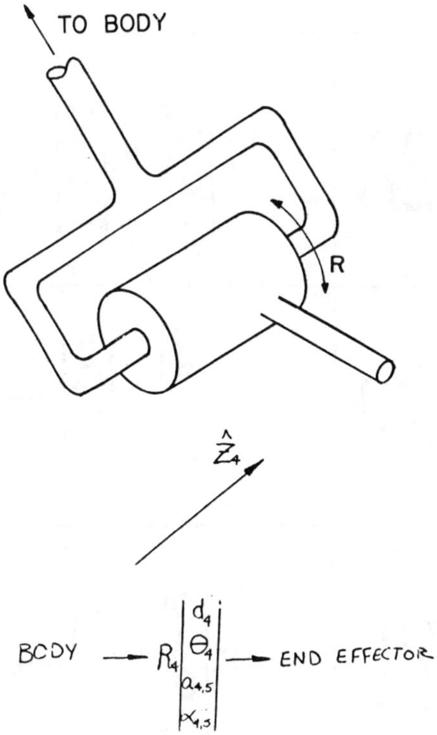

Figure 9b Wrist 1B

The most common wrists in pick and place applications consist of two degrees of freedom. These wrists are shown in Figures 10a, b, and c. Wrist 2A is a combination used to obtain pitch and roll and is the most common two degree of freedom wrist used. Wrist 2B is used to obtain yaw and pitch. Wrist 2C is similar to 1A except that an additional translational degree of freedom is available.

Three degrees of freedom wrists are probably the most common of all wrist categories and will be even more populat as applications become more complex. These wrists are shown in Figures 11a and b. Wrist 3A is used to obtain roll, pitch, and yaw. This design was originally used in the Cincinnati Milicron robots and is now used in many systems.

TABLE 2

Kinematic Arrangements of Manipulator Wrists Used in Industrial Works

	1x4 Matrix Equiv.	$\alpha_{4,5}$	Vector Equiv.
1A	$\rightarrow R_4 \begin{vmatrix} d_4 \\ \theta_4 \\ a_{4,5} \\ \alpha_{4,5} \end{vmatrix} \rightarrow$		
1B	$\rightarrow R_4 \begin{vmatrix} d_4 \\ \theta_4 \\ a_{4,5} \\ \alpha_{4,5} \end{vmatrix} \rightarrow$		
2A	$\rightarrow R_4 \begin{vmatrix} 0 \\ \theta_4 \\ 0 \\ \alpha_{4,5} \end{vmatrix} R_5 \begin{vmatrix} d_5 \\ \theta_5 \\ a_{5,6} \\ \alpha_{5,6} \end{vmatrix} \rightarrow$	IF $P_3 \Rightarrow \alpha_{4,5} = 0°, 180°$ IF $R_3 \Rightarrow \alpha_{4,5} = \pm 90°$	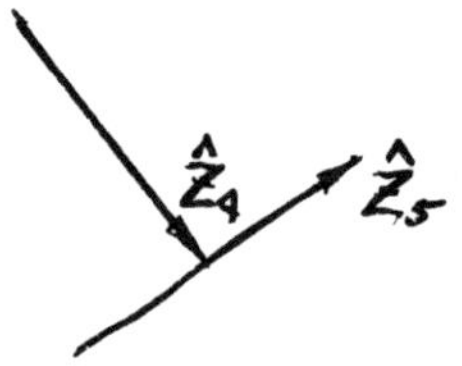
2B	$\rightarrow R_4 \begin{vmatrix} 0 \\ \theta_4 \\ 0 \\ \alpha_{4,5} \end{vmatrix} R_5 \begin{vmatrix} S_5 \\ \theta_5 \\ a_{5,6} \\ \alpha_{5,6} \end{vmatrix} \rightarrow$	IF $P_3 \Rightarrow \alpha_{4,5} = \pm 90°$ IF $R_3 \Rightarrow \alpha_{4,5} = 0, 180°$	
2C	$\rightarrow R_4 \begin{vmatrix} 0 \\ \theta_4 \\ 0 \\ 0 \end{vmatrix} P_5 \begin{vmatrix} d_5 \\ \theta_5 \\ a_{5,6} \\ \alpha_{5,6} \end{vmatrix} \rightarrow$		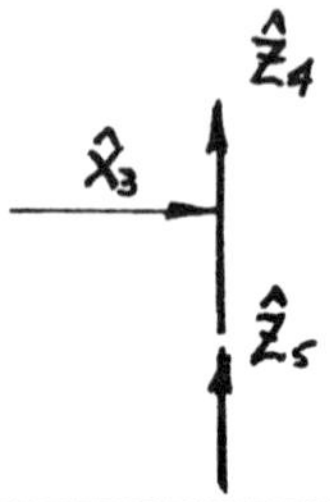
3A	$\rightarrow R_4 \begin{vmatrix} 0 \\ \theta_4 \\ 0 \\ \alpha_{4,5} \end{vmatrix} R_5 \begin{vmatrix} 0 \\ \theta_5 \\ 0 \\ 90° \end{vmatrix} R_6 \begin{vmatrix} S_6 \\ \theta_6 \\ a_{6,7} \\ \alpha_{6,7} \end{vmatrix} \rightarrow$	IF $P_3 \Rightarrow \alpha_{4,5} = 0°, 180°$ IF $R_3 \Rightarrow \alpha_{4,5} = \pm 90°$	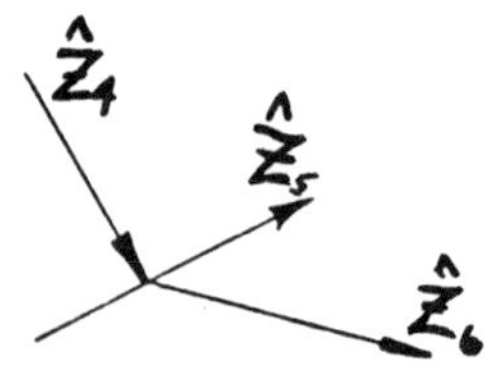
4A	$\rightarrow R_4 \begin{vmatrix} 0 \\ \theta_4 \\ a_{4,5} \\ 90° \end{vmatrix} R_5 \begin{vmatrix} 0 \\ \theta_5 \\ 0 \\ 90° \end{vmatrix} R_6 \begin{vmatrix} S_6 \\ \theta_6 \\ a_{6,7} \\ \alpha_{6,7} \end{vmatrix} \rightarrow$		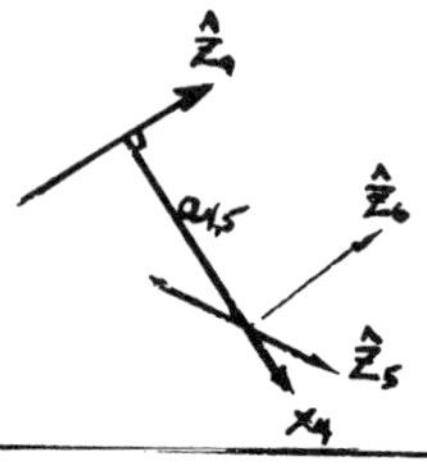

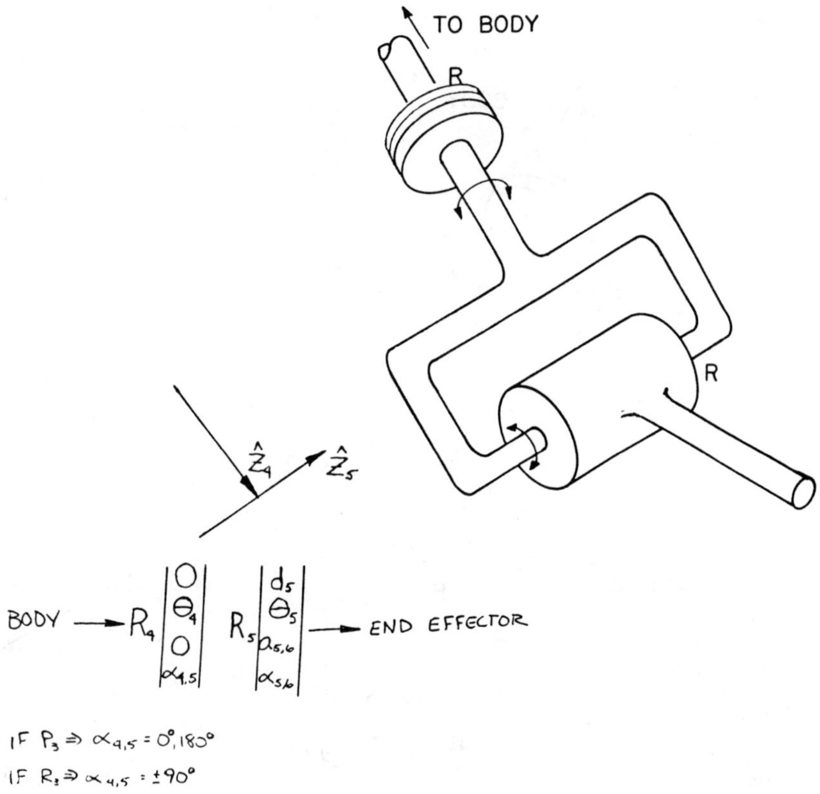

Figure 10a   Wrist 2A

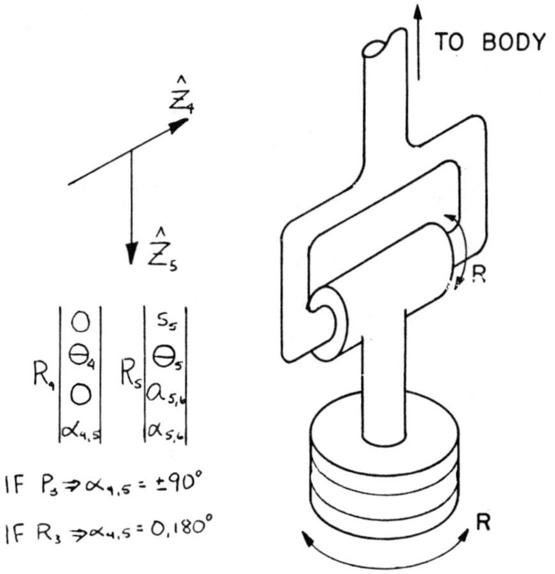

Figure 10b   Wrist 2B

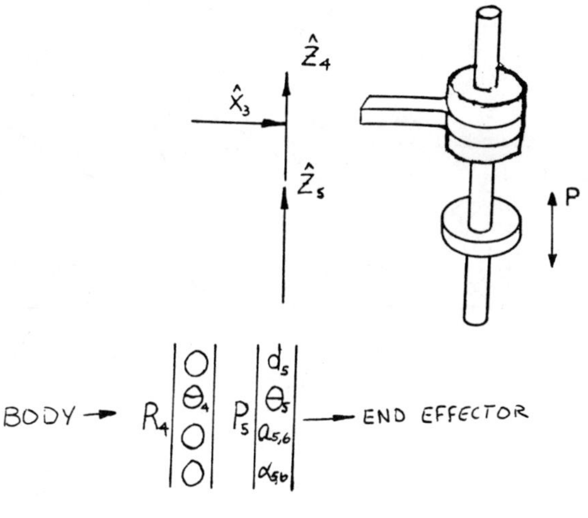

Figure 10c   Wrist 2C

The orientation of the last link relative to the first link in Wrist 3B is usually described in Euler angles (Rot z; Rot x; Rot z'). It is used by Unimation and others whose design is Stanford based. It is probably the most common three degree of freedom wrist.

Table 3 contains a listing of many of the currently available robotic manipulators. Tabulated are the Body and Wrist configurations.

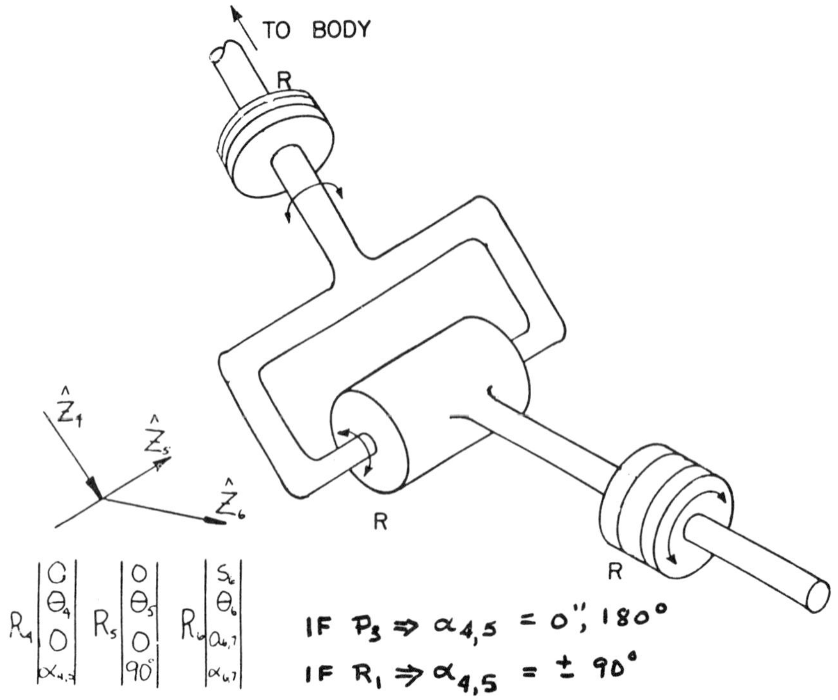

Figure 11a   Wrist 3A

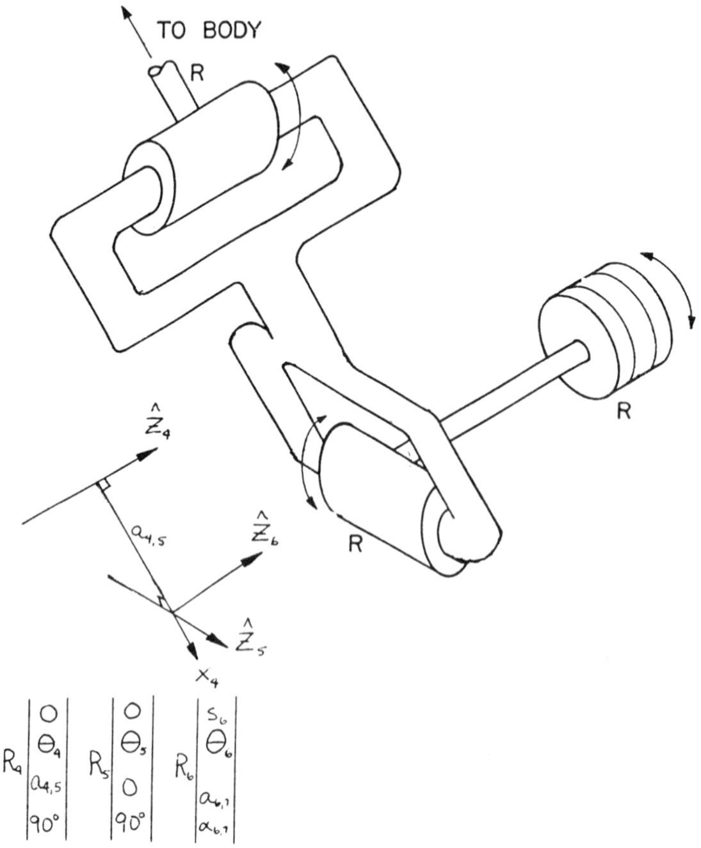

Figure 11b   Wrist 3B

TABLE 3

KINEMATIC ARRANGEMENTS OF BODY AND WRISTS IN INDUSTRIAL MANIPULATORS

MANUFACTURING / MODEL	MODEL NUMBER	TYPE BODY, WRIST
Accumatic Machine Co.		
Nu-Man		RRR, 2A
Advanced Robotics Co.		
Cyro	750	PPP, 2A
	820	RRR, 2B
	2000	PPP, 2A
AKR		
Spray Painting Robot		RRR, 3A
Anorad		
Anorobot		PPP, 2A
Armax Robotics, Inc.		
Armax Robot		RPP, 3A
Asea		
	IRb-6	RRR, 2B
	IRb-60	RRR, 2B
Automatix		
Robovision II		RRR, 2B
Bendix		
AA Series CNC		RRP, 2B or 3A
ML Series CNC		RRR, 3A
Cincinnati Milicron		
	T3-566	RRR, 3B
	T3-586	RRR, 3B
Copperweld		
	CR10	RPP, 1A
	CR50	RPP, 2B
	CR100	PPP, None
Cybotech	H8	RPP, 1B or 2B

Continued . . .

TABLE 3. Continued

	G80	PPP, 3A
	P15	RRR, 3A
	V80	RRR, 3A
	V30	RRR, 3A
Devilbiss		
	TR3003S	RRR, 3A
	TR3500W	RRR, 3A
GCA/PAR Systems		
	PARXR6100	PPP, 3A
General Electric		
	A12 Allegro	PPP, 3A
	AW-7 Arc Welding	PPP, 2B
	P5	RRR, 2B
	S6	RRR, 3A
	GP 132	RRR, 3A
	GP 66	RRP, 3A
	MH 33	RRP, None
	SN 220	RRP, 2A
General Numeric		
Series M	A-0	RPP, 2B
	0	RRP, 2A
	1	RPP, 2A
	GN3	RPP, 2A
Hitachi		
"Mr. Aros"		RPP, 1B
Process robot		RRR, 2B
Spray paint		RRR, 2B
Hodges Robotics International		
MRZ-5		RRR and RRP, 2A
Industrial Automates		
Industrial Automates		RRP, 1A
Intarm		
Intarm 100-S		RPP, 1A
International Business Machines		
RS1		PPP, 3A
7535		PRR, 2C
International Robomation/Intelligence		
IRI-M50		RRR, 2B

Continued . . .

TABLE 3. Continued

Mack Co.	
B-A-S-E Series II	PPP, None
Microbot	
Minimover 5	RRR, 2B
Nordson Co.	
Nordson Robot System	RRR, 3A
Positech Co.	
Probot	RPP, 2B
Prab	
Model E	RPP, 2A
Model FA	RPP, 2A
Model FB	RPP, None
Model FC	RPP, None
Model 4200	RRP, 2A
Model 5800	RRP, 2A
Precision Robots, Inc.	
PRI-1000	RRR, 3A
Sandhu Design, Inc.	
Rhino XR-1	RRR, 2B
Schrader-Bellows	
Motion Mate	RP, None
Seiko Instruments U.S.A.	
Model 100	RPP, None
Model 200	RPP, None
Model 400	RP, 1B
Model 700	RP, 1A
Thermwood Co.	
Series Six	RRR, 3A
Series Seven	RRR, 3B
Series Three	RRR, 2B
Unimation, Inc.	
Apprentice	RRP, 2A
250	RRR, 3A
500	RRR, 3A
600	RRR, 3A
750	RRR, 3A
1000	RRP, 2A
2000	RRP, 3A
4000	RRP, 3A

Continued . . .

TABLE 3. Continued

United States Robots	
Maker I	RRP, 2B
Westinghouse	
2000	RRR, None
4000	RRR, 2B
5000	PPP, 3A
7000	PPP, 2A
Yaskawa Electric America	
Motoman L3	RRR, 2B
Motoman L3L	RRR, 2B
Motoman L10G	RRR, 2B
Motoman S30	RRR, 2B

BIBLIOGRAPHY

1. Tesar, D., "A Review of Modeling, Control and Design of Manipulators in Terms of the Planar N DOF System", Jan. 1978, Unpublished Notes, University of Florida, Gainesville, Florida.

2. Roth, B., J. Rastegar, and V. Scheinman, "On the Design of Computer Controlled Manipulators, 2nd IFTOMM Symposium on Robots and Manipulators, Udine, Springer-Verlag, 1974, Volume 1, pp 93-113.

3. Duffy, J., and J. Rooney, "On the Closures of Spatial Mechanisms," ASME Paper No. 72-Mech-77, Mechanisms Conference, 1972.

4. Denavit, J. and R.S. Hartenburg, "A Kinematic Notation for Lower Pair Mechanisms Based on Matrices," Jour. of Appl. Mech., Vol. 22, Trans. ASME, Vol. 77, 1955, pp. 215-221.

5. Paul, R.P., "Modeling, Trajectory Calculation and Servoing of a Computer Controlled Arm," Stanford Intelligence Laboratory, Stanford University, AIM 177, 1972.

6. Duffy, J., Analysis of Mechanisms and Robot Manipulators, John Wiley, N.Y., N.Y., 1980.

7. 1982 Robotics Industry Directory, Technical Data Base Corp., P.C. Flora Publ., 1982.

Traveling Assembly Robot

H. Hirabayashi
Hitachi Ltd.
Yokohama, Japan

T. Hamada
Hitachi Ltd.

M. Akaiwa
Hitachi Ltd.

K. Kikuchi
Hitachi Ltd.

Abstract

An experimental travelling assembly robot system has been developed for flexible assembly. The pilot system is featured by reduction of cycle time due to simultaneous assembly and parts transfer. The articulated robot can assemble parts of various shapes and sizes with its versatile close-fitting table. In the pilot system there are stations along the track for special functions not performed by the robot. Each robot assembles one product in one trip around the track. By increasing or decreasing the number of robots, the system can be effectively adjusted to changes in rate or mode of production.

ABSTRACT

An experimental travelling assembly robot system has been developed for flexible assembly. The pilot system is featured by reduction of cycle time due to simultaneous assembly and parts transfer. The articulated robot can assemble parts of various shapes and sizes with its versatile close-fitting table.
In the pilot system there are stations along the track for special functions not performed by the robot.

Each robot assembles one product in one trip around the track. By increasing or decreasing the number of robots, the system can be effectively adjusted to changes in rate or mode of production.

INTRODUCTION

Today, the production lines of manufacturing industries which produce products with little variation in style are highly automated, however, industries in which there are many variations of a product are still not highly automated.
In the future, batch-production with medium sized batches is expected to account for more production lines, but as yet batch production is much less efficient than mass-production.
This paper will discuss the outline of the travelling assembly robot system as it is applied for production of a certain complicated product.

This system will increase productivity of batch-production with medium sized batches. In this system, the robot produces 8 styles of a product. The product is made of high accuracy parts for which the measuring error is less than 10 to 30 microns. The style of the product is changed one to five times a week. Several hundred finished products are produced per day on this production line.

1. SYSTEM CONFIGURATION

The parts of the product are assembled in the order shown in Fig. 1. Since high accuracy assembly is required, the measuring and adjusting operations are included in the assembly process. In addition, shrinkage fit operations are performed.

The 5 concepts are evaluated in this paper for application to an assembly system for a certain product requiring complicated assembly. The peripheral machine (machine which performs operations robots can not perform such as measuring and shrinkage fit operation) is shown in Fig. 2.

Concept 1 is composed of a free flow conveyor, simple parts-feeders and low degrees of freedom arms (pick & place) without any robots.

This concept is useful for short time assembly, long product life and limited style production.
Introducing of this concept to assembly system in Fig. 1 requires complicated parts-feeders with many parts-stockers, because the assembly system has a predetermined product length requirement that must be met. Therefore, a wide variety of parts are necessary in order to assure this requirement will be met. It must be noted that pick & place is inadequate for products requiring complicated assembly due to its low degree of freedom.

The tray-setting method was adopted for concepts 2, 3, 4 and 5. In this method, the parts for one product are prepared on the parts-tray.

Concept 2 is the same as Concept 1 except that some pick & places are replaced by robots. This concept is good only when the groups of pick & places and robots are segregated.
However, pick & places and robots must be randomly arranged for products requiring complicated assembly. Therefore, Concept 2 cannot be effectively applied.

In Concept 3 and 4, peripheral machines are not effectively utilized since the robot performs operations at one peripheral machine at a time. Furthermore, if production is increased, the peripheral machines with low rate of operation must also be increased.

Concept 5 makes effective use of both the robots and the peripheral machines. This concept is featured by the travelling robot which moves along the circular track. Several robots are able to travel along the rail-track while simultaneously assembling parts.

A diagram and a photograph of this system are shown in Figs. 3 and 4, respectively. There are several peripheral machines along the rail-track which perform the operations not performed by the robots. The robots, each of which is followed by a controller, stop at the peripheral machines and pick up already processed subassemblies in exchange for their unprocessed subassemblies. In this way the robots proceed along the rail-track picking up parts, assembling them and delivering and receiving subassemblies from the peripheral machines, and after one trip around the rail-track completely assemble one product.

Several robotic technologies chiefly concerning travelling functions, have been developed to realize Concept 5.

2. ROBOT WITH TRAVELLING FUNCTION

Robots are composed of two bogie trucks. The first truck carries the robot arm,a close-fitting table and the driving motor. The rear truck carries the robot controller and the power inlet. A diagram and a photograph of this robot are shown in Figs. 5 and 6, respectively.

The robot arm has an articulated structure with the specifications shown in table 1. The electrically driven arm comprises the forearm, the upper arm,the rotation and the arm supporter. In contrast to the ordinary industrial robot, the arm of the travelling robot is suspended from the top of the arm supporter.
Also, this travelling robot replaces some of the linkage with chains. Adoption of these two technologies enables a wider work area.

The wrist has two degrees of freedom (necessary for products requiring complicated assembly) and it uses the differential gears. The posture of the hand can be kept regardless of the motion of the upper arm and the forearm by means of the parallel linkage and chain adopted for power delivery.

A bogie truck comprises two pairs of wheels. Each pair turns on an axis fixed to the base of the truck.
The motor delivers power to the rack gear fixed to the rail-track via the pinion gear of the truck. The wheels are made of urethane in order to decrease vibration that occurs while travelling. The side-rollers of the bogie truck roll along the side of the rail for stable travel.

Since the track is circular, the power for the robot controller can not be supplied via electric wire. Therefore, the rail-track is fitted with a power cable which supplies the robot with electricity via power inlet.

3. CLOSE-FITTING TABLE

The robots must grasp and assemble various kinds of parts for products requiring complicated assembly. An example here has the precise insertion of a peg into a tight hole.

In order to perform this, the multipurpose hand shown in Fig. 7 was developed for grasping many parts of various dimensions and weights. The close-fitting table shown in Fig. 8 was developed for ease of insertion which can correct positioning error. With the aid of this table and the multipurpose hand, the robot can easily assemble many kinds of parts. The close-fitting table is installed under the hand of the robot as seen in Fig. 5.

4. CONTROL

It is necessary to synchronize control of both robots and peripheral machines in the assembly system. The host microprocessor is in change of supervising and controlling all robots and peripheral machines. The command of the host microprocessor simultaneously starts the peripheral machine and robot operations at an interval of one command cycle.

Fig. 9 shows the control configuration of this system.
Since the robot controller travels along a circular track, some effective means of communicating between the robot controller and the host microprocessor must be adopted.
Therefore, the rail-track at a peripheral machine is equiped with an optical signal transmitter/receiver, and the same transmitter/receiver is installed in each robot. Every time the robot stops at a peripheral machine, it communicates with the host microprocessor through the transmitters/receivers.

4.1 Host Microprocessor (HMP)

The block diagram of the HMP is shown in Fig. 10.
The HMP employs an 8-bit microprocessor (HMC 6800) to supervise and control all robots and peripheral machines through digital input/output interfaces. The processing of the HMP comprises transmitting and receiving necessary signals to and from the robots and peripheral machines.

The following is an example of process performed by the HMP at Station II in Fig. 3. When a robot arrives at Station II, it transmits the arrival signal to the HMP through a digital signal transmitter/receiver. After receiving the arrival signal from the robot, the HMP confirms that the previous operation has been completed at Station II.
Then the HMP transmits the code number of the operation to the robot.

The code number of the operation denotes a series of operations the robot should perform. Since these operations are taught to the robot in advance, the robot will perform the operation according to the code number. Depending on the state of the robots and peripheral machines such as the supply of parts, malfunction etc, the HMP can transmit a different code number to the robot even at the same station.

The robot starts the operation according to the code number, and then transmits the name of the operation it is performing to the HMP. When the robot finishes the operation at the station, it transmits the end-of-operation signal. After receiving the end-of-operation signal, the HMP confirms that there is no robot at the next station. Then the HMP transmits the travel command and code number of the operation the robot should perform during travelling, and simultaneously transmits the start signal to the peripheral machine.

To summarize, the HMP receives the state of the robots and transmits the operation commands to all robots. To prevent miscommunication between the transmitter and receiver, every time the receiver receives a signal from the transmitter,it sends back the same signal as confirmation.

The process performed by the HMP at one station has been described. In actual operation, the HMP supervises and controls several robots and peripheral machines simultaneously.
The HMP manages this process by time-sharing.

4.2 Robot Controller

The block diagram of the robot controller is shown in Fig. 11. The robot controller which accompanies the robot arm, travels along the rail-track. When the robot stops at a peripheral machine, it communicates to the HMP. The robot controller employs an 8-bit microprocessor (HMC 6800) to control the arm and the bogie truck and to communicate with the HMP through digital input/output interfaces, analog input/output interfaces and counters.

4.3 Optical Signal Transmitter/Receiver (Optical Transceiver)

In order for the robot controller to communicate with the HMP, the rear bogie truck is equipped with an optical transceiver. A diagram of optical transceiver is shown in Fig. 5, and details are in Fig. 12. The optical system, shown in Fig. 12, comprises a pair of optical transceivers which are identical in their structures. One is mounted on the robot side, and the other on the rail-track side.

An optical transceiver is composed of 8 light emitting diodes (LEDs) as the transmitter and 8 phototransistors as the receiver and works as an 8-bit transceiver between the robot controller and the HMP. Information is thus exchanged between a robot and the HMP, when the robot stops at a peripheral machine and the two optical transceivers face to each other.

5. SOFTWARE

The configuration of software of the system is shown in Fig. 13. There are two loops in the configuration.
The first loop includes teaching and manual operation and the second loop includes automatic operation.

First, teaching of the robot is performed using the first loop. After teaching is completed and confirmed, the robot is automatically operated, in which a cycle of the second loop is executed for one step of the robot motion sequences.

When the robot receives an operation command from the HMP, it analyzes the command, and performs the operation according to the analysis result.
When the robot finishes the operation, it requests the next operation command to the HMP. Some operations are performed at the station, and some others are performed while the robot is travelling.

The sampling cycle time of this system is 20 ms.
The first half of which is used for the arm control and the latter half for the travel control. The travelling can, thus, be controlled independently of the arm.

A new method for smoothing the robot trajectory was developed as shown in Fig. 14.
The trajectory of conventional robots exactly pass through all intermediate points between the origin and destination, which resulted increased cycle time duc to acceleration and deceleration at each point. In many practical cases, however, these intermediate points are not necessarily followed precisely.

In this system, a trajectory portion forming a corner is changed to a parabolic trajectory in order to save operation time and to smooth the robot motion.

Fig. 14 shows an example which employs a group of points, Q_1, Q_2, ... , Q_7, instead of a group of points, P_2, P_3, ..., P_5. Following is the algorithm to get points, Q_1 and Q_2. Point P_2 is on the corner which is formed by lines, $\overline{P_1P_2}$ and $\overline{P_2P_3}$. Since $\overline{P_1P_2}$ is shorter than $\overline{P_2P_3}$, Q_1 is decided on the middle point between P_1 and P_2, and Q_2 is decided in order that $\overline{Q_1P_2}$ is equal to $\overline{P_2Q_2}$. All other Points, Q_3, Q_4, ..., Q_7, are calculated according to the same algorithm.

Fig. 15 shows that the robot motion with parabolic interpolation at points B and C requires less time than with linear interpolation.

CONCLUSION

The travelling assembly robot system was described as a pilot system for medium-sized batch-production.
In this system, the robot assembles the product while travelling along the circular track, and it completes one product per revolution.

To realize this system, several technologies were developed, including the travelling mechanism, the optical signal transmitter/receiver for communication between the host microprocessor and the robot arm, and the close-fitting table for ease of assembly and the trajectory generating method. These technologies are fundamental to robotics and efforts are being made to apply them to actual production lines in order to increase productivity.

BIBLIOGRAPHY

[1] Abraham, R.G. "Programmable automation of batch assembly operations", The Industrial Robot, Vol. 4, No. 3, pp. 119-131 (September 1977).

[2] Takahashi, M. "An Assembly Robot System with Twin Arms and Vision", Proc. of 12th. I.S.I.R., pp. 111-120 (June 1982).

[3] Sugimoto, K. "Trajectory Interpolation of Robot Hand", Trans. JSME, Vol. 48, No. 431, pp. 990-997 (July 1982) (in Japanese).

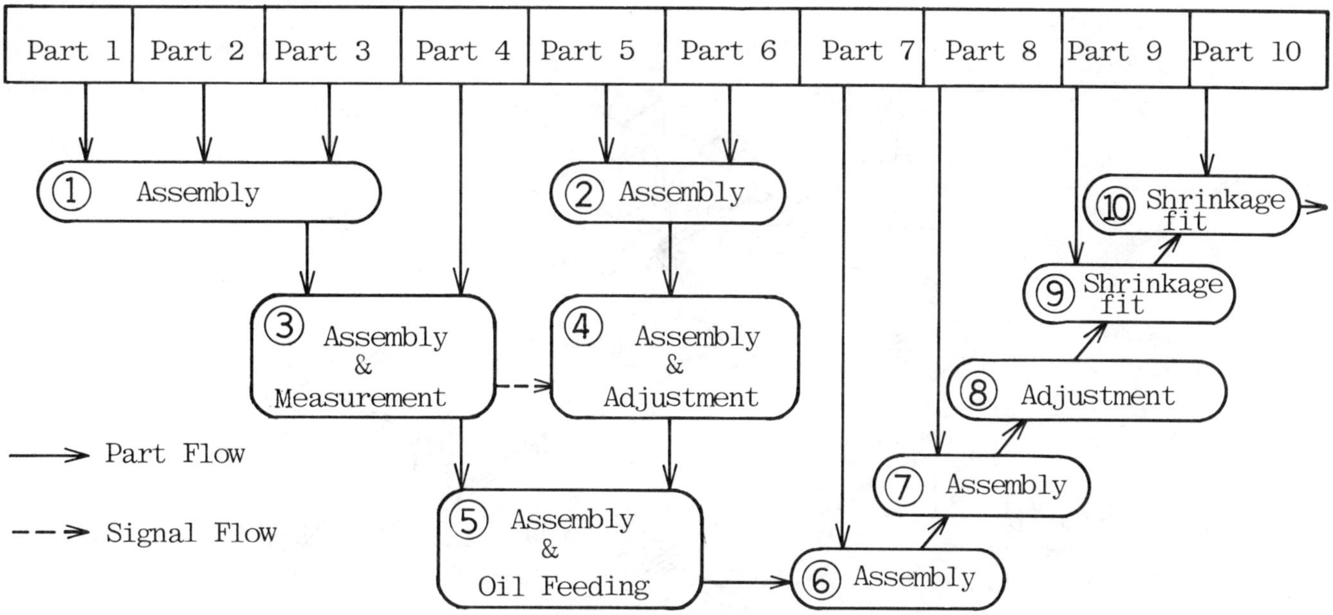

Fig. 1 Assembly Process

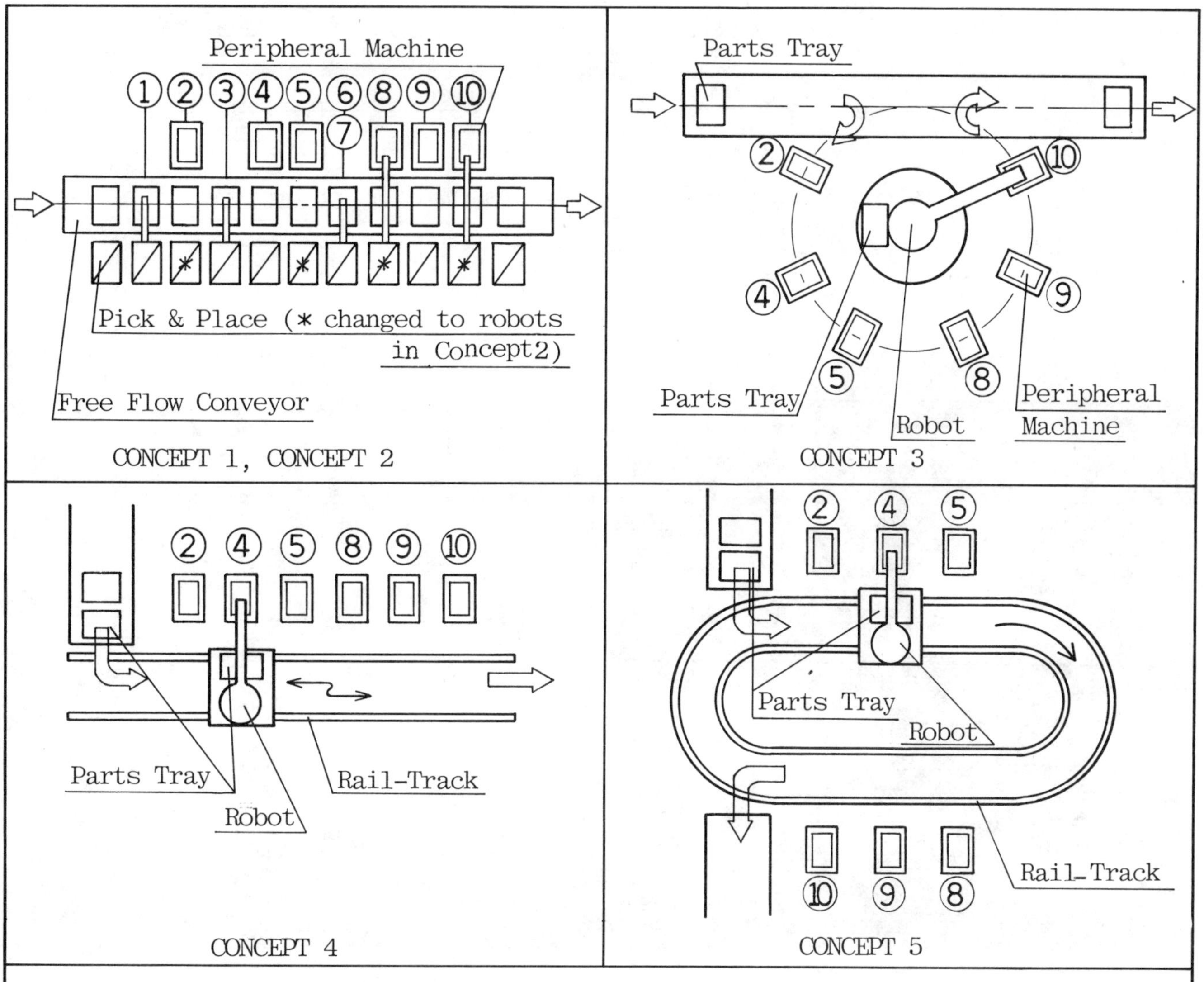

Fig. 2 System Configuration

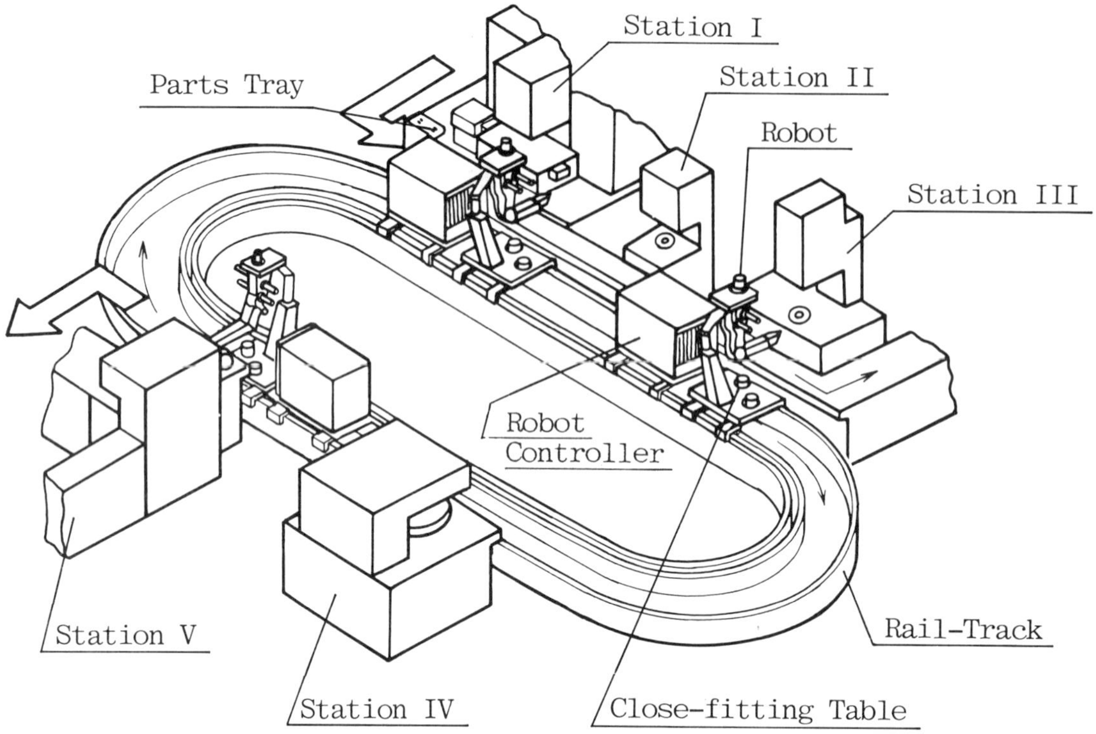

Fig. 3 Diagram of Travelling Assembly Robot System

Fig. 4 Outlook of System

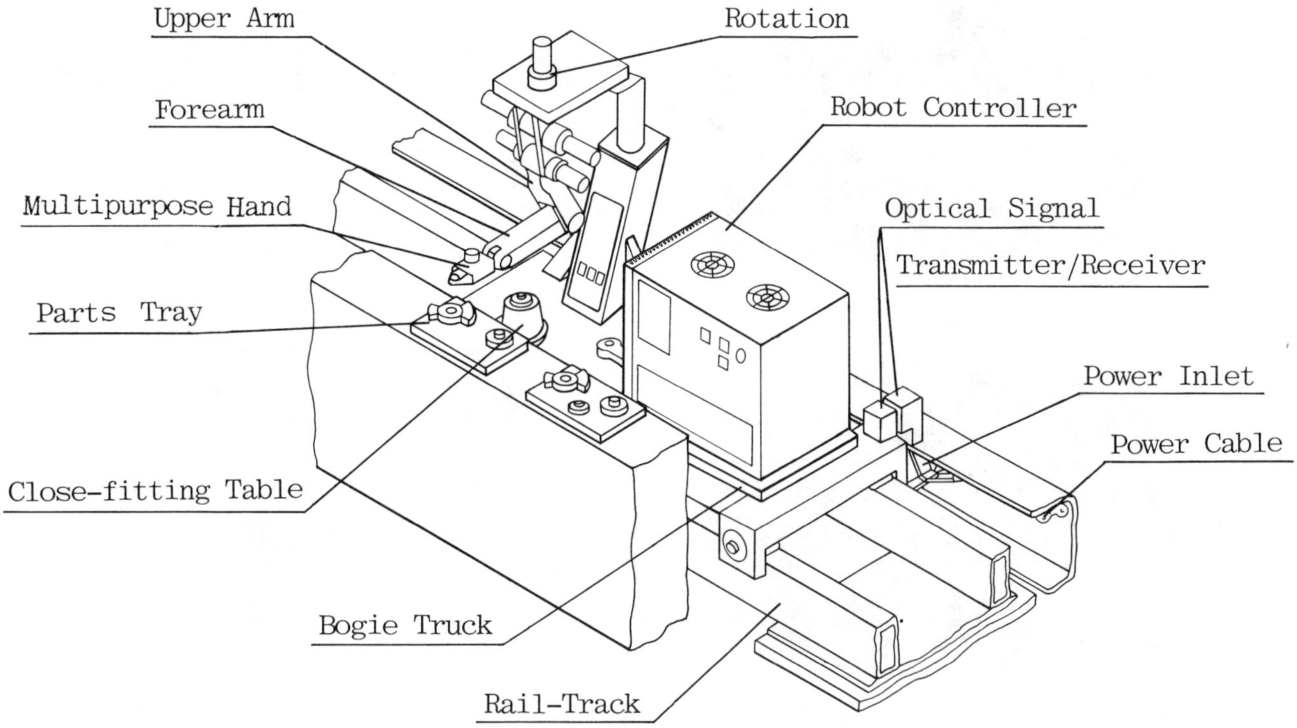

Fig. 5 Diagram of Robot

Table 1 SPECIFICATIONS

Robot			
Structure			Electric Articulated Arm with Travel Function
Degree of Freedom			6
Maximum Velocity	Arm	Rotation	120°/s
		Upper Arm & Forearm	800 mm/s
	Wrist	Bend	120°/s
		Roll	180°/s
	Travel		1000 mm/s
Repeatability			± 0.3mm
Load Capacity			7 kgf (incl. Hand Weight)
Weight			150 kgf (incl. Bogie Truck)
Controller			
Programming Method			Teach-in / Playback
CPU			HMC 6800
Position Control			Software Servo
Weight			50 kgf

Fig. 6 Outlook of Robot

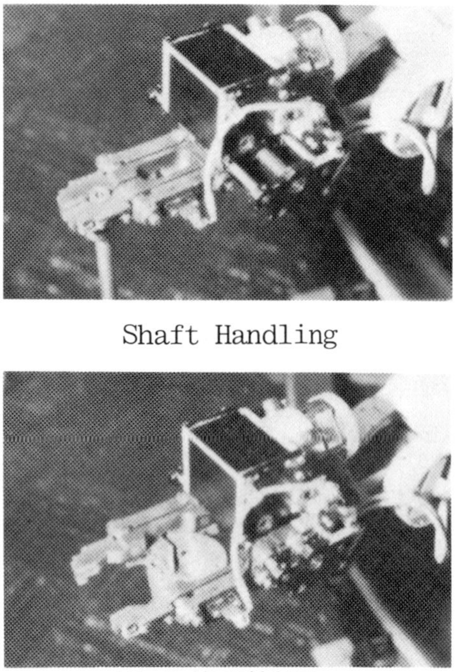

Shaft Handling

Ring Handling

Fig. 7 Multipurpose Hand

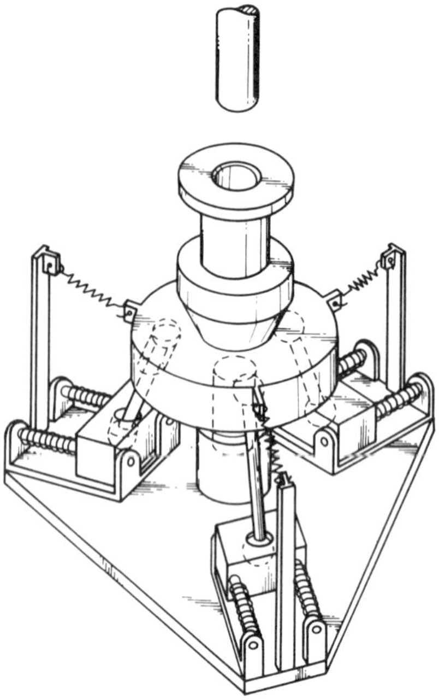

Fig. 8 Close-fitting Table

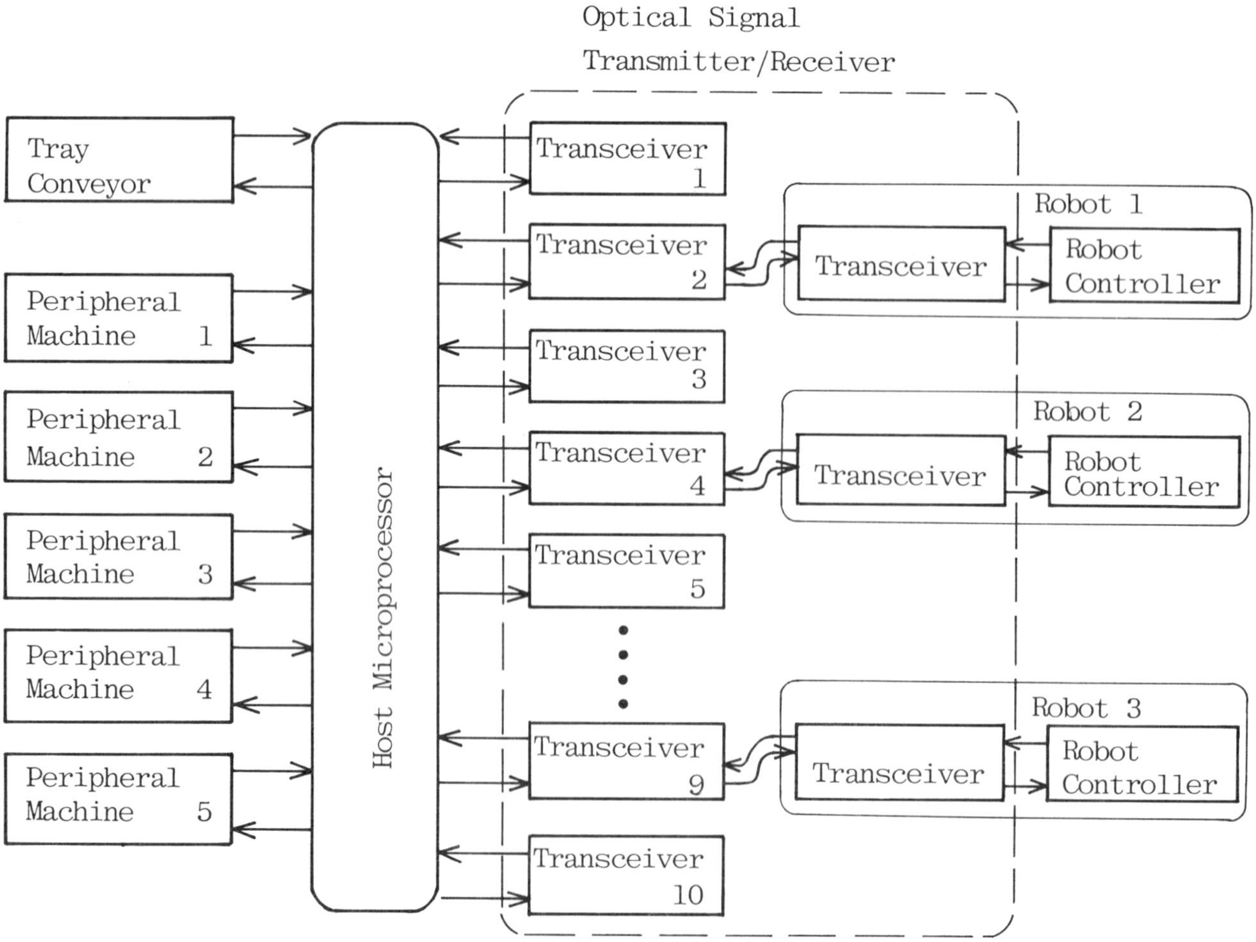

Fig. 9 Control Configuration of System

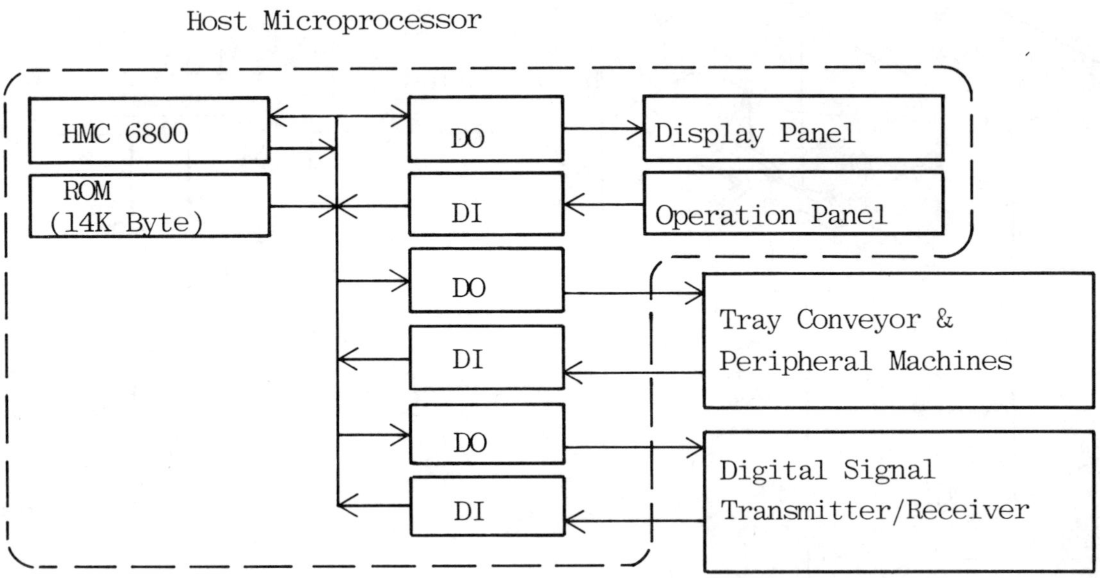

Fig. 10 Block Diagram of Host Microprocessor

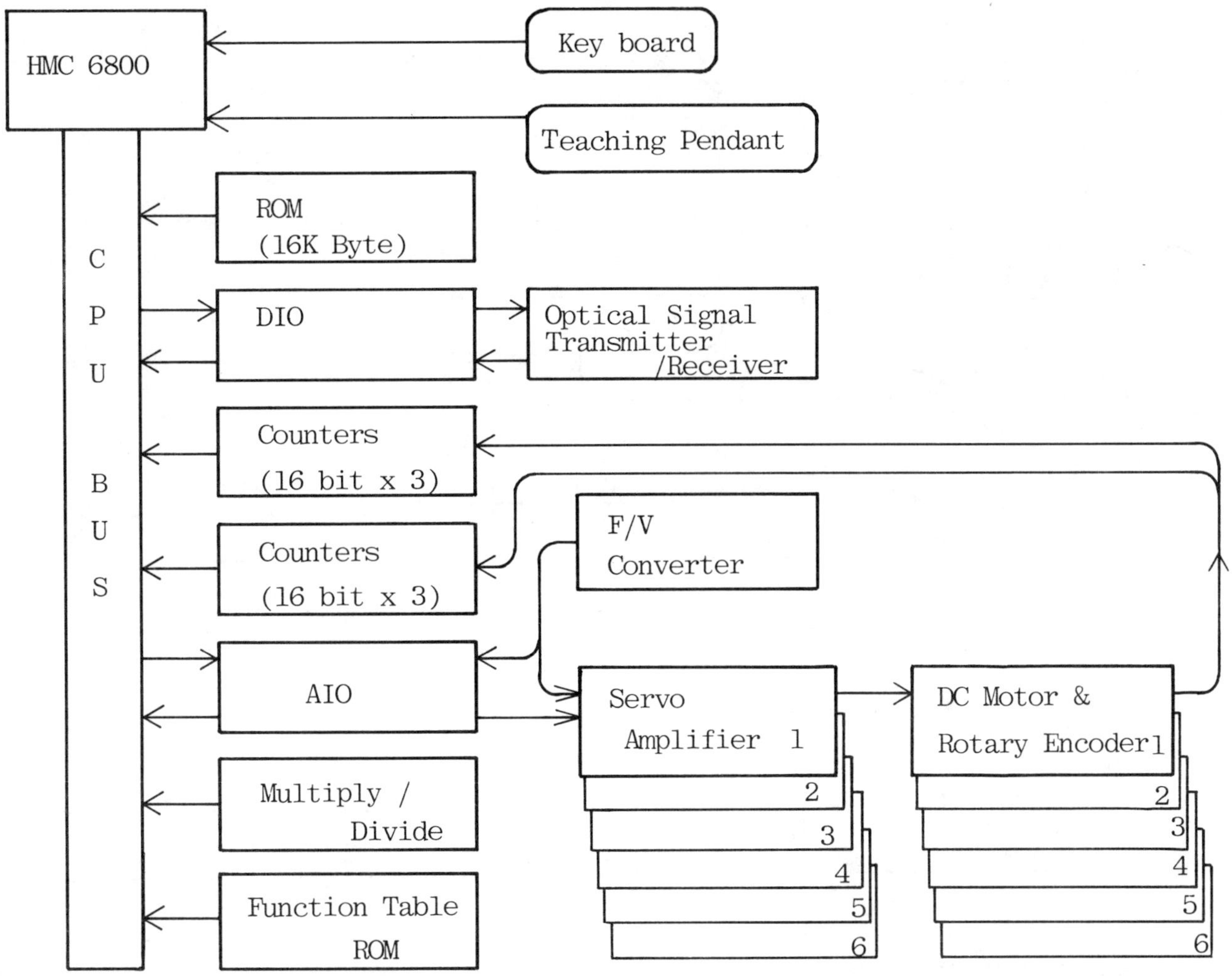

Fig. 11 Block Diagram of Robot Controller

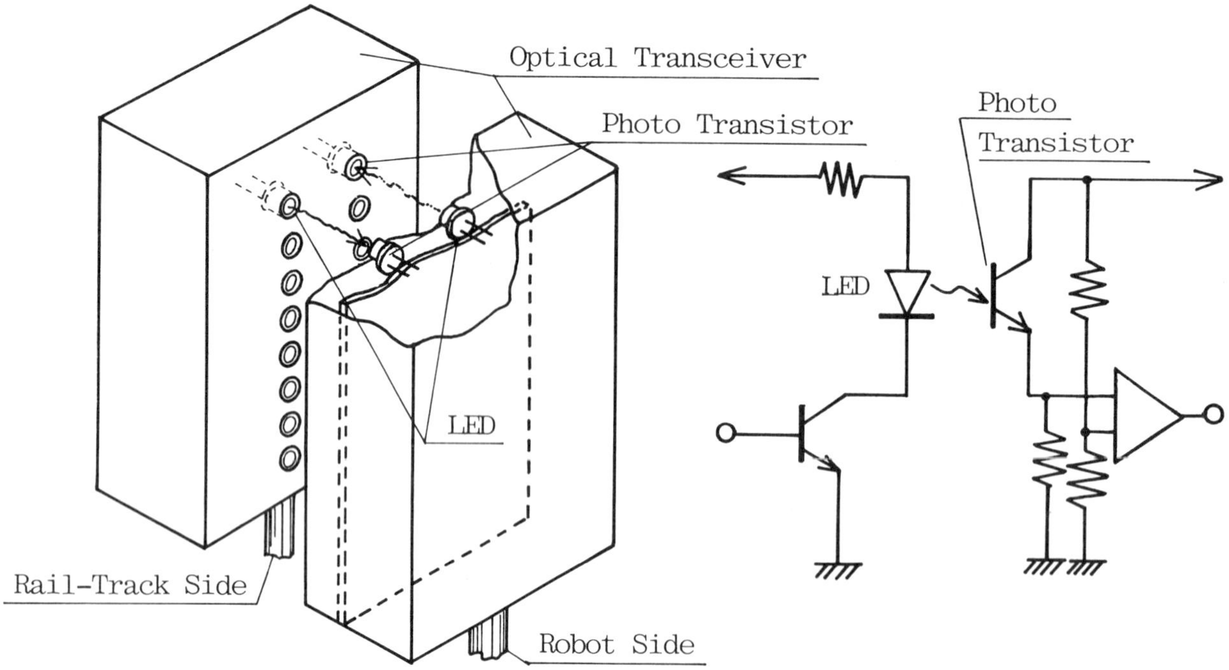

Fig. 12 Optical Signal Transmitter/Receiver System

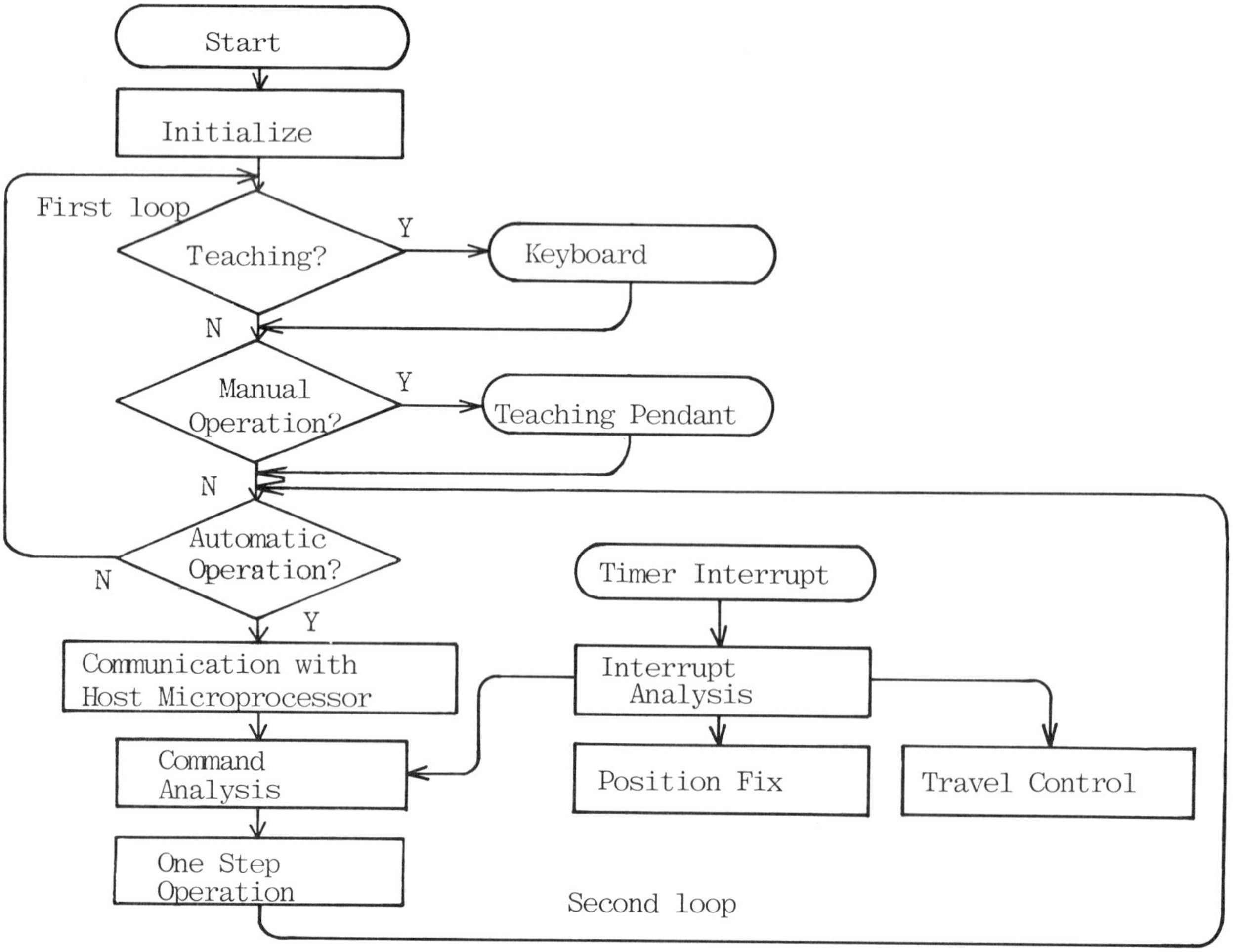

Fig. 13 Configuration of Software

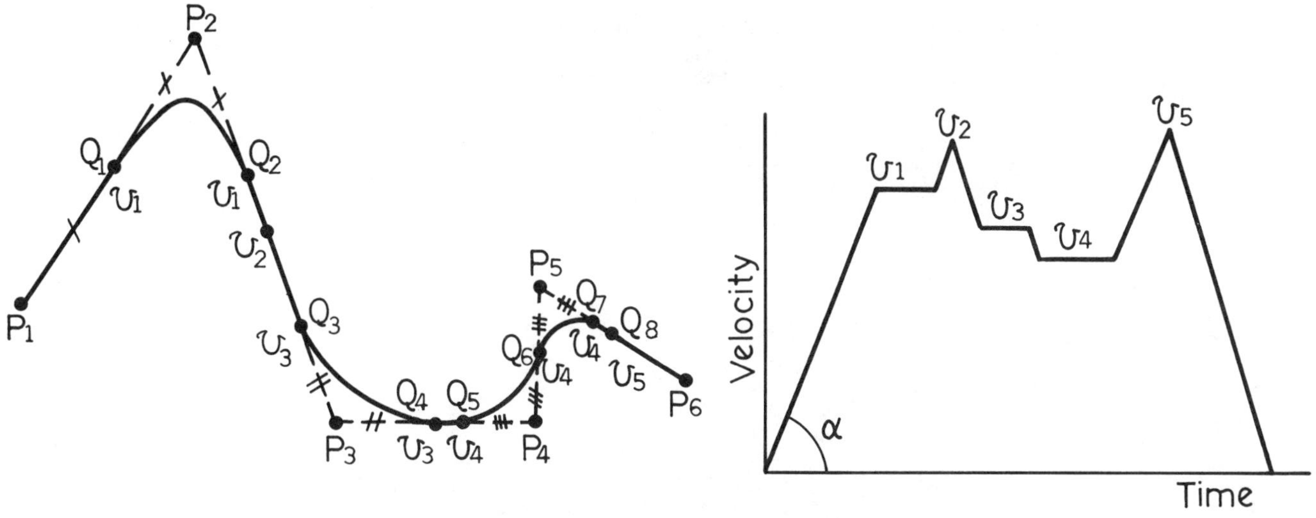

Fig. 14   Smoothing of Robot Trajectory

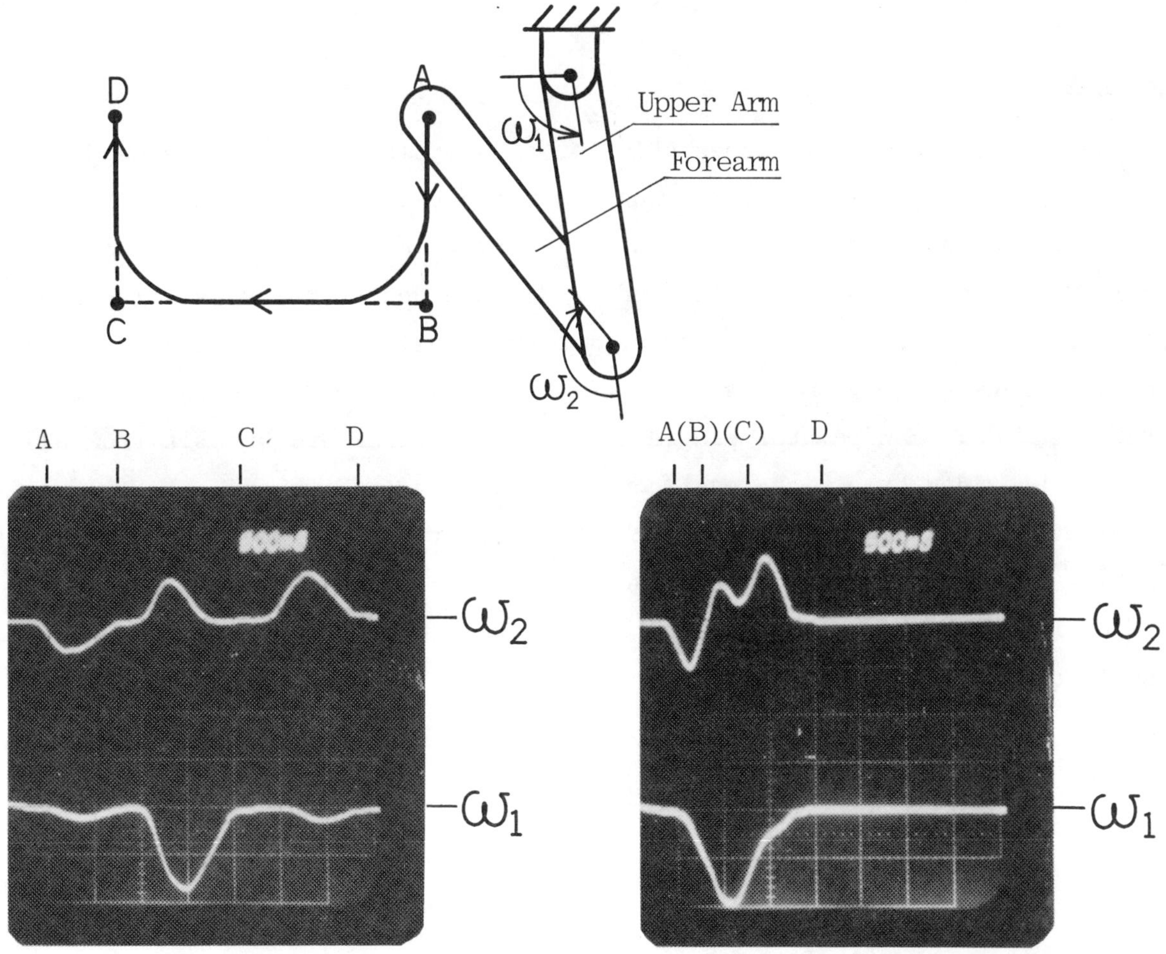

Linear Interpolation

Parabolic Interpolation

Fig. 15   Comparison of Two Interpolations

Robot Manipulator Control Under UNIX

Vincent Hayward
Purdue University
West Lafayette, Indiana

Richard P. Paul
Purdue University

Abstract

The paper describes the implementation of a manipulator language embedded in the "C" computer programming language running under the UNIX time-sharing system. World representation and motion primitives necessary to describe a manipulator task are provided by a set of system calls and predefined data-structures. The language allows for Cartesian path generation, mathematically described trajectories, and conveyor tracking (e.g. moving coordinate frames). The design takes into account force feedback. Manipulator programs are "C" programs and benefit from all the features provided by a high-level computer language. However, from the manipulation stand point, they are comparable with programs written in dedicated manipulator languages. The user also can take advantage of all the features provided by the UNIX operating system. Input-output operations and user interaction are thus readily available.

1. Introduction

It is generally agreed that the present decade will witness the introduction of the so-called second generation robot systems into industry [Paul,79, Shimano,80, Coiffet,82]. This new generation is characterized by the association of internal and external sensors with one or more articulated computer controlled mechanical arms. The underlying idea is to give the robot a high degree of flexibility or in other terms, the robot is expected to be adaptive and versatile.

Programmability has been shown to be the central idea to achieve such a result. Within a single system, the ability to specify motions and to acquire information about the external world for decision making must be provided. The ability to record information for later use is also important. Furthermore, new generation robots tend to minimize the taught part of the task description and emphasize the programmed part.

Sensors are likely to provide a broad range of types of information [Rosen,77]. A rough classification shows that a wide variety of data types and data structures are required to represent the sensory information. Low level sensors deal with simple notions that can be represented by boolean or numerical scalar values. Typical representatives of this class of sensors are touch sensors and position encoders. Higher level sensors like tactile sensors or computer vision systems require more complex data types to represent the spatial information [Rainer 82, Hasegawa 82].

Programming a robot manipulator consists of describing the different positions the manipulator must reach in space in order to grasp and move objects from place to place and to exert forces and torques. Decisions must be made during the execution of the task according to the information acquired from the physical environment by sensors or from the robot's mechanical parameters. Consequently, it is not possible to plan trajectories ahead of time as the motion must be calculated while the task is being executed. In other words, the task should be executed or, from the manipulator stand point, interpreted asynchronously with the arm motion in order to provide for decision making while the arm is moving [Paul 81]. It also appears that complete manipulator applications require far more computation, data processing, and user interaction programming than actual motion control programming.

All these reasons put together lead us to believe that a flexible and powerful manipulator language should present similar features to those of a high level computer programming language. This present research effort aims at providing a robot system that combines extensive capabilities while not limiting future expansion. It should constitute a core system permitting further exploration in many areas under study such as plan generation, sensor integration, learning, task structuring, and cooperation.

2. Design Approach

There are two different approaches to the design of a new manipulator language. One of them is to consider the design of a new dedicated language focusing on the robot control primitives. Representatives of these languages, PAL [Takase 81], AL [Mujtaba 79] , VAL [Unimation 80], AML [Taylor 82], have been developed in universities and industries. This approach allows the syntax to be tailored to describe manipulator tasks as conveniently and concisely as possible. The creation of dedicated languages also involves the creation of specific tools as scanners and parsers whose implementation is irrelevant to the actual robotic work. Any modification is rendered difficult. Modularity, manipulator independence and language mobility are then impaired. On the other hand, conventional high-level computer languages seem to be suitable to express robotic tasks, provided that a well chosen set of data structures and manipulation primitives are defined.

The robot control system presented here and referred to as RCCL (1) makes use of the "C" programming language [Kernighan 78] as a base language. The robot language consists of a set of

(1) RCCL stands for Robot Control C Library.

This work is partially supported by a Grant from the CNRS project ARA (Automatique et Robotique Avancée), France. Facilities to perform this research are provided by the Purdue University CIDMAC project. Richard Paul is the Ransburg Professor of Robotics. This material is also based on work supported by the National Science Foundation under the Grant No. MEA-8119884. Any opinions, findings, conclusions, or recommendations expressed in this publication are those of the authors and do not necessarily reflect the views of the National Science Foundation.

user callable functions and predefined data-structures suitable to describe a manipulator task. RCCL, entirely written in "C", runs under the UNIX operating system itself written in "C". A great consistency throughout the system is thus insured. All the features offered by UNIX for maintaining and debugging computer programs are readily available during the development of the system itself or while writing manipulator programs.

3. System description

The system consists of three parts as shown by figure 1.

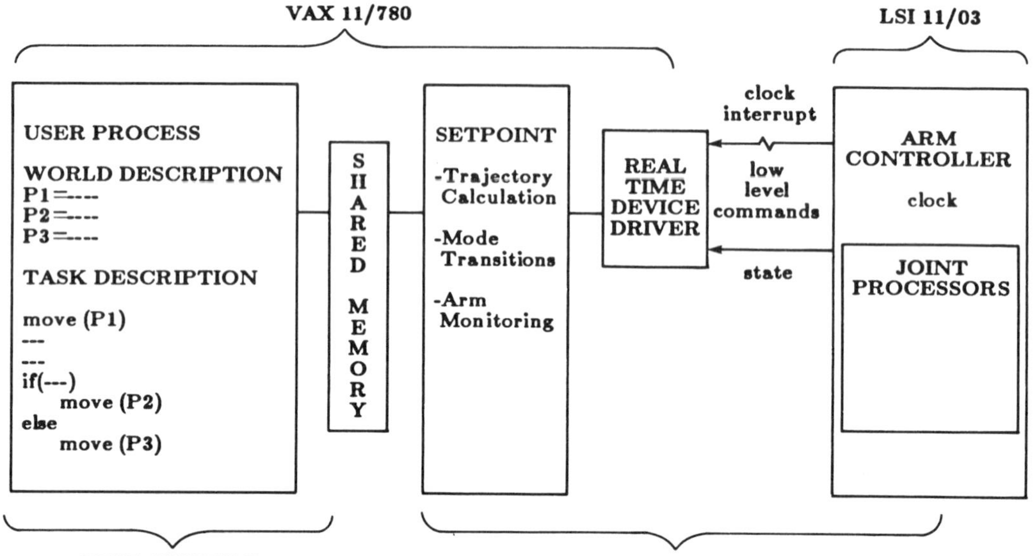

Fig. 1

The higher level part or "user" process is the world model description and the specification of the task. It does not differ from an ordinary UNIX process and is executed under time sharing. Motions are specified by calls to the "move" primitive taking two argument pointers. The first argument refers to a data structure describing the position to be reached in terms of a coordinate frame transformation equation. The second argument refers to a data structure containing time and load requirements for the current path segment.

A call to the "move" primitive can be considered by the user as instantaneous. Consequently, an invocation of the "move" primitive can be followed by any sequence of "C" statements executed while the arm is moving toward the goal position. During motion, all the data related to the state of the manipulator as well as the the data related to the motion and world description are accessible. This allows us to include in the manipulator program any types of computation and input-output operations to external devices. This processing can be performed either asynchronously or synchronously at sample time intervals.

The trajectory, which is continually being evaluated, can be updated or modified while the arm is running. If the arm approaches a position and there is no pending call to the "move" primitive, the arm is brought to rest. This organization provides for two important features :

- Tracking (A task can be described in respect with a moving coordinate frame. A special case is functionally defined trajectories).

- Environmental interaction (It is possible to include in the manipulator programs any input-output operation. A special case is user interaction).

The second part of the system called the "setpoint" process performs the trajectory calculation according to information obtained from the "move" primitive. This process also monitors the arm and transmits to the robot controller the joint values as they are computed. The "setpoint" process is interrupt driven and is executed at the servo clock rate. This process communicates with the "move" primitive via a block of shared data and synchronizes with it via an event flag. The last part of the system consists of the servo code, low level arm control code, and sensory information collection.

The "user" and "setpoint" processes are implemented in a VAX 11/780 minicomputer. As the system is now implemented, the "setpoint" process is executed in kernel mode for efficiency.

4. Language Description

4.1. Basic constructs

Homogeneous transformations are mathematical constructs used to describe the position and orientation of coordinate frames. They have been traditionally used in manipulation and computer vision because they are easy to interpret and are convenient for use in computer implementations [Paul 80]. A homogeneous transformation can either be interpreted as the description of the position and orientation of a coordinate frame with respect to another, or as a transformation performed on the first coordinate frame. Homogeneous transformations are a very general tool [Roberts 65]; however in manipulation the transformations are restricted to the following :

$$\begin{vmatrix} nx & ox & ax & px \\ ny & oy & ay & py \\ nz & oz & az & pz \\ 0 & 0 & 0 & 1 \end{vmatrix}$$

The left upper 3 X 3 matrix is a rotation matrix made with three orthogonal vectors "n", "o", and "a". When the corresponding coordinate frame is attached to the end-effector of a manipulator, the three vectors "n", "o", and "a" can be interpreted as the "normal" vector, the "orientation" vector, and the "approach" vector. The "p" vector is a translation vector or "position" vector.

Relative positions of objects can be described in terms of transformations products. For example, let "OBJ", a transformation, describe the position of an object relative to a reference coordinate frame. Let "HOLE" represent the position of a hole in respect with the frame "OBJ". The matrix product "OBJ HOLE" which is also a homogeneous transformation, describes the position of the hole relative to the reference coordinate frame.

One important property of orthogonal homogeneous transformation is that the inverse transformation can be obtained at reduced computational costs.

4.2. World Representation

One dedicated transformation "T6", is used to represent the position of the end-effector with respect to the reference coordinate frame located at the base of the manipulator. A given manipulator position can be specified in absolute coordinates by writing:

T6 = POS

However, such a description is usually insufficient. For instance, one might need to express that a tool attached to the arm end-effector must reach the position "POS". This is achieved by writing:

T6 TOOL = POS

A more complete description of a motion to a goal position is often written as:

TBL T6 TOOL = CONV OBJ PG

Where:

TBL is the position of the manipulator with respect to reference coordinate frame.

T6 describes the position of the manipulator end-effector with respect to the reference coordinate frame attached to the shoulder or to the base of the manipulator.

TOOL expresses the position of a tool attached to the end-effector.

CONV represents a conveyor belt, it can be a moving coordinate frame with respect to the reference coordinate frame.

OBJ is the position of the object to be grasped lying on the conveyor belt.

PG is the position of the end-effector, relative to "OBJ", where the object is to be grasped.

Position equations are solved for "T6" to obtain the desired position of the manipulator with respect to the reference coordinate frame:

$$T6 = TBL^{-1} \text{ CONV OBJ PG TOOL}^{-1}$$

In the RCCL language, the transformations used in position equations fall into two categories. Constant transforms which are numerically specified at initialization time. Functionally defined transforms which are associated to a "C" function, will be evaluated at sample time intervals. In both cases, their values can be accessed or modified during the execution of the task. The values are determined either by computation or with values obtained from external sensors.

The "makeposition" system call allows the user to set up a dynamic data structure suitable to describe a position equation and returns a pointer to this data structure. The "makeposition" function performs an optimization by pre-multipling pairs of constant transformations.

Several system calls allow the user to specify constant transformations in numerical form or to read them from files of previously recorded transformations. The current implementation allows us to teach one unknown transform per position equation. However, it is planned to embed in the system the structured teaching methods described in [Takase 81].

Numerical specifications can be performed using rotational systems such as Euler angles, roll, pitch, and yaw angles, or rotation of a given angle about a vector in space. It is also possible to use the two orthogonal "approach" and "orientation" vectors.

4.3. Motion Control

A complete task specification is split into a succession of path segments, each of them corresponding to an instance of the "move" primitive. While the first argument to "move" specifies the position to be reached, the second argument is a pointer to a data-structure describing the manner the motion is to be performed.

4.3.1. Path Generation

Two motion schemes are currently available. Straight line motion is obtained by controlling the joint angles of the manipulator such that the coordinate frame attached to the tool moves along straight line trajectories in the Cartesian space. In this case, the "makeposition" function expects an additional argument to specify which transformation of the equation is to be considered as the "tool" transformation. The change in orientation is performed in terms of two simultaneous rotations. The first rotation is conducted about a vector parallel to the cross product of the initial and final "approach" vectors. The "approach" vector is attached to the "tool" coordinate frame in the direction in which the tool is pointing. The second rotation is conducted around the approach vector and brings the orientation to its final value. This rotation scheme has the advantage of causing no change in the motion of the first five links of the manipulator when, for example, picking up objects randomly lying on a conveyor belt. For many manipulators, the second rotation corresponds to a rotation of the final joint.

During a path segment motion, the trajectory is computed with respect to the goal position coordinate frame. This coordinate frame is represented by the transformation which left multiplies the "TOOL" transformation when the equation is solved for "T6". The position equation is solved for a goal position coordinate frame at the beginning of the path segment. For example, the equation :

$$\text{TBL T6 TOOL} = \text{CONV OBJ PG}$$

is first reduced to:

$$\text{TBL T6 TOOL} = \text{CONV TEMP1}$$

in which "OBJ PG" is reduced to "TEMP1". The new goal position coordinate frame "TEMP1" is computed as:

$$\text{TEMP1} = \text{CONV}^{-1}\ \text{TBL T6 TOOL}$$

The position equation is temporarily modified by the addition of a functionally defined transformation "DRIVE" on the right hand side of the goal position. The value of the "DRIVE" transformation is dependent upon the time. At the beginning of the path segment, its value describes the differential transformation between the two successive positions, at each end of the path. At the end of the path, its value is the identity transformation.

It must be noted that the optimization routine has knowledge of the structure of the task and avoids to pre-multiply "TEMP1" by the inverse of "TOOL" in order to obtain further simplification. The equation can then be solved for "T6" :

$$\text{T6} = \text{TBL}^{-1}\ \text{CONV TEMP1 DRIVE TOOL}^{-1}$$

During the execution of the path segment, "T6" is repeatedly evaluated as the coefficients of the functionally defined transformation "DRIVE" are linearly interpolated. The joint values are obtained by the analytical kinematic solution for the particular manipulator [Paul 81].

This type of motion, although imposing a heavy computational load, presents several advantages :

- The motions are simple and easily predictable.
- Provisions are made for conveyor tracking and functionally defined motions.
- For a manipulator with the first two or three joints rotational, the inertial load is reduced as the load is moved closer to its base.

Unfortunately, the joint rates will become infinite if the manipulator goes into degeneracy [Paul 83].

The second type of motion is a joint interpolated motion and is linear in the joint coordinate space. The position equation is solved and the joint angles obtained only once at the beginning of a path segment for the goal position. Intermediate values are obtained by linear interpolation. This type of motion offers the following advantages :

- Manipulator degeneracies do not cause a problem.
- The joint velocities are easily predictable and the motion speed is only limited by individual maximum joint velocities.

Efficient manipulator tasks require a combination of these two types of motion.

4.3.2. Transitions generation

Discontinuities occur when the manipulator moves from one path segment to another. Transitions must be provided between each path segment. Smooth transitions are obtained by specifying continuity of position and velocity at both ends of the transitions. Figure 2 depicts the evolution of one coordinate in the coordinate system of the corresponding motion, Cartesian space for straight line motions, joint space for joint interpolated motion.

A quadratic polynomial function provides for continuity of position and velocity, and zero acceleration at both ends of a transition. The quadratic function is used to compute the coefficients of the "DRIVE" transformation in the case of a straight line motion or the joint angles in case of joint interpolated motion. When the manipulator is transitioning between two path segments of different types, the transition is performed in the motion mode of the subsequent path segment. While the trajectory plotted in the coordinate system of the subsequent path segment is linear, it appears curved in the coordinate system of the previous path segment. In order to avoid a velocity discontinuity, we modify the position of the point B as shown in

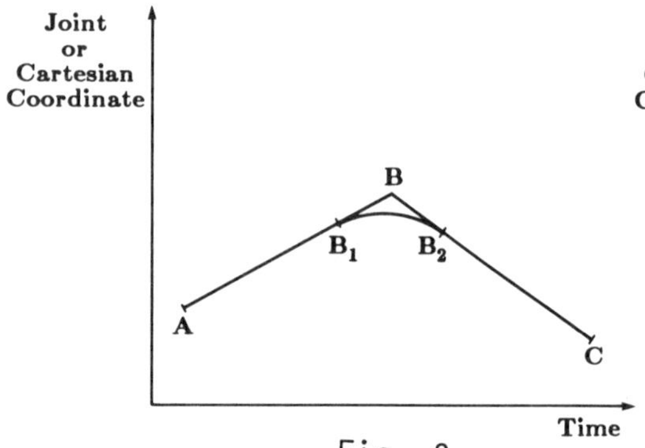

Fig. 2

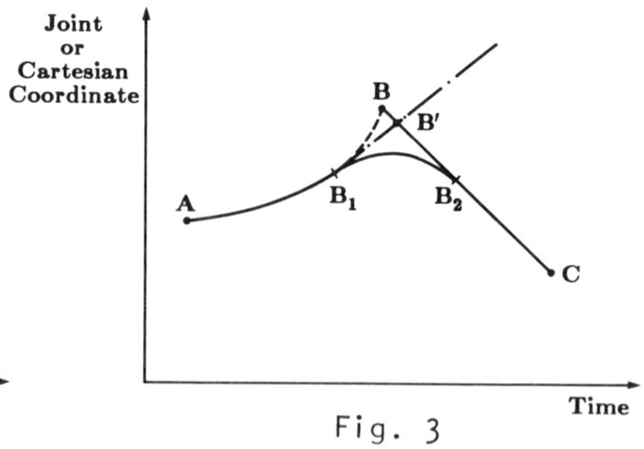

Fig. 3

figure 3. The corresponding joint or Cartesian velocity is computed to estimate a new position B'.
The resulting position error is negligible except when the arm is moving close to a degenerated configuration. For both motion types, the manipulator never moves through a specified position but close to it and it is necessary to stop the arm to obtain a perfect position setting. For example, in order to move along a perfectly square trajectory we are required to stop the arm at each corner.

4.3.3. Moving Coordinate Frames

In the case of joint interpolated motion, tracking is obtained by estimating the position of point C when the manipulator will reach it. Better accuracy is obtained when the moving coordinate frame is functionally defined as opposed to being determined from an external sensor. In this case the position must then be estimated linearly. In the case of straight line motion, the set of joint values is obtained at each time sample interval, and the arm will automatically track the moving coordinate frame.

4.4. Force Control

Force control must be used when the manipulator is to come into contact with rigid objects of the environment such as the mating parts. The motion must be controlled in terms of desired forces and moments when the contact occurs or has occurred. These contact forces and moments are conveniently expressed in the tool coordinate frame [Shimano 78]. Three cases must be taken into account:

- The arm is moving in free space and is position servoed. It must encounter an obstacle and certain joints must then be force servoed when the contact occurs.

- The arm is touching an object and moves along a constrained trajectory while exerting forces and torques.

- The arm is moving along a constrained trajectory and must leave the contact for a free space trajectory. All the joints must then be position servoed.

Whenever the arm comes into contact with a solid object and is exerting forces on it, the arm is said to enter a comply mode which must be specified in the manipulator program. The RCCL language offers three primitives to handle the compliance specifications.

The "untilf" primitive causes a transition into comply mode and terminates the current motion when a force condition is met. The "untilp" primitive causes a transition into position servo mode if a position error is detected in the specified directions; the current motion is also terminated. The motion can also terminate if the transition does not occur within a specified distance in the direction of the motion. This abnormal condition can be tested with the "getdist" function returning the distance when the manipulator actually stopped.

The "comply" primitive explicitly brings the arm in comply mode. This primitive is used to alter the current compliance state. It is often used before issuing a "move" statement in order to obtain a constrained motion. The arm will remain in its current compliance state until a call to the "lock" function brings the arm back to position servoing.

For any motion terminated on a condition, the world model may have to be updated to take into account the actual position where the manipulator stopped. This new position information might be very useful in any subsequent motion related to this location. For example, consider the case of a manipulator picking up an object which it had previously placed on a surface whose height is only approximatively known. If the final position of the object was updated, then the manipulator is able to retrieve the object immediately. The "update" function takes a position equation pointer and a transformation pointer as arguments. It causes the transformation to be updated in the given position equation in order to reflect the new position. The same "update" function can be used to teach positions when one transformation is left unspecified.

4.5. Semantics of the "move" primitive

When a "move" statement is specified in a program, the user must keep in mind that the call is executed when the arm is approaching the position specified by the previous "move" statement. All the code following a "move" statement is executed asynchronously while the arm is moving toward the next position. The execution of the code segment begins when the manipulator is in position B1 (Figure 2), at the beginning of the transition. The system computes at that instant the next position and determines the parameters of the transition.

When a "move" statement is followed by a "until" statement, a call to the "getdist" function immediately before the next "move" statement allows to decide if the transition from position servo mode to comply mode (or vice versa) was successful.

5. Summary

The following list summarizes the different system calls available to the user.

- Gentr_eul, gentr_rpy, gentr_nao, gentr_rot : a family of calls to create constant transformations.

- Gettr(name) : read a transformation from a file.

- Newtrans(name, fn) : create a transformation, attached to the function fn.

- Makeposition(name, lefthandside, EQ, righthandside, TOOL, tool) : set up the position equation data structure and return a pointer to it.

- Movecart(p, m) : move the manipulator in Cartesian mode according to the position equation "p" with mode "m" (time and load specifications).

- Movejnts(p, m) : move the manipulator in joint mode according to the position equation "p"

- Movestop(t) : bring the manipulator to rest for time "t". If the last motion occurred in Cartesian mode, the solution will continue to be repeatedly evaluated. Movestop(0) brings the velocity to zero at the end of the current motion.

- Movewait(e) : brings the manipulator to rest until event "e" occurs. Useful for synchronizing the execution with operations of undetermined durations such as grasping an object.

- Moveconf(p, m, c) : move the manipulator in joint mode according to position equation "p" with mode "m" to the manipulator configuration "c".

- Update(p, t) : Update the transformation "t" in the position equation "p" according to the current position of the arm.

- Untilf(directions, forces, dist) : monitor the forces and the torques along or around the specified directions in the tool coordinate frame. Transition to comply mode when the force conditions are met and terminate the current motion. Terminate the motion if the distance "dist" is exceeded in the direction of the motion.

- Untilp(directions, limits, dist) : monitor the position error in the tool coordinate frame. Transition to position servo mode when the limits are reached or distance "dist" is exceeded in the direction of the motion.

- Getdist() : delay the transition to the next motion and return the distance when the manipulator actually stopped.

- Comply(directions, forces) : go into comply mode and exert force and torques along or around specified directions.

- Lock(directions) : go into position servo mode along or around specified directions.

6. Example

The following program demonstrates the task description for grasping a washer laying on a working table and inserting it on a vertical pin. The program makes use of a computer vision system able to give the X and Y positions of the washer and the pin. The thickness of the washer is assumed to be small and the manipulator must first make contact with the table and then grasp it while sliding the fingers along the table. This strategy insures that the manipulator neither stops too high and misses the washer nor smashes the table position servoing. The manipulator will then carry the washer above the pin and will start a motion in the Z direction attempting to insert the washer. Most of the time, the manipulator will not be able to fit the washer around the pin due to positioning inaccuracies. This condition must be checked and an appropriate action executed. One among the possible strategies [Inoue 74] is to perform a spiraling motion in the X Y plane while monitoring the displacement in the Z direction. Sooner or later, the manipulator while exerting a vertical force will start a descending motion and must then transition to another comply mode before reaching the final position.

The example is given in "C" language. The syntax is sufficiently close to other structured programming languages to make it easily understandable. A system file included at the very beginning, contains all the specialized type declarations.

```
/*
 * Washer grasping and insertion on a pin
 */

#include "../i/manip.h"          /* manipulation types declarations */

TRSF_PTR spire;                  /* global functionnally defined transform */

pumatask()
{
        int spirefn();           /* spiral generating function */
/*
 * Constant Transformations
 */

        TRSF_PTR
         /* arm with respect to table    */
         base = gentr_rot("BASE", -200., 300., 200., yunit, 0.),

         /* end effector */
         tool = gentr_rot("TOOL", 0., 0., 180., yunit, 0.),

         /* washer with respect to table (initialized by the camera) */
         washer  = newtrans("WASHER", const),
```

```
        /* washer approach with respect to washer */
        washer_appro = gentr_rot("WASHER_APPRO", 0., 0., 10., yunit, 180.),

        /* where to grasp with respect to washer */
        washer_grasp = gentr_rot("WASHER_GRASP", 0., 0., 0., yunit, 180.),

        /* final position with respect to table (initialized by the camera) */
        pin = newtrans("PIN", const),        /* value set by camera */

        /* final position with respect to pin */
        pinbase = gentr_rot("PINBASE", 0. ,0. , 0., yunit, 180.),

        /* where to land with respect to pin */
        pintip = gentr_rot("PINTIP", 0., 0., 50., yunit, 180.),

        /* pin approach with respect to pin */
        pin_appro = gentr_rot("PIN_APPRO", 0., 0., 55., yunit, 180.),

        /* move free with respect to pinbase */
        pin_away = gentr_eul("PIN_AWAY", 0., 10., 20., 20., -20., 0.);

/*
 * Positions
 */
        POS_PTR
        approach = makeposition("APPROACH",      /* approach the washer */
                base, t6, tool, EQ, washer, washer_appro, TOOL, tool),

        getwasher = makeposition("GETWASH",      /* grasp the washer */
                base, t6, tool, EQ, washer, washer_grasp, TOOL, tool),

        above = makeposition("ABOVE",            /* above the pin */
                base, t6, tool, EQ, pin, pin_appro, TOOL, tool),

        touch = makeposition("TOUCH",            /* touch the top of the pin */
                base, t6, tool , EQ, pin, pintip, TOOL, tool),

        spiral = makeposition("SPIRAL", /* touch the top and move spiraling */
                base, t6, tool, EQ, pin, pintip, spire,TOOL, tool),

        down = makeposition("DOWN",              /* final position */
                base, t6, tool, EQ, pin, pinbase, TOOL, tool),

        away = makeposition("AWAY",              /* all done */
                base, t6, tool, EQ, pin, pinbase, pin_away, TOOL, tool);
/*
 * Modes
 */
        MODE_PTR
                towash = setmode(200, 2000, 0.),
                totable = setmode(150, 500, 0.),
                topin = setmode(150, 1000, 0.),
                totop = setmode(150, 500, 0.),
                searching = setmode(150,2000),
                tobottom = setmode(200, 1500, 0.),
                frompin = setmode(200, 1500, 0.),
                torest = setmode(200, 2000, 0.);
```

```
/*
 * initialize functionaly defined transform
 */
        spire = newtrans("SPIRE", spirefn);      /* attach function to    */
                                                 /* transform             */
/*
 * task
 */

#define TOLERANCE       5

        readcamera(washer);
        movejnts(approach, towash);
        readcamera(pin);
        movecart(getwasher, totable);
        untilf("fz", 100, TOLERANCE);    /* no check is needed since */
                                         /* the arm cannot miss the table */
                grasp();
                if (movewait(grasped) == NO)
                        giveup("washer not found");
        lock("fz");
        movejnts(above, topin);
        movecart(touch, totop);
        untilf("fz", 100, TOLERANCE);
        if (getdist() < TOLERANCE) {  /* hole not found at first attempt */
                movecart(spiral, searching);
                untilp("dz", SMALL, TOLERANCE);
        }
        lock("fz");
        comply("fx fy",0 ,0 );  /* constrained motion   */
        movecart(down, tobottom);
        untilf("fz", 100, TOLERANCE);
        lock("fz");
        release();
        movewait(released);
        movecart(away, frompin);
        movejnts(park, torest);
}

/*
 * generate a spiraling motion in the x y plan
 */

spirefn()
{
        extern int timesegment;          /* time spend during current motion */
        double time,
               omega = 1.,               /* one turn per second */
               growth = .5;              /* .5 mm per second */

        if (timesegment <= 0)
                return;
        time = timesegment / 1000.;      /* in seconds */
        spire->p.x = time * growth * cos(omega * timesegment * PIT2);
        spire->p.y = time * growth * sin(omega * timesegment * PIT2);
}
```

7. Conclusion

One of the goals of this work was to demonstrate that a powerful and efficient manipulator language can be implemented in terms of function calls. This idea matches conveniently one of the design options of the "C" language which does not provide any syntactical construct to directly handle formated or unformated input-output, string processing, dynamic memory allocation, and other high-level constructs. These constructs are related to the capabilities of a given installation rather than to the capabilities of computers in general. Our system is merely the addition to the "C" language of system calls concerned with the control of a manipulator. It therefore benefits from the advantages of the UNIX system as far as maintenance, portability and modularity are concerned. The macro capabilities provided by the "C" language may also be used to improve the cosmetic aspects and the readability of the manipulator programs. The three layer organization mentioned earlier, allows an easy interface with other languages supported by UNIX. Interpreters like LISP or APL are attractive options. The system is currently running two different manipulators, a Scheinman Stanford Arm and a Unimation Puma 600. In the future, it will be necessary to develop debugging tools such as step by step execution, trace, and replay facilities.

8. Acknowledgments

I would like to thank Bill Fisher who programmed the servo code and is responsible for the modifications to the arm controllers and the initial development of the control software. The specialized real-time device driver and the additions to the UNIX kernel code are the work of George Gobble whose help has been most valuable.

9. References

Coiffet, P., "La robotique: vers une nouvelle révolution industrielle," Le courrier du CNRS, Septembre 1982.

Hasegawa, T., "A New Approach To Teaching Object Description For A Manipulator Environment," Proceeding of the 12th ISIR, Paris, June 1982.

Inoue, H., "Force Feedback in Precise Assembly tasks," MIT Artificial Intelligence Laboratory, Memo No. 308, Aug. 1974.

Kernighan, B. K., Ritchie, D. M., "The "C" Programming Language," Prentice-Hall Software Series, 1978.

Mujtaba, S., Goldmann, R., "AL User's Manual," AIM-323, 1979, Stanford Calif.: Stanford University Artificial Intelligence Laboratory.

Paul, R. P., "Manipulator Language," Workshop On The Research Needed to Advance The State Of Knowledge In Robotics, April 15-17, 1980, organized by J. Birk and R. Kelley, supported by N.S.F.

Paul, R. P., Shimano, B. E., Mayer, E. G., "Kinematic Control Equations for Simple Manipulator," IEEE Transactions on Systems, Man, and Cybernetics, Vol SMC-11, No 6, June 1981.

Paul, R. P., Stevenson, C., "Kinematics of Robot Wrists," International Journal of Robotic Research, Vol 2, No 1, 1983.

Paul, R. P., "Robot Manipulators: Mathematics, Programming, and Control,"MIT Press 1981.

Paul, R. P., "Robots, Models, and Automation," Computer Magazine, July 1979.

Raibert, M. H., Tanner, E. J., "Design and Implementation of a VLSI Tactile Sensing Computer," International Journal of Robotic Research, Vol 1, No 3, 1982.

Roberts, L. G.,"Homogeneous Matrix Representation and Manipulation of N-Dimensional Constructs," MIT, Lincoln Labs., Document MS-1045, May 1965.

Rosen, C. A.,Nitzan, D., "Use of Sensors In Programmable Automation," Computer Magazine, December 1977.

Shimano, B. E., "The Kinematic Design and Force Control of Computer Controlled Manipulators," Stanford Artificial Laboratory, Stanford University, AIM 313, 1978.

Shimano, B., "Levels of Communication and Programming Methods," Workshop On The Research Needed to Advance The State Of Knowledge In Robotics, April 15-17, 1980, organized by J. Birk and R. Kelley, supported by N.S.F.

Takase, K., Paul, R. P., Berg E., "A Structured Approach to Robot Programming and Teaching," Proc. COMPSAC79. New York: IEEE Computer Society, 1981.

Taylor, R. H., Summers, P. D., Meyer, J. M., "AML: A Manufacturing Language," International Journal of Robotic Research, Vol 1, No 3, 1982.

Unimation Company, "User's Guide to VAL," 1980.

Designing Custom Robots for In-Plant Use

Robert A. Brown
Acushnet Company
New Bedford, Massachusetts

Abstract

Most industrial efforts involve sorting, arranging, and combining, best done by computers which can rapidly generate a wide variety of control instructions. Robots can respond in various ways since they work within a pseudo volume determined by the products of the range and resolution of their axes. Yet the speed and accuracy of Robots are severely limited by their inertia $I=MR^2$ and stiffness. A solution is to custom design Robots small, light, and stiff as possible for your applications. The paper describes the principles and equipment used to develop a highly successful in-house Robot system.

INTRODUCTION

In 1977 the Acushnet Company decided to make a major commitment to improved product quality and lower manufacturing costs by designing new automatic equipment for our Titleist golf ball plant.

We considered applying conventional machine design approaches using cams and air cylinders, as well as a Robotic approach. For various reasons, we divided the automation tasks between conventional and Robotic design groups.

Both approaches have been highly successful. The conventional approach has resulted in the complete automation of the ball cover and core assembly, automatic buffing of the flash caused by the cover molding operation, and automatic orientation and stamping of the name and number on the finished ball.

The Robotic approach was applied to the tasks of automatically winding golf balls and then automatically removing freshly painted golf balls from a moving conveyor following spray paint operations.

I will discuss the paint spray Robot application in this paper as an example of Designing Custom Robots for In-Plant Use.

The Robotic approach is a method of generating positional Variety (V) using the minimum number of actuators. Variety is defined as the number of different things, considered as a whole, that can happen.

When X = number of distinct stopping positions of the Robot on each axis,

and n = number of axes of the Robot,

then $V = X^n$.

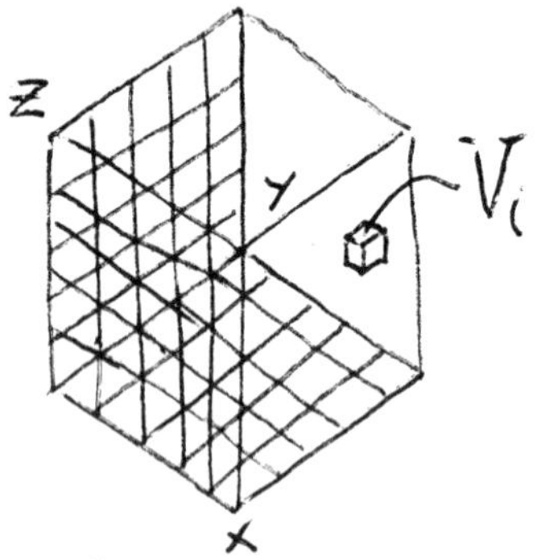

Assuming a resolution of .2% of the range of each axis and a three-axis Robot, then X=250, n=3 and

$$V = 500^3 = 125{,}000{,}000 \text{ possible positions.}$$

A conventional machine capable of generating the same variety using 2 position actuators would require

$$2^n = 125{,}000{,}000$$

or $n = 27$ actuators.

Since a simple three-axis Robotic type machine has greater potential for generating Variety than a conventional machine, we hoped the Robotic approach would help us deal with the bewildering complexity involved in our automation projects.

DISCUSSION

Our first Robotic machine is shown in Figure 1. Initially we used small synchronous motors as actuators which were manually controlled with forward, reverse, and off switches. This allowed us to learn how a Robot deals with its world. These low-power motors also revealed where excessive loads existed and where to adjust gearing to optimize Robot performance.

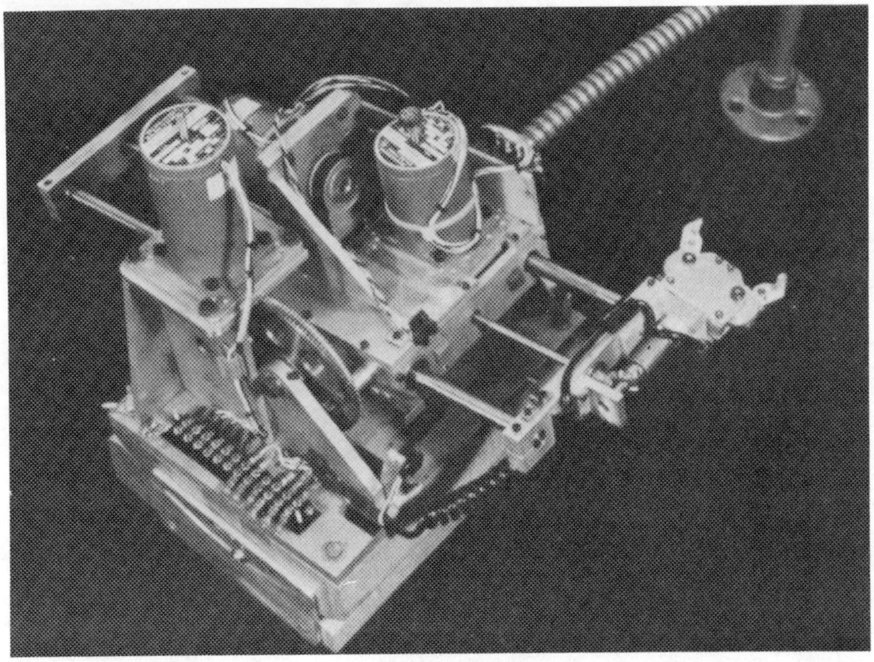

Figure 1

After determining the mechanical structure and sequence of operations needed to perform specific tasks using a Robot, we sat down to design the control system.

High on our list of priorities was a means of controlling and limiting acceleration, while maximizing displacement with minimum cycle times. We were required to perform a 90° trajectory over a 33" radius in less than .75 seconds. This required machine motions comparable to that produced by the finest cams.

We were also required to keep the Robot cost below $25,000 for adequate payback. This mandated the use of the simplest step-motor/micro-processor system possible. Knowing we were facing a very difficult problem, we decided to take a good look at the kinematics involved in the machine motions.

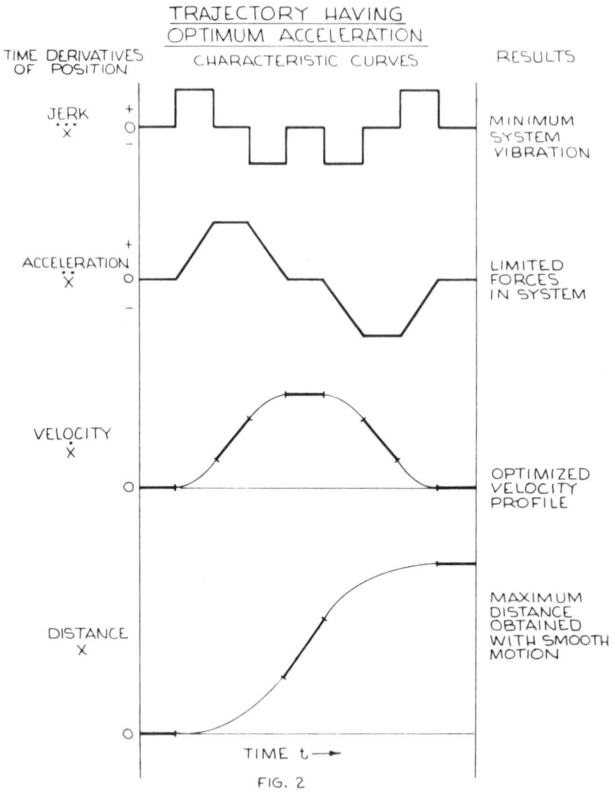

Figure 2

Figure 2 shows the relation between Position, Velocity, Acceleration, and Jerk which is needed to obtain smooth motions. Using this representation, we established an Algorithm which generated the machine Positions from the Jerk values through a smooth process of Digital Integration. In a cam-driven system, the Positions cut in the cam generate the Velocity, Acceleration, and Jerk through a process of Differentiation which accentuates any roughness and vibration.

The results of this control system exceeded expectation. The machine operated smoothly at speeds which you would expect from a well designed mechanical system and we successfully demonstrated that we could meet the cycle time requirements of the paint spray operation.

Historically, painted balls were manually unloaded from the painted ball conveyor onto drying trays (as shown in Figure 3) and loaded into trucks for holding in a drying room.

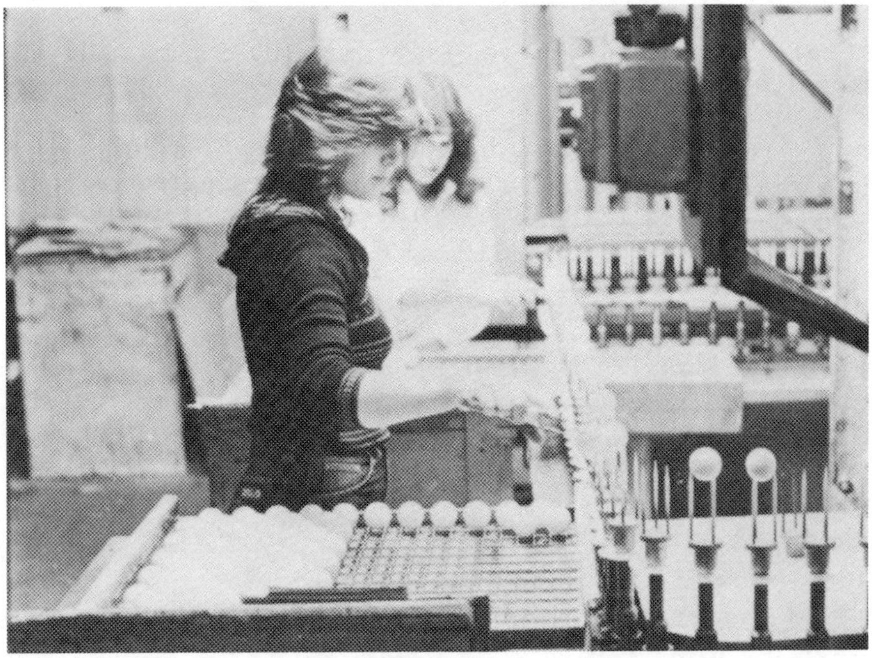

Figure 3

It took three operators to unload balls from each spray booth conveyor. We had seven booths operating three shifts for a total of 63 operators. Our goal was to automatically load unpainted balls, unload painted balls onto trays, and then use a single operator to move loaded trays and trucks from several booths. The balls are first loaded onto the ball conveyor (as shown in Figure 4) from a hopper which stores enough balls for several hours of painting.

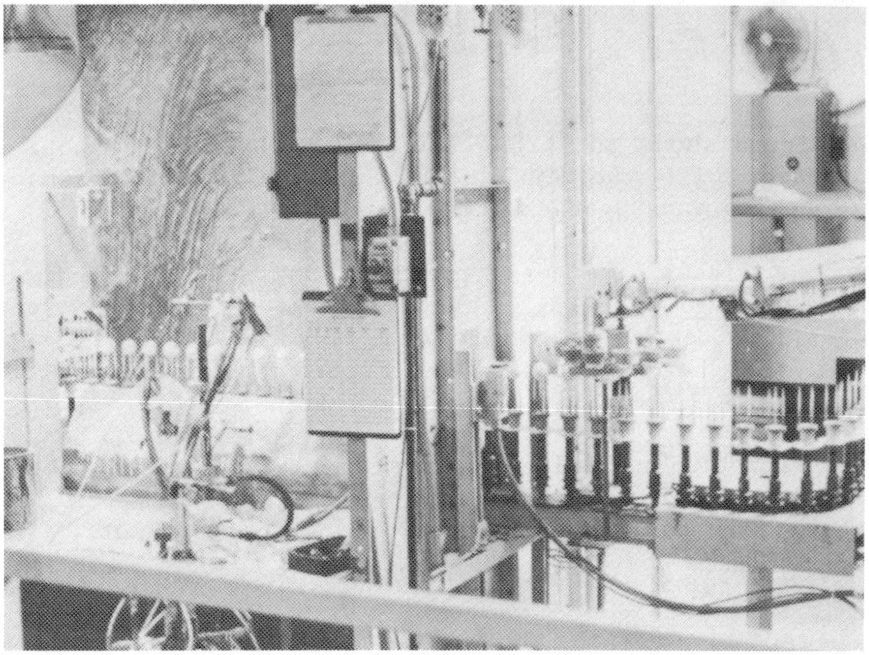

Figure 4

The balls are then sprayed at over 100 balls per minute in a conventional spray booth.

Figure 5 shows the Robot unloader picking and placing four balls into specially designed trays having a staggered 2" spacing for maximum packing density. The total cycle time, including returning to the ball conveyor ready for the next pick, is less than two seconds.

Figure 5

This Figure shows the tray conveyor which indexes and feeds trays in conjunction with the Robot. It automatically "downstacks" empty trays at one end and "upstacks" the loaded trays at the other end.

Figure 6 shows the main control cabinet which monitors and controls the whole paint spray system, including Robot, tray conveyor, spray booth, and stack vent sensors. Note the programmable controller and the five custom fabricated motor drive cards in the control cabinet.

The programmable controller "supervises" the system operation, but due to its relatively low operating speed, custom-built, high-speed controllers were required for each of the stepping motors.

Figure 6

As shown in Figure 7, the Robot itself consists of a set of stepping drive motors located near the rotary axis.

Figure 7

Power is transmitted to a gripper capable of picking up four balls at a time. Each finger is individually geared and spring loaded so that the picked balls are precisely lined up prior to placement in the trays.

The motor and limit switch leads are carried to a connector box through a long flexible conduit so as to minimize wire fatigue.

The Robot can be disconnected and removed from the tray conveyor and a replacement installed into pinned bushings within a few minutes. The ability to quickly exchange Robots seems indispensable in a three-shift production environment. The spare Robots are stored on a Robot transporter shown in Figure 8.

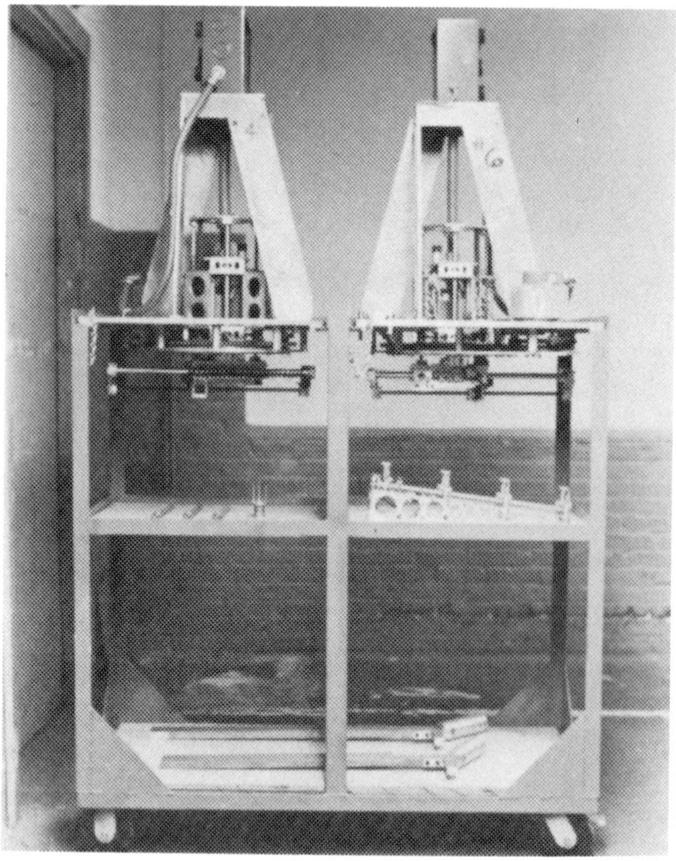

Figure 8

The gripper is also modular and readily removed by loosening two captive screws. Installing a replacement automatically makes the electrical connections needed for the limit switches on the gripper allowing the gripper to be replaced in about 15 seconds.

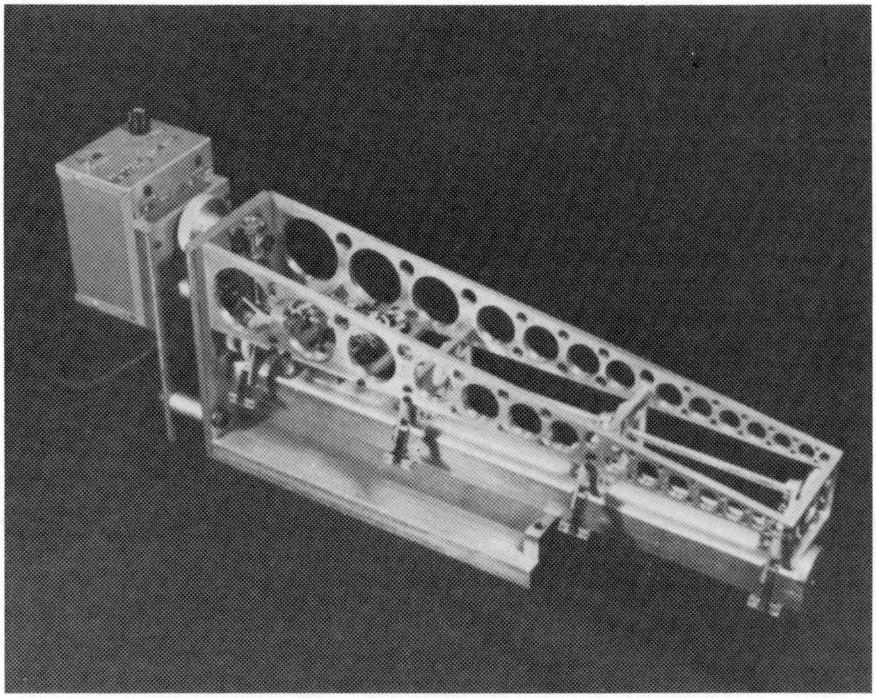

Figure 9

The gripper and its limit switches are adjusted on a calibration stand (shown in Figure 9).

The Robot and its limit switches are adjusted on a calibration table (shown in Figure 10).

Figure 10

A ball box representing four golf balls is placed in the gripper. The rotation, vertical position, and extension for the 12 trajectories are then checked against scribed lines on the calibration table.

We attempt to adjust all axes to within ±.010" which is equivalent to ± one pulse of the stepping motor. However, this is not always possible due to flex within the Robot. This is an inherent problem with all Robot systems. If you try to establish its position when the Robot is wound up either through deceleration or vibration, the gripper will be in a slightly different position when it is at rest. This problem can be dealt with in our system by tailoring the trajectories to minimize acceleration and jerk during critical Robot homing periods in the program.

These Robots, like all machines, occasionally experience malfunctions. These incidents can be classified in terms of "Crashes" or "Crunches." Pushing the reset button gets you out of a Crash as it does with most computer Crashes. In a Crunch, something gets bent or broken. These can be avoided by laying out the work space whenever possible so as to tolerate the "loss" of an axis. The motors, gearing, and shafting should be sized so that the motors simply stall if something jams. Step motors are excellent for this purpose.

It sometimes helps to make the system flexible so that it will also bend over obstacles. Sometimes these overload precautions hamper the operation of the system. A motor sized to stall may stall due to a momentary peak force due to vibration or backlash. Also, a limber system will have larger, lower frequency oscillations at the end effector. However, some unnecessary shutdowns and more time for homing may be less costly than one big Crunch.

Whenever there is a system failure, the question arises: "Is it a mechanical, electrical, or computer problem?" This clearly defines the three team members required to get a Robot system going and keep it going. It has been our experience that a very high level of performance is required in all three disciplines in order to obtain a Robot system upon which the whole organization can depend.

RESULTS

One of the greatest advantages that came out of the paint spray automation effort was a large reduction in the number of dropped wet balls. In spite of the skill and diligence of the manual operators, thousands of freshly painted balls were dropped each week. These have to be stripped and repainted at great expense. When properly aligned, the Robot system does not drop balls.

Pick marks on the balls are practically eliminated due to the precise pursuit and capture method used by the Robot. Production was increased due to the ability of the Robot system to operate through lunch and coffee breaks. The payback was less than one year based upon direct charges made for machine fabrication and installation.

The automatic painting system operates approximately 23 hours a day, 5-6 days a week. The first system has been operating for two years. Two other systems are in operation and four more are being fabricated at this time. Once a day, while the booth is shut down to refill the spray gun and clean the ball support spindles, parts of the Robot are lightly oiled by the machine operator. Then once a week, the Robot is wiped down and reoiled. Aside from this, our operating uptime has exceeded 97%, and preventative maintenance requirements have been minimal. We intend to overhaul each Robot every year or so to replace drive belts, spline and high helix drive shafts, and flex cabling.

CONCLUSION

The most important lesson learned from this effort is the relation of Robot size to performance. We learned from the very beginning that we could not obtain rapid cycle times if we reached too far or made the Robot too big.

The following formula shows the relation of the power required to rotate a solid round arm of length L through an angle θ in time t having a density D with a modulus of elasticity of E, and a flex deflection of d_{max}.

$$P = \frac{4 \pi L^7 D^2 \theta^2}{3 E d_{max} t^3}$$

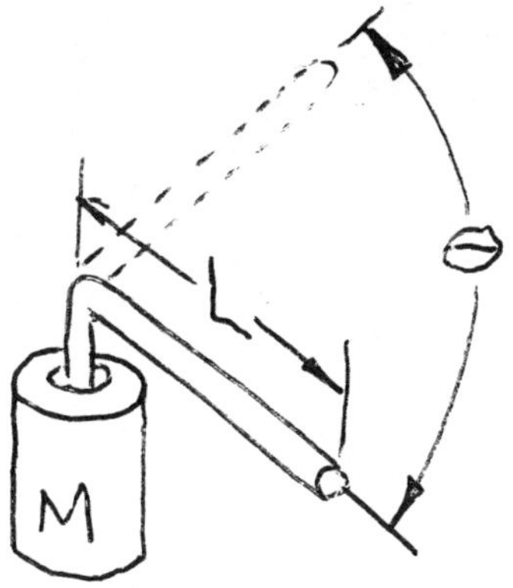

From a practical point of view, $\Delta P/\Delta T$, the rate of change of power, is the limiting factor and is determined by the torque to inertia ratio of the drive motor. Thus, a key equation in Robotics is:

$$\frac{dP}{dt} = \frac{4 \pi L^7 D^2 \theta^2}{3 E d_{max} t^4}$$

These limitations are made more severe by the fact that the torque to inertia ratio drops dramatically for higher power stepping motors. This equation shows that a small increase in Robot arm length or small decrease in cycle time greatly increases drive motor performance requirements.

This equation emphasizes the need to design the Robot system to minimize the size of the Robot itself. This may involve bringing piece parts close to the Robot and taking them away with some independent means. It also requires keeping any and all unnecessary weight out of all parts in the Robot's extremities. This requires machining out sections and drilling lightening holes whenever possible. It also requires the use of materials of high specific stiffness arranged for maximum rigidity. It requires getting used to designing the stationary parts of the Robot to be massive in contrast to tiny finger parts which may be held together with #0-80 jeweler size screws.

Due to the tremendous penalties of size, it has been my experience that the potential gains in performance more than offset the costs of designing your own Robot to meet your exact needs.

Testing and Simulation Techniques for Industrial Robots

Vladimir Cop
The Research Institute of Metalworking Industry
Sidlisko, Czechoslovakia

Abstract

The paper analyses the testing problems of selected characteristics, especially those concerning dynamics and accuracy. Simulation methods for checking of interactions between a robot and the particular machine and auxiliary device functions are introduced. Special stands designed for checking of these functions and interactions enable simulation of machining, forming, die casting and related technology conditions. These arrangements make it possible to diagnose concurrently 18 functions. Testing and simulation techniques effectively contribute to improving of quality and reliability of tasks solved in the stages of development, production and utilization.

1. REQUIREMENTS ANALYSIS FOR FUNCTION AND VERIFICATION IN AUTOMATED SYSTEMS.

Quality improvement in automated technological workplaces /ATW/ with industrial robots and manipulators /IRaM/ requires high quality of the research and production results as well as that of special service conditions at users´ plants.

The research stage includes the application of the design methods with prescribed reliability. In addition there are methods of extensive subassembly testing regarding functionality, operating life and reliability as well as complex methods of functional model and prototype testing.

In the production stage there is a set of methods and means used for the production quality control which are usually defined in coincidence with the technical conditions of the particular subassemblies and product itself.

It is already the research stage in which it is necessary - from the point of view of application - to verify the particular functional and operational characteristics of the IRaM especially regarding the automatic cycle of a machine, the requirements resulting from the technology itself and the interaction of the IRaM with the other technological and auxilliary devices at a workplace. Similar situation appears in the stage of designing the ATW with IRaM at the user or through the special designing-and-supplying organization. These are the areas where methods of production and automated devices simulation prove to be useful.

Finally, in the application stage it is necessary to know the instantaneous technical state of the automated devices in the ATW as well as to provide a quick trouble shooting.

This short analysis makes it clear that providing high quality

ATW with IRaM requires to develope the methods of simulation, diagnostics and autodiagnostics for the IRaM and for the particular auxilliary devices operating in an automatic cycle. Losses resulting from unsuitable coordination of an automatic cycle of the IRaM, machines and other auxilliary devices and those resulting from idle time and repairs still represent a big effectiveness potential.

As a theoretical basis for examining the functional and reliability characteristics can be regarded interactions between two or more elements, subassemblies or products functioning in an automatic cycle.

The interactions are characterized as follows:

- time duration of the interaction between two or more elements, subassemblies or products functioning in an automatic cycle,
- information on the physical aspects of the interaction between two or more elements in an automatic cycle as well as on the degree of utilization of their nominal characteristics,
- feature range of the elements, subassemblies and products entering the automated system.

On the basis of these characteristics it is possible to determine the critical interactions enabling to prove the presence of the dependence between the interaction characteristics and functional and reliability ones.

This theoretical basis is important mainly in the research and development stages where the features of the IRaM and the ATW with IRaM respectively cannot be experimentally verified.

From the point of view of an experimental verification this theoretical basisserves for determining the diagnostic points as well as for building the equipment based on the theoretical conclusions resulting from the interactional analysis of the functions and reliability.

2. SIMULATION AND DIAGNOSTICS OF INTERACTIONS AND BEHAVIOUR OF FUNCTIONS.

Practical application of the theoretical conclusions concerning the interaction and function examination requires the appropriate equipment.

Chronometers are used for examining the time interaction between two or more elements or products in an automatic cycle. The data concerning the time loading of an element in an autómatic cycle enable to assess the probability of the failure occurrance, i. e. the characteristic of the subassembly and system reliability. Chronometers are also used for trouble shooting in a pilot plant or a production system.

Simulators of operating conditions, especially those of subassemblies, serve for data detection concerning the physical behaviour of the interaction between the elements and the degree of their loading in an automatic cycle. For a continuous verification of automated workplace or industrial robot system functions it is necessary to combine the methods of measuring the particular physical features /pressure, friction, temperature, oscillation, etc./ including a time analysis of a cycle. Such analyses yielded the data for defining the critical load values for the particular subassemblies in an automated equipment.

Simulators of functions of the individual elements, subassemblies and of the whole ATW with IRaM are also used for determining the range of the elements and subassemblies entering the implemented automated system. The results of the measured values of the feature range represent important data for determining the failure occurrance probability, i. e. the device reliability. These methods are also used in the input check within the IRaM and the appropriate automated equipment manufacture.

Next we introduce some practical methods that we apply in the development and manufacture of the IRaM.

2.1 Simulation of the damper operation.

The simulation of the damper operation functions is performed on a testing stand /Fig. 1/ which enables the simulation of the following parameters:

- reduced mass acting on the damper /m_{red}/,

- speed of the reduced mass /v/,
- working stroke of the damper /s/,
- damper looping frequency /f/.

The testing stand utilizes the mobility of a rotary arm acting on the damper through a stop and a sliding unit. The simulation of the variation of the reduced mass moving towards the damper is achieved by changing the stop position in the arm or by increasing the moment of inertia of the rotary arm by adding a simulating load to it.

The simulation of the damper load parameters /m_{red}, v, s, f/ enables to experimentally verify the working diagram of the damper /functional dependance of an axial force of the damper on its stroke/ and at the same time the damping energy at various operational speeds.

2.2 Simulation of positional sensor operating conditions.

The simulation of positional sensor /magnetic and induction sensors/ operating conditions is performed on the testing stand /Fig. 2/ which enables simulation of the following parameters /operating conditions/:

- switching frequency /by velocity change of the rotary arm provided with a switching element of the sensor - shutter, magnet, etc./ /f, v/,
- sensor feeding /Un/,
- positional orientation of the sensor /P/,
- distance between the functional part and the switching element of the sensor /s/.

Multiple combinations of the simulated conditions at the testing stand enable to verify the reliability and the range of the positional sensor.

2.3 Simulation of the rotary drive operating mode.

The operating mode simulation of the rotary servodrive /electric, hydraulic/ is performed on the testing stand /Fig. 3/ including a hydraulically operated brake, a power supply and a control circuit.

The simulated conditions of the drive operation are as follows:

- rotation speed /v/,
- moment of load /M_k /,
- sense of rotation.

The simulation of the servodrive operating mode results in the amplitude-frequency characteristics, the effectiveness and the contact interactions between the transmission unit teeth.

2.4 Simulator of the industrial robot, machine and auxiliary equipment operating functions.

The simulation of the technological devices operating functions within the modelling of the ATW with IRaM utilizes a unit-built stand enabling simulation of a wide range of production machines and peripheral equipment /NC production machines with a horizontal and a vertical axes of operation, forming and casting machines, simple handling, etc./.

The unit-bulit stand structure enables to compose simulators of the production machines operating functions and of interactions between a simulated machine and an industrial robot from a number of subassemblies /functional units/. Fig. 4a and 4b represent examples of a simulated model workplace arrangement including 3 NC machines and an industrial robot.

The above mentioned model workplace makes it possible to simulate

- a complex automated cycle and its time optimization,
- the functions of machine loading and unloading as well as the particular positional interactions between a gripper, a workpiece and a machine fixture,
- checking and interlocking functions.

Readjustment purposed at modelling another workplace with different both technology and layout takes two shifts.

The working environment simulator consisting of an air-conditioning booth and the stand for vibration simulation is designed for IRaM testing in different conditions.

2.5 Chronometer

Measuring of time control structures of an operating cycle, structures of control-informational signals and interlocking reliability in the ATW requires a multichannel chronometer which digitally records the pulses determining the beginning and the end of the observed action.

Therefore a multichannel chronometer has been developed whose input circuits including buffer stores are multiplex-tested. Each electric signal state change at the particular input has its corresponding time of day. The time of day, the input number and the kind of change /rising or falling electric signals/ are recorded in the chronometer store. The chronometer can operate in two modes:

1. the mode of temporary record and data selection /chronometry/,
2. the mode of permanent data record with storing the latest recorded data /diagnostics/.

Data selection is performed manually or automatically on a display or a recording device /printing unit, punchint unit etc./. The basic technical parameters:

- input number: 16
- input voltage level: 24 V, 48 V = 24 V, 48 V, 110 V, 220 V
- resolution: 0,01 s
- method of representation:

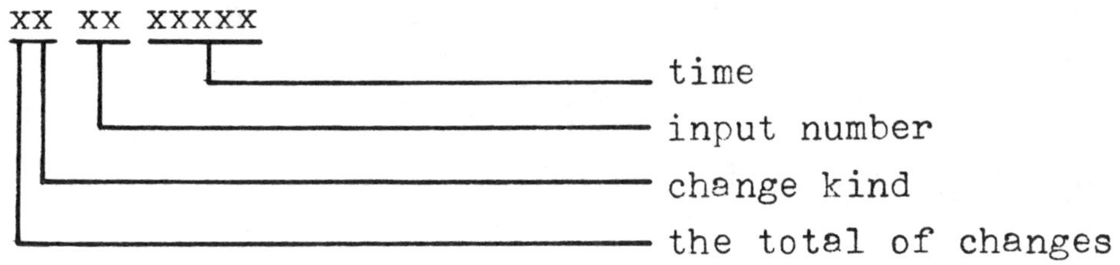

- the input is in BCD code.

2.6 Chronometric and diagnostic device

A combined chronometric and diagnostic device /Fig. 5/ represents an important means for complex evaluation of functions and interactions of IRaM. The device is based on the combination of measurements of the interaction behaviour within the particular physical quantities of oscillation, temperature, pressure etc. as well as of the cycle time characteristic. This combination enables a simultaneous evaluation of time and functional loading of an element, subassembly and an assessment of its reliability characteristics.

2.7 Control system diagnostics for IRaM

A separate problem of the control system quality improvements is the extanding of their diagnostic functions:

- diagnostics of the control system interactions with other subsystems of the mechanical checking-and-interlocking parts of the IRaM,
- autodiagnostics of the control system itself.

The new developed RS 1 C control system ic characterized by the following basic diagnostic capabilities:

1. In the interaction diagnostics of the IRaM subsystems with the control systems:
   - drive function failure indication,
   - sensor malfunction indication,
   - overrunning of the time limit for the programmed movement.
2. In the autodiagnostics of the control system itself:
   - hardware complexity test with the control system switched on,
   - text editor - instruction correctness test or continuously carried out instruction parameter range test,
   - RAM service test.

3. CONCLUSIONS

Modern methods of IRaM and ATW designing make full use of the simulation and diagnostics methods. These methods substantially influence the effectivness of the practical application of the IRaM in the ATW.

The complexity of the technological tasks and the corresponding complexity of the technical implementation of the IRaM as well as their control systems and auxiliary devices necessirely lead to use of these methods.

In addition to the comprehension of the simulation methods for operating functions and interactions in the particular subassemblies of the IRaM, a simulation of functions at the automated workplaces is increasingly used. It concerns predominantly positioning of technological heads in the welding and assembly processes, workpiece fixturing and removing in the process of machine, conveyor and pallet loading and unloading. Cycle optimization enables to cut time by more than 15%.

The diagnostics has gradually mastered the methods of autodiagnostics of mechanical, hydraulic, pneumatic and electrical elements and subassemblies has been gradually worked out too. In addition practical methods of diagnostics of the complex automated manufacturing systems representing the basis for establishing of automated robotized lines, workshops and plants has been developed.

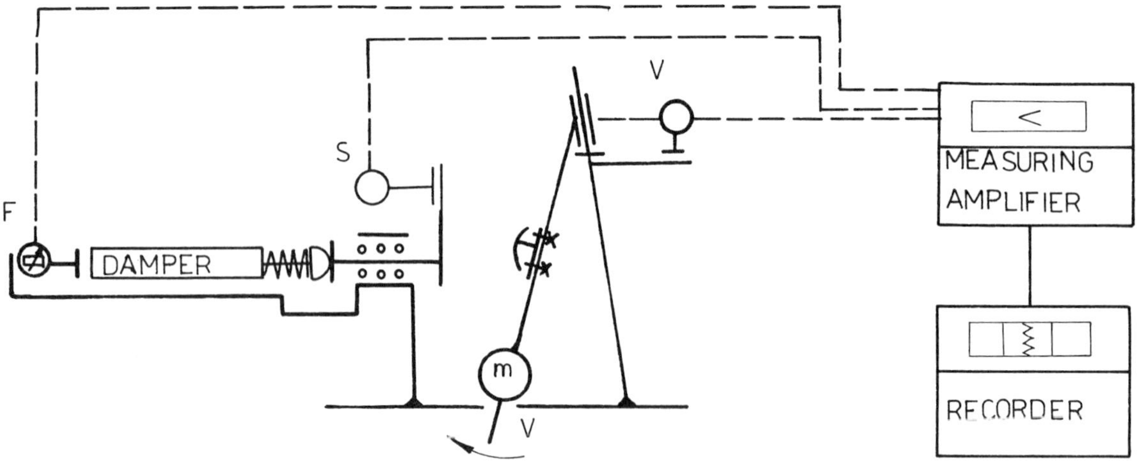

FIGURE 1

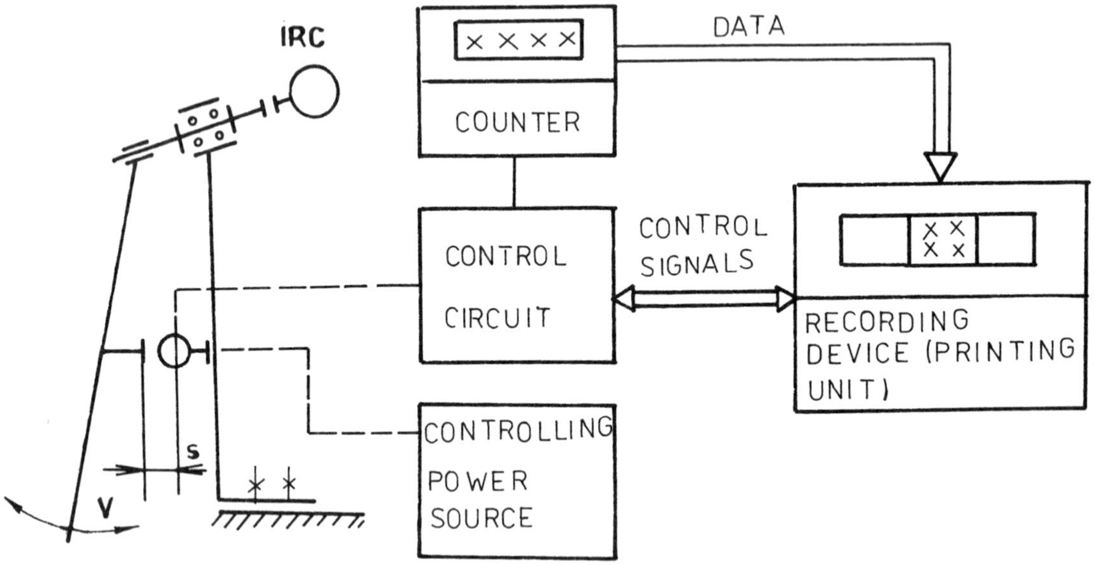

FIGURE 2

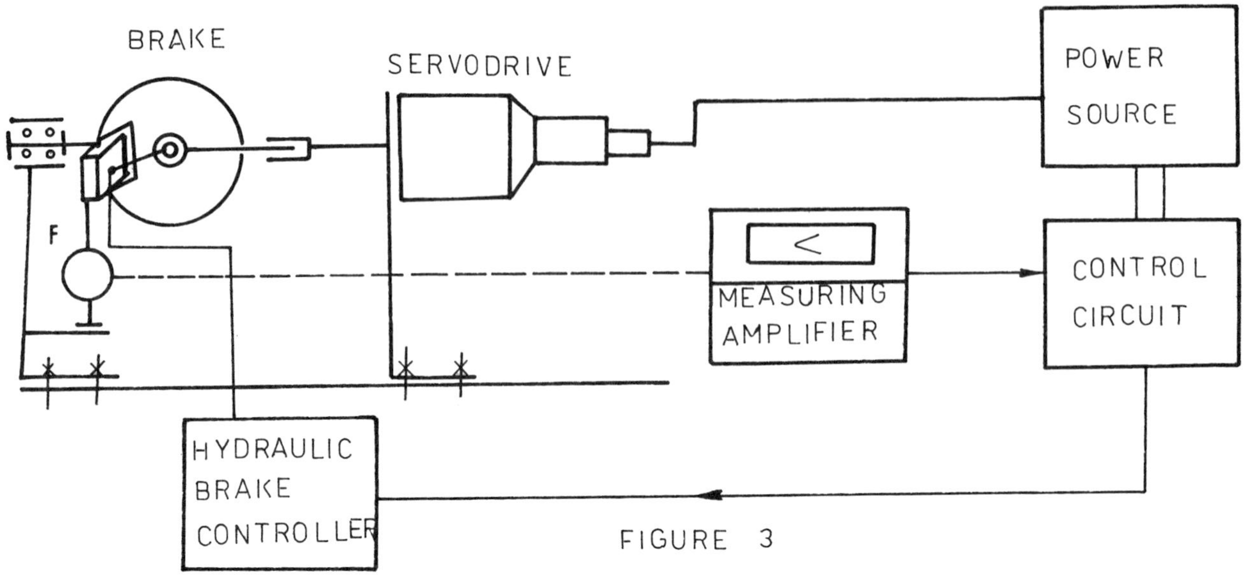

FIGURE 3

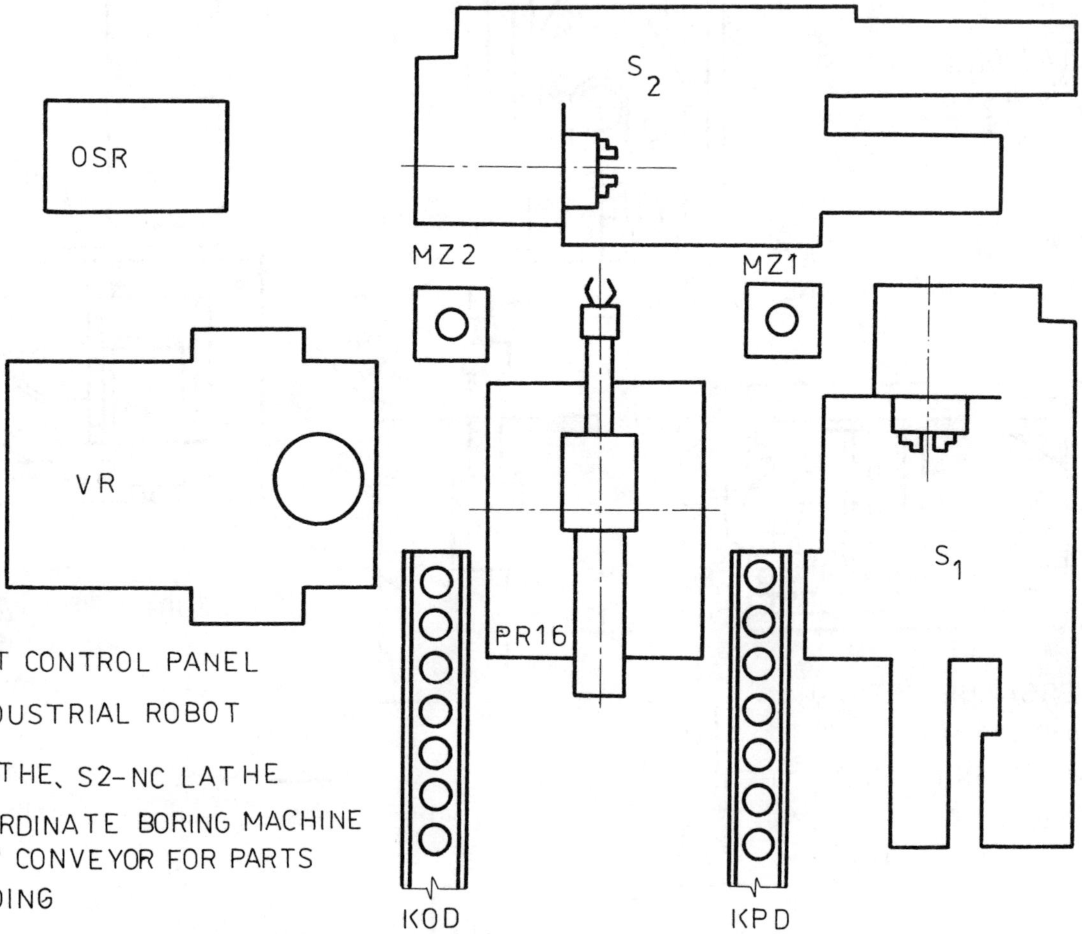

OSR-ROBOT CONTROL PANEL

PR-16 INDUSTRIAL ROBOT

S1-NC LATHE, S2-NC LATHE

VR-NC COORDINATE BORING MACHINE

KPD-STEP CONVEYOR FOR PARTS FEEDING

KOP-STEP CONVEYOR FOR PARTS REMOVING

MZ1 AND MZ2-IN-PROCESS MAGAZINES

FIGURE 4A LAYOUT OF THE WORKPLACE INCLUDING 3 NC MACHINES AND A ROBOT

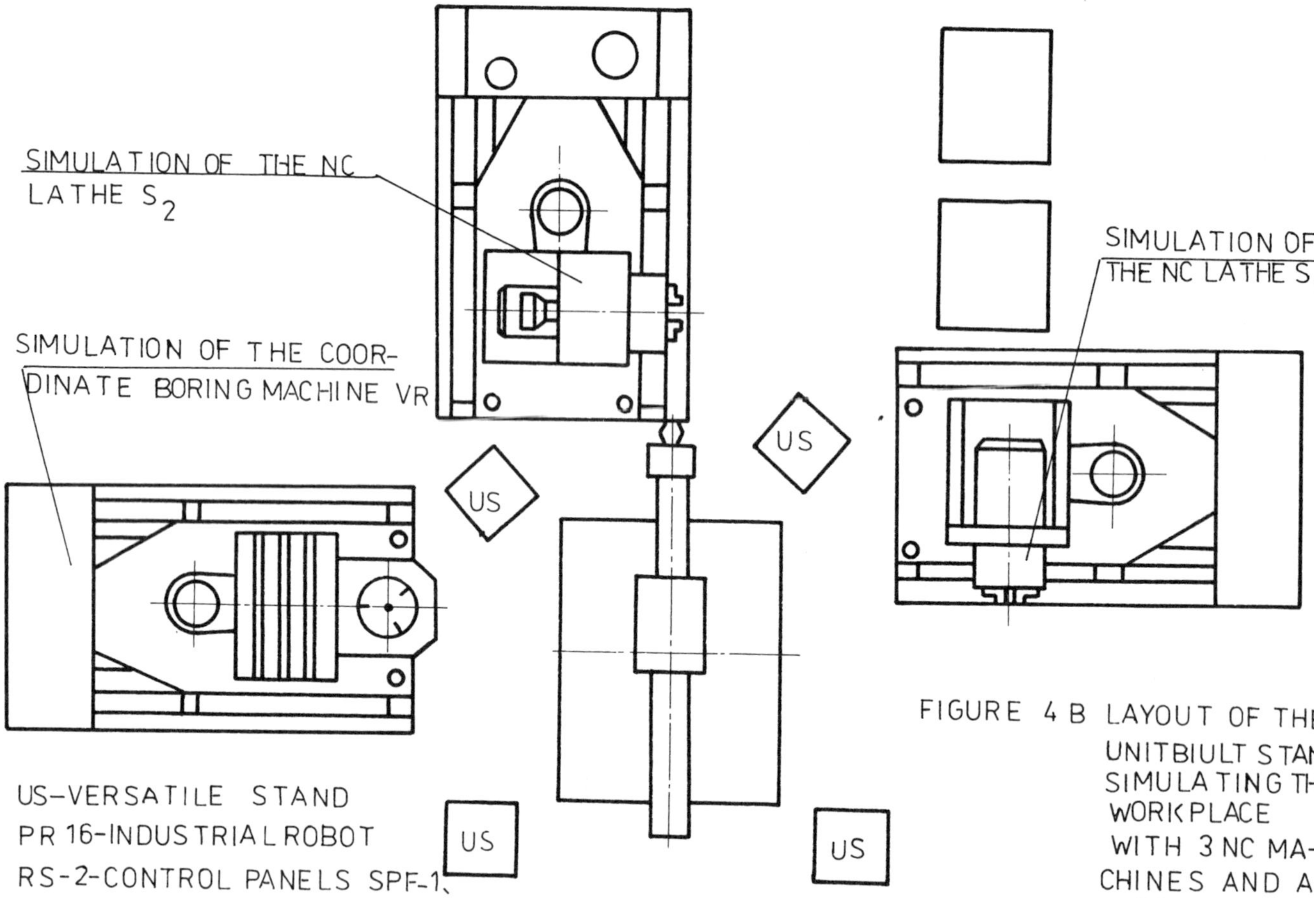

FIGURE 4 B LAYOUT OF THE UNITBIULT STAND SIMULATING THE WORKPLACE WITH 3 NC MACHINES AND A ROBOT

US-VERSATILE STAND
PR 16-INDUSTRIAL ROBOT
RS-2-CONTROL PANELS SPF-1, PR

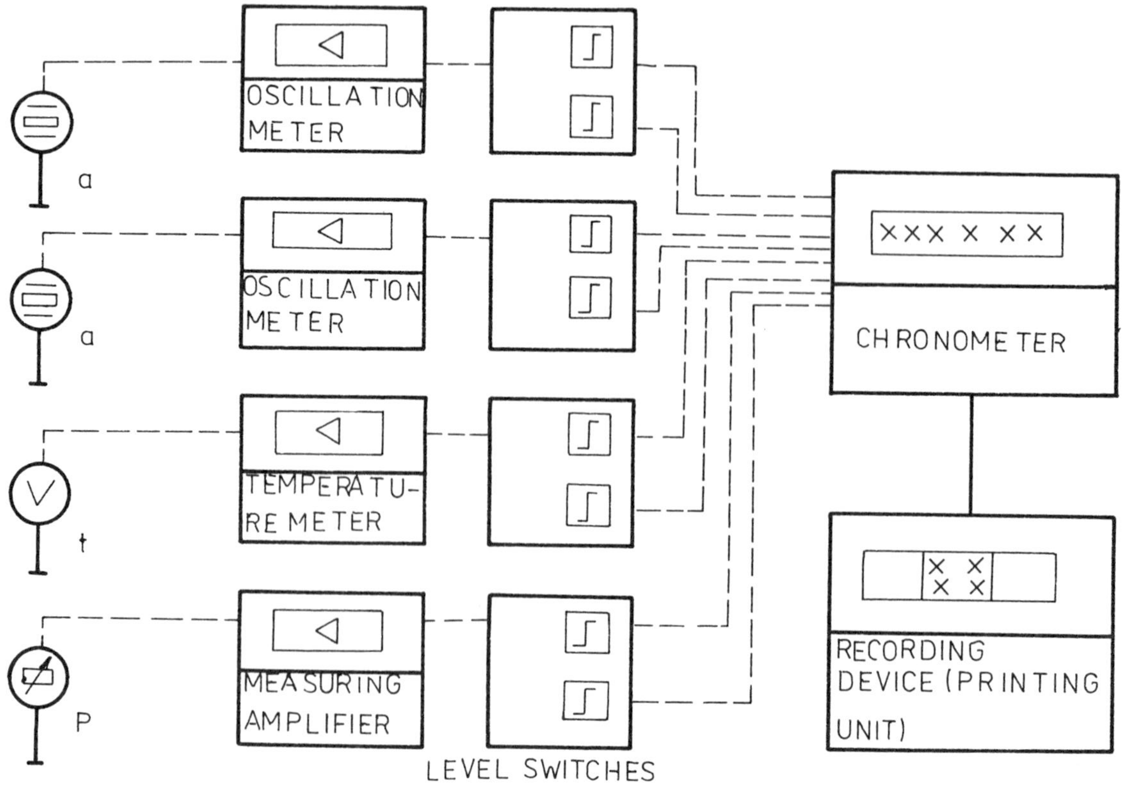

FIGURE 5

Straightline Motion of a Manipulator through Hybrid Computation

J.F. Steinman
Drexel University
Philadelphia, Pennsylvania

R.D. Klafter
Drexel University

Abstract

The possibility of using a combination of analog and digital techniques for performing the computations required to produce straightline motion of a three-degree of freedom cylindrical coordinate robot is proposed. A scheme, based on generating quadratic functions of time with sets of cascaded integrators, so that a small number of non-linear computing elements may be used, is developed. Advantages and disadvantages of the scheme are discussed.

1. INTRODUCTION

A central issue in the continuing advancement of robot technology is computational load. In order for the next generation of robots to attain a high degree of autonomy from their human operators they will have to spend more time on sensing and less time on the control of manipulator motion. A case in point is the movement of an arm along a straightline trajectory.

Various routine tasks call for the execution of straightline trajectories: approaching an object in preparation for grasping, painting, and welding are a few examples. However, controlling the arm so that it moves along such trajectories is non-trivial. The difficulty arises from the fact that linear motion of the joint variables does not not produce linear motion of the end effector.

In world coordinates, straightline motion can be formulated in terms of homogeneous transforms (1,2). If the starting and ending positions of a trajectory are given, then motion between these positions is specified by a time dependent "drive transform" (1). This transform, computed from the trajectory end points, gives not only position but also orientation of the end effector as a function of time. In order to move the end effector along the trajectory, a command signal is computed using the drive transform and then sent to the various joint controllers. For digital controllers this command signal takes the form of a sequence of set points, each one indicating the next position to which the joints must move. Unfortunately the computations involved in generating a set point are considerable.

First the host (or main) processor must compute the desired position in world coordinates by evaluating the drive transform. Then it must convert this postion from world coordinates to joint coordinates. The process of conversion invariably involves many multiplications and several evaluations of transcendental functions.

Generally robot manufacturers reduce the number of set points required by programming joint controllers to interpolate between successive points. However, there is a limit to how much interpolation can accomplish. If the time interal between the update of successive set points is too large, it is possible to excite structural resonances of the manipulator (1). Thus, even with interpolation the computational burden posed by straightline motion is large. In fact, for industrial robots presently in use, the maximum speed of motion is set by computational rather than mechanical constraints.

Various methods for solving this problem might be investigated. The use of special purpose processors or parallel processing techniques are possibilities. Another approach migh be to devise a hybrid computing scheme in which digital calculations are supplemented with analog calculations. This approach is attractive because of the effectively unlimited speed capabilities of an analog system. In what follows we discuss the feasibility of implementing such a hybrid scheme for one robot geometry.

2. MANIPULATOR GEOMETRY AND MATHEMATICAL DESCRIPTION OF STRAIGHTLINE MOTION

A method of implementation was examined for the three degree of freedom cylindrical coordinate robot shown schematically in Figure 1. The manipulator geometry consists of a waist axis with a joint range of $\pm 170^{\circ}$, a prismatic axis with non-zero minimal extension, and a wrist axis with a joint range of $\pm 90^{\circ}$. Since the manipulator possesses only three degrees of freedom, orientation of the end effector can not be varied independently of position. This characteristic would make the robot useless in any practical application of straightline motion. It was purposely chosen, however, so that the important issues surrounding a hybrid computing scheme would not be obscured by mathematical complexity.

Due to the lack of orientational variability, the transformation from world to joint coordinates can be obtained without the use of homogeneous transforms. Inspection of the geometry yields the following relationships.

$$\theta_1 = \tan^{-1}(-P_x/P_y)$$

$$d_2 = \sqrt{P_x^{\,2}+P_y^{\,2}} - \sqrt{a_3^2 - (P_z-d_1)^2} \tag{1}$$

$$\theta_3 = \tan^{-1}\left[(P_z-d_1)/\sqrt{a_3^2 - (P_z-d_1)^2}\right]$$

This transformation is seen to be extremely simple. In addition, it indicates that the robot has no multiple joint solutions. That is, every world position is mapped into a single joint position. The uniqueness of the mapping is a direct result of the joint constraints and would disappear if any one of the constraints were relaxed. For purposes of examining a hybrid computation scheme, uniqueness is advantageous since it eliminates the need for decision making during trajectory execution.

Notwithstanding its simplicity, the world to joint transformation poses a significant computational burden. Two square roots and two arctangents must be evaluated for each set point.

Straightline motion of the end effector is produced by sending an appropriate command signal to each joint controller. To specify the command signal, a functional representation must be chosen for position in world coordinates. This function can then be substituted into the world to joint transformation, yielding the time dependence of each joint. For robots with no orientational variability, the appropriate function is simply the vector representation of a line given by equation 2.

$$\underline{P}(t) = \begin{bmatrix} P_x \\ P_y \\ P_z \end{bmatrix} = \begin{bmatrix} a_x \\ a_y \\ a_z \end{bmatrix} s(t) + \begin{bmatrix} b_x \\ b_y \\ b_z \end{bmatrix} = \underline{a}\, s(t) + \underline{b} \tag{2}$$

This vector funtion is completely specified once the vectors $\underline{a}$ and $\underline{b}$, and the scalar function s(t), have been chosen. $\underline{a}$ and $\underline{b}$ depend on the starting and end points as well as the speed of motion along the trajectory. In choosing s(t) considerable freedom is available. The only restrictions are that it must give rise to physically realizable motion and that it must lend itself to some sort of analog implementation. In accordance with the first of these restrictions s(t) is chosen to be a monotonically increasing funtion of time whose derivatives vanish at the end points of the motion. This choice produces a trajectory veloctity $\underline{v} = \underline{a}\dot{s}$ which starts and ends with the value zero. For convenience we also specify that s(0)=0 so that the vector $\underline{b}$ acquires the straightforward interpretation of starting position. Each joint variable can be obtained as a function of time by substituting equations 2 into equations 1. It is these functions which we wish to evaluate through a combination of digital and analog techniques.

3. CHOOSING AN IMPLEMENTATION SCHEME

Although more than one mode of implementation may be possible, not all choices are equally practicable. As a first possibility, we considered a scheme in which the world to joint transformation is performed through entirely analog means.(Figure. 2)

The position, $\underline{p}(t)$, could be computed by the host processor and output to an analog computer via a multiplexed D/A converter. This implementation would completely free the host processor of the computations involved in the world to joint transformation. It has the disadvantage, however, of requiring quite a large number of non-linear computing elements i.e., two dividing, 3 squaring, 2 square roots, and 2 arc-tangent elements. While analog integrated circuits for each of these functions exist, their accuracy is limited. Thus it is desirable to keep their number to a minimum. Furthermore, the arctangent amplifier poses a particular problem. Analog realizations of this function generally have an output range of 0 to 90^{0}. As can be seen by inspection of the geometry, many straightline trajectories will require a change in θ_1 or θ_3 of greater than 90^{0}. Although techniques are available whereby the range of output angle can be increased to 180^{0} (3), they introduce added complexity.

An alternative approach becomes possible when the function s(t) is specified by a set of piecewise connected polynomials. This is because the two functions which form the square root arguments in the world to joint transformation are quadratic functions of the position vector components. After using equations 2 to substitute in the expression for the position vector compents it becomes clear that $x_1(t)$ and $x_2(t)$ are also quadratic in s(t).

$$x_1(t) = P_x^2(t) + P_y^2(t) = (a_x^2 + a_y^2)\, s^2(t) + 2(a_x b_x + a_y b_y)\, s(t) + (b_x^2 + b_y^2)$$

$$x_2(t) = a_3^2 - (P_z - d_1)^2 = -a_z^2 s^2(t) - 2a_z(b_z - d_1)s(t) + [a_3^2 - (b_z - d_1)^2] \qquad (3)$$

Given that s(t) is polynomial during any finite time interval, $x_1(t)$ and $x_2(t)$ will also be polynomial functions of time during that interval. This makes available the option of generating $x_1(t)$ and $x_2(t)$ with a sequence of cascaded integrators. It is to be expected that an implentation based on generating quadratic functions with linear computing elements will offer certain advantages. It is this scheme which we have adopted.

4. DESCRIPTION OF ADOPTED SCHEME

s(t) is chosen to be a sequence of 5 connected polynomials, each of degree three or less. This coice entails a linear manipulator motion which accelerates smoothy from zero velocity and which has a finite third derivative. (Although not a sufficient condition, a finite 3rd derivative of the position function is certainly necessary if large jerk in the joint variables is to be avoided.) Motion profiles of the first three derivatives of s(t) are given in Figure 3. Note that for time in any one of the 5 indicated regions the functional specification of s(t) remains the same.

4.1 COMPUTATION OF d_2

First let us consider the implementation of the joint variable d_2. Implementation begins with the generation of the polynomials $x_1(t)$ and $x_2(t)$. Since s(t) is a polynomial of first order or higher (in region 3 of Figure 3 it is simply a linear function of time), $x_1(x_2)$ will be of order two or higher.Such a function can be generated by a set of two cascaded integrators fed with a time varying input as shown in Figure 4a. The functional form of the input $u_1(u_2)$, will change depending on the region into which the time instant falls. For regions 1, 2, 4, or 5 the input is given by a fourth order polynomial. However, in region 3, the largest interval of the trajectory, it is constant valued. Leaving aside the question of how the input $u_1(u_2)$ will be produced (this will be discussed in section 4.4) we see that all the host processor has to do is compute and set the initial values of integrators. Despite the fact that the functional description of $x_1(x_2)$ changes between regions, initial conditions need be set only once since the function and its first derivative are continuous.

Having made available $x_1(t)$ and $x_2(t)$ all that is needed to complete generation of the joint variable d_2 is to introduce two square root amplifiers and a summer as shown in Figure 4a.

4.2 COMPUTATION OF θ_2

The realizations of joint variables θ_1 and θ_2 proceed along similar lines but are complicated by the need to evaluate the arctangent functions that appear in equations 1. Consider first the realization of θ_1. As previously mentioned, computing elements are available which directly evaluate the arctangent function. In this application, however, the form of the arctangent argument is known explicity. This suggest the possibility of evaluating the arctangent through direct integration. That is,

$$\theta_1(t) = \theta_1(0) + \int_o^t \frac{\frac{d}{d\tau}[-P_x(\tau)/P_y(\tau)]}{1 + [P_x(\tau)/P_y(\tau)]^2} d\tau \tag{4}$$

Whether this approach will solve more problems than it creates is not clear at first. However, after substituting in the explicit representations of p_x and p_y given by equation 2 and after performing some albebraic reduction, it becomes apparent that this mode of implementation is exceedingly simple.

$$\theta_1(t) = \theta_1(0) + \int_o^t \frac{(b_x a_y - b_y a_x)\dot{s}}{(a_x^2+a_y^2)s^2+2(a_x b_x+a_y b_y)s+(b_x^2+b_y^2)} d\tau \tag{5}$$

The integrand numerator, a function proportional to $\dot{s}$, is simply a polymonial. This can be treated as an input (u_4) to be generated by the same method used to obtain u_1 and u_2. The denominator is none other than the function $x_1(t)$, already available from the implementation of d_2. Thus this method of arctangent implementation requires nothing more than the introduction of a divider followed by an integrator.

In addition to its simplicity, this method has the advantage of partially resolving the problem of limited range in output angle. Obtaining the arctangent through direct integration automatically produces an output angle in the range $-\pi/2$ to $\pi/2$. Because no straightline trajectory involves a change in θ_1 greater than 180° this range is nearly sufficient. However, due to the fact the 180° change can take place anywhere withing the $\pm 170^o$ range allowed by the joint constraints it is not completely sufficient. This difficulty is easily remedied by adding an offset (u_5) to the output of the arctangent integral. The offset value will change depending on the range of values assumed by θ_1 during a trajectory. However, it need be set only once during any given trajectory. The implementation of θ_3 is shown in Figure 4b.

4.3 COMPUTATION OF θ_3

The method for realizing θ_3 is nearly identical to that used for θ_1. Substituting the position vector components of equation 2 into the integral formula for the arctangent leads to the following expression.

$$\theta_3(t) = \theta_3(0) + \int_o^t \frac{a_z \dot{s}}{\sqrt{-a_z^2 s^2 - 2a_z(b_z-d_1)s+[a_3^2-(b_z-d_1)^2]}} \tag{6}$$

The integrand numerator is once again proportional to $\dot{s}$. The denominator is equivalent to the function $\sqrt{x_2(t)}$, available from the implementation of d_2. Generation of θ_3 is thus accomplished by the addition of a divider and an integrator. To θ_3 no offset need be added, since the mechanical configuration constrains this angle to the range $\pm 90^o$.

4.4 COMPLETE SCHEME OF IMPLEMENTATION

The complete scheme of implementation is presented in the block diagram of Figure 5. Of the 11 voltages input to the analog computer, 7 are time independent and are output directly by the host processor (via a D/A converter and miltiplexer). The remaining 4 voltages are time dependent. The signal generator which produces these time dependent voltages is controlled by the host, but must be able to operate with a high degree of independence.

Consider first the time independent inputs. Six of them ($\dot{x}_{10}$, x_{10}, $\dot{x}_{20}$, x_{20}, x_{50}, x_{60}) are integrator initial conditions. Their values are computed by the host processor just before the start of trajectory execution. The remaining time independent input, u_5, serves an an offset to joint variable θ_1. As previously indicated, this offset is necessitated by the fact that mechanical constraints allow a greater range of variation than the 180° needed for any straightline trajectory. In order for the computer variable x_6 to take on a set of angle values symmetrically placed about zero, u_5 is chosen according to

$$u_5 = [\theta_1(t_c) - \theta_1(0)]/2 \tag{7}$$

$\theta_1(t_c)$ and $\theta_1(\theta)$ are the final and initial values of θ_1 respectively.

It is important to note that all 7 of the time independent inputs share the characteristic that that the host processor need compute their values only once for a given trajectory.

Turning now to the time dependent inputs u_1, u_2, u_3, and u_4, it becomes clear that the method used to generate these functions strongly affects the ultimate success of the computing scheme. The inputs are given by the following expressions.

$$u_1 = 2(a_x^2+a_y^2)(s\ddot{s}+\dot{s}^2)+2(a_xb_x+a_yb_y)\ddot{s} \qquad u_3 = a_z\dot{s}$$

$$u_2 = -2a_z^2(s\ddot{s}+\dot{s}^2)-2a_z(b_z-d_1)\ddot{s} \qquad u_4 = (b_xa_y-b_ya_x)\dot{s}$$

It can be seen that in time region 3, where $\dddot{s} = \ddot{s} = 0$ and $\dot{s} = 1$, all four of these expressions become constant valued. This is advantageous in that generating the functions poses a computational burden only during the acceleration (regions 1 and 2) and deceleration (regions 4 and 5) intervals of the trajectory. Nevertheless, it is desirable that computational burden be minimized throughout the duration of the trajectory. This is true not only so that the host can be freed up for activities such as sensing. Assuming that the length of the acceleration interval (t_a) is fixed, the minimum trajectory execution time occurs when the constant velocity portion of the trajectory is eliminated. Such a trajectory will last $2t_a$ seconds. If the generation of u_1, u_2, u_3, and u_4 involves heavy computations then t_a cannot be set at a value low enough to obtain high trajectory speeds. It is for this reason that we seek a method of signal generation which maintains a high degree of autonomy from the host processor.

We are considering several methods of implementing the signal generator. One obvious possibility is to have the host processor generate a set of samples for each of the functions, outputting them sequentially during acceleration and deceleration of the arm. It seems likely that this approach will be disadvantageous due to the large number of samples required to obtain sufficient accuracy. A more attractive option is to generate the functions using 4 sets of cascaded integrators. This is possible since the inputs are piecewise connected polynomials. Although this alternative has not been thoroughly investigated it appears that, during acceleration and deceleration, the only tasks to be performed by the host will be a small number of arithmetic computations and the switching of integrator initial conditions.

In Figure 5 we have shown the joint controllers on the output side of the analog computer. These controllers must, of course, be capable of accepting analog input.

In order to verify that the block diagram formulation shown in Figure 5 is correct, the analog portion of the system was simulated using the IBM Continuous Systems Modeling Program (CSMP). As expected it was found that the analog computer specified generates the correct command signals for producing straightline motion of the end effector. As part of an error analysis, work has been started on modelling the effects on accuracy of non-ideal computing elements. More will be said about solution accuracy in section 6.

5. SCALING OF THE ANALOG COMPUTER

The block diagram of Figure 5 details the physical quantities to be represented and the sequence of mathematical operations to be performed by the analog computer. What remains is to associate with each computing amplifier, a gain chosen so that the amplifier's output voltage stays within a prescribed range. These gains are determined through the process of scaling, which is usually performed in the two stages of time scaling and magnitude scaling (4,5).

For a typical analog computing application, the gains are calculated anew each time the parameter values of the problem are changed significantly. This is done to insure that during most of the run, voltages representing mathematical quantities are large compared to offset voltages. In this application, parameter values change frequently. They change every time the trajectory end points change or the speed of motion changes. Having the capability of adjusting the gains between each trajectory execution would undoubtedly be desirable. It could be accomplished through the introduction of digitally controlled gain selectors. However, for simplicity, we confine our attention to an analog computer with amplifier gains set once for all trajectories and all speeds, bearing in mind that the quality of performance will be correspondingly reduced.

The desired speed of motion along a trajectory determines the time scaling of the analog system. Whenever the speed is changed the system must be rescaled. For the analog computer shown in Figure 5, the process of time scaling is quite simple. In fact, once the maxiumum trajectory speed has been chosen it simply consists of scaling the magnitudes of 4 inputs and two initial conditions. The reason for this is that, of the operations performed by the computer, those not purely algebraic in nature are simple integrations. No feedback of any integrator output or any quantity dependent on an integrator output occurs. This can be contrasted with the situation that arises in the analog solution of differential equations. There, feedback of integrator outputs becomes necessary and gives rise to characteristic time constants in the system. That the selected system has no characteristic time constants is of course not surprising given that the original problem formulation was purely algebraic. However, it has the important consequence that amplifier gains need not be changed when the time scale (i.e., the trajectory duration time t_c) changes. Instead, as has been indicated, a change in time scale is accomplished by altering the magnitude of those inputs and initial conditions whose magnitudes depend on the quantity

t_c, e.g., u_1, u_2, u_3, u_4, $\dot{x}_{10}$, and $\dot{x}_{20}$.

Magnitude scaling must be performed in a very conservative way for the system under discussion. It must provide for the largest value taken on by each analog computer variable at any time during any straightline trajectory feasible for the robot geometry. Of course, it is unlikely that more than 2 or 3 of the computer variables will pass through their maxima during any one trajectory. Those which do not will have higher sensitivity to voltage offset errors. Such restrictive scaling is nevertheless necessary as no gain adjustments will be attemped between trajectories.

Before scale factors can be computed, the maximum magnitude of each variable must be identified. This entails solving several geometric maximization problems. For the robot under discussion, sufficiently tight bounds were obtained for most variables without great difficulty. The amplifier gains that resulted all fall withing the range of .2 to 3.

One consequence of the analog approach used here is that it results in angle joint constraints (constraints on θ_1 and θ_3) which are somewhat reduced from those imposed by purely mechanical considerations. The cause for this is the limited dynamic range of the dividing elements. For the divider considered in this design (Analog Devices 535), the denominator must always be greater than approximately one volt, a value one tenth of the output maximum. Since the scale factors relating geometric variables to their voltage representations are completely determined by the magnitude scaling, these denominator constraints can only be met by restricting the range of affected variables. For the computer configuration shown in Figure 5 a minimum restriction on the divider donominators results in a restriction in the range of the output variables θ_1 and θ_3. For θ_1 the range of variations was originally 154° (less than 180° due to the minimal radial extension). After taking divider constraints into account, this becomes 143°. The range of θ_3 is reduced from 180° to 168°.

Though not large these reductions are not insignificant. However, they are similar in kind to the allowances that must be made for manipulator singularities in a digital approach. As can be seen from observation of many commercial robots, joint speeds become infinite as the manipulator approaches a singularity. In the computing scheme presented here high joint speeds manifest themselves as small denominator inputs to the analog dividers.

6. SUMMARY AND CONCLUSION

The possibility of generating the command signal needed in straightline motion of a robotic arm through the use of hybrid computations has been proposed and discussed. In particular, the feasibility of implementing a hybrid computing scheme for a 3 degree of freedom, cylindrical coordinate robot was examined and benefits and drawbacks were identified.

It was found that, theoretically at least, a scheme of implementation could be devised. The scheme avoids the explicit computation of quadratic functions and can, therefore, be realized with a small number of non-linear computing elements. Significant advantages both in reducing the overall computational load on the host processor and in increasing the maximum possible speed of motion are projected to be provided by the scheme. All that reamins to complete the theoretical specification of the scheme is to devise an efficient method for generating, under the control of the host processor, the four polynomial functions of time u_1, u_2, u_3, and u_4. Although it has yet to be verified, it seems likely that such a method will consit of using 4 sets of cascaded integrators to generate the polynomials with the host processor serving to set constant valued inputs and initial conditions.

Together with its advantages the hybrid computing scheme here examined has inherent disadvantages. The first of these is a reduction below the range of variation allowed by mechanical constraints of the range of joint angles θ_1 and θ_3. This reduction was seen to be associated with the limited dynamic range of analog dividers. More serious than this is the likelihood that the analog section of the computing system will introduce large error. Analog computers generally have solution errors considerably larger than those associated with digital computations. There are several ways that distrubances can be introduced to the signal carrying voltages of an analog computer. For the analog computer discussed here finite precision of nonlinear computing elements (.5% of full range), voltage quantization associated with D/A conversions, and d.c. voltage offsets at the outputs of active elements, may all have appreciable effects on the solution accuracy. Furthermore, sensitivity to these sources of error is increased in the proposed scheme by the inability to rescale between successive changes in parameter values. Although an error analysis is as yet imcomplete, presently available results indicate that the scheme introduced here may not provide sufficient accuracy to be applicable using presently available technology.

We condlude that further work should be done to identify more precisely the effects of the error sources listed above and to investigate methods whereby accuracy can be improved, either now, or after the improvement of presently available operational amplifiers and nonlinear computing elements. Subsequent to this, attention should be directed to the question of whether computing methods of the type described here could be applied to robots with more complicated geometry.

REFERENCES:

1. Paul, R.P., Robot Manipulators: Mathematics, Programming, and Control, The MIT Press, Cambridge, 1981.

2. Paul, R.P., "Manipulator Cartesian Path Control", IEEE Trans. on Systems, Man, and Cybernetics SMC-9, 11 (Nov. 1979), 702-711.

3. Sheingold, D.H., editor, Nonlinear Circuits Handbook, Analog Devices, Inc., Norwood, Massachusetts, 1976.

4. Rekoff, M.G., Analog Computer Programming, Charles E. Merrill Books, Inc., Columbus, 1967.

5. Korn, G.A., Korn, T.M., Electronic Analog and Hybrid Computers, McGraw-Hill Inc., 1964.

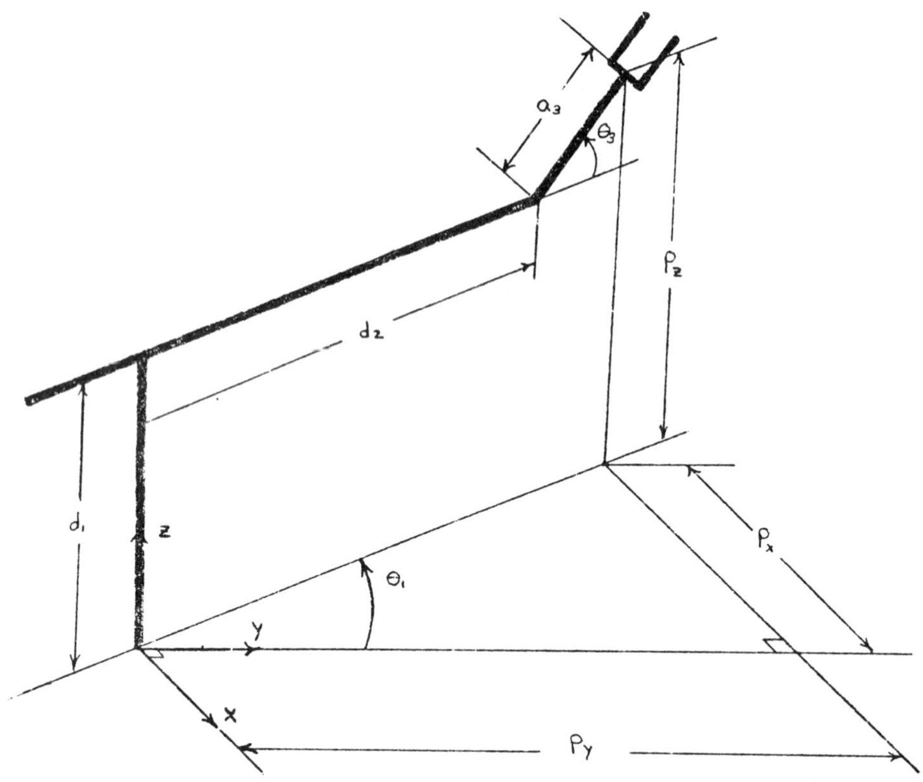

Figure 1. Robot geometry for which a hybrid computing scheme was devised. Joint variables are θ_1, d_2, and θ_3.

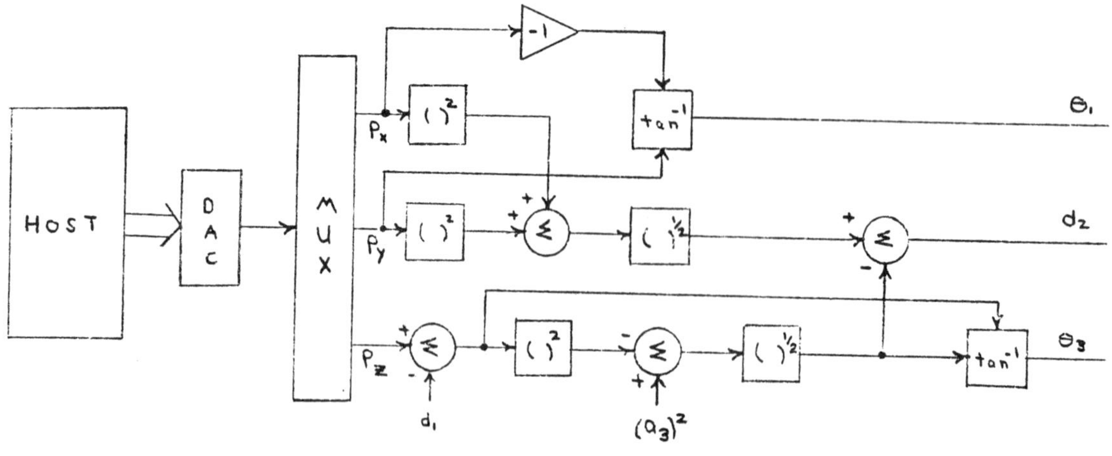

Figure 2. Initial attempt at implementation scheme. World position of trajectory points are computed by host and output to an analog computer.

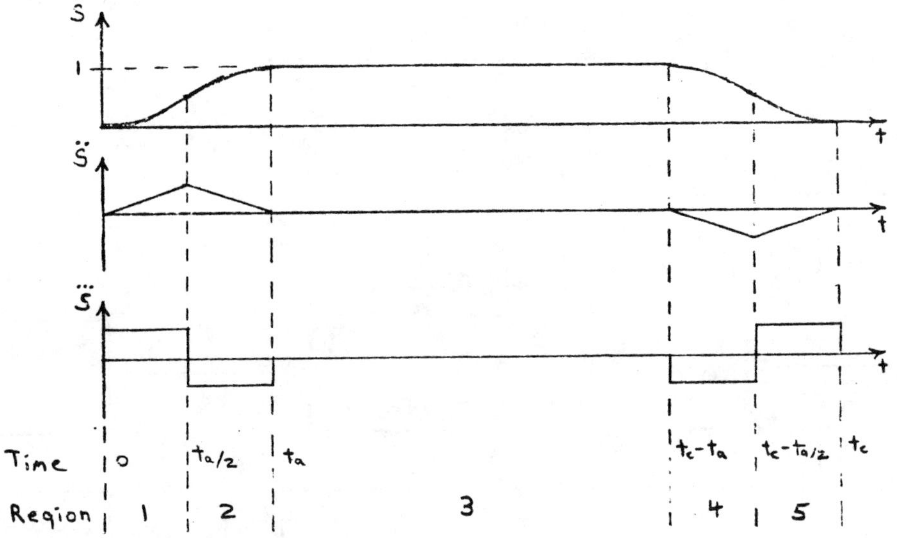

Figure 3. Derivatives of the position parameter function s(t). Trajectory speed is proportional to $\dot{s}$.

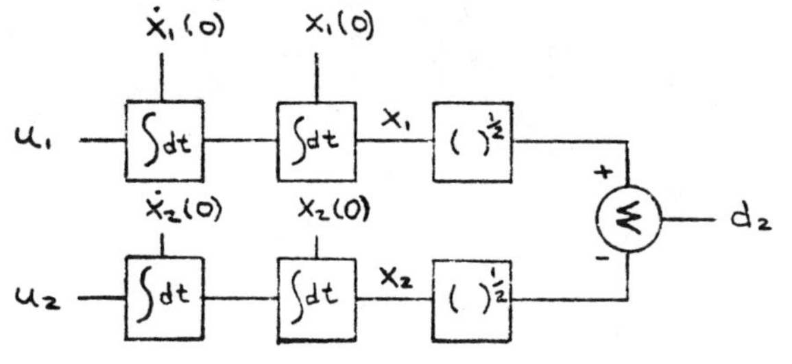

Figure 4a. Configuration of analog elements for computation of joint variable d_2.

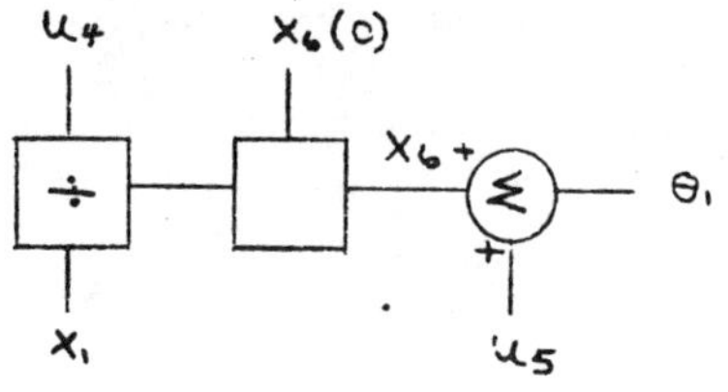

Figure 4b. Configuration of elements for computation of θ_1.

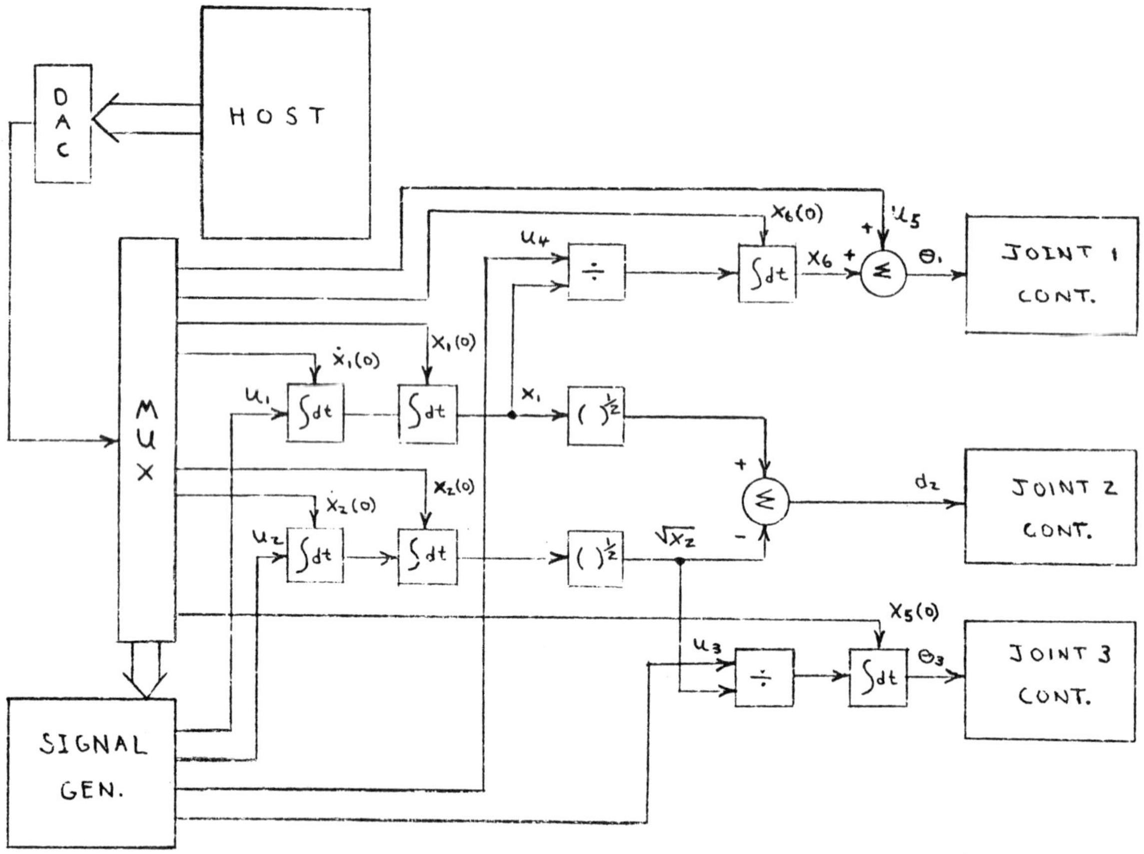

Figure 5. Complete block diagram of chosen scheme for implementing hybrid computation.

Development of New Type of Mobile Robot TO-ROVER

Masaharu Takano
University of Tokyo
Tokyo, Japan

Gotaro Odawara
University of Tokyo

Abstract

A new type of mobile robot, TO-ROVER, developed here, has the ability to run on the floor, and go up and down stairs, over obstacles, and turn in narrow spaces. Controlled manually or by a micro computer with simple control procedure, the mechanism of TO-ROVER consists of two pairs of three-arm units. Experiments show the stable motion of going up and down stairs.

1. INTRODUCTION

This study relates to the development of mobile robot "TO-ROVER" which is used for maintenance robot system "MOOTY". MOOTY is now being developed by our research group* which works mainly in nuclear power plant for automatic inspections and repairs or those of remote manual operation. The maintenance robot system has to possess the following functions as shown in Fig. 1: Sensory system to recognize the environment and defects of machines, arm-hand system to work for repairs, mobile system to walk around the machines and equipments, and intelligence system to comprehend his circumstance and to decide his action. The function needed for mobile system is that it can not only run on floor but go up and down stairs, go over obstacles and turn in narrow space, as the space and paths of nuclear power plant, for example, are so narrow and complicated.

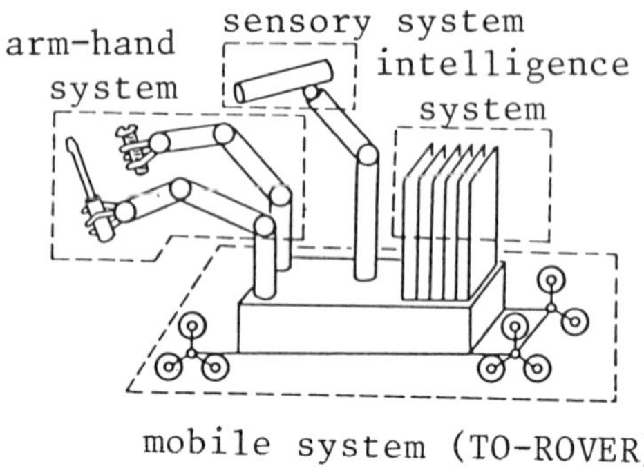

Fig. 1 Concept of maintenance robot system MOOTY

Many types of mechanism of mobile robot as wheel type, 2,3,4,5,6 and 8-legged type or crawler type are studied, but wheel type doesn't serve our purpose because it cannot go up stairs, and other types have difficulties in stability and complexity of automatic control. TO-ROVER developed here improves these points and it has the feature that it can automatically and stably go up and down stairs detecting steps without any particular sensor, turning in narrow space, and the software of its computer control of these motion is very simple. In the following the mechanism and control of TO-ROVER and the experiments of the motion are described.

2. MECHANISM OF TO-ROVER

The mechanism and motion of TO-ROVER are very simple as shown in Fig. 2. It has seven degrees of freedom; ①-④ drive wheels of front and rear, right and left sides independently, ⑤, ⑥ drive arms of front and rear sides, ⑦ bends these drive unit stands. Wheels are mainly driven for running or turning on floor, and arms are driven for going up and down stairs. Bending of unit stands is used for turning in narrow space (Fig. 3) or helps the arm-hand system to work in high position.

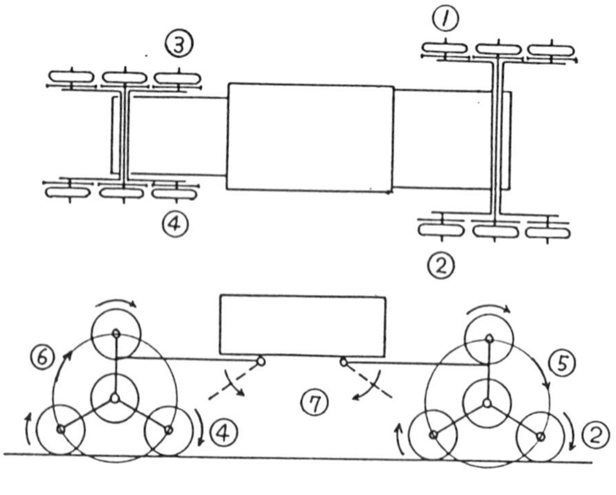

Fig. 2 Mechanism of TO-ROVER

---

*A MOOTY Committee: H.Yoshikawa(chairman), M.Takano, S.Ozono, G.Odawara, T.Miyoshi, T.Arai(Univ. of Tokyo); K.Shimo(Tsukuba Univ.); members of Nuclear Eng. Lab. of Toshiba Elec. Co. Ltd.

Fig. 4 shows the prototype of TO-ROVER. The specification is as follows:

wheelbase × width	300mm × 210mm
arm length	50mm
drive motors	7 × 11watts DC motors
sensors	2 encoders 4 tachometers
max. speed	580mm/s
max. arm speed	0.75rps (2.25step/s)
max. step height	75mm
min. turn space	250mm

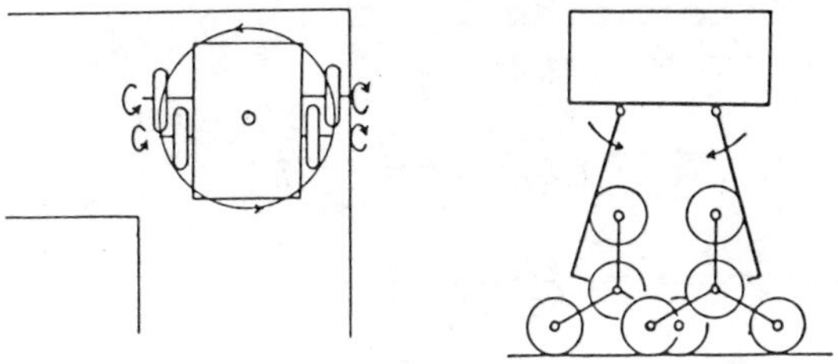

Fig. 3 Turn in narrow space

Fig. 4 Prototype of TO-ROVER (TO-ROVER-I)

3. MOTION CONTROL AND EXPERIMENTAL RESULTS

3.1. Turning on plane

As TO-ROVER has no steering mechanism, it turns only by the difference of the velocity between right and left wheels. If there is no slip in the direction of wheel rotation (longitudinal direction), the turning radius is,

$$R = \frac{i+1}{i-1}\,\frac{B_f + B_r}{4}, \qquad (1)$$

$$(\, i = v_o / v_i \,)$$

but when the slip occurs in both longitudinal and lateral directions, shown in Fug. 5, it becomes[4)]

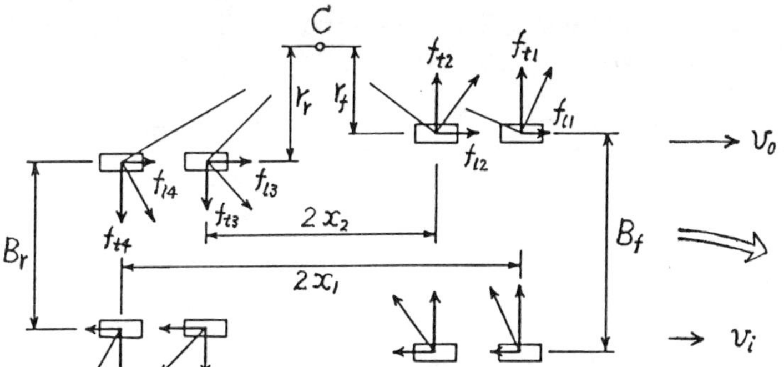

C : relative turning center of left wheels
f_{li} : longitudinal friction resistance
f_{ti} : lateral friction resistance

Fig. 5 Friction resistance of wheels

$$R_e = \frac{i+1}{i-1}\,\frac{B_e}{2}, \qquad (2)$$

B_e is determined by the following equations.

$$f_{l1} = \frac{r_f}{\sqrt{r_f^2 + x_1^2}} f, \quad f_{l2} = \frac{r_f}{\sqrt{r_f^2 + x_2^2}} f$$

$$f_{l3} = \frac{r_r}{\sqrt{r_r^2 + x_1^2}} f, \quad f_{l4} = \frac{r_r}{\sqrt{r_r^2 + x_2^2}} f$$

$$f_{t1} = \frac{x_1}{\sqrt{r_f^2 + x_1^2}} f, \quad f_{t2} = \frac{x_2}{\sqrt{r_f^2 + x_2^2}} f$$

$$f_{t3} = \frac{x_1}{\sqrt{r_r^2 + x_1^2}} f, \quad f_{t4} = \frac{x_2}{\sqrt{r_r^2 + x_2^2}} f$$

$$f = \frac{\mu W}{8},$$

$$B_f(f_{l1} + f_{l2}) + B_r(f_{l3} + f_{l4}) = 2(f_{t1}x_1 + f_{t2}x_2 + f_{t3}x_1 + f_{t4}x_2)$$

$$B_e = B_f + 2r_f, \qquad (3)$$

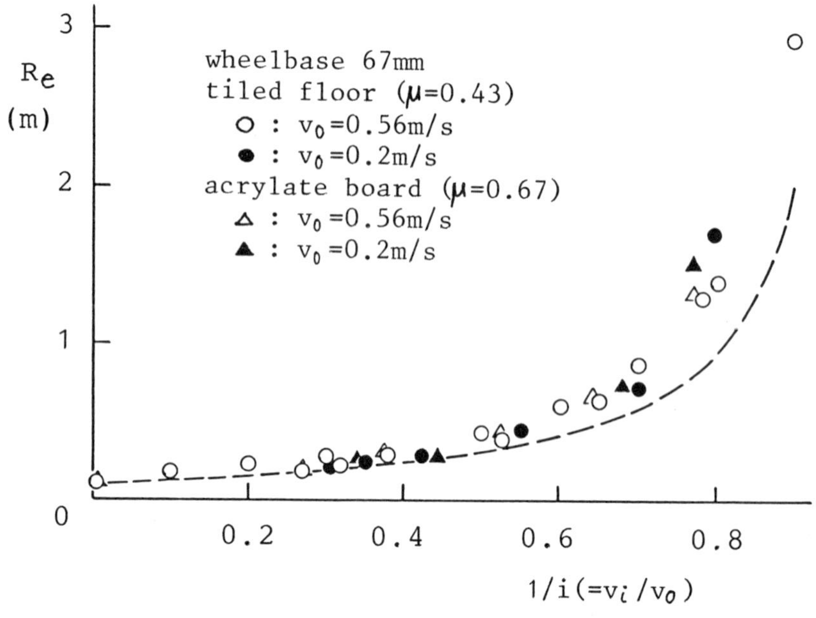

Fig. 6 An experimental result of turning

where the friction coefficient μ to both directions is assumed to be the same and the weight W to be uniformly distributed to all wheels. In our case $(B_f+B_r)/2=170$ mm and $B_e=200$mm. Fig. 6 shows the relation of turning radius R_e to velocity ratio i obtained by experiment. R_e shows a tendency to become larger than theoretical value as 1/i increases. R_e varies largely caused by floor condition, therefore some other sensory systems are needed to detect his situation in order that it could run exactly on a given way.

The driving force and torque of front and rear wheel are respectively,

$$F_f = f_{l1} + f_{l2}, \qquad F_r = f_{l3} + f_{l4}, \qquad (4)$$

$$T_f = F_f r_w, \qquad T_r = F_r r_w, \qquad (r_w = \text{wheel radius}) \qquad (5)$$

where the specific sliding of wheel is assumed not to influence the friction force. In case of tiled floor (μ=0.43) $F_f \fallingdotseq F_r$ =0.4kgf, $T_f \fallingdotseq T_r$ = 0.8kg·cm, which is sufficiently small compared with motor capacity.

3.2. Control of Going Up Stairs

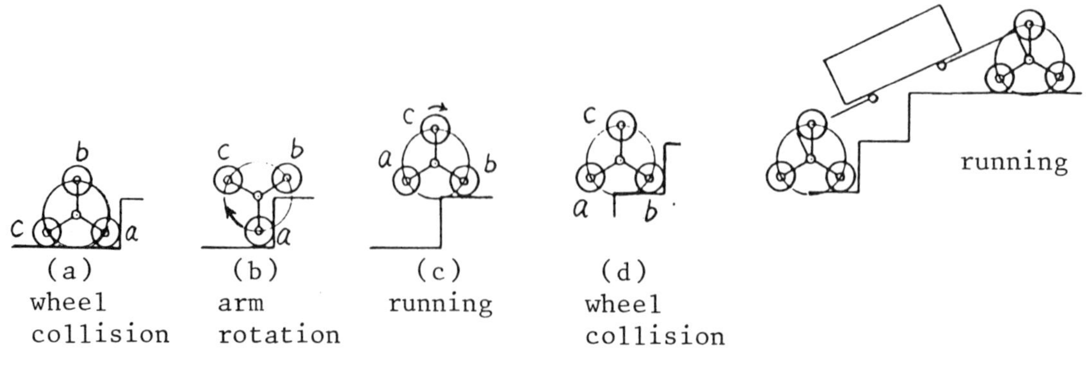

Fig. 7 Going up mode control

The control procedure of going up stairs is shown in Fig. 7. (a) In running mode the condition whether the front wheel collides with vertical board of the first step is examined. The wheel rotation becomes zero by collision and it is detected by tachometer generator. (b) The collision means the existence of the step, and the arm is rotated by 120° and climb the first step. (c) Holding the arm angle 120° wheel is rotated and runs forwards. (d) Again the collision of the front wheel is detected. TO-ROVER can go up stairs by the repetition of this procedure in front and rear drive units without missing steps. In this control the arm rotation angle 120° means that it is the value of the angle measured from horizontal plane and the inclination of main body has to be detected.

There may be the case that the wheel collides with the wall of the next step while arm is rotated when the pitch of the step is much smaller than arm length. In this case the wheel inversely rotates along the wall and no danger occurs.

If step number n of the stairs is previously known the simpler control procedure of going up stairs can be adopted. After the collision of the first step it is sufficient to rotate steadily the arm and wheel until arm rotation angle becomes n times 120°. In this case the wheel rotation helps to go forwards and arm not to miss the step.

The driving torque needed to go up stairs is, assuming all weight is put on an arm, $T_a = WR$. In TO-ROVER $T_a = 19\text{kg}\cdot\text{cm}$ and needed power is 4 watts by 1 step/s speed.

3.3. Control of Going Down Stairs

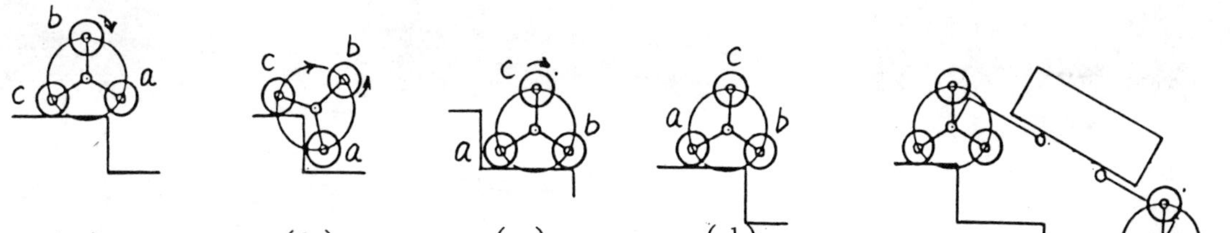

Fig. 8 Going down mode control

Fig. 8 shows the control procedure. (a) The condition whether the front wheel floats from the floor in running. If it floats the arm gives the inverse torque to motor which rotates inversely and this rotation can easily detected by rotary encoder. (b) The floating of front wheel means that it comes to the first down step. Arm is rotated by 120° and at the same time wheel is driven inversely by small torque. This inverse drive has an effect that the wheel to be contacted with the next step is forced to approach the vertical board of the previous step and it secures the stable going down motion. (c) After 120° rotation of the arm, wheel is rotated and runs forwards. (d) Again the floating of front wheel is detected. Thus TO-ROVER can go down stairs stably by the repetition of this procedure in front and rear drive units without missing his step.

If step number of stairs is previously known and the step pitch is not so large, the simpler control procedure of going down stairs can be adopted. After detecting the floating of front wheel on the first step, it is sufficient to rotate steadily the arm and to drive the wheel inversely until it goes down n steps.

3.4. Experiments of Going Up and Down and Discussion

Fig. 9 shows the motion of TO-ROVER controlled by a micro computer. By the simple control procedure it can go up and down stably the stairs with 45° average slope and 70mm step height. For the smooth motion the constant velosity control of arm rotation is needed, while the wheel must be controlled by constant driving torque in going up and down mode and by constant velocity control in running mode.

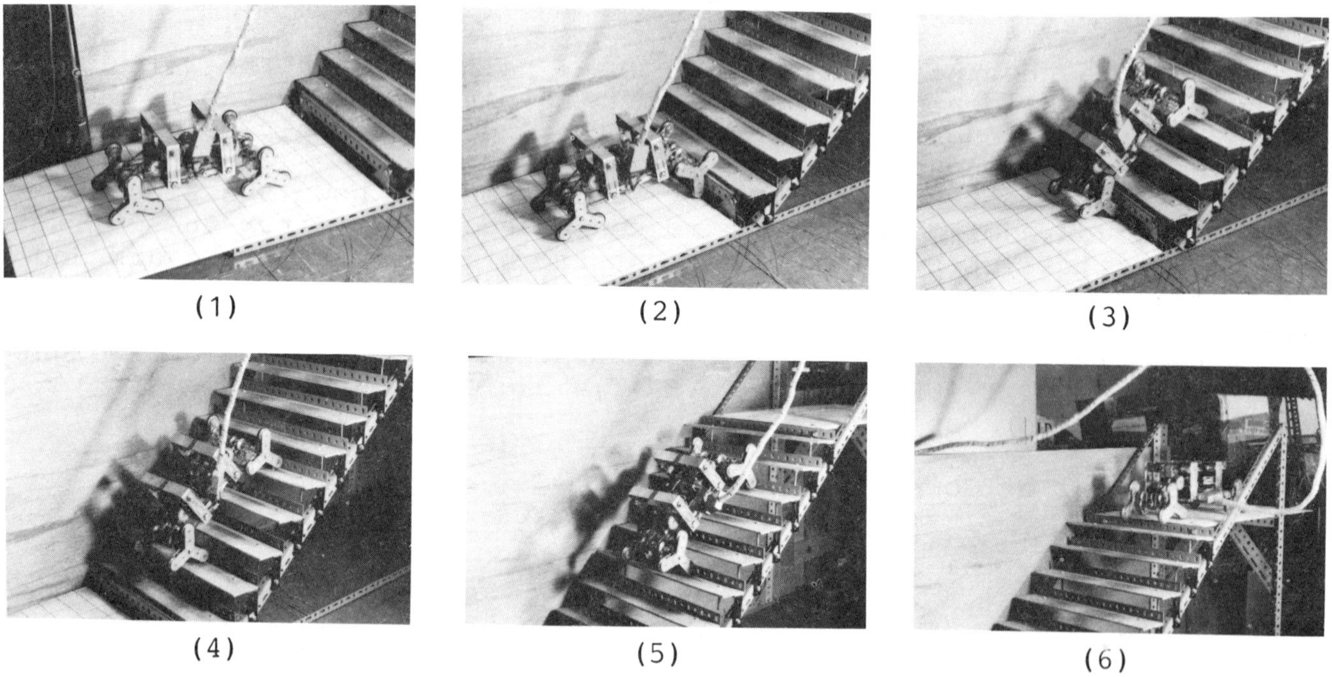

Fig. 9 An example of going up stairs controlled by simple procedure

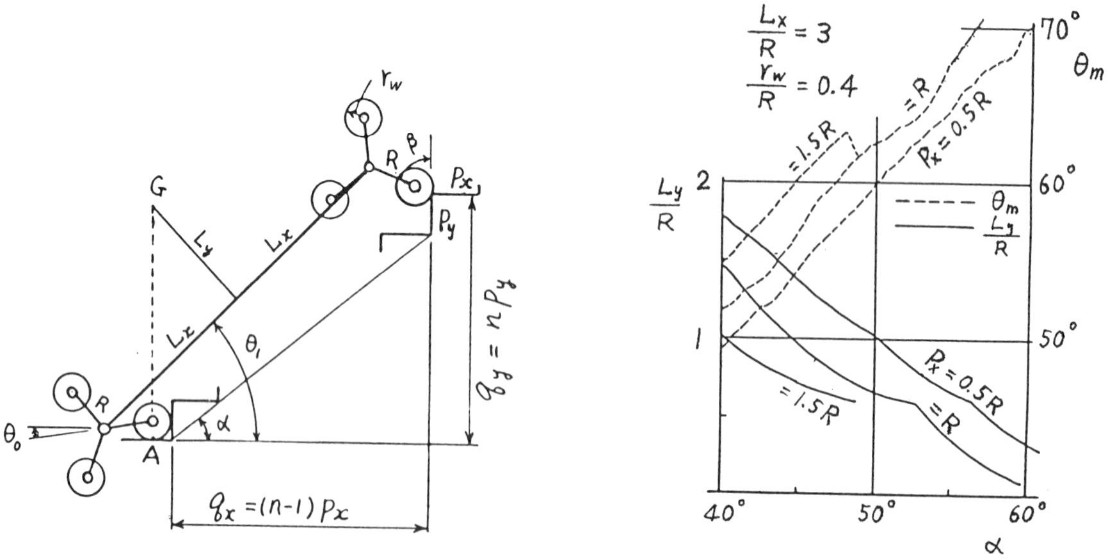

Fig. 10 Relation of permissible height of center of gravity L_y and maximum inclination θ_m to average slope α of stairs

There are some limits for the stairs that TO-ROVER can go up and down stably. First the step must not be so high that the arm cannot reach it. The permissible maximum height is theoretically $\sqrt{3}R$, R is the length of arm, or practically 1.5R. Second the average slope α of stairs must be less than maximum inclination θ_m of TO-ROVER. It goes up and down stairs varying its inclination in every step. Supposing the worst case of the state of going up as is shown in Fig. 10 (a), the limit value of the height of center of gravity L_y and inclination θ_1 are calculated from the following equations;

$$q_x = (n-1)p_x = 2L_x\cos\theta_1 - R\cos\theta_0 + (R+r_w)\sin\beta - r_w, \qquad (6)$$

$$q_y = np_y = 2L_x\sin\theta_1 - R\sin\theta_0 - (R+r_w)\cos\beta + r_w, \qquad (7)$$

$$L_y\sin\theta_1 = L_x\cos\theta_1 - R\cos\theta_0,$$

where

$$\theta_0 \leqq \theta_s = \sin^{-1}(p_y/\sqrt{3}R) - 30^\circ.$$

For a certain value of β, θ_1 becomes maximum θ_m. Obtained from these equations, the relation of L_y and θ_m to average slope α of stairs is shown in Fig. 10 (b). It depends on step pitch p_x and in short, θ_m is 10°- 15° greater than α in this case.

4. APPLICATION TO MOOTY

Fig. 11 is a maintenance robot system MOOTY developed by our research group applied by TO-ROVER mechanism. The dimension is 1150mm wheelbase, 900mm width 1600mm height and 400kg weight.

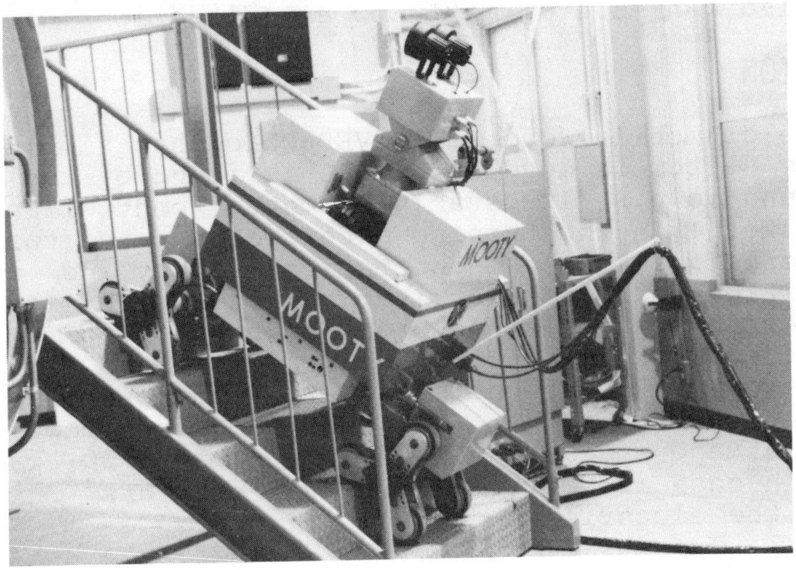

Fig. 11 Maintenance robot system MOOTY

Four kinds of control mode are possible; manual control mode, composite manual control mode, supervisory control mode and automatic control mode. In composite manual control the operator can switch on and off some arranged motor drives at the same time, which makes the operation very easy to go up and down stairs. This system carries three micro computers used for the supervisory control, where the operator gives a running or turning command by a joistic lever and supervises the automatic going up and down motion controlled by the computers. For the complete automatic rovering the vision and higher intelligence systems are needed in addition to this system.

The stable going up and down of 35° slope stairs is confirmed for the time being.

5. FURTHER DEVELOPMENT

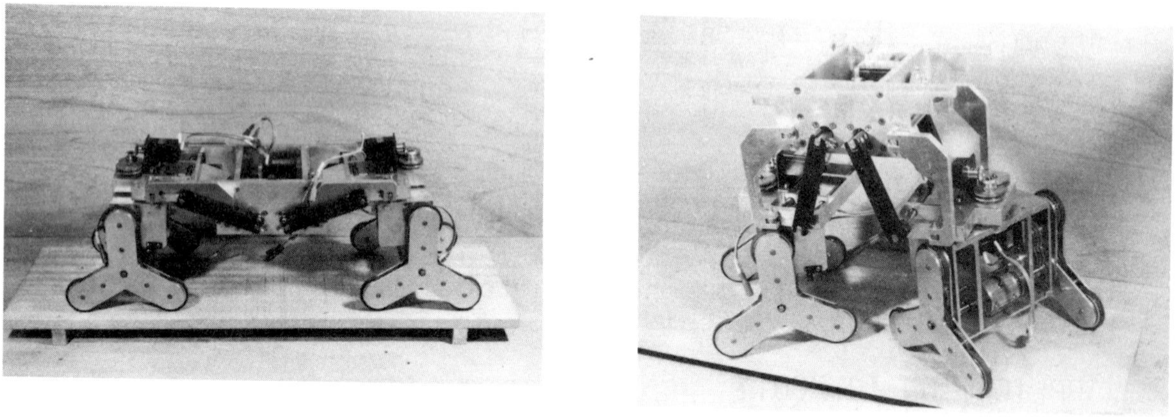

Fig. 12 TO-ROVER-II with steering mechanism

Fig. 12 is the model TO-ROVER-II which has steering mechanism with seven degrees of freedom. It remarkably improves the turning characteristics and also can move directly to any directions.

6. CONCLUSION

New type of mobile robot TO-ROVER which has simple mechanism and simple sensory system is developed. It is found by the experiment that it can go up and down stairs stably by manual control and by the simple control software. In the turning experiments, the turning radius is found a little larger than the theoretical value and it varies caused by floor condition and load distribution to wheels. It would be better to add the steering mechanism in order to have a good controllability.

As for the power to drive to go up stairs and to turn on floor, the samll size of TO-ROVER mentioned here needs only several watts in each motor, but we can estimate that the necessary power is proportional to the weight and speed, and it becomes about 300 watts for 0.1 step/s speed in case of MOOTY. This fact prevents the mobile robot from carrying the electric power source.

ACKNOWLEDGEMENT

The authors are greatly thankful for Mr. Takao MIKAMI of Nuclear Engineering Laboratory, Toshiba Electric Co., who has helped me in designing and manufacturing TO-ROVER's and MOOTY.

REFERENCE

1) Internationale Wehr-Revue, 1/1969, 63.
2) W. Kohler: Stand der Fernbedienungs-Technik, 8th ISIR (1978), 805.
3) R. B. McGhee and G. I. Iswandhi: Adaptive Locomotion of a Multilegged Robot over Rough Terrain, IEEE Trans., Vol. SMC-9, No. 4 (1979), 176.
4) D. E. Okhotsimski et al.: Integrated Walking Robot Simulation and Modelling, World Congr. IFAC, Vol. 7, No. 2 (1979), 917.
5) S. Hirose and Y. Umetani: Control System of 4-legged Walking Vehicle and Experiments of Adaptive Walking to Terrain, Preprint of 18th Meeting of SICE (1979), 259.
6) I. Hayashi: Practical Analysis of Tracked Vehicle Steering Depending on Track Slippages, Trans. JSME, Vol. 41, No. 352 (1975), 3470.
7) T. Harama et al.: Development of Robot for Inspection and Maintenance in Nuclear Power Plants, Hitachi Revue Vol. 60, No. 2 (1978), 49.
8) H. Yoshikawa: Nuclear Power Plant and Robot, Nuclear Engineering, Vol. 28, No. 12 (1982), 25.
9) M. Takano: A Trend of The Development of Maintenance Robot, Nuclear Engineering, Vol. 28, No. 12 (1982), 36.

AUTHOR'S INDEX

A

B

C

D

E

F

G

H

I

J

K

L

M

N

O

P

R

S

T